River Dynamics

Rivers are important agents of change that shape the Earth's surface and evolve through time in response to fluctuations in climate and other environmental conditions. They are fundamental in landscape development and essential for water supply, irrigation, and transportation. This book provides a comprehensive overview of the geomorphological processes that shape rivers and that produce change in the form of rivers. It explores how the dynamics of rivers are being affected by anthropogenic change, including climate change, dam construction, and modification of rivers for flood control and land drainage. It discusses how concern about environmental degradation of rivers has led to the emergence of management strategies to restore and naturalize these systems, and how river management techniques work best when coordinated with the natural dynamics of rivers. This textbook provides an excellent resource for students, researchers, and professionals in fluvial geomorphology, hydrology, river science, and environmental policy.

Bruce L. Rhoads is Professor of Geography and Geographic Information Science at the University of Illinois, Urbana-Champaign. He has been actively engaged in research on river dynamics for over 35 years. He has been a Guggenheim fellow and has received awards for research excellence and a distinguished research career from the American Association of Geographers. He is a fellow of the American Association of Geographers and the American Association for the Advancement of Science.

"*River Dynamics* represents a comprehensive and concise overview of contemporary knowledge of river process and form. The text is thoroughly illustrated with relevant photographs, drawings, and graphs, and the referencing is thorough and up to date. Rhoads is skilled at providing clear explanations of complex phenomena and the book includes just enough historical perspective to allow the reader to understand the background for contemporary conceptualizations of rivers. The book starts with a thorough discussion of basic information on river process and form, and then discusses river management in a manner that effectively integrates basic and applied science. This book will provide an engaging textbook for advanced undergraduate and graduate courses, and a resource for river-management professionals."
Ellen Wohl, *Colorado State University*

"This is a superb textbook on fluvial geomorphology – the book we have long been waiting for in this field. It provides a clear framework for understanding and is written in such a way, with the headings posed as questions, that we can immediately identify the key issues and content of each section. It provides detailed analysis and information on all the main topics in fluvial geomorphology, as well as a few not usually covered well. It can be used at various levels of information, from the main principles and concepts to detailed analysis of, for example, hydraulics and processes. It provides a rich source of information for undergraduates, for more specialist courses, and for fluvial management. I shall immediately recommend it as the main text for my own taught course in fluvial geomorphology. It has a copious reference list, with an excellent balance between classic references that provided the foundation of our subject and the more recent research developments, allowing the reader to appreciate how ideas and knowledge have developed to our present state of understanding, while also highlighting ongoing debates and questions. It is clearly illustrated with numerous figures to complement the explanations. It provides a firm scientific basis for sustainable management of rivers and catchments that builds on a thorough understanding of principles, processes, and dynamics."
Janet Hooke, *University of Liverpool*

"Bruce Rhoads draws on his unique experience to comprehensively review concepts in river dynamics. This book will be of great benefit to students, scientists, and managers of rivers by blending field and laboratory studies with theory and application."
Marwan Hassan, *University of British Columbia*

River Dynamics

Geomorphology to Support Management

Bruce L. Rhoads
University of Illinois

CAMBRIDGE
UNIVERSITY PRESS

CAMBRIDGE
UNIVERSITY PRESS

University Printing House, Cambridge CB2 8BS, United Kingdom

One Liberty Plaza, 20th Floor, New York, NY 10006, USA

477 Williamstown Road, Port Melbourne, VIC 3207, Australia

314–321, 3rd Floor, Plot 3, Splendor Forum, Jasola District Centre, New Delhi – 110025, India

79 Anson Road, #06–04/06, Singapore 079906

Cambridge University Press is part of the University of Cambridge.

It furthers the University's mission by disseminating knowledge in the pursuit of education, learning, and research at the highest international levels of excellence.

www.cambridge.org
Information on this title: www.cambridge.org/9781107195424
DOI: 10.1017/9781108164108

First published 2020

Printed in the United Kingdom by TJ International Ltd, Padstow Cornwall

A catalogue record for this publication is available from the British Library.

ISBN 978-1-107-19542-4 Hardback

Header image: photograph by Sam Riche for the *Indianapolis Star*

Contents

Preface   *page* ix

1

Introduction   1

1.1   Why Are Rivers Important?   1
1.2   What Is a River?   2
1.3   What Is a River System?   3
1.4   What Is Fluvial Geomorphology?   4
1.5   How Do the Dynamics of River Systems
      Vary over Time and Space?   10
1.6   What Is the Role of Humans in River
      Dynamics?   14

2

The Dynamics of Drainage Basins and Stream Networks   15

2.1   Why Is a Focus on River Systems over
      Geologic Timespans Important?   15
2.2   How Is the Formation of River Channels
      Related to Hillslope Processes?   15
2.3   What Are Some Limitations of Existing
      Models of Channel Initiation?   20
2.4   What Are the Characteristics of Rills?   22
2.5   How Do Gullies Differ from Rills?   22
2.6   How Is the Development of Stream
      Channels Related to Drainage Basin
      Characteristics?   24
2.7   How Do River Networks Form, Grow, and
      Evolve?   29
2.8   How Are Channel Networks Organized
      Geometrically, Hierarchically, and
      Topologically?   39
2.9   What Are the Scaling Characteristics
      of River Networks?   44
2.10 Is the Organization of River Networks
      Governed by Optimization Principles?   46

3

Sediment Dynamics at Global and Drainage-Basin Scales   47

3.1   How Important Are Rivers in Transporting
      Sediment and Dissolved Material
      at the Global Scale?   47
3.2   What Factors Control the Total Sediment
      Flux from Drainage Basins?   49
3.3   How Have Humans Affected Sediment
      Dynamics at Global Scales?   57
3.4   What Is the Relative Importance of
      Different Controlling Variables for
      Influencing River Sediment Flux?   59
3.5   What Factors Influence Landscape
      Denudation?   60
3.6   What Is the Mass Balance of a Watershed?   63
3.7   What Is the Sediment Delivery Problem
      and Its Relation to the Sediment Delivery Ratio?   63
3.8   What Is a Sediment Budget and How Is
      It Useful for Evaluating Watershed Sediment
      Dynamics?   65

4

Flow Dynamics in Rivers   72

4.1   Why Are Flow Dynamics in Rivers
      Important?   72
4.2   What Is the Basic Metric of Flow in
      Rivers?   72
4.3   How Are Flows in Rivers Classified
      According to Changes in Depth and
      Velocity?   73
4.4   How Is the Dimensionality of Fluid
      Motion in a River Determined?   74
4.5   What Is the Relation of Flow Classification
      and Dimensionality of Fluid Motion to
      Hydraulic Complexity?   75

4.6   What Are the Components of Flow
       Energy in Rivers?   75
4.7   How Can River Flows Be Characterized
       in Terms of Force, Stress, and Power?   76
4.8   Are River Flows Laminar or Turbulent?   78
4.9   What Is the Relation between Inertial and
       Gravitational Forces in Rivers?   80
4.10   How Is Flow Resistance in Rivers Related to
       Velocity?   82
4.11   How Is Flow Resistance in Rivers Related
       to Bed Shear Stress?   83
4.12   What Factors Determine the Reach-
       Averaged Bed Shear Stress?   83
4.13   What Is Boundary Layer Theory and
       How Is It Related to Flow in Rivers?   86
4.14   How Can Fluid Motion in Rivers Be Described
       Mathematically in Three Dimensions?   95

5

Sediment Transport Dynamics in Rivers   97

5.1   Why Is Sediment Transport Important in
       River Dynamics?   97
5.2   What Types of Material Flux Occur
       in Rivers?   97
5.3   What Are the Major Sources of Sediment
       Transported by Rivers?   97
5.4   What Are the Major Types of Sediment
       Transport in Rivers?   98
5.5   What Factors Influence the Transport
       of Fine Suspended Sediment?   100
5.6   What Factors Control the Mobilization
       of Bed Material?   105
5.7   How Is the Shear Stress Acting on Grains
       Determined?   115
5.8   How Is Bed-Material Load Transported in
       Suspension?   116
5.9   How Is Bedload or Bed-Material Transport
       Related to Shear Stress and Stream Power? 119
5.10   What Factors Influence the Fractional
       Transport of Bed Material?   122
5.11   How Is Bed-Material Transport Related
       to Particle Motion in Rivers?   129
5.12   How Is Bed-Material Transport
       in Rivers Linked to Changes in Channel
       Morphology?   131

6

Magnitude-Frequency Concepts and the Dynamics of Channel-Forming Events   134

6.1   Why Are Magnitude and Frequency
       Concepts Important for Understanding
       River Dynamics?   134
6.2   How Is Water Delivered to Rivers from
       Drainage Basins?   134
6.3   How Does River Flow Vary over Time?   135
6.4   How Is the Frequency of Floods Determined? 138
6.5   How Is the Frequency of River Flows
       Determined?   142
6.6   What Is a Channel-Formative Event?   144
6.7   What Is the Concept of Dominant
       Discharge and How Is It Related to River
       Equilibrium?   144
6.8   What Is the Concept of Geomorphic
       Effectiveness and How Does It Relate to
       Channel-Formative Events?   157

7

The Shaping of Channel Geometry   164

7.1   How Is Channel Geometry Related to the
       Three-Dimensionality of River Form?   164
7.2   How Is Channel Form Related to the
       Geometry of River Flow?   164
7.3   How Has Physically Based Analysis Been
       Used to Examine Channel Geometry?   178
7.4   How Have Empirical and Physical
       Approaches Been Integrated in the Study
       of Channel Geometry?   183
7.5   How Does Channel Geometry Change
       through Time?   183

8

Channel Planform – Controls on Development and Change   186

8.1   What Is Channel Planform and Why Is It
       Important?   186
8.2   What Are the Major Types of Channel
       Planforms?   186
8.3   What Environmental Factors Are
       Associated with Differences in Channel
       Planform?   188

8.4   How Do Environmental Conditions Differ for Meandering and Braided Rivers?   188

8.5   Under What Environmental Conditions Do Straight Channels Occur?   192

8.6   What Is the Environmental Domain of Anabranching of Rivers?   194

8.7   What Are the Implications of Changes in Environmental Conditions for Channel Planform Change?   195

9

The Dynamics of Meandering Rivers   197

9.1   Why Is the Meandering of Rivers Important?   197

9.2   Why Do Rivers Meander?   197

9.3   What Are the Major Planform Characteristics of Meandering Rivers?   205

9.4   What Are the Major Morphological Components of Meandering Rivers at the Bar-Unit Scale?   207

9.5   What Are the Patterns of Flow in a Meandering River?   213

9.6   How Is Bed Material Transported in a Meandering River?   218

9.7   How Do Riverbanks Erode?   221

9.8   How Do Meandering Rivers Migrate over Time and Space?   225

10

The Dynamics of Braided Rivers   234

10.1   Why Are the Dynamics of Braided Rivers Important?   234

10.2   How Does Braiding Occur?   234

10.3   What Are the Basic Morphological Attributes of Braided Rivers?   237

10.4   What Process–Form Interactions Occur at the Bar-Unit and Bar-Element Scales in Braided Rivers?   241

10.5   What Are the Dynamics of Braided Rivers at the Planform Scale?   246

11

The Dynamics of Anabranching Rivers   252

11.1   Why Are Anabranching Channels Important?   252

11.2   Why and How Do Rivers Anabranch?   252

11.3   What Are the Dynamics of Wandering Gravel-Bed Rivers?   256

11.4   What Are the Dynamics of Anastomosing Rivers?   259

11.5   What Are the Characteristics of Other Anabranching Rivers?   261

11.6   How Is Anabranching Related to the Dynamics of the World's Largest Rivers?   266

12

The Dynamics of River Confluences   269

12.1   Why Are Confluences Important?   269

12.2   What Are the Planform Characteristics of Confluences?   269

12.3   How Does Channel Form Change at Confluences?   271

12.4   What Are the Characteristics of Flow at Confluences?   272

12.5   What Are the Dynamics of Bed Morphology at Confluences?   281

12.6   What Are the Dynamics of Sediment Transport at Confluences?   285

12.7   What Are the Dynamics of Confluent Meander Bends?   287

12.8   How Do Confluences Change over Time?   290

12.9   What Are the Dynamics of Mixing at Confluences?   290

13

The Vertical Dimension of Rivers: Longitudinal Profiles, Profile Adjustments, and Step-Pool Morphology   294

13.1   Why Is the Longitudinal Profile of Rivers Important?   294

13.2   What Are the Characteristics of the Longitudinal Profile of a River?   294

13.3   What Factors Influence Downstream Change in the Size of Bed Material?   297

13.4   What Factors Influence Equilibrium Longitudinal Profiles of Alluvial Rivers?   305

13.5   What Factors Govern the Equilibrium Longitudinal Profiles of Bedrock Rivers?   306

13.6   How Do Longitudinal Profiles Adjust to Changes in External Forcing?   308

13.7 How Does Vertical Adjustment Influence the Morphology of Steep Channels?   313

14

The Dynamics of Floodplains   319

14.1 Why Are Floodplains Important?   319
14.2 What Is a Floodplain?   319
14.3 Why Do Floodplains Develop along Rivers?   321
14.4 What Are the Major Depositional Processes on Floodplains?   321
14.5 What Are the Major Erosional Processes on Floodplains?   331
14.6 How Are Types of Floodplains Related to Stream Power and Material Properties?   333
14.7 How Is Floodplain Sedimentology Related to River Planform Type?   335

15

Human Impacts on River Dynamics   343

15.1 Why Are Human Impacts on River Dynamics Important?   343
15.2 What Are the Major Impacts of Humans on Rivers?   343
15.3 How Does Human-Induced Change in Land Cover/Land Use Affect River Systems?   344
15.4 How Might Anthropogenic Climate Change Affect River Dynamics?   356
15.5 How Does Channelization Affect River Dynamics?   358
15.6 How Do Dams Affect River Dynamics?   361

15.7 How Does Instream Mining Affect River Dynamics?   366

16

River Dynamics and Management   369

16.1 Why Is an Understanding of River Dynamics Important in River Management?   369
16.2 How Is Scientific Inquiry on River Dynamics Related to Management?   369
16.3 What Are the General Goals of Environmental Management of Rivers?   371
16.4 What Are the Main Environmentally Oriented River Management Strategies?   373
16.5 How Can Geomorphological Assessments Contribute to River Management?   378
16.6 What Is the Role of Geomorphology in Implementation of Management Strategies?   389

Appendix A Power Functions in Fluvial Geomorphology   404
Appendix B Characterization of Fluvial Sediment   406
Appendix C Measuring Discharge and Velocities in Rivers   411
Appendix D Measurement of Sediment Transport in Rivers   414
Symbols   416
References   432
Index   507

Color plates are to be found between pp. 342 and 343.

Preface

Flow river flow, past the shady trees,
Go river go, go to the sea
Flow to the sea
"Ballad of Easy Rider," The Byrds, 1969

I have always been fascinated by flowing water. At a very young age, I remember wading into small streams near my home in eastern Pennsylvania to catch salamanders and assemble piles of rocks and sticks in vain efforts to stem the flow. Along with my sisters and cousins, I went fishing in my grandfather's jon boat on the local creek. Rather foolishly (in hindsight), we would row the boat right up to the face of the local abandoned mill dam and walk across its crest to feel the rush of cascading water across our feet. Only later, when I became aware that these dams are drowning machines, did I realize how dangerous this was.

When I went to college, I found that I could combine my love of science with my interest in flowing water to pursue the scientific study of rivers. As I became aware of how important rivers are in shaping landscapes on our planet, and just how unusual flowing water is on the surface of most planets, my fascination with the dynamics of rivers grew. I still remember clearly the excitement accompanying my personal discovery that I wanted to become a river scientist. Somewhat ironically, my PhD research examined dryland mountain streams in Arizona, where, despite several years of intensive field work, I never saw water flowing in the streams I studied.

Since receiving my PhD in 1986, I have been a professor at the University of Illinois, where I have engaged in basic and applied research on river dynamics. Teaching students has been remarkably rewarding, especially when I see these students become fascinated, as I am, by the dynamics of rivers. Working with graduate students who are passionately interested in rivers has been particularly gratifying. I have been fortunate to have had the opportunity to work with many talented graduate students. It has been fulfilling to see them succeed as professionals in academia, government agencies, and private industry.

Teaching is one of the major motivations behind the writing of this book. Over the past 20 years, the scope of knowledge related to river dynamics has advanced rapidly. In my own teaching at the advanced level, I have found it increasingly difficult to refer students to a concise text that includes the breadth of material I cover in my courses. These students also have diverse backgrounds, ranging from those with extensive coursework in college-level mathematics and science to those with minimal math and science coursework. This diversity necessitates an approach to teaching that allows those with advanced math and science backgrounds to feel challenged, while not overwhelming those without advanced training in math and science. I also feel strongly that the material should not be dumbed down to the extent that it does not include content commensurate with what a student at the advanced level should learn, given the current state of knowledge. This book represents an attempt to capture the balance between depth and breadth, at least as I see it. I have tried to at least touch on most major topics, but my personal predilections show through in that some topics are covered in greater detail than others. I have organized each chapter in the book using headings and subheadings in the form of questions. This approach emphasizes that science proceeds by asking questions and seeking answers to these questions. Not all of the headings represent scientific questions, but many of them do.

Many students are innately attracted to the relevance of river science to environmental issues, particularly human impacts on rivers and efforts to manage rivers in ways that enhance environmental quality. I have examined these types of issues in my own research and, in a world in which the footprint of humans is becoming increasingly large, understand why they are appealing to students. The last two chapters of the book explicitly address these issues, but given the complexity of rivers, I strongly feel that basic understanding of river dynamics, the focus of the majority of the book, is important for effective management. Attempts to manage rivers based on a rudimentary understanding of their complexity are unlikely to be successful. The title of the book, which includes the phrase "geomorphology to support management," is meant to emphasize that the book is not directed explicitly at management but is intended to provide information relevant to management. In this sense, I hope the book will be of value to river managers seeking to learn more about how rivers function. Not all aspects of river dynamics, particularly those that occur over long timescales, are relevant to management. Nevertheless, these long-term dynamics are important for understanding how rivers contribute to erosional and depositional processes that shape the surface of the Earth.

The other motivation behind writing this book is largely personal. Throughout my career, I have always thought that someday I would write a book on river dynamics. When actively engaged in research, accomplishing this task is difficult, given constant demands on one's time. My career as a professor now exceeds 30 years, and the time had come when it was now or never. As a lifelong learner, I have found the task of writing this book to be enormously fulfilling. Early in my career, I discovered that topics I thought I knew thoroughly, I did not know nearly as well as I thought I did once I had to teach those topics to students. Teaching improved my learning. The same discovery has occurred in producing this book. I learned much that was new about topics I thought I knew well. My appreciation of the talents of my fellow river scientists, as well as of the brilliance of their ideas, has deepened greatly through this experience. I have done my best to represent their efforts correctly, but take full responsibility for any omissions, errors, or misrepresentations. I thank those who took the time to read through drafts of chapters, including Mike Church, Tom Dunne, Brett Eaton, Karen Gran, Marwan Hassan, Janet Hooke, Kory Konsoer, and Jonathan Phillips. Your willingness to devote your valuable time on my behalf is greatly appreciated. Thanks also to Allison Goodwell for her help in producing an important figure and to Tanya Shukla for her expert camera work on some of the images in the book.

As a first-generation college student, I am blessed to have parents who emphasized the importance of education, who encouraged me to pursue my interests, whatever they might be (including fluvial geomorphology!), and who contributed to my education financially. Although my father is gone, I have stimulating conversations with my mother on an almost daily basis. At 90-plus years old, she continues to inspire me to maintain a proper perspective on life. I fell short as a teacher in getting my children interested in science and rivers; both pursued career paths more closely akin to that of my wife, a social worker. Jamie and Steven, you are the joy of my life, and I am immensely proud of you both. To my wife, Kathy, I cannot begin to express how much your love and support has meant to me over the years. Without it, this book would not have been possible. You are my rock.

1 Introduction

1.1 Why Are Rivers Important?

Rivers are integral to Earth's hydrological cycle as well as major agents of landscape change. One of the distinguishing characteristics of Earth is that much of its surface consists of water. Oceans cover 71% of Earth's surface, glacial ice about 10% of the total land surface, and lakes about 2.0% to 3.5% of the total nonglaciated land area (Verpoorter et al., 2014; Messager et al., 2016). The global surface area of rivers is only about 0.58% of Earth's nonglaciated land surface (Allen and Pavelsky, 2018). In addition to surface water, shallow groundwater supports the growth of terrestrial plants, which consist mainly of water. Through evapotranspiration, sublimation, advection, condensation, and precipitation, water is transferred into the atmosphere, moved laterally to new locations, and redelivered to Earth's surface. Water supplied to land surfaces moves back to the oceans through runoff, lateral flow of groundwater, and atmospheric processes. Within this hydrological cycle, rivers are the main natural features that convey water from the land surfaces to the oceans. The movement of water over and through landscapes induces solution and erosion of earth materials, the products of which are flushed into rivers. Thus, rivers convey not only water but also sediment and a host of biogeochemical constituents.

Rivers are important for a variety of reasons. From a human standpoint, rivers are a major source of water. Although they contain only 0.006% of the freshwater on Earth (Shiklomanov, 1993), rivers supply drinking water for many communities around the world. Water from rivers also is used in industrial operations and for irrigation of agricultural land. Rivers afford opportunities for recreational activities and aesthetic enjoyment. Through the construction of dams for the generation of hydropower, rivers contribute substantially to energy production. As of 2016, hydropower accounted for about 16% of total global electricity generation and supplied 71% of all renewable electricity (World Energy Council, 2019). Large navigable rivers serve as vital transportation corridors for the movement of material goods of economic value. Historically, humans have exploited rivers as convenient disposal receptacles for wastewater from domestic, industrial, commercial, and agricultural activities. Recognition of the adverse effects of wastewater disposal on water quality has led to the development of management strategies that seek to prevent or mitigate pollution of rivers and protect clean water. The intersection between rivers and humans also can have negative consequences for society. Flooding is the most expensive natural hazard in the United States, generating billions of dollars in losses each year. Erosion of riverbanks can threaten structures located along rivers and lead to loss of property. In some areas of the world, rivers have been established as political boundaries, and changes in the courses of rivers over time may result in disputes between governments. From an ecological perspective, rivers are important components of ecosystems, supporting diverse animal and plant communities. Alteration or pollution of rivers through human action can disrupt these communities, generating environmental concern about ecosystem degradation. Finally, from a geomorphological perspective, rivers are primary agents of landscape change on Earth, delivering the majority of sediment eroded from terrestrial landscapes to the ocean basins. Clearly, rivers are vital to human existence and integral components of Earth's environment.

1.1.1 Why an Emphasis on River Dynamics?

A key emphasis of this book is that rivers are inherently dynamic features. The flow of water in rivers is ever changing in response to variations in precipitation and runoff. Hydrological fluctuations occur over the short term with changes in weather and over the long term with changes in climate. Anthropogenic modifications of climate associated with greenhouse gas emissions have the potential to change the hydrology of rivers on a global scale (Nijssen et al., 2001; Arnell and Gosling, 2016). As flow varies, so does the amount of transported sediment, which responds to changes in the delivery of eroded material to the river from the surrounding landscape and to changes in the capacity of the river to move sediment derived from its bed and banks. Changes in flow and the amount of transported sediment occur frequently and are conspicuous to anyone who observes a river regularly. Less obvious in many cases, even to a regular observer, is change in the form of a river. Although many rivers appear, at least from a human perspective, to change little over time, this lack of evident change merely reflects the long timescales over which change occurs. A view of almost any river in which timespans of decades to centuries were compressed into a few minutes would reveal considerable change. Over such timescales, rivers move and shift across landscapes through processes of erosion and deposition. Over timescales of thousands to millions of years, rivers develop and are eradicated in conjunction with the evolution of entire landscapes.

1.1.2 How Is Fluvial Geomorphology Related to the Study of Rivers?

The study of rivers as natural components of the Earth system falls within the science of fluvial geomorphology – an interdisciplinary field that is embedded within parent disciplines of geography and geology, but that also draws upon and intersects with concepts from fluid mechanics, hydraulics, hydrology, and ecology. Growth in scholarship in fluvial geomorphology over the past 60 to 70 years has exploded as Earth scientists have fully recognized the important role that rivers play in Earth surface systems. Moreover, throughout human history, settlements ranging from small villages to large cities have developed near rivers to take advantage of access to water for human consumption, for agriculture, for industry, for power generation, and for recreation. Increasing societal concern about management of rivers, especially management aimed at protecting environmental values and sustaining natural functions, has greatly enhanced public awareness of the relevance of fluvial geomorphology for generating useful knowledge to guide environmental policy and decision-making. Thus, fluvial geomorphology – once a rather small, esoteric branch of science – has blossomed into a field of considerable scholarly and societal importance.

1.1.3 What Is the Purpose of This Book?

This book presents foundational principles in fluvial geomorphology and explains why these principles are important for understanding rivers as dynamic agents of change in the Earth

system. It also relates scientific understanding of rivers to important societal concerns, including the response of rivers to global change; impacts on rivers of land-use change, dams, channelization, and other human activities; and efforts to manage rivers to balance considerations of natural hazards versus environmental quality. It presents essential information on the current understanding of river dynamics and, at least to some extent, relates this understanding to management concerns. The goal is to provide a resource both for scientists interested in fluvial geomorphology and for practitioners dealing with river management issues.

The book is organized around questions related to topics encompassed by fluvial geomorphology. As it is a scientific field, research within fluvial geomorphology is driven by questions. Scientists, including fluvial geomorphologists, are curious about the world and seek knowledge by asking questions and then engaging in research to answer those questions. In some cases, answers to the questions are concrete, whereas in many cases, definite answers have yet to emerge. The search for definitive answers to research questions fuels the process of scientific inquiry.

1.2 What Is a River?

At first glance, the answer to the question "What is a river?" may seem obvious. At the broadest level, a river can be thought of as a body of flowing water that follows a distinct course. This general view, while not inaccurate, does not fully capture the complexity and, to some extent, uncertainty of what a river is or, for that matter, is not. A river is a stream of water in the sense that stream refers to flow. Stream also is a term typically used to describe a small river, but no absolute scientific criteria exist for distinguishing a river from a stream. The form and dynamics of rivers are generally similar over a large range of scales, indicating that the distinction between a stream and a river is mainly colloquial.

Rivers are commonly characterized as watercourses where flow occurs within a channel with well-defined banks. This characterization is generally consistent with the geomorphic perspective of a river as a channeled flow of water. Nevertheless, complications abound. Some rivers, those referred to as intermittent or ephemeral, flow only occasionally and are identified as such in the absence of flow based on the existence of a dry riverbed. Others have multiple channels or poorly defined channel banks. Thus, not all rivers flow all the time, well-defined banks may not always exist, and the number of channels can vary.

Rivers also do not occur in isolation but are components of river networks. The identification of the path of any particular river within a network can be somewhat subjective, based on human preferences. For example, if average amount of flow provided the basis for making such decisions, the Ohio River would be designated as the Mississippi River (or, alternatively, the lower Mississippi River would be renamed the Ohio River), because the amount of flow in the Ohio River typically exceeds that of the upper Mississippi River on an annual basis. The arrangement of rivers in networks can also lead to debate about the length of a particular river. For many years, the Rio Apurimac basin in Peru was considered the source of the Amazon River; however, in 2014, geographers claimed that the Rio Mantaro basin is the most headward source, adding 75–92 km to the maximum length of the river (Contos and Tripcevich, 2014).

Another complicating issue is whether features produced by contemporary river processes should be included as part of the river. Rivers often are associated with flow within a channel or set of channels; however, in alluvial rivers, or those carved within sediment deposited by the river itself, the development of depositional areas of land known as floodplains (discussed in detail in Chapter 14) is linked closely with processes that shape and maintain the channel. In such cases, an appropriate perspective is to view the channel and floodplain as integrated components of the river, rather than identifying the river with the channel only and viewing the floodplain as a separate feature. By contrast, bedrock rivers, or those carved into rock, may not develop floodplains. Also, rivers that are actively incising may become disconnected from their floodplains, so that this feature no longer is an integral component of the river system.

Such complications show that an answer to the question "What is a river?" is more nuanced than it may first appear. The contents of this book inform this question comprehensively, at least from a geomorphological perspective. Rivers are also vital components of ecosystems, an issue that will be touched upon briefly but not treated in detail.

1.3 What Is a River System?

The concept of a system is useful for examining rivers as geomorphological features. A system is a group of interacting components that constitute a unified whole. It is delineated by temporal or spatial boundaries and situated within an environmental setting. In open systems, energy and matter can cross system boundaries to influence internal interactions among system components, which determine system dynamics.

Drainage basins, also known as watersheds and catchments, provide natural geomorphological units for defining river systems within terrestrial landscapes (Leopold et al., 1964). A drainage basin delimits a portion of the Earth's surface that contributes runoff, sediment, and dissolved constituents to a river system. It consists of two basic components: hillslopes and the river network. The drainage divide, a topographic boundary separating runoff between adjacent watersheds,

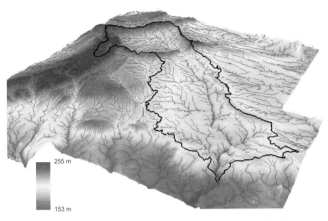

Figure 1.1. Three-dimensional view of a drainage basin for the upper part of the Sangamon River in Illinois, United States, showing drainage divide (black line), basin outlet (red dot), and stream network (blue lines). (A black and white version of this figure will appear in some formats. For the color version, please refer to the plate section.)

defines the boundary of a drainage basin (Figure 1.1). Runoff within the boundary of the drainage basin will move over or through hillslopes into the river network within the watershed. The downstream locus, or mouth, of the watershed provides a common outlet for all water and sediment exiting the watershed (Figure 1.1). A drainage basin can be defined upstream of any particular location along a river network, ranging from the downstream limit, where a large river flows into the ocean, to the most headward locations, where the smallest streams begin. Thus, drainage basins are arranged in a hierarchical, nested configuration.

Rivers develop within drainage basins and are influenced by inputs of precipitation to these basins, which produce runoff on and through hillslopes, supplying water to rivers. Runoff on hillslopes also erodes sediment that moves into the rivers. Thus, rivers are open systems that receive fluxes of material from hillslopes within drainage basins.

Three interrelated components – flow, sediment transport, and morphology – characterize a river system (Figure 1.2). Flow varies with changes in runoff and provides the mechanism for erosion, transportation, and deposition of sediment within the river system. Under certain circumstances, sediment transported by the river can affect hydraulic characteristics of the flow, leading to feedback between the flow and sediment transport. The movement of sediment shapes the morphology of the river, especially material mobilized along the boundaries of channelized flow – the bed and banks of the river. In many cases, the main morphological feature associated with the river system is a channel, but morphology includes any component of form that is part of the river system. Some morphological features, such as floodplains, are influenced by deposition of fine material delivered to the river from hillslopes. River morphology constrains hydraulic conditions, thereby influencing characteristics of the flow. According to this view, river dynamics involve inputs of water and sediment from drainage basins that drive interaction among flow, sediment transport, and morphology (Figure 1.2). This structure, which appears simple, in reality leads to complex dynamics, in part because the interdependence is typically highly nonlinear. For example, the existence of thresholds in river systems can result in an abrupt change in the state or morphology of a system when the threshold is attained or surpassed.

1.4 What Is Fluvial Geomorphology?

The field of fluvial geomorphology derives its name from *fluvius* – the Latin word for river, stream, or running water; *gé* (Γή) – the Greek word for land or earth; and *morphé* (μορφή) – the Greek word for form or shape. This branch of science not only studies how rivers shape Earth's surface; it also examines the dynamics of rivers. It is a subfield of geomorphology, a relatively young science rooted in the development of the earth sciences, particularly geology and physical geography (Tinkler, 1985).

1.4.1 What Is the History of Fluvial Geomorphology?

Recognition of the importance of fluvial processes in shaping Earth's surface emerged in the eighteenth century and was

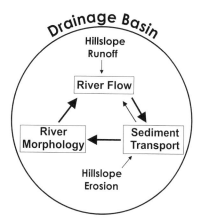

Figure 1.2. Basic structure of a river system. Runoff and erosion on hillslopes within drainage basins deliver water and sediment to rivers, the dynamics of which are characterized by interaction among flow, sediment transport, and morphology.

closely associated with the advent of uniformitarianism. This principle holds that natural processes operating over long time-spans govern the dynamics of the Earth system. The foundation of uniformitarian thought was established by James Hutton (1795), who has been referred to as the founder of modern geology (Bailey, 1967), the first great fluvialist (Chorley and Beckinsdale, 1964), and even the founder of geomorphology (Orme, 2013). Hutton's ideas emphasized the importance of fluvial processes in ceaselessly reshaping Earth's surface. Although Hutton died shortly after publishing his work, his ideas were championed by John Playfair (1802), who expanded on Hutton's scheme and expressed its central tenets with a degree of inspirational clarity that greatly surpassed the convoluted prose of Hutton. What has become known as Playfair's Law is a shining example of his eloquent style and a beautiful articulation of the way in which rivers, arranged in networks, carve the landscape into a system of valleys (Box 1.1). At the time it was written, the statement countered the prevailing notion that Earth's valleys were carved by one or more divinely instigated catastrophic deluges (Orme, 2013).

The term "geomorphology" was introduced in the late 1800s, probably by geologist W.J. McGee (Tinkler, 1985), at a time when the nascent field of landform studies was growing rapidly through insights gained from explorations of the dramatic landscapes of the American West (Chorley and Beckinsdale, 1964; Sack, 2013). Fluvial processes stood at the center of ideas emerging from these explorations. John Wesley Powell (1875, p. 208), who undertook the first organized scientific expedition of the Colorado River through the Grand Canyon, noted in his report: "All the mountain forms of this region are due to erosion; all the cañons, channels of living rivers and intermittent streams, were carved by the running waters, and they represent an amount of corrasion difficult to comprehend." Powell introduced the seminal concept of base level, the idea that erosion by rivers has a vertical limit. In the case of a tributary, it is the level of the main river it joins, and in the case of a main river that flows into the ocean, it is sea level.

Grove Karl Gilbert, who worked under the direction of John Wesley Powell, set an example for much contemporary research in geomorphology, including fluvial studies (Pyne, 1980). Through his training in geology and mechanics, Gilbert approached problems of landform development by

---

BOX 1.1 | PLAYFAIR'S (1802) LAW

Every river appears to consist of a main trunk, fed from a variety of branches, each running in a valley proportioned to its size, and all of them together forming a system of vallies, communicating with one another, and having such a nice adjustment of their declivities, that none of them joins the principal valley, either on too high or too low a level; a circumstance which would be infinitely improbable, if each of these vallies were not the work of the stream that flows in it.

integrating field observations with physical theory. His work on the Henry Mountains in Utah invoked systems concepts, equilibrium thinking, and force-balance relations to explain how running water acts to shape landscapes through erosion and deposition (Gilbert, 1877). He introduced the concept of grade, a state of dynamic adjustment whereby a river attains a capacity of transport equivalent to the amount of sediment supplied to it. Later in his career, he conducted seminal experimental research on the transportation of sediment by running water (Gilbert, 1914) and tried to use knowledge he gained from these experiments to understand how the introduction of vast amounts of sediment by humans from hydraulic mining for gold affected rivers of the Sierra Nevada (Gilbert, 1917) (see Chapter 15).

The development of geomorphology accelerated between the late 1800s and the early 1900s when William Morris Davis, a geologist at Harvard who also championed the development of geography as a formal academic discipline in the United States, proposed his theory of the geographical cycle, or cycle of erosion (Davis, 1889, 1899). This theory held that landscapes uplifted above sea level and eroded by fluvial action evolve through a systematic sequence of distinctive stages (Figure 1.3). An uplifted landscape progresses through stages of youth, maturity, and old age as rivers incise into it and hillslopes produced by river incision are worn down by surface runoff. Eventually, the entire landscape, if not uplifted beforehand, is beveled to a flat surface standing just above sea level – a feature Davis referred to as a peneplain. The cyclic aspect of the theory comes into play when uplift eventually recurs, reinitiating the sequential stages of development. The scope of the theory addressed the evolution of landscapes of regional extent over timespans involving millions or tens of millions of years. The impact of Davis and his theory was enormous and dominated geomorphological inquiry from the late 1800s to the middle of the twentieth century (Chorley et al., 1973). This method of inquiry mainly involved trying to classify landscapes into stages of the cycle through visual observations or map-based analysis of landscape characteristics combined with descriptive geological investigations that focused on defining the relative timing of different stages of landscape evolution. Although cyclic models were developed for landscapes other than those dominated by fluvial erosion in humid-temperate environments (e.g. Hobbs, 1921), and alternative perspectives on landscape evolution emerged in response to the Davisian view (e.g. King, 1953; Penck, 1972), all these models also were qualitative. Thus, geomorphology textbooks published between 1900 and 1950 were largely descriptive treatises on how landforms, including rivers, develop and change (Rhoads, 2013).

Table 1.1. Elements of a dynamic basis of geomorphology.

- Study of geomorphological processes and landforms as various kinds of responses to gravitational and molecular stresses acting on earth materials
- Quantitative determination of landform characteristics and causative factors
- Formulation of empirical equations by mathematical statistics
- Building concepts of open dynamic systems and steady states for all geomorphological processes
- Deduction of general mathematical models to serve as natural quantitative laws

(from Strahler, 1952a)

Seminal work by Robert E. Horton (1945), a hydraulic engineer, on the erosional development of drainage basins and stream networks based on physical reasoning planted a seed of change amongst a new generation of geomorphologists who were increasingly dissatisfied with the qualitative Davisian approach. A subsequent landmark paper by Arthur Strahler (1952a) called for a dynamic basis of geomorphology focusing on a quantitative approach to the study of landforms and the processes that shape landforms (Table 1.1). This new perspective, which supplanted the Davisian view and has persisted since the 1950s, emphasizes the importance of understanding processes and process–form interactions in producing knowledge of geomorphic systems (Rhoads, 2013). It is consistent with Gilbert's approach to geomorphological inquiry, which was long overshadowed by Davis's ideas. The conception of geomorphic processes, including those related to rivers, as the manifestation of mechanistic action has led to the infusion of principles from classical (Newtonian) mechanics into geomorphology. Foundational work in fluid dynamics and hydraulics by Daniel Bernoulli (1700–1782), Antoine de Chezy (1718–1792), Robert Manning (1816–1897), William Froude (1810–1879), and Osbourne Reynolds (1842–1912) has become relevant to process-based studies of river dynamics. Thus, contemporary fluvial geomorphology has twin historical roots: one planted in the earth sciences (geology and physical geography) and the other in the development and application of classical mechanics to topics related to rivers (hydraulics and fluid dynamics) (Orme, 2013).

A major change in fluvial geomorphology that accompanied the shift toward process-based inquiry has been a focus on issues with time and space scales that coincide with societal concerns related to rivers. Whereas the Davisian perspective tended to examine how rivers shaped landscapes over regional scales and millions of years, the process approach embraces studies examining how individual rivers or sections of rivers

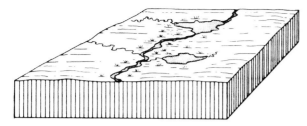

A In the initial stage, relief is slight, drainage poor.

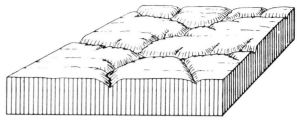

B In early youth, stream valleys are narrow, uplands broad and flat.

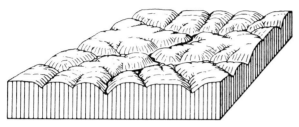

C In late youth, valley slopes predominate but some interstream uplands remain.

D In maturity, the region consists of valley slopes and narrow divides.

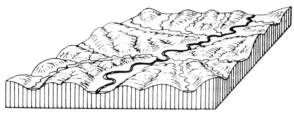

E In late maturity, relief is subdued, valley floors broad.

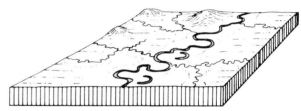

F In old age, a peneplain with monadnocks is formed.

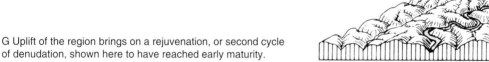

G Uplift of the region brings on a rejuvenation, or second cycle of denudation, shown here to have reached early maturity.

Figure 1.3. Stages in the cycle of erosion (from Strahler, 1965).

are influenced by individual formative events or changes in the frequency of these events. It also explores the role of humans as geomorphic agents who are capable of changing rivers directly and triggering responses to these changes. The relevance of fluvial geomorphology to river management and to attempts to protect, preserve, or improve the environmental quality of river systems has therefore increased, resulting in rapid growth in research in this field of science (Figure 1.4) and in its visibility within the public domain.

1.4.2 What Are the Different Styles of Scientific Inquiry in Fluvial Geomorphology?

Scientific inquiry in fluvial geomorphology is diverse. Any scheme that attempts to capture the full range of diversity will almost certainly be incomplete. With that in mind, at least seven different styles of inquiry can be recognized (Table 1.2). These styles are not necessarily mutually exclusive and often intersect to some extent within individual investigations.

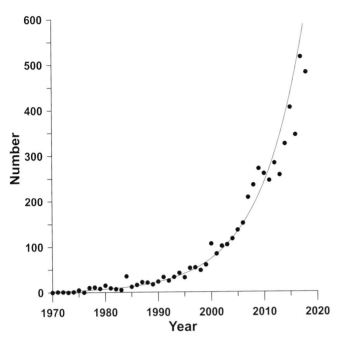

Figure 1.4. Increase in number of publications containing topic words *fluvial* and *geomorphology* between 1970 and 2018 based on a Web of Science© search.

Many theoretical principles in fluvial geomorphology derive from the basic sciences such as physics, chemistry, and biology. Examples from physics include conservation of mass and momentum as well as force–resistance relations. Gravitational forces and fluid forces are the principal types of forces acting in river systems, and expressions for these forces typically are derived from principles of mechanics, at least at the highest level of formalism. Developing foundational principles within fluvial geomorphology that have the same level of certainty as principles from foundational sciences has proven challenging. For example, sediment transport is an essential fluvial process, and considerable effort has been devoted to the search for universal geomorphic transport laws (Dietrich et al., 2003; Hicks and Gomez, 2016). Despite this effort, the development of such universal relations remains elusive.

Other principles, sometimes referred to as regulative principles, are qualitative. Regulative principles constrain possibilities related to the structure or dynamics of a fluvial system and therefore provide a basis for the development of explanatory frameworks (Rhoads and Thorn, 1993). Examples of such principles include optimality conditions, such as a fluvial system tending toward a steady state or toward a state that minimizes or maximizes a system property, and nonlinear dynamical behavior, such as evolution of the system toward an attractor state or along a distinct trajectory. Whether qualitative or quantitative, sets of principles provide guidance for the development of models, either mathematical or conceptual, to represent the structure and dynamics of fluvial systems.

Table 1.2. Styles of inquiry in fluvial geomorphology.

Theoretical	Development of conceptual, logical, or mathematical principles that constitute theoretical constructs
Experimental	Production of information through constrained measurements under controlled conditions
Hypothetical (modeling)	Development and evaluation of simple to sophisticated models developed through analogical reasoning based on theoretical knowledge Exploration of model implications through simulations and sensitivity testing
Taxonomic	Ordering of variety through comparisons of similarities and differences among individual objects Discovery of natural affinities to establish classes and categories
Statistical/ probabilistic	Assessment of expectations versus outcomes based on probabilities Documenting/ discovering statistical regularities and associations
Historical	Analysis of development or change over time. Postulation of causes that led to this development
Case study	Analysis of a single or small number of cases to establish detailed case-specific knowledge that, through informal inferences, can inform theoretical knowledge

(after Rhoads, 1999)

Contemporary perspectives within geomorphology embrace the model-theoretic view (MTV) of scientific theory (Rhoads and Thorn, 2011). According to MTV, models are the primary constituents of theory structure. Models connect theoretical and regulative principles to testable hypotheses about the real world (theoretical models) and also provide data-based representations of these models (data models). Empirical testing involves comparing hypotheses derived from theoretical models with evidence embedded in data models (Figure 1.5). Theoretical models are representational devices that facilitate intellectual access to real-world phenomena; in the case of fluvial geomorphology, representation focuses on river systems. The manner of representation includes descriptions, mathematical equations, probabilistic functions, computer algorithms, diagrams, pictorial displays, images, and physical artifacts (i.e., "hardware") (Odoni and Lane, 2011; Grant et al., 2013; van de Wiel et al., 2016). However, no single model fully captures the content of

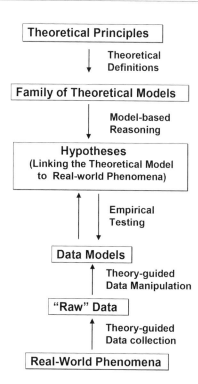

Figure 1.5. The model-theoretic view of scientific theory (adapted from Rhoads and Thorn, 2011).

a theory. Because many different discrete representations of theoretical principles are possible, these principles provide support for an interconnected set, or family, of models. Thus, according to MTV, a theory comprises a set of theoretical principles and a family of models that embodies these principles (Giere, 1988).

Data models also are representational, but these models represent real-world phenomena through information, or data, collected about the phenomena. The collection and processing of data are guided by theory and conducted with a theoretical objective in mind – usually a hypothesis or set of hypotheses derived from a theoretical model. The result of data processing is a data model. The processed information embodied in the data model provides the basis for testing the hypotheses derived from the theoretical model. In cases where theoretical understanding of a phenomenon is uncertain, data models may be developed relatively autonomously from theoretical models. However, scientists continuously strive to link data models that yield intriguing patterns or outcomes to explanatory theoretical principles through the development of theoretical models. Thus, models and theory are intertwined at all levels of scientific practice.

The MTV also provides a basis for embracing different styles of inquiry (Table 1.2). Theoretical, hypothetical, and taxonomic styles provide the basis for the formulation of theoretical models, whereas experimental, historical, and case study approaches relate closely to the production of data models. Statistical/probabilistic styles of inquiry may contribute to either type of model, depending on the specific way in which the model is applied.

Over the past several decades, the development of analytical and numerical models of river systems has increased dramatically (Coulthard and Van de Wiel, 2013a; Nelson et al, 2016; Pizzuto, 2016). Such models provide the basis for the development of sets of hypotheses about the dynamics of fluvial systems. Predictions about some aspect of a river system produced by a model represent hypotheses. In using the model to generate predictions, it is assumed that the underlying governing equations of the model represent valid representations of a river system. Hypothesis testing involves comparing the model predictions with results of data models derived from a systematic measurement program. Validated models can then be used to develop additional predictions about river systems that extend beyond the domain of the information and setting on which the data model was based. This use of validated mathematical models to explore potential real-world implications can be thought of as a form of experimentation (Kirkby, 1996; Church, 2011); however, the outcomes of numerical experiments, to be connected to the real world, must be validated through empirical testing using appropriate data models. Thus, numerical experimentation is a method of hypothesis generation rather than a method of generating observations for model testing, as is the case with empirical experimentation. In this sense, predictions or forecasts generated by mathematical models represent sophisticated thought experiments (Kirkby, 1996).

Taxonomy, or classification, as in any scientific endeavor, has played and continues to play an important role in fluvial geomorphology (Buffington and Montgomery, 2013; Kondolf et al., 2016). The stages in Davis's cycle of erosion (Figure 1.3) represent categories of landscape development. Today, fluvial geomorphologists continue to try to determine general characteristics of river systems to provide the basis for identifying distinct kinds of rivers. This effort to classify constitutes a basic form of theorizing, in the sense that generalization is accomplished by hypothesizing that different kinds of rivers exist and that each kind of river shares common properties compared with other kinds. Classes, once established, also provide the basis for further theorizing about causal mechanisms that lead to the development of different kinds of rivers. Over time, however, classes may change as theoretical understanding evolves (Church, 2011).

Empirical experimental research in geomorphology, which plays an important role in the development of data models, is not restricted solely to the use of scaled physical (hardware) models (e.g. Peakall et al., 1996) but can be viewed more

Table 1.3. Types of experimental research in geomorphology.

Scaled Physical Experiments	Use of a scaled physical model of a fluvial feature, usually in a laboratory setting, where an attempt is made to ensure similarity of geometrical, kinematic, and dynamical criteria between the model and a field prototype
Classic Field Experiments	Intentional direct interference with natural conditions of a fluvial landscape to obtain controlled results on processes that shape the landscape or on the conditions that govern landscape change
Unscaled (Analog) Experiments	Use of unscaled physical models or small-scale natural features to explore the possible behavior of full-scale field prototypes
Paired Field Experiments	Comparison of two or more similar fluvial landscapes in which at least one of the features is deliberately manipulated and at least one is not manipulated
Statistical Field Experiments	Use of multiple cases or plots to evaluate the effects of different treatments or case-specific conditions
Inadvertent Field Experiments	Manipulation of a fluvial feature for a purpose other than experimentation that provides an extraordinary opportunity to document the response of the feature to a specific type of change

(adapted from Church, 2011)

generously as encompassing a variety of field investigations (Table 1.3). Unscaled physical models have been widely used in fluvial geomorphology to explore the dynamics of river systems (Schumm et al., 1987; Metivier et al., 2016). Insights provided by such models, despite the lack of strict scaling with field prototypes, have been substantial (Paola et al., 2009). Inadvertent experimental opportunities to study river responses to human-induced change are not uncommon, given that humans have modified river channels in a variety of ways for purposes other than scientific experimentation. Intentional field experiments for a scientific purpose are less common, but recent efforts have sought to attain at least partial experimental control at field scales (Wohl, 2013a; Sukhodolov, 2015).

Process-based field investigations in fluvial geomorphology commonly implement rigorous measurement protocols designed explicitly to test hypotheses derived from theoretical models at a single site or a small number of sites. Measurement protocols for such investigations are based on a theory-guided experimental design, but in the field it is typically not possible to control boundary conditions, such as the inputs of flow and sediment to the reach under study or the composition and morphology of the channel bed and banks within this reach. It is possible, however, to document the boundary conditions in detail. This information on boundary conditions provides the basis for calibrating theoretical models to predict river dynamics in the specific field situation of interest. Testing involves comparing model predictions with outcomes of data models generated by field measurements. These process-based case studies (Table 1.2) are pseudo-experimental in the sense that complete experimental control is not achieved, but data are collected using an experimental design aimed at testing theoretical models within specific field contexts (Richards, 1996).

Statistical analysis became prominent in fluvial geomorphology after the transition to process-based research in the 1950s. Many field studies in fluvial geomorphology involve collection of data sets with large sample sizes and analysis of the collected data using statistical methods to try to isolate the independent covariance between variables of interest (Piegay and Vaudor, 2016). Some of this work is highly exploratory with only a weak connection to explanatory theoretical principles. Other work, such as statistical modeling, intersects with the hypothetical style of inquiry (Rhoads, 1992); statistical models are often formulated a priori based on theoretical reasoning, and then hypotheses embedded in the models are evaluated through significance testing by statistically fitting the model to empirical data. Many bivariate and multivariate relations in fluvial geomorphology are expressed in the form of power functions, given that variables related to rivers typically have log-normal probability distributions (Appendix A). Familiarity with power functions and the method by which such functions are derived statistically is therefore important for understanding statistical associations related to river systems.

Historical studies in fluvial geomorphology rely on two types of information: historical records produced intentionally by direct measurement or monitoring programs and geohistorical data derived from the artifacts of human or biophysical activities. Systematic attempts to collect scientific data of relevance to understanding rivers are relatively recent. Relevant sources that can provide information about past conditions of rivers include newspaper articles, old maps, land survey records, stream gaging data, sediment discharge information, ground-based photographs, bridge surveys, and travel accounts (Trimble and Cooke, 1991; Trimble, 2008, 2013; Grabowski and Gurnell, 2016). Aerial photography, available for areas in the United States since the 1930s, and satellite remote-sensing imagery, available for most areas of the world since the 1960s, now afford remarkable opportunities to examine dynamic change in the characteristics of rivers over time (Gilvear and Bryant, 2016), particularly given that most of these data are in the public domain. Geohistorical data typically extend beyond the limited temporal domain of human-generated information on river systems. Such data include sedimentological information, methods of absolute and relative dating, biogeochemical analyses, and archeological artifacts (Brown et al., 2016; Jacobson et al., 2016).

1.4.3 What Are the Basic Types of Scientific Reasoning in Fluvial Geomorphology?

Two distinct types of scientific reasoning are common in fluvial geomorphology: deductive reasoning and abductive reasoning. Both these types of reasoning are informed by theory, but in different ways (Rhoads and Thorn, 1993). Deductive reasoning is common in process-based approaches to inquiry, which typically involve testing of theoretical hypotheses about general process or process–form relations. Theoretical hypotheses are derived deductively from theoretical models. Such hypotheses typically have the form if A, then B, where commonly A is a cause and B is an effect. Testing of deductive theoretical hypotheses in process-based studies involves comparing the claims of these hypotheses with results embodied in a data model. Ideally, the test should involve generation of A and confirmation that A causes B through data-based evidence embodied in a data model. Of course, if B does not occur, or occurs because of a cause other than A, the outcome of the test would not support the hypothesis. In some instances (e.g., nonexperimental conditions), it may not be possible to generate A directly; in such cases, process-based studies typically rely on abductive reasoning.

Many studies based on historical sources of information, but particularly those that rely on geohistorical data, are fundamentally reconstructive in nature in that the goal is to determine the event or series of events (A) that caused the development of a contemporary fluvial feature. In such cases, data on the cause (A) are not directly or even indirectly available (e.g., a large flood that occurred in the distant past cannot be "remeasured" if it was not measured at the time it occurred). As a result, abductive, rather than deductive, reasoning is common in geohistorical investigations. This reasoning involves first observing a feature B (an effect) for which one seeks an explanatory cause (A). By consulting background knowledge, a potential explanation of the type "if A, then B" is identified as providing a possible explanation for B. In other words, A becomes a hypothetical cause of B. Abductive reasoning based on geohistorical information has an inherently higher level of uncertainty than deductive reasoning based on direct documentation of cause–effect relations because the documentation of a particular effect (B) does not guarantee the occurrence of a specific cause (A) (Rhoads and Thorn, 1993). Another cause (C) may also account for the particular effect of interest (B). Typically, further work is needed to evaluate the hypothesis. For example, if A causes D in addition to B, and evidence for D can be confirmed in conjunction with B, the hypothesis gains further support.

A complication of simple cause–effect reasoning in fluvial geomorphology is that rivers are situated within complex natural environments, where a variety of factors can affect these systems. In particular, contingency is often a major consideration in determining how rivers are structured and how they change through time. This contingency includes attributes of the particular contemporary environmental setting in which the river is located as well as the particular historical sequence of circumstances that have shaped the characteristics of the river. In this sense, every river is unique. Although this inherent contingency confounds determinations of the extent to which general physical principles govern river dynamics, it also highlights how manifestations of processes governed by general principles develop in particular instances. From a scientific perspective, the variety of different methods of inquiry (Table 1.2) provide tools for trying to unravel the interrelated roles of generality and specificity in river dynamics. From a practical standpoint of river management, consideration of the influence of contingent factors on river dynamics is important, particularly when the focus of management is on a particular river.

1.5 How Do the Dynamics of River Systems Vary over Time and Space?

The dynamics of river systems, which are related to the hydrological and hydraulic characteristics of river flow, vary over a wide range of temporal and spatial scales. River flows are turbulent (see Chapter 4); thus, the smallest relevant scales of dynamic variability of these flows are defined by the smallest length (ℓ_{emin}) and time (t_{tmin}) scales of turbulent flow, known as Kolmogorov microscales:

$$\ell_{emin} = \left(\frac{v^3}{\varepsilon_d}\right)^{0.25} \tag{1.1}$$

$$t_{tmin} = \left(\frac{v}{\varepsilon_d}\right)^{0.5} \tag{1.2}$$

where v is the kinematic viscosity of water and ε_d is the turbulent dissipation rate (Tennekes and Lumley, 1972). Assuming a value of v for water at 20 °C (1.00×10^{-6} m^2 s^{-1}) and a value of ε_d for turbulent river flow of 0.0005 m^2 s^{-3} (Sukhodolov et al., 1998) yields Kolmogorov microscales of $\ell_{emin} = 0.0002$ m and $t_{tmin} = 0.04$ s. The time-space domain

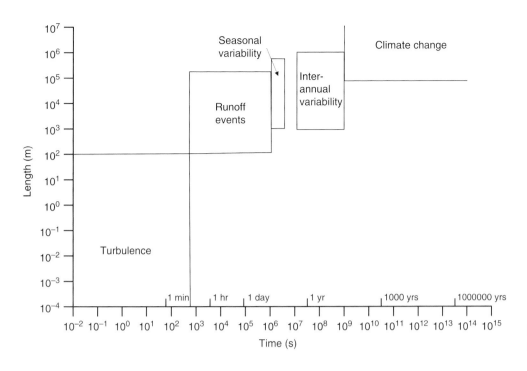

Figure 1.6. Time-space domains of flow dynamics in rivers.

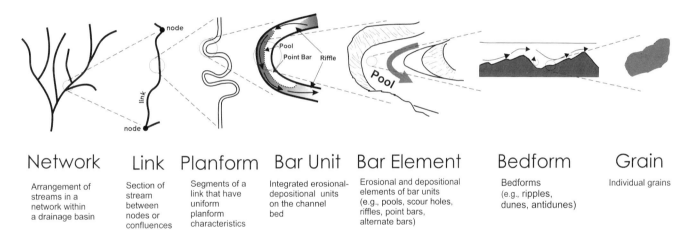

Network
Arrangement of streams in a network within a drainage basin

Link
Section of stream between nodes or confluences

Planform
Segments of a link that have uniform planform characteristics

Bar Unit
Integrated erosional-depositional units on the channel bed

Bar Element
Erosional and depositional elements of bar units (e.g., pools, scour holes, riffles, point bars, alternate bars)

Bedform
Bedforms (e.g., ripples, dunes, antidunes)

Grain
Individual grains

Figure 1.7. Hierarchical morphological structure of alluvial river systems.

of turbulence extends up to coherent turbulent structures that in the world's largest rivers may be a hundred meters in diameter and evolve over timescales of many minutes (Figure 1.6).

Variations in runoff associated with changing weather conditions are a major source of flow variability in rivers. The duration of individual hydrological events can vary from a few minutes in the smallest streams to several weeks in large rivers. Highly localized events may affect only a few hundreds of meters of small streams, whereas large storms can produce variations in flow that extend over hundreds of kilometers of river length. Seasonal variability is another factor that can produce spatial and temporal variation in river flow. The

timescale of this variability, which may affect broad geographic areas, is typically several months. Weather conditions also vary from year to year or over periods of years, resulting in inter-annual variability in river flow. Over timescales of decades to millennia, changes in climate can contribute to flow variability. The spatial domain of climate change extends to the global scale. Changes in river flow at this scale are of growing concern in relation to human-induced climate change.

The morphological structure of alluvial river systems within drainage basins can be viewed hierarchically (Figures 1.7 and 1.8). At the largest scales, rivers form branching networks that extend throughout drainage basins. Discrete segments of rivers between nodes where streams join, also known as confluences,

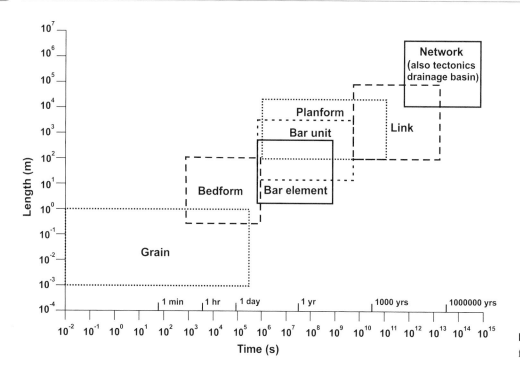

Figure 1.8. Time-space domains of fluvial morphodynamics.

constitute links. Within links, the pattern or planform of the river – what it looks like when viewed from above – becomes evident. Within rivers with different types of planforms, distinctive patterns of erosion and deposition on the channel bed produce bar units. Parts of bar units are associated with discrete morphological features, or bar elements, that differ in substrate composition and elevation of the channel bed. At still smaller scales, different types of bedforms develop within the river system. Individual grains represent the smallest morphological units within river systems, but the absolute size of individual grains can vary over several orders of magnitude, ranging from large boulders a meter or more in diameter to microscopic clay particles.

The morphodynamics of river systems vary depending on the morphological features of interest. Three relevant timescales can be identified: geologic, modern, and event (Figure 1.9) (Schumm and Lichty, 1965). Large drainage basins and river networks evolve over thousands to millions or even tens of millions of years. The evolution of watershed size, shape, and relief; of vegetation throughout watersheds; and of the characteristics of the stream networks within these watersheds generally occurs over geologic timescales. These aspects of river system morphology are dependent mainly on the climatic and geological conditions that exist over these long time intervals (Figure 1.9). Geological conditions include the type of rock into which the watershed is carved, structural characteristics of this rock, such as folding and faulting, and the spatial extent and rate of tectonic activity. Over geologic timescales, drainage basins and river networks

come into existence, coevolve, and are eradicated as the Earth's terrestrial surfaces are affected by global changes in climate and tectonism. Climate influences runoff and erosion, which carve drainage basins and stream networks into the geological framework of the landscape. It also plays a major role in determining the type of vegetation on the landscape. The properties of the drainage basin, river network, and vegetation, along with climate and geology, in turn shape the form of bedrock channels, which typically evolve over geologic timescales. Moreover, these factors determine the past (paleo-) hydrology, hydraulics, and morphology of rivers, evidence of which is sometimes preserved in the sedimentary record.

Although the properties of drainage basins, stream networks, and river longitudinal profiles change continuously over time through the interplay of erosional, depositional, and tectonic activity, amounts of change usually are negligible over timescales of decades to centuries. These properties, along with characteristics of vegetation, are, in many instances, relatively constant over modern timescales. The hydrological and sediment regimes of a river system within a drainage basin consist of flows and sediment fluxes of different magnitudes and frequencies occurring through time and over space in conjunction with characteristics of climate, vegetation, and drainage-basin morphology. Depending on the properties of these regimes and the extent to which the regimes are stationary, characteristic forms can develop in alluvial rivers. These aspects of river morphology exhibit constancy over time in the sense that they vary not at

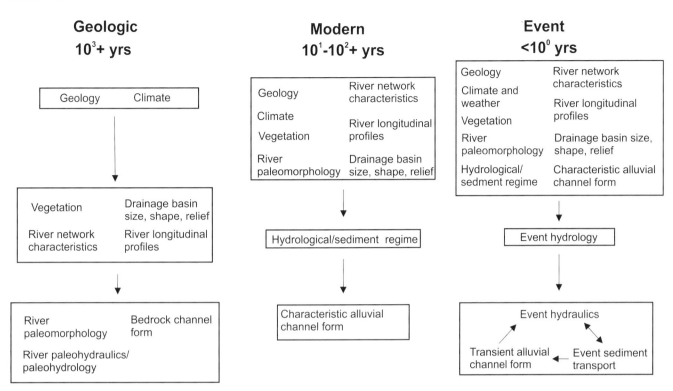

Figure 1.9. Hierarchical structure of river morphodynamics and controlling factors over different timescales. Absolute times in years associated with geologic, modern, and event timescales are approximate. Arrows indicate direction of causality (adapted from Schumm and Lichty, 1965).

all or fluctuate only a small amount about a constant average state over modern timescales, even though the hydrologic and sediment regimes encompass considerable variability in the magnitudes of flows and sediment fluxes produced by discrete hydrological events. The concept of characteristic forms often is equated with the regulative principle of a steady or equilibrium state; however, considerable confusion surrounds the use of equilibrium terminology in geomorphology (Thorn and Welford, 1994). Despite this confusion, the notion of equilibrium states in river systems often provides the basis for contemporary environmental management of rivers.

Over timescales of individual hydrological events or series of events, i.e., event timescales, many aspects of channel form exhibit transient dynamics (Figure 1.9). For many fluvial systems, event dynamics occur within the context of characteristic forms, which constrain these dynamics (Figure 1.9). Transiency in rivers that develop characteristic forms is limited because unchanging aspects of channel form regulate interaction among hydraulic conditions, sediment transport, and channel form. Nevertheless, in many rivers, even those that remain fairly constant in form, channel position can change through event-driven avulsion or migration. Bar forms and bed forms can be rearranged as flow varies within an event or between events. Sorting of sediment on the surface of bar elements may be altered as the flow rises and falls.

Not all rivers develop characteristic forms; some exhibit marked transient morphodynamics, even over modern timescales. In rivers that are highly sensitive to change, virtually all aspects of river morphology can be rearranged by discrete events. The form of such rivers does not vary closely about a constant average morphological state. For these rivers, event dynamics may occur over timescales much greater than 10^0 years.

The hierarchical structure of morphodynamics over space and time provides two important lessons for understanding river systems. First, processes that occur elsewhere along a river, within the network of which the river is part, and within the watershed within which the river is situated influence the form and dynamics of a river at any particular location. This lesson emphasizes the importance of spatial connectivity within river systems (Czuba and Foufoula-Georgiou, 2015), even though the understanding of this connectivity is far from complete (Fryirs, 2013). Second, not all aspects of the river system change dynamically at the same rates or are influenced by the same temporal scales of flow variability. Thus, rivers have morphological "memories" related to historical change. A river is a "physical system with a history" (Schumm, 1977, p. 10). Although it is clear that river dynamics encompass an immense range of temporal scales, understanding of interconnections among processes and forms across this range of scales is far from complete.

1.6 What Is the Role of Humans in River Dynamics?

Missing from the time-space conceptual scheme (Figure 1.9) is the role of human agency. Increasingly, humans have become agents of change in river systems. An abundance of geomorphological research over the past several decades shows that humans are substantially affecting river systems at the event and modern timescales (see Chapter 15). In some cases, humans also appear to be producing long-lasting effects that may persist over geologic time. Through watershed-scale modifications of vegetation cover and climate, hydrological and sediment regimes are altered, resulting in changes in channel form. These same modifications can affect the characteristics of individual hydrological events and the responses of rivers to these events. Moreover, humans have directly altered rivers by reconfiguring the form of channels, by reshaping floodplains, and by constructing barriers, such as dams, along rivers. These activities have been pursued through attempts to manage rivers to achieve specific societal goals. Over the past several decades, management strategies have arisen that focus explicitly on environmental goals. Although such strategies seek to enhance environmental quality, in many cases they still involve active manipulation of rivers. Proper understanding of river dynamics is essential for effective river management. Past management efforts that have failed to fully consider how rivers respond to human intervention have often led to unanticipated and undesired consequences. Only by understanding how rivers respond to change can management attain the intended goals while avoiding negative consequences.

2 The Dynamics of Drainage Basins and Stream Networks

2.1 Why Is a Focus on River Systems over Geologic Timespans Important?

Over geologic timespans, landscapes sculpted by fluvial processes exhibit considerable change. Drainage basins and river networks develop, evolve, and even cease to exist as the interplay between fluvial dissection of the landscape, changes in climate, and endogenic geological processes transform Earth's surface. Over these timespans, land masses move around on Earth's surface through the mechanism of plate tectonics. Orogenic activity at plate margins raises mountain ranges high above sea level, deforms Earth's crust through folding and faulting, and generates volcanism. By contrast, fluvial erosion opposes uplift of Earth's surface produced by endogenic processes. In the absence of endogenic activity, erosion would eventually wear Earth's surface down to a nearly level plain standing just above sea level (Davis, 1899). The lack of these types of beveled erosional plains on Earth attests to the important role that endogenic processes play in landscape evolution (Phillips, 2002). Over the past 40 years, growing recognition of the significance of these processes has given rise to tectonic geomorphology, which seeks to link landscape form, especially in mountainous terrain, to tectonic activity (Anderson and Anderson, 2010; Owen, 2013). For portions of Earth's surface that have been affected less by tectonism, for example, landscapes associated with stable cratons, the effects of fluvial erosion over geologic timespans have varied primarily in relation to changes in climate. Such changes can produce distinct topographic signatures of past climatic regimes or reset the clock on fluvial erosion entirely, as has been the case for portions of North America and Europe influenced by extensive glaciation during the Pleistocene Epoch (2.58 million to 11,700 years before present (BP)). In any case, few, if any, landforms on the surface of the Earth are believed to be more than 70 million years old (Vasconselos et al., 2019), and many are much younger (Selby, 1985, p. 4). Although 70 million years may seem long, it constitutes only about 1.5% of the total age of Earth. Thus, relative to the length of time Earth has existed, terrestrial surfaces are highly dynamic and ever changing. Fluvial action is a key agent contributing to landscape dynamics over geological timespans.

2.2 How Is the Formation of River Channels Related to Hillslope Processes?

The question of how rivers form is a fundamental one in fluvial geomorphology. A river needs to exist before any aspect of river dynamics can be considered. Attempts to address this question have focused mainly on the development of channels based on the perspective that streams and rivers are fundamentally a type of channelized flow. This type of flow differs from unchannelized flow that occurs on hillslopes – the other major component of drainage basins. The key issue in the formative dynamics of rivers is to determine how unchannelized flow transforms into channelized flow. The starting point for analysis of this problem is typically a sloping surface that represents a hillslope unmodified by erosional processes. Analysis then focuses on how water flowing over this unchanneled hillslope erodes channels.

Before considering the problem of channel development, the main geomorphic processes acting on hillslopes must be identified. These processes include diffusive processes and advective processes. Diffusive processes are slope dependent and do not involve transport of earth material, or eroded sediment, by water. Soil creep and rainsplash are classic examples of diffusive hillslope processes. Soil creep involves slow downslope movement of soil under the influence of gravity, and rainsplash consists of net downslope displacement of soil particles by the impact of raindrops. In both cases, the flux of sediment by these processes increases with increasing slope, and sediment tends to move from areas of high slope to low slope, thereby eroding elevated areas and filling in depressions. The result is a reduction of relief and topographic irregularities. Advective processes involve transport of earth material, or eroded sediment, by flowing water and are dependent on the hydrodynamic characteristics of the flow. Advective transport is capable of removing sediment from a location and transporting it far from the site of removal. The transport of sediment in channelized flows is advective. Sheetwash, the movement of soil particles by overland flow, or water flowing over the hillslope as irregular sheet, also is advective, but can behave diffusively depending on the characteristics of the slope.

The problem of channel initiation by overland flow on hillslopes has been examined using two approaches. The first approach employs a force–resistance relation to define a threshold for initiation of erosion. The second applies mathematical stability analysis to sediment-flux relations to determine the conditions that lead to growth of topographic perturbations into channels. A third approach focuses on erosion by subsurface processes.

2.1.1 How Is the Formation of Channels on Hillslopes Related to Overland Flow?

2.1.1.1 How Has Channel Initiation Been Related to Force–Resistance Relationships?

Classic work on channel initiation has focused on erosion by advective processes (Horton, 1945). Precipitation falling on a hillslope produces runoff, water flowing over the surface of the hillslope as an irregular sheet, when the intensity of rainfall exceeds the infiltration capacity of the soil – the rate at which the soil can absorb the rainfall. This type of runoff is referred to as infiltration excess overland flow. Assuming that the precipitation affects the entire hillslope, the depth of overland flow (h) increases as a power function of distance from the top of the slope x (Figure 2.1):

$$h = f(x^b) \tag{2.1}$$

where $b < 1$ (Horton, 1945; Anderson and Anderson, 2010, p. 388). The shear stress (τ_b) of this flow acting on the hillslope surface at any location along the path of the flow is

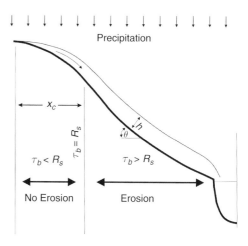

Figure 2.1. Conceptual model of channel initiation by overland flow (based on Horton, 1945).

$$\tau_b = \rho g h \sin \theta \tag{2.2}$$

where ρ is the density of the flowing water, g is gravitational acceleration, and θ is the local slope angle in degrees. This formulation assumes that the flow does not vary greatly in depth or velocity over space or over time as it moves down the slope. The boundary shear stress represents the force per unit area of the water in the downslope direction. It acts on the soil or weathered material on the surface of the hillslope and has the potential to mobilize this material. In this threshold approach, erosion occurs where τ_b exceeds the resistance of soil or weathered material to mobilization (R_s). This resistance depends on properties of the slope materials, such as texture and cohesion, as well as on vegetation cover. Resistance can be expressed as shear strength to provide a dimensionally balanced relation between shear stress and resistance. Upslope of the location where $\tau_b = R_s$ is the belt of no erosion (x_c) (Figure 2.1). Downstream of this location, erosion leads to the development of small parallel channels known as rills (Figure 2.2). Deposition may occur near the base of the slope where the hillslope gradient decreases, especially if this part of the slope has a pronounced concave-upward form. Deposition at the lower end of the hillslope presumably would fill the lower parts of the rills, or diminish the size of these features, producing discontinuous channels.

With the zone of erosion, random microscale irregularities in the topography of the initial hillslope surface concentrate erosive overland flow, producing feedback between erosion and flow concentration that excavates rills (Figure 2.3). Repetition of this process produces a set of parallel rills across a hillslope. The transformation of parallel rills into a stream channel involves enlargement of one of the rills more rapidly than the others through the positive feedback mechanism of flow

Figure 2.2. A rill on a hillslope in Israel.

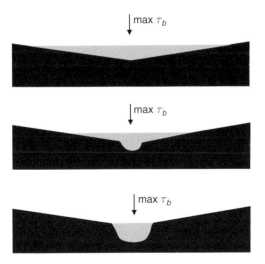

Figure 2.3. Concentration of flow (shaded) in a linear depression that progressively leads to the development of a rill.

concentration. Dominance of a particular rill may be caused by random variations in rill size and by capture of flow in adjacent rills by the dominant rill during storm events that break down or overtop rill divides. As the dominant rill captures flow, it incises into the hillslope, producing strong lateral topographic gradients that redirect flow from the adjacent rills toward the

dominant rill – a process referred to as cross grading. Eventually, the dominant rill evolves into a stream channel, and strong lateral flow toward the stream on emerging valley sides eliminates any vestiges of the adjacent rills (Figure 2.4).

2.1.1.2 How Has Channel Initiation Been Related to Sediment Fluxes?

Whereas the force–resistance approach considers only mobilization of material on the hillslope as the mechanism of channel initiation, linear stability analysis includes mass fluxes. The method considers an initial unchanneled hillslope of constant form and specifies a system of governing equations, i.e., a theoretical model, that defines the transport of sediment and continuity of sediment flux over this surface. Small perturbations are introduced to the mathematical representation of the hillslope system in the form of systematic infinitesimal wave-like forms (ridges and troughs) distributed across the hillslope surface (Figure 2.5). The troughs of these infinitesimal perturbations can be viewed as topographic features that can potentially develop into incipient channels. Stability analysis is then used to determine whether various wavelengths of perturbations, which have an influence on the sediment flux, are stable or unstable. Stability indicates that patterns of sediment flux on the perturbed surface either do not cause the perturbations

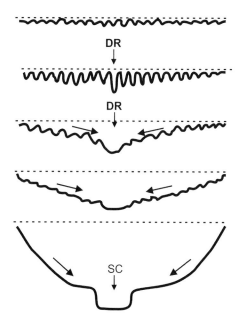

Figure 2.4. Cross-sectional view of hillslope showing progressive development of parallel rills into a dominant rill (DR) and evolution of the dominant rill into a stream channel (SC) through cross grading. Arrows indicate convergence of flow. Dashed line is original hillslope surface.

to grow or eliminate the perturbations; i.e., incipient channels do not form. Instability results in growth of the perturbations; i.e., incipient channels develop on the hillslope.

Advective transport of sediment by sheetwash on a hillslope can be expressed in general form as

$$q_s = f(q, S) \tag{2.3}$$

where q_s is the volumetric flux of sediment per unit width of slope, q is the volumetric flux of water per unit width, and S is the gradient of the hillslope ($\tan \theta$). For this type of transport function, stability conditions are related to the rate of change in sediment-transport flux with respect to change in discharge ($\partial q_s / \partial q$) relative to the magnitude of sediment flux per unit discharge (q_s / q). Instability occurs when $\partial q_s / \partial q > q_s / q$, i.e., when perturbations produce a rate of increase in sediment flux per unit increase in discharge greater than the local ratio of sediment flux to discharge. Assuming discharge increases downslope, if $\partial q_s / \partial q > q_s / q$ the sediment flux per unit discharge also increases downslope, leading to net erosion and channel growth. This condition is characteristic of perturbed concave-upward slopes. Stability occurs when $\partial q_s / \partial q < q_s / q$ and $\partial q_s / \partial q = q_s / q$, which are characteristic of perturbed convex-upward and straight slopes, respectively (Smith and Bretherton, 1972). Many natural hillslopes exhibit a convexo-concave form whereby the top part of the hillslope is convex and the lower part of the slope is concave (Figure 2.1). The two

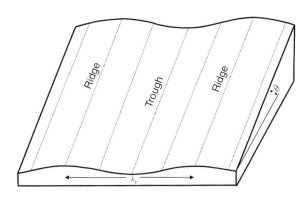

Figure 2.5. Idealized hillslope surface for stability analysis with lateral perturbations of ridges and troughs (λ_p = perturbation wavelength).

components join at a short straight section or inflexion point. Thus, the results of the stability analysis indicate that channel development, as indicated by instability of perturbations, will occur downslope of the straight segment or inflection point. The zone from the hillslope crest to the upper end of the concave segment is similar to a belt of no erosion (Horton, 1945), but differs from it because erosion can occur in this region (i.e., it is not below the threshold for motion); however, transport on the convex part of the slope is dominated by diffusive effects that eliminate initial perturbations.

Initial stability analysis based on Eq. (2.3) indicates that perturbations with the shortest wavelengths are the most unstable and grow fastest (Smith and Bretherton, 1972). Thus, the spacing of emerging channels occurs at the highest spatial frequencies. In other words, the model predicts a lateral spacing of channels that tends to zero – a condition that is not consistent with the finite spacing of parallel rills on hillslopes. By introducing into linear stability analysis a transport function that accounts for micro-scale nonlocal transfer of eroded material, the growth of short-wavelength perturbations is suppressed (Loewenherz, 1991). Instead, the most unstable perturbations are associated with intermediate wavelengths. This refined stability analysis suggests that initial incisions of channels into a hillslope will have a finite spacing – a characteristic of developing rills. It also indicates that incision in some cases extends upstream of the inflection point between concave and convex portions of the hillslope, with the upstream extent of incision governed by the efficacy of diffusive transport mechanisms and the wavelength of the disturbance causing the incision. Additional refinements of stability analysis include explicit consideration of transverse pressure gradients and advective versus diffusive sediment fluxes on channel initiation (Loewenherz-Lawrence, 1994).

Channel initiation has also been examined by combining linear stability analysis of advective overland flow with the threshold approach in which erosion does not occur until an erosional threshold is exceeded at a critical distance down the slope (x_c) (Izumi and Parker, 1995). An important contribution of this work is that linear stability based on Eq. (2.3) (Smith and Bretherton, 1972) does not yield a finite characteristic wavelength of instability because it is based on a normal flow approximation. Consideration of spatial gradients in hydrodynamic conditions, ignored in the normal flow approximation, yields a solution in the form of an expression for the characteristic wavelength of channel spacing (λ_{cs}):

$$\lambda_{cs} = 0.000181 c_f^{-0.25} \tau_{bc}^{1.125} S^{-1.25} I^{-0.25} \tag{2.4}$$

where c_f is a friction coefficient, τ_{bc} is the critical bed shear stress of the hillslope material, and I is rainfall intensity. This

Figure 2.6. Convex hillslope with lateral perturbations (from Izumi and Parker, 2000).

equation predicts the spacing of incipient channels on hillslopes, providing a basis for testing of the theory. It does not, however, predict geomorphic evolution of the channels to a mature, mutually stable configuration.

Traditional stability analysis regards channel initiation as an "upstream-driven" phenomenon resulting from changes in sediment-transport capacity associated with perturbations on hillslopes. Channel development can also occur by headward erosion from the base of a hillslope through "downstream-driven" processes. Stability analysis of such a case shows that convex-upward hillslopes are unstable to transverse perturbations under certain hydrodynamic conditions (Izumi and Parker, 2000) (Figure 2.6). The analysis generally predicts characteristic spacing of incipient channels of between 30 and 100 meters.

A family of theoretical models provides a framework for addressing channelization of hillslopes in transport-limited and detachment-limited environments (Box 2.1) (Smith, 2010). Linearized analysis and numerical solutions of nonlinear formulations of this family of models show that four mechanisms influence the domain of channel-cutting instability: advective transport of sediment by water, variations in the configuration of the free water surface, advection-driven diffusion, and slope-driven diffusion. All the models, except for those with advective transport and no free water surface, result in the emergence of well-defined channels. Nonlinearities of transport processes amplify instability, leading to strong

---

BOX 2.1 | TRANSPORT- VERSUS DETACHMENT-LIMITED CONDITIONS IN MODELS OF FLUVIAL SYSTEMS.

Two types of conditions are invoked to characterize morphological change in quantitative models of fluvial systems: transport-limited and detachment-limited (Pelletier, 2012a).

Transport-limited conditions are governed by spatial gradients in the capacity of the flow to transport sediment. If transport capacity remains constant over space, neither erosion nor deposition will occur. If transport capacity increases over space, erosion will occur, and if transport capacity decreases over space, deposition will occur. Transport-limited conditions assume that actual transport always equals potential (capacity) transport and that changes in capacity are responsible for morphological change through erosion and deposition. Changes in capacity are governed by changes in hydraulic properties of the flow. Transport-limited conditions typically apply to abundant coarse clastic material (sand size and larger), the movement of which is dependent on spatial variations in hydraulic conditions of the flow. Formulations based on transport-limited conditions can include an entrainment threshold, below which no movement of sediment on the surface under consideration occurs.

Detachment-limited conditions are dependent on the force exerted by the flow on an erodible surface. The magnitude of the force determines the rate of erosion: the larger the force, the greater the erosion rate. Any eroded material moves completely out of the system under consideration. Deposition of transported material within the system does not occur, as is often the case in transport-limited conditions. Detachment-limited conditions typically apply to erosion of bedrock or soil, where the eroded material is small (silt or clay) and transport of this eroded material does not depend strongly on hydraulic conditions. Formulations based on detachment-limited conditions can include a detachment threshold, below which no detachment of sediment on the surface under consideration occurs.

selection of the fastest-growing perturbations, which set the spacing of incipient channels. The introduction of diffusive transport mechanisms stabilizes perturbations of high lateral frequencies (i.e., small lateral spacing). Detachment-limited environments exhibit a greater degree of instability than analogous transport-limited environments. For transport laws of the form $q_s = q^n S^m$, instabilities emerge for values of $n > 0$ in detachment-limited environments, but only for $n > 1$ in transport-limited environments (Smith, 2010).

2.2.1 How Is the Formation of Channels on Hillslopes Related to Subsurface Flow?

All the models discussed thus far invoke overland flow as the primary advective transport mechanism for incipient erosion of channels. The classic model (Figure 2.1; Horton, 1945) explicitly specifies infiltration excess overland flow as the type of runoff for channel erosion, but stability analyses implicitly adopt a similar perspective, given that runoff is assumed to originate at the crest of the slope and move downslope from that location. Infiltration excess overland flow is most common in arid environments with thin soils or exposed bedrock and a lack of vegetation cover. Under these conditions, rainfall intensity can exceed infiltration capacity, leading to surface runoff under unsaturated subsurface conditions. Infiltration excess overland flow is less common in humid environments with thick soils and abundant vegetation cover. Interception of rainfall by leaf canopies, branches, and blades of grass increases evaporation losses and converts rainfall into stemflow and throughfall. Thus, interception reduces the total amount of precipitation reaching the soil as well as the intensity of water delivery to the soil surface. Moreover, the roots of flora and borrowing activities of fauna within the soil prevent soil compaction and provide abundant, interconnected macropores that promote an open soil structure and enhanced rates of infiltration.

In humid environments, overland flow generally occurs only when soil is saturated, either through water emerging onto the surface from the subsurface as return flow or through direct precipitation onto saturated areas. The zones of saturation expand upslope away from valley bottoms and contract downslope toward valley bottoms in response to variations in precipitation inputs (Dunne and Leopold, 1978, pp. 265–271). Thus, the spatial extent of saturation-excess overland flow in humid temperate environments varies spatially with antecedent moisture conditions. Rarely, if ever, will overland flow extend from the crest to the bottom of a hillslope in a humid environment. Thick vegetation cover increases the resistance of the hillslope surface to erosion and impedes the flow of water down the slope, thereby diminishing its erosional potential. For these reasons, the carving of channels through erosion by overland flow is more difficult to accomplish in humid environments than in arid or semiarid environments.

Subsurface processes represent an alternative mechanism to overland flow for initiating the formation of channels (Dunne, 1980). This alternative conceptualization attributes channel initiation to seepage erosion. Analogously to piping in soils, seepage erosion occurs where groundwater moving through bedding planes, pores, fractures, and joints in underlying bedrock emerges at the surface (Dunne, 1990). Weakening of the rock through frost action and chemical weathering reduces cohesion and tensile strength while increasing porosity and hydraulic conductivity. If the seepage flow has sufficient force to entrain and remove weathered rock material, seepage erosion at the site leads to the development of an overhanging springhead scarp – a mechanism referred to as groundwater sapping. The magnitude of the seepage force is directly related to the discharge per unit area of outflow directed normal to the scarp face and inversely related to the saturated hydraulic conductivity of the rock. Field observations in Vermont indicate that groundwater sapping is capable of producing springhead scarp retreat of 10 to 30 cm over a period of several weeks of strong outflow during the snowmelt season (Dunne, 1980).

The conceptual model of channel initiation based on seepage erosion starts with uplift of a rock mass with an unchanneled tilted surface above sea level (Figure 2.7). The uplifted mass is composed of rock of uniform permeability. The water table within the uplifted mass slopes toward the base level at the edge of the mass, providing a hydraulic gradient that drives groundwater flow. Heterogeneities in characteristics of the rock or in rates of chemical weathering lead to spatial variation in permeability and the development of a local perturbation of high permeability that concentrates groundwater flow. Initial sapping at this location produces a springhead scarp. Lowering of the ground surface by erosion further concentrates groundwater flow in the springhead, which enhances chemical weathering and sapping. Through this positive feedback between topographic deformation and concentration of groundwater flow, the scarp migrates headward over time, resulting in the development of a channel.

2.3 What Are Some Limitations of Existing Models of Channel Initiation?

Models of channel initiation provide insight into possible mechanisms by which stream channels develop on terrestrial landscapes, yet none of these models has been tested rigorously at field scale, at least in the sense of directly observing the initiation of stream channels on an unchanneled hillslope and relating this initiation to formative mechanisms embodied in the models. The classic overland flow model (Horton, 1945) and stability analyses focus mainly on the development of

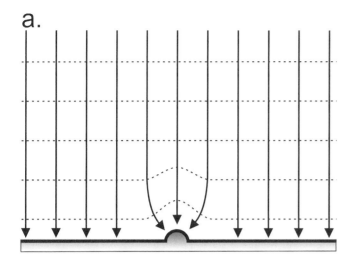

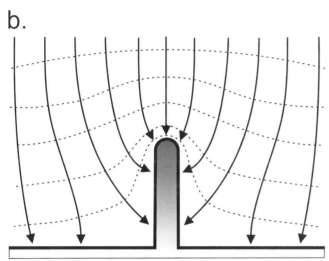

Figure 2.7. Planview of channel initiation by groundwater flow and seepage erosion through the development of a springhead scarp in an uplifted surface. (a) Scarp develops at a site of high permeability that concentrates groundwater flow (dark lines with arrows). (b) Headward migration of springhead by groundwater sapping (shaded). Dashed lines show contours of equipotential (after Dunne, 1980).

incipient channels rather than of full-size stream channels. These incipient channels typically are identified as rills, because hillslopes constitute the formative environment and rills are hillslope erosional features (Horton, 1945). Thus, these models largely pertain to the development of rills rather than streams.

Complete transformation of a system of parallel rills into a stream channel through micropiracy, incision, and cross grading has yet to be observed at field scale. Rill systems often develop as parallel networks of relatively similar-sized channels on hillslopes. Such parallel networks form when advective transport is small compared with diffusive transport or when surface roughness is small (McGuire et al., 2013). Experimental studies and numerical simulations indicate that parallel networks of channels can remain stable and do not evolve by

capturing flow from adjacent channels to form a dominant rill that subsequently incises to form a stream channel (Gomez et al., 2003; McGuire et al., 2013; Schneider et al., 2013; Bennett et al., 2015; Shen et al., 2015a; Bennett and Liu, 2016). Rills are not "baby" stream channels that mature through the evolutionary sequence of positive feedback (e.g., Figure 2.4).

Linear stability analysis can only determine the initial response of the hillslope system to the perturbations; in the case of instability, it cannot predict the time evolution, or growth, of the perturbations, because such growth will influence sediment dynamics in ways not addressed by linear analysis. Thus, the connection between the incipient channels predicted by linear stability analysis and fully formed channels and channel networks is incomplete. Linear stability theory does not capture the evolution of a rill by knickpoint development and migration (Slattery and Bryan, 1992; Bennett, 1999). It also generally predicts that the development of rills occurs only on concave-upward surfaces and that channels extend slightly upstream of the inflection point between concave and convex surfaces. By contrast, natural rills can extend through convex surfaces to positions close to drainage divides (Figure 2.8). Experimental studies and nonlinear simulation models are needed to evaluate the feedbacks between form and flow in evolving rills. Nonlinear simulation models, in contrast to linear analytical models, can capture interaction among flow, sediment transport, and evolving channel form.

Channel initiation by seepage erosion (Dunne, 1980) faces the problem that this process occurs over geologic timescales and is difficult to document directly in the laboratory or in the field. Experimental studies have examined the problem of seepage erosion in noncohesive or weakly consolidated sediments, confirming that channel initiation through groundwater sapping can occur in these materials (Howard and McLane, 1988; Kochel et al., 1988; Lobkovsky et al., 2004, 2007; Pornprommin and Izumi, 2010). Generally, these studies show that sapping produces scarps with amphitheater shapes that retreat headward over time. Such studies, however, do not directly address the issue of whether seepage erosion is important for the initiation and development of bedrock channels and valleys. This issue is relevant for determining the origin of fluvial features not only on Earth (Laity and Malin, 1985; Kochel and Piper, 1986; Onda, 1994; Schumm et al., 1995), but also on Mars (Aharonson et al., 2002). Nevertheless, mechanistic evidence supporting erosion of channels in bedrock by seepage erosion and sapping is sparse (Lamb et al., 2006); weathering concentrated at the seepage face usually is necessary to reduce the cohesion of intact rock before sapping can occur by seepage erosion (Dunne, 1990). The capacity of sapping and seepage erosion to initiate channels remains uncertain. Some large amphitheater-headed canyons on Earth with substantial seepage outflow from vertical headwalls have been carved by megafloods rather than by groundwater sapping (Lamb et al., 2008a).

Figure 2.8. Rill extending downslope from near crest through convex upper part of an eroding hillslope in Georgia.

2.4 What Are the Characteristics of Rills?

Rills develop on soil-mantled hillslopes with gradients between 2 and 12° (Bull and Kirkby, 1997) and have dimensions on the order of 0.5 to 30 cm in width and depths that do not exceed about 30 cm (Knighton, 1998) (Figure 2.2). From a practical standpoint, the depth of rills is not great enough to impede normal tillage practices, and normally these features are obliterated by tillage (Grissinger, 1996). Rills typically are ephemeral features that form, fill in, and reform, often in different locations, in response to individual runoff events or seasonal variations in these events.

Negative feedback exists in rill systems that can lead to stabilization or eradication of individual channels over time. Channelization of flow can concentrate coarse soil particles on the bottom of the rill, inhibiting further incision. Resistant subsurface soil layers, such as clay hardpans, can also impede incision. Experimental work indicates that infiltration of runoff into the bed of a rill can lead to decreases in volumetric flow rate, inhibiting erosion (Slattery and Bryan, 1992). Rainsplash is an effective process for counteracting rill incision (Dunne and Aubry, 1986). Elevated areas between rills that slope toward the rills promote diffusion of rain-splashed sediment into these small channels (Dunne, 1980). Moreover, rainsplash can occur in the absence of overland flow, leading to infilling of channels during times when flow does not occur within rills to counter infilling.

2.5 How Do Gullies Differ from Rills?

Although rills sometimes evolve into gullies – erosional features much larger than rills – in most cases they do not. Instead, most gullies are not just enlarged rills but independent erosional

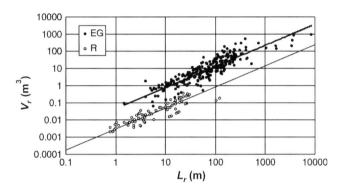

Figure 2.9. Relation between volume (V_r) and length (L_r) of rills (R) and ephemeral gullies (EG). Parallel lines indicate similarity in scaling but offset between data for rills and gullies indicate distinct differences in absolute size (from Capra et al., 2009).

features that tend to occur on steeper hillslopes than rills (Gao, 2013). The dimensions of gullies vary widely, but these features are large enough that they disrupt normal tillage operations. Rills and gullies have similar morphological scaling characteristics but are distinctly different in absolute size (Capra et al., 2009) (Figure 2.9). The lack of transitional forms can lead to abrupt transitions between rills and gullies when these erosional features occur together on hillslopes (Figure 2.10).

Gullies commonly indicate accelerated erosion on hillslopes, developing in response to natural or human-induced environmental change. Changes in climate may affect vegetation cover and rainfall intensity, enhancing landscape erodibility. Human activities such as farming, deforestation, or urbanization can decrease the resistance of the soil to erosion or increase rates of runoff. The formation of gullies is a common symptom of land degradation (Castillo and Gomez, 2016).

Figure 2.10. A meandering rill (foreground) transitioning abruptly into a large gully on an eroding hillslope in Georgia.

Figure 2.11. Headwall of a gully in Georgia.

The initiation of gullies, in contrast to rills, involves a positive feedback mechanism in which the concentration of flow within an incipient channel is sufficiently great to exceed substantially the resistance of the material on the bed of the channel (Gao, 2013). Moreover, the material into which the channel incises is of sufficient depth and has either uniform or decreasing resistance over depth so that incision can proceed unimpeded. As incision occurs, flow is further concentrated, promoting additional erosion (Kirkby and Bracken, 2009). Incision often does not happen uniformly along the length of a gully but develops through upslope migration of a gully headwall (Figure 2.11). As the headwall migrates upslope, it produces a steep-walled trench much larger in cross-sectional area than the dimensions of the flow conveyed within the bottom of the trench. Flow within most gullies is ephemeral, occurring only in response to storm events. The rate of migration of the headwall varies with local conditions and may occur during a single event or over many years. Upstream retreat occurs until dissipation of erosional energy in a plunge pool at the base of the headwall and

the reduction in the contributing drainage area above the headwall are sufficient to prevent further erosion. Gully development may also involve the subsurface process of soil piping (Bull and Kirkby, 1997; Gao, 2013), the erosion of earth material by flow through well-defined conduits within the soil (Bryan and Jones, 1997). Collapse of soil pipes can initiate gully development, and piping may be involved in headwall retreat or sidewall collapse (Swanson et al., 1989; Wilson, 2011; Nichols et al., 2016). Mass movement, including failure of steep, high channel banks or the gully headwall, also contributes to gully development (Kirkby and Bracken, 2009; Gao, 2013).

Gullies typically are hillslope erosional features rather than incipient stream channels; however, no definitive criteria exist for distinguishing a gully from a stream channel. Some gullies, especially large permanent ones, share morphological characteristics and fluvial processes with certain types of stream channels, especially those of incised channels. On the other hand, many stream channels are fundamentally different from gullies in terms of both morphological characteristics and formative processes. Large, permanent gullies can be discontinuous and not connect downslope with rivers or streams (Bull and Kirkby, 1997). In some cases, entire gully networks confined to hillslopes may connect poorly or not at all with stream systems within drainage basins in which these networks develop. Some small gullies are ephemeral, filling in and reforming seasonally much like rills (Capra et al., 2009).

2.6 How Is the Development of Stream Channels Related to Drainage-Basin Characteristics?

A key issue that has become of considerable interest in fluvial geomorphology over the past few decades is the determination of how much hillslope area and length are needed to support the development of a stream channel. If drainage basins consist of channels and hillslopes, points of transition from hillslopes to channels mark locations of change not only in landscape form but also in process mechanisms. The problem under consideration here is the morphological complement of channel initiation. The concern focuses on the extent to which the landscape within a watershed consists of channels versus how much of the landscape remains unchanneled, or, alternatively, how much hillslope contributing area or length is required to initiate and maintain a channel.

2.6.1 What Factors Control the Density of Drainage?

Early attempts to address this issue did so holistically by considering the relation between the total length of streams within a drainage basin and the total area of the basin. Dividing the total length of all streams within a drainage network (L_t) by the total area of the watershed (A_d) yields the drainage density (D_d):

$$D_d = L_t/A_d \tag{2.5}$$

This metric indicates the extent to which streams dissect a drainage basin. The higher the drainage density, the more dissected the basin is by the stream network. Values of D_d vary from about 2 km km^{-2} to as great as 800 km km^{-2} (Chorley et al., 1984, p. 319; Ritter, 1986, p. 186). Differences in D_d among drainage basins reflect variations in the relative effectiveness of factors that control the development of channelized flow. These factors include climate (Gregory, 1976), vegetation (Melton, 1957), lithology (Wilson, 1971), relief (Schumm, 1956), and slope angle (Lin and Oguchi, 2004). Interaction among these factors in different environmental settings makes it difficult to determine independent effects. In general, drainage densities increase with increasing precipitation, decreasing vegetation

cover, increasing relief, increasing slope angle, and decreasing rock resistance to erosion. However, exceptions to these relations occur. For example, drainage density can decrease with increasing slope angle if hillslope erosion is dominated by mass wasting (Talling and Sowter, 1999). Drainage density also increases over time in landscapes evolving from an initial, unchanneled state (Ruhe, 1952; Schumm, 1956; Talling and Sowter, 1999).

When using data from a digital elevation model (DEM), drainage density can be defined in a dimensionless form as the ratio of channelized pixels to the total of number of pixels in the watershed (D_{dd}) (Sangireddy et al., 2016a) (Figure 2.12). Analysis of D_{dd} for 101 drainage basins in the United States shows that dimensionless drainage density tends to decrease with increasing mean annual precipitation (MAP) in arid and semiarid environments (MAP < 1050 mm yr^{-1}) and increase slightly with increasing MAP for humid environments. Drainage density exhibits a complex relationship with relief, decreasing marginally with increasing relief in arid and semiarid environments, but increasing with increasing relief in humid environments. The relation of drainage density with relief also varies with total relief, increasing with increasing relief for low and high values of relief, but decreasing with increasing relief for intermediate values of relief. Increasing vegetation cover leads to a reduction in drainage density in arid and semiarid environments but corresponds to a slight increase in drainage density in humid environments. Overall, drainage density tends to decline with increasing vegetation cover. The analysis indicates that no significant relationship exists between rock type or strength and D_{dd}, suggesting that climate, vegetation, and topography are more important than geology in determining drainage density.

2.6.2 What Is the Constant of Channel Maintenance?

The inverse of D_d defines the average amount of hillslope drainage area required to maintain a unit length of stream at the scale of the entire drainage network:

$$C_m = A_d/L_t = 1/D_d \tag{2.6}$$

where C_m is the constant of channel maintenance (Schumm, 1956). If drainage density is high, less hillslope drainage area is required, on average, to support a unit length of stream. Assuming that streams bisect, or divide in half, the average drainage area needed to maintain a unit length of the streams yields

$$l_o = (0.5)(1/D_d) = 0.5/D_d \tag{2.7}$$

where l_o can be considered an estimate of the length of overland flow if infiltration excess overland flow begins at the crest of a hillslope (Figure 2.1) (Horton, 1945). Because l_o is derived

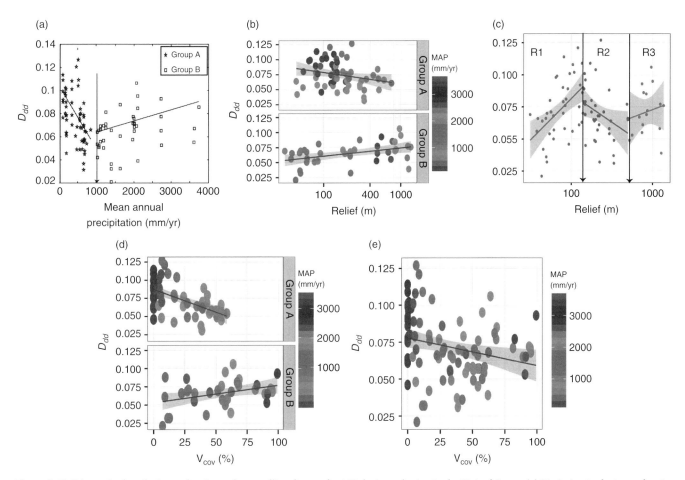

Figure 2.12. Dimensionless drainage density and controlling factors for 101 drainage basins in the United States. (a) Variation in drainage density with mean annual precipitation (MAP) for two groups: A: MAP < 1050 mm yr^{-1} and B: MAP ≥ 1050 mm yr^{-1}. (b) Variation in drainage density for groups A and B with relief. Colors of dots correspond to MAP. (c) Variation in drainage density for three classes of relief. (d) Variation in drainage density with percent vegetation cover (V_{cov}) for groups A and B and (e) for all data. Colors of dots correspond to MAP (from Sangireddy et al., 2016a). (A black and white version of this figure will appear in some formats. For the color version, please refer to the plate section.)

from morphological data on drainage area and stream length, and overland flow does not necessarily originate at hillslope crests, l_o is actually an estimate of the average length of hillslope needed to support a unit length of channel. Given that drainage densities for stream networks range from 2 to 800 km km^{-2}, values of C_m range from 0.5 km^2 km^{-1} to 0.00125 km^2 km^{-1} and values of l_o from 250 m to 0.63 m. These ranges of C_m and l_o provide first-order approximations of the amounts of hillslope area and hillslope length needed to support a unit length of stream. However, it must be emphasized that the metrics average information over entire networks from the largest rivers to the smallest headwater streams, including interior portions of drainage networks, where the maintenance of stream channels is largely a function of flow and sediment entering a particular reach of channel from upstream rather than from adjacent hillslopes. Thus, the network-scale averaging method does not provide detailed information on the length of hillslope or size

of drainage area required to initiate a channel in the headwaters of a drainage basin.

2.6.3 Where Do Channels Begin?

A fruitful approach to determining the amount of drainage area or hillslope length needed to initiate and sustain a stream channel is to ask the question: where do channels begin on the landscape (Montgomery and Dietrich, 1988)? This approach focuses on identifying where channels start to form relative to the position of drainage divides and on evaluating the factors that control these locations of channel initiation. It emphasizes the importance of channel heads, or the upstream boundary of concentrated water flow and sediment transport between definable banks (Dietrich and Dunne, 1993). Channel heads are crucial fluvial forms that demarcate the transition from hillslope to channel processes (Kirkby, 1980). Under conditions where the hillslopes and stream network are mutually adjusted to one

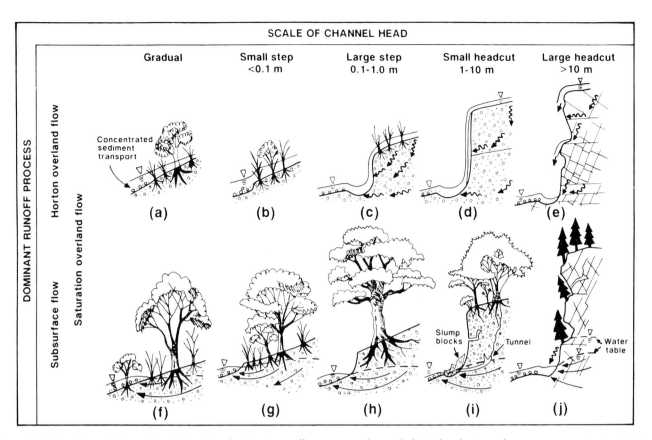

Figure 2.13. Typology of channel heads based on dominant runoff processes and morphological scale. Smooth arrows represent saturation subsurface flow and wiggly arrows represent percolation through unsaturated zones (from Dietrich and Dunne, 1993).

another, i.e., in a steady state, the distance from the drainage divide to the channel head controls the drainage density and corresponds to the hillslope length needed to generate a channel. The drainage area upstream of the channel head defines the hillslope area required to produce and sustain a channel.

The study of channel heads has led to the development of a classification scheme for these geomorphic features based on their morphological characteristics and dominant runoff processes (Figure 2.13). The morphological characteristics describe the transition from the channel to the upstream hillslope. Gradual heads transition without a break in the longitudinal profile between the channel and the hillslope. These types of heads are hardest to identify in the field and may extend over long distances. Stepped heads and headcuts both involve abrupt breaks in slope and the existence of a vertical or steeply sloping transition between the channel floor downstream and the surface of the hillslope above the head. Consistently with models of channel initiation, the dominant runoff processes that influence the characteristics of channel heads include infiltration excess overland flow, saturation-excess overland flow, and subsurface flow. In many cases, a variety of mechanisms shape channel heads, including erosion by overland flow, plunge pool scour at

the base of a step or headcut, seepage erosion by sapping and piping, and mass failure of steps or headcuts (Figure 2.14).

In coastal mountain ranges of California and Oregon, the position of channel heads, defined by the length (L_d) and area (A_d) of the drainage basin between the channel head and the drainage divide, relates strongly to the valley slope (S_v) (Montgomery and Dietrich, 1989):

$$L_d = L_{do}S_v^{-0.83} \tag{2.8}$$

$$A_d = A_{do}S_v^{-1.65} \tag{2.9}$$

where $L_{do} = 67$ m and $A_{do} = 1978$ m^2. The lengths and areas of contributing basins upslope of the channel heads decrease as the valley slope increases. The first relation conforms to the hypothesis that channels begin at a threshold distance from the drainage divide because of increasing boundary shear stress (Horton, 1945) or a transition in hillslope form that promotes incision by advective processes (Smith and Bretherton, 1972). The second relation encapsulates, in general form, how advective processes lead to channel initiation. Drainage area is strongly related to discharge; thus, this equation represents the two major variables q and S in the general

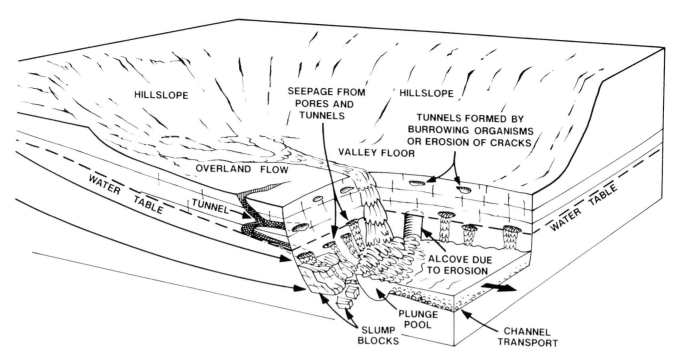

Figure 2.14. Runoff and erosion processes at a steep, high channel head (from Dietrich and Dunne, 1993).

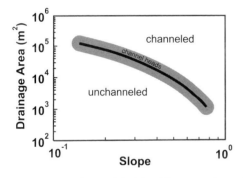

Figure 2.15. Threshold relation (black line) between channeled and unchanneled portions of drainage basins in Oregon and California. Shading indicates domain of variability in the relation. (from Montgomery and Dietrich, 1992).

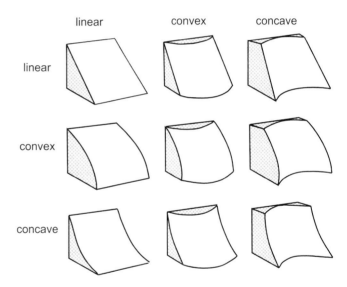

Figure 2.16. Typology of hillslope forms based on longitudinal profile (top to bottom) and planform shape (left to right). Flow-concentrating hillslopes have concave planforms.

transport function used in channel-initiation analyses (Eq. (2.3)). It defines a morphological threshold between channeled and unchanneled conditions within these landscapes (Montgomery and Dietrich, 1992) (Figure 2.15). Channeled portions of the landscape tend to have larger products of slope and drainage area compared with unchanneled areas. The slope-area threshold for channelization corresponds to landscape elements where the boundary shear stress for overland flow exceeds the critical value for erosion (Dietrich et al., 1992, 1993)). Such elements typically are hillslopes with a convergent planform, which promotes the concentration of overland flow (Figure 2.16). For hillslope elements that lie below the threshold, the shear stress of saturation-excess overland flow does

not exceed the critical value and channelization typically does not occur; however, seepage erosion may be active at channel heads or produce channelization where the land surface has been disturbed, for example, by cattle grazing. On steep hillslopes, landslides and debris flows are primary mechanisms for initiating channels, but these features are not strictly fluvial in origin. Hillslope elements with divergent planforms tend to preclude the development of saturation-excess overland flow,

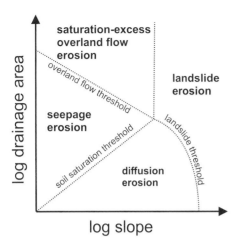

Figure 2.17. Process regimes defined by threshold relations for landscape erosion by different mechanisms (adapted from Montgomery and Dietrich, 1994).

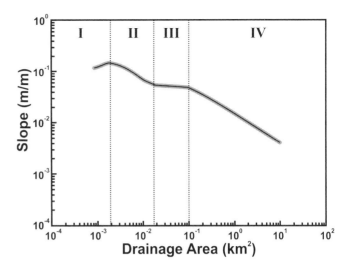

Figure 2.18. Scaling relation between drainage area (A_d) and slope (S) for drainage basins. Roman numerals correspond to domains defined in the text (from Ijjasz-Vasquez and Bras, 1995).

and here the landscape is dominated by diffusive transport (e.g. soil creep and rainsplash) (Figure 2.17).

Climate is an important factor influencing the location of channel heads. In semiarid environments, channel heads have longer hillslope lengths and larger contributing areas than those in humid environments (Henkle et al., 2011). Changes in elevation within mountainous environments can alter precipitation regimes, modifying slope-area thresholds for channel-head location (Henkle et al., 2011). Characteristics of hillslope sediment affect the channel-head threshold by changing the resistance to entrainment (Istanbulluoglu et al., 2002). Over the short term, wildfire reduces resistance to erosion on vegetated hillslopes, decreasing the slope-area threshold for channel-head development (Hyde et al., 2014) and producing upslope migration of channel heads (Wohl, 2013b). Terrain roughness, characterized by the degree of valley incision and hillslope gradients, is another potential factor influencing the slope-area relation for channel-head locations in mountainous environments; this relation tends to be well-defined where roughness is high but poorly defined where roughness is low (Imaizumi et al., 2010). In some cases, the relationship between slope and drainage area at channel heads is weak or poorly defined (Julian et al., 2012; Placzkowska et al., 2015), confirming the complexity of the channel-initiation process. Strong bedrock control on patterns of subsurface flow may disrupt the relation between topographic characteristics and channel-head locations (Jaeger et al., 2007). In other cases, simple threshold relations are evident. Channel heads produced by overland flow and landslides in a small watershed in Thailand exceed a slope threshold independently of drainage area, whereas channel heads produced by seepage erosion exceed an area threshold independently of slope (McNamara et al., 2006). Moreover, channeling of hillslopes appears in different forms,

especially when landscapes are disturbed. Ephemeral channels and gullies, common in disturbed landscapes, tend to have heads located at lower slopes and smaller drainage areas than permanent channels (Vandaele et al., 1996; Jefferson and McGee, 2013). These conditions reflect changes in channel-initiation thresholds associated with landscape disturbances that increase runoff or reduce soil resistance to erosion.

The use of DEMs to examine the topographic characteristics of landscapes has led to speculation about the location of channel heads in relation to the cumulative frequency distribution of slope and contributing area throughout entire watersheds (Montgomery and Foufoula-Georgiou, 1993; Ijjasz-Vasquez and Bras, 1995). Four distinct regions/process regimes can be identified on such plots: region I, where the slope-area relation has a positive gradient and diffusive hillslope transport dominates; region II, where the relation has a negative gradient and unchannelized advective transport dominates; region III, where the relation is transitional between regions III and IV; and region IV, where the relation has a negative gradient and channelized advective transport dominates (Figure 2.18). According to this scheme, channel heads should be located within regions III and IV. Empirical data on channel-head locations and slope-area distributions indeed show that channel heads often are positioned in regions III and IV of these distributions (McNamara et al., 2006; Henkle et al., 2011; Jefferson and McGee, 2013).

Although most work on channel heads has identified and mapped these features in the field, increasing effort is being devoted to automated extraction of channel heads from DEMs using geospatial analysis techniques (e.g., Hancock and Evans, 2006; Julian et al., 2012; Clubb et al., 2014). Such efforts have the potential to increase substantially the availability of information on channel-head locations in relation to landscape characteristics. Because these methods rely on process-based relations of where channels should start in relation to topographic characteristics, they also provide the basis for exploring factors leading to discrepancies between expected and actual channel-head locations.

2.7 How Do River Networks Form, Grow, and Evolve?

Once stream channels begin to develop within a drainage basin, these channels organize into river networks. The problem of how networks form, grow, and evolve is a fundamental one in fluvial geomorphology. This problem has been addressed using a variety of styles of inquiry, including conceptual, experimental, and numerical approaches.

Drawing on the mode of landscape evolution depicted in the cycle of erosion model (Davis, 1899; see Chapter 1), drainage-network development on an uplifted, initially unchanneled surface was conceived as progressing through an extension stage involving 1) initiation of channels, 2) elongation of streams into the uplifted surface to the maximum extent supported by the coevolving drainage area, 3) elaboration, or the addition of tributaries to the elongated streams, and 4) maximum extension – the attainment of complete elaboration (Figure 2.19) (Glock, 1931). After the extension stage, an integration, or reduction, stage occurs, characterized by elimination of channels from the network within the interior part of the network through reduction of relief, valley widening, and floodplain development (Abrahams, 1977).

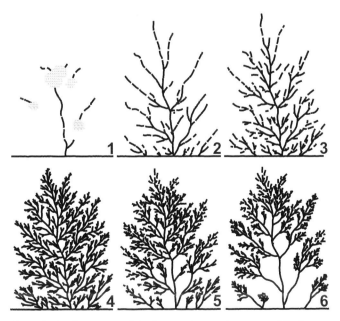

Figure 2.19. Planview model of network growth on an uplifted surface (Glock, 1931). Sequence 1–4 is part of the extension stage and sequence 5–6 represents the integration, or reduction, stage.

In the classic overland flow model (Horton, 1945), the development of a stream channel within a well-defined valley (Figure 2.4) is followed by the formation of rills on side slopes of the main valley (Figure 2.20). Through cross grading, dominant rills become tributary streams. Rills and master rills then form on the valley sides of these tributary streams. This process

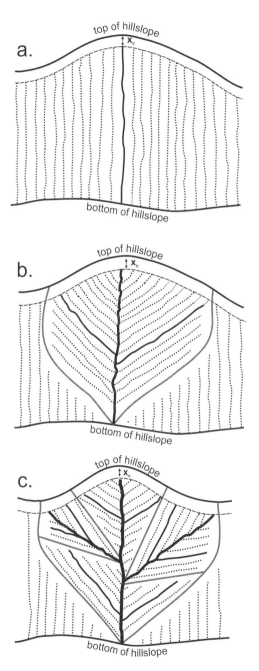

Figure 2.20. Planview model of network growth by overland flow (after Horton, 1945). (a) Parallel rills (dotted lines) flanking an evolving dominant rill. (b) Rills draining into a stream with two dominant tributary rills on valley slopes. (c) Development of tributary rills on valley slopes to tributary streams. Dark gray lines are drainage-basin and subbasin boundaries.

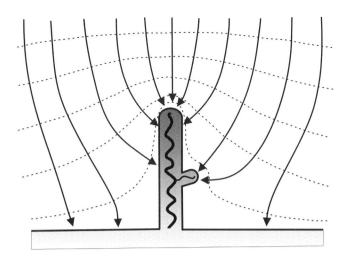

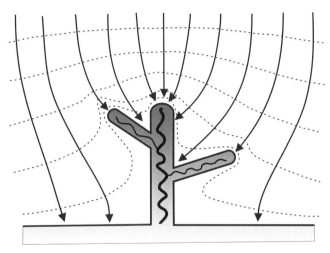

Figure 2.21. Planview model of network growth based on groundwater sapping and seepage erosion (adapted from Dunne, 1980).

repeats itself until the network expands to the point where channel initiation no longer occurs; i.e., the remaining hillslope lengths are less than the threshold for channel initiation (x_c).

The model of network growth based on seepage erosion and groundwater sapping adopts a scheme based on repetition of springhead sapping (Dunne, 1980). Once a channel has formed, heterogeneities in permeability along the valley wall lead to headward growth of a new channel head, producing a tributary to the original channel. The process of channel-tip initiation and extension recurs until seepage along existing channels is insufficient to sustain erosion of additional channels (Figure 2.21).

2.7.1 How Have Physical Models Been Used to Explore Network Development?

Given that most stream and river networks evolve over geologic timescales, the process of drainage-network development can only be observed firsthand in exceptional circumstances, and

even then only at small scales. The formation of networks has been partly documented on newly exposed unchanneled landscapes produced either artificially (Schumm, 1956) or by natural processes (Morisawa, 1964). Additions and losses of channels occurred within these evolving small-scale natural networks, but the period of observation did not allow complete documentation of the entire evolutionary process of drainage-network development. This process has, however, been documented for miniature drainage networks generated in the laboratory. Although details differ, experimental models generally consist of a large rectangular box filled with sediment. Rainfall is applied to the sloping surface of sediment to generate saturation-excess overland flow, which exits the box through an outlet at the downslope end. The elevation of the outlet can be varied to change base level, and, in some cases, the slope of the surface can be adjusted.

2.7.1.1 How Do Networks Evolve under the Influence of Saturation-Excess Overland Flow?

Channel development in saturation-excess overland flow experiments typically occurs through rapid headward erosion from the downstream end of the sloping surface rather than by channel initiation at a critical distance down the sloping surface (Flint, 1973). Initial microtopography of the sediment surface has a strong effect on network growth. In experiments in the Runoff Erosion Facility (REF) at Colorado State University, small ponds formed on the surface of the sediment and became connected to one another by streams of surface runoff to form a broad, shallow drainage network (Schumm et al., 1987). Surface topography directed runoff toward the center of the surface, where the longest and deepest connection in pond-link drainage developed. Erosion proceeded rapidly headward from the outlet along this line of concentrated flow to produce a wide main channel with vertical banks. Locations where runoff from interconnected pond systems flowed over the banks of the main channel became headcuts that migrated along the path of interconnected ponds to produce tributary channels. Thus, initial microtopography, which influences the concentration of surface water, controlled the emerging network pattern.

Experiments in the REF demonstrate that the mode of growth of evolving networks is sensitive to surface slope and changes in baselevel. On a relatively mild slope with baselevel lowering, the network grows slowly headward with frequent addition of tributaries, producing a high drainage density within the network (expansion mode) (Figure 2.22). On a relatively steeply sloping surface without a lowering of baselevel, the network extends rapidly headward into the available unchanneled area (extension mode). The evolving networks conform to some extent with the stages of Glock's (1931) model, but overlap exists between the extension and reduction stages. The interior part of the network attains a condition of maximum extension, followed by reduction in

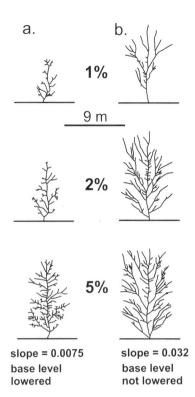

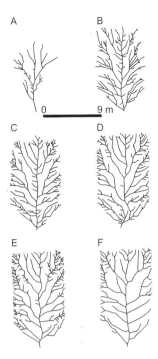

Figure 2.23. Network development in the REF showing extension followed by reduction: (a) 2 h, (b) 10 h, (c) 33 h, (d) 96 h, (e) 106 h, (f) 216 h (from Schumm et al., 1987).

Figure 2.22. Network growth in REF experiments showing (a) expansion mode and (b) extension mode. Percentages indicate the percentage of total flow volume in the experiments (after Parker and Schumm, 1982).

channels, particularly within the interior of the network (Figure 2.23). However, at the same time as reduction is occurring, the network is still expanding into unchanneled areas through the formation of new channels on the network periphery. This dynamic prolongs the period of maximum drainage density. Channels, once formed, are fixed in place and do not change through stream capture, whereby one stream erodes headward into another and captures its flow. No evidence was observed of tributary formation through micropiracy and cross grading as envisioned by Horton (1945).

Experimental work using an experimental facility similar to the REF corroborates that initial microtopography strongly influences the final form of networks, in the sense that initial flow paths determine channel locations (Hancock and Willgoose, 2001a). Channels incise, advance headward, and branch along these flow paths through migrating headcuts. Network growth does not exhibit evidence of stream capture or of tributary development through micropiracy and cross grading as envisioned by Horton (1945). The pattern of network development over time generally corresponds to Glock's (1931) model with the network expanding to a full extent and then stabilizing (Figure 2.24). It also is similar for both high and low slopes, with growth rate varying directly with rainfall intensity.

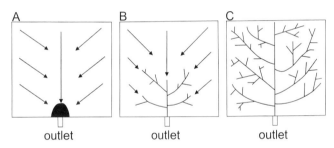

Figure 2.24. Basic pattern of network growth for low and high slope conditions in an experimental erosional basin. (a) Stage 1 – unchannelized runoff concentrated in broad flows (arrows) toward area of incision and erosion (black shading) near outlet. (b) Stage 2 – channels grow headward from area of incision near outlet as headcuts migrate along broad flow paths. (c) Stage 3 – channel network expands to full extent and stabilizes. Time to full network development occurs 3.33 times faster for the low slope, which had higher rainfall intensities than the high slope (from Hancock and Willgoose, 2001a).

Experimental work has also examined stream-network development in the REF on sloping planar surfaces with a uniform base level (Pelletier, 2003). Four different initial conditions result in four different patterns of network evolution (Figure 2.25). None of these patterns conforms to the classic conceptual models of drainage-basin evolution. Notably, stream capture plays an important role in network development. Capture occurs as channels migrate laterally into one another and flow from one of the channels is diverted into the other channel. For a simple sloping planar surface, rates of extension of subparallel channels increase with increasing slope; however, channels begin to form upslope from base level, rather than through headcut migration at base level, in accordance with the concept of a threshold

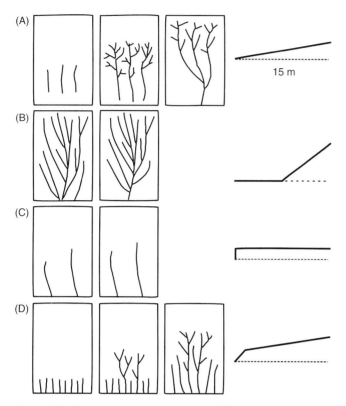

Figure 2.25. Patterns of network growth at different times (left) for different initial longitudinal profiles (right). (a) uniform slope, (b) flat surface in front of slope segment, (c) raised platform, and (d) segmented slope (after Pelletier, 2003).

distance for channel initiation. These experiments demonstrate that boundary conditions (outlet versus uniform base-level) and initial conditions (morphological configuration of the unchanneled slope) can have a strong influence on the mode of network growth.

2.7.1.2 How Do Networks Evolve under the Influence of Subsurface Flow?

Experimental studies indicate that headward growth and bifurcation in drainage networks that grow through sapping generally conform to the seepage erosion conceptual model (Figure 2.21) (Kochel et al., 1985). In experiments with a single, discrete basin outlet, initial headward growth occurs largely through episodic headwall slumping (Gomez and Mullen, 1992; Berhanu et al., 2012), and rates of headward growth are directly related to the hydraulic gradient of the subsurface flow (Howard and McLane, 1988). The stage of headward extension is followed by a phase of enhanced tributary development through bifurcation and slow headward growth of tributaries. In unconsolidated sand, the network area expands largely through lateral valley widening and elimination of divides between adjacent tributaries, whereas the network perimeter increases through headward extension of tributary channels (Gomez and Mullen, 1992). Basin circularity (C_d) defines the relation between drainage-basin area (A_d) and the length of the drainage-basin perimeter (P_d):

$$C_d = A_d/\{\pi[(P_d/\pi)/2]^2\} \tag{2.10}$$

Circularity decreases during initial stages of network growth when the trunk stream elongates headward, increasing the perimeter of the network relative to the drainage area (Figure 2.26). Extension of the network following elongation involves valley widening, divide decay, and the addition of

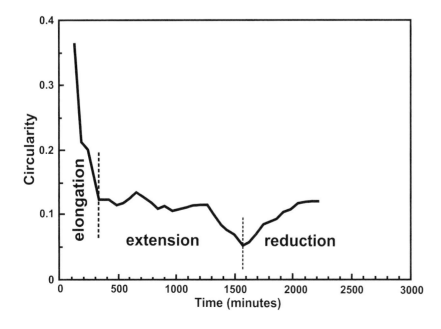

Figure 2.26. Changes in basin circularity index over time in experiments on network growth by sapping (after Gomez and Mullen, 1992).

tributaries, which tend to increase both the drainage area and the network perimeter. Thus, circularity remains relatively constant with a slight decline toward the end of the extension phase when tributary development dominates. Reduction in tributaries through valley widening and divide decay results in a slight increase in circularity in late stages of network development as the network area increases and the length of the network perimeter decreases.

Channels in seepage experiments without a single, discrete basin outlet indicate that two main factors influence the mode of network growth: the source of groundwater and the configuration of the sediment surface (sloping versus horizontal) (Marra et al., 2015) (Figure 2.27). Groundwater fed into the upstream end of a sloping surface of sediment (distant source) leads to channel formation by mass wasting at the downstream end of the sediment body. Channels grow headward through seepage erosion by mass wasting at the channel head and incise and enlarge by fluvial processes within the channels. Rapidly growing channels capture groundwater from adjacent channels, inhibiting or stopping the growth of small channels (Schorghofer et al., 2004). Bifurcation is limited, leading to the development of a few large parallel channels. Precipitation directly onto the surface of a sloping body of sediment (local source) produces saturation-excess overland flow toward the downslope end of the surface and initiation of parallel channels through erosion by surface runoff (Marra et al., 2015). Initial v-shaped channels are transformed into u-shaped channels once subsurface flow enters the channels and causes seepage erosion and mass wasting at the channel head. Channels grow headward in parallel fashion but are relatively shallow compared with the distant source case where subsurface flow is deep and groundwater capture occurs. Although minor branching may occur, especially near channel heads (Schorghofer et al., 2004), overall the channels remain parallel to one another and develop a uniform lateral spacing. The degree to which channels bifurcate appears to depend on the difference between the depth of flow within the sediment layer at the head of an elongating channel and the depth of flow within the channel; large differences between these depths tend to promote strong flow into the channel head from different directions, leading to bifurcation (Pornprommin et al., 2010). Such conditions are favored by a horizontal, rather than a sloping, sediment surface. Local sources allow inputs of seepage flow to channels on a horizontal surface from multiple directions, promoting more varied branching than when a distant source supplies seepage flow from a specific direction to channels (Figure 2.27).

2.7.2 How Have Mathematical Models Been Used to Explore Drainage-Basin and Network Development?

Over the last 30 years, mathematical modeling of the dynamics of river systems at the drainage-basin/network scales has

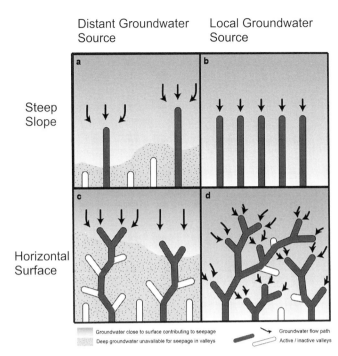

Figure 2.27. Channel systems formed by groundwater seepage from local or distant sources on steeply sloping and horizontal surfaces. A distant source (a, c) results in valley abandonment due to upstream capture of groundwater, whereas a local groundwater source (b, d) is less prone to flow capture. Horizontal surfaces (c, d) have a strong tendency to form valley bifurcations in contrast to steep slopes (a, b). Valleys emerging from a distant groundwater source result in an open landscape as no valleys develop downstream of large valleys (from Marra et al., 2015).

increased dramatically. Initial efforts to simulate the development of river networks involved two-dimensional models governed by rules about how networks form on a flat, planar surface that neglects variations in elevation throughout watersheds (Scheidegger, 1967; Howard, 1971a; Stark, 1991; Masek and Turcotte, 1993). Such models produce networks that look realistic; however, the neglect of gradient-driven erosional dynamics causes the networks generated by these models to deviate from scaling relations for natural river networks (Niemann et al., 2001). Thus, the realistic simulation of river networks requires three-dimensional models that account for the gravitational effects that drive erosion mechanics.

2.7.2.1 What Are Landscape-Evolution Models?

Three-dimensional models intended to simulate the evolution of landscape topography and river networks are known as landscape-evolution models (LEMs) (Martin and Church, 2004: Willgoose, 2005; Tucker and Hancock, 2010). An LEM consists of governing equations for conservation of mass; the movement of sediment and possibly solutes on hillslopes; runoff generation and flow routing; and erosion, transport, and deposition of sediment by rivers (Tucker and Hancock, 2010). Diffusive and advective processes are incorporated, and many models can simulate detachment-limited and transport-limited conditions. Functions for conversion of bedrock to weathered material (regolith or soil) and tectonic uplift also are included (Chen et al., 2014). Solutions to the governing equations for a set of boundary and initial conditions are obtained using numerical methods that produce time-stepped solutions throughout a spatially discretized computational domain representing the landscape.

2.7.2.2 How Can LEMs Be Used to Explore Landscape Evolution?

A variety of LEMs are available for simulating landscape evolution, including the evolution of river networks (Table 2.1). A main use of these models is to perform numerical experiments to explore hypothetically how various factors influence landscape evolution. Rigorous testing of model predictions of landscape evolution over geologic timescales is challenging and requires for the landscape of interest information on driving forces (e.g. climate or changes in baselevel), the timing of changes in landscape form, and initial and boundary conditions (Bishop, 2007). Alternatively, models can be tested against landforms, including stream networks, created in small-scale landscape-evolution experiments (Hancock and Willgoose, 2001b, 2002). Yet another approach is to determine how well statistical scaling characteristics of simulated topography and stream networks generated by such models conform with statistical scaling characteristics for natural drainage basins and river networks (Pelletier, 1999). From a practical perspective, LEMs are proving useful in evaluating landscape change over short timescales (decades to centuries) relevant to land management. Processes that contribute to land degradation, such as gully, rill, and

Table 2.1. List of some landscape-evolution models.

Model	Source
SIBERIA	Willgoose et al. (1991a,b)
DELIM	Howard (1994)
GOLEM	Tucker and Slingerland (1994)
CHILD	Tucker et al. (2001)
CASCADE	Braun and Sambridge (1997)
CAESAR	Coulthard et al. (2002)
CIDRE	Carretier et al. (2009)
SIGNUM	Refice et al. (2012)
LE-PHIM	Zhang et al. (2016a)

sheetwash erosion, are now being simulated using LEMs (Hancock et al., 2008; Coulthard et al., 2012; Hancock et al., 2014a).

LEMs have been used to explore the influence of tectonism, climate, and even vegetation on the morphologic evolution of landscapes dominated by fluvial erosion (Ibbitt et al., 1999; Beaumont et al., 2000; Istanbulluoglu and Bras, 2005; Bishop, 2007; Colberg and Anders, 2014). Under specific tectonic forcing conditions, simulations suggest that classic conceptual models of landscape evolution are consistent with predicted patterns of evolution (Kooi and Beaumont, 1996) (Figure 2.28) Classic modes of evolution, except for parallel slope retreat (King, 1953), require relatively simplistic initial landscape geometries. In simulations, landscapes with different initial forcing mechanisms that disrupt the balance between tectonic mass input and denudational mass output exhibit nonlinear modes of evolution (Kooi and Beaumont, 1996), implying that initial conditions strongly influence the dynamics of landscape evolution.

2.7.2.3 What Is Model Nonlinearity and What Are Its Implications for Landscape Evolution?

Nonlinearity in system models arises when change in predicted output is not proportional to change in input. Nonlinear mathematical models are characterized by an equation or set of equations to be solved that cannot be expressed as a linear combination of the unknown variables or functions within the equation or set of equations. Typically, the dynamics of nonlinear system models are much more complex than those of linear system models. Predicted dynamics can involve multiple stable states and chaotic behavior, in which small differences in initial conditions lead to large differences in system outputs over time or space. Whereas linear systems always reach the same equilibrium state regardless of initial conditions, nonlinear systems may have several different stable states that vary depending on the initial conditions of the system. Moreover, the evolution toward a particular stable state from a particular set of initial conditions may be highly sensitive to perturbations, so this evolution may shift from one stable state to another stable state in response to perturbations.

If nonlinearity in LEMs captures the dynamics of natural landscapes, it implies that variability in initial conditions can strongly influence landscape evolution (Perron and Fagherazzi, 2012). Morphological characteristics of landscapes, including the structure of river networks, can vary either because landforms evolving from different initial conditions reach different equilibrium states or because landscapes evolving from different initial conditions toward a single state converge too slowly to reach equilibrium under natural conditions. Moreover, perturbations or changes in

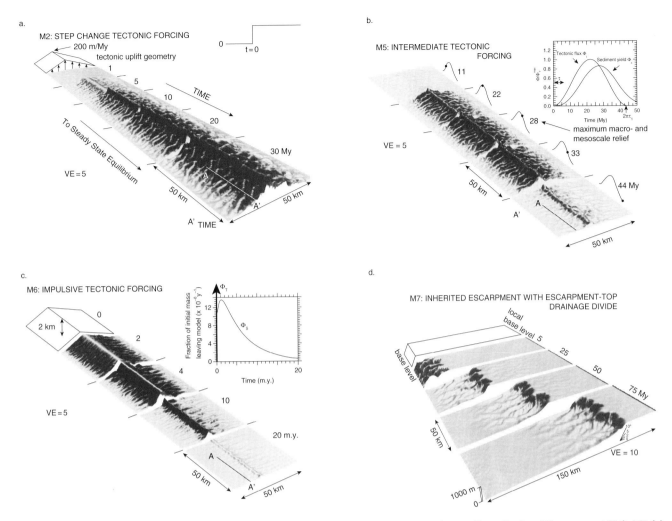

Figure 2.28. Simulated modes of landscape evolution for different initial and boundary conditions (from Kooi and Beaumont, 1996). M2 (a): step change in tectonic uplift and development of steady-state topography as erosion balances uplift (Hack, 1960; Willett and Brandon, 2002). M5 (b): episodic, cyclical tectonism with phase lag between uplift and erosion à la Penck (1972). M6 (c): impulsive tectonic uplift followed by quiescence à la Davis (1899). M7 (d): inherited elevated surface with escarpment à la King (1953).

process regimes may lead to evolution of landscapes toward different equilibrium states even though initial conditions of the landscapes are similar.

2.7.2.4 How Do Models Provide Insight into the Factors That Control Valley Spacing?

The spacing of valleys in linear mountain ranges is a problem that has received considerable attention within the context of landscape-evolution modeling. Valleys draining linear ranges exhibit rather uniform spacing (Hovius, 1996; Perron et al., 2008a). Although the development of uniform valley spacing has been attributed to the imposition of this pattern, which formed outside the mountain range, onto the range as it widens (Castelltort and Simpson, 2006), other work indicates that it develops through an evolutionary competition between narrow adjacent drainage basins for drainage area (Perron et al., 2008b, 2009). Valley spacing scales with a Peclet number

(Pe) – a dimensionless ratio normally used to quantify relative timescales of advective and diffusive processes associated with mass or heat transfer in fluids:

$$Pe = (k_i \ell^{2m+1})/D_e \qquad (2.11)$$

where k_i is an advective erodibility coefficient for river incision into bedrock, m is an exponent relating the rate of river incision to drainage-basin area (see Chapter 13), ℓ is a length scale, and D_e is soil diffusivity (Perron et al., 2009). Values of $Pe \approx 1$ correspond to the transition from hillslope (diffusion-dominated) to valley (advection-dominated) morphology. As Pe increases above 1, advective processes become more dominant over diffusive processes, leading to more active valley development and a decrease in valley spacing. This effect also can be characterized in terms of length scale. Setting $Pe = 1$ in Eq. (2.11) and solving for ℓ yields a critical length scale (ℓ_c) for valley formation:

$$\ell_c = (D_e/k_i)^{[1/(2m+1)]} \qquad (2.12)$$

An increase in the importance of advective processes (k_i and m) reduces ℓ_c, which decreases valley spacing (λ_v), whereas an increase in the importance of diffusive processes (D_e) increases ℓ_c and λ_v. The theoretical framework based on Pe can also be used to characterize valley branching (Perron et al., 2012). Numerical simulations indicate that when $Pe <$ 250 a series of parallel, uniformly sized valleys develop along a linear mountain range with the spacing of these valleys related directly to the magnitude of Pe. However, when $Pe >$ 300 an array of parallel, uniform valleys is unstable, and a perturbation will cause the drainage area of the perturbed valley to enlarge at the expense of its neighbors. Once $Pe > 60$ on side slopes of the enlarging valley, tributary valleys develop, resulting in a branching valley network. The instability caused by perturbing one valley propagates through the rest of the valleys until all valleys have stabilized in either branching or nonbranching configurations. Spatial variation in bedrock erodibility (Giachetta et al., 2014) or changes in baselevel (McGuire and Pelletier, 2016) can influence competition between evolving valleys and the resulting steady-state spacing of valleys.

2.7.2.5 How Have LEMs Been Used to Examine Stream-Network Development?

The use of LEMs to analyze in detail the growth of river networks has been somewhat limited. Simulations show that channel networks expand headward into an unchanneled, uplifted surface with random variations in topography (Willgoose et al., 1991c) (Figure 2.29). The rate of growth in drainage density over time in simulations is fairly linear with the rate decreasing as network expands toward the boundaries of the computational domain (Willgoose et al., 1991b). Drainage density under constant tectonic forcing (steady state) reaches a constant maximum value, and no reduction in density is evident at late stages of network evolution such as that in a landscape undergoing a progressive reduction in relief (Schumm et al., 1987). The details of network growth are sensitive to initial conditions, with variations in network characteristics developing for identical simulated erosional histories and values of model constants (Willgoose et al, 1991b; Howard, 1994). Given only differences in initial unchanneled random topography, otherwise identical simulations generate dissimilar arrangements of drainage basins, stream-network patterns, and divides, but similar average landscape morphology. High initial roughness favors increased branching of evolving networks, as flow moving down the rough surface is forced to converge locally by roughness variations, whereas low roughness favors less branching, as flow paths can move unimpeded down the slope parallel to one another (Simpson and Schlunegger,

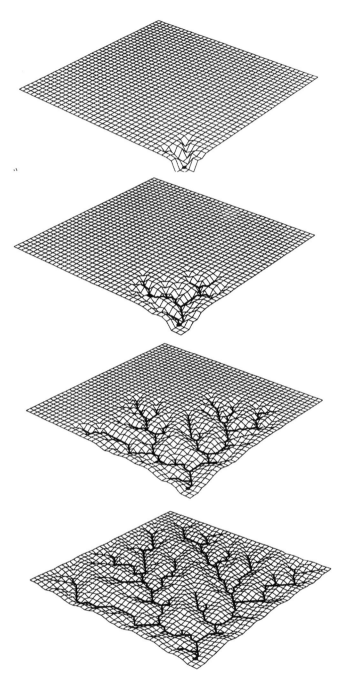

Figure 2.29. Simulated headward growth of a channel network (from Willgoose et al, 1991c).

2003) (Figure 2.30). These simulation results are consistent with laboratory experiments that have demonstrated the importance of initial conditions in network growth (e.g., Pelletier, 2003). The findings also point out the difficulty in determining precisely the evolutionary history of natural river networks, given that initial conditions are unlikely to be as simplistic as relatively flat, unchanneled, sloping surfaces with random microtopography.

In developed networks, simulations indicate that temporal variations in the threshold for channel initiation associated with changes in land cover or climate can produce hysteresis in drainage density over time (Rinaldo et al., 1995; Perron et al., 2012). Even though critical shear stresses for channel initiation return to previous values over time, the drainage density is not the same for a given value of critical shear stress as it was at the earlier time. This relation seems to reflect the nonlinear response of the landscape to channel formation. In essence, it is easier to form a channel than to eliminate one. Once a channel forms, it creates convergent topography in the form of a small valley. When the threshold for channel formation increases, this convergent topography sustains enough drainage area to maintain the channel.

2.7.2.6 How Do Stream Networks Compete with One Another over Geologic Timescales?

Over geologic timescales, dynamic reorganization of established river networks occurs, as migrating divides change the shapes of drainage basins and the structure of networks by capturing river channels in adjacent basins. The capacity of numerical models to simulate stream capture becomes important over these timescales (Howard, 1971b; Benaichouche et al., 2016). Modeling of divide migration and stream-network reorganization indicates that a characteristic metric of river profile form predicts patterns of divide movement amongst adjacent drainage basins (Willett et al., 2014). Based on the assumption that erosion rate is proportional to the rate of energy expenditure of flowing water as it moves downslope within a fluvially eroded landscape with uniform uplift rate (U_r) and rock erodibility (k_i), the relationship between elevation (Z) at any point (x) along a river in a network and the drainage area above that point in the network is

$$Z(x) = Z_b + \left(\frac{U_r}{k_i A_o^m}\right)^{1/n} \chi \qquad (2.13)$$

where Z_b is the elevation at the mouth of the river (i.e., at base level), and m and n are exponents in the stream power model of bedrock incision for a specific river system (Perron and Royden, 2013) (see Chapter 13). The quantity χ is an integral function of drainage area (A_d):

$$\chi = \int_{x_{bl}}^{x} \left(\frac{A_o}{A_d(x)}\right)^{m/n} dx \qquad (2.14)$$

where A_o is a reference drainage area (1 or 10 km^2) and x_{bl} is the distance at baselevel. Values of χ have units of distance and, because uplift rate and erodibility are assumed to be constant, serve as metrics for steady-state bed elevations at various locations (x) upstream from the mouth of a river (Eq. (2.14)). If basins separated by a divide are in steady state, i.e., no divide migration is occurring, plots of Z versus χ for rivers in both basins will be linear, and values of χ at channel heads

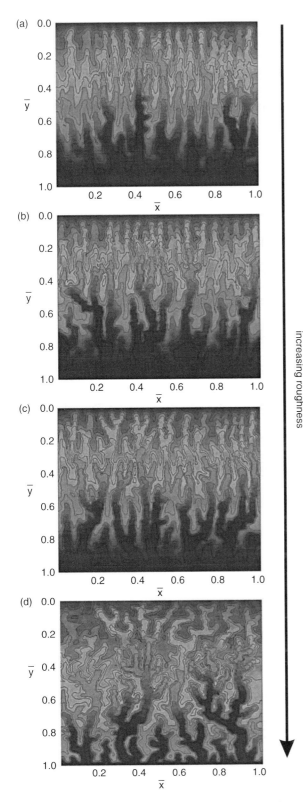

increasing roughness

Figure 2.30. Influence of roughness on stream dissection of an initial slope. As the roughness of the initial surface increases, the branching of streams and valleys dissecting the surface also increases (red corresponds to high elevations, blue to low elevations) (from Simpson and Schlunegger, 2003). (A black and white version of this figure will appear in some formats. For the color version, please refer to the plate section.)

near the divide will be identical for the two basins (Figure 2.31). If not, the divide for the basin with lower χ values near the divide will migrate toward the basin with higher χ values near the divide until conditions for steady state are achieved. Mapping of the spatial distribution of χ for numerical simulations of drainage-basin evolution in which divides actively migrate confirms this dynamic: divides more away from channel heads with low χ toward channel heads with high χ, divides do not stabilize until values of χ on both sides of the divides are equal, and at steady state all points in the simulated fluvial landscape have elevations predicted by values of χ (Willett et al., 2014). Thus, by mapping the spatial distribution of χ for natural landscapes, the extent to which landscapes are in equilibrium can be determined and, for those that are not in equilibrium, patterns of future divide migration can be assessed. Moreover, pronounced nonlinearities in plots of Z versus χ provide evidence of stream capture in cases where spatial patterns of χ suggest that capture may have occurred. The assumption underlying this scheme is that divides reach a steady state whereby no further migration is possible because river incision rates on both sides of a divide balance one another.

Predictions of the χ mapping approach contrast to some degree with the dynamics of landscape evolution predicted by LEMs, which exhibit dependence on the flow routing algorithm used in the LEM (Shelef and Hilley, 2013). Simulations in which surface runoff within an LEM moves downslope in multiple directions, rather than simply following the path of steepest descent, result in continuous divide migration even

under conditions of uniform vertical uplift, bedrock erodibility, and precipitation (Pelletier, 2004). Also, alternative metrics, such as contrasts in mean local relief and mean topographic gradient across divides, may provide more information on divide mobility than χ mapping when uplift rate and rock erodibility are not uniform (Forte and Whipple, 2018).

2.7.2.7 How Have LEMs Been Used to Model Stream-Network Development by Seepage Erosion?

The modeling of landscape evolution and stream-network development by seepage erosion has received much less attention than the modeling of landscape evolution by surface runoff. Since the proposal that stream networks in Vermont form largely through subsurface processes (Dunne, 1980), other stream networks formed by such processes have been identified. In particular, networks incised into Citronelle Formation in the Panhandle of Florida have formed by the action of seepage erosion (Schumm et al., 1995). The Citronelle Formation consists of highly permeable nonmarine quartz sands with discontinuous layers of clay and gravel. Streams incised into this material have valleys with steep walls and flat bottoms. Valley heads are amphitheater shaped, and springs emerge from the base of valley headwalls. This landscape has inspired several efforts to model the dynamics of its development. Use of an analytical model to simulate network expansion over time indicates that the ratio of headward growth to branching in this network has a length scale of

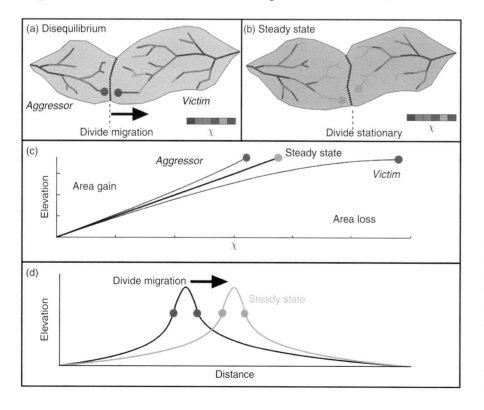

Figure 2.31. Mapping of χ across drainage divides indicates disequilibrium and a tendency for the divide migration (a) or steady state and a stationary divide (b). Plots of elevation versus χ show basin conditions at the divide relative to the steady-state condition (c). Channel locations (aggressor) that plot above (to the left, green dot) of the steady-state curve have high erosion rates relative to channel locations (victim) that plot below (to the right, red dot) of the curve, leading to divide migration (d) until steady state (yellow dots) is achieved (from Willett et al., 2014). (A black and white version of this figure will appear in some formats. For the color version, please refer to the plate section.)

about 460 m, indicating that branching on average occurs when the channels grow headward by this distance (Abrams et al., 2009). Coupling of a simple groundwater model with the hypothesis that streams grow in the direction that groundwater enters channel heads (Cohen et al., 2015) indicates that channels within the network should branch at a characteristic angle of 72°, a value that conforms well to field data on branching angles for sapped networks on the Florida Panhandle (Devauchelle et al., 2012; Petroff et al., 2013).

Many LEMS simulate erosion either by surface runoff or by seepage. In humid-temperate and tropical landscapes of the world, both surface and subsurface flow can play roles in erosion and the development of stream channels. Full hydrological coupling of subsurface flow and surface flow is the first step toward development of an integrated model that includes erosion by seepage, overland flow, and channelized flow. An LEM that incorporates this type of coupling indicates that subsurface flow can strongly influence the pattern of landscape evolution as the groundwater flow field evolves with changing topography (Zhang et al., 2016a). Simulated steady-state landscapes with subsurface flow have steeper slopes and higher relief than steady-state landscapes without this flow. When subsurface flow occurs, overland flow and erosion by this flow are constrained mainly to downslope locations near valley bottoms. To balance rock uplift rates, hillslopes steepen and diffusive processes increase to maintain appropriate rates of downslope movement of sediment.

2.8 How Are Channel Networks Organized Geometrically, Hierarchically, and Topologically?

River networks display remarkable patterns of spatial organization. Moving headward, these networks branch into smaller and smaller streams, whereas moving in the opposite direction, small streams flow together to eventually form great rivers. Quantitative characterization of the branching structure of river networks is essential for conducting analyses aimed at defining general characteristics of network spatial structure.

2.8.1 How Are Stream Networks Defined?

The first step in network analysis is to define a network. In the past, this effort relied mainly on topographic maps, aerial imagery, or, in cases of small networks, by mapping the extent of stream channels in the field. Streams depicted as blue lines on topographic maps often provided the basis for network delineations. These lines represent a cartographer's interpretation of stream locations based mainly on stereoscopic inspection of aerial imagery. Blue lines on topographic maps often fail to accurately represent the smallest streams in networks, especially

when these maps are compiled at fixed scales. Blue lines can be extended manually based on patterns of contour crenulations, but this practice is rather subjective. The blue-line approach has been superseded by the development of algorithms to automatically extract channel networks from DEMs. These automated methods, like the blue-line approach, are dependent on data resolution, but the growing availability of high-resolution topographic information, such as light detection and ranging (LiDAR) data (Persendt and Gomez, 2015), provides the basis for computer-based algorithms to ascertain channel networks at unprecedented levels of detail and accuracy (Pelletier, 2013; de Azeredo Freitas et al., 2016; Sangireddy et al., 2016b).

2.8.2 What Are the Major Geometric Patterns of Stream Networks?

River networks often exhibit distinct differences in the geometric pattern of streams within the network (Howard, 1967; Morisawa, 1985) (Figure 2.32). Distinctions among different network geometries are based largely on visual inspection. These distinctions are somewhat subjective and scale dependent, but differences in scaling characteristics and junction angles have been identified among geometrical patterns that can provide the basis for objective classification (Mejia and Niemann, 2008; Jung et al., 2015). In the absence of strong geological control, networks develop a uniform pattern of branching similar to the pattern associated with trees. This classic type of drainage pattern is referred to as dendritic. Increases in regional slope in excess of about 3° lead to the development of parallel networks (Jung et al., 2015). Easily erodible rocks, such as those in badlands, often display fine-grained pinnate patterns. Rectangular and trellis patterns are indicative of geological control, by angular configurations of joints and faults in rectangular patterns and by spatially alternating resistant ridges and erodible valleys in trellis patterns. Other less common network patterns include radial, annular, and contorted geometries. Distributary geometries, where a single main channel diverges into multiple subparallel channels each with its own mouth, are common on alluvial fans and deltas (Figure 2.33).

2.8.3 What Methods Are Used to Define the Hierarchical Structure of Stream Networks?

Two primary methods are used to characterize the hierarchical structure of stream networks: the stream ordering system (Horton, 1945; Strahler, 1952b) and the link-magnitude system (Shreve, 1967) (Figure 2.34). The rules for stream ordering are: 1) source tributaries not joined by another tributary are assigned a value of one; 2) the joining of two streams with the same order γ increases the order by one; and 3) if a stream of

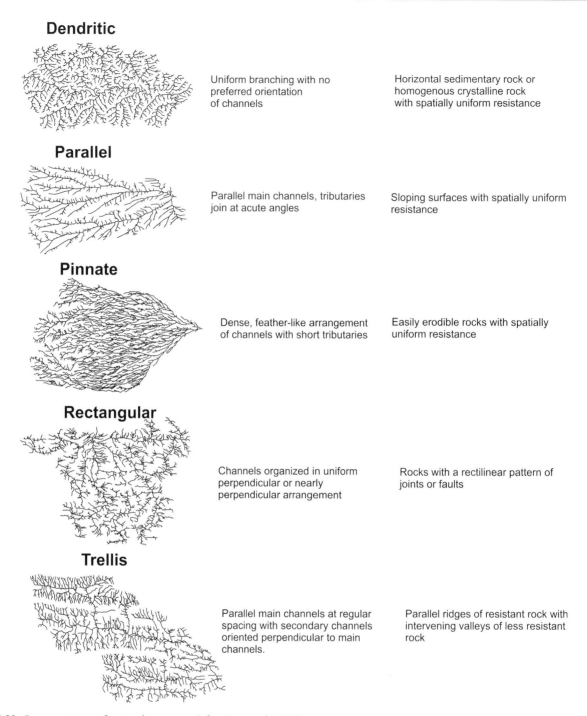

Dendritic

Uniform branching with no preferred orientation of channels

Horizontal sedimentary rock or homogenous crystalline rock with spatially uniform resistance

Parallel

Parallel main channels, tributaries join at acute angles

Sloping surfaces with spatially uniform resistance

Pinnate

Dense, feather-like arrangement of channels with short tributaries

Easily erodible rocks with spatially uniform resistance

Rectangular

Channels organized in uniform perpendicular or nearly perpendicular arrangement

Rocks with a rectilinear pattern of joints or faults

Trellis

Parallel main channels at regular spacing with secondary channels oriented perpendicular to main channels.

Parallel ridges of resistant rock with intervening valleys of less resistant rock

Figure 2.32. Common types of network geometry (after Jung et al., 2015).

lower order (Υ) joins a stream of higher order ($\Upsilon + 1$), the stream segment downstream of the junction becomes part of the upstream segment with the higher order. A well-known shortcoming of the stream ordering system is that it fails to account quantitatively for the influence of every tributary on the structural organization of the network. This shortcoming led to the development of the link-magnitude system, which treats the network as a set of links and nodes: links are segments between nodes, and junctions correspond to nodes (Shreve, 1966, 1967). Step one of the link-magnitude system is the same as that of the stream ordering system; thus, magnitude one links, also known as exterior links (M), are identical to first-order streams. In step two, when two links join, the magnitude of the downstream link is equal to the sum of the magnitudes of the two joining links. Thus, in contrast to the stream ordering system, magnitude changes at every junction in the network in direct proportion

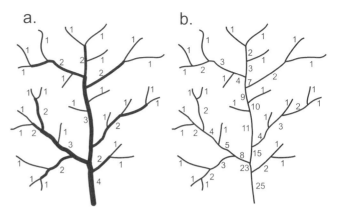

Figure 2.34. Stream ordering (a) and link magnitudes (b).

$$R_{Ai} = \overline{A}_d(\Upsilon + 1)/\overline{A}_d(\Upsilon) \tag{2.17}$$

where R_{bi} is the bifurcation ratio, R_{Li} is the stream length ratio, and R_{Ai} is the drainage area ratio, and these ratios are determined for individual successive pairs of orders i (e.g., order 1 versus order 2, order 3 versus order 4). Values of R_b, R_L, and R_A for entire stream networks, commonly referred to as Horton ratios, sometimes are approximated by averaging the individual ratios for all orders (e.g. Leopold et al., 1964), but precise values are obtained from statistical analysis. Plots of stream numbers versus order and of stream lengths versus order indicate that the number of streams decreases exponentially with order, whereas the length and drainage areas of streams increase exponentially with order (Figure 2.36):

$$N_\Upsilon = a_1 e^{\beta_1 \Upsilon} \tag{2.18}$$

$$\overline{L}_\Upsilon = a_2 e^{\beta_2 \Upsilon} \tag{2.19}$$

$$\overline{A}_{d\Upsilon} = a_3 e^{\beta_3 \Upsilon} \tag{2.20}$$

where a_1, a_2, and a_3 are constants, $\beta_1 = -\ln R_b$, $\beta_2 = \ln R_L$, and $\beta_3 = \ln R_A$. Ordinary least squares regression analysis of $\ln(N_\Upsilon)$, $\ln(\overline{L}_\Upsilon)$, and $\ln(\overline{A}_{d\Upsilon})$ versus Υ yields estimates of β_1, β_2, and β_3, from which estimates of the Horton ratios R_b, R_L, and R_A are derived as

$$R_b = e^{-\beta_1}, R_L = e^{\beta_2}, R_A = e^{\beta_3} \tag{2.21}$$

The statistical nature of the laws of drainage composition (Eq. (2.18)–(2.20)) means that these relations do not provide an exact description of network structure, but characterize average tendencies (Smart, 1968): for example,

$$N_\Upsilon \approx R_b^{\Upsilon_m - \Upsilon}; \overline{L}_\Upsilon \approx \overline{L}_1 R_L^{\Upsilon-1}; \overline{A}_{d\Upsilon} \approx \overline{A}_{d1} R_A^{\Upsilon-1} \quad \Upsilon = 1, 2, \ldots \Upsilon_m \tag{2.22}$$

Figure 2.33. Landsat 7 image of a distributary drainage pattern on the Lena River Delta, Siberia, 2–27-2000. (A black and white version of this figure will appear in some formats. For the color version, please refer to the plate section.)

to the number of exterior links upstream of the junction on each tributary. Correspondingly, the magnitude of the network at the outlet equals the total number of exterior links in the network. The total number of links in the network, both exterior and interior, is $2M - 1$. Although both the order and the magnitude of the river at the outlet of a network increase with network size, these measures are scale dependent and thus are not absolute indices of size. Order and magnitude at the outlet of a basin can change as the resolution of information on network extent increases and additional headwater streams are added to the network (Figure 2.35).

2.8.4 How Has Stream Ordering Been Used to Quantitatively Analyze the Branching Structure of Stream Networks?

The stream ordering system provides the basis for laws of drainage-network composition (Horton, 1945) that define how the number of streams in each order (N_Υ), the average length of streams in each order ($\overline{L}_\Upsilon$), and the average drainage area of streams in each order ($\overline{A}_{d\Upsilon}$) change with order (Υ). Three ratios reflect these changes (Horton, 1945; Schumm, 1956):

$$R_{bi} = N(\Upsilon)/N(\Upsilon + 1) \tag{2.15}$$

$$R_{Li} = \overline{L}(\Upsilon + 1)/\overline{L}(\Upsilon) \tag{2.16}$$

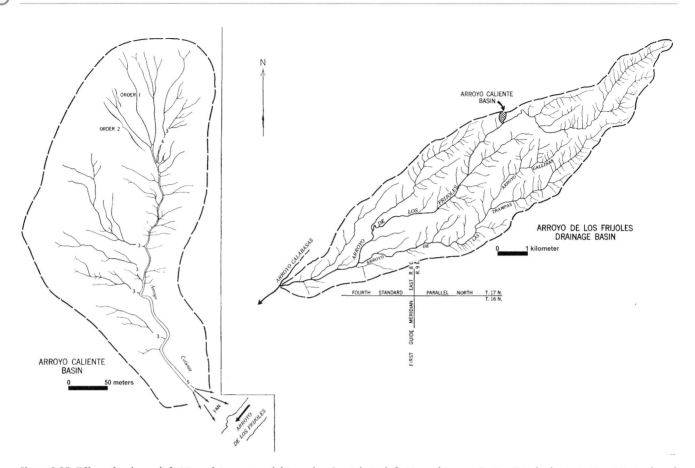

Figure 2.35. Effect of scale on definition of river network hierarchy. On right is definition of Arroyo De Los Frijoles basin in New Mexico based on a 1:24000 scale map. On left is field mapping of Arroyo Caliente, a first-order tributary on the 1:24,000 map (from Leopold and Miller, 1956).

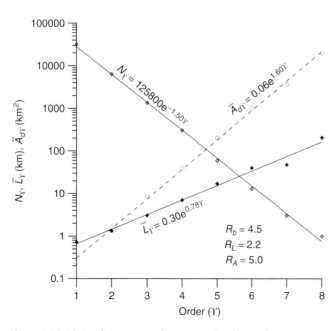

Figure 2.36. Plots of stream numbers, mean lengths and mean areas versus order, Powder River basin, WY (data from Peckham, 1995).

where Υ_m is the order of the network at the outlet stream. Values of the three ratios generally conform to rather narrow ranges; R_b varies from 3 to 5 with a modal value of 4, R_L varies from 1.5 to 3.5 with a modal value close to 2, and R_A varies from 3 to 6 (Abrahams, 1984; Kirchner, 1993).

Although the regularity of the ratios has been viewed as indicative of the uniformity of drainage-network composition despite variations in climate, lithology, and a host of other factors that might influence this composition, criticism has been leveled against this interpretation. In part, the relations seem to be at least partly an artifact of the ordering system itself. Because changes in order are not additive as in the link-magnitude system, increases in order are restricted by the rules of ordering. In networks in which the pattern of branching results in progressive joining of segments toward a defined endpoint or outlet, increases in order will become progressively more difficult to accomplish using the rules of the stream ordering system. This attribute contributes to the exponential form of the laws of drainage composition. The implication is that these laws are insensitive to a wide range of variation in the geometric or topological characteristics of stream networks (Kirchner, 1993). In this sense, the laws have limited value for identifying similarities or differences in network characteristics and, in turn, for determining how variations in possible controlling factors, such as differences in geology or climate, produce differences in these characteristics. Another shortcoming of the method is that it violates several assumptions of ordinary least squares regression, leading to error in estimation of Horton ratios (Furey and Troutman, 2008).

2.8.5 How Has the Link-Magnitude Method Been Used to Analyze Network Topology?

The link-magnitude method of characterizing stream-network structure provides the foundation for topological analysis of networks. Topological analysis focuses on the arrangement of stream segments within a river network independently of the geometric characteristics of these segments (e.g., the lengths of segments or the angles at which segments join at nodes). Foundations for topological analysis rely on the depiction of networks in topological form (Figure 2.37). For any network of magnitude M, the total number of possible arrangements of the links within the network $N(M)$ equals (Knighton, 1998)

$$N(M) = \frac{(2M - 2)!}{M!(M - 1)!} \qquad (2.23)$$

Each different arrangement of a network is referred to as a topologically distinct channel network (TCDN) (Figure 2.37). Attributes of networks depicted in topological form include the diameter of the network, which is the length of the longest path through the network expressed in terms of the number of links from the outlet to the most distant magnitude one link, and network width, which is the number of links at successive link distances from the outlet to the most distant magnitude one link. Network width has proven to be especially useful in relating the hydrological response of a drainage basin to the spatial structure of its drainage network (Chapter 6).

The random topology model of stream networks proposes two hypotheses: 1) in the absence of strong geological controls, natural river networks are topologically random, i.e., all TCDNs are equally likely; and 2) in uniform environments, interior and exterior link lengths have separate statistical distributions that are independent of location within drainage basins. Considerable effort was devoted to testing these two hypotheses from the late 1960s to the early 1980s. Results indicate that small networks conform with topological randomness to a greater extent than large networks, that link length distributions are complex and do not seem to conform well to

Figure 2.37. All possible topologically distinct channel networks for $M = 6$.

any particular probability distribution, and that exterior versus interior link lengths generally differ from one another, but in rather a nonsystematic manner (Abrahams, 1984). Related work based on consideration of probability distributions of lower-order tributaries to higher-order streams supports the conclusion that stream segments are distributed randomly throughout drainage networks (Dodds and Rothman, 2001c).

Representation of stream networks topologically also has highlighted the geometrical arrangement of tributary entry and tributary development in relation to branching angles. Links bounded by tributaries that enter the link from the same side are *cis* links and those bounded by tributaries that enter the link from opposite sides are *trans* links (James and Krumbein, 1969) (Figure 2.38). Although the random model predicts that *cis* and *trans* links should occur with equal frequency, *trans* links tend to be more abundant than *cis* links, presumably because tributaries on the same side of a stream segment compete for drainage area, whereas those on opposite sides do not (Abrahams, 1984). Moreover, if a stream branches from a main river at an acute angle (<90°), a common phenomenon in most natural river networks (De Serres and Roy, 1990), tributaries along the branching stream develop preferentially toward the obtuse-angle (>90°) side of this stream (Flint, 1980). The asymmetry of junction angles produces differences in the availability of drainage area to support channel formation on the two sides of the branching stream (Figure 2.38). More drainage area is available on the outer (obtuse-angle) side of the branching stream than along its inner (acute-angle) side.

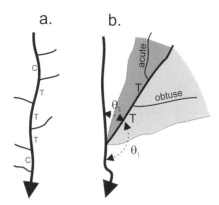

Figure 2.38. (a) Main stream with *cis* (C) and *trans* (T) links. (b) Obtuse (θ_1) and acute (θ_2) angles at a branching stream to a main river. The obtuse side of a branching stream has a greater drainage area (light shading) than the acute side (dark shading), leading to preferential development both of an obtuse tributary to the branching stream and of *trans* links upstream of the confluence of the branching stream with the main river. Arrows indicate direction of flow.

This effect favors the development of *trans* links immediately upstream of the junction of a branching stream with a main river. It also demonstrates that network structure within the interior portions of these networks is constrained by spatial relations among tributaries in relation to the availability of drainage area to support channel development and maintenance.

2.9 What Are the Scaling Characteristics of River Networks?

The issue of scaling examines the extent to which the characteristics of river networks change or remain similar as the scale of the network increases or decreases. In other words, do large networks have the same characteristics as small networks, or do the characteristics of these different-size networks differ? Horton ratios represent scaling laws in the sense that if the individual ratios remain constant over successive orders, the characteristics of bifurcation, length, and area remain similar with increasing stream size. Scaling, however, is particular to the type of metric used to examine scaling. One type of metric may indicate that the attributes measured by the metric remain similar with changes in size, whereas another metric may show that other attributes of a stream system change with size. Thus, different types of metrics and analyses have been employed to explore scaling issues related to stream networks and drainage basins.

Scaling is tied closely to the concept of fractals and fractal geometry (Mandelbrot, 1982). Fractal scaling refers to objects of different sizes that exhibit identical deterministic or statistical

characteristics. In the study of river systems, the analysis is statistical, and fractal scaling relates to invariance in statistical properties describing some aspect of the system with changes in geometry or size (Rodriguez-Iturbe and Rinaldo, 1997). An object that conforms to fractal scaling exhibits self-similarity. If one zooms in on or out on the object, its overall geometry looks similar; it exhibits infinite nesting of structure at all scales. Without a scale of reference, the size of such an object cannot be ascertained from its appearance, which is scale invariant. The fractal dimension of an object is a measure of its complexity that defines how the detail of the object's geometry changes with the scale at which it is measured. It can also be interpreted in the context of space filling. A line that is so complex it completely fills a two-dimensional surface (i.e., a plane) has a fractal dimension of 2.

In the case of the laws of drainage composition, the relative constancy of Horton ratios with stream order is only a general indicator of self-similarity. Rigorous evaluations should be based on analysis of probability distributions of drainage areas or stream lengths for different orders (Peckham and Gupta, 1999; Dodds and Rothman, 2001b). This type of analysis shows that probability distributions match closely for different orders, supporting the idea that river networks exhibit self-similarity (Peckham and Gupta, 1999).

One of the most fundamental scaling relations identified in the study of river systems is the relation between the length of a stream, measured from a given location along its path to the drainage divide, to the drainage area of the stream, measured upstream of the given location. The relationship between stream length (L) and drainage area (A_d) was initially examined empirically for streams in the Appalachian Mountains of Virginia, yielding the power function (Hack, 1957)

$$L = 1.4 A_d^{0.6} \qquad (2.24)$$

Geometrical similarity of basins over all scales implies that the exponent in Eq. (2.24) should have a value of 0.5. This value of the exponent would dimensionally balance the units of the two sides of the equation, resulting in an isometric relationship between stream length and drainage area. Under this condition, basin shape would remain constant over size. The exponent of 0.6 suggests that large basins have elongated shapes relative to small basins. Some empirical studies support elongation (Rigon et al., 1996), but others do not (Willemin, 2000; Dodds and Rothman, 2001a). Another possible explanation is that basin shape remains constant, but lengths of rivers tend to increase in relation to the total drainage area in large drainage basins (Peckham, 1995). This relation also may reflect the fractal nature of the measurement of river lengths at different scales (Robert and Roy, 1990). Related work has examined the relationship between basin length (L_d), the length of the basin along the axis of the main valley, and basin area for stream systems ranging over nearly 12 orders of magnitude in basin size (Montgomery and Dietrich, 1992):

$$L_d = 1.78\,A_d^{0.49} \tag{2.25}$$

The value of the exponent close to 0.5 suggests that the relation between basin length and drainage area is isometric. Support for an isometric relation between stream length and drainage area has been reported for 37 of the largest river basins in the world (Dodds and Rothman, 2000).

The analysis of fractal scaling in river systems is extensive (Rodriguez-Iturbe and Rinaldo, 1997). Approaches based on Horton ratios assume that river networks are inherently fractal, and under this assumption the use of the ratios becomes a way to estimate fractal dimensions (Kirchner, 1993). Using Horton ratios, a relation for the fractal dimension D_f of stream networks can be defined as (Tarboton et al., 1988; La Barbera and Rosso, 1989)

$$D_f = \frac{\log R_b}{\log R_L} \tag{2.26}$$

This relation is valid for $R_b > R_L$, which is the case with most natural stream networks. Data for R_b and R_L (Morisawa, 1962) indicate that D_f varies from 1.7 to 2.5 (Tarboton et al., 1988). Topologically random networks have values of $R_b = 4$ and $R_L = 2$, yielding $D_f = 2$ (Shreve, 1967). These findings suggest that channel networks are space filling. This conclusion conflicts with work on channel-initiation thresholds that suggests channel networks are not space filling because a scale limit exists for landscape dissection by streams. Other empirical work indicates that D_f varies from 1.5 to 2 with an average of about 1.6 to 1.7 – values that are not space filling (La Barbera and Rosso, 1989). Including the fractal dimension of river lengths (D_l) in Eq. (2.26) leads to (Tarboton et al., 1990)

$$D_f = D_l\frac{\log R_b}{\log R_L} \tag{2.27}$$

Since $D_l \approx 1.11$, inclusion of this factor increases reported values of D_f from less than two (La Barbara and Rosso, 1989) to close to two; however, it also has the consequence of producing values of D_f in excess of two for both natural and topologically random networks – a result inconsistent with 2D planar representations of river networks (Phillips, 1993). An alternative formulation (La Barbera and Rosso, 1990)

$$D_f = \frac{1}{(2 - D_l)}\frac{\log R_b}{\log R_L} \tag{2.28}$$

has the same disadvantage. The use of Horton ratios to examine fractal characteristics of river systems has been extended to define fractal dimensions of: 1) the length–area relationship of subbasins within a drainage basin, 2) the total length of streams in a network as a function of scale, 3) the mainstream length–area relationship (Hack's law), and 4) the total drainage area of streams as a function of areal scale (Beer and Borgas, 1993).

An alternative approach to addressing the question of whether river networks are fractal is to use a box counting procedure whereby boxes of different sizes are superimposed onto a depiction of a drainage network. A power-law relation between box size (s_b) and the number of boxes intersecting a stream segment (n_b) defines the fractal dimension:

$$n_b = as_b^{-D_f} \tag{2.29}$$

This type of analysis for natural stream networks shows that systematic deviations occur from the power-law relation; thus, a single, well-defined fractal dimension cannot be identified and the networks are not self-similar (Beauvais and Montgomery, 1997). The analysis does indicate that channel networks are space filling up to limits of dissection defined by channel-initiation thresholds.

Another approach to the exploration of self-similarity in river networks draws upon stream ordering and tree graphs (Peckham, 1995). This method focuses on how tributaries within a branching network enter rivers of larger size as defined by stream order. Streams of lower order joining a stream of higher order are *side* streams, and streams of higher order that a side stream joins are *absorbing* streams (Dodds and Rothman, 2001c). In a self-similar branching network, a lower triangular matrix defines the relation between side streams and absorbing streams:

$$\begin{bmatrix} \overline{N}_{2,1} & & & & \\ \overline{N}_{3,1} & \overline{N}_{3,2} & & & \\ \overline{N}_{4,1} & \overline{N}_{4,2} & \overline{N}_{4,3} & & \\ \vdots & \vdots & \vdots & \ddots & \\ \overline{N}_{\Upsilon,1} & \overline{N}_{\Upsilon,2} & \overline{N}_{\Upsilon,4} & \cdots & \overline{N}_{\Upsilon,\Psi} \end{bmatrix}$$

Each entry in the matrix ($\overline{N}_{\Upsilon,\Psi}$) represents the number of side streams of a specific order Ψ joining an absorbing stream of a specific order Υ. Absorbing streams of order one are not included because first-order streams cannot be joined by a stream of lower order. Self-similar branching trees are characterized by constant values on matrix diagonals (Figure 2.39). Matrices for natural networks can be evaluated to determine the extent to which matrix values conform to those for self-similar trees and to explore reasons for deviations from self-similarity (Peckham, 1995; Dodds and Rothman, 2001c).

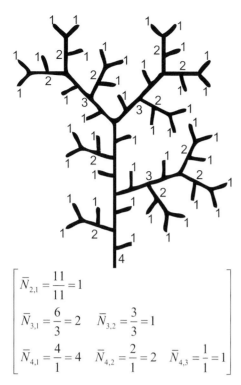

$$\begin{bmatrix} \bar{N}_{2,1} = \dfrac{11}{11} = 1 \\[2mm] \bar{N}_{3,1} = \dfrac{6}{3} = 2 \quad \bar{N}_{3,2} = \dfrac{3}{3} = 1 \\[2mm] \bar{N}_{4,1} = \dfrac{4}{1} = 4 \quad \bar{N}_{4,2} = \dfrac{2}{1} = 2 \quad \bar{N}_{4,3} = \dfrac{1}{1} = 1 \end{bmatrix}$$

Figure 2.39. Example of a self-similar fourth-order drainage network and its lower triangular matrix showing number of side streams of lower order (numerator) joining absorbing streams of higher order (denominator). Average values of the ratios are constant along diagonals of the matrix.

2.10 Is the Organization of River Networks Governed by Optimization Principles?

A different method of examining the structure of stream networks is to develop computer algorithms that generate networks by optimizing energy expenditure within a drainage system (Howard, 1990; Rodriguez-Iturbe and Rinaldo, 1997). Energy expenditure corresponds to the power of the flow it moves along the river (see Chapter 4). Optimization principles can be local or global. Local principles include minimization of energy expenditure within any link of the network and equal energy expenditure per unit length of channel anywhere within the network. A global principle is minimum energy expenditure in the network as a whole. An initial network pattern generated randomly or by a headward growth model is rearranged iteratively until the network configuration satisfies the optimization principles. Networks generated by optimization algorithms are known as optimal channel networks (OCNs) (Rodriguez-Iturbe and Rinaldo, 1997).

Comparisons between scaling properties of natural river networks and those of OCNs provide insight into the degree to which natural networks conform to the structure of OCNs. For example, OCNs predict that networks will elongate with increasing network size, or drainage-basin area, in conformance with Hack's law (Ijjasz-Vasquez et al., 1993). In many cases, OCNs and real networks display remarkable similarity in geometrical characteristics and in scaling properties, suggesting that optimality principles govern the development of natural drainage networks (Rinaldo et al., 2014). The view of natural river networks as OCNs has been linked to the general scientific concept of self-organized criticality, whereby complex systems with many interacting components evolve into stable configurations (Bak, 1996). OCNs are spatial models of self-organized criticality, with self-organization governed by optimality principles (Rodriguez-Iturbe and Rinaldo, 1997).

Although the similarity between OCNs and natural channel networks has been proposed as an explanation for the structure of natural river networks, causality is difficult to identify in optimization approaches. Confirming that river systems actually minimize rates of energy expenditure at both local and global scales is difficult to achieve empirically, especially throughout entire networks. Even if minimization does occur, the extent to which the evolution of drainage networks is caused by minimization remains unclear. The attainment of minimization states may be a consequence of network evolution that involves nonlinear dynamics of myriad interacting fluvial processes. Optimality approaches imply that ideal conditions exist in nature toward which a system is inherently attracted – a teleological form of explanation. Such explanations do not specify the fluvial processes that generate network structure.

3 Sediment Dynamics at Global and Drainage-Basin Scales

3.1 How Important Are Rivers in Transporting Sediment and Dissolved Material at the Global Scale?

The primary job of rivers is to redistribute water across terrestrial surfaces of the Earth and to deliver much of this water to oceans (Phillips, 2010). In this sense, rivers are fundamentally hydrological entities. Sediment dynamics are a by-product of hydrological processes, but the role of rivers in redistributing sediment is of primary interest to geomorphologists. This redistribution of sediment maintains or changes the morphology of Earth's land surfaces. It also delivers clastic material to shallow ocean environments through delta formation and deposition on continental shelves. The material load of rivers includes products of chemical weathering processes that decompose earth materials into dissolved solids. Delivery of these dissolved solids to streams and rivers through surface and subsurface hydrological pathways contributes to the material flux of fluvial systems.

Geomorphological activity at the global scale involves the movement of eroded material from the terrestrial land surfaces to the ocean basins. Besides rivers, other geomorphic agents, such as glaciers, wind, and wave action, also move eroded material. Both fluvial and glacial processes depend on gravitational gradients, which are directed toward sea level – the ultimate baselevel for these processes. Coastal and eolian processes, although less constrained directionally, contribute to the material flux from land surfaces to the oceans. Global estimates of the average total annual material flux to the oceans by sediment suspended in the water column of rivers ranges from 13.5 to about 26.7 gigatons (Gt) per year, with the most recent estimates around 18–20 Gt yr^{-1} (Table 3.1). The amount of sediment moving on the bed of rivers is difficult to ascertain, and estimates typically assume that this material is a small

Table 3.1. Estimates of global material flux delivered to oceans (Gt yr^{-1}).

Rivers		
Sediment in suspension	18.3	Holeman (1968)
	26.7	Jansen and Painter (1974)
	13.5	Milliman and Meade (1983)
	15	Walling and Webb (1983)
	20	Milliman and Syvitski (1992)
	15.5	Dedkov and Gusarov (2006)
	16	Ludwig and Probst (1998); Syvitski et al. (2005)[a]
	14	Syvitski et al. (2005)[b]
	18	Syvitski (2003); Syvitski et al. (2003)
	19	Milliman and Farnsworth (2011)
Sediment moving on bed	2	Syvitski (2003); Syvitski et al. (2003)
Dissolved solids	5	Syvitski (2003); Syvitski et al. (2003)
	3.8	Milliman and Farnsworth (2011)
	4.2	Garrels and Mackenzie (1971)
Wind	0.3–0.9	Various references in Mahowald et al. (2005)
	0.7	Syvitski (2003); Syvitski et al (2003)
	0.06	Garrels and MacKenzie (1971)
Glaciers, sea ice, icebergs	2	Garrels and MacKenzie (1971)
	4.5	Syvitski et al. (2019)
Coastal erosion	0.4	Syvitski (2003); Syvitski et al. (2003)
	0.5	Syvitski et al. (2019)
	0.2–0.9	Garrels and MacKenzie (1971)

[a] Without reservoirs, [b] with reservoirs.

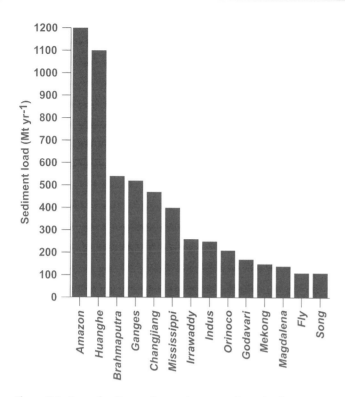

Figure 3.1. Annual sediment flux to the oceans from the fourteen rivers with the largest fluxes (data from Milliman and Farnsworth, 2011).

proportion ($\approx$ 10%) of the suspended load. Rivers also transport a substantial amount of dissolved solids (Table 3.1). Based on the most recent estimates, the total flux of solid and dissolved material to oceans from rivers is about 25 Gt yr^{-1}. By contrast, estimated material fluxes to the oceans by glacial, eolian (wind), and coastal processes are much less than those produced by fluvial action (Table 3.1). Collectively, these processes deliver about 3 to 6 Gt yr^{-1} of sediment to the oceans each year. Rivers clearly are the dominant agents of material transport on Earth's surface, accounting for about 80% to 90% of the total flux to the oceans. This global comparison does not provide information on the local effectiveness of each process at producing erosion of landscapes. In some locales, especially high mountain environments, rates of glacial erosion exceed rates of fluvial erosion (Koppes and Montgomery, 2009). However, globally, the effectiveness of glaciers as erosional agents is constrained by the limited extent of ice cover on Earth's surface.

At a global scale, fourteen major rivers account for about 30% of the total sediment flux to oceans (Figure 3.1). The Amazon River has the largest sediment flux (1200 Mt yr^{-1}) followed closely by the Huanghe River in China (1100 Mt yr^{-1}). All these rivers, except the Godavari River in India, drain high mountain environments (> 3000 m above mean sea level). Rivers in Asia and Oceania produce the greatest flux of sediment to the oceans, accounting for nearly two-thirds of the total flux (Figure 3.2). Interior drainage by

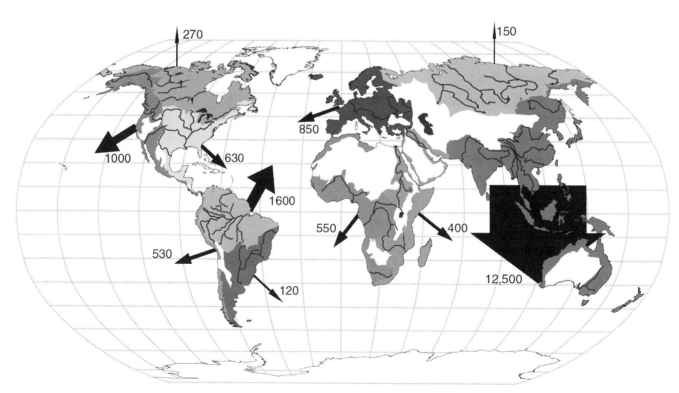

Figure 3.2. River fluxes of suspended sediment to the oceans from different geographic regions. Values are in Mt yr^{-1} (from Milliman and Farnsworth, 2011).

rivers that do not reach oceans encompasses large parts of central Africa, Eurasia, and Australia. In these regions, redistribution of sediment by rivers can produce changes in landscape morphology but does not contribute to the global flux of sediment to the oceans.

Estimates of sediment fluxes of rivers to oceans must be tempered with the caveat that the most downstream measurement locations on which these estimates are based often are located far upstream of where rivers flow into the ocean (Phillips and Slattery, 2006). Deposition of transported sediment along lower reaches of rivers downstream of measurement locations is particularly likely on passive continental margins characterized by low-gradient coastal plains and abundant estuaries and deltas. Along the Atlantic coast of the United States, less than 10% and perhaps less than 5% of river-transported sediment reaches the continental shelf or deep ocean over millennial timescales (Meade, 1982). Thus, fluxes to oceans are likely to be overestimated in such environments.

3.2 What Factors Control the Total Sediment Flux from Drainage Basins?

The flux, or discharge, of sediment by a river at the mouth of a basin is referred to as the sediment load (Q_s). It typically has units of mass per unit time, or in metric units, tons per year for annual values. Sediment yield (Q_{sy}) is obtained by dividing the total sediment load for a drainage basin by the basin area (A_d):

$$Q_{sy} = \frac{Q_s}{A_d} \tag{3.1}$$

This metric provides the basis for comparisons of relative fluxes among basins of different sizes. The relation between sediment flux and landscape erosion is expressed as denudation rate, the rate of surface lowering averaged over the entire watershed area. Given data on Q_{sy} in units of t km^{-2} yr^{-1}, the denudation rate (δ_d) in millimeters per thousand years (mm kyr^{-1}) is

$$\delta_d = \frac{1000 Q_{sy}}{\rho_s} \tag{3.2}$$

where ρ_s is the sediment density (kg m^{-3}).

Factors that influence mobilization of sediment on hillslopes and within stream channels as well as the movement of eroded material across hillslopes and through the stream network ultimately influence the amount of sediment transported out of a drainage basin by its main river. These factors include the amount of flow, the erosive effectiveness of the flow, and the capacity of the flow to transport eroded sediment (see Chapter 5). The amount, effectiveness, and capacity of flow can vary temporally and spatially, which will affect spatial

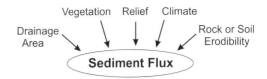

Figure 3.3. Major factors influencing sediment flux from drainage basins.

and temporal patterns of mobilization and transport. Once material is mobilized, spatial continuity of transport is required for the sediment in transport to remain in transport and not be redeposited before reaching the basin outlet. Variations in flow over time produce temporal variability in mobilization and transport at particular locations within a drainage basin. The effects of spatial and temporal variability on river sediment fluxes are most pronounced at the event timescale; averaging over events diminishes these effects at modern and geologic timescales.

Studies of fluxes out of drainage basins have focused on variables that influence sediment regimes of rivers over modern timescales (Figure 1.9), including climate, vegetation, drainage basin size, drainage basin topography, and geology, particularly rock or soil characteristics (Figure 3.3). Hypotheses related to these variables derive from basic physical reasoning about how each variable should influence sediment flux through its effect on runoff and erosion. The reasoning considers the independent effects of each variable, assuming all other variables are held constant.

Although climate and vegetation can be identified as separate factors influencing sediment flux, these two variables are strongly interrelated. Climate is characterized mainly in terms of mean annual precipitation and temperature. As mean annual precipitation increases, runoff should increase, resulting in increased hillslope and channel erosion. Therefore, sediment flux should be greater in wet environments compared with dry environments. Temperature will affect sediment flux mainly through its influence on the susceptibility of the landscape to erosion. Excessively cold or hot conditions will limit vegetation growth, exposing soil to runoff and erosion. Where temperatures are consistently below freezing, permafrost will develop, limiting the erodibility of frozen soil. Moreover, precipitation will occur mainly in the form of snow rather than runoff-generating rain.

Vegetation impedes hillslope erosion by reducing rates of runoff through interception of rainfall and evapotranspiration, by increasing hydraulic resistance to overland flow, and by increasing the shear strength of soil and weathered material. Wet environments with thick vegetation cover are less susceptible to erosion than drylands where soil and weathered material are exposed directly to runoff. Increases in vegetation should therefore decrease sediment flux.

Basin topography affects sediment flux mainly through its influence on hillslope gradients. Increases in relief, or more specifically the relief ratio, which is the maximum relief in a drainage basin divided by the length of the basin, should promote increases in sediment flux. The relief ratio is a general index of the average steepness of hillslopes within a drainage basin: basins with high relief ratios usually have steeper hillslopes than basins with low relief ratios. Increases in hillslope steepness will enhance diffusive and advective hillslope erosion processes, both of which are dependent on slope angles (see Chapter 2). Advective transport of sediment by streams in basins with high relief ratios also will tend to be greater than sediment transport by streams in basins with low relief ratios.

If all other variables are constant, the amount of sediment eroded in a large basin will usually exceed the amount eroded in a small basin based purely on the absolute differences in basin size. Thus, total sediment loads (Q_s) should increase with basin size. However, the relation for sediment yields, which standardize loads by basin size, is less clear. The form of the relation between sediment yield and basin area depends on the extent to which continuity of transport occurs throughout the basin over the timescale at which the flux is considered, usually an annual basis. In other words, if sediment is mobilized by erosion in the headwaters of a small basin versus a large basin, how likely is it that this mobilized sediment will reach the outlet of the basin within a year? The potential for disruption of continuity of transport from source location to the basin outlet and for temporary storage of sediment seems more likely in a large basin than in a small basin. Under such conditions, sediment yield should decrease as basin size increases.

The erodibility of rock and soil has an obvious connection to sediment flux. Drainage basins formed in weak, highly erodible rocks, such as shale or mudstones, will generate more sediment than those developed in strong, resistant rocks, such as quartzite or schist. Other factors besides lithology, such as jointing and weathering, can also affect rock strength and erodibility. Overall, however, sediment flux should increase as rock erodibility increases. Similarly, basins with highly erodible soils should produce more sediment than basins with soils resistant to erosion.

3.2.1 How Is Sediment Flux Related to Precipitation and Runoff?

Attempts to evaluate hypothetical arguments relating sediment flux to controlling variables have been confounded to some extent by complex interaction among these variables. Initial attention focused on the relation between climate and sediment flux under the assumption that climate, by influencing runoff, should strongly influence the amount of sediment in rivers. Data on sediment yield based on both sediment loads for streams in the central United States and sedimentation rates in reservoirs throughout the United States have been related to effective precipitation, the precipitation required to produce measured annual runoff (Langbein and Schumm, 1958). These data consist of group averages of sediment yield computed from numerous individual cases for discrete ranges of effective precipitation; no variances or ranges are reported for the data sets used to compute the group averages. Curves drawn on plots of sediment yield versus effective precipitation represent the interpreted relationship between these two variables (Figure 3.4). A key constraint is to have the curves extend to a sediment yield of zero for zero precipitation. These curves suggest that sediment yield peaks around an effective mean annual precipitation of about 300 mm – a value that corresponds roughly with the transition between desert shrub and grasslands. Because vegetation is dependent on precipitation, and vegetation was not controlled for in the analysis, the interpreted curves indicate that in extremely dry environments (< 300 mm) sediment yield becomes increasingly limited by lack of runoff as vegetation density decreases, whereas in wet environments (> 300 mm) sediment yield is limited by increasing vegetation cover despite increasing amounts of precipitation to produce runoff.

If the constraint of zero sediment yield at zero effective precipitation is relaxed, lines of best fit estimated using ordinary least squares regression indicate that sediment yield declines with increasing effective precipitation over the range of available data, reflecting the mitigating effect of increasing vegetation cover on sediment flux (Figure 3.4). Although it is reasonable to assume that sediment yield becomes zero for zero effective runoff, the lack of data for the interval 0 to 250 mm of effective precipitation calls into question the nature of the relation between sediment yield and effective precipitation for dry environments. In particular, it is not clear that yield declines progressively over this interval. Predictive modeling of sediment yield for different precipitation and vegetation regimes supports the interpretation that sediment yield peaks at around 250–300 mm of mean annual precipitation, with sediment yield decreasing above and below this peak value (Istanbulluoglu and Bras, 2006). However, these predictions may reflect the rather simplistic structure of relations among soil erosion, runoff, and vegetation cover embedded in the model, while not accounting for complicating effects of other variables on sediment yield.

Other work suggests that the relation between precipitation and sediment yield is complex, especially in the light of confounding effects of vegetation and other governing factors (Douglas, 1967; Wilson, 1973). A comprehensive analysis of data for 1246 measurement locations throughout the world reveals no relation between sediment yield and mean annual precipitation (Walling and Kleo, 1979) (Figure 3.5). An attempt to average these data by intervals of precipitation

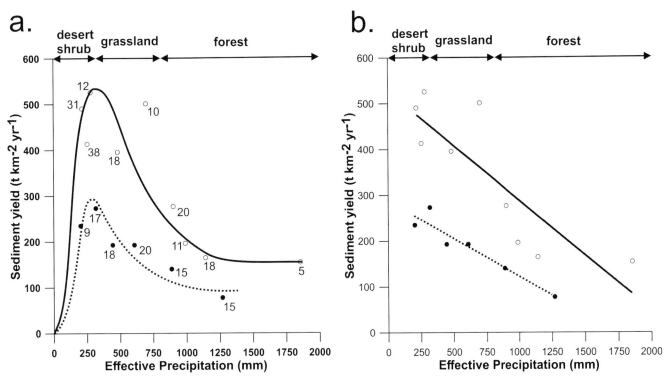

Figure 3.4. (a) Sediment yield versus effective mean annual precipitation for measured stream data (solid dots, dashed curve) and reservoir data (open circles, solid curve). Values beside data points indicate the number of cases from which averages constituting each point were computed (after Langbein and Schumm, 1958). (b) Best-fit lines for the same data determined from least squares linear regression.

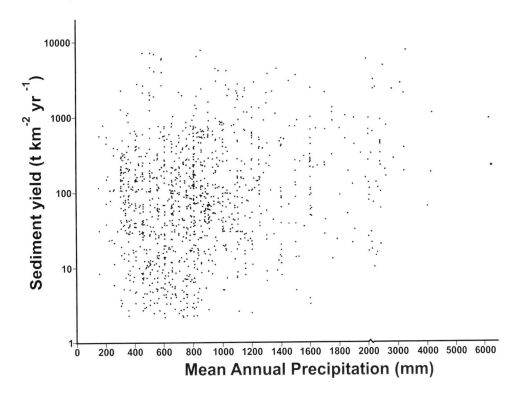

Figure 3.5. Sediment yield versus mean annual precipitation for 1246 rivers around the world (from Walling and Kleo, 1979).

values produces multiple peaks, including one around 400 mm (Figure 3.6), which is somewhat similar to the peak of the curve for the data from the United States (Figure 3.4). This complicated curve through averaged data also indicates that sediment yield may increase at high rates of mean annual precipitation and, presumably, runoff – a finding consistent

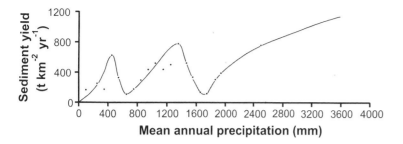

Figure 3.6. Variation in sediment yield with mean annual precipitation for group averages data for individual river basins (see Figure 3.5) (from Walling and Kleo, 1979).

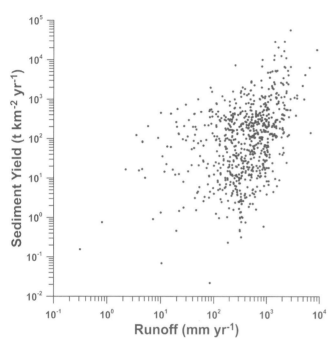

Figure 3.7. Relation between annual sediment yield and runoff for 735 rivers throughout the world (from Milliman and Farnsworth, 2011).

thereby limiting erosion by abundant runoff. On the other hand, erosion is also limited in sparsely vegetated dry environments by the lack of precipitation. The high degree of variance in the data also suggests that local factors besides precipitation and runoff play important roles in sediment yield when examined at a global scale. Most analysis has focused on mean annual precipitation, even though other aspects of precipitation, such as intensity and seasonality, may contribute to variability in sediment yield.

3.2.2 How Is Sediment Flux Related to Relief?

Denudation rate and sediment load exhibit a strong positive relationship with relief or relief ratio (Figure 3.8). In fact, relief provides the basis for organizing the relationship between drainage area and sediment load into different functional groupings (Milliman and Syvitski, 1992; Milliman and Farnsworth, 2011). Although sediment loads for rivers around the world that drain to oceans tend to increase with drainage area, the scatter is quite pronounced (Figure 3.9). For a set of 280 rivers draining into the ocean, this scatter is reduced by defining relations between sediment load and drainage area for different categories of river basins based on maximum elevation of the basins above sea level (Milliman and Syvitski, 1992). Because the rivers under consideration all flow into the sea, the elevation categories are, in effect, categories of total relief. Relations between load and basin area generally conform to power functions of the form

$$Q_s = aA_d{}^b \tag{3.3}$$

Values of b do not differ greatly among the groups, indicating that the rate of increase in sediment load with drainage area is not strongly dependent on total relief. On the other hand, values of a differ markedly among the groups and serve as scaling parameters defining differences in sediment loads with differences in relief. Rivers draining to the oceans from high mountains (> 3000) have the highest sediment loads (largest values of a). Rivers draining mountains with maximum

with some prior studies (Fournier, 1960; Douglas, 1967). However, given that individual values of sediment yield vary over three to four orders of magnitude for the entire range of precipitation values, defining meaningful differences between group averages, which are all roughly the same order of magnitude, is problematic. Adding or subtracting one extreme value within a group can greatly alter the group average (Walling and Webb, 1983). Direct analysis of the relation between sediment load and runoff confirms that, on a global basis, little or no distinct relationship exists between these two variables (Milliman and Farnsworth, 2011) (Figure 3.7).

The lack of a strong relation between precipitation or runoff and sediment flux indicates that the opposing effects of vegetation cover and total runoff preclude the emergence of a distinct trend in sediment yield with increasing precipitation or runoff. Wet landscapes are well vegetated,

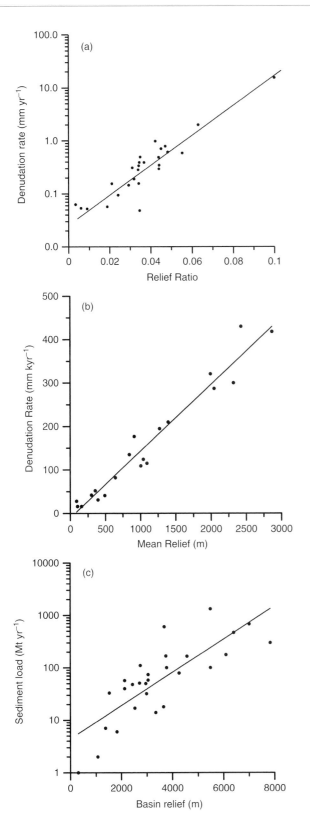

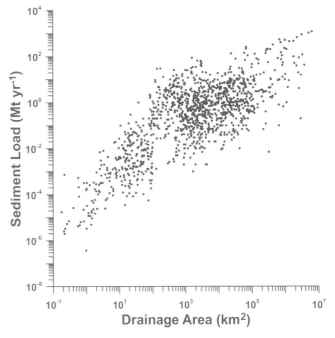

Figure 3.9. Sediment load versus drainage area for 1241 rivers around the world (from data compiled by Covault et al., 2013).

elevations between 1000 and 3000 m above sea level separate into geographic groups, with rivers in Africa, North America, and Northern Europe having higher sediment loads than mountains of similar elevation in other parts of the world. These geographic differences suggest that other factors, such as tectonics, rock type, and climate, influence the sediment loads of mountain rivers. Lowland (100–500 m) and coastal plain (< 100 m) rivers generally have sediment loads one to three orders of magnitude less than those for high mountain rivers for a given drainage area.

Analysis of sediment load–drainage area relations for four elevation categories greater than 100 m using an expanded dataset for 747 rivers that drain to oceans confirms that average sediment load relative to drainage area increases with total relief, but the scatter of data is much larger than for the smaller set of 280 rivers (Figure 3.10). Exponents of Eq. (3.3) are still similar for the four categories, but values of *a* differ by an order of magnitude between lowland and high mountain rivers. The influence of relief, or elevation, on sediment loads seems to reflect steep hillslope gradients in mountain environments, production of sediment for fluvial erosion by glacial action and mechanical weathering, and orographic effects of mountains on precipitation.

3.2.3 How Is Sediment Flux Related to Drainage Area?

The relation between sediment yield and drainage area is more complicated than the relation between load and

Figure 3.8. (a) Denudation rate versus relief ratio for small watersheds in the upper Cheyenne Basin, USA (Hadley and Schumm, 1961). (b) Denudation rate versus mean relief for 20 river basins in the United States and Europe (Ahnert, 1970). (c) Sediment load versus basin relief for 32 rivers around the world (Summerfield and Hulton, 1994).

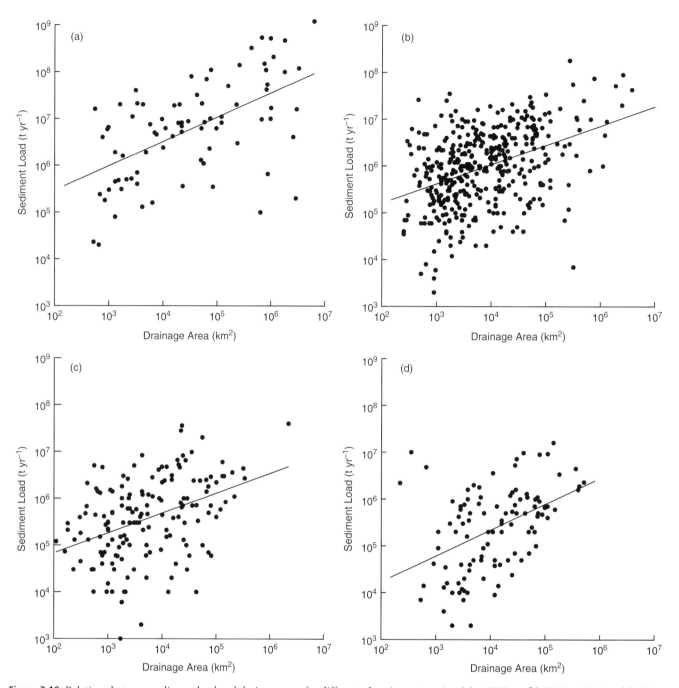

Figure 3.10. Relations between sediment load and drainage area for different elevation categories: (a) > 3000 m, (b) 1000 to 3000 m, (c) 500 to 1000 m, (d) 100 to 500 m (data from Milliman and Farnsworth, 2011).

drainage area. Decreases in sediment yield with increasing drainage area have been well documented for a wide range of environmental settings (Walling and Kleo, 1979; Milliman and Syvitski, 1992; de Vente et al., 2007) (Figure 3.11). This inverse relation between sediment yield and drainage area seems to reflect increased opportunities for sediment storage to occur in large basins compared with small basins. Analysis of this relation, however, is prone to the possibility of spurious

correlation because drainage area occurs in the denominator of Q_{sy}, leading to a strong inherent tendency for sediment yield and drainage area to be inversely related to one another statistically (Waythomas and Williams, 1988; Worrall et al., 2014). Despite this concern, the storage interpretation persists. Large drainage basins are more likely than small drainage basins to have rivers with large, expansive floodplains that store substantial amounts of sediment, reducing the flux of sediment per

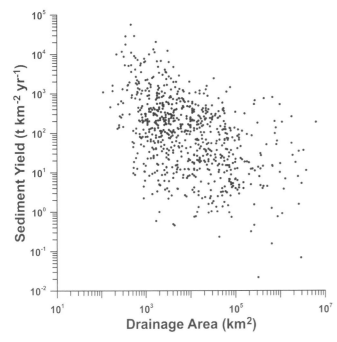

Figure 3.11. Sediment yield versus basin area for 774 rivers around the world (data from Milliman and Farnsworth, 2011 compiled by Covault et al., 2013).

unit area of the watershed (Milliman and Farnsworth, 2011). In large drainage basins that include high mountains, only a small fraction of the total basin area may be characterized by high relief. An example is the Amazon River basin, which drains the Andes Mountains with peak elevations above 3000 m but has less than 10% of its total drainage area with elevations above 500 m. Within expansive low-relief portions of the drainage network, the main stem and tributaries are flanked by extensive floodplains. Not only is the potential for sediment storage great in such systems, but sediment production is limited by low relief throughout much of the basin (Walling and Webb, 1996).

A comprehensive analysis of the sediment yield–drainage area relation indicates that exceptions to the negative trend between sediment yield and drainage area are common and that in many cases this relation can be positive or exhibit no trend (de Vente et al., 2007). The sediment-yield relation based on the sediment discharge equation (Eq. (3.3)) is

$$Q_{sy} = \frac{Q_s}{A_d} = a \frac{A_d{}^b}{A_d} = a A_d{}^{b-1} \qquad (3.4)$$

Thus, the value of the exponent in the sediment yield–drainage area relation will be negative when $b < 1$, equal to zero when $b = 1$, and positive when $b > 1$. The principle of mass conservation

provides a basis for examining the form of the yield–area relationship. Sediment yield is the difference between the total amount of erosion per unit drainage area (Q_{sp}) and the total flux of sediment into storage per unit drainage area (Q_{st}) within the drainage basin upstream of the point at which yield is determined:

$$Q_{sy} = Q_{sp} - Q_{st} \qquad (3.5)$$

Erosion and storage can occur on hillslopes and within the channel network. If sediment yield exists at all scales ($Q_{sy} > 0$), $Q_{sp} > Q_{st}$ for all drainage areas. The relationship between sediment yield and basin area depends on relative amounts of change in supply (ΔQ_{sp}) and storage (ΔQ_{st}) with changes in drainage area (Worrall et al., 2014) (Figure 3.12). If both supply and storage change by the same amount, sediment yield remains constant as area increases (Figure 3.12c). The lack of a trend in sediment yield with area indicates that the fluvial system is maintaining a balance between changes in supply and storage per unit area over the range in scale of drainage-basin size. Sediment yield declines with either constant supply and increasing storage (Figure 3.12b) or decreasing supply and increasing storage (Figure 3.12e). Negative trends between sediment yield and drainage area typically develop when upland hillslopes supply abundant amounts of sediment to headwater streams and favorable locations exist within downstream channels and valley bottoms for deposition and storage of sediment, a typical situation in many agricultural drainage basins with enhanced rates of soil erosion (Walling and Webb, 1996; Dedkov, 2004). Increases in storage with increasing basin area relative to supply reduce sediment yield with increasing area. Decreases in connectivity between hillslopes and floodplain-buffered rivers may contribute to the negative trend by decreasing the supply of sediment to rivers as drainage area increases. Sediment yield increases with increasing supply and constant storage (Figure 3.12a) or increases in supply that exceed increases in storage (Figure 3.12d). Positive trends develop when channel erosion and excavation of stored sediment within valley bottoms are occurring in addition to delivery of sediment from hillslopes, so that large increases in supply overwhelm available storage as drainage area increases (Church and Slaymaker, 1989; Dedkov and Moszherin, 1992; Church et al., 1999).

In some cases, the relation between sediment yield and drainage area is scale dependent. Changes in erosional and depositional processes on hillslopes and within channels at different spatial scales can produce peaked relations between sediment yield and drainage marked by a positive relationship at small scales and a negative relationship at large scales (Church and Slaymaker, 1989; Xu and Yan, 2005) (Figure 3.12f). Negative trends tend to occur as the

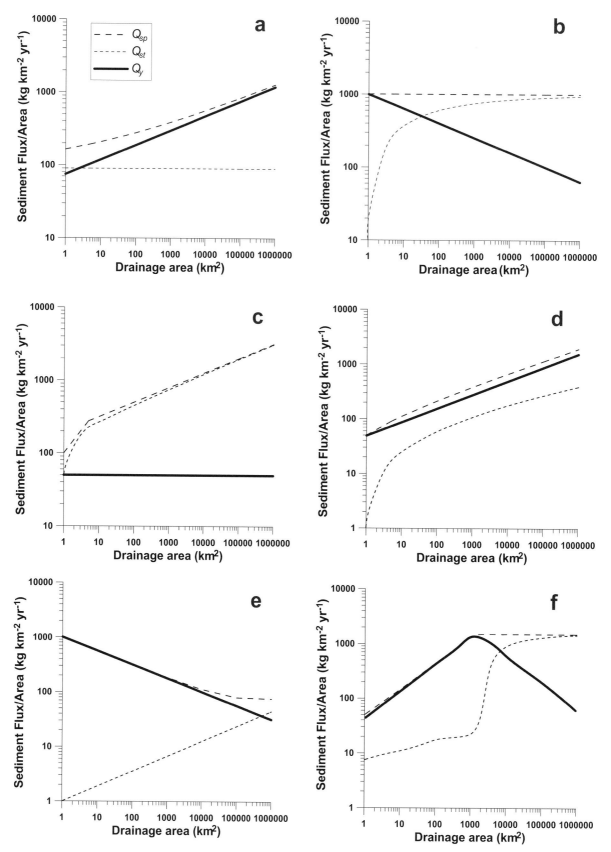

Figure 3.12. Hypothetical patterns of sediment yield (Q_{sy}) for different relations between supply (Q_{sp}) and storage (Q_{st}) with increasing drainage area: (a) constant Q_{st}, increasing Q_{sp}, increasing Q_{sy}, (b) constant Q_{sp}, increasing Q_{st}, decreasing Q_{sy}, (c) increasing Q_{sp} and Q_{st}, constant Q_{sy}, (d) increasing Q_{sp} and Q_{st}, increasing Q_{sy}, (e) decreasing Q_{sp}, increasing Q_{st}, decreasing Q_{sy}, (f) increasing then steady Q_{sp}, slowly increasing Q_{st} with abrupt increase at threshold for floodplain formation, increasing then decreasing Q_{sy}. For all cases, Q_s increases with drainage area.

Table 3.2. Variations of sediment yield with rock type for drainage basins in New Zealand, including ratio of average a and range of a in the relation $Q_{sy} = aP_{ma}^{2.3}$ for all rock types.

Lithology	Sediment yield (t km^{-2} yr^{-1})	Precipitation (mm)	Ratio of average a relative to granite, gneiss, marble	Range of a
Granite, gneiss, marble	17–350	3700–8400	1	0.0000001–0.00000035
Greywacke, argillite	44–1740	1060–8100	15	0.0000021–0.000033
Schist, semi-schist	22–29,600	660–11,200	25	0.000006–0.000067
Weak marine sedimentary	1200–20,000	1200–2400	900	0.000052–0.0004

(after Hicks et al., 1996)

Table 3.3. Relation of sediment yield to basin slope and local basin relief for different rock types in 47 Andean drainage basins.

Functional relations	Variation in lithologic erodibility (ψ_{LE}) by rock type		
	Igneous	Metasedimentary	Weak sedimentary
$Q_{sy} = 0.45\psi_{LE}S_d^{3.36}$	1	7.2	46
$Q_{sy} = 2.6 \times 10^{-7}\psi_{LE}R_{ra}^{3.02}$	1	6.9	33

(based on data in Aalto et al., 2006)

basin scale exceeds a threshold for floodplain development, which will vary from basin to basin, because the development of floodplains provides storage that decreases yields. The relation between sediment yield and drainage area can also change over time depending on changes in land cover or land management practices (Renwick and Andereck, 2006).

3.2.4 How Is Sediment Flux Related to Rock Type?

The influence of rock type on sediment yield has been demonstrated clearly within particular environments. In New Zealand, a strong relation exists between sediment yield and rainfall for 203 drainage basins sorted into categories based on rock strength (Hicks et al., 1996). Basins in four different rock types all exhibit relations between yield and mean annual precipitation (P_{ma}) of the form

$$Q_{sy} = aP_{ma}^{2.3} \tag{3.6}$$

where the coefficient *a* provides a scaling index differentiating rock erodibility (Table 3.2). Ratios of average values of *a* for four different rock types relative to the average value of *a* for the most resistant rock type

(crystalline and metamorphic rocks) show that sediment yields for the most erodible rocks are on average about 900 times greater than those for the most resistant rocks (Table 3.2). In a similar analysis, power-function relations between sediment yield and average basin slope and between sediment yield and relief for drainage basins developed on different rock types in the Andes Mountains of Bolivia exhibit similar exponents but varying coefficients (Aalto et al., 2006). The ratio of coefficients provides an index of erodibility, with sediment yields for weak sedimentary rocks exhibiting rates of erosion an order of magnitude greater for given average basin slopes or basin relief than rates for igneous rocks (Table 3.3). Rock age often is also related to erodibility, with sediment yields for rivers draining landscapes formed of young rocks exhibiting a tendency to have higher sediment yields than rivers draining landscapes formed of old rocks (Milliman and Farnsworth, 2011).

3.3 How Have Humans Affected Sediment Dynamics at Global Scales?

Most measured sediment data on which analyses of sediment flux rely have been obtained within the past several decades – a period during which human activities have strongly influenced sediment dynamics. Thus, efforts to identify natural factors controlling sediment dynamics are complicated by possible bias in data associated with human effects. Land disturbances, such as agriculture, grazing, deforestation, mining, and construction activities, tend to enhance delivery of sediment to rivers and streams (see Chapter 15). Sediment loads for many rivers throughout the world may be higher during the period of record than were loads prior to anthropogenic disturbance. The problem of accelerated soil erosion, which began thousands of years ago with the advent of agriculture (Syvitski and Kettner, 2011), has been well documented (Montgomery, 2007a; Dotterweich, 2013). Industrial mechanization of farming, deforestation, mining, and construction accelerated soil erosion in the early twentieth century until recognition of this problem led to widespread implementation of soil conservation practices. Some soil-erosion rates have increased by an order of magnitude, which certainly has had an impact on sediment loads of rivers (Walling, 2006), especially loads of fine-grained sediment (Owens et al., 2005). However, caution must be exercised in relating soil erosion to sediment loads because this relation is strongly influenced by connectivity between hillslopes and channels – a problem explored in detail in subsequent sections of this chapter.

3.3.1 How Have Dams Influenced Global Sediment Fluxes?

Trapping of sediment in reservoirs behind dams is the dominant human-induced mechanism reducing sediment loads (Walling, 2006). Although the construction of small dams has occurred for centuries, most major dams in the world were constructed between the early 1900s and the present, with a period of major construction between 1950 and 1980 (Walling, 2006) (Figure 3.13). Large reservoirs ($\geq$ 0.5 km^3 storage capacity) now intercept more than 40% of the total river discharge to the oceans (Vorosmarty et al., 2003). The local trap efficiency of large reservoirs – the ratio of total outflux of sediment from a reservoir to total influx to the reservoir – globally is nearly 90%. Global mean trap efficiency, the percentage reduction in sediment load at the mouth of major basins of the world, including contributions of tributaries downstream of reservoirs as well as contributions from rivers in undammed basins, increased dramatically between 1950 and 1980 (Figure 3.13).

Determining the counteracting effects of soil erosion and reservoir retention on fluvial sediment fluxes at a global scale is a difficult task. A comprehensive analysis of recent trends in sediment loads for 145 major rivers throughout the

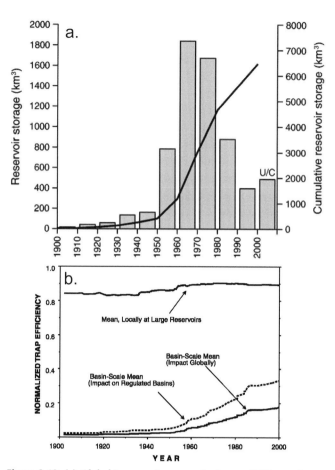

Figure 3.13. (a) Global increase in reservoir storage (U/C - under construction) (Walling, 2006). (b) Changes in trap efficiency of the world's largest reservoirs, including mean local trap efficiency, mean basin-scale trap efficiency in regulated basins, and mean basin-scale efficiency for all basins (Vorosmarty et al., 2003).

world indicates that about half of these rivers exhibit statistically significant increases or decreases in load over the past few decades (Walling and Fang, 2003). In some rivers, such as the Mississippi, the Danube, and the Yellow, loads have declined because of reservoir construction or implementation of soil and water conservation practices. In others, such as the Hongshuihe River, China and the Yazgulum River, Kazakhstan, sediment loads have increased as a result of expansion and intensification of agriculture or other land disturbances. Reductions in sediment load are more common than increases, indicating that the effect of reservoir retention surpasses that of enhanced soil erosion. Half of the historical sediment records are nonstationary, exhibiting a trend over time, suggesting that humans are altering sediment fluxes. However, the other half are stationary, or constant over time, suggesting either that humans are not altering sediment loads in these basins or that counteracting effects of increases in retention and erosion rate maintain stationarity.

Although human-induced soil erosion has increased the global flux of sediment within river systems by an estimated 2.6 Gt yr^{-1} ($\pm$ 0.6 Gt yr^{-1}), trapping of sediment in reservoirs has more than offset this effect (Syvitski et al., 2005). Estimates of sediment retention by reservoirs are about 4.0 Gt yr^{-1} (Syvitski et al., 2005) and may exceed 5 Gt yr^{-1} (Milliman and Farnsworth, 2011, p. 140). As a result, the total sediment flux to the oceans has decreased by about 1.4 Gt yr^{-1} ($\pm$ 0.3 Gt yr^{-1}). Reservoirs trap an estimated 16–20% of the total global sediment flux to the oceans (Vorosmarty et al., 2003; Syvitski et al., 2005). Overall, current evidence indicates that dam construction by humans now is having a major influence on the global flux of sediment by rivers. The problem of reduced sediment flux to deltas is of particular concern, given that this flux is central to delta morphodynamics (Syvitski and Saito, 2007) and that the sustainability of delta environments is increasingly at risk (Syvitski et al., 2009; Tessler et al., 2015).

3.4 What Is the Relative Importance of Different Controlling Variables for Influencing River Sediment Flux?

Multivariate analysis is required to unravel the independent influences of different controlling variables on sediment flux. The development of multivariate models has focused for the most part on determining which combinations of possible controlling variables yield the best predictive equations for estimating global sediment yields (Jansen and Painter, 1974; Ludwig and Probst, 1998) or evaluating changes in sediment delivery to the oceans in response to sea level variations (Mulder and Syvitski, 1996). Such studies highlight variables that are useful for prediction, but do not identify the relative importance of the variables. A series of investigations (Syvitski and Morehead, 1999; Syvitski et al., 2003, 2005) has examined the factors controlling the flux of sediment from river basins, leading to the development of a comprehensive multivariate predictor of fluvial sediment flux (Syvitski and Milliman, 2007). This model has the form:

$$Q_s = 0.0006\eta_s Q^{0.31} A_d^{0.5} R_{rm} T_{da} \qquad T_{da} \geq 2\ °C \qquad (3.7)$$

$$Q_s = 0.0012\eta_s Q^{0.31} A_d^{0.5} R_{rm} \qquad T_{da} < 2\ °C \qquad (3.8)$$

where Q_s is in units of Mt yr^{-1}, Q is river discharge in km^3 yr^{-1}, A_d is drainage area (km^2), R_{rm} is maximum relief (km), and T_{da} is basin-averaged temperature (°C). The inclusion of T_{da} as a factor influencing sediment load is rather novel. In the model, T_{da} serves as a surrogate for climate and its relation to weathering, vegetation, and precipitation-generating mechanisms (i.e., storm types and precipitation intensity). The variable η_s is defined as

$$\eta_s = e_g L_f (1 - T_e) e_{se} \qquad (3.9)$$

where e_g is a glacial erosion factor ($e_g \geq 1$), L_f is an average basin-wide lithology factor, T_e is the trapping efficiency of lakes and reservoirs ($1 - T_e \leq 1$), and e_{se} is a human-induced soil-erosion factor (Table 3.4). The model accounts for 96% of the variance in sediment loads for a database of 294 rivers and 95% of the variance in an independent set of load data for an additional 194 rivers. Fitting the model to data by sequentially adding variables to it reveals the relative importance of these variables. Basin area and relief together account for 57% of the total variance in the sediment load data, whereas temperature accounts for 10%, lithology for 8%, and discharge for 3%. The ice-cover factor is important in glaciated basins but does not contribute significantly to a reduction in unexplained variance for the entire data set. Collectively, the geological and geographic variables (Q, A_d, R_{rm}, T_{da}, L_f, and e_g) account for 79% of the total variance in sediment loads. The anthropogenic factors of reservoir trapping and human influence on landscape erosion individually contribute to a 9% reduction in unexplained variance and together to a 16% reduction in this variance. The statistical model provides insight into the relative importance of natural geologic and geographic factors versus anthropogenic factors on global sediment fluxes. It clearly shows that the effects of anthropogenic factors on sediment fluxes, while not as dominant as natural factors, are detectable at a global scale and that these factors contribute substantially to the variance of these fluxes.

A qualitative approach to evaluating the effects of individual factors on sediment fluxes involves classifying the data by groups and then examining flux relationships within these groups. The analysis of sediment flux by drainage areas for different elevation categories (Figure 3.10) or by slope for different rock types (Table

Table 3.4. Values of factors for the variable η_s in Eq. (3.9).

Factor	Value
e_g	$e_g = (1 + 0.09A_i)$ where A_i is percentage of basin that is ice covered ($e_g = 1$, no ice cover, $e_g = 10$, 100% ice cover)
L_f	0.5 to 3.0 with low values for strong, resistant rocks and high values for erodible materials
T_e	0.9 (entire basin affected by large reservoirs) to 0 (no reservoirs)
e_{se}	0.3 high population density and high per capita Gross National Product (GNP) 1.0 low population density 2.0 high population density and low per capita GNP

(from Syvitski and Milliman, 2007)

3.3) represents examples of this approach. Analyses of this type have explored how sediment fluxes vary for river basins of similar elevations within the same climatic region, but with different ages of rocks and for drainage basins of similar elevations and ages of rock, but with different climates. Sediment fluxes for rivers draining wet tropical mountains with old rocks are one to two orders of magnitude less than fluxes for rivers draining wet tropical mountains with young rocks, which tend to be less resistant than old rocks (Milliman and Farnsworth, 2011). Moreover, sediment loads for rivers draining young arid mountains are 5 to 10 times less for the same drainage area than loads for rivers draining young tropical mountains. Thus, holding other factors constant, increasing aridity reduces sediment yields – a finding that is consistent with Eq. (3.7) and (3.8) if Q is viewed as an index of wetness. Rivers draining young, often tectonically active mountains with high precipitation and runoff ($> 750 \, \text{mm yr}^{-1}$) contribute disproportionately to the global mass flux to the oceans, accounting for 62% of the suspended-sediment flux and 38% of the dissolved solid flux, even though these environments constitute only 14% of the total land area (Milliman and Farnsworth, 2011).

The focus on variables related to climate, topography, geology, and human influence does not completely capture the full complexity of fluvial sediment fluxes. Extreme events, such as earthquake-generated landslides that deliver exceptional amounts of sediment to rivers (Chuang et al., 2009; Hovius et al., 2011) or hydrologic conditions that result in extraordinary amounts of sediment mobilization (Farnsworth and Milliman, 2003; Nearing et al., 2007; Schiefer et al., 2010), can in some instances have a pronounced influence on total sediment flux within a river system. Extrapolations based on short-term instrumented records that fail to properly account for the influence of rare, extreme events on sediment fluxes can lead to biased estimates of total sediment loads over decadal to centennial timescales.

Seismicity, or earthquake activity, is emerging as a potentially important factor influencing sediment yield that generally has been overlooked in the past. Large earthquakes can generate landslides that serve as a major source of sediment to rivers in tectonically active environments (Dadson et al., 2004). However, seismicity also seems to influence sediment yield more generally. For both large and small drainage basins in Europe, seismicity, as measured by peak ground acceleration (PGA), is significantly correlated with sediment yield, even after accounting for the influence of topography and lithology (Vanmaercke et al., 2014). This effect on sediment yield appears to involve long-term enhancement of mass movement and mechanical weathering by moderate seismic events rather than delivery of large amounts of sediment to rivers by landslides in the immediate aftermath of extreme seismic events (Vanmaercke et al., 2014, 2017).

3.5 What Factors Influence Landscape Denudation?

3.5.1 What Are the Shortcomings of Using Sediment Flux Data to Estimate Landscape Denudation?

Studies of landscape denudation focus on rates of removal of mass from continental surfaces through erosional processes, including fluvial erosion. The timescale of interest is geological, i.e., thousands to millions of years. One way to approach such studies is to use instrumented records of sediment flux and dissolved solids to determine rates of denudation based on these records (Eq. (3.2)) and then extrapolate these rates over geologic timescales (Table 3.5). This method is potentially subject to biases associated with anthropogenic impacts on sediment fluxes and with extreme events, such as volcanic eruptions, natural dam breaks, and major landslide events, that may not be adequately captured by short-term records yet have a pronounced influence on sediment fluxes (Korup, 2012). Moreover, vegetation and climate characteristics, and even rates of tectonic deformation, can vary over geologic timescales, introducing nonstationarity into long-term time series of sediment fluxes (Schumm and Rea, 1995). Thus, estimates of denudation rates based on instrumented records should be viewed as first-order approximations. Spatially averaged rates (e.g., Table 3.5) also are somewhat deceptive, because denudation will vary spatially within these regions in conjunction with the factors that control sediment flux. The Amazon River basin again provides an instructive example: most production of sediment for river transport occurs in the Andes Mountains in the headwaters of the basin, whereas the interior part of the basin contains numerous large rivers with well-developed floodplains where net storage of sediment occurs (Warrick et al., 2014). Basin-averaged rates do not capture this pronounced spatial variability in denudation rates.

Table 3.5. Regional denudation rates estimated from suspended sediment and dissolved solid loads for rivers.

Continent	Denudation rate (mm kyr^{-1})
Africa	17
European Arctic	20
North America	70
Europe	110
South America	100
Asia/Oceania	310
Australia	25

(Milliman and Farnsworth, 2011)

3.5.2 How Are Denudation Rates Estimated from Cosmogenic Radionuclides Related to Controlling Factors?

Another method to estimate long-term denudation rates is to use cosmogenic radionuclides (CRNs), particularly ^{10}Be

(Box 3.1). Based on concentrations of CRNs in sediments sampled at the mouths of drainage basins, basin-averaged rates of denudation can be determined over timescales of thousands to hundreds of thousands of years. A compilation of CRN-derived denudation rates for non-glaciated landscapes indicates that rates are not related to

---

BOX 3.1 | DETERMINING BASIN-WIDE EROSION RATES USING COSMOGENIC RADIONUCLIDES

Determinations of erosion rates using cosmogenic nuclides (CRNs) have revolutionized efforts to evaluate long-term denudation of Earth's surface. This method has become the tool of choice for estimating erosion rates at a variety of spatial scales ranging from a single outcrop to large watersheds (Granger and Schaller, 2014). CRNs are produced through cascades of nuclear reactions initiated by high-energy particles from primary cosmic rays that constantly bombard the Earth (Anderson and Anderson, 2010). Interaction of these high-energy particles with molecules within Earth's atmosphere produces secondary cosmic rays consisting of neutrons, protons, and muons. Nuclear reactions of these secondary rays with minerals near Earth's surface result in the production of CRNs. The surface concentration of a CRN (C_0 atoms g^{-1}) is directly proportional to the production rate of the CRN at the surface (P_0 atoms g^{-1} yr^{-1}) and inversely proportional to the radioactive decay constant of the radionuclide (λ_c yr^{-1}) and to the denudation rate of the surface (von Blanckenburg 2005):

$$C_0 = P_0/(\lambda_c + \delta_d/z^*) \tag{3.1.1}$$

where z^* (cm) is the absorption depth scale of the rock or soil, which is directly related to the cosmic ray absorption free path (Λ g cm^{-2}) and inversely related to the density of the rock or soil (ρ_s g cm^{-3}). Values of z^* determine the depth to which production of CRNs takes place beneath the surface. The CRN of interest in erosion studies is typically ^{10}Be, which is produced in minerals such as quartz. The depth of production of ^{10}Be is typically about 0.6 m for rock and 1 m for soil (Granger and Schaller, 2014). The increase in concentration of ^{10}Be within a grain is indicative of the residence time of the grain beneath the eroding surface. Grains beneath slowly eroding surfaces will have long residence times and approach the surface slowly, resulting in high ^{10}Be concentrations. Grains beneath rapidly eroding surfaces will have short residence times and approach the surface rapidly, resulting in low ^{10}Be concentrations. Thus, ^{10}Be concentrations serve as "clocks" that provide information on the rate at which surface material is stripped away by erosion. The timescale of this clock is 10^3 to 10^5 years.

Remarkably, a handful of sand near the mouth of a river that representatively integrates sources of eroding sediment from throughout a drainage basin can provide information on average erosion rates throughout the basin (von Blanckenburg, 2005; Granger and Schaller, 2014). Such a sample will include grains from rapidly eroding areas of the basin with low ^{10}Be concentrations and grains from slowly eroding portions of the basin with high ^{10}Be concentrations. Because in this representative, integrated sample the abundance of grains of each type occurs in proportion to areas of the watershed that are eroding at different rates, the average ^{10}Be concentration provides a basin-averaged rate of denudation. Over periods of 10^3 to 10^5 years, the timescale of denudation typically is much shorter than the timescale of radioactive decay of ^{10}Be (i.e., $\delta_d/z^* \gg \lambda_c$), which has a half-life of 1.39×10^6 years. Thus, λ_c can be ignored in Eq. (3.1.1), and solving for δ_d yields (Covault et al., 2013)

$$\delta_d = \frac{P_0 z^*}{C_0} \tag{3.1.2}$$

Rates estimated from Eq. (3.1.2) integrate the combined effects of physical erosion and removal of mass through chemical weathering (production of solutes). The method assumes that sediment storage in the catchment over the timescale of denudation is minimal. The estimated denudation rate can be converted to a basin-averaged mass yield using Eq. (3.2) and to a mass load using Eq. (3.1).

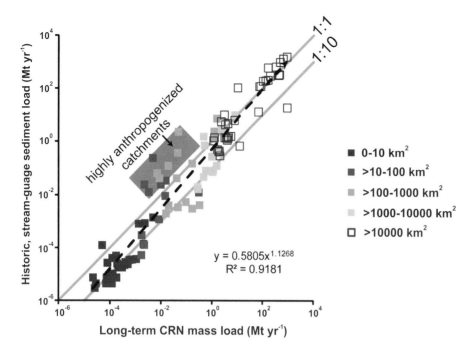

Figure 3.14. Stream sediment loads versus CRN mass loads for 103 matched locations. Sediment loads are colored according to drainage-basin area. Dashed line represents best-fit power function. Light lines show ratios of CRN to stream loads. Highly anthropogenized watersheds exhibit increase in stream loads relative to CRN loads (from Covault et al., 2013). (A black and white version of this figure will appear in some formats. For the color version, please refer to the plate section.)

precipitation or temperature, that some areas show a strong relationship between relief and denudation and others do not, and that denudation rates are typically high in areas with active tectonism, independently of relief (von Blanckenburg, 2005). Moreover, in areas with intensive land use, denudation rates derived from measured sediment load data greatly exceed CRN-derived rates, indicating that anthropogenic disturbance has enhanced erosion rates in these areas (Covault et al., 2013).

Comparison of historic stream-derived sediment loads with CRN-estimated mass loads for 103 sites sampled at approximately the same locations indicate that about two-thirds of the CRN-derived mass loads exceed the stream-derived loads (Covault et al. 2013) (Figure 3.14). Part of this disparity reflects the inclusion of chemical weathering denudation in the CRN estimates. For the most part, however, the difference is attributed to the occurrence of infrequent, large-magnitude erosive events that recur over centennial to millennial timescales and that are captured by the CRN-derived estimates but not by the historical data (Kirchner et al., 2001). In seventeen cases the stream-derived loads exceed the CRN-derived loads, signaling the influence of anthropogenic effects on contemporary sediment fluxes (Figure 3.14). The difference between the two estimates of load diminishes with increasing load, which generally is indicative of basin size. Buffering of sediment fluxes in large watersheds may account for the increasing similarity of mass load estimates with increasing load. Small basins lack storage areas for sediment that can buffer changes in erosion rates elsewhere in the watershed, whereas large basins often contain abundant storage sites, particularly

floodplains, that buffer sediment dynamics within rivers. Small transport-reactive basins are more likely to be prone to the influence of intermittent, high-magnitude transport events where the signals of these events are transferred rapidly to sediment deposited at the basin outlet. Historical records will often exclude these events. By contrast, large well-buffered drainage basins with abundant storage will effectively dampen the signal of extreme events or changes in environmental conditions. If upstream erosion increases, downstream deposition may increase to balance this change; conversely, if upstream erosion decreases, downstream evacuation of stored sediment will increase. Such responses will tend to maintain a relatively constant flux of sediment from the basin over time (Metivier and Gaudemer, 1999; Dearing and Jones, 2003; Phillips, 2003). As a result, denudation rates will exhibit broad similarity over different timescales.

Multivariate analysis shows that ^{10}Be-derived drainage-basin denudation rates are most strongly related to slope and relief, whereas climate (precipitation and temperature) as well as vegetation contribute to only a minor extent to reducing unexplained variance (Portenga and Bierman, 2011). Attempts to extend this finding to examine the geographic distribution of denudation and the relative importance of different environments to sediment fluxes have produced mixed results. The exponential relation

$$\delta_d = 11.9e^{0.0065S_d} \qquad (3.10)$$

where S_d is the mean slope computed over a 5×5 km moving window for 990 river basins accounts for about half of the variance in denudation rate (Willenbring et al.,

2013). Below an average drainage-basin slope value of 200 m km^{-1} the relationship breaks down and denudation rate is viewed as equivalent to the constant in Eq. (3.10), i.e., 11.9 mm kyr^{-1}. Using Eq. (3.10) to compute denudation rates on a global scale shows that although denudation rates are highest in mountainous environments, low-sloping regions of the world, which encompass a much greater proportion of the total land area, dominate the global sediment flux. This finding contradicts results of previous studies indicating that numerous relatively small mountain rivers supply the greatest amount of sediment to the oceans (Milliman and Syvitski, 1992; Milliman and Farnsworth, 2011). Concern about the possible influence of scale effects on averaging of drainage-basin slopes (Warrick et al., 2014) led to reestimation of global denudation rates accounting for these effects (Larsen et al., 2014). Slope averaging has a strong influence on denudation rates, and fine-grained slope data indicate that land areas with mean slopes > 15° account for 52% of global denudation (Larsen et al. 2014). Moreover, estimated global sediment flux to the oceans is 19 Gt yr^{-1} and estimated global solute flux is 3.9 Gt yr^{-1} – values that closely match estimates based on stream measurement data (Table 3.1).

3.6 What Is the Mass Balance of a Watershed?

Whereas sediment loads and yields provide information on the amount of sediment transported out of a drainage basin, these metrics do not link sediment flux to erosion within the basin. The sediment mass balance analysis represented by Eq. (3.5), which considers only the stream system, can be extended to the entire drainage basin:

$$Q_s = Q_{si} - \Delta M_{st} \qquad (3.11)$$

where Q_{si} is annual influx of material mass to the basin (t yr^{-1}) and ΔM_{st} (t yr^{-1}) is the annual change in the total material mass of the basin above the baselevel elevation of the outlet. The main sources of mass influx (Q_{si}) are uplift produced by endogenic geological processes, such as tectonism or isostasy, and deposition of atmospheric dust. Assuming that both these influxes are minor over the modern timescale yields

$$Q_s = -\Delta M_{st} \qquad (3.12)$$

Thus, over periods of a few decades or centuries, many drainage-basin sediment systems, particularly those in tectonically inactive areas, are characterized by net erosional fluxes of material out of the basin that reduce the total mass of the basin. This formulation emphasizes the connection between sediment flux and erosion of drainage basins. It does not account for the contribution of dissolved solids to mass flux out of the basin and of chemical weathering to change in mass storage, which can be important in some environments. The focus here is on the flux of sediment, ignoring the contribution of dissolved materials.

3.7 What Is the Sediment Delivery Problem and Its Relation to the Sediment Delivery Ratio?

Soil erosion is a primary process that contributes to mass flux within a soil-mantled drainage basin. As soil is eroded it moves downslope under the influence of gravity through advective and diffusive processes (see Chapter 2). Eventually, it reaches streams, where it is transported through the drainage network toward the basin outlet. Concern about how much eroded soil is transported out of a drainage basin over a period of many years, decades, or even centuries is a fundamental problem of scientific interest to geomorphologists and of practical interest to soil conservation specialists (Trimble, 1975). This issue is referred to as the sediment delivery problem (Walling, 1983). A key metric that has been used in analysis of this problem is the sediment delivery ratio (SDR) (Glymph, 1954; Roehl, 1962):

$$\text{SDR} = \frac{Q_{sy}}{Q_{sp}} \qquad (3.13)$$

The SDR originated largely within the context of soil management concerns, and in this context the denominator of Eq. (3.13) has typically been viewed as the production of sediment associated with soil erosion on hillslopes (Q_{se}) rather than from all possible sources of eroded sediment. In other words, it is often assumed that $Q_{sp} = Q_{se}$. Extension of the concept to drainage-basin scales recognizes that the river network, particularly sediment stored on valley bottoms, on floodplains, and within channels, can at times become a source of eroded sediment, leading to a channel erosion source term (Q_{sc}). Thus, $Q_{sp} = Q_{se} + Q_{sc}$. Assuming that all sediment leaving the basin comes from hillslopes or the channel system, the maximum value of the SDR using Q_{sp} in Eq. (3.13) is one.

The SDR for most drainage basins is less than one; in many cases, much less than one (Roehl, 1962; Walling and Kleo, 1979; Walling, 1983). To account for this discrepancy between erosion and yield, which will have a value of one if the relation between erosion and yield is balanced (Trimble, 1975), storage of eroded sediment must occur within the basin. Sediment eroded from hillslopes and channels does not exit the basin over the same timescale as this erosion but instead is deposited before reaching the basin outlet.

Although the SDR is a simple way to characterize the aggregate connection between sediment mobilization and sediment output, in practice it is difficult to operationalize. Actual measurements of sediment mobilization by soil erosion typically are limited to small plots. Results of such experiments must be extrapolated empirically to scales of hillslopes, sets of hillslopes, and entire drainage basins, typically using models. A commonly used empirical model for estimating soil erosion is the universal soil loss equation (USLE) and its revision, the revised universal soil loss equation (RUSLE). The basic USLE has the form

$$Q_{se} = R_F K_F L_F S_F C_F P_F \tag{3.14}$$

where R_F is a rainfall erosivity factor (erosivity unit per area per time), K_F is a soil erodibility factor (mass per erosivity unit), S_F is a slope gradient factor (dimensionless), L_F is a slope length factor (dimensionless), C_F is a crop management factor (dimensionless), and P_F is an erosion control practice factor (dimensionless). The RUSLE retains the same factors but includes various refinements for estimating these factors (Renard et al., 1991). A standard practice for determining sediment yield to water bodies from eroding hillslopes within drainage basins is to estimate soil loss using the USLE-based method and then compute sediment yield using the SDR (U.S. Department of Agriculture, 1983):

$$Q_{sy} = Q_{se} * \text{SDR} \tag{3.15}$$

The use of SDRs to estimate yields reflects the recognition that gross soil-erosion rates neglect storage and lead to overestimates of sediment delivery to streams (Kinnell, 2004).

The SDR approach to computing sediment yield resulting from agricultural soil erosion has motivated efforts to relate the SDR to controlling factors to provide the basis for estimating appropriate values of the delivery ratio for specific applications. Much like sediment yield, the SDR tends to be inversely related to drainage area (Glymph, 1954; Roehl, 1962) and positively related to relief ratio (Roehl, 1962) (Figure 3.15). SDR, like sediment yield, can increase with basin area if watershed conditions reduce sediment production as area increases (e.g., hillslope erosion declines), if storage of sediment within the river system decreases as area increases, or if channel erosion increasingly contributes to sediment yield as area increases. SDRs computed from Eq. (3.13) with Q_{se} used in the denominator can exceed one if contributions of sediment to Q_{sy} from channel erosion are large relative to those from hillslope erosion (Lu et al., 2005).

3.7.1 What Are the Limitations of the Sediment Delivery Ratio Concept?

At the hillslope scale, the SDR method fails to consider the influences of upslope sediment contributions on the SDR

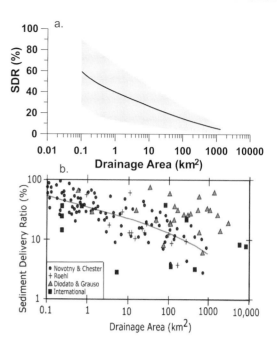

Figure 3.15. (a) Sediment delivery ratio versus drainage area. Dark line is the median value and shading shows range of data from numerous studies (U.S. Department of Agriculture, 1983). (b) Data on sediment delivery ratio versus drainage area from several studies throughout the world (from James, 2018).

(Kinnell, 2004, 2008a). Moreover, the problem of soil loss measurement is highly scale dependent (Parsons et al., 2006, 2008a). Soil loss over a fixed experimental plot, whatever its size, is not necessarily equivalent to erosion within the plot, with erosion consisting of the detachment and transport of sediment particles. Soil loss from a plot is typically measured by the mass flux at the downslope end of the plot (Kinnell, 2008b). However, some particles within the plot may be eroded, i.e., detached and transported, but not leave the plot over the timescale of the experiment. Technically, the movement of such particles contributes to soil erosion (Q_{se}), but at a length scale smaller than the scale under consideration. As the length scale of an experimental plot or of a hillslope increases, the number of particles traveling the distance associated with the new, longer length scale over a certain period of time will decrease, thereby decreasing the flux of sediment out of the area in relation to the amount mobilized. Thus, the SDR will decline with increasing length scale based on consideration of particle travel distances. Because of this scale dependency of gross erosion, the SDR is difficult to characterize in an absolute sense.

The use of a simple SDR to characterize the sediment flux out of a drainage basin in relation to the total amount of erosion occurring upstream of the basin outlet has been viewed as a black-box method that fails to illuminate the sediment delivery process (Walling, 1983; Richards, 1993; Trimble and Crosson, 2000). If changes in storage represent sediment input and the sediment yield represents the output (Eq. (3.12)), a variety of processes within the drainage basin contribute to transformation of the input into output. Peering inside the black box to better understand the hillslope and channel processes that transform sediment fluxes within a drainage basin into stream-transported fluxes leaving the basin has become a major area of research interest in watershed-scale fluvial geomorphology over the past several decades.

3.8 What Is a Sediment Budget and How Is It Useful for Evaluating Watershed Sediment Dynamics?

The concept of a sediment budget provides the foundation for a methodological framework within which to examine watershed-scale sediment dynamics. Sediment budgets represent accounting schemes to evaluate spatial patterns of sources of sediment, storage areas for sediment, and connectivity between sources and sinks (Reid and Dunne, 2016). In essence, a sediment budget extends the conservation of mass framework to explore the transfer of sediment between discrete geomorphological components of an integrated hillslope-river network system. It attempts to document how sediment fluxes interact with sediment storage zones within the system to generate a flux of sediment out of the drainage basin in which this system is embedded.

Sediment budgets can be constructed over a variety of time and space scales (Hinderer, 2012), but fluvial research has focused mainly on watershed-scale budgets over timescales of decades to centuries. Although the sediment-budget concept has a long history in fluvial studies (Reid and Dunne, 2016), work on watershed-scale sediment budgets began in earnest in the 1970s (Dietrich and Dunne, 1978). Since then, sediment budgets have been developed for a variety of different watersheds throughout the world for a variety of different purposes, both scientific and applied (Reid and Dunne, 2016).

Seminal work on historical erosion and sedimentation in the Hill Country of western Wisconsin (Trimble, 2013), especially within the Coon Creek drainage basin (Trimble, 1981, 1983, 1999, 2009), illustrates the value of the sediment-budget approach (Figure 3.16). The budget for the Coon Creek basin was constructed using the USLE to estimate amounts of upland sheet and rill erosion. Limited measurements of stream suspended-sediment loads, along with reservoir sedimentation rates and regional sediment-yield relations,

provided estimates of sediment yield. A key component of the budget computations was to determine changes in sediment storage in valley bottoms using repeat surveys of river-channel cross sections, borings into floodplain deposits, exposed stratigraphic sequences, and evidence of buried soils. Gully erosion was estimated by subtracting sedimentation volumes in valley bottoms from erosion volumes produced by sheet and rill erosion. Colluvial storage, the least precise component of the budgets, is the difference between the sum of all deposition plus sediment yield and total rill, sheet, and gully erosion. The valley system within the watershed was divided into four sections: upland valleys, tributary valleys, the upper main valley, and the lower main valley. Budgets were constructed for three periods: 1853–1938 – a period of intensive agricultural development in the watershed, 1938–1975 – a period when soil conservation practices were implemented on uplands, and 1975–1993 – continuation of soil conservation.

The main lessons emerging from this work are that 1) sediment discharge and yield have remained relatively constant through time despite large changes in net upland erosion rates; 2) the SDR, when considering total, as opposed to net, upland erosion, is about 0.05 to 0.06 (Trimble, 1983); 3) a large percentage of eroded sediment (90–95%) is deposited on hillslopes and in valley bottoms within the basin; 4) changes in upland soil erosion have been accompanied by corresponding changes in storage; 5) erosion sources have changed through time as stored sediment has been excavated by channel erosion in some parts of the watershed, for example, the upper main valley after 1938; and 6) the lower main valley is a major sediment sink within the river network. Between 1853 and 1938, extensive agricultural development mobilized large amounts of sediment on uplands, but much of this sediment was stored on hillslopes and in valley bottoms, which acted as filters to regulate the sediment flux out of the basin. Since 1938, conservation practices have reduced rates of upland erosion by sheet, rill, and gully erosion, and the net flux of sediment into the stream system from the uplands has decreased. During this period, corresponding decreases in total storage related to channel erosion have maintained a constant flux of sediment out of the watershed. Overall, storage of sediment in valley bottoms and subsequent remobilization of some of this stored sediment has produced profound changes in the morphology of streams and floodplains over time (Trimble, 2013) (see Chapter 15).

The example of Coon Creek illustrates one of the most important contributions of the sediment-budget approach: the emphasis on sediment storage as a major component of fluvial systems. Sediment mobilized on uplands by soil erosion often is redeposited on hillslopes before entering streams, and some of the sediment that enters streams is deposited on

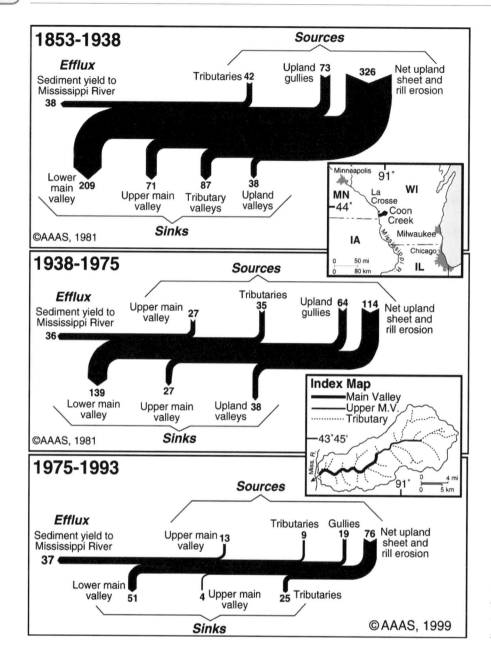

Figure 3.16. Sediment budget for Coon Creek drainage basin, Wisconsin, USA (Trimble, 1999). Numbers are annual averages for the periods in 10^3 Mg yr^{-1}.

floodplains or along channel margins before reaching the watershed outlet. In watersheds dominated by these storage effects, sediment delivery ratios are often less than 25% of total erosion and in some cases less than 10% (Phillips, 1991; Walling et al., 2002; Delmas et al., 2012).

Sediment budgets compartmentalize watershed-scale sediment dynamics into sources and sinks. For transfers of sediment to occur, these compartments must be connected. Connectivity, or the lack thereof, disconnectivity, is therefore fundamental to understanding the sediment dynamics exemplified by sediment budgets. Connectivity refers both to the spatial continuity of sediment transport pathways between different components of a fluvial system and to the efficacy

of transport processes in maintaining continuity of mass between these components. To some extent, connectivity is a scale issue. Given differences in process domains at different spatial scales (e.g., the difference between diffusive and advective sediment-transport processes on hillslopes versus advective processes only in streams) and differences in pathway continuity (e.g., discontinuous pathways of sediment transport on hillslopes versus continuous pathways in streams), it is unlikely that sediment-transfer processes and, thus, sediment budgets exhibit scale invariance (Slaymaker, 2006). Consequently, sediment budgets for hillslopes will differ from sediment budgets for entire watersheds (Delmas et al., 2012).

Disconnectivity in sediment cascades refers to factors that constrain the efficiency of sediment transfers (Fryirs et al., 2007; Fryirs, 2013). Such constraints can occur longitudinally (barriers), laterally (buffers), and vertically (blankets). Longitudinal linkages are defined in the context of upstream to downstream sediment transfers within the channel network. Longitudinal barriers to sediment movement include expansions in channel or valley dimensions; natural dams associated with large woody debris, landslides, or beaver activity; and artificial dams constructed by humans (see Chapter 16). Lateral linkages connect the channel network to the surrounding landscape and include hillslope–channel and channel–floodplain relationships (see Chapters 2 and 14). Lateral buffers promote demobilization of sediment transported laterally toward or away from river channels. Examples include natural or artificial levees that prevent hillslope material from entering channels and floodplains that promote deposition of suspended sediment transported by overbank flows. Vertical linkages focus mainly on within-channel deposition. Vertical blankets involve storage of sediment within channels through bar formation and through the infilling of gravel substrates by fine sediments. All these processes disrupt the cascade of sediment through fluvial systems and influence the spatial structure of sediment budgets.

The temporal scale of coupling among different compartments of a watershed-scale fluvial system is also important in understanding sediment cascades (Harvey, 2002). The relevance of timescale in watershed sediment dynamics can be illustrated by considering the SDR. An SDR less than one (Eq. (3.13)) indicates that sediment production exceeds sediment yield. If this situation persisted indefinitely, the drainage basin would progressively fill with sediment (Graf, 1988a, p. 134). Thus, SDRs and sediment budgets are dependent on the timescales over which they are constructed. SDRs vary over different timescales, and over long periods, a balance between erosion and yield requires that the SDR equal one (Parsons et al., 2006; Hoffmann, 2015). The prevalence of alluvial deposits in the rock record suggests that SDRs less than one are not uncommon, even over geologic timescales.

3.8.1 What Factors Control Sediment Residence Time?

The problem of temporal scale focuses on residence times; i.e., the length of time sediment resides within a particular part of the system. The residence time (t_{sr}) is a function of the volume of sediment (V_s) in a component of the system and the long-term average volumetric flux of material within that component (Q_{sv}) (Slaymaker, 2003):

$$t_{sr} = \frac{V_s}{Q_{sv}} \qquad (3.16)$$

Parts of a fluvial system where the long-term transport rates are high and volumes of material are small will have short residence times, whereas parts that have large volumes and low transport rates will have long residence times. Fine sediment suspended within the water column of a river that maintains a constant, relatively high material flux rate will have a short residence time within the drainage basin. On the other hand, fine sediment deposited by overbank flow on a floodplain that is only remobilized after many decades or centuries by migration of the river channel into the location of deposition will have a long residence time. Factors that promote disconnectivity typically increase residence times, establishing a relationship between the spatial and temporal dynamics of sediment (Fryirs, 2013). If spatial disconnectivity decreases the flux of sediment (Q_s), residence time will increase. Moreover, if disconnectivity promotes deposition, the volume of stored material will increase (V_s), which also increases residence time. Direct determination of residence times of stored sediment typically involves estimating the age of the material using dating techniques (Brown et al., 2009). Common methods include radiocarbon dating if stored sediments contain organic material (Törnqvist and van Dijk, 1993) and optically stimulated luminescence (OSL) dating for stored sediments containing quartz minerals (Wallinga, 2002). Nevertheless, the lack of reliable dating of sediments is a shortcoming of many sediment budgets (Brown et al., 2009).

3.8.2 How Can Sediment Tracing Contribute to Understanding of Sediment Dynamics?

Fine sediment tracing, sometimes referred to as sediment fingerprinting, provides a method for identifying connections between sources of sediment within a drainage basin and the flux of material out of the basin (Collins and Walling, 2004; Walling, 2005). The purpose of fine sediment tracing is to determine the spatial provenance, or sources, within a watershed of sediment transported in suspension out of the drainage basin. A variety of potential tracers can be used to determine sources of sediment (Table 3.6). Although the tracer method can be applied to coarse particles (d'Haen et al., 2012), most tracer studies at watershed scales focus on fine sediment (silt, clay, and fine sand). Evaluations of coarse sediment connectivity

Table 3.6. Potential tracers for fine sediment.

Physical Properties	Mineralogical Properties	Mineral Magnetic Properties	Geochemical Properties	Biogeochemical Properties	Fallout Radionuclides	Cosmogenic Radionuclides	Stable Isotope Ratios	Biogenic Markers
Color	Bulk mineralogy	Magnetic susceptibility	Major elements	Nitrogen	^{137}Cs	^{10}Be	δ^{15}N	Pollen
Grain Shape	Heavy minerals	Anhysteretic remanent magnetization	Trace elements	Phosphorus	^{210}Pb	^{26}Al	δ^{13}C	Spores
	Clay minerals	Isothermic remanent magnetization	Elemental composition of major minerals	Carbon	^{7}Be			
	Quartz cathodoluminescence							

(adapted from D'Haen et al., 2012)

typically concentrate on stream channels (Hooke, 2003a); such evaluations have been the focus of some sediment budgets (Fryirs and Brierley, 2001).

To conduct fine-sediment tracer analysis, samples must be collected both at a target location, usually within the stream at the basin outlet, and at a variety of locations within each type of source, which can include forest, pasture, cultivated land, or any set of land-cover categories for which distinctive sets of tracer signatures can be identified. If a stream sample and a particular source sample both have the same highly distinctive characteristics, such as color, degree of magnetization, or mineralogy, it may be possible to identify the source of the stream sediment qualitatively. In most cases, such obvious associations do not occur. Instead, statistical approaches are used when the selection of a tracer is not obvious and when the stream sample is a composite of multiple sources.

Two steps are involved in the statistical approach to fine-sediment tracing analysis (Collins et al., 1997). The first step identifies the most useful geochemical properties for distinguishing among the different sources. Various chemical, mineralogical, biological, or magnetic analyses of the sediment samples are conducted to generate information on potential tracers (Table 3.6). Statistical tests are then conducted to determine which tracers differ significantly among the different sources. Once tracers distinctive to each source are identified, the second step employs an unmixing model to ascertain the extent to which the stream sediment consists of sediment from the different sources (Collins et al., 1998, 2010; Fox and Papanicolaou, 2008; Abban et al., 2016). Results of the unmixing analysis yield estimates of the percentage contribution of each source to the sediment flux at the basin outlet. This type of analysis

also can be used to evaluate contributions of sources to floodplain storage (D'Haen et al., 2012), to assess channel–floodplain sediment exchange (Belmont et al., 2014; Stout et al., 2014), to distinguish between contributions from surface and subsurface erosion (Hancock et al. 2014b), and to determine changes in sediment flux following wildfire (Wilkinson et al., 2009a).

Tracer information is complementary to the development of sediment budgets because it helps to identify where sediment reaching the basin outlet or accumulating on floodplains has come from, thereby allowing precise apportionment of sediment fluxes among various sources (Figure 3.17). Whereas budgets allocate sediment fluxes by process (erosion, deposition, or yield), sediment tracing, or fingerprinting, links these fluxes to specific types of sediment sources within the watershed. Incorporating fine sediment tracing into sediment-budget analysis leads to enhanced understanding of watershed sediment dynamics.

An integrated approach combining the use of ^{137}Cs measurements, sediment source fingerprinting, bed sediment surveys, and conventional river monitoring was successfully employed to establish the fine-grained sediment budgets of two lowland agricultural watersheds in the United Kingdom (Walling et al., 2006). The ^{137}Ce measurements generated information on upland soil-erosion rates and patterns of redistribution on uplands. This artificial fallout radionuclide, released into the atmosphere between the early 1950s and the 1970s by testing of thermonuclear weapons, accumulated readily in soils during this time, and its subsequent redistribution across hillslopes can be related through conversion models to soil erosion, transport, and deposition (Walling and He, 1999). Identification of distinct tracer signatures for soils in cultivated fields versus pasture provided the basis for developing separate sediment budgets illustrating the contributions of these two types of land use to the overall sediment flux of the two watersheds (Figure 3.18). These budgets show that upland erosion is the dominant source of sediment in this system. Both cultivated land and pasture contribute to erosion in proportion to the percentage of land cover each occupies in these cultivated-dominated watersheds. Most eroded sediment is deposited within the field of origin or between the field and the stream channel network. Erosion and deposition within the stream network are relatively minor. The SDRs for the two catchments are both about 1%, indicating that little of the eroded sediment leaves the watersheds on an annual basis.

Although sediment tracing has become a major tool for developing sediment budgets, it is not without shortcomings. A key assumption underlying the method is that the tracer characteristics of sediment remain constant as the material moves through the landscape. In other words, the technique assumes that the tracers exhibit conservative behavior; i.e., tracer characteristics are preserved in transport, and sediment particles have the same characteristics when sampled at the basin outlet that these particles had when located in the source area. Similarity between signatures of these unchanging characteristics provides the basis for statistical evaluation of how much each source contributes to the basin sediment flux. Various types of physical, chemical, and biological transformations of sediment can occur as it moves through the landscape, introducing nonconservative behavior into properties of this sediment. Such changes must be taken into consideration when conducting tracer studies and

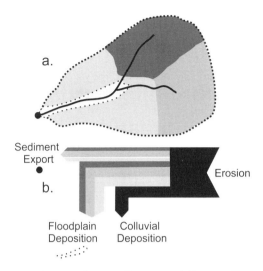

Figure 3.17. Schematic diagram illustrating different sediment sources within a watershed (a, colors) and apportionment of sediment fluxes by sources at watershed outlet (dot) and floodplain deposition (yellow area) based on fine-sediment tracing analysis (b). (A black and white version of this figure will appear in some formats. For the color version, please refer to the plate section.)

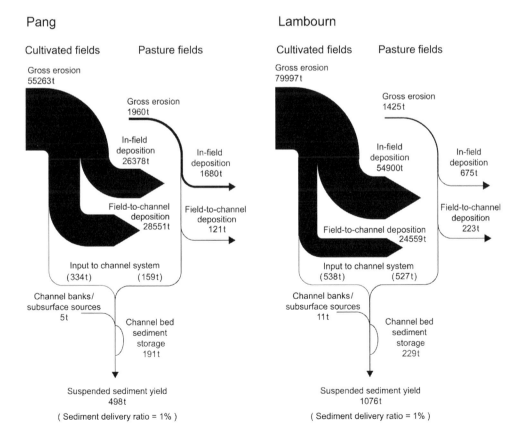

Figure 3.18. Sediment budgets for the Pang and Lambourn watersheds in the United Kingdom (from Walling et al., 2006).

evaluating the outcomes of these studies (Koiter et al., 2013). Another issue of concern is that the results of tracer studies do not shed light on the processes that link sources of sediment to samples collected at a target location, such as the flux of sediment at the basin outlet. Even if a tracer behaves conservatively, the method identifies the original source of the sediment, not the proximal source. For example, sediment from a pasture, the original source, could move onto a floodplain, a proximal source, before it is remobilized by floodplain erosion and transported to the basin outlet. In this sense, the method is a black-box approach (Koiter et al., 2013). Attempts to address this concern have explored the use of tracers, such as radionuclides, that exhibit known nonconservative behavior. Knowledge of this behavior can be drawn upon to enhance tracer studies. Differences in concentrations between radionuclide tracers with long half-lives (^{10}Be) and those with short half-lives (^{210}Pb and ^{137}Ce) and in apportionment estimates of upland erosion based on these concentrations provide the basis for determining the amount of upland sediment at the stream sampling location derived from long-term floodplain storage (Belmont et al., 2014). Preferably, tracing studies should be coordinated with detailed studies of erosional and depositional processes within drainage basins so that fluxes of sediment at target locations as determined by tracing results can be linked to the fluvial processes that move sediment from particular sources to this location.

3.8.3 How Are Models Used to Evaluate Sediment Dynamics and Sediment Budgets?

Another set of tools to support the development of sediment budgets is the use of models. A variety of models exist for predicting soil erosion and sediment transport at the watershed scale (Aksoy and Kavvas, 2005). Prediction of soil erosion often is necessary in sediment budgets unless the watershed is small enough to evaluate erosion from field-based assessments. RUSLE2 is a computer program that includes the capability to account for soil detachment, transport, and deposition over defined hillslope overland flow paths (U.S. Department of Agriculture, 2013). It uses an empirical-equation form of the USLE to compute detachment, and process-based equations provide estimates of transport and deposition. The Soil and Water Assessment Tool (SWAT) is a widely used basin-scale model for quantifying and predicting the impacts of land management practices on water, sediment, and agricultural chemical yields in large complex watersheds with varying soils, land use, and management conditions over long periods of time (Neitsch et al., 2009). It predicts soil erosion using the modified universal soil loss equation (MUSLE), which replaces the rainfall erosivity factor in the USLE with a runoff erosivity factor, thereby avoiding the need to use SDRs to predict sediment yields (Williams and

Berndt, 1977). MUSLE also allows predictions of sediment yield for individual storm events. Companion models EPIC, a crop and soil productivity model, and APEX, a watershed and land management simulation model, also incorporate the MUSLE. Another model commonly used to assess the effect of land management decisions on water, sediment, and chemical loadings within a watershed system is the Agricultural Nonpoint Source Pollution Model (AGNPS), which incorporates the RUSLE (Bingner et al., 2015). The Watershed Erosion and Prediction Project (WEPP) model, a physically based, spatially distributed continuous simulation erosion prediction model, has been developed by the U.S. Department of Agriculture as an alternative to the USLE and RUSLE2 (U.S. Department of Agriculture, 1995). This program also is available in a geographic-information system format (GeoWEPP) (Renschler, 2003). The model is capable of simulating sheet and rill erosion on hillslopes as well as hydrologic and erosion processes in small watersheds. It also produces estimates of sediment yield and SDRs at the watershed scale.

Other physically based models for estimating soil erosion include the European Soil Erosion Model (EUROSEM) (Morgan et al., 1998), OPENLISEM (Baartman et al., 2012), and SHESED (Wicks and Bathurst, 1996). Many landscape evolution models (Table 2.1) are being adapted to predict sheet, rill, and gully erosion on uplands and to route eroded sediment into and through the channel network, accounting for deposition along the way, and predicting the resultant fluxes of sediment out of watersheds. These models also can be used to examine interactions among climate, tectonics, and sediment yield, thereby providing insight into sediment budgets over geologic timescales (Coulthard and van de Wiel, 2013b). Spatially distributed models targeted specifically at predicting sediment budgets of watersheds (Wilkinson et al., 2008; Patil et al., 2012) have proven valuable for exploring connections among erosion, transport, and deposition on hillslopes and within channel systems as well as the influence of these connections on sediment yield (Wilkinson et al., 2006, 2009a, 2014).

At global scales, WBMsed, a distributed suspended-sediment flux model based in part on the BQART relations (Eq. (3.7) and (3.8)) (Syvitski and Milliman, 2007) provides a tool for exploring variability in water and sediment fluxes both over time and between continents (Cohen et al., 2014). A simple global-scale model incorporating the effects of slope, soil texture, mean monthly rainfall, and mean monthly leaf area index on sediment detachment, as well as a transport component that accommodates the effects of slope and soil texture, reproduces the long-term sediment yield of 128 global rivers with a Pearson correlation coefficient (R value) of 0.79 (Pelletier, 2012b). It also predicts sediment delivery ratios that decrease with basin area over six orders of magnitude in basin size.

Comprehensive analysis of the predictive capabilities of fourteen different soil-erosion and sediment-yield models indicates that appropriate application of most models is scale dependent and that simple spatially lumped models often perform better than physically based, spatially distributed models for predicting sediment yield, even though lumped models do not provide insight into sediment sources and sinks (de Vente et al., 2013). Given the need for spatially distributed information in sediment budgets, refinement and testing of spatially distributed models is necessary. Finally, in steep terrain, mass movement processes, including soil creep, debris flows, dry ravel, and landslides, may deliver substantial amounts of sediment to streams (Dietrich and Dunne, 1978). Accounting for the contributions of these processes to stream transport and storage is an important component of sediment budgets in such environments (Reid and Dunne, 2016).

3.8.4 What Is the Value of Sediment Budgets for River and Watershed Management?

Sediment budgets provide a useful framework for management of sediment dynamics within watersheds and the possible effects of these dynamics on stream habitat, sediment pollution, and river-channel stability (Slaymaker, 2003; Walling and Collins, 2008). The Coon Creek example clearly shows that sediment fluxes, as measured by sediment yields, may in some cases be fairly insensitive to changes in land management practices (Walling, 1999), especially if the system is well buffered against change through the availability of abundant storage locations. On the other hand, sediment budgets illustrate how land-use changes, particularly the development of modern agricultural practices and urbanization, have modified erosion rates and the spatial distribution of erosion and storage (Phillips, 1991; Beach, 1994; Brierley and Fryirs, 1998; Fryirs and Brierley, 2001; Walling et al., 2002; Jackson et al., 2005; Allmendinger et al., 2007; Gellis et al., 2017). In some cases, small areas of watersheds can contribute substantially to the total sediment load of a river, and the construction of a sediment budget can help in identifying these locations of exceptional erosion (Stubblefield et al., 2009; Day et al., 2013; Nichols et al., 2013). Assessment of erosional heterogeneity and the contributions of erosional hotspots, areas of high erosional activity, to basin-scale sediment dynamics can guide management by determining where the implementation of conservation practices aimed at regulating erosion and sediment transfer will achieve maximum benefit.

The utility of sediment budgets in a management context has led to the development of a variety of methods for developing these budgets (Reid and Dunne, 2016). Budgets need not be complex to be useful, but the information used to construct the budget should be appropriate to the level of accuracy and precision required. Fully quantifying all aspects of a sediment budget is difficult to accomplish, especially as the size of the watershed and the timescale of the budget increase. Many budgets contain an unestimated component computed solely through subtraction of estimated quantities under the assumption of conservation of mass. This practice tends to hide accumulated error in the estimated components within the unestimated component (Kondolf and Matthews, 1991). For example, in the sediment budget for the Coon Creek watershed, the quantity of colluvial deposition was obtained by subtracting the sum of estimated deposition within channels and sediment discharge at the basin outlet from estimated quantities of erosion under the assumption that total deposition plus sediment export must equal total erosion (Trimble, 1983). The budget did not include a direct estimate of colluvial deposition based on field measurements or modeling. This method of determining the quantity of colluvial deposition assumes that all estimated components have no error, which almost certainly is not the case (Kondolf and Matthews, 1991). It represents a problem of budget closure (Parsons, 2012). Rarely do budgets include estimates of uncertainty of individual estimated components, and most budgets focus on sediment suspended within the flow and neglect sediment transported on the bed of rivers. The timescale of budgets often is poorly specified, and estimates of budget components may represent dramatically different timescales. Ideally, sediment budgets should specify clearly the timescale over which they are applicable, strive to produce estimates of budget components consistent with the chosen timescale, not include unestimated components obtained by subtraction, and provide estimates of uncertainty for all reported components of the budget (Parsons, 2012).

4 Flow Dynamics in Rivers

4.1 Why Are Flow Dynamics in Rivers Important?

Rivers consist fundamentally of channeled flows of water. Although the flow of water in rivers typically is not of primary interest to fluvial geomorphologists, it is integral to other processes of interest, such as sediment transport and river morphodynamics (Figure 1.2). Therefore, knowledge of flow in rivers is vital to the understanding of river dynamics. The study of fluid flow in channels and pipes is referred to as hydraulics, a branch of applied science and technology encompassed by civil, environmental, and mechanical engineering. The field of fluid mechanics provides the theoretical underpinnings of hydraulics. Fluid mechanics is part of mechanics, the subfield of physics concerned with the behavior of bodies when subjected to forces. The fluid of interest in rivers is, of course, water, and the type of channel of interest is an open channel – a channel that has solid boundaries on the bottom and two sides and is open to the atmosphere at the top. River channels are examples of open channels with forms that typically are much more complex than artificial open channels designed by engineers. Open-channel hydraulics is the field that examines the mechanics of flow in open channels, including rivers.

This chapter presents basic principles of open-channel hydraulics, providing the foundation for subsequent chapters that relate these principles to the fluvial processes involved in the dynamics of river channels. The focus throughout this chapter is on a reach of river or a section several multiples of the channel width in length. In the scheme presented in Chapter 1, a reach corresponds to the planform scale. The emphasis here is on relatively straight channels of fairly uniform shape, size, and composition of sediment on the channel bed, rather than on more complex configurations. Within a reach, no major tributaries join the channel. Although these assumed characteristics are somewhat simplistic compared with the planform-scale characteristics of many natural rivers, they provide a starting point for defining the basic hydraulic principles that govern river flow. Consequences of deviations from these assumptions are clearly noted both in this chapter and in subsequent chapters.

4.2 What Is the Basic Metric of Flow in Rivers?

Discharge, the volume of water passing along the river per unit time ($L^3 T^{-1}$), is the basic metric of flow in open channels. This metric is fundamental not only to hydraulics but also to hydrology (see Chapter 6). Thus, it serves as a variable connecting the mechanics of the flow and the quantity of flow.

From a hydrological perspective, changes in discharge reflect changes in the delivery of water to streams from the surrounding landscape. These changes, in turn, are related to variations in amounts of runoff from hillslopes by mechanisms of infiltration excess overland flow, saturation excess overland flow, and subsurface flow through soils (throughflow) or through underlying parent material (groundwater flow) (see Chapter 6). Thus, in hydrology, discharge is treated as a dependent variable.

In hydraulics, the topic of this chapter, discharge represents an amount of water that the river must convey. Therefore, it is an independent variable imposed on the river system. In an open channel, the discharge (Q) of water through a cross section is defined as

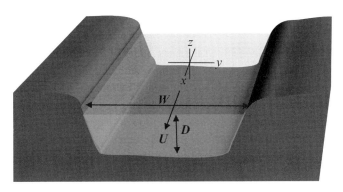

Figure 4.1. Idealized diagram of a channel reach illustrating water-surface width (W), mean flow depth (D), and mean flow velocity (U) at a cross section. Three-dimensional coordinate system is also shown. (A black and white version of this figure will appear in some formats. For the color version, please refer to the plate section.)

$$Q = UWD \qquad (4.1)$$

where W is the width of water surface at the cross section, D is the mean water depth over the cross section, and U is the mean velocity of the flowing water over the cross section (Figure 4.1). Because $WD = A$, the cross-sectional area of the flow, discharge also equals

$$Q = UA \qquad (4.2)$$

From a hydraulic perspective, changes in discharge are accommodated by changes in W, D, and U – the component variables of discharge. If discharge is constant, continuity of mass requires that WDU remains constant:

$$W_1 D_1 U_1 = W_2 D_2 U_2 = W_3 D_3 U_{3...} \qquad (4.3)$$

where 1,2,3 ... are successive cross sections along the reach. The magnitude of each individual variable may change, but the product of these three variables does not change. In this formulation, the density of the fluid (ρ), which must be multiplied by WDU to obtain volumetric mass flux, is assumed to be constant.

4.3 How Are Flows in Rivers Classified According to Changes in Depth and Velocity?

The science of hydraulics has developed a scheme for classifying flows in open channels into different types based on spatial and temporal variations in D and U (Chow, 1959) (Figure 4.2). Water-surface width traditionally is ignored because many artificial open channels have constant widths. In natural channels, W often varies, but this variation has the same effect as variation in depth for the purposes of the classification scheme. The discussion of classification assumes constant width, but variations in W must be considered when ascertaining which types of flow within the scheme occur most

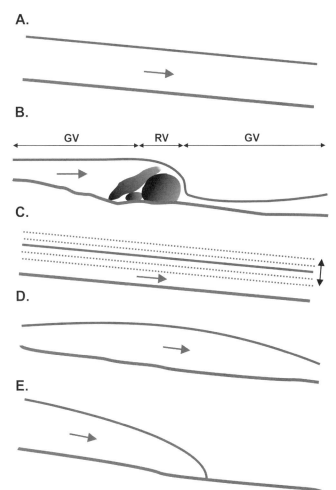

Figure 4.2. Classification of flow types: (a) steady, uniform flow, (b) gradually varied (GV) and rapidly varied (RV) flows, (c) unsteady, uniform flow (rare), (d) gradually varied, unsteady flow (flood wave), (e) rapidly varied unsteady flow (flood bore). Brown lines are channel bed and blue lines are water surface. (A black and white version of this figure will appear in some formats. For the color version, please refer to the plate section.)

commonly in natural channels. The idea behind the scheme is that U and D (and/or W in a natural channel) are interdependent for a constant discharge and also mutually accommodate changes in discharge associated with increases or decreases in the supply of water to the channel.

Flows in which D and U do not change over time (t) or over distance along the channel (x) are uniform, steady flows (Figure 4.2). This type of flow occurs when discharge is constant over time, no additions of flow occur along the reach, and channel geometry does not change along the reach. Such conditions are rare in natural rivers, but at times can be closely approximated in streams that have been artificially modified by human action. Varied flows occur when discharge is constant but U and D vary over distance because of changes in channel form. If these changes occur abruptly over a short distance, the flow is rapidly varied; otherwise, it is gradually varied. Unsteady flow develops when discharge through the reach changes over time. Uniform, unsteady flow requires that D and U remain constant over the length of the reach but change over time. To produce this type of flow, water would have to be added instantaneously in identical amounts at uniform increments of length along a channel of unvarying form so that D and U do not change along the channel even though these two variables change through time at all locations along the channel (Figure 4.2). Such conditions are hard, if not impossible, to achieve either artificially or naturally; thus, this type of flow is included for completeness of the scheme but is not likely to occur in rivers. Unsteady rapidly varied and gradually varied flows are characterized by changes in U and D over time and over distance along the reach of an open channel. The distinction between these two types of flow again is based on whether changes occur gradually or abruptly. In deserts, heavy rainfall can produce flash floods in which a flood bore at the front of the flood wave propagates down the dry bed of an irregular channel – an example of rapidly varied, unsteady flow (Figure 4.2). The incremental rise of water levels associated with the downstream movement of flood waves in many perennial rivers exemplifies gradually varied unsteady flow.

4.4 How Is the Dimensionality of Fluid Motion in a River Determined?

The dimensionality of fluid motion in a river refers to the number of directions in three-dimensional space in which persistent motion of the fluid occurs. In a river, water typically flows mainly in the downstream or streamwise (x) direction, but it may also move in the lateral (y) or vertical (z) direction (Figure 4.1). An appropriate way to think about this issue is in terms of vectors, which are physical quantities that define the magnitude and direction of motion. In an open channel, velocity vectors specify the local magnitude and direction of the flow. Vectors typically are depicted as arrows where the length of an arrow is proportional to the velocity magnitude and the orientation of an arrow indicates the direction of motion. In three-dimensional space, the magnitude of a velocity vector ($\mathbf{u}$) and its direction at any location within the cross section of an open-channel flow can be specified as

$$\mathbf{u} = \sqrt{u^2 + v^2 + w^2} \tag{4.4}$$

$$\alpha_{\mathbf{u}1} = \tan^{-1}\left(\frac{v}{u}\right) \tag{4.5}$$

$$\alpha_{\mathbf{u}2} = 90 - \cos^{-1}\left(\frac{w}{\mathbf{u}}\right) = \tan^{-1}\left(\frac{w}{\sqrt{u^2 + v^2}}\right) \tag{4.6}$$

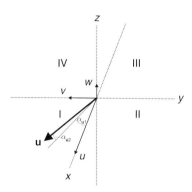

Figure 4.3. Cartesian coordinate system for a three-dimensional velocity vector ($\mathbf{u}$) and vector components (u,v,w). Quadrant numbers (I–IV) correspond to x-y plane.

where u is the mean local velocity component in the x direction, v is the mean local velocity component in the y direction, w is the mean local velocity component in the vertical direction, $\alpha_{\mathbf{u}1}$ is the angle between the x-z plane and the projection of the vector onto the x-y plane, and $\alpha_{\mathbf{u}2}$ is the angle between the vector and the x-y plane (Figure 4.3). Generally, water in a river flows downstream, or in the positive x direction ($u > 0$). In a right-handed coordinate system looking downstream, positive values of $\alpha_{\mathbf{u}1}$ correspond to $v > 0$ or clockwise rotation of the vector relative to the x axis (vector in x-y quadrant I) (Figure 4.3). Conversely, negative values of $\alpha_{\mathbf{u}1}$ correspond to $v < 0$ or counterclockwise rotation of the vector relative to the x axis (vector in x-y quadrant II). In rare cases, pockets of flow can move upstream ($u < 0$), resulting in vectors directed toward x-y quadrant III ($v < 0$) or quadrant IV ($v > 0$). Positive values of $\alpha_{\mathbf{u}1}$ in these cases represent clockwise rotation from 180° (vector in x-y quadrant III), whereas negative values correspond to counterclockwise rotation from 180° (vector in x-y quadrant IV). Fluid can also move vertically, with positive values of w corresponding to upward motion and negative values corresponding to downward motion. This motion will produce an upward (positive w) or downward (negative w) orientation of the velocity vector relative to the x-y plane.

Now that velocity vectors have been defined, it is straightforward to define the dimensionality of fluid motion. If all velocity vectors throughout a reach of stream are persistently aligned in the streamwise direction of the channel (parallel to the x axis), the flow is unidirectional, or one-dimensional with respect to the orientation of fluid motion. It is important to emphasize that the properties of a unidirectional flow can still vary in multiple dimensions. If the magnitudes, but not the orientations, of velocity vectors change over depth and over width in a unidirectional open-channel flow, which is typically the case, the rate of flow varies in three dimensions. Two-dimensional fluid motion in an open channel typically is characterized by $u > 0$ and $v \neq 0$. In three-dimensional fluid motion, all three mean velocity components differ from zero.

4.5 What Is the Relation of Flow Classification and Dimensionality of Fluid Motion to Hydraulic Complexity?

The schemes of flow classification and dimensionality of fluid motion provide a qualitative framework for examining the complexity of river flows. As the flow becomes more varied and unsteady, complexity increases. Likewise, as the dimensionality of fluid motion increases, complexity increases. Steady, uniform, unidirectional flows are the simplest, whereas rapidly varied, unsteady flows characterized by three-dimensional fluid motion are the most complex. Where do most rivers fall in this range of complexity? Most natural rivers exhibit enough morphological variability to produce at least gradually varied flow, and rapidly varied flow is not unusual, especially in rivers influenced strongly by large substrate materials or by abrupt changes in channel bed elevation (Figure 4.2b). Virtually all rivers are unsteady to some extent, given variations in flow associated with runoff from weather events (see Chapter 6). The issue of dimensionality of fluid motion hinges in part on the time and space scales over which dimensionality is of interest. Over short timescales of less than a second to perhaps several seconds, virtually all river flows exhibit three-dimensional fluid motion. This high-frequency three-dimensionality may not be important when considering channel morphodynamics over long timescales. Nevertheless, persistent two- and three-dimensional fluid motion in rivers is common and occurs in response to forces acting on flow associated with spatial variations in channel form. In reality, then, many rivers exhibit spatially varied, unsteady flow characterized by two- or three-dimensional fluid motion. Nevertheless, many geomorphological studies have used and continue to use simple models founded on steady, uniform flow or gradually varied, unidirectional flow to try to link flow in rivers to sediment transport and to channel morphodynamics (Figure 1.2). The validity of characterizing complex river flows using simple models hinges on the extent to which deviations from assumed conditions compromise results generated by the application of these models. The study of compromising effects is the focus of ongoing research. In the treatment here, it is sufficient to acknowledge potential limitations of simple approaches while also embracing the need to start with basic concepts.

4.6 What Are the Components of Flow Energy in Rivers?

Natural rivers flow downhill over distance from high elevations to low elevations. In other words, the water flows down a gravitational gradient, and the source of energy for this downhill movement is gravitational energy. This type of energy is potential energy – energy that underlies the potential for displacement based on the position of an object, in this case a body of water, in relation to a difference in elevation. As the water flows, it also has kinetic energy, or energy related to motion. These two types of energy, gravitational potential energy and kinetic energy, provide the basis for characterizing the energy of river flows. The treatment here focuses on uniform steady and gradually varied flows.

An idealized analysis of energy considers steady flow in a rectangular open channel flowing down a sloping bed (Figure 4.4), but the relations derived from the analysis apply to natural channels with nonrectangular geometries. Energy as a physical quantity has dimensions $M\,L^2\,T^{-2}$. In the study of energy in open channels, this quantity is typically divided by the weight of the water $(M\,L\,T^{-2})$ to yield a measure of energy called *head*, defined in the dimension of length (L). The total head (H_t) of the flow at any particular cross section along the open channel is (Figures 4.4 and 4.5)

$$H_{ti} = Z_i + D_i \cos\theta + \alpha_{ei}\frac{U_i^2}{2g} \qquad (4.7)$$

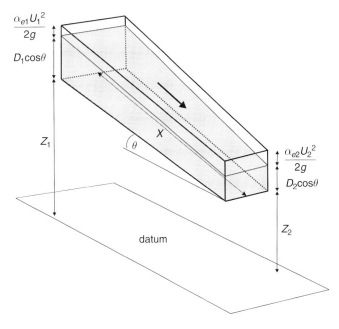

Figure 4.4. Flow (gray) in a rectangular open channel flowing downhill between cross sections 1 (upstream) and 2 (downstream) where energy conditions are defined.

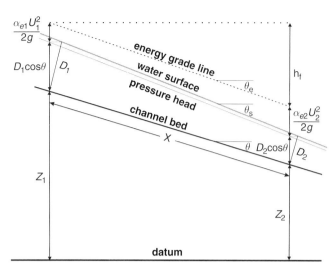

Figure 4.5. Energy relations for a gradually varied flow between two cross sections in an open channel.

where Z is the elevation head, or potential energy related to the elevation of the channel bed above an arbitrary datum, $D_i \cos\theta$ is the pressure head, or potential energy related to the water depth (D) above the channel bottom, $\alpha_{ei}\frac{U_i^2}{2g}$ is the velocity head, or kinetic energy related to the velocity (U) of flow, θ is the slope angle of the channel bed, α_e is the energy coefficient, g is gravitational acceleration, and the subscript i is a cross-section designator. The energy coefficient corrects the velocity head for nonuniform velocity distributions throughout a cross-section of the flow. It is the ratio of the actual kinetic energy of the flow to the kinetic energy corresponding to the mean velocity. If all local velocities throughout the entire cross section equal the mean velocity, $\alpha_e = 1$; otherwise, $\alpha_e > 1$ (Dingman, 2009, pp. 298–299). Values of the coefficient generally fall between 1 and 2, with higher values for natural rivers than for straight, artificial channels (Chow, 1959, p. 28).

Conservation of energy requires that the total energy head at an upstream location equal the total energy head at the downstream location plus any loss of head (h_f) between the two locations (Figure 4.5):

$$Z_1 + D_1\cos\theta + \alpha_{e1}\frac{U_1^2}{2g} = Z_2 + D_2\cos\theta + \alpha_{e2}\frac{U_2^2}{2g} + h_f$$

(4.8)

The head loss, or decrease in total energy, therefore is equal to the difference between the total head at the two locations:

$$h_f = \left(Z_1 + D_1\cos\theta + \alpha_{e1}\frac{U_1^2}{2g}\right) - \left(Z_2 + D_2\cos\theta + \alpha_{e2}\frac{U_2^2}{2g}\right)$$

(4.9)

In many rivers, the slope angle (θ) is small ($< 6°$ or 0.105 radians) and $\cos\theta \approx 1$. Under these conditions, Eq. (4.9) becomes

$$h_f = \left(Z_1 + D_1 + \alpha_{e1}\frac{U_1^2}{2g}\right) - \left(Z_2 + D_2 + \alpha_{e2}\frac{U_2^2}{2g}\right)$$

(4.10)

The rate of decrease in energy is defined by the slope of the energy grade line (S_e) along the length of the reach (X):

$$S_e = \frac{h_f}{X} = \sin\theta_e$$

(4.11)

where θ_e is the slope angle of the energy grade line. For steady, uniform flow, by definition $D_1 = D_2$ and $U_1^2 = U_2^2$, and Eq. (4.10) simplifies to

$$h_f = Z_1 - Z_2$$

(4.12)

In other words, the total head loss for a uniform steady flow is equal to the change in bed elevation from upstream to downstream. Moreover, the energy gradient for this flow is

$$S_e = \sin\theta$$

(4.13)

For low slope angles ($\theta < 6°$), $\sin\theta$ is nearly identical to $\tan\theta$, the slope of the channel bed (S). The change in energy, although referred to as head loss, is not lost in an absolute sense, because energy must be conserved. Energy is lost, or expended, only in the sense that it is no longer available to propel water along the river. Much of this expended energy is converted into heat by friction with the channel boundary, but some of it may become involved in mechanical processes, such as the transport of sediment.

4.7 How Can River Flows Be Characterized in Terms of Force, Stress, and Power?

A uniform steady flow can be viewed as a block of water moving downstream within an open channel (Figure 4.6). The force of this block of water on the channel boundary is defined by basic physical principles. The weight (Wt) of a solid object resting on an inclined surface can be resolved into different vector components (Figure 4.7). Both the weight of the block and its component vectors have dimensions of force (MLT^{-2}). The component of the object's weight oriented perpendicular to the inclined surface ($Wt\cos\theta$) represents a normal force, and the component oriented parallel to the inclined surface ($Wt\sin\theta$) is a shear force. In the study of rivers, the shear force is of particular interest, because it represents the component of the total force acting in the downstream direction. This component can potentially

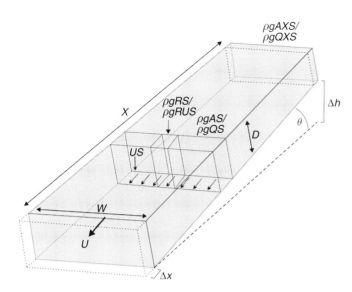

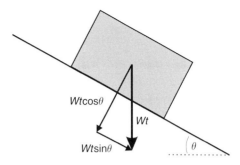

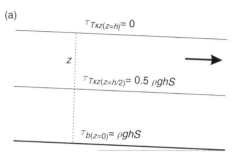

Figure 4.7. Force components of the weight (Wt) of a solid object on an inclined surface.

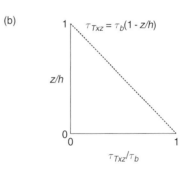

Figure 4.6. A body of water flowing downhill in an open channel. Arrows on bed indicate direction of shear force per unit length, stream power per unit length, shear stress, and stream power per unit area.

initiate and sustain the downstream movement of sediment on the channel bed (see Chapter 5).

The total weight of the block of water under consideration (Figure 4.6) is its volume times its specific weight (weight per unit volume), where the specific weight is the product of the density of water (ρ) and gravitational acceleration (g):

$$Wt = \rho g W D X \tag{4.14}$$

and the shear force (F_b) of the block acting on the bottom of the inclined channel is

$$F_b = \rho g W D X \sin\theta \tag{4.15}$$

or, for a river channel of relatively mild slope,

$$F_b = \rho g W D X S \tag{4.16}$$

In the SI system, force has units of Newtons (N). Dividing by the length of the reach yields the force per unit length (Figure 4.6):

$$F_b / X = \rho g W D S \tag{4.17}$$

and dividing Eq. (4.17) by the width equals the shear force per unit area, or the average shear stress on the channel bed:

$$\tau_b = \rho g D S \tag{4.18}$$

For flow in an infinitely wide channel with a flat bed, where the local depth (h) is nearly equivalent to the mean cross-sectional depth, the local shear stress on the bed equals

$$\tau_b = \rho g h S \tag{4.19}$$

Figure 4.8. (a) Shear stress at the surface, mid-depth, and the bed in an infinitely wide steady, uniform flow; (b) linear variation of shear stress above the bed in this type of flow.

Because the local shear stress at any position in the flow above the bed varies with the thickness of the overlying water, it decreases linearly in the x-z plane, reaching a value of zero at the water surface (Figure 4.8). Eq. (4.18) ignores the effects of the channel sides, or banks, on the reach-averaged shear stress, and commonly an alternative expression is used to account for these effects (Figure 4.6):

$$\tau_b = \rho g R S \tag{4.20}$$

where R is the hydraulic radius. In an open channel, $R = A/P$, where P is the length of the submerged portion of the channel boundary, the wetted perimeter. In a channel with a flat bottom and vertical banks, $P = 2D + W$, but in most natural channels the wetted perimeter must be determined by survey-ing the geometry of the flow (see Appendix C). In deep, narrow flows, P substantially exceeds W, but the proportional difference between these two variables diminishes as W increases in relation to D.

Physical metrics of power are commonly used to character-ize river flows, especially with regard to the capacity of the flow to transport sediment or erode the channel boundary (Rhoads, 1987a). As the block of water moves downstream, the application of a shear force on the channel boundary over distance equals work or energy (M L^2 T^{-2}). If the movement occurs at a mean velocity U, the result is energy expenditure or stream power:

$$P_b = \rho g WDUXS \tag{4.21}$$

where P_b has SI units of Watts (W) (M L^2 T^{-3}). Dividing P_b by the length of the reach yields the stream power per unit length (Figure 4.6) (W m^{-1}):

$$\Omega = \rho g WDUS = \rho g QS \tag{4.22}$$

and dividing Ω by the wetted perimeter results in the stream power per unit area (W m^{-2}) (Figure 4.6):

$$\omega = \rho g RUS = \tau_b U \tag{4.23}$$

Total power can also be divided by the weight of the water to yield

$$US = P_b/Wt \tag{4.24}$$

the power per unit weight of water, which is a vector quantity defining the vertical rate of decrease in elevation of the water (Figure 4.6).

4.8 Are River Flows Laminar or Turbulent?

Flowing water can exhibit laminar or turbulent behavior. Both types of behavior are characteristic of the flow itself, not inher-ent properties of water as a fluid. Examination of the conditions that generate laminar versus turbulent flows is important for understanding the hydraulic characteristics of rivers.

4.8.1 What Is a Laminar Flow?

Laminar flow consists of thin layers of fluid moving parallel to one another at different velocities over depth in response to an applied shear stress (τ) (Figure 4.9). The fastest layers are located near the surface of the flow, where it is bounded by overlying air, and no movement at all occurs where molecules of water are attached directly to the bed. Between the bed and the surface, the velocity of the layers progressively increases

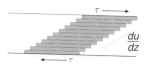

Figure 4.9. Layered vertical structure of a laminar flow.

upward from the bed. An analogy for this type of flow is the pattern a deck of cards sitting on a solid surface makes if pushed forward by hand from the top of the deck (Dingman, 2009). The forward force per unit area on the top of the deck repre-sents an applied shear stress. In response to the application of this shear stress, the card at the bottom of the deck remains in its position, whereas the top card moves forward by the greatest amount. In between, cards move forward by amounts propor-tional to their distances from the solid surface.

In laminar flow, viscosity characterizes friction between adjacent layers and is a measure of internal resistance of the fluid to movement. It is related both to the strength of cohe-sion between molecules moving at different speeds in adjacent layers and to exchanges of molecules moving at different speeds between the layers. The relation between the viscous shear stress of a laminar flow at any position z above the bed in the x-z plane (τ_{vxz}) (Figure 4.3) and the velocity gradient in this plane (du/dz) is defined by Newton's Law of Viscosity:

$$\tau_{vxz} = \mu \frac{du}{dz} \tag{4.25}$$

where μ is the dynamic viscosity (M L^{-1} T^{-1}). Values of μ define how resistant a fluid is to deformation, or flow. A fluid with a large value of μ requires a larger shear stress to produce a specific gradient of velocity (fluid deformation) than does a fluid with a small value of μ (Figure 4.10). Conversely, for a specific shear stress, the fluid with the higher dynamic visc-osity will deform less (smaller du/dz) than a fluid with a low dynamic viscosity. Exchanges of molecules between layers result in a vertical transfer of momentum, or a momentum flux, within the flow. This flux, which has the same dimension as shear stress (M L^{-1} T^{-2}), is a diffusive process that acts down the gradient of velocity from regions of high velocity near the surface to regions of low velocity near the bed. Thus, Newton's Law of Viscosity can be viewed not only as an expression of shear stress between adjacent layers moving at different velo-cities, but also as an expression of momentum flux associated with molecular-scale diffusion.

4.8.2 What Is Turbulent Flow?

Turbulent flow, in contrast to laminar flow, is characterized by various scales of fluid motion that transport small discrete amounts of water, or fluid parcels, from one part of the fluid

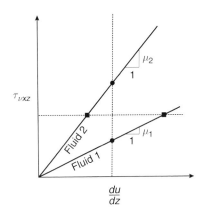

Figure 4.10. Two fluids with different dynamic viscosities. A larger shear stress is needed for fluid 2 (high viscosity) compared with fluid 1 (low viscosity) to produce the same velocity gradient. Conversely, for a given shear stress, fluid 1 deforms (flows) more readily than fluid 2.

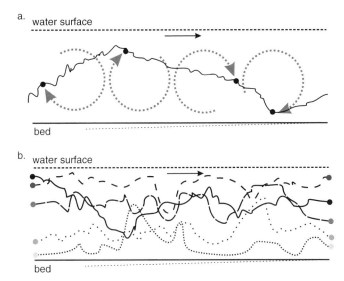

Figure 4.11. Examples of turbulent flow in rivers. (a) Path (solid line) of a discrete fluid parcel (black dots) in relation to a large rotating eddy (gray dotted circles) moving with the flow. (b) Complex paths (lines) of fluid parcels (dots) within a turbulent flow containing an array of eddies of different sizes, shapes, and rates of rotation.

column to another. Distinct turbulent structures that influence the transport of fluid parcels are referred to as eddies. Although turbulent fluid motion can occur in different directions, the vertical transfer of fluid parcels between the water surface and the channel bed is of particular interest (Figure 4.11). Because of turbulent motion, fast-moving water near the surface can be moved rapidly into regions of slow-moving water near the bed. Alternatively, slow-moving water near the bed suddenly may be transported upward into a region of otherwise high-velocity flow near the surface. The motion of fluid parcels in turbulent flow is not in distinct layers; instead, these parcels, influenced by eddies of various sizes, shapes, and rates of rotation, move irregularly throughout the water column (Figure 4.11).

The net effect of turbulence is to produce internal resistance to flow through exchange of momentum within the water column. Friction between distinct layers, as occurs in laminar flow, is not important in turbulent flow, because such layers do not exist. Instead, resistance to motion is related primarily to the exchange of momentum as fluid parcels with specific velocity characteristics move to locations in the water column where the surrounding flow has velocity characteristics different from those of the parcels. Viscous stresses still exist in a turbulent flow but are small in relation to turbulent stresses.

The shear stress, or momentum flux, in a turbulent flow (τ_t) has been characterized in a manner analogous to that of laminar flow:

$$\tau_{txz} = \varepsilon \frac{du}{dz} \tag{4.26}$$

where ε is the eddy viscosity. In contrast to the dynamic viscosity, which is associated with molecular-scale diffusion of momentum in a laminar flow, the eddy viscosity represents

exchange of momentum within the flow produced by large-scale turbulent eddies. As will be shown in Section 4.13.4, the characterization of ε is an important problem in the analysis of turbulent flows.

4.8.3 What Factors Influence Whether the Flow is Laminar or Turbulent?

Comparison of Eq. (4.25) and Eq. (4.26) provides the basis for determining whether a flow is dominated by viscous or turbulent shear stresses. As a first-order approximation, the velocity profile can be characterized as linear over the depth of the flow, allowing U/D to be substituted for du/dz in the two equations. As will be shown in Section 4.13.4, such an approximation is quite crude, but it is useful for comparative purposes. An approximation of the eddy viscosity can be formulated by assuming that its magnitude is proportional to the momentum of the flow (ρU – momentum per unit volume) and a characteristic vertical length scale of turbulent eddies (ℓ_e), which in the case of an open channel should be proportional to the mean depth of flow (D):

$$\varepsilon \propto \rho UD \tag{4.27}$$

Substituting these approximations into Eq. (4.25) and Eq. (4.26) and expressing the relation between the turbulent and viscous shear stresses as a ratio yields

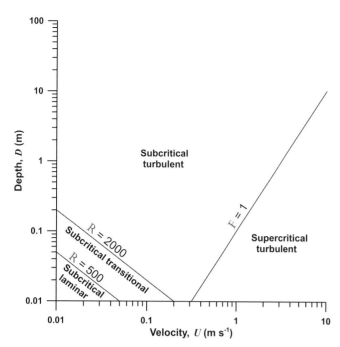

Figure 4.12. Domains of subcritical, supercritical, turbulent, laminar, and transitional flows in relation to the range of hydraulic conditions (depth and velocity) typically found in streams and rivers.

$$\frac{\tau_{txz}}{\tau_{vxz}} \propto \frac{\rho UD\left(\frac{U}{D}\right)}{\mu\left(\frac{U}{D}\right)} = \frac{\rho UD}{\mu} = \frac{UD}{\nu} = \mathbf{R} \qquad (4.28)$$

where $\nu = \mu/\rho$ is the kinematic viscosity and $\mathbf{R}$ is the Reynolds number, named after British hydraulician Osbourne Reynolds, who first recognized its importance (Dingman, 2009). Sometimes the hydraulic radius (R) is substituted for D in Eq. (4.28) to better account for the effects of the banks on flow in open channels, including rivers (Dingman, 1984; Knighton, 1998). For $\mathbf{R} < 500$, flow in open channels is dominated by viscous forces. The domain $500 < \mathbf{R} < 2000$ is transitional, and disturbances to the flow may or may not trigger the development of turbulence depending on the frequency, amplitude, and persistence of a disturbance. For $\mathbf{R} > 2000$, open-channel flows are fully turbulent. The value of ν depends on water temperature but is on the order of 1×10^{-6} m^2 s^{-1} (20 °C), ranging from 1.6736×10^{-6} m^2 s^{-1} at 2 °C to 0.8007×10^{-6} m^2 s^{-1} at 30 °C.

Given the small magnitude of ν, virtually all rivers, including the smallest streams, are fully turbulent (Figure 4.12). Consider a flow only 0.1 m deep flowing at 0.1 m s^{-1} with a water temperature of 20 °C. This flow will have a Reynolds number of 10,000, still well above the threshold for fully turbulent flow. In medium to large rivers, Reynolds numbers often exceed 1 million and can be greater than 5 million.

Recognizing that rivers are fully turbulent is of considerable importance. It means that river dynamics, including the movement of sediment and patterns of erosion and deposition, are governed by complex turbulent flows rather than less complex laminar flows. The Reynolds number can be thought of as an approximation of the ratio of the timescale of turbulent diffusion of momentum to the timescale of molecular diffusion of momentum in a flow with the same length scale (ℓ). The large values of $\mathbf{R}$ for natural rivers clearly emphasize that turbulence is several of orders of magnitude more effective at redistributing momentum within the flow than is molecular exchange. The flux of momentum association with turbulence influences the distribution of flow momentum near the bed and banks of a river, which, in turn, determines the shear stresses acting on the channel boundary.

4.9 What Is the Relation between Inertial and Gravitational Forces in Rivers?

A comparison similar to that performed for turbulent versus viscous forces can be conducted to examine the relationship between inertial and gravitational forces in rivers. The inertial force is represented by the approximation of the turbulent shear stress (numerator of Eq. (4.28)), whereas the gravitational force is simply the force per unit area of the water on the channel bed, which is assumed for simplicity to be horizontal (i.e., $\cos \theta \approx 1$):

$$\frac{\tau_{txz}}{\tau_g} \alpha \frac{\rho UD\left(\frac{U}{D}\right)}{\rho gD} = \frac{U^2}{gD} \qquad (4.29)$$

This relation has conventionally been expressed in square root form:

$$\sqrt{\frac{U^2}{gD}} = \frac{U}{\sqrt{gD}} = \mathbf{F} \qquad (4.30)$$

where $\mathbf{F}$ is the Froude number, named after William Froude, an English hydraulic engineer. Note that the denominator of this dimensionless metric has units of velocity and corresponds to the celerity, or rate of propagation, of gravity waves on the surface of an open-channel flow. Subcritical flow occurs when $\mathbf{F} < 1$. These relatively deep, slow-moving flows are described as tranquil and streaming (Chow, 1959, p. 13) (Figure 4.13). Waves produced locally on the surface of such flows can propagate upstream because the wave celerity exceeds the velocity of the flow. Moreover, the surface of the flow is not affected by submerged morphological features on the riverbed because of the strong damping effects of gravitational forces. When $\mathbf{F} = 1$, the flow is in a critical state where the gravitational and inertial forces are balanced. When $\mathbf{F} > 1$, the flow is supercritical, and waves produced by disturbance of the water surface are swept downstream. These relatively shallow, fast-moving flows are described as shooting, rapid,

Figure 4.13. (a) Subcritical turbulent flow, Wabash River, USA. (b) Supercritical flow with migrating surface waves above underlying antidunes, Wabash River, USA.

or torrential (Chow, 1959, p. 13). Strong inertial forces can deform the water surface into downstream migrating waves as supercritical flow moves over an erodible or morphologically varied riverbed (Figure 4.13). Most river flows are subcritical except for high-velocity torrents in mountain rivers or floods in channels of modest gradient and low flow resistance (Figure 4.12). As an example, a flow 1 meter deep must have a velocity in excess of 3 meters per second to attain supercritical conditions.

Abrupt spatial transitions between subcritical and super-critical flow are common in some rivers, especially those with steep channels and highly irregular beds, such as mountain streams (Vallé and Pasternack, 2006). When flow transitions over distance from a subcritical to a supercritical state, the velocity increases and the depth decreases. This type of transition is a hydraulic drop. The abrupt increase in velocity and decrease in depth as water flows over a steep ledge or waterfall is an example of a hydraulic drop. When the opposite transition occurs, from a supercritical to a subcritical state, the velocity decreases and the flow depth increases in the form of a hydraulic jump – a stationary wave on the water surface with a celerity equal, but opposite, to the flow velocity. Such transitions are common where a steep channel containing supercritical flow abruptly decreases in gradient. Hydraulic jumps and drops represent forms of steady rapidly varied flow. The frequent juxtaposition of hydraulic drops and jumps within relatively steep rivers with large obstacles, such as boulders or rock ledges, generates white water for recreational activities such as rafting and kayaking (Figure 4.14). Hydraulic jumps also play an important role in energy dissipation in river systems. Flow within a hydraulic jump is often highly turbulent, and considerable energy loss by turbulent dissipation, the conversion of kinetic energy to thermal energy,

Figure 4.14. Whitewater rafting over a hydraulic jump downstream of a hydraulic drop.

occurs within the jump (Wyrick and Pasternack, 2008). As a result, the energy of the flow downstream of an abrupt jump is considerably less than the energy upstream of the jump (Chow, 1959, p. 45).

4.10 How Is Flow Resistance in Rivers Related to Velocity?

In rivers and streams, friction along the channel boundary resists the tendency for water to flow downhill under the influence of gravity. In a steady uniform flow, the balance between the driving force of gravity and the resisting forces generated by friction along the channel boundary determines the mean velocity (U). The driving gravitational shear stress is described by Eq. (4.20). As indicated in Eq. (4.28), the resisting shear stress associated with boundary friction in a turbulent flow is proportional to $\rho UD(U/D)$. Therefore, the bed shear stress associated with frictional resistance can be expressed as

$$\tau_b = c_f \rho U^2 \tag{4.31}$$

where c_f is a constant of proportionality known as the friction coefficient. Because the driving shear stress (Eq. (4.20)) and resisting shear stress (Eq. (4.31)) are balanced,

$$\rho g R S = c_f \rho U^2 \tag{4.32}$$

Solving for U:

$$U = \sqrt{gRS/c_f} \tag{4.33}$$

or

$$U = C\sqrt{RS} \tag{4.34}$$

where $C = \sqrt{g/c_f}$. This equation, known as the Chezy equation, is named after Antoine de Chezy, a French hydraulic engineer. The coefficient C is the Chezy coefficient – a resistance coefficient inversely related to c_f. Note that C is not dimensionless; therefore, values of C depend on the system of units employed (Dingman, 2009, p. 221). The Chezy equation is an example of a reach-averaged resistance equation. Given values of R and S, along with an estimate of C, the formula can be used to determine the mean velocity (U) within an open channel.

Despite being grounded in a balance of forces framework, the Chezy equation has not commonly been used to estimate velocities of uniform, steady flows in open channels, mainly because no widespread tabulations of values of C exist (Yen, 2002). Instead, a similar function, the Manning equation, derived empirically by Robert Manning, an Irish engineer, has been widely adopted (Dooge, 1991):

$$U = \frac{R^{0.67}S^{0.5}}{n} \tag{4.35}$$

where n is Manning's roughness coefficient and all units of measurement are metric. Manning's n, like the Chezy coefficient, is not dimensionless, and the right-hand side of the equation must be multiplied by an adjustment factor if units other than metric are used (Dingman, 2009, p. 243). Given the widespread use of the Manning equation by engineers in the United States, numerous sources are available on methods to estimate Manning's n for rivers and streams, including tables of values with descriptions of stream conditions (Chow, 1959, Table 5.6), values of n corresponding to photographs of stream channels (Chow, 1959, figure 5.5; Barnes, 1967; Arcement and Schneider, 1989; Hicks and Mason, 1991; Soong et al., 2012; Yochum et al., 2014), guidance for partitioning the total resistance into separate components associated with different types of roughness (Cowan, 1956; Arcement and Schneider, 1989; Coon, 1998), and use of empirical relations to estimate n (e.g., Limerinos, 1970; Jarrett, 1984). Values of Manning's n for natural lowland rivers typically fall within the range of 0.02 to 0.08, depending on obstructions to flow (Chow, 1959, table 5.6; Barnes, 1967). In mountain rivers with step-pool morphology and high amounts of large wood, values of Manning's n can be quite large, consistently exceeding 0.08 and in some cases approaching 1.00 (Yochum et al., 2014).

Another commonly used reach-averaged flow resistance formula for rivers is the Darcy–Weisbach equation, originally developed for determining the head loss (h_f) of flow through pipes:

$$h_f = f_f \left(\frac{X}{d_{pd}} \right) \left(\frac{U^2}{2g} \right) \tag{4.36}$$

where X is the length of the pipe, d_{pd} is the pipe diameter, and f_f is the dimensionless Darcy–Weisbach friction factor. For a pipe filled with water, the relation between the hydraulic radius and the pipe diameter is

$$R = \frac{A}{P} = \frac{\pi \left(\frac{d_{pd}}{2} \right)^2}{2\pi \left(\frac{d_{pd}}{2} \right)} = \frac{d_{pd}}{4} \tag{4.37}$$

In other words, the pipe diameter is equal to 4 times the hydraulic radius. Substituting $d_{pd} = 4R$ into Eq. (4.36), recognizing that for low-gradient channels $h_f/X \approx S$, and solving for U yields the Darcy–Weisbach resistance equation for open-channel flow:

$$U = \sqrt{\frac{8gRS}{f_f}} \tag{4.38}$$

This equation has been used extensively to examine flow resistance in relatively steep natural rivers with gravel and boulder beds (Bathurst, 1985; Comiti et al., 2007; Ferguson, 2007; Rickenmann and Recking, 2011; Yochum et al., 2012). Often, the expression $\sqrt{8g/f_f}$ is used to represent resistance, in which case the Darcy–Weisbach equation is virtually identical to the Chezy equation if $\sqrt{g/c_f}$ is substituted for C, where $f_f = 8c_f$.

All three reach-averaged resistance equations have similar forms and include R and S as variables for predicting velocity in open channels. The Chezy and Darcy–Weisbach equations relate U to the square root of R and S, whereas the Manning equation assigns slightly greater weight to R compared with S. The similar forms of the equations allow the different coefficients to be easily related to one another (Yen, 2002; Dingman, 2009). Given the widespread abundance of information on Manning's n, values of n commonly are converted into values of C or f_f:

$$C = \frac{R^{0.167}}{n} \tag{4.39}$$

$$f_f = \frac{8gn^2}{R^{0.33}} \tag{4.40}$$

$$c_f = \frac{gn^2}{R^{0.33}} \tag{4.41}$$

where all conversions are based on the metric system of units.

4.11 How Is Flow Resistance in Rivers Related to Bed Shear Stress?

The force-balance derivation of the generic reach-averaged resistance relation (Eq. (4.32)) shows that the resistance coefficient (c_f) can be related to the bed shear stress, thereby providing a link among flow resistance, mean velocity, and forces acting on the bed of the river. Solving Eq. (4.33) for c_f yields

$$c_f = \left(\frac{\sqrt{gRS}}{U}\right)^2 = \frac{u_*^2}{U^2} \tag{4.42}$$

where $u_* = \sqrt{gRS}$ is the shear velocity. Note that u_* is an expression of the bed shear stress in the dimensions of velocity:

$$u_* = \sqrt{\frac{\rho g RS}{\rho}} = \sqrt{\frac{\tau_b}{\rho}} \tag{4.43}$$

or

$$\tau_b = \rho u_*^2 \tag{4.44}$$

According to Eq. (4.42), the friction coefficient increases as the bed shear stress increases for a constant mean velocity, or as the mean velocity decreases for a constant bed shear stress. If the gravitational shear force per unit area driving the flow increases, yet velocity remains constant, friction must increase to maintain a constant velocity. Conversely, if the velocity decreases, but the driving shear stress remains constant, friction must have increased to slow the flow down. If the driving and resisting forces are balanced, as is usually assumed in a steady flow, the bed shear stress is the product of U^2 and c_f (Eq. (4.31)). In other words, the bed shear stress embodies the effects of flow resistance and the inertia of the flow on the channel boundary.

4.12 What Factors Determine the Reach-Averaged Bed Shear Stress?

The bed shear stress for a uniform, steady flow over a river reach typically is determined using Eq. (4.20) rather than Eq. (4.31) because this formulation avoids the need to estimate the value of a resistance coefficient. Only data on S, R, and ρ are required to compute the bed shear stress. In a gradually varied steady flow measured in the field, the energy gradient upstream of a cross section S_e replaces the channel gradient and the bed shear stress applies only to specific cross sections of interest, rather than the entire reach, because the hydraulic radius varies over the reach. Reach-scale resistance for gradually varied flows also can be estimated directly, although the method is quite cumbersome (Dingman, 2009, pp. 241–244). In numerical models of gradually varied steady flow that employ resistance coefficients, such as HEC-RAS (U.S. Army Corps of Engineers, 2016), the energy slope is computed directly at a cross section using the Manning equation, and this slope can be used to compute the bed shear stress. The physical interpretation of such a bed shear stress is ambiguous, given that the estimated energy slope occurs only at the cross section and does not extend spatially along the river. Alternatively, the mean bed shear stress over a reach bounded by two cross sections can be estimated using the average of the energy slopes and the hydraulic radii for the two bounding cross sections.

4.12.1 How Do Scales of Roughness Relate to Flow Resistance and Bed Shear Stress?

In any case, the average bed shear stress incorporates the effects of frictional resistance on the flow at a range of spatial scales within the length of channel over which the bed or energy slope extends. This range typically includes the grain, bedform, bar-element, bar-unit, and planform scales (Figure

1.7). Morphological elements at each of these scales contribute to reach-scale resistance and therefore to the bed shear stress. Determining the individual effects of different scales of morphological elements on flow resistance is a challenging problem.

4.12.2 What Is Grain Resistance?

Considerable attention has focused on the influence of sediment grains on frictional resistance, especially in gravel-bed streams where large particles constitute a major component of total resistance. A common approach is to relate a metric of flow resistance to the ratio D/d_i, where d_i is a characteristic particle diameter representing the roughness of the channel bed. An example for f_f is (Ferguson, 2007)

$$\sqrt{\frac{8}{f_f}} = a\left(\frac{D}{d_i}\right)^b \tag{4.45}$$

Such relations clearly indicate that resistance decreases as either flow depth increases or the size of particles on the bed decreases. Furthermore, for high relative roughness ($D/d_i < 1$) the relation between resistance and roughness tends to be linear ($b = 1$), whereas for low relative roughness ($D/d_i \gg 1$) resistance changes nonlinearly ($b = 0.167$) (Ferguson, 2007). In steep mountain streams characterized by major variations in the morphology of the channel bed, Eq. (4.45) has been applied successfully to characterize flow resistance by replacing d_i with σ_z, the detrended standard deviation of bed elevations (Yochum et al., 2012).

4.12.3 What Is Bedform Resistance?

Whereas grain resistance is important in rivers with beds covered by gravel, cobbles, and boulders, bedform resistance has a strong influence on the flow in rivers with sand beds.

Bedforms consist of aggregations of individual grains (Figure 4.15, Table 4.1) (Venditti, 2013). The development of different types of bedforms has been associated with changes in Froude number (**F**) and stream power per unit area (ω). Ripples and lower-stage plane beds develop at the lowest values of **F** and ω, with ripples forming in finer bed material than plane beds (Figure 4.16). As power continues to increase, dunes appear (Figure 4.16), and these bedforms eventually are washed out and replaced by upper-stage plane beds. At Froude numbers close to or greater than 1, antidunes become the dominant bedform. Flow over these bedforms is characterized by distinctive waves on the water surface (Figure 4.13). Chutes and pools occur at Froude numbers from about 1.1 to greater than 2 (Cartigny et al., 2014).

Sequential changes in bedforms with increasing flow levels lead to a highly nonlinear pattern of flow resistance (Figure 4.17). In particular, the development of dunes leads to enhanced resistance. Two basic approaches have been used to account for the influence of bedforms on reach-averaged flow resistance (Yen, 2002). The linear separation approach divides the total resistance into the sum of two components: one related to resistance generated by the channel boundary without bedforms (grain resistance) and the other related to resistance produced by the bedforms (form resistance). Various formulations of this approach include

$$f_f = f_f' + f_f'', n = n' + n'', D = D' + D'', \tau_b = \tau_b' + \tau_b'' \tag{4.46}$$

where the single primes represent the component of the variable associated with grain resistance and the double primes indicate the component related to bedform resistance. The inclusion of τ_b emphasizes that the total bed shear stress can be strongly influenced by bedform resistance – an issue of importance in determining mobilization of sediment on the bed (see Chapter 5). The nonlinear approach treats form and grain resistance as a single, composite factor and does not

Table 4.1. Characteristics of alluvial bedforms.

Bedform	Height	Wavelength	Height/ Wavelength	Migration velocity (m s^{-1})
Ripple	0.01 to 0.04 m	< 0.06 m	< 0.1	1×10^{-4} to 1×10^{-3}
Dune	0.1 to 10 m Mainly between 0.05D and 0.33D with a mean of 0.167D	0.1 to 100 m Mainly between 2D and 16D with a mean of 5D	< 0.06	1×10^{-4} to 1×10^{-3}
Antidunes	0.03 to 1 m	$2\pi D$	<0.06	Stationary to 0.01 to 0.05

(after Bridge, 2003)

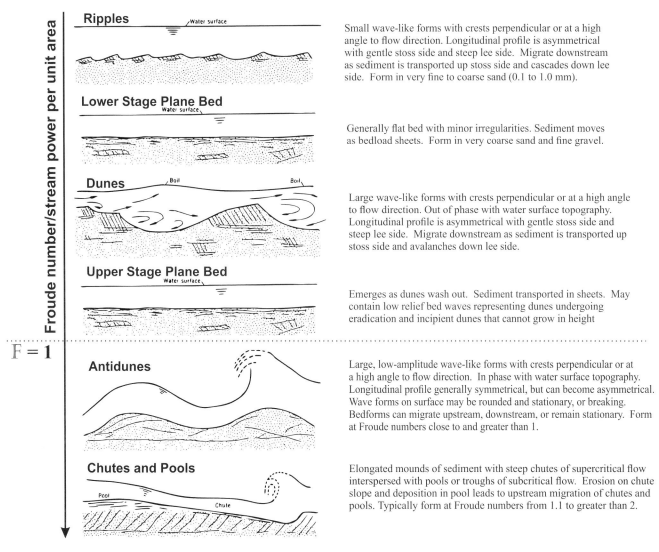

Ripples
Small wave-like forms with crests perpendicular or at a high angle to flow direction. Longitudinal profile is asymmetrical with gentle stoss side and steep lee side. Migrate downstream as sediment is transported up stoss side and cascades down lee side. Form in very fine to coarse sand (0.1 to 1.0 mm).

Lower Stage Plane Bed
Generally flat bed with minor irregularities. Sediment moves as bedload sheets. Form in very coarse sand and fine gravel.

Dunes
Large wave-like forms with crests perpendicular or at a high angle to flow direction. Out of phase with water surface topography. Longitudinal profile is asymmetrical with gentle stoss side and steep lee side. Migrate downstream as sediment is transported up stoss side and avalanches down lee side.

Upper Stage Plane Bed
Emerges as dunes wash out. Sediment transported in sheets. May contain low relief bed waves representing dunes undergoing eradication and incipient dunes that cannot grow in height

$F = 1$

Antidunes
Large, low-amplitude wave-like forms with crests perpendicular or at a high angle to flow direction. In phase with water surface topography. Longitudinal profile generally symmetrical, but can become asymmetrical. Wave forms on surface may be rounded and stationary, or breaking. Bedforms can migrate upstream, downstream, or remain stationary. Form at Froude numbers close to and greater than 1.

Chutes and Pools
Elongated mounds of sediment with steep chutes of supercritical flow interspersed with pools or troughs of subcritical flow. Erosion on chute slope and deposition in pool leads to upstream migration of chutes and pools. Typically form at Froude numbers from 1.1 to greater than 2.

Figure 4.15. Types of alluvial bedforms in sand-bed rivers (adapted from Simons and Richardson, 1966).

require direct knowledge of the bed configuration. Instead, separate total resistance relations are developed for different ranges of Froude number (Camacho and Yen, 1991; Yen, 2002).

Although gravel-bed rivers do not exhibit the classic suite of bedforms that occur in sand-bed rivers, local accumulations of particles may form pebble clusters (Brayshaw, 1984; Strom and Papanicolaou, 2008) and, under certain conditions, dune-like features can develop (Pitlick, 1992; Dinehart, 1992; Carling, 1999). These bedforms influence flow resistance (Hassan and Reid, 1990), but less systematically than bedforms in sand-bed rivers. Other factors that affect flow resistance include variations in the geometry of the stream channel, vegetation growing on the banks or bed, the movement of sediment on the bed or within the water column, bar forms, and irregularities in channel planform, such as

meandering. Methods to partition resistance caused by these factors are for the most part fairly qualitative and subjective (Cowan, 1956).

4.12.4 What Are the Advantages and Disadvantages of Reach-Averaged Shear Stress?

The advantage of the reach-averaged shear stress (Eq. (4.20)) is that it embodies the net effect of all scales of flow resistance on the force acting on the channel boundary. This advantage can also be seen as a disadvantage if one is interested in spatial variability in the force acting on the boundary at scales less than the planform scale; for example, at bar-unit, bar-element, bedform, or grain scales. Moreover, the movement of sediment is the result of the component of the reach-averaged shear stress acting on sediment grains, not the total shear stress. Thus, the reach-averaged shear stress is not necessarily a good index of the competence of the stream to mobilize sediment on the bed or in the banks. Localized analysis of the driving and resisting forces of turbulent flow is required to address this issue.

Figure 4.16. (a) Subaqueous ripples on a riverbed (note leaf at bottom center for scale). (b) Dunes on an exposed bar.

4.13 What Is Boundary-Layer Theory and How Is It Related to Flow in Rivers?

Boundary-layer theory is a body of principles in fluid mechanics related to the flow of a fluid over a solid surface or around solid obstacles (Schlichting and Gersten, 2016). This theory applies to any type of fluid, both liquids and gases, and deals with laminar and turbulent flows. It has a wide range of applicability, including aerodynamics (e.g., flow of air over the wings of an airplane), atmospheric science (e.g., the flow of air over mountains), and hydrodynamics (the flow of water through pipes). Rivers flow in channels with solid boundaries, and interaction of the flow with these boundaries drives the erosional and deposition processes involved in river-channel dynamics. Boundary-layer theory therefore provides a theoretical context for understanding how water flowing over a riverbed or past river banks produces forces that lead to channel change.

A boundary layer in a flowing fluid refers to the portion of the flow that is affected by friction at a boundary. It is a region of the flow in proximity to the boundary in which velocity changes with distance from the boundary. A freestream is a flow that remains unaffected by a boundary, such as wind high above Earth's surface. In a freestream, the mean velocity of the flow is constant over height. The lack of a velocity gradient indicates that neither viscous nor turbulent shear stresses are affecting the flow, whereas the development of a velocity gradient signals the presence of these stresses.

If a freestream encounters a boundary, friction begins to affect the flow (Figure 4.18). Viscous stresses act immediately, transmitting the retarding effect of the boundary upward into the flow. The vertical extent over which the flow is affected by frictional effects determines the thickness of the boundary layer (δ_L), which is defined as the height at which $u_z = 0.99u_o$, where

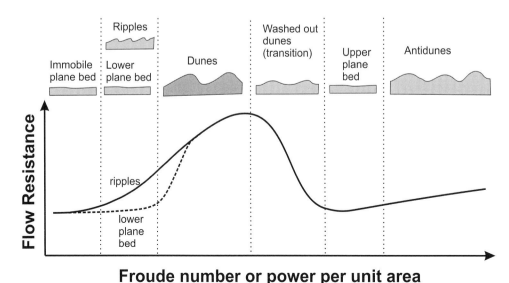

Figure 4.17. Variation in flow resistance with sequential changes in bedforms as Froude number and stream power per unit area increase.

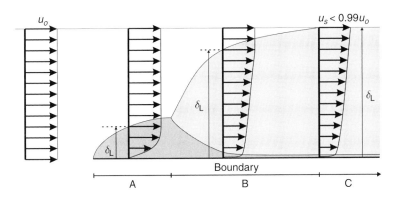

Figure 4.18. Development of a turbulent boundary layer when a freestream with a uniform velocity profile (left) encounters a boundary. In region A, a gradient in velocity develops near the boundary as viscous stresses affect the flow (dark shading). In region B, the onset of turbulence occurs and the region of turbulent flow (light shading) expands throughout the water column. In region C, turbulent flow has become fully developed and the boundary-layer thickness (δ_L) extends from the bed to the surface. The velocity (u_s) at the surface is now less than 99% of the freestream velocity (u_o).

u_z is the mean velocity at height z and u_o is the freestream velocity. If the Reynolds number of the flow exceeds the threshold for turbulent behavior, turbulence will develop a short distance from the start of the boundary. The onset of turbulence results in rapid vertical expansion of the boundary layer. When the thickness of the boundary layer stabilizes, a fully developed turbulent boundary layer has formed. Even in the deepest natural rivers, turbulent stresses generated by friction at the channel bed and banks operate over the entire flow depth. In other words, the turbulent boundary layer extends from the bed to the surface (Figure 4.18).

Boundary-layer theory is important in studies of rivers because it provides the basis for understanding the vertical structure of velocity profiles of turbulent river flows in relation to frictional resistance at the channel bed and the relation of frictional resistance, which determines bed shear stresses, to the structure of velocity profiles. By focusing on velocity profiles in turbulent flows, and the relation of these profiles to frictional resistance and bed shear stress, the theory is useful for determining how shear stresses associated with the flow act locally on the channel bed rather than over an entire river reach. It can be used to determine spatial variation in bed shear stresses at subreach scales.

4.13.1 What Is the Structure of a Turbulent Boundary Layer in a Channel with a Smooth Bed?

The turbulent boundary layer of flow over the smooth bed of a wide open channel consists of several different components (Figure 4.19). The issue of smoothness (or roughness) is important in mobilization of particles on the bed, a topic examined in Chapter 5, but here the bed is considered to be perfectly smooth or containing only microscopic imperfections. Immediately adjacent to the bed, velocities are low because of strong frictional effects. Local Reynolds numbers fall below the laminar-flow threshold and viscous forces dominate over turbulent forces, producing a thin layer known as the viscous sublayer. The dimensionless thickness of this sublayer (z_v^+) is

$$z_v^+ = \frac{z_v u_*}{v} \tag{4.47}$$

where z_v is the actual thickness of the sublayer. Although estimates of z_v^+ are as high as 12 (Middleton and Wilcock, 1994, p. 387; Bridge, 2003, pp. 26–27), the most commonly cited value in the fluid mechanics literature is 5 (Tennekes and Lumley, 1972, p. 160; Nezu and Nakagawa, 1993, p. 16; Furbish, 1997, p. 378; Dingman, 2009, p. 186; Schlichting and Gersten, 2016, p. 525), or

$$z_v = \frac{5v}{u_*} \tag{4.48}$$

Values of z_v for many smooth open-channel flows, including those characteristic of natural rivers, are quite small – often less than 0.1 mm (Figure 4.19). Eq. (4.48) indicates that the thickness of the sublayer increases with increasing kinematic viscosity and decreases with increasing shear velocity, which can be viewed as an index of turbulence. Intense turbulence decreases the thickness of the viscous region near the bed. Flow within the viscous sublayer is not completely laminar because of interaction with overlying turbulent flow (Nezu and Nakagawa, 1993, pp. 166–169). The profile of mean velocity (u_z) within the sublayer is linear:

$$u_z = \frac{u_*^2 z}{v} \tag{4.49}$$

As distance from the boundary increases, frictional effects diminish and velocities increase, producing a buffer layer with local Reynolds numbers in the transitional range between laminar and turbulent flow. The upper limit of this buffer layer is

$$z_t = \frac{z_t^+ v}{u_*} \tag{4.50}$$

with values of z_t^+ ranging from 30 (Tennekes and Lumley, 1972, figure 5.7; Nezu and Nakagawa, 1993, p. 16; Furbish, 1997, p. 378) to as high as 70 (Schlichting and Gersten, 2016,

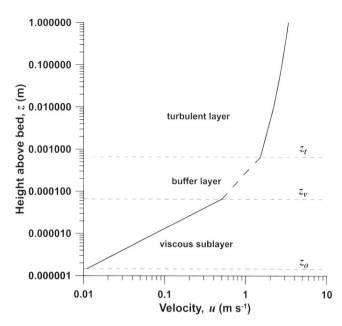

Figure 4.19. Structure of a turbulent boundary layer in a smooth open channel. Values of z_v and z_t are computed for a flow with $R = 1$ m, $S = 0.001$ m/m, and $v = 1.33 \times 10^{-6}$ m^2 s^{-1}) with $z_v^+ = 5$ and $z_t^+ = 50$.

p. 525). Regardless of the specific value, the upper limit of the buffer layer in many flows in smooth open channels is still less than 1 mm from the bed (Figure 4.19). No simple formula exists for predicting the velocity profile within the buffer layer (Nezu and Nakagawa, 1993, p. 16). Above the buffer layer, the flow is fully turbulent. Detailed analysis of the velocity profile within this thick turbulent layer, which extends over the vast majority of the total flow depth, must be based on theory concerning the characteristics of turbulent stresses in flowing water.

4.13.2 What Are Turbulent Stresses?

When a flow is turbulent, instantaneous velocities at any particular level within the water column ($\hat{u}$) will vary over time. This variation in velocities is caused by redistribution of momentum within the flow by turbulent eddies. As fluid parcels traveling at different speeds move within the water column and interact with one another, momentum exchange among the parcels leads to local accelerations and decelerations of the flow.

At any level in the flow, the velocity signals over time in x, y, and z directions can be decomposed into mean and fluctuating components (Figure 4.20):

$$\hat{u} = u + u' \ , \ \hat{v} = v + v' \ , \ \hat{w} = w + w' \qquad (4.51)$$

where the values with the primes represent the fluctuations (Figure 4.20). Although the means of the fluctuations

equal zero, the means of the products of the fluctuations usually do not. Multiplying the mean products of the fluctuations by water density (ρ) yields values of turbulent stress (M L^{-1} T^{-2}):

$$\tau_{txx} = -\rho\overline{u'u'}, \quad \tau_{tyy} = -\rho\overline{v'v'}, \quad \tau_{tzz} = -\rho\overline{w'w'} \qquad (4.52)$$

$$\tau_{txy} = -\rho\overline{u'v'}, \quad \tau_{txz} = -\rho\overline{u'w'}, \quad \tau_{tyz} = -\rho\overline{v'w'} \qquad (4.53)$$

where the overbars indicate averages of products of the fluctuations over the time series of a velocity record. For example, for a time series of data on u', v', and w' consisting of n measurements of each velocity component (Figure 4.20), $\overline{u'w'}$ is computed as

$$\overline{u'w'} = \frac{\sum_{1}^{n} u'w'}{n} \qquad (4.54)$$

The various terms in Eq. (4.52) and (4.53) are referred to as Reynolds stresses. The first set (Eq. 4.52) is the normal stresses and the second set (Eq. 4.53) defines the shear stresses. The negative sign in front of the terms represents a sign convention so that negative products (e.g., $-u'w'$) are positive stresses and positive products (e.g., $u'w'$) are negative stresses. In this sense, the Reynolds stresses characterize resisting forces per unit area produced by fluctuating motions within the flow. In a turbulent flow, the Reynolds stresses will not be zero, even if mean fluid motion is one-dimensional, for example, $u > 0$, $v = w = 0$. Of particular importance is τ_{txz}, the turbulent shear stress related to the vertical redistribution of streamwise momentum, which can be related to the development of the profile of streamwise velocity (u) in a one-dimensional turbulent open-channel flow.

The Reynolds stresses are equal in magnitude to the momentum flux associated with turbulent fluctuations, but act in opposition to it (Furbish, 1997, pp. 353–354; Dingman, 2009, p. 118). When viewed as momentum fluxes, means of products of turbulent fluctuations define the net transport of momentum by turbulence. The velocity fluctuations represent advection terms. Consider a positive vertical velocity fluctuation w'. This fluctuation has the capacity to transport some property of the fluid, such as heat, or of the flow, such as momentum. The momentum per unit volume of the streamwise flow associated with turbulence is $\rho u'$; thus, $\rho u'w'$ is the instantaneous turbulent vertical flux of streamwise momentum. It is also the instantaneous turbulent streamwise flux of vertical momentum. Averaging values of $u'w'$ over time and multiplying by density results in mean momentum fluxes. For example, $\rho\overline{u'w'}$ represents the mean turbulent vertical flux of streamwise momentum.

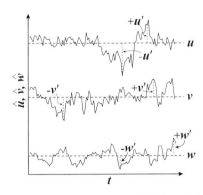

Figure 4.20. Time series of $\hat{u}$, $\hat{v}$, and $\hat{w}$ (solid lines) showing mean (dashed horizontal lines) and fluctuating (primes) components in an open-channel flow.

4.13.3 How Are Turbulent Stresses Related to Change in Velocity over Depth?

The total shear stress in the x-z plane at any location within a turbulent flow (τ_{Txz}) is equal to the sum of the viscous and turbulent stresses, or, drawing upon Eq. 4.25, 4.26, and 4.53,

$$\tau_{Txz} = \tau_{vxz} + \tau_{txz} = (\mu + \varepsilon)\frac{du}{dz} = \mu\frac{du}{dz} + \varepsilon\frac{du}{dz} = \mu\frac{du}{dz} - \rho\overline{u'w'}$$

(4.55)

Because viscous stresses are only important extremely close to the boundary, over the majority of the flow only the turbulent shear stress needs to be considered:

$$\tau_{txz} = -\rho\overline{u'w'} = \varepsilon\frac{du}{dz}$$

(4.56)

This equation shows that turbulent shear stresses in the x-z plane are related to the magnitude of the eddy viscosity and to the rate of change in velocity over depth.

4.13.4 What Is the Law of the Wall in Turbulent Boundary Layer Flows?

Integration of Eq. (4.56) over depth is necessary to determine the relationship between the velocity profile, frictional resistance, and shear stress at the channel bed. To develop an analytical solution for this equation, the factors that control the velocity fluctuations u' and w' must be ascertained. As parcels of fluid are transported vertically over the water column by turbulent eddies, u' and w' vary over time (Figure 4.21). The positive gradient in the streamwise velocity over depth tends to produce eddies that rotate clockwise, which, in turn, tends to generate fluctuations u' and w' with opposite signs. As an eddy of length scale ℓ_e

transports fluid from below ($z - \ell_e$) upward to a particular position z above the bed, it produces a positive vertical velocity fluctuation ($+w'$). The arriving mass of fluid, because it originates close to the bed where friction is high, will typically have less streamwise momentum than the flow at z, resulting in negative streamwise velocity fluctuation ($-u'$). Conversely, if an eddy of length scale ℓ_e transports fluid downward to z, it generates a negative vertical velocity fluctuation ($-w'$), and this downward-moving fluid will typically have more momentum than the flow at z, resulting in a positive streamwise velocity fluctuation ($+u'$).

The magnitude of velocity fluctuations should be related to 1) the local gradient of mean velocity (du/dz) and 2) the length scale, or size, of turbulent eddies (ℓ_e). Strong local gradients in u indicate that as fluid moves vertically it encounters over short distances fluid moving at different velocities, resulting in pronounced momentum exchange. The length scale ℓ_e can be viewed as the distance over which a parcel of fluid travels before this parcel loses its momentum to the moving fluid surrounding it. Large length scales result in the transport of parcels of water moving at a certain velocity into water at a different level moving either much more slowly or much faster than the arriving parcels. Both these effects should enhance turbulent fluctuations. The structure of turbulence is assumed to be isotropic, or varying uniformly over all three dimensions of the flow, so that u' and w' have approximately the same absolute magnitudes. This assumption is approximate; in straight, natural rivers, $|\overline{w'}| \approx 0.65|\overline{u'}|$ over most of the flow depth (Sukhodolov et al., 1998).

Based on the reasoning that velocity fluctuations are related to the velocity gradients and to eddy length scale, the magnitudes of fluctuations can be expressed as

$$u' = \ell_e\frac{du}{dz}, \quad w' = \ell_e\frac{du}{dz}$$

(4.57)

where the expressions on the right-hand sides of the two equations have opposing signs because positive fluctuations of one velocity component tend to be accompanied by negative fluctuations of the other component. According to this assumption,

$$\overline{u'w'} = -\ell_e^2\left(\frac{du}{dz}\right)^2$$

(4.58)

This quantity represents a kinematic momentum flux, and the corresponding shear stress (opposite in sign) is

$$\tau_{txz} = -\rho\overline{u'w'} = \rho\ell_e^2\left(\frac{du}{dz}\right)^2$$

(4.59)

To allow negative shear stresses with negative velocity gradients, Eq. (4.59) is commonly written as

$$\tau_{txz} = \rho \ell_e^{2} \left| \frac{du}{dz} \right| \left(\frac{du}{dz} \right)$$ (4.60)

This slight modification is not of great consequence in open channels, where velocity gradients are predominantly positive.

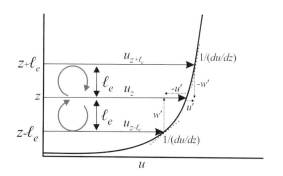

Figure 4.21. Velocity profile in a turbulent open-channel flow showing how eddies of length scale ℓ_e produce fluctuations in w' and u' at level z.

As a first-order approximation, values of ℓ_e can be related linearly to the distance from the channel bottom:

$$\ell_e = \kappa z$$ (4.61)

where κ is von Karman's constant. This relation proposes that eddies increase in size linearly as distance from the channel bottom increases. Substituting this expression into Eq. (4.59) yields

$$\tau_{txz} = \rho \kappa^2 z^2 \left(\frac{du}{dz} \right)^2$$ (4.62)

Comparison of Eq. (4.62) with Eq. (4.26) shows that

$$\varepsilon = \rho \kappa^2 z^2 \left(\frac{du}{dz} \right)$$ (4.63)

Equating the resisting stress in Eq. 4.62 to the driving shear stress of the flow (Eq. 4.19) and integrating this equation over depth results in (Boxes 4.1 and 4.2)

BOX 4.1   LAW OF THE WALL: INTEGRATION.

1) Set the driving shear stress (Eq. 4.19) equal to the resisting shear stress (Eq. 4.62):

$$\rho ghS = \rho \kappa^2 z^2 (du/dz)^2$$ (4.1.1)

2) Rearrange to isolate du:

$$du^2 = (1/\rho)(1/\kappa^2)(1/z^2)(\rho ghS)dz^2$$ (4.1.2)

3) Simplify and take the square root of each side:

$$du = (1/\kappa)(1/z)(ghS)^{0.5}dz$$ (4.1.3)

4) Given $u_* = \sqrt{ghS}$:

$$du = (1/\kappa)(1/z)(u_*)dz$$ (4.1.4)

5) Integrating Eq. (4.1.4) with respect to z:

$$u = (1/\kappa)(u_*)\ln z + c$$ (4.1.5)

6) Cannot eliminate the constant of integration, c, by setting $u = 0$ and $z = 0$ because $\ln(0)$ is undefined. Instead, assume $u = 0$ when z is very small (z_o):

$$0 = (1/\kappa)u_*\ln(z_o) + c$$ (4.1.6)

$$c = -(1/\kappa)u_*\ln(z_o)$$ (4.1.7)

7) Substitute for c and rearrange:

$$u = (1/\kappa)u_*\ln(z) - (1/\kappa)u_*\ln(z_o)$$ (4.1.8)

$$u = (u_*/\kappa)[\ln(z) - \ln(z_o)]$$ (4.1.9)

$$u = (u_*/\kappa)\ln(z)/\ln(z_o)$$ (4.1.10)

BOX 4.2 | LAW OF THE WALL – AN ALTERNATIVE DERIVATION.

The traditional derivation of the Law of the Wall equation assumes that $\ell_e = \kappa z$ and the resisting shear stress can be related to the driving shear stress at the channel bed (Eq. 4.32). Two problems arise with these assumptions (Dingman, 2009, pp. 196–197). First, the formulation for ℓ_e implies that eddies should be largest at the water surface, ignoring the constraining effect of the water–air interface on eddy size. Second, the velocity profile should reflect the distribution of shear stress throughout the water column, not just at the bed. To address the first issue, an alternative expression can be introduced for ℓ_e (Henderson, 1966):

$$\ell_e = \kappa z \left(1 - \frac{z}{h}\right)^{0.5} \tag{4.2.1}$$

This equation has a parabolic form with $\ell_e = 0$ at the channel bed and water surface and a maximum value around $z/h = 0.65$. The second issue involves consideration of the distribution of the driving shear stress over depth in a turbulent, steady, uniform open-channel flow. This distribution is simply a linear relation governed by the increasing weight of the water from the water surface to the bottom (Figure 4.8). For a channel of low gradient,

$$\tau_{txz} = \rho g (h - z) S \tag{4.2.2}$$

Using these relations, the force balance between driving and resisting shear stresses is (see Eq. (4.1.1))

$$\rho g (h - z) S = \rho \kappa^2 z^2 \left(1 - \frac{z}{h}\right)\left(\frac{du}{dz}\right)^2 \tag{4.2.3}$$

where

$$\varepsilon = \rho \kappa^2 z^2 \left(1 - \frac{z}{h}\right)\left(\frac{du}{dz}\right) \tag{4.2.4}$$

or

$$\rho g (h - z) S = \rho \kappa^2 z^2 \left(\frac{1}{h}\right)(h - z)\left(\frac{du}{dz}\right)^2 \tag{4.2.5}$$

Simplifying, solving for du, and taking the square root of both sides:

$$du = (1/\kappa)(1/z)(u_*)dz \tag{4.2.6}$$

This equation is identical to Eq. (4.1.4), and integration proceeds in the same manner.

$$u = \frac{u_*}{\kappa} \ln\left(\frac{z}{z_o}\right) \tag{4.64}$$

where z_o is the roughness height or roughness length, the height above the bed at which $u = 0$. Eq. 4.64 is known as the Prandtl–von Karman Law of the Wall for turbulent wall-bounded shear flows. It indicates that the velocity profile in the turbulent part of a flow in an open channel with a smooth bed, the region above the buffer layer, has a logarithmic form (Figure 4.19). An important caveat of the log law is that it assumes the channel is wide and not affected substantially by effects from side walls or, in the case of natural rivers, by friction along the channel banks.

The velocity profiles of many natural rivers are logarithmic, especially in the lower 15–20% of the total flow depth ($z/h \leq 0.15$–0.20). Debate exists about the extent to which the law of the wall applies to the entire region of turbulent flow within rivers ($z/h > 0.20$) (Bridge, 2003, pp. 33–34; Dingman, 2009, p. 196). Far from the bed, the form of the velocity profile may no longer strongly reflect the influence of the boundary on the flow as manifested by the inclusion of the roughness length in Eq. (4.64). Instead, the profile may become a function of distance above the bed (z/h) rather than z/z_o. Relations of this form are known as velocity-defect laws. Another approach is to add a wake parameter to a velocity-defect law to account for the deviation of some velocity profiles from the logarithmic law in the outer portion of the turbulent boundary layer (Bridge, 2003, p. 33). The logarithmic law predicts that the maximum velocity occurs at the water surface, but both experimental results and some field observations indicate that the maximum velocity, even in straight, smooth open channels, can be located beneath the surface – a phenomenon known as velocity dip. Under these conditions, the logarithmic law clearly is inadequate to define the velocity profile. Attempts to characterize such profiles have involved adding velocity-defect or wake terms to the logarithmic law (Nezu and Nakagawa, 1993, pp. 16–17; Yang et al., 2004; Guo, 2013). The velocity-dip phenomenon is most pronounced in straight narrow open channels with $W/D < 7$ where secondary currents generated at side walls affect the middle of the flow (Yang et al., 2004; Guo, 2013). Many natural rivers have $W/D > 7$. The extent to which velocity profiles in natural rivers deviate from the law of the wall has not been extensively evaluated, but at least some rivers exhibit velocity profiles that conform to the law of the wall over the entire flow depth.

4.13.5 How Can the Law of the Wall Be Used to Determine the Local Bed Shear Stress?

From a geomorphic perspective, the goal of the derivation of the law of the wall is not to predict the velocity profile over depth, but instead to develop a relation that can be used to predict the shear stress locally on the channel boundary. As long as some portion of the lower part of the turbulent boundary layer exhibits a logarithmic profile, shear stresses acting on the boundary locally can be determined based on Eq. (4.64). A common method involves measuring mean velocities (u) at several different heights above the bed (z) over the lower 15–20% of the flow depth and using the resulting data to fit a linear relation between u and z. Eq. (4.64) can be expanded as

$$u = \frac{u_*}{\kappa}\ln(z) - \frac{u_*}{\kappa}\ln(z_o) \tag{4.65}$$

or

$$u = a(\ln z) + b \tag{4.66}$$

where $a = u_*/\kappa$ and $b = -(u_*/\kappa)\ln z_o = -a(\ln z_o)$. Linear regression analysis of u (dependent variable) on $\ln(z)$ (independent variable) yields estimates of a and b (Bauer et al., 1992; Bergeron and Abrahams, 1992), from which estimates of the local shear velocity, bed shear stress, and roughness length can be derived as

$$u_* = \kappa a \tag{4.67}$$

$$\tau_b = \rho u_*^2 \tag{4.68}$$

$$z_o = e^{-b/a} \tag{4.69}$$

When conducting such analysis, it is important to verify that the relation between measured values of u and $\ln(z)$ is indeed logarithmic and to report statistical confidence intervals for estimates of u^*, τ_b, and z_o (Bauer et al.,1992). Also, the expression for shear velocity (Eq. (4.67)) requires an estimate of von Karman's constant. The value of κ is not necessarily constant, especially when sediment transport occurs (Gaudio et al., 2010; Ferreira, 2015). As a constant, it is considered to have an exact value of 0.41 for clear-water flows (Furbish, 1997, p. 374; Bridge, 2003, p. 31) but often is assigned a value of 0.4 (Dingman, 2009, pp. 184–185).

The roughness height z_o provides an estimate of the height above the bed at which the velocity goes to zero. For smooth beds, this value is typically quite small and lies within the viscous sublayer (Figure 4.19). For natural rivers with beds covered by coarse particles, the value of z_o tends to vary directly with the size of these particles (Whiting and Dietrich, 1990; Wiberg and Smith, 1991).

Data on u and z may exhibit considerable scatter over depth, even when the overall form of the relation is logarithmic. High degrees of scatter in the data lead to increased levels of imprecision in estimates of shear velocity, shear stress, and roughness height. An alternative approach is to try to eliminate the need for regression analysis by defining a relation between z_o and particle size for streams in which grain resistance is the primary factor influencing the roughness height. Such a relation can then be substituted for z_o in Eq. (4.64) and the equation solved for shear stress. A relation of the general form

$$z_o = ad_i/30 \tag{4.70}$$

is often used to estimate the roughness height in gravel-bed rivers (Whiting and Dietrich, 1990; Wiberg and Smith, 1991; Wilcock, 1996). In this equation, d_i is a characteristic particle size and a is an empirical constant. Typical characteristic particle sizes include d_{84} or d_{90}, which correspond to sizes of particles (L) larger than 84% or 90% of the material by weight on the channel bed (see Appendix B). In other words, for d_{84}, 84% of the material by weight on the bed of the river is smaller than a particle of this size. The value of a generally is around 3, so that $z_o \approx 0.1d_i$. Substituting Eq. (4.70) into Eq. (4.64) and solving for τ_b yields

$$\tau_b = \rho(\kappa u_z)^2 \left(\ln\frac{30z}{ad_i}\right)^{-2} \tag{4.71}$$

This method requires sampling and analysis of bed material to determine d_i. However, once d_i is determined, only a single measurement of velocity (u_z) at height z above the bed somewhere within the lower 20% of the flow is needed to estimate τ_b. A version of Eq. (4.71) that uses depth-integrated values of velocity and depth is (Wilcock, 1996)

$$\tau_b = \rho(\kappa\overline{u})^2 \left(\ln \frac{h}{e(ad_i/30)} \right)^{-2} \tag{4.72}$$

where $\overline{u}$ is the depth-averaged velocity (average velocity over h, the local depth) and e is the base of natural logarithms (Wilcock, 1996).

Comparison of estimates of τ_b derived from Eq. (4.64), (4.71), and (4.72) with $a = 2.85$ and $d_i = d_{90}$ indicates that estimates derived from Eq. (4.72) exhibit the highest degree of replicability; i.e., repeat estimates of bed shear stress at a specific location vary least using this method (Wilcock, 1996). Repeat estimates from Eq. (4.71) are three times more variable than those from Eq. (4.72) but are still considerably less variable than repeat estimates derived from linear regression analysis using Eq. (4.64). Although this comparison highlights differences in levels of precision of the different methods, it does not highlight differences in accuracy, which requires independent knowledge of the actual bed shear stress against which estimates of bed shear stress produced from each approach can be compared.

4.13.6 What Other Approaches Are Used to Determine the Local Bed Shear Stress?

Besides the method based on the law of the wall, several other approaches have been used to estimate the local bed shear stress in natural rivers (Biron et al., 2004a; Bagherimiyab and Lemmin, 2013).

4.13.6.1 What Is the Reynolds Stress Approach?
If an instrument is available to measure velocities at a high sampling rate in two dimensions, time series of $\hat{u}$ and $\hat{w}$ can be obtained to estimate the near-bed Reynolds shear stress from velocity fluctuations:

$$\tau_b = -\rho\overline{u'_b w'_b} \tag{4.73}$$

where the subscript b indicates values of u' and w' obtained at a single location in the water column close to the bed. An important issue related to this method is the proximity to the bed at which $\hat{u}$ and $\hat{w}$ should be measured. The Reynolds shear stress generally varies above both smooth and rough beds in a characteristic pattern with peak values occurring at about $0.10z/h$ and values decreasing from this peak both toward the bed and toward the surface (Nikora and Goring, 2000; Biron et al., 2004a). Measurement within the peak provides a maximum near-bed shear stress, but the relation of this maximum to sediment transport, which, as will be seen in Chapter 5, is often of the goal of estimating bed shear stress, has yet to be fully evaluated.

As discussed in Box 4.2, the shear stress within a uniform, steady flow should vary linearly from the surface to the bed. Another approach to evaluating the bed shear stress is to use measurements of time series of $\hat{u}$ and $\hat{w}$ at numerous positions over depth and determine the pattern of density-normalized Reynolds stresses $(-\overline{u'w'})$ over depth (Nikora and Goring, 2000). Typically, this pattern will be linear from the surface to the peak at around $0.10z/D$ (Figure 4.22). When multiplied by ρ, the intercept of a linear regression between $-\overline{u'w'}$ and z/h provides an estimate of τ_b.

4.13.6.2 What Is the Turbulent Kinetic Energy Approach?
The total turbulent kinetic energy (TKE) of a turbulent flow is related to the sum of the variances of velocity fluctuations in three dimensions:

$$\text{TKE} = 0.5(\overline{u'^2} + \overline{v'^2} + \overline{w'^2}) \tag{4.74}$$

In oceanographic studies, this quantity has been related to bed shear stress through the simple function (Kim et al., 2000)

$$\tau_b = 0.19\rho\text{TKE} \tag{4.75}$$

The TKE approach has been used to estimate bed shear stress in several fluvial studies (Biron et al., 2004a; MacVicar and Roy, 2007a; Bagherimiyab and Lemmin, 2013). A variant on this method uses only the vertical velocity fluctuations to estimate bed shear stress:

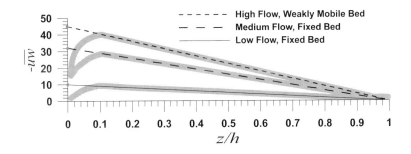

Figure 4.22. Linear profiles (dashed and solid lines) of density-normalized Reynolds stresses (cm² s⁻²) over depth for different flow conditions in a straight, gravel-bed channel. Shaded areas indicate pattern of data scatter in relation to fitted trends. Gray dots show intercepts of linear profiles corresponding to bed shear stresses (adapted from Nikora and Goring, 2000).

$$\tau_b = 0.9\rho\overline{w'^2} \tag{4.76}$$

The TKE method requires an instrument capable of measuring $\hat{u}$, $\hat{v}$, and $\hat{w}$. Both methods encounter the same issue as the Reynolds stress method, i.e., the appropriate level at which to obtain velocity measurements, since both TKE and w' tend to vary with height above the bed. Typically, the peak levels occur at about $0.10z/h$ (Biron et al., 2004a).

A variant on the TKE method is based on the wall similarity concept, which holds that under uniform flow at high Reynolds numbers, an extended depth range exists where turbulent energy production and dissipation are in equilibrium, so that turbulent diffusion is negligible (Bagherimiyab and Lemmin, 2013). Under these conditions the bed shear velocity can be expressed as a vertical flux of TKE (Lopez and Garcia, 1999; Hurther and Lemmin, 2000):

$$u_* = \left(1.667\overline{(u'^2 + v'^2 + w'^2)w'}\right)^{0.33} \tag{4.77}$$

or

$$\tau_b = \rho\left(1.667\overline{(u'^2 + v'^2 + w'^2)w'}\right)^{0.67} \tag{4.78}$$

for the range $z/h = 0.25$ to 0.60. If values of $\overline{(u'^2 + v'^2 + w'^2)w'}$ are relatively constant over this range, the constant quantity can be incorporated into Eq. (4.78) to compute the bed shear stress. This method has not been widely applied to river flows but may have considerable potential for estimating bed shear stress.

4.13.6.3 What Is the Ray-Isovel Approach?

Methods for determining the local bed shear stress assume the channel is infinitely wide and that the channel banks do not severely disrupt the hydraulics or turbulence characteristics of the flow. Thus, these methods are best applied within the central parts of rivers, where the influence of the banks on the flow are negligible. The channel banks represent another source of frictional resistance, and near the interface between the bed and the banks, both of these boundaries may affect the flow. Within a straight natural river channel with a fairly symmetrical cross-sectional form, the distribution of streamwise velocities in a steady, uniform flow typically decreases both away from the banks and away from the bed, reaching a maximum within the center of the flow near the surface (Figure 4.23). Profile (law of the wall, linear Reynolds stress extrapolation) and nonprofile (Reynolds stress, TKE) methods of estimating bed shear

(a)

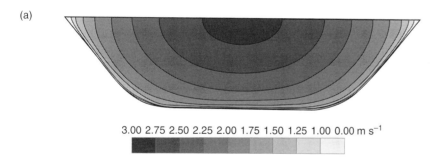

3.00 2.75 2.50 2.25 2.00 1.75 1.50 1.25 1.00 0.00 m s^{-1}

(b)

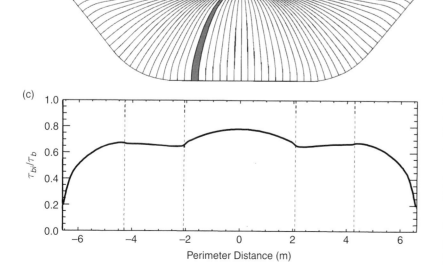

(c)

Figure 4.23. (a) Typical velocity distribution for flow in a straight channel with a symmetrical cross section. (b) Pattern of rays orthogonal to isovels for the velocity distribution above. Gray shading shows a subarea corresponding to adjacent rays. (c) Pattern of the local boundary shear stress (τ_{bi}) predicted by the ray-isovel method relative to the mean bed shear stress over the entire cross section ($\tau_b = \rho gDS$). Dashed lines indicate curving portions of the channel perimeter at the transition from the bed to the banks (after Kean and Smith, 2004).

stress are likely to encounter problems near the channel banks where the velocity contours and channel boundary are sloping in the cross-stream direction and secondary currents develop (Hopkinson and Wynn-Thompson, 2016). The robustness of these methods has not been rigorously evaluated for such conditions.

One approach to partitioning boundary shear stress within an open channel is to view the boundary as consisting of two domains, the bed versus the two channel banks, and analyze the stresses separately within these two domains. In channels of simple geometry, such as rectangular or trapezoidal forms, the distinction between the bed and the banks (or sides) is rather straightforward. The problem reduces to isolating the portions of the flow cross section that contribute to shear stresses acting on the two banks from the portion that contributes to stresses acting on the bed (Khodashenas and Paquier, 1999; Guo and Julien, 2005; De Cacqueray et al., 2009; Ansari et al., 2011; Kabiri-Samani et al., 2013). Such approaches are difficult to apply to many natural rivers, where the transition between the banks and the bed may be gradual rather than abrupt.

Since the 1930s, attempts have been made to use information on the pattern of velocity within river cross sections to estimate spatial variation at the channel boundary (Leighly, 1932). The pattern of mean streamwise velocity (u) within the cross section commonly is depicted as isovels, or lines of equal velocity (Figure 4.23). These isovels are produced from detailed measurements of u throughout the cross section. The distribution of boundary shear stresses acting over the transverse profile of a channel cross section is related to subareas of the flow bounded by rays aligned orthogonal to the pattern of the isovels (Leighly, 1932):

$$\tau_{bi} = \rho g S \frac{A_i}{P_i} \tag{4.79}$$

where τ_{bi} is the boundary shear stress associated with subarea A_i and P_i is the length of the wetted perimeter associated with A_i. Because the rays represent surfaces of zero shear, the driving shear stress associated with the mass of water within a subarea must be balanced by the resisting shear stress along the solid channel boundary of the subarea. If the zone of maximum velocity is submerged below the water surface, the rays originating at the bed extend only to the position of maximum velocity; above this position rays extend to the water surface, indicating that this part of the flow does not contribute to boundary shear stress (Leighly, 1932; Chiu and Lin, 1983).

The ray-isovel method has not been applied widely to natural rivers, but incorporation of it into hydraulic modeling has demonstrated the potential of the method for exploring patterns of shear stress along channel boundaries, including channel banks (Kean and Smith, 2004) (Figure 4.23). Key

issues that must be fully reconciled include whether interfacial shear stresses between the ray-defined subareas are indeed negligible and whether secondary currents contribute substantially to boundary stresses (Ansari et al., 2011; Hopkinson and Wynn-Thompson, 2016). Also, where patterns of isovels are highly complex, delineation of rays may be difficult, if not impossible.

4.14 How Can Fluid Motion in Rivers Be Described Mathematically in Three Dimensions?

Treatments of river hydraulics in this chapter have focused mainly on uniform, steady, unidirectional flow. In Section 4.5 it was noted that fluid motion in many rivers is three-dimensional. Some aspects of this three-dimensionality will be examined in subsequent chapters, especially in relation to spatial variations in channel planform (see Chapter 9). The physical characterization of such flows based on theoretical principles from fluid mechanics is mathematically complex and represents a major reason why the study of flow in rivers now involves the use of sophisticated numerical models. For example, the simple cross-sectional continuity equation for uniform, steady, unidirectional flow (Eq. (4.1)) becomes an elemental partial differential equation when dealing with steady, three-dimensional fluid motion:

$$\frac{\partial u}{\partial x} + \frac{\partial v}{\partial y} + \frac{\partial w}{\partial z} = 0 \tag{4.80}$$

This equation, while perhaps imposing to those with limited mathematical backgrounds, simply indicates that a change in mass flux of water in one direction must be accommodated by a change in mass flux of water in another direction to ensure that no net increase or decrease in mass flux occurs. The basic equations describing the three-dimensional motion of a fluid are known as the Navier–Stokes equations. A full derivation of these equations is beyond the scope of this book, but when considering turbulent, steady, open-channel flow the equations have the form (Williams, 1996)

$$\rho \left(u \frac{\partial u}{\partial x} + v \frac{\partial u}{\partial y} + w \frac{\partial u}{\partial z} \right)$$
$$= F_x - \frac{\partial p}{\partial x} + \mu \nabla^2 u - \rho \left(\frac{\partial \overline{u'^2}}{\partial x} + \frac{\partial \overline{u'v'}}{\partial y} + \frac{\partial \overline{u'w'}}{\partial z} \right) \tag{4.81}$$

$$\rho \left(u \frac{\partial v}{\partial x} + v \frac{\partial v}{\partial y} + w \frac{\partial v}{\partial z} \right)$$
$$= F_y - \frac{\partial p}{\partial y} + \mu \nabla^2 v - \rho \left(\frac{\partial \overline{u'v'}}{\partial x} + \frac{\partial \overline{v'^2}}{\partial y} + \frac{\partial \overline{v'w'}}{\partial z} \right) \tag{4.82}$$

$$\rho\left(u\frac{\partial w}{\partial x} + v\frac{\partial w}{\partial y} + w\frac{\partial w}{\partial z}\right)$$

$$= F_z - \frac{\partial \overline{p}}{\partial z} + \mu\nabla^2 w - \rho\left(\frac{\partial \overline{u'w'}}{\partial x} + \frac{\partial \overline{v'w'}}{\partial y} + \frac{\partial \overline{w'^2}}{\partial z}\right) \qquad (4.83)$$

where p is pressure, F_x, F_y, and F_z are body forces in the x, y, and z directions, respectively, ∇^2 is LaPlace's operator, and the overbars indicate time averaging. Eq. (4.81)–(4.83) are known as the Reynolds-averaged Navier–Stokes equations. This set of equations cannot be solved analytically and requires the use of numerical methods to develop predictive models of three-dimensional turbulent fluid

flow in rivers and open channels. Note that the Reynolds stresses appear as terms on the right-hand side of the equations within the parentheses. Also, viscous effects are captured by terms associated with the Laplace operator. In open-channel flows, the body force F_x represents the force per unit volume acting in the streamwise direction ($F_x = \rho g \sin\theta$) (Nezu and Nakagawa, 1993, p. 132). Gravitational body forces per unit volume in the y direction can be important when the alignment of the river is not straight and centrifugal forces act on the flow, producing spatial variation in water-surface elevations and the development of lateral gravitational gradients.

CHAPTER

5 Sediment Transport Dynamics in Rivers

5.1 Why Is Sediment Transport Important in River Dynamics?

The movement of sediment within rivers is the fundamental process linking the form of river systems to the dynamics of flow. In alluvial rivers, channels are carved into material transported and deposited by the river. This accumulated sediment is referred to as the floodplain (see Chapter 14). Together, erosion and deposition, processes directly related to sediment transport, shape the form of alluvial river channels and floodplains. In bedrock rivers, transported sediment plays an instrumental role in abrading the bed of the channel, contributing to erosion and channel change over time (Sklar and Dietrich, 2004, 2012; Lamb et al., 2008b).

Given the vital link between sediment transport and river morphodynamics, attainment of an in-depth understanding of the processes by which sediment moves in rivers has become a goal of paramount importance in fluvial geomorphology. A voluminous body of research has emerged on this topic over the past several decades. In some respects, the search for answers has led river scientists down a rabbit hole, where the complexity of the problem seems to grow under increasingly intense scrutiny. Sediment transport in rivers is complicated, and ready solutions to this knotty problem have not been forthcoming. Nevertheless, a wide range of useful concepts and models have been developed through attempts to better understand how rivers transport sediment. These concepts and models provide a foundation for ongoing scientific inquiry.

5.2 What Types of Material Flux Occur in Rivers?

The material flux of rivers consists of dissolved and solid loads. Chemical weathering of rock and soil produces solutes that are carried into rivers by surface runoff (overland flow), by flow of water through soil (throughflow), and by groundwater flow. This dissolved load contributes substantially to the material flux of rivers (Table 3.1). Although it is not a major factor influencing river morphodynamics, the transport of dissolved material is important for evaluating the evolution of landscapes over geologic timescales and for determining mass budgets of river systems, especially with regard to mass fluxes from continents to ocean basins (see Chapter 3).

The solid load of a river consists of rock and mineral particles, also known as clastic material. The term "solid load" is used most frequently in water-quality investigations, whereas river scientists focusing on research related to material flux of rivers typically use the term "sediment load." Although both terms refer to the amount of clastic material transported by rivers, differences exist between data on solid concentrations and on sediment concentrations because of differences in the analytical methods used to determine these values (Gray et al., 2000).

5.3 What Are the Major Sources of Sediment Transported by Rivers?

Sources of sediment transported by rivers can be located either outside the river channel or within the channel

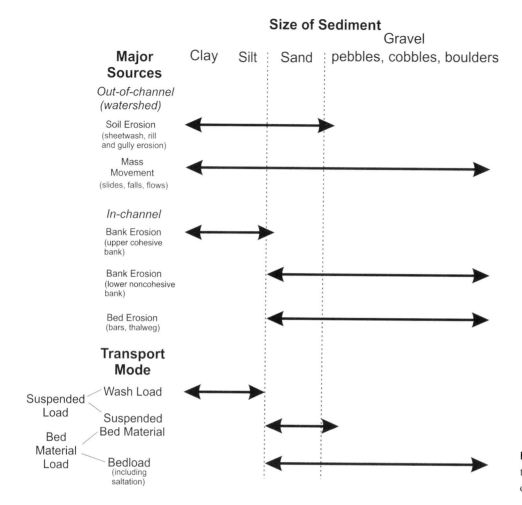

Figure 5.1. Major sources and transport modes of different sizes of particles transported by rivers.

(Figure 5.1). Soil erosion and mass movement are the two major out-of-channel processes that deliver sediment to rivers. Soil erosion occurs in a wide range of environmental settings, whereas mass movement is most common in steep terrain, particularly mountain environments. Because most soils consist mainly of sand, silt, and clay, erosion of soil predominantly supplies particles of these size classes to rivers. Even if coarse particles are available, soil-erosion processes generally are not capable of transporting particles larger than fine pebbles into a river. Mass movement, on the other hand, can deliver particles of all sizes to rivers, including large cobbles and boulders.

In-channel sources of sediment include the channel banks and the channel bed. In many cases, channel banks of alluvial rivers consist of a cohesive upper part containing silt, clay, and fine sand and a lower noncohesive part containing particles ranging in size from sand to boulders, depending on the range of grain sizes found on the bed of the river. Erosion of a channel bank over its full height results in both fine and coarse particles entering the river. The texture of the channel bed largely depends on what sizes of particles the river is

capable of transporting and the size range of particles supplied to the river by out-of-channel sources. Typically, the coarsest particles available within an alluvial river system are located on the channel bed. Characteristics of the bed include bars, or local accumulations of sediment, and the thalweg, or the part of the bed where the deepest and often the fastest flow occurs. Particles on the bed can be mobilized by flow within the channel, thereby becoming a source of sediment for transport.

5.4 What Are the Major Types of Sediment Transport in Rivers?

Sediment transport in rivers can be classified into types based on the mode of transport and the source of the transported material (Figures 5.1 and 5.2). Sediment within rivers is transported either in suspension within the water column or by rolling, sliding, or hopping along the channel bed. This sediment is delivered to the river system from hillslopes or is generated within the system through erosion of the channel bed or banks. The various types of sediment transport are not mutually exclusive and often occur simultaneously.

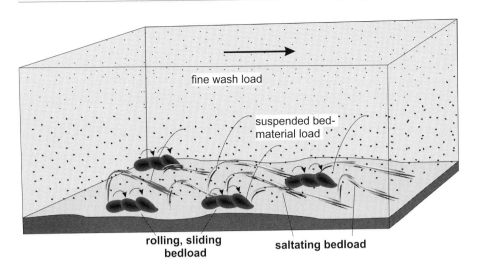

Figure 5.2. Diagram illustrating the different types of sediment transport in rivers. (A black and white version of this figure will appear in some formats. For the color version, please refer to the plate section.)

5.4.1 What Is Wash Load?

Wash load consists of the finest particles transported in suspension by rivers, typically silts and clays (Figure 5.1). Because it is carried in suspension, wash load contributes to the suspended load of rivers (Figure 5.2). Fine silts and clays do not exhibit strong dependence on hydraulic conditions, and the amount of this material within the flow is typically well below the capacity for transport. In fact, the transport capacity of wash load cannot be readily computed from hydraulic characteristics (Einstein and Chien, 1953). Colloidal particles less than 1 μm in size remain in suspension indefinitely through Brownian motion (Haw, 2002), and the influence of molecular-scale forces cannot be ignored for clays and perhaps fine silts. In a flowing river with $R > 2000$, turbulent eddies generally produce sufficient upward lift to maintain uniform suspension of silts and clays throughout the water column. Thus, large amounts of silt and clay can be transported by rivers, and the amount suspended within the flow is largely dependent on the amount supplied to the river from various sources.

Wash load originates from out-of-channel and within-channel sources (Figure 5.1). Soil erosion is a primary source of fine suspended sediment consisting of silt and clay, i.e., material with a diameter less than 0.063 mm. In steep terrain, mass movement can be a source of fine sediment, particularly if this movement involves failure of soil material on hillslopes into channels. The term wash load reflects the notion that much fine sediment is washed into the river channels by runoff from the surrounding watershed and moves through channels without substantial amounts of it being deposited along the way. Given the ease with which silt and clay remain in suspension, particles of this size typically do not accumulate on the beds of rivers. Fine suspended sediment may be deposited on floodplains during overbank flows and can play an important role in floodplain development (see Chapter 14).

Although wash load is often associated with inputs of sediment to a river from the surrounding watershed, it may include material introduced to the flow from bank erosion within the channel. Floodplains form the banks of alluvial channels, and erosion of these banks can deliver floodplain sediment to within-channel flows. In particular, the upper parts of channel banks along many alluvial rivers consist of cohesive silt and clay layers deposited by floods that inundate the floodplain. As banks fail, this cohesive material collapses into the river channel and over time disaggregates into fine suspended particles. Rivers that are actively eroding their banks may contribute substantial amounts of sediment to wash load. Wash load also affects water quality both by changing the turbidity characteristics of rivers and by transporting sediment-adsorbed contaminants.

5.4.1.1 What Is a Hyperconcentrated Flow?

Although wash load can be carried in virtually unlimited quantities by rivers, high concentrations of fine material lead to a transition to non-Newtonian behavior of the flow. At volumetric concentrations of fines between 3% and 10%, depending on the grain-size characteristics of these fines, the rheological properties of the flow change through the development of a yield stress for deformation and abrupt increases in viscosity of the fluid–sediment mixture. The result is the onset of hyperconcentrated flow – a type of flow transitional between fluvial processes and debris flows or mud flows (Pierson, 2005). Hyperconcentrated flows are capable of transporting large quantities of coarse sediment, particularly sand, in suspension. Some river systems can attain hyperconcentrated conditions. Well-known examples are rivers draining the Loess Plateau of China, where sediment concentrations by volume sometimes exceed 40% (Zhang et al., 2016b), and the main river into which streams from the Loess Plateau drain, the Yellow River, which has maximum sediment concentrations by volume exceeding 10% (Xu, 2002; van Maren et al., 2009; Kong et al., 2017).

5.4.2 What Is Bed-Material Load?

Bed-material load has traditionally been viewed as sediment transported by rivers that comes from the channel bed. The view that the source of bed-material load is strictly the channel bed has broadened with expanding knowledge of sediment transport in rivers, but the label persists. Bed-material load can be classified into two types: suspended bed-material load and bedload (Figures 5.1 and 5.2).

5.4.2.1 What Is Suspended Bed-Material Load?

Suspended bed-material load, like wash load, is carried in suspension within the flow (Figure 5.2). It typically consists of sand but in exceptionally turbulent flows may include small pebbles. Unlike wash load, the transport of suspended bed-material load is strongly dependent on hydraulic conditions. Thus, the amount of suspended bed-material load varies with changing flow characteristics, such as strength of turbulent eddies that can prevent suspended particles from falling to the bed.

Major sources of suspended bed-material load outside the channel include soil erosion and mass movement. If sand-size material from these sources enters a channel during a high flow it may be immediately suspended. However, as flow declines, the material is deposited on the channel bed, where it becomes an in-channel source for a subsequent high flow. Besides the channel bed, the lower parts of eroding channel banks, which often consist of noncohesive coarse sediment, including sand, can also be an in-channel source of suspended bed-material load. The suspended bed-material load, together with wash load, constitutes the suspended-sediment load of a river (Figure 5.1).

5.4.2.2 What Is Bedload?

Bedload is sediment that rolls, hops, or slides along the bed of the river (Figure 5.2). As with bed-material load, the transport of bedload is strongly dependent on hydraulic conditions. The amount of bedload therefore varies greatly with changes in flow. A wide range of particles can be transported as bedload, ranging from fine sand to large boulders. Sediment on the bed of a river is the major source of bedload, but bank erosion and mass failures on hillslopes during high flows may also contribute to bedload (Figure 5.1).

Saltation is a common hopping motion characterized by asymmetrical, low-angle trajectories of sand grains and, for highly turbulent flows, small gravel (Figure 5.2). The phenomenon of saltation involves not only brief suspension of sediment grains but also repetitious contact of particles with the bed. In this sense, it includes aspects of suspension and of

movement completely in contact with the bed. Although saltation is sometimes viewed as a suspension mode of transport (Einstein, 1942), more commonly it is categorized as a type of bedload transport (Knighton, 1998; Garcia, 2008). Large particles, such as pebbles, cobbles, or boulders, typically roll or slide along the bed. This type of transport is sometimes referred to as traction load to distinguish it from saltating bedload.

5.5 What Factors Influence the Transport of Fine Suspended Sediment?

The bulk of sediment transported by most rivers consists of suspended material, and the vast majority of this suspended material in many cases consists of wash load (sediment less than 63 μm in size). Attempts to characterize the transport of fine suspended sediment, which for the most part is independent of hydraulic conditions, have focused on factors that influence the delivery of fine sediment to river systems.

5.5.1 How Is the Transport of Fine Suspended Sediment Related to Discharge?

The relation between the concentration of suspended sediment within a river (C_s, mg L^{-1}) and discharge (Q) often is expressed as a power function:

$$C_s = aQ^b \tag{5.1}$$

The rationale underlying this relation is that increases in C_s reflect increases in runoff, which, in turn, enhance erosion of fine sediment from land surfaces throughout the watershed. In other words, large amounts of runoff, or Q, flush more sediment into the river system than events with small amounts of runoff. Thus, discharge is viewed as a proxy variable that represents the net effects of the many factors that influence the erosion, delivery, and transport of fine sediment within the river system. In many, but not all, cases, plots of measured C_s versus measured Q form linear patterns on graphs with logarithmic axes, indicating that the relation between these two variables conforms to Eq. (5.1) (Figure 5.3).

Because Eq. (5.1) is a statistical relation, the coefficient a and exponent b lack physical meaning. Nevertheless, a can be viewed as a scaling parameter that defines the general level of erosion of fine sediment within a watershed. For constant b, increases in a correspond to increases in suspended concentration for a given discharge. The exponent b specifies the responsiveness of suspended-sediment concentration to changes in discharge. When $b > 1$, the rate of increase in sediment concentration with discharge becomes greater as discharge increases. Such a condition typically arises when

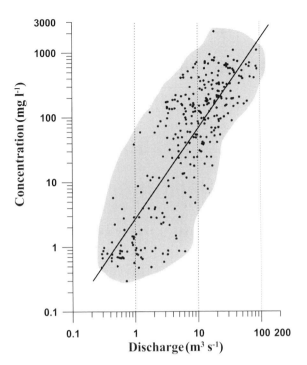

Figure 5.3. Suspended-sediment concentration versus discharge for the River Creedy, UK. Shading indicates the domain of scatter and vertical lines illustrate range of variability in C_s for specific values of Q (adapted from Walling, 1977).

new sources of sediment throughout the watershed increasingly become available with increasing runoff, including sources within the river system, such as bank erosion. Conversely, when $b < 1$, the rate of increase in concentration declines with increasing discharge, signaling perhaps exhaustion of sediment available for delivery to the river system.

Establishing relations between measured C_s and measured Q, such as Eq. 5.1, forms the basis for the development of sediment rating curves. Direct determination of C_s for every Q that occurs over time within a river would require continuous sampling and analysis of water-sediment samples, an effort prohibited in all but exceptional cases by logistics and cost. The rating-curve approach allows sediment concentrations for discharges to be estimated based on an established relationship between C_s and Q. In this approach, data collected in the field for a limited set of flow events in a river provide information on values of C_s and Q. Statistical analysis is then used to fit a function to these data, which may take the form of a power function. Using Eq. 5.1 as an example, once values of a and b are determined from statistical analysis, this equation can be used to estimate values of C_s for any measured value of Q. Estimates of suspended-sediment concentrations can be used to estimate suspended-sediment loads. Given a value of C_s for a particular Q, the suspended-sediment load (Q_{ss}) in kg s^{-1} is

$$Q_{ss} = 0.001 C_s Q \tag{5.2}$$

or in tons per day,

$$Q_{ss} = 0.0864 C_s Q \tag{5.3}$$

Although the rating-curve method is commonly used to estimate suspended-sediment loads in rivers, it must be applied cautiously. Power-function rating curves estimated by applying ordinary least squares regression analysis to log-transformed data are subject to transformation bias that affects the accuracy of concentrations estimated directly from Eq. (5.1). In particular, this approach tends to underestimate sediment loads for large discharges (Gao, 2008). Corrections are required to account for this bias (Ferguson, 1986a, 1987a). An alternative is to use other statistical methods and models to estimate relations between C_s and Q (Jansson, 1985, 1996; Asselman, 2000; Horowitz, 2003; Cox et al., 2008). The use of alternative models is particularly important when the relation between C_s and Q does not display a linear trend on plots with logarithmic axes (Crowder et al., 2007).

5.5.2 Why Is Relating Suspended-Sediment Concentration to Discharge Overly Simplistic?

Relating sediment concentrations to discharge only is a highly simplistic approximation of a complex set of processes involving erosion of soil and weathered material within a watershed, delivery of this eroded material to the river, and transport of the material within the river. Concentration–discharge relations often exhibit considerable scatter, with values of C_s varying over several orders of magnitude for specific values of Q (Figure 5.3). Detailed analysis of relations between C_s and Q reveals several reasons for such scatter. First and foremost, if Q is viewed as a hydraulic variable, the large amount of scatter is not surprising, because the transport of fine sediment, particularly wash load, is governed predominantly by the supply of this material to the river, not by a hydraulically determined sediment-transport capacity. Variations in sediment supply from the watershed will depend on a host of factors driving erosion and delivery of this material to the river during particular runoff events, rather than on the amount of water flowing in the river.

5.5.2.1 Why Are Differences in Travel Times of Water and Sediment Important?
Differential spatial translation of sediment waves and discharge waves along a river system can contribute to scatter in relations between C_s and Q. The velocity of a discharge wave or sediment wave relative to the velocity of the water or sediment is known as its celerity (Dingman, 2009). The velocity of a discharge wave or sediment wave can be determined

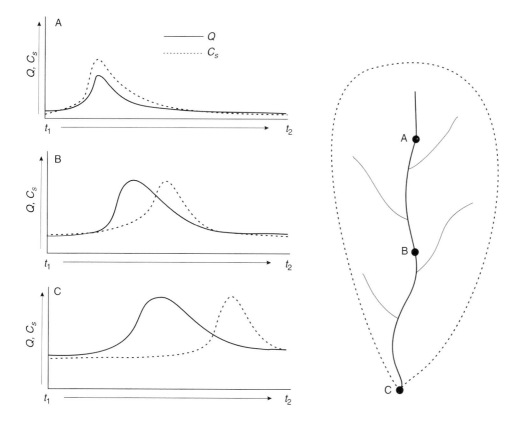

Figure 5.4. Translation of discharge waves (Q) and sediment-concentration waves (C_s) along a river system during the time interval t_1 to t_2 showing the faster downstream movement of the flood wave relative to the concentration wave.

by plotting positions of the peaks of variations in Q and C_s over time at different locations along a river. If waves of discharge and suspended-sediment concentration have different velocities, i.e., these waves move at different rates, the peaks of the waves will increasingly separate over distance (Figure 5.4). Early work suggested that sediment-concentration waves tend to move more slowly than discharge waves (Heidel, 1956), but other investigations indicate that sediment waves sometimes have greater celerities than discharge waves (Bull et al., 1995; Bull, 1997). Separation of the waves complicates simple relations between C_s and Q because the magnitudes of the two variables do not covary directly, which is necessary to produce a linear relationship on logarithmic plots.

5.5.2.2 What Is the First Flush Phenomenon and How Does It Influence Hysteresis in Relations between Suspended-Sediment Concentration and Discharge?

Differences between patterns of variation in sediment concentrations and discharge can also occur over time because of differential response times of sediment and water delivery mechanisms to the river system. One well-known example of this type of behavior is the first flush phenomenon. Often, available sediment that has accumulated between runoff events through sustained weathering of soils or hillslopes, through accumulation of sediment on impervious surfaces and in storm sewers, or by deposition within stream channels is

flushed quickly into and through the stream system during the early part of a runoff event. Under these circumstances, sediment concentrations peak before the discharge peaks. Such an effect can also occur annually if seasonal variations in runoff promote enhanced availability of sediment prior to a period of wet weather (Bussi et al., 2017). The occurrence of peak sediment concentrations before a peak in discharge may also reflect the predominance of sediment sources close to the sampling location; the travel time of sediment from such a source will be short relative to the travel time of water arriving at the sampling location from throughout the basin at peak discharge. In either case, C_s and Q do not covary proportionally throughout the event. Instead, C_s is large for relatively small values of Q during the rising limb of the hydrograph, whereas it is small for these same values of Q during the falling limb. Moreover, peak values of C_s do not correspond to peak values of Q. Under these conditions, a plot of C_s versus Q throughout the event forms a looped pattern, known as clockwise hysteresis (Figure 5.5).

In some cases, C_s may peak after the peak in Q, forming anticlockwise patterns of hysteresis (Figure 5.5). One possible interpretation of such patterns is that sediment is supplied from distant sources that require considerable travel times to reach the sampling location (Klein, 1984). Alternatively, the pattern may reflect activation of an erosional mechanism late in the event, such as initiation of bank erosion. Spatial variability in rainfall amounts and intensities within a watershed also can produce

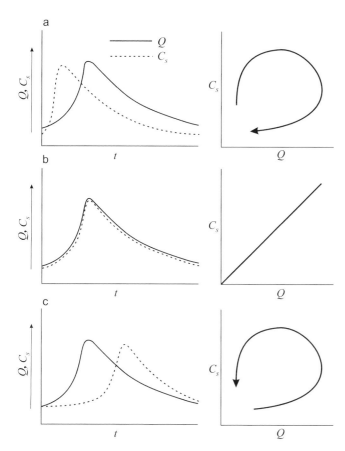

Figure 5.5. (a) Peak in C_s preceding peak in Q (left) results in clockwise hysteresis (right). (b) Similar timing of peaks results in linear relation. (c) Peak in Q preceding peak in C_s results in counterclockwise hysteresis. All axes for plots of C_s and Q are logarithmic.

hysteresis by influencing the spatial distribution of erosion and delivery of sediment to the river system.

Complex forms of hysteresis, such as figure eight patterns with clockwise and anticlockwise loops, lead to complicated explanations invoking multiple interacting processes of sediment flushing, exhaustion, storage, delayed activation, and remobilization throughout an event (Vercruysse et al., 2017). Hysteresis complicates the relationship between C_s and Q because only in cases where the two variables covary directly through time will this relationship plot as linear on a graph with logarithmic axes. Separate power functions for rising and falling limbs of hydrographs may be needed to accurately capture the relation between suspended-sediment concentration and discharge for clockwise and counterclockwise patterns of hysteresis. Such functions may be inadequate when hysteresis exhibits more complex patterns.

5.5.2.3   How Can Sediment Fingerprinting Contribute to the Understanding of Sediment Hysteresis?

Sediment fingerprinting (Chapter 3) serves as an important complement to interpretations of fine sediment dynamics founded on temporal relations between C_s and Q, particularly when this fingerprinting is conducted at the same timescale as sediment sampling (Wilson et al., 2012). Suppose clockwise hysteresis is observed in the relation between C_s and Q. If fingerprinting analysis of sediment sampled during the early part of the event indicates that the bulk of this sediment came from a source close to the sampling location, the results provide support for the interpretation that the pattern of hysteresis is caused by the spatial proximity of the sediment source to the sampling location. Conversely, if fingerprinting indicates that during the latter part of an event with anti-clockwise hysteresis, sources far from the sampling location contribute large amounts of material to suspended sediment, this result supports the conclusion that the pattern of hysteresis is caused by long travel distances between the sampling location and the primary sediment source. Coordinating sediment fingerprinting and suspended-sediment sampling over multiple timescales can help to identify how intra-event, seasonal, and interannual variations in suspended-sediment concentrations are linked to variability in dominant sources of sediment (Vercruysse et al., 2017).

5.5.3   How Do Controlling Factors of Suspended-Sediment Dynamics Vary with Timescale?

Given that power-function relations between C_s and Q are simplistic, often prone to high degrees of scatter, and misrepresentative of relations characterized by hysteresis, attention has increasingly turned toward multivariate analysis of the factors controlling the production, delivery, and transport of fine sediment in rivers. Controlling factors vary with the timescale under consideration (Table 5.1).

5.5.3.1   What Are the Most Important Controlling Factors at Interannual Timescales?

Over interannual timescales, climatic conditions and land-cover characteristics strongly influence suspended-sediment dynamics (Table 5.1). Changes in these characteristics commonly result in changes in suspended-sediment concentrations and loads. The connectivity of hydrological pathways determines how water and sediment move through a watershed into the river system and through this system. If these pathways change or are disrupted over time, the delivery of fines to river systems can be enhanced or impeded. Humans often play a role in changes to these pathways. In forested areas, unpaved roads often generate runoff, yield fine sediment, and serve as runoff pathways, thereby enhancing sediment delivery to streams (Reid and Dunne, 1984; Reid et al., 2016). The construction of dams, a topic treated in more detail in Chapter 15, can substantially disrupt within-river pathways of sediment movement.

Table 5.1. Variables related to the dynamics of suspended sediment in rivers at different timescales.

Event	Seasonal	Interannual
Discharge	Discharge	Discharge
Peak discharge		
Runoff duration		
Total water volume		
Rate of change in discharge		
Precipitation intensity, pattern, duration, and amount	Precipitation intensity, pattern, and amount	Precipitation intensity, pattern, and amount
Antecedent runoff	Soil moisture	Land-cover characteristics
Antecedent precipitation		
Antecedent soil moisture		
Time between sediment-mobilizing events	Weather or farming-related changes in land cover	Connectivity of hydrological pathways within the watershed
Spatial extent of runoff	Within-channel sediment storage	Human impacts, including dam construction, soil tillage, soil conservation practices, deforestation, urbanization, and mining
Connectivity of pathways of runoff in relation to source areas of sediment		
Within-channel sediment storage		

5.5.3.2 How Does Seasonality Affect Suspended-Sediment Dynamics?

Seasonal variations in suspended-sediment concentration largely reflect seasonal variations in precipitation and land-cover characteristics. Storm types, convective versus frontal, often vary seasonally and are associated with different precipitation intensities and patterns. Vegetation characteristics also differ seasonally in many climates, especially in conjunction with farming activities. Seasonal differences in controlling factors can produce distinct seasonal contrasts in the relation between discharge and sediment concentration (Walling, 1977; Lecce et al., 2006). Seasonal variations may reflect intra-annual storage and release cycles, whereby certain parts of the year (e.g., dryer months) promote sediment storage and others (e.g., wetter months) promote flushing of this stored sediment (Vercruysse et al., 2017).

5.5.3.3 What Are the Characteristics of Fine–Suspended Sediment Dynamics at the Event Scale?

The most complex suspended-sediment dynamics occur at the event scale. Multiple interacting factors influence the response of the sediment system at this scale (Table 5.1), producing myriad types of event dynamics. In particular, simple relationships between concentration and discharge over the course of an event typically do not occur; instead, hysteresis relations are common. Detailed analysis of hysteresis patterns indicates that temporal variations in suspended-sediment concentrations at a particular sampling location are strongly influenced by place- and time-specific conditions (Gao and Josefson, 2012a; Gellis, 2013; Aich et al., 2014). Precipitation characteristics, such as rainfall intensity, total precipitation, and rainfall duration, often are correlated strongly with sediment concentrations, indicating the importance of storm characteristics for runoff and sediment delivery, independently of discharge (Lana-Renault and Regues, 2009; Oeurng et al., 2010; Lopez-Tarazon and Estrany, 2017). The occurrence of phenomena that deliver large of amounts of sediment to a river, such as landslides or episodic erosion of channel banks, may dramatically increase sediment concentrations relative to discharge during an individual hydrological event. Antecedent conditions, especially the magnitude of antecedent runoff or precipitation and the length of time since a prior sediment-mobilizing event, can influence sediment concentration by determining the amount of sediment available for delivery from the watershed and the extent to which the occurrence of prior events has exhausted the supply of sediment (Gray et al., 2014). Connectivity can vary with storm magnitude and spatial extent; local runoff events of small magnitude may only mobilize sediment within a small number of transport pathways between hillslopes and the river system, whereas large regional events may generate widespread runoff and corresponding sediment production, activating a large number of sediment-transport pathways (Sherriff et al., 2016).

Although C_s–Q relations often exhibit marked hysteresis at the intra-event scale, aggregation of suspended-sediment transport over an event has a filtering effect that reveals strong relations between the transport of suspended material and the size of an event. In small drainage basins in New York and Arizona, rating curves are difficult to develop at the intra-event scale because of intra-event variability in the relation between suspended-sediment concentration and discharge. Nevertheless, total sediment load per event (tons) is strongly related to event peak discharge for the New York watersheds (Gao and Josefson, 2012b) and event-specific sediment yield (tons km^{-2}) is strongly related to event runoff depth (mm) for the Arizona watersheds (Gao et al., 2013). Even though intra-event variation in suspended-sediment concentrations exhibits complicated patterns of hysteresis, magnitudes of suspended-sediment transport per event in these two contrasting environments are well-defined functions of basic metrics of flow size (peak discharge and runoff depth).

5.6 What Factors Control the Mobilization of Bed Material?

The initiation of motion of particles on the channel bed is a key component of the morphodynamics of alluvial rivers. Once particles are mobilized, the channel bed is set in motion and can deform through erosion and deposition. Mobilization of material on the channel bed can also lead to erosion and deposition along the channel banks (see Chapter 9). Ultimately, channel change begins with mobilization of the channel boundary.

5.6.1 What Are the Forces Acting on Particles on the Bed of a River?

In a steady, one-dimensional, uniform flow, the forces acting on a grain resting on the sloping bed of a river include the gravitational force (F_g), the drag force (F_D), and the lift force (F_L) (Figure 5.6). The normal component of the gravitational force ($F_g \cos\theta$) resists grain motion along with friction associated with interlocking of grains with one another as determined by the pivoting or friction angle (ϕ_o). The drag force and the downslope component of the gravitational force ($F_g \sin\theta$) are driving forces acting in the downstream direction along the slope of the channel bed, whereas the lift force acts in opposition to the normal component of the gravitational force. The

most basic approach to grain entrainment considers a flat bed, so that $F_g \cos\theta = F_g$ and $F_g \sin\theta = 0$. Moreover, friction associated with interlocking grains is neglected. Under these conditions, resistance to motion is associated with the gravitational force, and only the drag force acts parallel to the bed.

The gravitational force consists of the submerged weight of the particle and is oriented normal to a flat bed:

$$F_g = (\rho_s - \rho)gV_p \tag{5.4}$$

where V_p is the volume of the particle, or, for a spherical particle,

$$F_g = (\rho_s - \rho)g\frac{4}{3}\pi r^3 = (\rho_s - \rho)g\frac{\pi}{6}d^3 \tag{5.5}$$

where ρ_s is sediment density, r is the particle radius, and d is the particle diameter. For particles of arbitrary shape, Eq. (5.5) can be generalized as

$$F_g \propto (\rho_s - \rho)gd^3 \tag{5.6}$$

The drag force of the mean (time-averaged) flow on the grain is (Bridge, 2003, p. 49; Middleton and Wilcock, 1994, p. 38; Garcia, 2008)

$$F_D = C_D\rho\frac{u_c^2}{2}A_s \tag{5.7}$$

where C_D is a dimensionless drag coefficient, A_s is the cross-sectional area of the grain, and u_c is a characteristic velocity. The lift force (F_L) develops as a result of velocity and pressure gradients between the bottom and the top of the grain. Typically, an expression identical to the drag force is used to characterize this force, with a dimensionless lift coefficient (C_L) replacing the drag coefficient:

$$F_L = C_L\rho\frac{u_c^2}{2}A_s \tag{5.8}$$

The resultant of the drag and lift forces (F_r) is

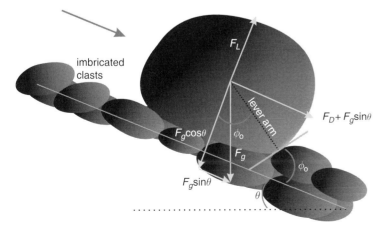

Figure 5.6. Components of the force balance acting on a grain on a sloping riverbed. (A black and white version of this figure will appear in some formats. For the color version, please refer to the plate section.)

$$F_r = C_r \rho \frac{u_c^2}{2} A_s \tag{5.9}$$

where

$$C_r = \sqrt{C_D^2 + C_L^2} \tag{5.10}$$

For a spherical grain,

$$F_r = C_r \rho \frac{u_c^2}{2} \pi r^2 = C_r \rho u_c^2 \frac{\pi}{8} d^2 \tag{5.11}$$

Generalizing for grains of arbitrary shape yields

$$F_r \propto \rho u_c^2 d^2 \tag{5.12}$$

It is important to note that the proportionality expressed in Eq. (5.12) does not include C_r and therefore assumes that this coefficient is a constant and can be eliminated from consideration in determining the resultant force. A constant value of C_r implies that C_D and C_L are also constants (Eq. (5.10)). In fact, C_D and C_L vary with grain shape and even for a given grain shape are not always constant (Middleton and Wilcock, 1994). Nevertheless, Eq. (5.12) serves as a starting point for a generalized consideration of drag and lift forces acting on a grain.

The characteristic velocity (u_c) commonly is associated with the mean velocity u acting at the level of the center of the particle (Middleton and Wilcock, 1994, pp. 38–39; Bridge, 2003, p. 49). A shortcoming of this representation is that it requires information on the near-bed velocity profile at the level of the grain. An alternative approach is to associate the characteristic velocity in Eq. (5.12) with the shear velocity u_*, which represents both the shear stress (τ_b) acting on the bed in units of velocity and the gradient of the velocity profile near the bed (Dingman, 2009, p. 480):

$$F_r \propto \rho u_*^2 d^2 \tag{5.13}$$

or, recalling that $\tau_b = \rho u_*^2$,

$$F_r \propto \tau_b d^2 \tag{5.14}$$

5.6.2 What Is the Dimensionless Bed Shear Stress?

The gravitational force resists motion, whereas the drag and lift forces act to move the grain. If the driving force exceeds the resisting force, the grain will be mobilized, resulting in particle entrainment. The ratio between generalized expressions of the driving (Eq. (5.14)) and resisting (Eq. (5.6)) forces is

$$\Theta = \frac{\tau_b d^2}{(\rho_s - \rho)g d^3} = \frac{\tau_b}{(\rho_s - \rho)g d} \tag{5.15}$$

where Θ is the dimensionless bed shear stress. If the dimensionless bed shear stress represented accurately the driving and resisting forces acting on particles of a specific size, the critical value at which incipient motion of the particles occurs (Θ_c) should be at or slightly greater than $\Theta_c = 1$. Eq. (5.15), however, is based on approximations of the driving and resisting forces and therefore is unlikely to conform to this critical value. Empirical investigation of particle mobilization is required to determine how values of Θ relate to incipient motion of particles of different sizes. Different methods have been developed for determining incipient motion (Box 5.1).

5.6.3 What Is the Critical Dimensionless Bed Shear Stress for Uniform Sediment?

In the 1930s, Albert F. Shields, an American PhD student studying in Berlin, conducted laboratory experiments to determine values of Θ_c for channel beds consisting of particles of relatively uniform size (Kennedy, 1995). This work related Θ_c to the boundary Reynolds number (R_b), a dimensionless metric characterizing particle size (d):

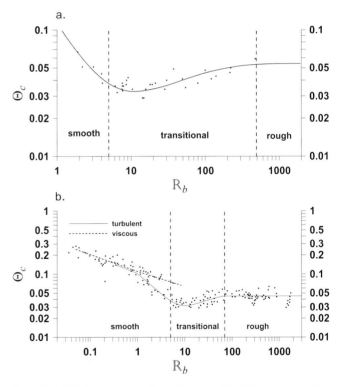

Figure 5.7. Relation between dimensionless critical shear stress and boundary Reynolds number. (a) Original data from Shields (from Buffington, 1999, table 3) with fitted curve (Guo, 2002). (b) Data from Yalin and Karahan (1979) showing difference between turbulent and viscous relations for low $\mathbf{R}_b$. Fitted curves from Garcia-Flores and Maza-Alvarez (1997) as presented in Garcia (2008). Dashed vertical lines separate smooth-, transitional-, and rough-boundary regimes.

BOX 5.1 | **HOW IS INCIPIENT MOTION OF BED MATERIAL DETERMINED?**

All studies of entrainment of bed material rely on determinations of incipient motion. Different methods have been used to assess incipient motion, and the choice of method can influence the results of the study. The most common methods for defining incipient motion are 1) extrapolating measured bedload transport rates either to zero or to a very small value considered to be a reference value, 2) visual observation of when grains begin to move on the bed, and 3) estimating the boundary shear stress required for the flow to move the largest transported particle and assuming that this shear stress represents the actual shear stress of the flow (Buffington and Montgomery, 1997; Petit et al., 2015). Visual observation was relied upon in early studies of incipient motion, such as the classic work of Gilbert (1914), but is rarely employed today, except perhaps in the form of analysis of high-speed videography of particle motion (e.g., Nino et al., 2003). Extrapolation of measured volumetric bedload transport rates per unit width of channel (q_{sbv}) to zero or nearly zero is commonly employed in experimental studies. For mixtures, volumetric transport rates per unit width of channel for each size fraction (q_{sbvi}) in nondimensionalized form are related to Θ or τ_b. Examples of nondimensionalized transport rates include (Wilcock and Crowe, 2003)

$$q^*_{sbvi} = \frac{\left(\frac{\rho_s}{\rho} - 1\right) g q_{sbvi}}{u_*^3 f_i} \tag{5.1.1}$$

and (Shvidchenko et al., 2001a)

$$q^*_{sbvi} = \frac{q_{sbvi}}{f_i \sqrt{\left(\frac{\rho_s}{\rho} - 1\right) g\, d_i^3}} \tag{5.1.2}$$

where f_i is the proportion of size fraction i in the surface bed material. Values of q^*_{sbvi} are plotted versus Θ or τ_b and the resulting relation extended either to $q^*_{sbvi} = 0$ or to where q^*_{sbvi} is equal to a very small number, such as 0.002. The corresponding value of Θ or τ_b is then designated as Θ_{ci} or τ_{bci}. This method is most commonly applied in laboratory studies where detailed measurements of transport rates by size fraction are possible (Wilcock and Crowe, 2003), although it has also been used in some field studies with high-quality bedload transport data (Wathen et al., 1995).

The competence method involves sampling the movement of sediment in a channel over numerous events and estimating the peak bed shear stress for these same events. The largest particle sampled during a particular flow is assumed to be the largest particle that the peak shear stress for that flow can move. In other words, the peak shear stress is the critical shear stress for that size of particle. The relation between particle size and critical shear stress defines a threshold function for incipient motion of bed material of different sizes. The competence method is clearly inappropriate for sediment with equal mobility, as it relies on selective entrainment of particles of different sizes. It also suffers from the limitation that the stream may have been able to transport a particle of larger size than the largest size sampled, but no particle of the largest possible size actually reached the bedload sampler during the duration of sampling. A variant of this method that avoids this problem is to mark particles over the full size range in a stream and then determine the largest marked particle mobilized by the flow (Petit et al., 2015).

$$R_b = \frac{u_* d}{v} \tag{5.16}$$

Experimental results indicated that values of Θ_c decline with increasing R_b for $R_b < 10$ and increase for $10 < R_b < 490$ (Figure 5.7a). Not only is $\Theta_c \neq 1$, but it varies with the size of the uniform particles on the channel bed. Although Shields's data set did not include information for $R_b > 500$, he proposed, based on the pattern of data and a single data point at $R_b = 473$, that $\Theta_c \approx 0.060$ for $R_b \geq 500$ (Buffington and Montgomery, 1997) – a value for Θ_c that has been widely cited (Rouse, 1939;

Vanoni et al., 1966; Wiberg and Smith, 1987). Subsequent work has attempted to refine the Shields relation by incorporating additional data from experimental and field investigations (Miller et al., 1977; Yalin and Karahan, 1979; Buffington and Montgomery, 1997).

5.6.3.1 How Is the Critical Dimensionless Shear Stress Related to Boundary Roughness?

To better understand the relationship between the dimensionless shear stress and the boundary Reynolds number, the value of R_b can be recast as a relation between particle diameter and the thickness of the viscous sublayer (z_v). Recalling that $z_v = 5v/u_*$ (Eq. (4.48)), rearranging this expression yields

$$u_* = \frac{5v}{z_v} \tag{5.17}$$

Substituting the right-hand side of this expression for u_* in Eq. (5.16) results in

$$R_b = 5\frac{d}{z_v} \tag{5.18}$$

Thus, when the diameter of the particles on the bed is equal to the thickness of the viscous sublayer ($d/z_v = 1$), $R_b = 5$. For a bed consisting of particles with a value of $R_b < 5$, these particles will be contained within the viscous sublayer and the boundary considered smooth. A rough boundary develops when the diameter of the particles is large enough to fully disrupt the viscous sublayer. Under this condition, which mainly pertains to gravels, entrainment of particles is fully dependent on turbulent stresses near the bed. Typically, the rough-boundary condition is equated with the zone where Θ_c becomes constant and no longer varies with R_b ($R_b > 500$, $d/z_v = 100$) (Miller et al., 1977; Buffington and Montgomery, 1997), although Yalin and Karahan (1979), in a widely cited revision of the original Shields diagram (Dingman, 2009), place this threshold as low as $R_b = 70$ ($d/z_v = 14$) (Figure 5.7b). A transitional region, usually dominated by entrainment of sandy bed material, occurs between the domains of smooth and rough boundaries.

The Shields curve can be defined by the equation (Guo, 2002)

$$\Theta_c = \frac{0.11}{R_b} + 0.054(1 - e^{-0.16R_b^{0.52}}) \tag{5.19}$$

which for large values of R_b converges asymptotically on $\Theta_c = 0.054$ in the rough-boundary region (Figure 5.7). This value is less than the traditional value of $\Theta_c = 0.06$ for $R_b > 490$, but is similar to the value of $\Theta_c = 0.056$ for rough boundaries recommended by Henderson (1966). Other work has suggested a rough-boundary value of 0.045 (Miller et al., 1977; Yalin and Karahan, 1979) (Figure 5.7).

5.6.3.2 How Is Incipient Motion of Particles Related to the Critical Bed Shear Stress?

Particle size (d) can be expressed in terms of Θ_c and R_b as (Dingman, 2009, p. 483)

$$d = \left(\frac{v^2 R_b^2}{g(\rho_s/\rho - 1)\Theta_c}\right)^{0.33} \tag{5.20}$$

Using Eq. (5.19) to determine values of Θ_c in Eq. (5.20), calculating d, and substituting values of Θ_c and d into

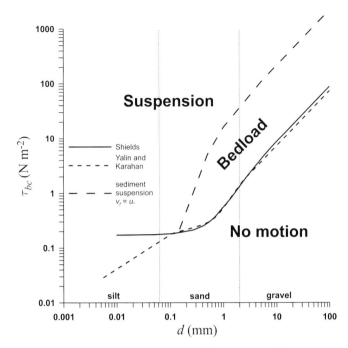

Figure 5.8. Relation between critical bed shear stress and grain size based on fitted lines in Figure 5.7. Curves based on $v = 1.31 \times 10^{-6}$ m^2 s^{-1}, $\rho_s = 2650$ kg m^{-3}, $\rho = 1000$ kg m^{-3}, $g = 9.81$ m s^{-2}. Also shown is the threshold for sediment suspension for coarse grains, which is approximated by $v_f = u_*$ (Wilcock and McArdell, 1993; Garcia, 2008).

$$\tau_{bc} = \Theta_c(\rho_s - \rho)gd \tag{5.21}$$

allows the Shields curve to be transformed into a relation between the critical dimensional bed shear stress, τ_{bc}, and the grain size of the bed material (d) (Figure 5.8). A similar analysis can be performed using relations for Yalin and Karahan's (1979) data developed by Garcia-Flores and Maza-Alvarez (1997) (Garcia, 2008, p. 52) (Figure 5.8). Note that for rough boundaries consisting of gravel, where Θ_c is essentially constant, the relation between the critical bed shear stress and grain size becomes linear on this logarithmic plot (Figure 5.8). In this domain, entrainment is strongly size dependent, with increases in grain size requiring corresponding increases in bed shear stress to entrain a particle.

5.6.4 Do Exact Thresholds for Motion of Particles Exist?

The extent to which relations between Θ_c and R_b or d neatly define the threshold for incipient motion of uniform sediment remains uncertain. Not all data included in analysis of these relations are for strictly uniform sediment. Moreover, reanalyses confined to data sets with small values of sediment sorting ($\sigma_s < 0.5$) still exhibit considerable amounts of scatter (Buffington and Montgomery, 1997) (Figure 5.9). Of particular interest has been the supposed constant value of Θ_c for $R_b > 500$, the domain of

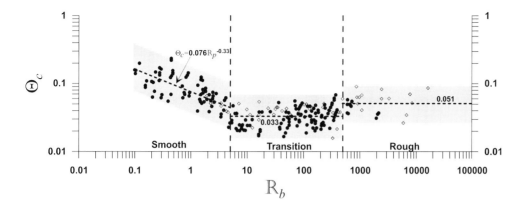

Figure 5.9. Relation between critical dimensionless shear stress and boundary Reynolds number (data from Buffington and Montgomery, 1997). Solid dots – incipient motion determined visually, open rhombi – incipient motion determined from reference-based method. Labeled dashed horizontal and sloping lines indicate approximate mean or trending values of Θ_c for smooth, transitional, and rough regimes (denoted by vertical dashed lines). Shaded areas represent range of variation of Θ_c.

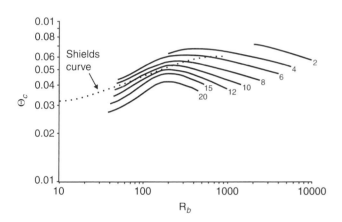

Figure 5.10. Relation between dimensionless critical shear stress and boundary Reynolds number for different relative roughness conditions (after Shvidchenko and Pender, 2000). Numbers by lines indicate R/d, which is inversely related to channel bed slope. Shields curve (Figure 5.7) shown for reference.

gravel-bed rivers. Even though the mean value may be constant (e.g., Figure 5.9), individual values for different rivers exhibit substantial variability (Petit et al., 2015). The threshold of motion for a given grain size (R_b or d) may therefore best be defined by a range of values rather than a single, discrete value. Such findings have led some to call into question the extent to which distinct thresholds for entrainment really exist (Lavelle and Mofjeld, 1987).

Incipient motion of uniform sizes appears to depend on the relative submergence, the ratio of hydraulic radius to particle diameter (R/d), and bed slope. Experimental work indicates that Θ_c, instead of becoming constant for $R_b > 500$, tends to decline (Figure 5.10). Increasing bed slopes produce shallower flows relative to particle size, increasing flow resistance and requiring greater Θ_c to mobilize bed material. The influence of slope and R/d on Θ_c has become a topic of considerable

interest because of its relevance to sediment dynamics in mountain streams (Gregoretti, 2008; Lamb et al., 2008c; Recking, 2009; Ferguson, 2012; Prancevic and Lamb, 2015a).

5.6.5 How Does Entrainment Occur in Particle Mixtures?

Entrainment relations (Figures 5.7 and 5.8) are presumed to apply to sediment of uniform size (constant d), even though the data used to develop the relations were not for sediment of completely uniform size (Buffington and Montgomery, 1997). Nevertheless, the mixtures used to develop the relations were, for the most part, much better sorted than bed material in natural rivers, the size of which can vary over several orders of magnitude (Buffington and Montgomery, 1997). Applying the Shields relation to predict entrainment of individual particles within a sediment mixture results in highly size-selective mobilization of all material coarser than medium sand (Figure 5.8). An important issue is whether this type of highly size-selective mobilization of bed material occurs in natural rivers. Since the 1980s, many different studies have attempted to address this question.

Entrainment of particles from a bed consisting of grains of different sizes is a complicated problem. Many factors affect entrainment in this type of situation, including the size and shape of the particles, the extent to which the particles are compacted, the protrusion of grains above the mean elevation of the bed, and the exposure of individual particles to the flow relative to upstream particles. As these factors vary, so will the critical shear stress required to move particular particles. Thus, the critical shear stress for entrainment is dependent not only on the size of particles, as indicated by relations developed for beds consisting of particles of relatively uniform size (Figure 5.8), but also on geometric relations among particles of different sizes and shapes that cover the channel bed.

5.6.5.1 What Is the Importance of Pivoting Angle in Understanding Entrainment of Sediment Mixtures?

Detailed analysis of the problem of grain entrainment from mixtures considers the forces acting on a grain positioned on other grains that must pivot out of position over the grains upon which it rests to move along a sloping bed (Figure 5.6). An important issue is how the size of an adjacent particle influences the entrainment of a clast of interest. If the adjacent particle differs in size from a clast of interest, this difference could affect the mobility of the clast compared with the situation where all particles are the same size.

Inclusion of the pivoting angle (ϕ_o) in the analysis of particle entrainment modifies slightly the formulation of the problem. Based on force-balance considerations, the relation between driving and resisting forces for a grain on a sloping bed is (Bridge, 2003, p. 2003)

$$l_a \cos \phi_o (F_D + F_g \sin \theta) = l_a \sin \phi_o (F_g \cos \theta - F_L) \qquad (5.22)$$

where l_a is the length of the lever arm (Figure 5.6). For a spherical grain, $l_a = d/2$. To isolate the effect of the pivoting angle, the bed is assumed to be flat or nearly flat, resulting in

$$l_a \cos \phi_o (F_D) = l_a \sin \phi_o (F_g - F_L) = l_a \sin \phi_o (F_g') \qquad (5.23)$$

Dividing both sides by $l_a \cos \phi_o$,

$$F_D = \tan \phi_o (F_g') \qquad (5.24)$$

$$\frac{F_D}{F_g'} = \tan \phi_o \qquad (5.25)$$

This equation indicates that at the moment of grain mobilization, the ratio of the driving to resisting forces equals the tangent of the pivoting angle, moving the grain up and over the grain against which it rests. It also shows that as the pivoting angle increases, a greater driving force is needed to mobilize the grain. Thus, grain resistance to motion increases with increasing pivoting angle. Note also that the dimensionless critical shear stress is an approximation of the ratio of driving to resisting forces acting on the bed (Eq. (5.15)). For a bed comprised of uniform grains, the pivoting angle and the dimensionless critical shear stress should be closely related. Indeed, Wiberg and Smith (1987) used a force-balance relation derived from Eq. (5.22) to show that the Shields curve (Figure 5.7) is closely approximated by a bed of uniform grains with $\phi_o = 60°$.

Numerous studies have examined the factors that influence the pivoting angle, sometimes referred to as the friction angle, of particles on the bed of a river (Miller and Byrne, 1966; Li and Komar, 1986; Komar and Li, 1986, 1988; Kirchner et al., 1990; Buffington et al., 1992; Prancevic and Lamb, 2015b). Detailed models of grain entrainment by pivoting have been developed based on physical consideration of forces acting on grains (Wiberg and Smith, 1987; James, 1990; Bridge and Bennett, 1992; McEwan and Heald, 2001; Vollmer and Kleinhans, 2007), but application of these models to entrainment of large numbers of particles situated on channel beds comprised of surrounding particles with a wide variety of geometric relations to one another has proven challenging. Although methods for measuring the pivoting angle in the field have been proposed (Johnston et al., 1998), conducting such measurements for extensive areas of channel beds is costly and time-consuming.

5.6.5.2 What Is the Relation between Pivoting Angle and Relative Grain Size?

Experimental work on pivoting angles has shown that these angles are dependent on particle size for both sand- (Miller and Byrne, 1966) and gravel-sized (Li and Komar, 1986) grains in the form

$$\phi_o = a(d_p/d_{bd})^b \qquad (5.26)$$

where d_p is the diameter of the intermediate axis of a pivoting grain, d_{bd} is the diameter of the intermediate axis of the base grains over which the pivoting grain moves, and a and b are empirical coefficients. The value of a is a scaling factor that depends on grain shape, with flat grains having higher values of a than grains with nearly equal short and intermediate axial diameters (Li and Komar, 1986). It represents the pivoting angle when pivoting grains are the same size as the base grains (uniform sediment). The value of b is negative, indicating that for base grains of constant size, pivoting angles decrease as the size of the pivoting grain increases. When considered in relation to Eq. (5.25), a decrease in ϕ_o with increasing d_p indicates that the force needed to move a grain of a particular size decreases as it becomes large relative to the grains over which it is moving. Values of b are -0.3 for sand grains (Miller and Byrne, 1966), -0.33 for angular gravel and flat pebbles, -0.36 for ellipsoidal gravel, and -0.75 for spheroidal gravel (Li and Komar, 1986).

5.6.5.3 What Is the Relation of Dimensionless Critical Shear Stress to Relative Grain Size?

The relation between the pivoting angle and grain size, along with the similarity between Θ_c and $\tan \phi_o$ as expressions of the ratio between the driving and resisting forces for grain entrainment, has led to the formulation of Θ_c for heterogeneous bed material as a function of grain-size ratio:

$$\Theta_{ci} = \Theta_{cr} \left(\frac{d_i}{d_r} \right)^b \qquad (5.27)$$

where Θ_{ci} is the critical dimensionless shear stress for a grain-size fraction (f_i) with mean diameter d_i, Θ_{cr} is the critical dimensionless shear stress for a reference grain size of diameter d_r, and b is an empirical coefficient. In most cases, d_{50} is the reference grain size, but in some cases the mean grain size (d_m) has been used. The effect of the exponent b on entrainment becomes clear by substituting Eq. (5.27) into Eq. (5.21) to determine the critical shear stress required to mobilize grain size d_i (τ_{bci}) in bed material consisting of a mixture of different grain sizes:

$$\tau_{bci} = \Theta_{cr}\left(\frac{d_i}{d_r}\right)^b (\rho_s - \rho)gd_i = \Theta_{cr}(\rho_s - \rho)gd_r^{-b}d_i^{1+b} \tag{5.28}$$

If $b = 0$, Eq. (5.28) reduces to

$$\tau_{bci} = \Theta_{cr}(\rho_s - \rho)gd_i \tag{5.29}$$

which is equivalent to Eq. (5.21). This relation defines size-selective entrainment for a sediment mixture identical to entrainment associated with a set of channel beds, each of which has a bed of uniform grain size d_i, but in which values of d_i differ among the beds. In other words, the presence of heterogeneous sediments does not influence entrainment of grains of different sizes. On the other hand, if $b = -1$, Eq. (5.28) equals

$$\tau_{bci} = \Theta_{cr}(\rho_s - \rho)gd_r \tag{5.30}$$

For this condition, particles of all sizes move at the same critical bed shear stress as the reference particle size d_r. All particles, regardless of size, are equally mobile and move at the same value of critical bed shear stress. Values of b between 0 and –1 define size-selective transport, but at levels less pronounced than that defined by Eq. (5.21). If $b > 0$, size-selective transport is even more pronounced than that predicted by Eq. (5.21). In this case, bed-material heterogeneity results in enhancement of size-selective grain mobilization. Finally, if $b < -1$, τ_{bci} is inversely related to d_i, and coarse grains are more mobile than fine grains.

5.6.5.4 To What Extent Is Entrainment of Bed Material in Gravel-Bed Rivers Size Selective?

Numerous studies have empirically evaluated Eq. (5.27) using field data on rivers with gravel beds or mixed sand and gravel beds (Table 5.2). At least one field study (Parker et al., 1982a) and one experimental study (Kuhnle, 1993a) report values of $b \approx -1$, suggesting that bed material in gravel-bed rivers is equally mobile. However, most studies have found values of b between –0.5 and –0.9. This range of values indicates that size-selective transport occurs in gravel-bed rivers, but that the relative mobility of grains of different sizes is tempered by the influence of other grains. Large grains tend to move more readily than if the bed consisted entirely of identical grains, whereas small grains tend to move less easily, presumably due to hiding effects by large grains. In fact,

Table 5.2. Reported values of coefficient and exponent in Eq. (5.27).

	Θ_{cr}	b	d_{50} (mm)	d_i/d_{50}
Parker et al. (1982a)	0.0876	−0.982	1.3–25	0.045–4.2
Andrews (1983)	0.0834	−0.872	54–74	0.3–4.2
Carling (1983)	0.045	−0.68	20	0.5–10
Ashworth and Ferguson (1989)	0.089	−0.74	23–98	0.1–2
Ashmore et al. (1992)	0.049	−0.69	18–32	0.15–3.2
Wathen et al. (1995)	0.05 0.086	−0.70 −0.90	21	0.3–3
Whitaker and Potts (2007)	0.044	−0.59	56	1.57–3.3

relations in the form of Eq. (5.27) are known as hiding functions, because it is generally argued that in mixtures of sediment, the influence of large grains on flow near the bed impedes the movement of fine grains. The form of hiding functions is generally consistent with the analysis of pivoting angles for particles of different sizes, which indicates that large particles perched on smaller particles will have a smaller $\tan\phi_o$ than small particles resting on relatively larger particles. In particular, protruding grains are highly susceptible to entrainment (Masteller and Finnegan, 2017). Interest in the properties of hiding functions has motivated detailed laboratory analysis of the factors influencing the value of b. These factors include the degree of sorting of the bed material, the median size of the bed material, the shear velocity, and whether d_i/d_r is greater than 1 or less than 1 (Petit, 1994; Shvidchenko et al., 2001a; Wilcock and Crowe, 2003; Buscombe and Conley, 2012).

5.6.5.5 To What Extent Is Entrainment of Bed Material in Sand-Bed Rivers Size Selective?

Less work has examined entrainment of heterogeneous sands. The limited range of grain sizes of sands compared with gravels indicates that river beds composed of sand move over a comparatively narrow range of critical bed shear stresses (Figure 5.8). The relation

$$\tau_{bc} = 0.00515(\rho_s - \rho)d_p^{0.57}\tan\phi_o \tag{5.31}$$

where $\tan\phi_o$ is determined by Eq. (5.26) with $a = 61.5$ and $b = -0.3$ predicts well the entrainment of a specific size of sand grains resting on a substrate of grains of a different size (Komar and Wang, 1984; Li and Komar, 1992). In particular, Eq. (5.31) predicts that as the ratio of d_p/d_{bd} increases τ_{bc} decreases, indicating that coarse sand grains resting on fine sand grains are entrained more readily than fine grains resting on coarse grains. The range of d_p/d_{bd} evaluated was rather small ($0.59 < d_p/d_{bd} < 1.41$) compared with the total size range of sand (0.063 to 2 mm), which can produce values of d_p/d_{bd} from less than 0.05 to greater than 30. Eq. (5.26) applied over this range yields pivoting angles greater than 90 degrees and negative values of $\tan\phi_o$ – an outcome that will inappropriately produce negative critical shear stresses (Eq. (5.31)).

Experimental work on the entrainment of sand directly into suspension shows that as d_p/d_{bd} decreases, higher values of Θ_c are required to entrain a particle of size d_p (Nino et al., 2003). For $0.1 < d_p/d_{bd} \geq 0.5$,

$$\frac{\Theta_{cpb}}{\Theta_{cps}} = 1.1 \left(\frac{d_p}{d_{bd}}\right)^{-0.82} \tag{5.32}$$

where Θ_{cpb} is the dimensionless critical bed shear stress to entrain a particle of size d_p from a bed of particle size d_{bd} and Θ_{cps} is the dimensionless critical shear stress to entrain a particle of size d_p from a smooth bed. This equation, which is similar to entrainment equations for heterogeneous gravel (Table 5.2), yields values of $7.3 > \Theta_{cpb}/\Theta_{cps} > 2$ for $0.1 < d_p/d_{bd} \geq 0.5$. For $0.5 < d_p/d_{bd} < 1$, $\Theta_{cpb}/\Theta_{cps} \approx 2$. Overall, the results show that a particle of size d_p on a smooth bed is more easily entrained into suspension than the same-size particle on a bed comprised of larger particles. The finest particles (0.112 mm) are not entrained into suspension at all but move only within the interstices of larger grains in the substrate.

A flume-based assessment of sand entrainment found that the exponent $b = -1.028$ in Eq. (5.27), which nearly conforms to equal mobility with perhaps a slight tendency for large grains to be entrained more readily than small grains (Kuhnle, 1993a). Experimental work on bed material of mixed sand and gravel shows that for $d_i/d_r < 1$, b can be less than -1, indicating that entrainment of sands becomes more difficult with decreasing particle size (Shvidchenko et al., 2001a). These findings support the conclusion that entrainment of heterogeneous sediment mixtures differs from that of homogeneous sand and that the net effect of heterogeneity is to impede the mobility of fine particles within the bed material relative to the mobility of coarse particles.

5.6.6 What Metrics Other than Bed Shear Stress Have Been Used to Determine Entrainment?

Alternatives exist to using the mean bed shear stress to evaluate entrainment of bed material. Other variables, such as critical unit discharge (q_c) (Bathurst, 1987, 2013; Ferguson, 1994) and critical stream power per unit area (ω_c) (Petit et al., 2005; Parker et al., 2011; Ferguson, 2005, 2012), have been related to incipient

motion of bed material. The use of these metrics is not as widespread as that of bed shear stress in either dimensional or dimensionless form, but q_c and ω_c do appear in some bed-material transport equations. Another alternative is the ratio of critical shear velocity (u_{*c}) to fall velocity (v_f), the rate at which a sediment particle falls through a body of water (Simoes, 2014). Because the shear velocity is the square root of the shear stress, this index of entrainment tends to have less scatter in a relation with R_b than does the dimensionless shear stress. Nevertheless, using the fall velocity to characterize resistance to movement may be meaningful for small particles prone to suspension, but it does not discriminate well amongst large particles, such as gravel, all of which have large fall velocities.

5.6.7 What Is the Concept of Impulse and How Is It Related to Entrainment?

The concept of impulse also has been related to sediment entrainment. Impulse (Im) refers to the application of force (F) over time:

$$Im = \int F \, dt \tag{5.33}$$

Experiments designed to examine the mobilization of a spherical particle resting on other spherical particles through the application of lift or drag forces show that particle mobilization depends not only the magnitude of an applied force but also on the duration over which the force is applied (Diplas et al., 2008). Large impulses can be achieved by applying either a high-magnitude force over a short duration or a low-magnitude force over a long duration (Figure 5.11). The integration of the magnitude and the duration of applied forces, the impulse, determines particle movement.

The concept of impulse calls attention to the importance of time in the problem of particle entrainment. The timescales over which forces are applied become an important aspect of this problem. Analysis of displacement of tracer particles in streams by individual flows shows that the path length of displacement is linearly related to impulse and that a power-function relation exists between the variance of displacement length and impulse (Phillips and Jerolmack, 2014). In this analysis, the timeframe of the impulse calculation (Eq. (5.33)) is the duration over which the shear velocity of the flow exceeds the critical shear velocity.

5.6.7.1 How Is the Concept of Impulse Related to Turbulence and the Role of Turbulence in Particle Entrainment?
The main value of the impulse conception of entrainment is that it has helped to motivate analysis of bed-material entrainment that fully accommodates the turbulent nature of river

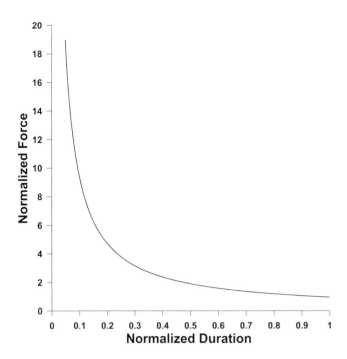

Figure 5.11. Curve of equal impulse in relation to normalized force and normalized duration. A large force acting over a short duration and a small force acting over a long duration have the same impulse. Normalized force = actual force/minimum force and normalized duration = actual duration/maximum duration.

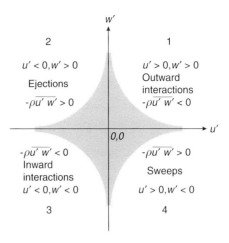

Figure 5.12. Quadrants for analysis of turbulent events defined by paired velocity fluctuations in the streamwise and vertical directions. Shaded area shows the hole size.

flows (Valyrakis et al., 2010, 2011, 2013; Celik et al., 2013). Interest in the role of turbulence in sediment movement has led to the call for a fluvial fluid mechanics that relates turbulence concepts in fluid mechanics to the study of interaction between flow and sediment entrainment and transport (Keylock, 2015). Thus far, the level of sophistication required to accurately measure time-varying characteristics of flow near the bed, especially when particles on the bed are actively moving, has restricted most studies to laboratory settings or small-scale field investigations. A complementary approach is to employ computational fluid dynamics (CFD) models to simulate turbulence characteristics of the flow near the bed, including the development and evolution of coherent turbulent structures (Keylock et al., 2012). Such simulations can provide a framework for exploring the role of turbulence in grain entrainment.

Quadrant analysis is an important tool for characterizing turbulence in rivers and its relation to sediment entrainment (Lu and Willmarth, 1973), particularly with regard to the magnitude and duration of turbulent forces acting on particles. This method provides the basis for relating turbulent fluid motion associated with coherent turbulent structures to the Reynolds shear stresses acting in the vertical-streamwise plane above the bed (τ_{txz}). The analysis uses paired time series for streamwise (u') and vertical (w') velocity fluctuations and

examines corresponding values of these fluctuations at each time in the series. Paired values are classified into four quadrants based on the signs of u' and w': 1) $u' > 0$, $w' > 0$, 2) $u' < 0$, $w' > 0$, 3) $u' < 0$, $w' < 0$, and 4) $u' > 0, w' < 0$. Typically, only sequences of products of u' and w' that exceed a certain absolute magnitude or hole size (H) are included in the classification:

$$H \geq \frac{|u'w'|}{\sigma_{u'}\sigma_{w'}} \tag{5.34}$$

where $\sigma_{u'}$ and $\sigma_{w'}$ are the root mean square values of the two fluctuating velocity components:

$$\sigma_{u'} = \sqrt{\frac{1}{n}\sum_{i=1}^{n}(\hat{u}_i - u)} \, , \quad \sigma_{w'} = \sqrt{\frac{1}{n}\sum_{i=1}^{n}(\hat{w}_i - w)} \tag{5.35}$$

The magnitude of H is arbitrary, but it commonly is set to a value between 1 and 5 (Robert, 2003, p. 42).

Based on the hole size criterion, turbulent events, defined by series of paired values of u' and w' that occur in the same quadrant and that equal or exceed H, can be identified. The mean of the product of velocity fluctuations over the events contributes to the Reynolds stress in the x-z plane (Eq. (4.53)). Events in quadrant two, known as ejections, and quadrant four, known as sweeps, generate positive values of τ_{txz}, whereas events in quadrants one (outward interactions) and three (inward interactions) produce negative values of τ_{txz} (Figure 5.12). Because values of τ_{txz} near the bed determine the turbulent shear stress acting on grains (see Chapter 4), ejections and sweeps contribute to large near-bed shear stresses. Moreover, because shear stresses equal forces per unit area, integrating values of near-bed $-\rho u'w'$ over each event provides information on the impulse per unit area.

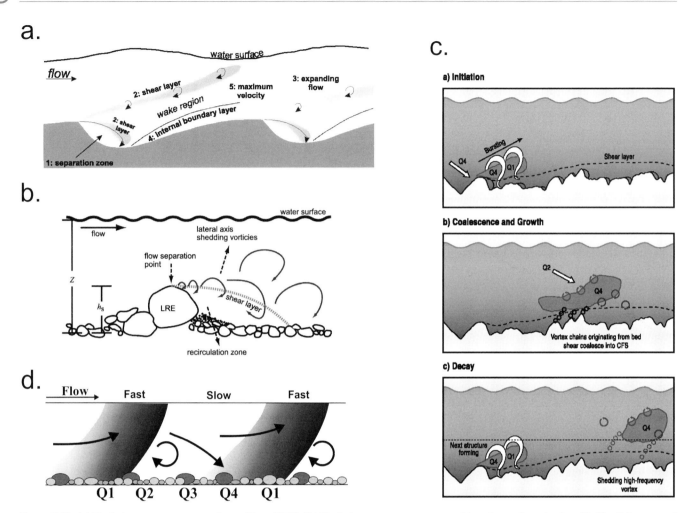

Figure 5.13. (a) Turbulent structures over dunes (Best, 2005). (b) Turbulent structures generated by a large obstacle clast (Buffin-Belanger and Roy, 1998). (c) The evolution of coherent turbulent structures over an irregular gravel bed (Hardy et al., 2016). (d) Large wedges of fast- and slow-moving flow in gravel-bed rivers (Numbered Qs refer to quadrants) (Buffin-Belanger et al., 2000).

5.6.7.2 What Are Coherent Turbulent Structures and How Are They Related to Sediment Entrainment?

Coherent turbulent structures generate events defined by the quadrant method. These structures are discrete, organized patterns of fluid motion that recur over time, and persist over time and cohere over space at scales characteristic of turbulence (Figure 1.6) (Venditti et al., 2013). Recognition of the importance of coherent structures emerged from research in fluid mechanics exploring turbulent flow over smooth boundaries with a well-developed viscous sublayer. A process known as bursting originates within the sublayer, whereby low-speed streaks, which alternate with high-speed streaks, oscillate and rise up from the bed to form horseshoe-shaped vortices that subsequently disintegrate or burst, ejecting high-speed fluid into the overlying turbulent flow (Nezu and Nakagawa, 1993, pp. 166–170). In rivers, the process of bursting, which has both ejection and sweep components, has been invoked as a possible mechanism for the initiation of

bedforms, particularly ripples, on hydraulically smooth sand beds (Best, 1992).

The beds of most natural rivers are hydraulically rough, and the interaction of turbulent flow with roughness elements on the bed is a primary mechanism generating coherent turbulent structures over gravel beds (Roy et al., 2004; Hardy et al., 2009, 2016; Franca and Lemmin, 2015; Mohajeri et al., 2016) and over bedforms (Best, 2005; Keylock et al., 2013, 2014a) (Figure 5.13). A suite of coherent turbulent structures develop in conjunction with flow over dunes (Figure 5.13). Shear layers that extend upward from the crests of dunes often lead to the development of visible boils on the surface of rivers. In gravel-bed rivers, vortices, or coherent rotating eddies, may be shed from individual roughness elements, such as large boulders or clusters of large particles (Buffin-Belanger and Roy, 1998; Lacey and Roy, 2008). Arrays of clasts of different sizes produce a complex amalgamation of vortices that collectively have a major influence on macroturbulence within gravel-bed rivers (Roy et al.,

1996; Shvidchenko and Pender, 2001b; Hardy et al., 2009, 2016). The structure of this macroturbulence is characterized by wedges of fast- and slow-moving flow that extend over the entire flow depth (Figure 5.13). These wedges have lengths 3 to 5 times the flow depth, have widths 0.5 to 1 times the flow depth, and are separated by interfaces inclined in the downstream direction at an average angle of 36° (Ferguson et al., 1996a; Buffin-Belanger et al., 2000; Roy et al., 2004).

The initiation, growth, passage, and dissipation of coherent turbulent structures generates turbulent events associated with different quadrants of u' and w' fluctuations, thereby influencing the near-bed shear stress. Coherent turbulent structures of different sizes and intensities not only generate different magnitudes of stresses acting on grains on the river bed, but also act over different durations. The development of these structures produces temporal fluctuations in near-bed velocities, pressures, and Reynolds shear stresses that lead to time-varying forces acting on particles on the bed (Schmeeckle et al., 2007; Vollmer and Kleinhans, 2007; Dey et al., 2011; Keylock et al., 2014b).

The extent to which impulses produced by different coherent turbulent structures are effective in mobilizing grains on the beds of rivers is an area of active research (Shih et al., 2017). Although the relation of impulses generated by distinct coherent structures to particle motion remains uncertain, some insight has been gained into the influence of turbulence on particle motion. Somewhat surprisingly, ejections, which generate positive near-bed shear stresses, are not necessarily strongly related to bed-material mobilization. Instead, the initiation of motion of bed material is correlated with sweeps and outward interactions (Nelson et al., 1995; Papanicolaou et al., 2001). The substantial role of outward interactions as opposed to ejections in particle motion suggests that the drag force per unit area, which is related to the absolute value of the near-bed streamwise turbulent normal stress ($|-\rho \overline{u'^2}|$), may be more important than the near-bed turbulent shear stress ($-\rho \overline{u'w'}$) in initiating particle motion. Strong sweeps and outward interactions have large streamwise turbulent normal stresses. The absolute magnitudes of near-bed normal stresses are often much greater than those of near-bed shear stresses (Paiement-Paradis et al., 2011). The movement of bed material has also been associated with abrupt accelerations and decelerations of near-bed flow, which may influence particle motion through effects on near-bed pressure gradients (Paiement-Paradis et al., 2011).

5.7 How Is the Shear Stress Acting on Grains Determined?

Another important issue related to bed-material entrainment is that the shear stress acting on grains on the bed often differs from the total bed shear stress (see Chapter 4). This issue focuses on the partitioning of flow resistance, or boundary shear stress, into components associated with different types of resistance. Although multiple scales of resistance exist in river systems (Figure 1.7), attention usually is focused on form versus grain resistance (Petit et al., 2015), where form resistance is predominantly related to bedforms or other large recurring resistance elements, such as gravel clusters. In many field studies, total boundary shear stress is computed from the product of hydraulic radius and slope (Eq. (4.20)). If substantial form resistance occurs in a stream, τ_b will overestimate the shear stress acting on the grains (τ_b') because a large proportion of the total stress will be associated with form resistance (τ_b''). In such cases, estimates of grain entrainment based on comparison of τ_b with τ_{bc} can be misleading. Mobilization requires $\tau_b' > \tau_{bc}$.

One way to estimate τ_b' is to compute it using a local determination of u_* derived from velocity profile measurements near the bed (see Chapter 4). Although in principle this method can yield an estimate of τ_b' at the exact location of a particle or set of particles of interest, in practice obtaining measurements of velocities precisely at the time of incipient motion or when particles on the bed are moving is a difficult task. More commonly, a grain resistance formula is used to estimate the grain shear stress. A variant of the law of the wall equation for turbulent flows over a rough boundary with a roughness length scale of k_s is (Garcia, 2008)

$$\frac{U}{u_*} = \frac{1}{\kappa} \ln\left(11\frac{D}{k_s}\right) \tag{5.36}$$

This equation is known as the Keulegan equation. A power function approximation of Eq. (5.36) used in some studies of grain resistance (e.g., Ferguson, 2012) is

$$\frac{U}{u_*} = 8.0\left(\frac{D}{k_s}\right)^{1/6} \tag{5.37}$$

The roughness length scale can be related to grain size as

$$k_s = a_c d \tag{5.38}$$

where a_c depends on the grain-size percentile, d_i, used in the equation (Garcia, 2008, table 2-1). Common values are $a_c = 1$ for d_{50} and $a_c = 3.5$ for d_{84}. Applying $c_f = u_*^2/U^2$ (Eq. (4.42)). to Eq. (5.36) with d_{50} substituted for k_s produces a relation for estimating the resistance associated with grain friction (c_f'):

$$c_f' = \left[\frac{1}{\kappa}\ln\left(11\frac{D'}{d_{50}}\right)\right]^{-2} \approx \left[8.1\left(\frac{D'}{d_{50}}\right)^{1/6}\right]^{-2} \tag{5.39}$$

Drawing upon Eq. (4.18) and (4.31), the bed shear stress associated with grain friction is

$$\tau_b' = \rho g D' S = \rho c_f' U^2 \tag{5.40}$$

Given information on D, d_{50}, and U, Eq. (5.39) and (5.40) can be used to estimate the component of the total bed shear stress associated with grain friction (τ_b'). In some studies, no adjustment is made for the fact that the flow depth D will be affected by form resistance components; in other words, it is assumed that $D' = D$, where D is the measured flow depth. A complete analysis should account for this effect. Substituting Eq. (5.39) into Eq. (5.40) and solving for D' results in

$$D' = \frac{U^2}{gS} \left[\frac{1}{\kappa} \ln \left(11 \frac{D'}{d_{50}} \right) \right]^{-2} \tag{5.41}$$

which, given additional information on S, can be solved iteratively for D'. This value can then be used with S in Eq. (5.40) to compute τ_b'. Commonly, the component of the total stress associated with form drag (τ_b'') is determined by subtracting τ_b' from τ_b, but methods do exist for determining resistance associated with form drag directly, especially for sand-bed streams with well-developed bedforms (Garcia, 2008). The critical dimensionless grain shear stress Θ_c' for rough turbulent flows tends to be less than the values of 0.030 to 0.60 reported for laboratory and field studies that do not fully account for the effects of form drag (Figure 5.7), ranging from 0.015 to 0.035 with an average of 0.019 (Petit et al., 2015).

5.8 How Is Bed-Material Load Transported in Suspension?

In rivers, particles finer than sand often are transported in suspension. For particles of sand size and coarser ($d > 0.063$ mm), the threshold for entrainment becomes an important factor determining particle mobility. For the most part these particles are transported as bedload, including saltating motion, once flow exceeds the threshold for entrainment (Figure 5.8). The threshold for suspension is approximated by $v_f = u_*$ (Wilcock and McArdell, 2003; Garcia, 2008) (Figure 5.8). For fine sands, the bed shear stress required to directly suspend this material exceeds the bed shear stress for mobilization as bedload by a factor of 2–3; however, as particle size increases, the difference between the bed shear stresses of the mobilization and suspension thresholds becomes an order of magnitude (Figure 5.8). Gravel-size particles typically are suspended only in flows with exceptionally large bed shear stresses.

The key to the transport of relatively coarse particles, such as sand or even fine gravel, in suspension is for the upward-

directed forces acting on the particles to exceed the downward-acting forces; otherwise, the particles will fall out of the flow and be deposited on the bed of the river. Assuming that the flow is one-dimensional in the streamwise direction and that the vertical velocity component of the mean flow is zero, the upward-directed force is related to turbulence. The downward-directed force is related primarily to the submerged weight of the particle. The interaction between turbulence and particle weight determines whether a particle of a particular size can be suspended as well as the concentration of suspended particles of a certain size above the bed of the river.

5.8.1 What Is the Balance of Forces of a Falling Particle?

In a column of still water, the balance of forces acting on a spherical particle falling toward the bed at constant velocity, known as the fall velocity (v_f), is

$$F_D = F_g \tag{5.42}$$

where F_g and F_D are defined by Eq. (5.5) and (5.7), respectively. Based on consideration of a spherical particle subject to pressure and viscous forces as it moves downward through the water column, English physicist G.G. Stokes in 1851 derived an expression for the drag force acting on the particle (Dingman, 2009, p. 475):

$$F_D = 6\pi \mu r v_f = 3\pi \mu d v_f \tag{5.43}$$

Equating Eq. (5.5) and (5.43) and solving for v_f yields Stokes' Law, which defines the fall velocity of fine particles through a column of still water:

$$3\pi \mu d v_f = (\rho_s - \rho) g \frac{\pi}{6} d^3 \tag{5.44}$$

$$v_f = \frac{(\rho_s - \rho) g d^2}{18 \mu} \tag{5.45}$$

Stokes' Law assumes that falling particles move slowly enough through the fluid that the boundary layers surrounding the moving particles are laminar. This condition generally applies to particles with $d \leq 0.1$ mm. For large rapidly falling clasts, boundary layers surrounding the particles become more complicated and include the development of laminar and turbulent wakes. The basic relation for a coarse spherical particle can be obtained by equating Eq. (5.5) and (5.7) with $u_c = v_f$ and $A_s = \pi r^2 = \pi d^2 / 4$ and solving for v_f:

$$v_f = \sqrt{\frac{4[(\rho_s - \rho)/\rho] g d}{3 C_D}} \tag{5.46}$$

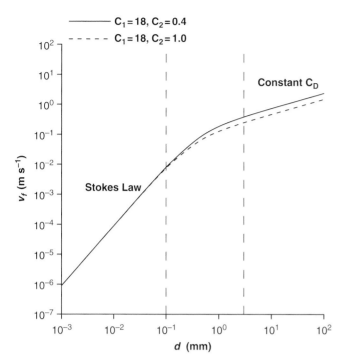

Figure 5.14. Fall velocity versus grain size as predicted by Eq. (5.47) for $\rho_s = 2650$ N m^{-3}, $\rho = 1000$ N m^{-3}, $v = 1 \times 10^{-6}$ m^2 s^{-1}, and C_1 and C_2 as shown in the legend. Lower ends of curves converge on Stokes' Law and upper ends converge on constant C_d in Eq. (5.46).

A variety of relations have been proposed for estimating v_f for particles greater than 0.1 mm in size (Rubey, 1933a; Watson, 1969; Dietrich, 1982; Cheng, 1997). The equation

$$v_f = \frac{[(\rho_s - \rho)/\rho]gd^2}{C_1 v + \{0.75 C_2 [(\rho_s - \rho)/\rho]gd^3\}^{0.5}} \tag{5.47}$$

applies over the entire range of viscous to turbulent conditions (Ferguson and Church, 2004). For smooth spheres, $C_1 = 18$ and $C_2 = 0.4$, which for small particle sizes ($d < 0.1$ mm) converges on Stokes' Law and for large particle sizes ($d > 3$ mm) converges on Eq. (5.46) with a constant drag coefficient of $C_2 = C_D = 0.4$. Values of $C_1 = 18$ and $C_2 = 1$ are recommended for irregularly shaped natural sands when d is determined by sieve diameter (Figure 5.14).

5.8.2 What Factors Influence Equilibrium Profiles of Suspended-Sediment Concentration?

The vertical turbulent transport of sediment through the water column is $\overline{w'C'_{svz}}$, where C'_{svz} is the local instantaneous fluctuation in volumetric sediment concentration at any level z in the water column and the overbar indicates averaging over time. The net vertical flux of suspended sediment (f_{ssz}) is

$$f_{ssz} = -v_f C_{svz} + \overline{w'C'_{svz}} \tag{5.48}$$

where C_{svz} is the local mean volumetric sediment concentration. Because the turbulence characteristics of the flow are not typically known, and because the action of turbulence is a diffusive process, the turbulent flux of sediment is commonly assumed to be proportional to the gradient in sediment concentration:

$$\overline{w'C'_{svz}} = -\varepsilon_{sv}\frac{dC_{svz}}{dz} \tag{5.49}$$

where ε_{sv} is the vertical sediment diffusivity coefficient. Substituting Eq. (5.49) into Eq. (5.48) and assuming equilibrium sediment concentrations within the water column in which the upward and downward fluxes of sediment are balanced yields

$$0 = -v_f C_{svz} - \varepsilon_{sv}\frac{dC_{svz}}{dz} \tag{5.50}$$

$$v_f C_{svz} = -\varepsilon_{sv}\frac{dC_{svz}}{dz} \tag{5.51}$$

$$\frac{dC_{svz}}{C_{svz}} = -\frac{v_f}{\varepsilon_{sv}}dz \tag{5.52}$$

The vertical turbulent diffusivity of sediment is often viewed as analogous to the vertical turbulent diffusivity of momentum (ε, or eddy viscosity) in the law of the wall derivation of the velocity profile relation (see Chapter 4). Setting $\varepsilon_{sv} = \varepsilon/\rho$, the kinematic eddy viscosity, and substituting $u_*/(\kappa z)$ for du/dz (see Eq. (4.1.4)) into Eq. (4.2.4) yields

$$\varepsilon_{sv} = \kappa u_* z\left(1 - \frac{z}{h}\right) \tag{5.53}$$

Combining this relation with Eq. (5.52), integrating, and evaluating the constant of integration at a reference level z_{rf} results in an equation for the variation in the volumetric sediment concentration over depth (Dingman, 2009, p. 494):

$$\frac{C_{svz}}{C_{svrf}} = \left[\left(\frac{h-z}{z}\right)\left(\frac{z_{rf}}{h-z_{rf}}\right)\right]^{R_o} = \left[\frac{(h-z)/z}{(h-z_{rf})/z_{rf}}\right]^{R_o}$$

$$= \left[\frac{(1-z/h)}{(1-z_{rf}/h)}\frac{z_{rf}}{z}\right]^{R_o} \tag{5.54}$$

where

$$R_o = \frac{v_f}{\kappa u_*} \tag{5.55}$$

and C_{svrf} is the volumetric concentration at z_{rf}. Eq. (5.54) is commonly referred to as the Rouse equation, and R_o is known as the Rouse number. Large values of R_o indicate pronounced gradients in concentration over depth, whereas small values

(a)

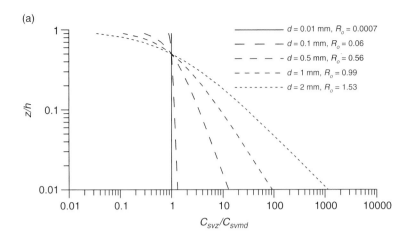

(b)

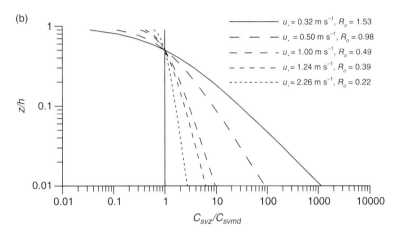

Figure 5.15. Changes in suspended-sediment concentration over depth predicted by Eq. (5.56). (a) Variation in concentration over depth with changes in particle size (d) for constant shear velocity ($u_* = 0.32$ m s^{-1}). (b) Variation in concentration over depth with changes in shear velocity (u_*) for constant $d = 2$ mm.

denote relatively uniform concentration gradients. If the reference level is equated to mid-depth ($h_{md} = h/2$), Eq. (5.54) reduces to

$$\frac{C_{svz}}{C_{svmd}} = \left[\left(\frac{h}{z} - 1 \right) \right]^{R_o} \tag{5.56}$$

Expressing the Rouse equation in this form is useful for illustrating how concentration over depth changes in relation to variation in the Rouse number.

Eq. (5.55) and (5.56) show that the sediment concentration at any depth relative to the concentration of sediment at mid-depth is directly related to the fall velocity of the sediment and inversely related to the shear velocity of the flow (Figure 5.15). For a constant shear velocity, concentrations will vary substantially over depth for large particles with large fall velocities; concentrations will be highest near the bed and diminish rapidly toward the surface. As particle size and fall velocity decrease, the gradient in concentration over depth will also decrease, becoming nearly uniform for the finest particles. Conversely, for a constant particle size, increases in shear velocity, which represent increases in turbulence, promote increasingly uniform concentrations of particles of that size over depth. Increases in the strength of turbulence allow particles of a particular size to be suspended at increasingly high concentrations at relatively large distances from the bed.

5.8.3 How Can the Effect of Complicating Factors on Sediment Suspension Be Accounted For?

Considerable effort has been devoted to evaluating the validity of the Rouse formulation of suspended-sediment concentration in open-channel flows (Garcia, 2008; Dingman, 2009, pp. 496–502). Potential complicating factors include the method used to determine the reference concentration in Eq. (5.54) (Pizzuto, 1984a), density stratification effects in sediment-laden flow with strong vertical gradients in sediment concentration (Wright and Parker, 2004; Yeh and Parker, 2013), and the influence of sediment on turbulence characteristics of the flow (Guo and Julien, 2001; Cao et al., 2003; Muste et al., 2005; Bennett et al., 2014). All these factors may lead to deviations of the basic Rouse model from observed gradients of sediment concentration. In general, the basic model tends to overpredict gradients of the concentration profiles, requiring the need for correction (Pizzuto, 1984a; Dingman, 2009, pp. 496–497).

The expression for ε_{sv} can be generalized as

$$\varepsilon_{sv} = \beta_c \kappa u_* z \left(1 - \frac{z}{h}\right) \tag{5.57}$$

where β_c is a diffusivity correction coefficient. Substituting this expression into Eq. (5.52) and integrating generates a modified Rouse equation:

$$\frac{C_{svz}}{C_{svrf}} = \left[\frac{(1 - z/h)}{(1 - z_{rf}/h)} \frac{z_{rf}}{z}\right]^{R_o{}'} \tag{5.58}$$

where $R_o{}' = R_o/\beta_c$. Note also that based on Eq. (5.57) and Eq. (4.2.4), $\beta_c = \varepsilon_{sv}/\varepsilon$ – the ratio of sediment diffusivity to eddy viscosity. Thus, increases in β_c imply that sediment diffusivity increases in relation to turbulence diffusivity. An increase in the value of β_c will decrease $R_o{}'$, which will, in turn, be associated with a decrease in the gradient of suspended-sediment concentration over depth. Several studies have developed empirical relations for β_c to refine estimates of $R_o{}'$ (van Rijn, 1984; Graf and Cellino, 2002; Jha and Bombardelli, 2009; Cheng et al., 2013; Pal and Ghoshal, 2016).

5.8.4 How Are Suspended-Sediment Loads Estimated from Sediment-Concentration Profiles?

Application of the Rouse equations (Eq. (5.54) or (5.58)) to determine sediment concentrations over depth requires estimates of C_{svrf} and z_{rf}. Several formulas have been developed to estimate equilibrium reference concentrations at reference heights (Garcia, 2008, table 2–6). The vast majority of these relations have been developed for uniform sediments. A useful equation for estimating C_{svrf} for nonuniform sediments in rivers at a reference height $z_{rf} = 0.05h$ is (Garcia and Parker, 1991)

$$C_{svrfi} = \frac{0.00000013(\lambda Z_{ci})^5}{1 + 0.000000043(\lambda Z_{ci})^5} \tag{5.59}$$

where

$$Z_{ci} = \frac{u_*{}'}{v_{fi}} R_{pi}^{0.6} \left(\frac{d_i}{d_{50}}\right)^{0.2} \tag{5.60}$$

$$\lambda = 1 - 0.288\sigma_\phi \tag{5.61}$$

$R_p = \left(\{[(\rho_s - \rho)/\rho]gd_i\}^{0.5} d_i\right)/v$, is the particle Reynolds number, $u_*{}'$ is the grain shear velocity, and i refers to specific grain-size fractions. Below the reference level, transport is assumed to occur as bedload. Eq. (5.59) applies to sediment in the size range 0.063 mm $< d <$ 0.5 mm.

Total suspended-sediment transport over local depth (h) can be determined from concentration profiles as (van Rijn, 1984)

$$q_{ss} = \int_{z_{rf}}^{h} C_{sz} u_z dz \tag{5.62}$$

where q_{ss} is the dry-mass suspended-sediment load per unit width (M L^{-1} T^{-1}), C_{sz} is the dry-mass concentration of suspended sediment (M L^{-3}), and transport below the reference level (z_{rf}) is treated as bedload (e.g., $z_{rf} = 0.05 h$). For suspended sediment–concentration profiles constructed from data collected in the field, the reference level represents the position closest to the bed at which suspended sediment was sampled. The inclusion of u_z, the streamwise velocity at different depths, requires measured data on variation of velocity over depth or a function to estimate this variation, such as the law of the wall (van Rijn, 1984). The method also assumes that suspended particles move at the same velocity as the flowing water and, in cases where the velocity profile is estimated, that sediment concentrations are not large enough to affect the shape of the velocity profile. When volumetric concentrations are estimated using a Rouse-based equation (Eq. (5.55) or (5.58)) and a reference-level function (Eq. (5.59)), these concentrations can be converted to dry-mass suspended-sediment concentrations by multiplying the volumetric concentration by the density of the sediment.

5.8.5 How Does the Rouse Number Provide Insight into Modes of Bed-Material Transport?

The Rouse number has been used to examine theoretically modes of bed-material transport in relation to the ratio of fall velocity to shear velocity (v_f/u_*) for sediment-transporting flows (Dade and Friend, 1998; Church, 2006). Logistic curves define relationships between the percentage of suspended load and v_f/u_* for different ratios of flow depth to bedload layer thickness. Three regimes of transport can be discriminated based on inflections in these curves: $v_f/u_* \leq 0.3$ – predominantly suspended load (suspended load >80–90% of total load), $0.3 > v_f/u_* > 3$ – predominantly mixed load, and $v_f/u_* \geq 3$ – predominantly bedload (suspended load <10–20% of total load). Mixed-load transport refers to conditions where fluxes of suspended bed material and bedload are approximately equal. Empirical analysis of sediment transport data for natural rivers also indicates that distinct domains of transport can be discerned on the basis of Θ and R_b, with the domain of predominantly suspended bed-material load corresponding to $\Theta \geq 1 - 3$ and $R_b < 100$ (Dade and Friend, 1998; Church, 2006).

5.9 How Is Bedload or Bed-Material Transport Related to Shear Stress and Stream Power?

5.9.1 How Is the Rate of Bedload or Bed-Material Transport Defined?

The traditional approach to determine the rate of bedload transport is to treat it from an Eulerian perspective. Given the large

Table 5.3. Definitions of volumetric sediment transport rate (q_{sbv}).

$q_{sbv} = C_{sbv} u_{sb} \delta_b$	C_{sbv} = volumetric bedload sediment concentration (volume of sediment/volume of sediment and water), u_{sb} = mean bedload grain velocity, δ_b = thickness of the bed involved in transport
$q_{sbv} = N_b \overline{V_p} u_{sb}$	N_b = number of particles moving in the bed per unit area, $\overline{V}_p$ = mean volume of particles in the moving bed
$q_{sbv} = E_{sv} \overline{L}_{pd}$	E_{sv} = volumetric rate of grain entrainment per unit area of the bed, $\overline{L}_{pd}$ = average total distance of grain transport

number of individual particles involved in transport by rolling, sliding, or saltating, the Eulerian approach focuses on the movement of particles over time past specific locations within the river system. Typically, these locations are river cross sections – planes perpendicular to the direction of the river flow.

The volumetric bedload transport rate is equal to the volumetric flux of sediment per unit width of the bed (q_{sbv}) ($L^2 T^{-1}$). This volumetric flux can be determined in various ways (Garcia, 2008; Haschenburger, 2013a) (Table 5.3). In geomorphological investigations, sediment transport rate often is expressed in dimensional form as the dry weight of sediment transport per unit width per unit time (q_{sbdw}) (kg m^{-1} s^{-1}):

$$q_{sbdw} = \rho_s q_{sbv} \tag{5.63}$$

or as the submerged weight per unit width per unit time (q_{sbsw}):

$$q_{sbsw} = (\rho_s - \rho) q_{sbv} \tag{5.64}$$

A common dimensionless formulation of sediment transport rate, also known as bedload transport intensity (Einstein, 1950), is

$$q_{sbv}^* = \frac{q_{sbv}}{d_r \sqrt{g \frac{(\rho_s - \rho)}{\rho} d_r}} \tag{5.65}$$

where d_r is a representative grain size.

5.9.2 What Are Common Functional Forms of Bedload and Bed-Material Transport Equations?

Eulerian approaches typically do not focus on the components of q_{sbv} but seek to relate bulk transport rates to controlling factors. Common functional forms of Eulerian bedload and bed-material transport equations are:

$$q_{sbv} \propto f(\Theta - \Theta_c)^b \tag{5.66}$$

$$q_{sbv} \propto f(\omega - \omega_c)^b \tag{5.67}$$

These functional forms show that the rate of bedload transport is a power function of either excess shear stress (Eq. (5.66)) or excess stream power per unit area (Eq. (5.67)). A multitude of equations have been developed for predicting bedload transport rates in rivers (Garcia, 2008; Dey, 2014), many of which derive from the basic functional formulations. A well-known and widely used version of Eq. (5.66) is (Meyer-Peter and Muller, 1948)

$$q_{svb}^* = k_{sb}[\eta_{sw} \eta_{fc}(\Theta - \Theta_c)]^b \tag{5.68}$$

where k_{sb} is an empirical coefficient, η_{sw} is a correction factor for sidewall effects, and η_{fc} is a correction factor for grain friction effects (Wong and Parker, 2006). The exponent (b) has a value of 1.5 to 1.67 (Table 5.4), indicating that bedload transport rate increases nonlinearly, and rapidly, with increasing excess shear stress. Eq. (5.68) works well for predicting the transport of uniform sediment on plane beds (Wong and Parker, 2006), but it also can be adapted, mainly through adjustment of η_{fc}, to provide reasonable predictions of total transport for experimental data on nonuniform sediment with bedforms (Huang, 2010). Many other bedload transport relations have been developed based on the concept of excess shear stress (Garcia, 2008; Dey, 2014).

A well-known empirical version of Eq. (5.67) is (Bagnold, 1980)

$$q_{sbsw} = 0.1\left[\frac{\omega - \omega_c}{0.5}\right]^{1.5}\left(\frac{D}{0.1}\right)^{-0.67}\left(\frac{d_r}{0.0011}\right)^{-0.5} \tag{5.69}$$

where $\omega_c = 5.75[0.04(\rho_s - \rho)d_r]^{1.5}(g/\rho)^{0.5}\log(12D/d_r)$ or, assuming $\rho_s - \rho = 1600$ kg m^{-3}, $g = 9.81$ m^2 s^{-1}, and $\rho = 1000$ kg m^{-3}, $\omega_c = 292d_r^{1.5}\log(12D/d_r)$. For unimodal sediments, $d_r = d_{50}$ to predict total transport rates (Bagnold, 1980). For bimodal sediments, separate values of critical stream power per unit area are estimated for values of d corresponding to each mode (ω_{c1} and ω_{c2}), and the geometric mean $\overline{\omega}_c = (\omega_{c1}\omega_{c2})^{0.5}$ is used in Eq. (5.69). The exponent corresponding to the excess stream power relation is similar to the exponent for the excess shear stress term in Eq. (5.68) (Table 5.4). Eq. (5.69) suggests that, for constant excess stream power, rates of bedload transport decrease as depths increase and decrease as the reference particle size increases.

The concept of stream power has also served as the basis for development of total bed-material load equations in rivers, which can be viewed as mixed-load formulas that predict transport of both suspended bed-material load and bedload. Dimensional analysis yields the following empirical equation for estimating bed-material concentration C_{sbm} in parts per million (ppm) for sand-bed rivers based on excess stream power per unit weight of water (US) (Yang, 1973):

$$\log C_{sbm} = a_1 + a_2 \log\left(\frac{US}{v_f} - \frac{U_c S}{v_f}\right) \qquad (5.70)$$

where

$$a_1 = 5.435 - 0.286\log\left(\frac{v_f d_{50}}{v}\right) - 0.457\left(\frac{u_*}{v_f}\right),$$

$$a_2 = 1.799 - 0.409\log\left(\frac{v_f d_{50}}{v}\right) - 0.314\left(\frac{u_*}{v_f}\right) \qquad (5.71)$$

and

$$\frac{U_c}{v_f} = 2.05 \quad \text{for} \quad \left(\frac{u_* d_{50}}{v}\right) \geq 70, \quad \text{or} \quad \frac{U_c}{v_f} = \left(\frac{2.5}{\log\left(\frac{u*d_{50}}{v}\right) - 0.06}\right)$$

$$+ 0.66 \quad \text{for} \quad 1.2 < \left(\frac{u_* d_{50}}{v}\right) < 70 \qquad (5.72)$$

Values of C_{sbm} can be converted to total sediment load (Q_s) in metric tons (t) per day using

$$Q_s = 0.0864 Q C_{sbm} \qquad (5.73)$$

A stream power equation for predicting bedload transport in gravel-bed rivers with the same functional form as Eq. (5.70), but different values of coefficients in Eq. (5.71) and (5.72), has also been developed (Yang, 1984).

Based on reasoning from fundamental physics, Bagnold (1966) derived the following total bed-material transport rate formula for sand-bed rivers based on stream power per unit area (ω):

$$q_{stsw} = q_{sbsw} + q_{sssw} = \omega\left[\frac{e_b}{\tan\phi_o} + \frac{e_s U_{ss}}{v_{ef}}(1 - e_b)\right] \qquad (5.74)$$

where q_{stsw} is the total bed-material transport rate by submerged weight, q_{sbsw} is the bedload transport rate, q_{sssw} is the suspended-sediment transport rate, e_b is the bedload transport efficiency – the fraction of stream power involved in bedload transport, e_s is the suspended-load transport efficiency – the fraction of stream power involved in suspended-sediment transport, and U_{ss} is the velocity of the suspended sediment. The effective fall velocity (v_{ef}) is the weighted fall velocity of the mixture of suspended sediment:

$$v_{ef} = \sum f_i v_{fi} \qquad (5.75)$$

where f_i is the proportion by weight of grain-size fraction i and v_{fi} is the fall velocity of this size fraction. Assuming that suspended sediment moves at the same mean velocity as the flow (U), an assumption challenged by experimental evaluations of suspended particle motion (Muste et al., 2005), and that $e_s = 0.01$, a value

Table 5.4. Examples of bedload transport formulas based on excess shear stress.

$q_{sbv}^* = 8(\Theta - 0.047)]^{1.5}$ (assumes η_{sw} and η_{fc} equal 1)	Meyer-Peter and Muller, 1948
$q_{sbv}^* = 12(\Theta - 0.047)]^{1.5}$	Wilson, 1966
$q_{sbv}^* = 5.7(\Theta - \Theta_c)]^{1.5}$ (Θ_c varies from 0.05 for $d = 9$ mm to 0.058 for $d = 3.3$ mm)	Fernandez Luque and van Beek, 1976
$q_{sbv}^* = 11.2\frac{(\Theta - 0.03)^{4.5}}{\Theta^3}$	Parker, 1978b
$q_{sbv}^* = 4.93(\Theta - 0.0470)]^{1.6}$ (best fit to reanalysis of Meyer-Peter and Muller relation)	Wong and Parker, 2006
$q_{sbv}^* = 3.97(\Theta - 0.0495)]^{1.5}$ (best fit to constrained reanalysis retaining exponent of 1.5)	
$q_{sbv}^* = 6(\Theta - 0.047)]^{1.67}$ (uniform and nonuniform sediments without bed forms) $q_{sbv}^* = 6(\eta_{fc}\Theta - 0.047)]^{1.67}$ (nonuniform sediments with bed-forms – relations provided for computing η_{fc})	Huang, 2010

determined from theoretical considerations of anisotropic momentum fluxes over the vertical column of the fluid, Eq. (5.74) becomes:

$$q_{stsw} = q_{sbsw} + q_{sssw} = \omega\left[\frac{e_b}{\tan\phi_o} + \frac{0.01U}{v_{ef}}\right] \qquad (5.76)$$

Values of e_b range from 0.11 to 0.15 depending on grain size and flow velocity.

Although Eq. (5.76) has not been used widely to predict total bed-material load, it provides insight into the extent to which the total rate of energy expenditure of the flow (stream power) is transformed into work performed in transporting sediment. The estimated values of e_b and e_s indicate that the fraction of total stream power that contributes to bedload transport greatly exceeds the fraction that contributes to suspended-sediment transport and that the percentage of total energy expenditure involved in sediment transport is on the order of 10–15%. If empirical data are available on ω, q_{sbsw}, and ϕ_o, the bedload efficiency can be computed as (Bagnold, 1973)

$$e_b = \frac{q_{sbsw}\tan\phi_o}{\omega} \qquad (5.77)$$

Comparison of bedload transport efficiencies for perennial, seasonal, and ephemeral streams indicate that these efficiencies vary by several orders of magnitude, with the highest

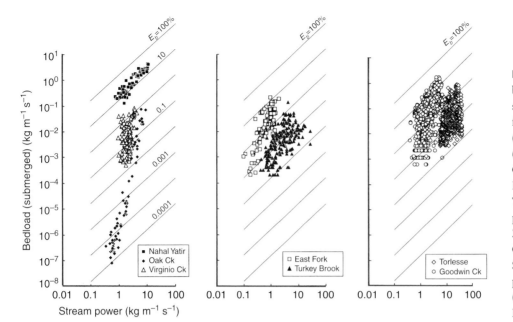

Figure 5.16. Relations among bedload transport rate (q_{sbsw}) and stream power per unit area (ω/ρ) for the ephemeral Nahal Yatir (Israel), the perennial Oak Creek (Oregon), the perennial Virginio Creek (Italy), the perennial East Fork (Wyoming), the perennial Turkey Brook (England), the perennial Torlesse Stream (New Zealand), and the seasonal Goodwin Creek (Mississippi). Slanted lines correspond to percentages of bedload efficiency ($E_b = 100e_b$) (from Reid and Laronne, 1995).

efficiencies occurring in the ephemeral stream, presumably because of the ready availability of poorly sorted material on the bed surface in this type of fluvial system (Reid and Laronne, 1995) (Figure 5.16).

5.9.3 Which Equations Predict Best?

The Eulerian approach has produced a wide array of formulas for predicting bedload and bed-material transport. While scientists and research engineers tend to value the theoretical content of formulas to gain insight into the physical mechanisms that underlie sediment transport, river engineers and managers often seek predictive accuracy. The ideal relation would be one that is grounded firmly on theoretical principles yet also has high predictive accuracy. Given the semi-empirical nature of most transport relations, many predict well for the domain of data used in model development. Several studies have compared the predictive accuracy of transport relations using independent data sets (Alonso, 1980; Gomez and Church, 1989; Nakato, 1990; Yang and Wan, 1991; Yang, 2005; Recking et al., 2012). No one formula seems to consistently outperform another for all cases, and all formulas may generate predictions with large errors in specific cases.

5.10 What Factors Influence the Fractional Transport of Bed Material?

While general bedload or bed-material transport relations predict bulk loads, even for sediment mixtures, river scientists often are interested in the transport of different size fractions. River beds typically are composed of particles of a variety of sizes. Exchange of particles between the active layer, the vertical portion of the bed involved in bed-material transport (Church and Haschenburger, 2017), and moving bedload can lead to changes in the size characteristics of the bed material, not only during a transport event but also when the event has ceased. Moreover, material entrained and mobilized locally may be transported far downstream. Thus, the transport of mixtures is also of importance in determining downstream trends in particle-size characteristics of river beds, a topic examined in detail in Chapter 13. Many benthic organisms in rivers require specific size characteristics of bed material for aquatic habitat. Understanding how fluvial processes affect habitat conditions requires insight into size-specific entrainment, transport, and deposition of bed-material load.

5.10.1 How Do Coarse Surface Layers Influence Fractional Transport of Bed Material in Gravel-Bed Rivers?

By far the majority of attention on size-specific transport has focused on gravel-bed rivers, where diameters of particles comprising the bed can vary over several orders of magnitude. Gravel-bed rivers typically contain bed material with median grain sizes (d_{50}) greater than 15 mm and values of sorting (σ_ϕ) greater than three (Garcia, 2008). The d_{50} of bed material in sand-bed rivers is usually between 0.1 and 1.0 mm, this material is well sorted ($\sigma_\phi = 1.0 - 1.5$), and the material at the surface has similar size characteristics to material below the

Figure 5.17. (a) Active slough across a gravel-armored point-bar surface. (b) Armored upstream end of slough and point bar a few days later (slough no longer active). (c) Distal part of slough and unarmored point bar showing transition to sandy deposits (foreground) that underlie the gravel armor. (d) Close-up of gravel armor.

surface. By contrast, the composition of surficial bed material in gravel-bed rivers often differs from material below the surface. In particular, an abrupt transition commonly occurs between a relatively coarse surface layer, exposed directly to fluid forces and involved most directly in bed-material transport, and finer material below this coarse surface layer (Figure 5.17). The presence of a coarse surface layer can strongly influence mobilization and transport of coarse particles by size fraction.

5.10.1.1 What Types of Coarse Surface Layers Occur in Gravel-Bed Rivers?

Classification of bedding for fluvial gravels includes at least two types, framework gravels and censored gravels, that have spatially continuous coarse surface layers (Figure 5.18). The structure of these layers, which includes interlocked and imbricated clasts, promotes stability of the streambed and increases resistance to transport. Imbrication occurs when elongated, relatively flat clasts overlap with one another and dip downward into the streambed with an upstream-orientated dip (Figure 5.6) (Qin et al., 2012). The grain-size distribution of surface material in matrix gravels is coarser than material in the subsurface, but the surface may be characterized by local clusters of coarse particles rather than a continuous layer of coarse material. The abundance of fine matrix material distinguishes this type of gravel deposit from framework or censored gravels. A large supply of fine sediment may lead to infilling of near-surface voids within framework or censored gravels, resulting in a filled gravel with a surface layer that is finer than the subsurface material (Carling,

1984). This condition, often referred to as embeddedness (Sennatt et al., 2006), is of ecological interest because it can limit the living space for benthic aquatic organisms.

A continuous coarse surface layer has been referred to variously as either armor or pavement. Some debates exist about the use of these two terms, mainly in relation to whether particles in the layer are in motion or not (Carling and Reader, 1982; Gomez, 1984; Jain, 1990). Here the term *armor* is adopted, since it has the broadest usage, but a distinction is made between static armor and mobile armor. Static armor, once formed, is immobile, whereas mobile armor involves the movement of particles of different sizes within the coarse surface layer.

The degree of development of an armor layer is measured by the ratio of the median grain size of the surface layer (d_{50srf}) to the median grain size of the subsurface (d_{50sub}). Typically this ratio ranges between 1 and 3. Values less than 1 can occur in filled gravels. The thickness of the surface layer is about one to two times the diameter of large grains in the layer, usually represented by d_{90} (Bunte and Abt, 2001; Parker, 2008).

5.10.1.2 What Is the Difference between Static and Mobile Armor?

The problem of armor development in gravel-bed streams has been examined within the context of the relation among bed material characteristics, flow competence, transport capacity, and sediment supply (Richards and Clifford, 1991; Gomez, 1995; Powell, 1998). Competence refers to the largest particles a river can transport for a given bed shear stress, whereas capacity is the largest bed-material load a stream can transport

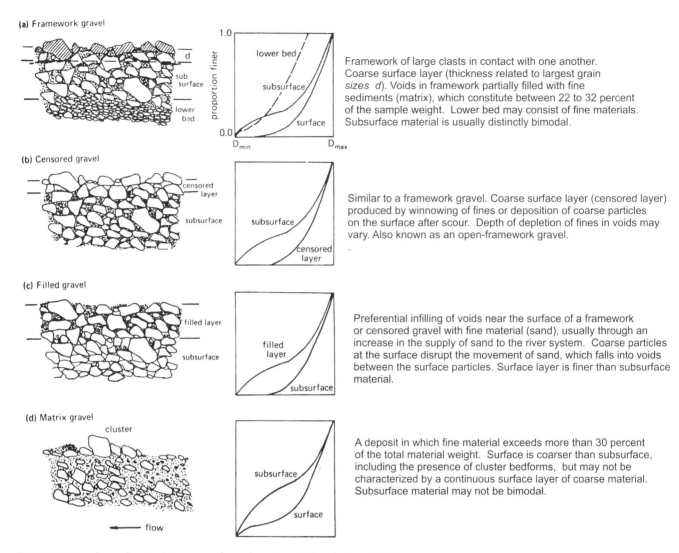

(a) Framework gravel

Framework of large clasts in contact with one another. Coarse surface layer (thickness related to largest grain *sizes d*). Voids in framework partially filled with fine sediments (matrix), which constitute between 22 to 32 percent of the sample weight. Lower bed may consist of fine materials. Subsurface material is usually distinctly bimodal.

(b) Censored gravel

Similar to a framework gravel. Coarse surface layer (censored layer) produced by winnowing of fines or deposition of coarse particles on the surface after scour. Depth of depletion of fines in voids may vary. Also known as an open-framework gravel.

(c) Filled gravel

Preferential infilling of voids near the surface of a framework or censored gravel with fine material (sand), usually through an increase in the supply of sand to the river system. Coarse particles at the surface disrupt the movement of sand, which falls into voids between the surface particles. Surface layer is finer than subsurface material.

(d) Matrix gravel

A deposit in which fine material exceeds more than 30 percent of the total material weight. Surface is coarser than subsurface, including the presence of cluster bedforms, but may not be characterized by a continuous surface layer of coarse material. Subsurface material may not be bimodal.

Figure 5.18. Typology of vertical structure of gravels in rivers (Church et al., 1987).

for a given bed shear stress. Competence relates to the size of material transported, whereas capacity relates to the amount of material transported. In general, armor layers develop in gravel-bed streams where sediment supply is limited in relation to flow competence and transport capacity. For both static and mobile armor layers, the effect of armor development is to regulate the rate of bed-material transport, particularly the rate of transport of different sizes of bed material.

Static armor develops when a bed composed of mixed sediment is sorted by the flow until the effect of sorting results in an immobile coarse surface layer. The most extreme example of this type of condition is a clear-water flow devoid of sediment flowing over a substrate of mixed sediment sizes. Such a situation may occur below a dam, which prevents downstream movement of bed-material load (see Chapter 15). Assuming the clear-water flow downstream of the dam is not competent to transport all sizes in the sediment mixture, fine bed material will be preferentially mobilized and transported downstream, resulting in degradation, or lowering, of the channel bed and progressive coarsening of the bed material. Eventually, only the coarsest immobile particles remain on the surface, the bed material discharge decreases to zero, and degradation of the bed ceases (Jain, 1990; Parker and Sutherland, 1990). Although a complete lack of coarse sediment supply is uncommon in natural rivers, the development of static armor tends to occurs in settings with limited sediment supply and restricted competence of the flow in relation to the full size range of the bed material (Gomez, 1983; Sutherland, 1987; Dietrich et al., 1989).

Both field and experimental studies indicate that mobile armor develops under conditions where sediment of bed-material size is supplied to the channel and the flow is competent to transport all sizes of material in the bed (Parker et al., 1982b; Andrews and Erman, 1986). In flume experiments with a supply of sediment equivalent in size characteristics to the initial unarmored bed-material sediment, preferential mobilization of fines in the unarmored bed leads to bed degradation, a decrease in bed gradient, and coarsening of surficial sediment (Parker et al., 1982b). Mobilization of fines is restricted by coarsening, which impedes the movement of fine grains and prevents exposure of fine grains beneath the armor layer to the flow. On the other hand, the concentration and exposure of coarse particles on the bed surface enhances the mobility of these particles. As the mobile coarse surface layer develops, bed-material discharge decreases relative to the incoming sediment supply, which promotes aggradation and an increase in bed gradient to an equilibrium configuration. At equilibrium, the bedload grain-size distribution (BSD) is equivalent to the grain-size distribution of the bed material (BMSD) (Jain, 1990). In other words, the fractions by volume or weight of particles of different sizes in the bedload (p_i) equal the fractions by volume or weight of corresponding sizes in the bed material (f_i). To attain this equivalency, particles of different sizes in the bed must be both entrained and transported equally. Thus, this condition corresponds to equal mobility (Parker and Klingeman, 1982; Parker and Toro-Escobar, 2002).

The relation between static and mobile armor layers can be characterized conceptually by considering how an initial unarmored bed evolves over time for different initial bed shear stresses (Figure 5.19). At low bed shear stresses, no particles are mobile and the bed does not evolve at all over time. In this case, no coarse surface layer develops. Static-armor layers form when the bed shear stress exceeds the threshold for mobilization of fine material in the initial unarmored bed, but the increased exposure of coarse grains on the bed surface once it has evolved into an armored equilibrium state is not sufficient to enable mobilization of these grains. Over time, the bedload transport rate declines to zero as all fine mobile particles are winnowed from the bed and only coarse immobile particles remain. The upper limit of static-armor development is defined by the lower limit of mobile armor development (Gomez, 1995). As the bed shear stress approaches this upper limit, the coarseness of static armor increases as winnowing becomes more pronounced (Figure 5.19).

Above the upper limit, the bed shear stress is great enough to mobilize all particles on the bed surface for the static-armor equilibrium state ($q_{sbv} = 0$) corresponding to the upper limit. In this domain, mobile armor develops. The formation of a mobile armor layer represents a mutual adjustment among the flow, sediment transport, and the textural characteristics of the bed surface to offset inherent differences in mobility of particles of different sizes (Figure 5.19). As a mobile armor layer develops from an initially unarmored bed, the bedload transport rate decreases as the coarsening surface regulates the mobility of both fine and coarse particles. For equilibrium conditions, particles are equally mobile, and the BSD equals the BMSD. As bed shear stress continues to increase within the domain of mobile armor development, the rate of bed-material transport increases and the coarseness of the layer progressively decreases. At the upper limit of the mobile armor domain, which corresponds to exceptionally high bed shear stresses, a coarse surface layer no longer exists (Jain, 1990; Gomez, 1995). The bed is unarmored, with the surface layer and subsurface having identical grain-size distributions (Parker, 2008). Bed-material transport rates increase dramatically because the mobility of particles in the subsurface is no longer constrained by a coarse surface layer. Under these conditions, regulation of transport of different size fractions by a coarse surface layer is not necessary for equal mobility. The bed shear stress is great enough to mobilize all particles in the active layer of the bed in proportion to their abundance in the subsurface without the regulating effect of a coarse surface layer (Laronne et al., 1994).

5.10.1.3 What Is the Relation between Equal Mobility and Partial Transport of Bed Material?

The concept of equal mobility has two components: 1) equal entrainment and 2) equal transport. As noted in Section 5.6, equal entrainment from a particle mixture occurs when the exponent in Eq. (5.27) equals −1, which results in a single value of critical shear stress for entrainment of particles of all sizes (Eq. (5.30)). Exponents for natural gravel-bed rivers are generally greater than −1 (Table 5.2), but less than zero, indicating that size-selective transport still occurs in the presence of mobile armor layers, but that such layers reduce differences in mobility compared with those expected if grains of other sizes were not present on the bed. Differences in grain mobility of different sizes of particles can also be assessed based on sediment transport data. Typically, such evaluations are performed by plotting fractional transport rates q_{sbdwi}/f_i, where f_i is the proportion of the grain-size fraction i on the bed surface, versus the diameter of particles in each fraction (d_i) for different transport events (Wilcock, 1997). The degree to which patterns of data deviate from a horizontal line on such plots indicates the extent to which the particle-size distribution of the bedload differs from that of the subsurface material, and thus from equal transport mobility. Deviation of the line from horizontal is indicative of partial transport; i.e., all sizes may be entrained and moving, but some size fractions are not being transported in amounts proportional to their presence in the bed material. Such fractions are being only partially transported. Both experimental work and field data for gravel-bed channels containing some sand indicate that the break point between fully mobile and partially mobile fractions tends to shift toward increasingly coarse fractions with increasing shear stress (Wilcock and McArdell, 1993, 1997) (Figure 5.20). Also, the bed shear stress required to fully mobilize a specific grain-size fraction is about twice the shear stress required for

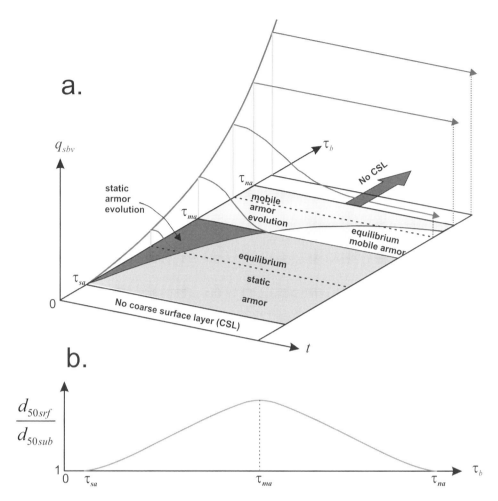

Figure 5.19. Diagram illustrating hypothesized, ideal model of armor layer development and coarseness for zero sediment influx but constant bed shear stress through time. (a) Evolution of equilibrium static armor and mobile armor over time (t) in relation to increasing bed shear stress (τ_b) and bedload transport rate (q_{sbv}). For $\tau_b < \tau_{sa}$, no bedload transport occurs and no coarse surface layer develops. For $\tau_{sa} < \tau_b < \tau_{ma}$, fines are progressively winnowed from the surface bed material until a static-armor layer develops and q_{sbv} goes to zero. For $\tau_{ma} < \tau_b < \tau_{na}$, the surface layer coarsens and transport rate decreases as fines in mixture are hidden beneath the mobile armor. All particles are mobile, and at equilibrium the grain-size distribution of the bedload approximates the grain-size distribution of the subsurface bed material. For $\tau_b > \tau_{na}$, mobility of coarse particles can be maintained without the regulating effect of mobile armor, and no coarse surface layer develops. Sediment influx will change the rate of armor evolution. (b) Coarseness of armor layer at equilibrium in relation to shear stress thresholds. Coarseness of surface layer reaches a maximum at the transition between static and mobile armor (τ_{ma}) and decreases with both increasing and decreasing τ_b (modified from Jain, 1990). (A black and white version of this figure will appear in some formats. For the color version, please refer to the plate section.)

incipient motion of that fraction (Wilcock and McArdell, 1993). Partial transport is favored when transport rates are low, the size range of the bed material is large, encompassing both sand and gravel, and when the mixture of sand and gravel is bimodal. Under these conditions, sand and possibly fine gravel in the mixture are readily mobilized and transported, whereas coarse grains may not move at all or only to a minor degree. At high transport rates, however, a tendency for equal mobility emerges even in bimodal sediments containing sand and gravel (Kuhnle, 1993b) (Figure 5.20).

In natural rivers, some evidence exists for equal mobility at high bed-material transport rates (Kuhnle, 1993b; Wilcock and

McArdell, 1993), but overall size-selective transport seems to occur under most flow conditions (Gomez, 1995). Given the wide array of factors that can complicate the development of armor layers and the effect of these layers on grain mobility in natural rivers compared with flumes, the difference between flume results and data for natural rivers is not surprising. Armor layers seem to persist as flow stage changes, but the characteristics of these layers probably evolve to some extent under changing flow conditions. Nevertheless, theoretical analysis suggests that mobile armor layers should persist as discharge changes (Wilcock and DeTemple, 2005; Parker et al., 2008), and experimental work indicates that mobile armor

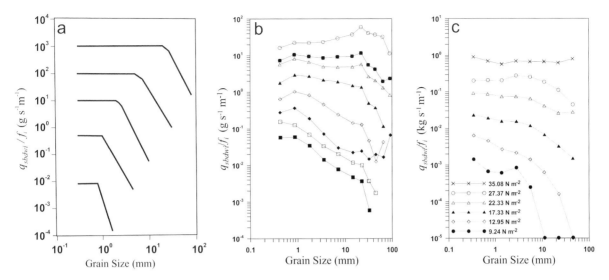

Figure 5.20. (a) Idealized representation of partial transport. Fractional transport rates scaled by the proportion of each fraction in the bed material remain constant for small grain sizes but decline for large grain sizes. The point of inflection between a horizontal and a declining relationship shifts toward larger grain sizes as transport rate increases (based on Wilcock and McArdell, 1993). (b, c) Fractional transport rates for Oak Creek, Oregon (b, from Wilcock and McArdell, 1993), and Goodwin Creek, Mississippi (c, from Kuhnle, 1993b) showing partial transport at low transport rates shifting toward equal mobility at high transport rates (equal mobility is represented by a horizontal relation between fractional transport rates and grain size across the full spectrum of grain size).

remains structurally stable over long flow durations (Powell et al., 2016). Another important factor in natural rivers that influences armor development is spatial variation in channel morphology at the bar element, bar unit, and planform scales (Figure 1.7). This variation affects the interaction among flow, sediment transport, and bed-material sorting, which, in turn, can produce spatial variability in the composition of bed-material characteristics at these different scales. Rarely, if ever, is variability at these spatial scales considered in experimental or theoretical evaluations of armor development.

5.10.2 What Models Have Been Developed to Predict Fractional Transport Rates?

A variety of models are available for predicting bedload transport by size fraction in rivers with nonuniform bed material (Parker, 2008). These models are capable of predicting transport rates in rivers with mobile armor layers. The Eulerian perspective still applies to these models in the sense that predictions focus on transport rates of each size fraction at a specific location in a river. Here, a few notable methods are presented to provide examples of these types of predictive relations.

5.10.2.1 What Is the Parker et al. (1982a) Method?

The Parker et al. (1982a) method is derived from empirical analysis of field data on bedload transport collected in Oak

Creek, Oregon. It applies to gravel transport only. The method includes two components (Parker, 2008):

1) a grain-size–based formulation for determining entrainment thresholds of different-size grains in the sediment mixture (see also Table 5.2):

$$\Theta_{ci} = 0.0876 \left(\frac{d_i}{d_{50}} \right)^{-0.982} \tag{5.78}$$

2) a transport function:

$$q_{sbvi} = f_i \frac{u_*^3}{[(\rho_s - \rho)/\rho]g} G(o_i) \tag{5.79}$$

where

$$o_i = \frac{\Theta_i}{\Theta_{ci}}$$

and

$$G(o_i) = \begin{cases} 0.0025 \exp[14.2(o_i - 1) - 9.28(o_i - 1)^2] \\ \quad \text{for } 0.95 < o_i \leq 1.65 \\ 11.2 \left(\frac{1 - 0.822}{o_i} \right)^{4.5} \\ \quad \text{for } o_i > 1.65 \end{cases} \tag{5.80}$$

The method uses data on the particle-size distribution of the subsurface bed material beneath the armor layer to determine values of d_i and d_{50} and on the bed shear stress, preferably the component acting on grains, to compute u_*

and Θ_i. Fractions by weight (f_i) of the subsurface bed material are determined on the basis of gravel-size material only. Transport rates are calculated for each grain-size fraction (i) and summed to obtain the total transport rate $q_{sbv} = \sum q_{sbvi}$. Procedures have been developed to calibrate the method to local conditions (Bakke et al., 1999, 2017) using a site-specific hiding function instead of Eq. (5.78) and a normalized version of the Parker (1978) transport function (Table 5.4) instead of Eq. (5.79):

$$G(o_i) = 11.2 \left(1 - \frac{0.853}{o_i}\right)^{4.5} \tag{5.81}$$

This normalized transport function, which was not derived from an analysis of the Oak Creek data, fits those data almost as well as Eq. (5.80).

A modified version of the model that relies on information on the grain-size characteristics of the surface (armor) layer of gravel-bed rivers has also been developed (Parker, 1990, 2008). Such information is difficult to obtain in natural rivers at transport-effective stages when the mobile armor is active. An advantage of this model is that it accounts for evolution of the coarse surface layer as the characteristics of this layer change in relation to changes in bed shear stress and bed-material transport. The surface-based transport relation has been incorporated into sediment-routing models for gravel-bed rivers to explore the problem of downstream fining, in which characteristics of the surface layer are of considerable interest (e.g., Cui et al., 1996; Hoey and Ferguson, 1997) (see Chapter 13).

5.10.2.2 What Is the Diplas (1987) Method?

Diplas (1987) revisited the analysis of Parker et al. (1982a) and developed a generalized relation that extends beyond the range of o_i specific to the Oak Creek data:

$$q_{sbvi} = f_i \frac{0.0025 u_*^3}{[(\rho_s - \rho)/\rho]g} (4) \times 17^{2.625b} \tag{5.82}$$

where

$$b = 1 - 1.205 \left[(\Theta_{50}/0.0873)^{(d_i/d_{50})^{-0.057}}\right]^{-1.843(d_i/d_{50})^{0.3214}} \tag{5.83}$$

Data requirements are the same as for the Parker et al. (1982a) method.

5.10.2.3 What Is the Wilcock and Crowe (2003) Method?

The method of Wilcock and Crowe (2003) is derived from laboratory data on bedload transport of sand and gravel mixtures. It uses information on the surficial grain-size distribution rather than the subsurface grain-size distribution. If surficial bed material in a natural river is sampled at low flow, when armor is static, the persistence of the character of the armor layer for mobile conditions is presumed in applying this method. The extent to which this assumption holds remains uncertain in many instances. The method is based on two components:

1) a grain size–based formulation for determining entrainment thresholds of different-sized grains in the sediment mixture:

$$\tau_{bci} = \tau_{bcm} \left(\frac{d_i}{d_m}\right)^b \tag{5.84}$$

where

$$\tau_{bcm} = (\rho_s - \rho)g d_m [0.021 + 0.015 \exp(-20 P_s)] \tag{5.85}$$

$$b = \frac{0.67}{1 + \exp\left(1.5 - \frac{d_i}{d_m}\right)} \tag{5.86}$$

and P_s is the proportion of sand in the surface layer.

2) a transport function:

$$q_{sbvi} = f_i \left\{\frac{u_*^3}{[(\rho_s - \rho)/\rho]g}\right\} 0.002 \left(\frac{\tau_b}{\tau_{bci}}\right)^{7.5} \quad \text{for } \left(\frac{\tau_b}{\tau_{bci}}\right) < 1.35 \tag{5.87}$$

$$q_{sbvi} = f_i \left\{\frac{u_*^3}{[(\rho_s - \rho)/\rho]g}\right\} 14 \left(1 - \frac{0.894}{\left(\frac{\tau_b}{\tau_{bci}}\right)^{0.5}}\right)^{4.5} \quad \text{for } \left(\frac{\tau_b}{\tau_{bci}}\right) \geq 1.35 \tag{5.88}$$

Data requirements include particle-size distributions for the surficial bed material to estimate d_i, d_m, f_i, and P_i as well as information on the bed shear stress (τ_b), again preferably the component acting on the grains (τ_b'). Because the critical bed shear stress of the mean grain size (τ_{bcm}) decreases with increasing sand content of the bed surface (Eq. (5.85)), increasing amounts of sand result in decreases of τ_{bc} for all fractions (i) (Eq. (5.84)), thereby increasing transport rates.

Bedload transport functions that predict fractional transport rates and the grain-size distributions of bedload have yet to be thoroughly tested. Additional data on bedload transport rates for sediment mixtures, especially mixtures that include gravel, are required to evaluate and refine existing models (Recking, 2016). In particular, field data on transport rates and bedload size distributions are needed to determine the extent to which relations that predict transport under experimental conditions can accurately predict transport in natural rivers.

5.11 How Is Bed-Material Transport Related to Particle Motion in Rivers?

An alternative to the Eulerian perspective on bedload transport is to adopt a Lagrangian approach that considers transport in terms of the rates of movement, transport distances, and transport pathways of moving grains. Work on bedload transport by Hans Albert Einstein (1942, 1950), Albert Einstein's son, considered the travel distance of particles, which was assumed to be a constant multiple of grain diameter, and cast entrainment and transport of size fractions of particles within a stochastic rather than a deterministic framework. This theoretical approach proposed that particles on the bed move in discrete steps with intervening periods of inactivity. Net bedload transport rates represent averages of the stochasticity of movement and resting of ensembles of individual particles. The method of computing dimensionless transport rates based on Einstein's (1950) bedload function is one of the most, if not the most, complex procedures for predicting sediment transport in rivers (Chang, 1988a; Gomez and Church, 1989). This seminal work called attention to the importance of the distance a particle travels during a transport episode and resting periods between transport episodes in defining bedload transport rates, with these rates being directly proportional to travel distances and inversely proportional to the duration of resting periods.

5.11.1 How Have Coarse-Particle Tracing Studies Contributed to the Understanding of Grain Kinematics?

Over the past several decades, considerable effort has been devoted to characterizing the movement of discrete particles, or grain kinematics (Haschenburger, 2013a), both within river systems and in experimental settings. A variety of methods exist for tracing coarse particles in natural rivers (Hassan and Roy, 2016), but most contemporary studies rely on magnets (Hassan et al., 1984) or passive integrated transponder (PIT) tags (Lamarre et al., 2005) inserted into tracer particles. The seeded tracers are then emplaced on the bed of a river and relocated after transport-effective events. Active tracers based on radio transmitters inserted into particles have also been used in a few settings (Ergenzinger et al., 1989; Habersack, 2001; McNamara and Borden, 2004) but are quite expensive to deploy. An advantage of active tracers is that particle locations and movement can be determined in real time, allowing tracing to occur during events, not just before and after an event as is the case with passive tracers.

The explosion of work on coarse-particle tracing in natural rivers has called attention to several important issues related to bedload kinematics, including the statistical characteristics of particle displacements, how displacement changes over time, differential displacement of particles of the same or different sizes over time, the thickness of the active bed layer, vertical exchange of transported and deposited particles within this active layer, residence times of particles in the bed through burial and removal from the active layer, and the role of different scales of river characteristics in tracer movement and deposition. The limited use of active tracers has provided some insight into the key issue of step length, the distance a particle travels during a transport episode, and rest periods of bedload particles (Ergenzinger and Schmidt, 1990; Schmidt and Ergenzinger, 1992; Habersack, 2001). Rest periods typically have an exponential distribution, but step-length distributions may conform to exponential (Schmidt and Ergenzinger, 1992) or gamma (Habersack, 2001) distributions. The length of steps increases with increasing discharge (Schmidt and Ergenzinger, 1992), and the probability distribution of step lengths also varies with discharge (Ergenzinger and Schmidt, 1990). Dimensionless step lengths (step length divided by grain size) are not constant, as proposed by Einstein (1950), but vary inversely with particle size (Habersack, 2001).

Step lengths cannot be determined from passive tracers, only the total amount of displacement during a particular event. This displacement (L_{pd}) is referred to as the path length. The path length presumably consists of multiple steps, unless a particle is in motion for the entire duration of an event, which active tracer studies indicate is highly unlikely (Habersack, 2001). For unconstrained clasts, or those exposed to the flow on the surface of the bed, path lengths tend to decline with increasing particle size, indicating that the distance of transport of clasts is size selective (Church and Hassan, 1992). The size dependency of path lengths decreases as transported clasts become buried within the bed, but some size dependency of total distance transported persists even when evaluated over many years (Haschenburger, 2013b). Path lengths of all particles during individual events increase with excess stream power (Hassan et al., 1992) or excess shear velocity (Phillips and Jerolmack, 2014). Independently of size, particle shape can have an influence on particle mobility, with spherical and rod-shaped clasts moving more readily than disc- or blade-shaped clasts (Carling et al., 1992).

5.11.2 What Is the Virtual Velocity of a Particle?

The virtual velocity (U_{vs}) of a sediment particle is defined as

$$U_{vs} = L_{pd}/\Delta t_c \qquad (5.89)$$

where Δt_c is the time during an event for which flow exceeds the threshold of motion for the tracer (Hassan et al., 1992). A variety of factors influence the virtual velocity. Methodologically, a single critical value for entrainment often is used for all tracers so that Δt_c is a constant for all particle sizes, but this approach can potentially bias the virtual velocities of grains of different sizes (Milan, 2013a). Ideally, a critical value of shear stress should be estimated for each size class, which will produce a different value of Δt_c for each class. Flow strength strongly affects virtual velocities, with values of U_{sv} increasing rapidly with increases in the excess stream power of peak discharge (Hassan et al., 1992) or excess grain dimensionless shear stress (Ferguson and Wathen, 1998).

As with path lengths, virtual velocities decrease with tracer size, indicating that downstream rates of transport are size selective (Ferguson and Wathen, 1998). Tracers seeded on the bed surface often are highly exposed to the flow and will initially tend to exhibit relatively large displacement lengths and virtual velocities (Church and Hassan, 1992). Over time, as the particles mix into the bed, virtual velocities decrease as particles become buried and resting periods increase (Ferguson et al., 2002; Ferguson and Hoey, 2002; Pelosi et al., 2016). This type of time-dependent behavior emphasizes the need to consider the vertical exchange of particles within the active layer of moving particles (Schick et al., 1987; Hassan and Church, 1994; Haschenburger, 2011) and through mechanisms of scour and fill (DeVries, 2003; Haschenburger, 2017), especially over long time scales (Haschenburger, 2013b).

Variations in channel form along a river, including variability at different scales (Papangelakis and Hassan, 2016; Hassan and Bradley, 2017), can produce spatial variations in the competence of the flow to transport bedload, resulting in preferential deposition of particles where competence is low (Lamarre and Roy, 2008; MacVicar and Roy, 2011; Haschenburger, 2013b; Kasprak et al., 2015). In bedrock rivers with small amounts of sediment cover on the channel bed, virtual velocities of tracer particles are relatively high compared with values for alluvial rivers because opportunities for exchange of tracers with underlying material, including burial, are limited (Hodge et al., 2013; Ferguson et al., 2017). The complex influence of multiscale spatial variability in channel characteristics and temporal variability in flow conditions on the transport of bedload particles in natural rivers results in distributions of path lengths that vary considerably among rivers and that do not conform to a common distributional form (Pyrce and Ashmore, 2003a).

5.11.3 How Can Bedload Transport Rates Be Estimated from Coarse-Particle Tracing?

Data on virtual velocities of tracer particles provide the basis for estimating rates of bedload transport in gravel-bed rivers, at least for the coarsest fractions that are amenable to tracing. The volumetric rate of transport (q_{sbv}) based on the virtual velocity is (Hassan et al., 1992)

$$q_{sbv} = U_{vs}\delta_b(1 - p_o) \qquad (5.90)$$

where δ_b is the thickness of the active layer of the bed and p_o is the porosity of the active layer. The thickness of the active layer can be determined either from measurements of scour and fill depths within a reach or from the vertical distribution of mobilized tracers within the subsurface (Haschenburger and Church, 1998; Church and Haschenburger, 2017). An advantage of this method over conventional techniques of measuring bedload transport at transects across a river using bedload samplers (see Appendix D) is that it provides insight into bedload transport throughout a reach and the connection of this transport to patterns of scour and fill. Importantly, such studies also yield information on the dynamics of the active layer, showing how the thickness of this layer tends to increase with increases in stream power (Haschenburger and Church, 1998) or dimensionless bed shear stress (Mao et al., 2017) – a finding supported by results of flume experiments (Wong et al., 2007).

5.11.3.1 What Is the Importance of Advection and Diffusion in Bedload Transport?

From a theoretical perspective, the bedload transport process as determined from the flux of individual particles involves both advection and diffusion. Advection is the net downstream movement of ensembles of particles starting at a particular location on the riverbed over time. Diffusion refers to the spreading of the spatial distribution of particles during transport. The emphasis of Eulerian approaches to bedload transport is on advection, whereas the Lagrangian perspective seeks also to understand the role of diffusion in bedload transport. Diffusion focuses on the mean squared displacement of an ensemble of particles σ_{dx}^2 over time, where

$$\sigma_{dx}^2 = \frac{1}{N}\sum_{i=1}^{N}[x(t) - x(t_0)]^2 \qquad (5.91)$$

$x(t)$ is the position of a particle at time t, $x(t_0)$ is the position of the particle at time $t = 0$, and N is the number of particles. Diffusion increases as the spread of the particles increases over time in relation to the initial position of the particles (Figure 5.21). In normal diffusion, the value of σ_{dx}^2 increases linearly

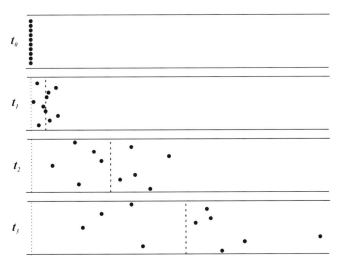

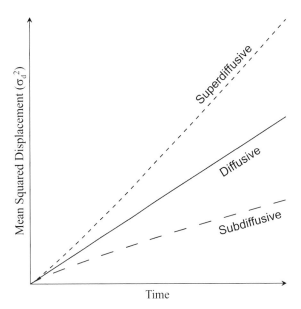

Figure 5.21. Planview of a stream channel showing at top the initial location of 10 tracer particles across the stream (black dots). Flow moves from left to right. Over time the particles are moved downstream. Mean displacement is indicated by the downstream shift of the heavy dashed line. This increase in mean displacement over time represents net advection of the group of particles. Diffusion refers to the increasing spread of the particles along the reach over time in relation to the initial position of the particles (light dashed line).

Figure 5.22. Plot of mean squared displacement versus time showing relation of subdiffusive and superdiffusive behavior to normal diffusive behavior.

over time. Diffusion of bed particles related to σ_{dx}^2 is characterized by a power function of the form

$$\sigma_{dx}^2 = \varepsilon_c t^b \tag{5.92}$$

where ε_c is a diffusion coefficient. Thus, normal diffusion occurs if $b = 1$. Anomalous diffusion is characterized by a nonlinear relation between σ_{dx}^2 and time, or $n \neq 1$. Processes where $b > 1.0$ are superdiffusive, whereas those where $b < 1.0$ are subdiffusive (Figure 5.22). The extent to which bedload grain kinematics in rivers conform to normal or anomalous diffusion is unclear, but some evidence indicates that tracer dispersal is superdiffusive (Liebault et al., 2012; Phillips et al., 2013).

Recognition of the importance of diffusive processes in bedload grain kinematics has led to theoretical explorations of the implications of anomalous versus normal diffusive processes for tracer dispersal in rivers (Ganti et al., 2010). It has also fueled efforts to determine the diffusive character of grain kinematics through detailed laboratory (Nino and Garcia, 1998; Lajeunesse et al., 2010; Martin et al., 2012; Fathel et al., 2016; Naqshband et al., 2017) and field studies (Drake et al., 1988; Nikora et al., 2002) of grain motion using ordinary or high-speed videography. To complement detailed experimental data, theoretical advances have emerged in the form of probabilistic, grain-kinematic

formulations of bedload sediment flux that incorporate both advective and diffusive components (Furbish et al., 2012, 2017). These two components of bedload transport may play an important role in the granular segregation process that leads to armoring in gravel-bed streams (Ferdowsi et al., 2017). Despite theoretical advances, semi-empirical Eulerian equations for predicting bedload transport derived in part from experimental studies of Lagrangian particle motion commonly take the form of excess shear stress formulations (Table 5.5).

5.12 How Is Bed-Material Transport in Rivers Linked to Changes in Channel Morphology?

5.12.1 What Is the Importance of Spatial Gradients in Transport for Channel Change?

In alluvial rivers, sediment transport has a strong process linkage with changes in channel form. These changes can be understood within the context of spatial gradients of transport and conservation of mass. For sediment of uniform density and porosity, the continuity equation for conservation of mass, or Exner equation, is (Garcia, 2008)

$$(1 - p_o)\frac{\partial Z}{\partial t} = -\frac{\partial q_{sbvs}}{\partial s} - \frac{\partial q_{sbvn}}{\partial n} + (D_{sv} - E_{sv}) \tag{5.93}$$

where Z is the bed elevation, D_{sv} is the volumetric rate of deposition of bed material from suspension onto the bed per unit area of the bed, E_{sv} is the volumetric rate of entrainment

Table 5.5. Examples of bedload transport equations based on particle motion studies.

$q_{sbv}^* = 43(\Theta - \Theta_c)(\Theta^{0.5} - 0.7\Theta_c^{0.5})$ (saltating sand and gravel particles)	Nino and Garcia, 1998
$q_{sbv}^* = 10.6(\Theta - \Theta_c)(\Theta^{0.5} - \Theta_c^{0.5} + 0.025)$	Lajeunesse et al., 2010

of bed material into suspension per unit area of the bed, s is a coordinate along the direction of the river, and n is a coordinate perpendicular to the direction of the river. This important equation provides fundamental insight into how morphological change occurs in rivers in relation to bed-material transport. In particular, it links sediment fluxes to changes in the elevation of the channel bed. Moreover, the differentials of bedload transport in the s and n directions emphasize that spatial gradients in this type of transport are mechanisms of morphological change. If the rate of transport in the s or n direction is increasing, net removal of sediment will occur along s or n and the elevation of the channel bed will decrease. In other words, a positive gradient in bed-material transport will produce a negative change in bed elevation. On the other hand, if the spatial gradients are negative, so that the flux decreases over s or n, deposition will occur and the bed elevation will increase. Similarly, if deposition from suspended bed-material load (D_{sv}) exceeds entrainment (E_{sv}), the bed elevation will increase, whereas if $E_{sv} > D_{sv}$, the bed elevation will decrease. Although not included explicitly in Eq. (5.93), the relation between D_{sv} and E_{sv} is influenced by spatial gradients in the transport capacity of the flow, which, in turn, are linked to spatial gradients in hydraulic conditions. If water transporting a large amount of entrained bed material in suspension decelerates as it flows along or across a river, transport capacity will decrease and deposition will occur. The opposite will take place if the flow accelerates spatially.

A vital lesson to be learned from this simple analysis is that it is not the absolute magnitude of the local rate of transport, or correspondingly, the bed shear stress above the threshold for sediment mobilization, that determines whether a particular portion of an alluvial river channel will be prone to change over time. Rather, it is the absolute magnitude of the spatial gradients in bed-material transport that governs channel change. Thus, knowledge of spatial gradients in bed-material transport is essential for understanding morphological change in rivers over time. These spatial gradients often vary with changing discharge and as the morphology of the river changes, leading to complex feedbacks among flow, sediment transport, and

channel form. The Exner equation is fundamental in attempts to model the morphodynamics of rivers (Mosselman, 2005). Numerical models of this type combine the physics of sediment transport with conservation of mass to simulate or predict dynamic change in river morphology over time and space (Church, 2006). A complete analysis must also consider spatial and temporal variations in the supply of sediment to the river system from the surrounding watershed that can modify existing spatial patterns of sediment fluxes within this system.

5.12.2 How Can Information on Channel Change Be Used to Estimate Bed-Material Transport Rates?

The relation between bed-material transport and channel change embodied in Eq. (5.93) provides the basis for studies that attempt to determine the rate of bed-material transport from changes in the morphology of alluvial channels (Ashmore and Church, 1998; Church, 2006). Such studies directly link spatial variation in sediment transport to changes in channel form – a central concern in fluvial geomorphology (Church, 2006). In a general sense, conservation of mass for any particular reach of a river system can be expressed as

$$\Delta V_s = V_{sin} - V_{sout} \tag{5.94}$$

where ΔV_s is the net change in sediment volume, V_{sin} is the volumetric input of sediment, and V_{sout} is the volumetric output of sediment. Changes in net volume of sediment will result in morphological change. The volumetric flux of sediment associated with morphological change (Q_{sbv}) over a time interval Δt is

$$Q_{sbv} = (1 - p_o)\frac{\Delta V_s}{\Delta t} \tag{5.95}$$

Values of ΔV_s can be determined through repeat topographical surveys of channel morphology. Ideally, survey frequency should be tied to the timing of events that produce morphological change (Church, 2006). If information also is available on V_{sin}, V_{sout}, or both, a sediment budget can be constructed for the reach of the river system under investigation.

The morphological method has been used to estimate bed-material fluxes associated with changes in channel form at a variety of spatial scales, ranging from the migration of bedforms (van den Berg, 1987; Claude et al., 2012) to the evolution of channel planform (Zinger et al., 2011), but the most common applications are at the bar-element and bar-unit scales (Vericat et al., 2017). Until recently, the method has been rather labor intensive, involving time-consuming topographic surveys, even when using fairly modern technology such as total stations, global positioning systems, and low-level aerial photogrammetry (Lane et al., 1994, 1995; Martin and Church, 1995). Advances in survey technology for obtaining repeated high-resolution topographic (HRT) data over different spatial scales underpin the growing use of the morphological method (Vericat et al., 2017). Through technological developments such as airborne and terrestrial laser scanning (TLS) (Heritage and Hetherington, 2007; Milan et al., 2007; Williams et al., 2014), structure from motion analysis of images obtained from unmanned aerial systems (Javemick et al., 2014; Woodget et al., 2015; Dietrich, 2016), and multibeam sonar (Parsons et al., 2005; Konsoer et al., 2016b), three-dimensional point clouds of fluvial features can be generated at high spatial resolutions over increasingly large spatial scales. These developments are accompanied by analytical improvements to facilitate both the production of accurate digital elevation models (DEMs) from point cloud data and the differencing of DEMs to estimate volumetric change in bed and bank material over time (Wheaton et al., 2010; Milan et al., 2011).

CHAPTER

6 Magnitude-Frequency Concepts and the Dynamics of Channel-Forming Events

6.1 Why Are Magnitude and Frequency Concepts Important for Understanding River Dynamics?

The mechanics of flow and sediment transport are fundamental mechanisms involved in the cycle of mutual interaction between process and form that shapes the morphology of river systems (Figure 1.2). Details of these mechanisms have been examined in the previous two chapters, but temporal variability in hydraulic conditions and sediment-transport processes play an important role in river dynamics. Flow in rivers, measured by discharge, changes over time, resulting in changing hydraulic conditions. These varying hydraulic conditions result in changes in sediment transport, which, in turn, influence channel morphology. Thus, the dynamics of rivers are integrally connected to temporal variability in river discharge. This variability is characterized by changes in the magnitude of discharge as well as by differences in the frequencies of discharges of various magnitudes. The study of temporal variation in river discharge falls within the domain of hydrology – the science of water supply, distribution, and movement on or near Earth's surface. This chapter examines fundamental hydrological principles and relates these principles to river morphology. In particular, it focuses on the magnitude and frequency of hydrological events as well as the relationship of magnitude and frequency concepts to the form and morphodynamics of rivers.

6.2 How Is Water Delivered to Rivers from Drainage Basins?

River flow is driven by the terrestrial component of the global hydrological cycle (Figure 6.1). Water is delivered to the river system through runoff from its watershed. Precipitation falling on hillslopes infiltrates into soil with rates of infiltration dependent mainly on soil permeability, which determines the hydraulic conductivity, or rate of movement of water through the soil, and antecedent moisture conditions. Soils with high permeability and hydraulic conductivity have high infiltration capacities. Moisture entering the soil at the surface gradually fills pore spaces and also moves downward into the subsurface as a wetting front. As pore spaces fill, rates of infiltration at the surface decrease because not all water delivered by precipitation can enter the moist soil. If the soil becomes saturated, water from additional rainfall accumulates on the surface, where it may move laterally down a hillslope as saturation excess overland flow. Rainfall rates that exceed the infiltration capacity of unsaturated soils generate infiltration excess overland flow (see Chapter 2). Interception of rainfall by vegetation reduces intensities at the soil surface, promoting infiltration; thus, infiltration excess overland flow typically develops where vegetation cover is sparse. Both saturation excess overland flow and infiltration excess overland flow contribute to surface stormflow during and immediately following precipitation events.

Water also moves as throughflow laterally and downslope within the soil under the influence of local pressure gradients.

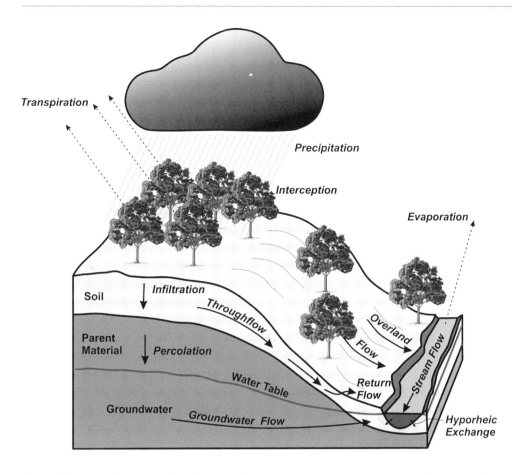

Figure 6.1. Pathways of the terrestrial water cycle. (A black and white version of this figure will appear in some formats. For the color version, please refer to the plate section.)

Throughflow contributes to subsurface stormflow. Accumulation of throughflow near the base of hillslopes may generate saturated conditions. At these locations, water within the soil emerges onto the surface as return flow that contributes to saturation excess overland flow.

Deep percolation of infiltrated water into underlying parent material leads to accumulation of water in saturated aquifers of groundwater. The water table marks the upper limit of the zone of saturation. Groundwater moves laterally along pressure gradients toward streams flowing along the bottoms of valleys. Inputs of groundwater as well as delayed shallow subsurface flow contribute to the baseflow of perennial streams. A lack of baseflow occurs seasonally in intermittent streams and permanently in ephemeral streams, which flow only through surface and shallow subsurface runoff from storms. In alluvial rivers, water within the flowing stream interacts with water within saturated sediment surrounding the submerged portion of the channel boundary. This process is known as hyporheic exchange, and the area within which interaction occurs is the hyporheic zone.

6.3 How Does River Flow Vary over Time?

Over time, the delivery of water to rivers fluctuates as amounts of runoff produced by different precipitation events varies. Variation of flow within a river caused by changes in delivery of water from the drainage basin is characterized by a hydrograph – a plot of change in discharge over time (Figure 6.2). Key components of the hydrograph are the rising limb, the peak discharge, and the falling, or recessional, limb. The time of rise is the difference between the initiation of the rising limb and the peak discharge. The lag to peak is the interval between the center of mass of rainfall and the peak discharge.

6.3.1 How Is Discharge Measured over Time in Rivers?

The production of hydrographs and evaluations of the magnitude and frequency of flow in rivers require data on changes in discharge over time. The measurement of discharge in rivers using the velocity-area method (Appendix C) is labor-intensive and expensive. Such measurements can only be accomplished occasionally, and continuous monitoring of streamflow using this method is impractical. Instead, the stage-discharge method is typically used to generate continuous records at specific locations along rivers. This method involves establishing gaging stations along rivers where the elevation of the water surface (stage) is measured over time (Turnipseed and Sauer, 2010). These measurements of stage are then related to occasional measurements of discharge at the gaging station using the velocity-area method through the development of a rating curve. In the past, measurements of stage involved the use of float and pulley systems placed in concrete stilling wells, which in turn were connected to the river through inlet pipes (Figure 6.3). Data from the pulley system were recorded on paper strip charts, which then had to be interpreted to determine variations in stage. Today, many gaging stations use modern technology such as radar, acoustic, optical, pressure, or gas-bubble sensors that do not require the installation of elaborate stilling wells (Figure 6.3). Data on stage from these devices are generated digitally and in many cases transmitted automatically to a central receiving location where the information is uploaded in real time to Internet-based information-delivery platforms.

In the United States, the U.S. Geological Survey is the primary organization that establishes and maintains stream gaging stations. Stage data at most stations typically are collected and reported at 15-minute intervals. Real-time and historical information for the gaging stations across the United States are available through the website USGS Water Data for the Nation (waterdata.usgs.gov).

To convert the stage data to estimates of discharge, a rating curve is constructed for the gaging station (Figure 6.4). Measurements of discharge are obtained using the velocity-area method at a variety of different stages. Statistical analysis is performed to generate a function relating discharge to stage. The resulting function can be used to estimate discharge for each measurement of stage. An online tool for automatically generating rating curves from field measurements of stage and discharge at gaging stations in the United States is available at the U.S. Geological Survey's WaterWatch website (https://water watch.usgs.gov). Once a rating curve is established, repeated measurements of discharge are required to confirm that the stage–discharge relation does not change over time. A stable stage–discharge relation depends to a large degree on the extent to which the form of the river channel at the gaging station

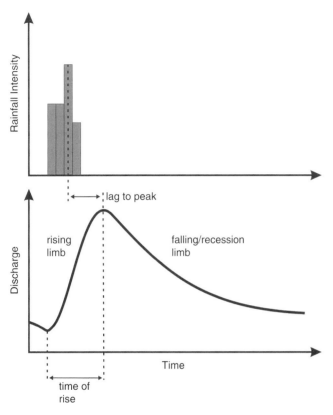

Figure 6.2. Characteristics of a hydrograph.

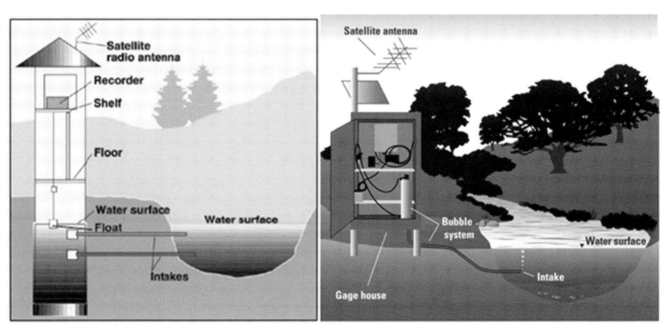

Figure 6.3. Examples of stream gaging operations for measuring water stage (from United States Geological Survey).

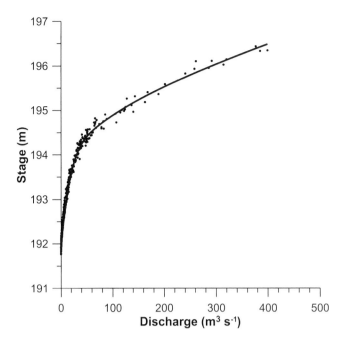

Figure 6.4. Rating curve for the Sangamon River at Monticello, Illinois (USGS Gaging Station 05572000).

remains constant over time. Systematic changes in the dimensions of a river channel will change the stage for a particular discharge, resulting in changes in the stage–discharge relation. If such changes are frequent and pronounced, the construction of a stable rating curve may not be possible. Therefore, considerable attention is given to the stability of a river channel over time when selecting sites for gaging stations. On the other hand, systematic shifts in stage–discharge relations over time have been used to identify progressive changes in stream-channel morphology (James, 1997; Juracek, 2004).

6.3.2 What Factors Influence the Shape of Hydrographs?

The shapes of hydrographs are related to the timing, amount, intensity, and spatial distribution of precipitation as well as to vegetation cover, all of which affect how quickly runoff is generated. High-intensity rainfall that produces copious amounts of runoff over a short period will produce peaked hydrographs characterized by short times of rise and lags to peak, as well as steep rising limbs. Similarly, runoff-generating rainfall that occurs mainly near the basin outlet will produce more peaked hydrographs than rainfall far from the basin outlet. Vegetation, by impeding runoff, can slow rates of rise of hydrographs. Hydrograph shapes are also affected by geomorphological characteristics of watersheds, including soil and parent material properties, basin relief, basin shape, and the structure of the channel network. These properties influence how fast water delivered to different parts of the basin moves over or through hillslopes to stream channels and then through stream channels to the outlet of the basin.

6.3.2.1 How Is Hydrograph Shape Related to Drainage-Basin Shape and Network Structure?

A considerable amount of research has explored the relation between drainage-network structure and hydrological response (Rodriguez-Iturbe and Rinaldo, 1997). At the most basic level, differences in the shape of a drainage basin and its associated network result in differences in hydrological response for a basin and network of constant size (drainage area and network magnitude), assuming that rainfall and runoff occur uniformly over the basin (Figure 6.5). The structure of the stream network primarily influences the travel-time distribution of water moving from the landscape to the basin outlet. The travel time is longer for a long path than for a short path.

Recognition of the influence of stream network structure on travel time distributions has led to sophisticated attempts to link watershed hydrology to the geomorphic characteristics of networks. The concept of the geomorphological instantaneous unit hydrograph (Rodriguez-Iturbe, 1993) connects the structure of networks as defined by stream order to basin hydrological response. The shape of an instantaneous unit hydrograph, which depicts how the basin responds to an instantaneous impulse of runoff, is a function of the bifurcation ratio (R_b), the drainage area ratio (R_A), and the stream length ratio (R_L) (Rodriguez-Iturbe and Valdes, 1979; Rodriguez-Iturbe and Rinaldo, 1997). For basins of fixed scale, the time to peak discharge is positively related to R_b and inversely related to R_A and R_L, whereas the magnitude of the peak discharge is positively related to R_L (Rodriguez-Iturbe and Valdes, 1979; Rosso, 1984). The width function, W_F, or the number of links that occur at a specific geometric distance along the channel network from the basin outlet, provides a useful metric for characterizing the influence of network structure on basin hydrological response (Kirkby, 1976; Rodriguez-Iturbe and Rinaldo, 1997). A plot of the width function over distance shows how the number of channels within the drainage network varies from the outlet toward the basin boundary (Figure 6.5). Because each link accounts for part of the total drainage area in the basin, the width function also depicts how drainage area varies with distance from the outlet. If water is delivered uniformly over the basin and moves at a constant velocity (U) everywhere within the river network, discharge at time t, $Q(t)$, at the watershed outlet is a function of the number of channels at a specific distance $x = Ut$ from the outlet, or

$$Q(t) \propto W_F(x) = W_F(Ut) \tag{6.1}$$

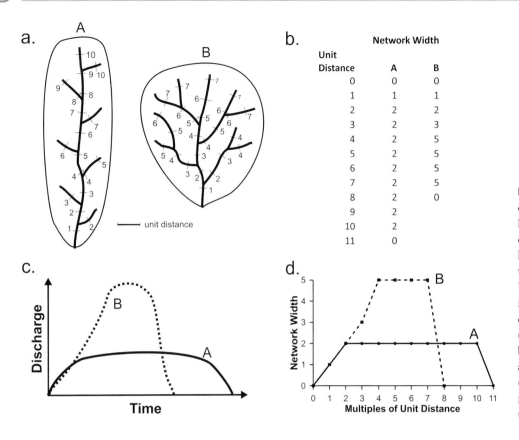

Figure. 6.5. (a) Elongated (A) and compact (B) networks showing locations of successive unit distances from the outlet to the headwaters along different paths through the drainage network. (b) Widths of the networks A and B in number of links at successive unit distances from the basin outlets. (c) Hypothetical hydrographs for basins A and B for uniform rainfall and runoff over the entire basins. (d) Plot of network width versus multiples of the unit distance from the network outlet.

In other words, the spatial variation in the number of channel links with distance from the outlet provides valuable information on the temporal pattern of hydrological response (Figure 6.5).

The characterization of network hydrological response as a direction function of the distance of channel segments from the basin outlet is often too simplistic to characterize hydrograph shapes for specific runoff events in natural river networks, but it does highlight the role of network structure in hydrological response. In natural networks, velocities tend to vary spatially throughout different river segments (links) within these networks, and rainfall rarely is distributed uniformly over drainage basins. Thus, considerable filtering of the hydrological signature of network structure occurs in many natural systems. Nevertheless, when combined with algorithms that account for movement of runoff into the channels from hillslopes and the flow of water through the channel network, approaches using the width function can produce reasonable estimates of flood hydrographs during specific events (Naden, 1992; Grimaldi et al., 2012).

6.4 How Is the Frequency of Floods Determined?

6.4.1 How Is Flood-Frequency Analysis Conducted Using Annual Maximum Series?

The frequencies of floods are determined based on peak discharges. The most common approach to flood-frequency analysis relies on data sets consisting of the largest peak discharge during each year of record for a gaging station. Such data sets are known as annual maximum series. The U.S. Geological Survey reports annual data on discharge for the 12-month period extending from October 1 to September 30 rather than the calendar year. This 12-month reporting period is known as a water year, and reporting for the period includes information on the peak discharge. Thus, in the United States, flood-frequency analysis typically is based on water years rather than calendar years. Once data for an annual maximum series are acquired, values of peak discharge in the series are ranked in order from highest to lowest. Any peak discharges of equal magnitude are ranked sequentially in arbitrary order. The plotting position (p_p) of each discharge value is determined as

$$p_p = \frac{m_r - a}{n_y + b} \tag{6.2}$$

where p_p is a probability estimate, m_r is the rank of discharge, n_y is the total number of years in the record, and a and b are constants. An alternative plotting position is

$$T_r = \frac{n_y + b}{m_r - a} \tag{6.3}$$

where T_r is an estimated recurrence interval, sometimes referred to as a return period. Most commonly, $a = 0$ and

$b = 1$, but plotting-position formulas have been proposed with values of a and b that differ from 0 and 1(Adamowski, 1981). Selection among alternatives is somewhat arbitrary. Nevertheless, the choice does influence the results of the flood-frequency analysis, particularly for tails of distributions.

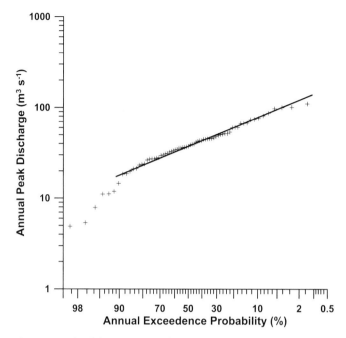

Figure 6.6. Flood-frequency plot for the Des Plaines near Gurnee, Illinois (normal probability axis) based on annual peak discharges from 1946 to 2017. Black line is flood-frequency curve fitted manually to the data, ignoring low-magnitude, high-frequency outliers.

Peak discharge data are plotted according to estimated probability (x axis) and magnitude (y axis), and a line of best fit is established through the data to produce a flood-frequency curve (Figure 6.6). A variety of probability distributions can be considered in determining flood frequencies, including normal, lognormal, extreme value type I (Gumbel), extreme value type III (Weibull), Pearson type III, and log-Pearson type III. The choice of distribution is arbitrary, and the goal is to find a curve that fits the data closely without major outliers from the curve. The term "curve" is somewhat misleading, because often the data will plot along a straight line when the distribution of the data conforms to the probability distribution under consideration.

Although models of probability distributions, as well as power-law and exponential functions (Malamud and Turcotte, 2006), can be fitted to data on estimated probabilities of peak discharges using statistical methods (Kidson and Richards, 2005), curves of best fit are often established manually to assess whether particular points, particularly extremes, may not conform to the general trend and constitute outliers (Dunne and Leopold, 1978). Such outliers can have a strong influence on quantitative methods of curve fitting, resulting in shifts of the fitted relation into an unrepresentative position on the plot. Some methods, such as the log-Pearson type III approach, necessitate quantitative estimation of a line of best fit (Figure 6.7). Log-Pearson type III analysis is now the recommended method for flood-frequency estimation in the United States (England et al., 2018), and the program PeakFQ

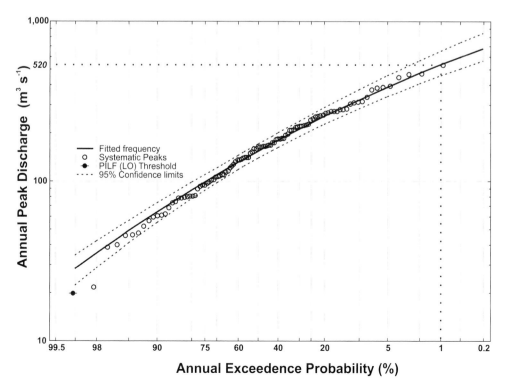

Figure 6.7. Log-Pearson type III flood-frequency curve for the Sangamon River at Monticello, IL, based on annual peak discharges from 1908 to 2016. Dotted lines designate the magnitude of the 100-year flood (Q_{100}). PILF (LO) Threshold defines threshold for potential influential low outliers.

has been developed to facilitate analysis using this method (Flynn et al., 2006).

Flood-frequency curves provide the basis for estimating exceedance probabilities or recurrence intervals of floods (Figures 6.6 and 6.7). For example, a flood with an exceedance probability of 0.01, as estimated from a flood-frequency curve, has a corresponding recurrence interval of 100 years. Thus, a flood of this magnitude has a 1% chance of being equaled or exceeded in any given year and, on average, is equaled or exceeded once every 100 years. Many practical applications of flood-frequency analysis adopt recurrence-interval terminology, such as the 100-year flood, the 50-year flood, or the 2-year flood. In geomorphology, discharges corresponding to different recurrence intervals are designated with appropriate subscripts; for example, Q_{100} for the 100-year flood or Q_2 for the two-year flood.

The concept of recurrence interval or return period can be confusing, because the two terms imply an interval of time between the occurrence of events. The inclusion of "on average" in the definition of flood recurrence is important. A flood of a particular size will be equaled or exceeded on average a particular number of times over the designated interval. However, the probability remains the same even if an event of this size has just occurred. If a 100-year flood occurs in a given year, the probability that it will be equaled or exceeded the following year is still 1%. The probability perspective of flood-frequency analysis provides no information on the timing of events. For example, three floods with peak discharges equal to the estimated 25-year flood could conceivably occur in three consecutive years and then not recur over the following 72 years.

The arithmetic mean of the annual series of peak discharges is known as the mean annual flood (Q_{maf}). Theoretically, this flood has a recurrence interval of 2.33 years if the data in the annual series conform to an extreme value type I (Gumbel) frequency distribution (Dunne and Leopold, 1978, p. 313). If the data do not conform to this distribution, the recurrence interval of the mean annual flood as determined graphically from the flood-frequency curve will deviate slightly from the theoretical value.

6.4.2 How Is the Accuracy of Flood-Frequency Analysis Related to Record Length?

The accuracy of flood-frequency analysis for determining the frequency of large floods depends to a major extent on the length of the hydrologic record. Considerable extrapolation will be involved in using a flood-frequency curve based on analysis of an annual maximum series 20 to 30 years in length to estimate the recurrence intervals of large floods, such as Q_{50} or Q_{100}. In some cases, historical information, such as an

individual discharge measurement of an extreme flood obtained well before the period of instrumented record or newspaper accounts of water stage during a flood of record, may be available. Inclusion of such information in a flood-frequency analysis can reduce uncertainty in extrapolation (Dalrymple, 1960; Halbert et al., 2016). Over even longer timescales, paleohydrological data on floods is valuable for placing floods of record within the context of extreme events that have occurred over timescales of centuries or even millennia (Blainey et al., 2002; Lam et al., 2017).

6.4.3 On What Assumptions Is Flood-Frequency Analysis Based?

Flood-frequency analysis is based on the assumptions that the discharge data represent a random sample drawn from a single, stationary (nontrending) population of floods and that the events corresponding to the peak discharges are independent of one another. The stationarity assumption is important, because if the hydrological regime of a river is trending over time, the frequency of a peak discharge of a given magnitude is not constant as is implied by a flood-frequency curve. Trends in climate, including human-induced global warming, may challenge the stationarity assumption for some rivers (Merz et al., 2014; Bloschl et al., 2015). If climate characteristics of a region are such that rainfall-runoff mechanisms constitute distinct populations, separate flood-frequency curves will be needed to adequately characterize these populations (Waylen and Woo, 1982; Barth et al., 2017).

Assuming that the data conform to stationarity and that the probability distribution selected for the analysis adequately represents the underlying flood population, outliers can be interpreted as events with actual probabilities that differ from the estimated probability based on plotting position. Outliers that plot above and to the left of the curve represent extreme floods with recurrence intervals greater than the length of record (n). Those that plot below and to the right of the curve probably occur more frequently than is indicated by the plotting position. Over time, outliers should progressively shift toward the curve as more discharge values are included in the data series. This interpretation of outliers should be exercised cautiously because other factors, such as nonstationarity, mixed populations of floods, and selection of an inappropriate probability distribution, can also produce outliers.

6.4.4 How Is Flood-Frequency Analysis Conducted Using Partial Duration Series?

Flood-frequency analysis based on annual maximum series considers only the largest peak discharges of each year. It ignores other large floods that may have occurred during any specific year. For example, if an event with a peak discharge equal in magnitude to an estimated two-year flood occurs during the same year as an event with a peak discharge equal in magnitude to an estimated 50-year flood, the smaller discharge will be ignored and only the larger discharge included in the annual maximum series. Flood-frequency analysis based on partial duration series attempts to overcome this limitation by including all peak flows over a certain magnitude threshold in the analysis, rather than restricting the number of included discharges to one per year. Because the method is based on a threshold, it is often referred to as the peaks over threshold method (Bezak et al., 2014).

The assumption that peak discharges are independent of one another also applies to the partial duration series and becomes of substantial concern when multiple discharges are included for a particular year. Decreasing the threshold includes more peak discharges in the analysis, but at the risk of the discharges not being independent of one another. Increasing the threshold enhances independence, but with a potential loss of valuable information. Methods exist for objectively determining thresholds based on statistical considerations (Lang et al., 1999), but the choice of a threshold remains somewhat subjective. Practical guidance suggests that peaks are independent if they are separated by a well-defined trough in the hydrograph and the discharge of the trough is at least 25% less than the discharge of the smaller of the two peaks (Dalrymple, 1960). Such a rule of thumb may not guarantee independence if the duration of the trough is short; in such instances, both peaks may still be part of the same runoff event.

Analysis for the peaks over threshold method using partial duration series is the same as analysis using the annual maximum series. The only difference is that n_y in Eq. (6.2) and (6.3) is replaced by n_t – the number of peak discharges that exceed the threshold. As a result, the exceedance probabilities and recurrence intervals generated by flood-frequency analysis of partial duration series differ from those produced by analysis based on annual maximum series.

6.4.5 How Do Flood Frequencies Differ for Partial Duration versus Annual Maximum Series?

The interpretation of recurrence interval differs between flood-frequency analyses based on annual maximum and partial duration series. For the annual maximum series, the recurrence interval specifies the average interval within which a flood of a given magnitude will occur as an annual maximum, i.e., as the largest peak discharge of the year in which it occurs. For the partial duration series, the recurrence interval specifies the average interval between floods of a given size irrespective of whether it is the annual maximum for

Table 6.1. Relations among exceedance probability and recurrence interval for partial duration series and annual maximum series.

Exceedance probability, partial duration	Recurrence interval, partial duration	Exceedance probability, annual series	Recurrence interval, annual series
4.54	0.22	0.99	1.01
3.33	0.33	0.95	1.05
2.44	0.41	0.91	1.1
2	0.5	0.86	1.16
1.33	0.75	0.74	1.36
1.1	0.91	0.67	1.5
1	1	0.63	1.58
0.5	2	0.39	2.54
0.2	5	0.18	5.52
0.1	10	0.095	10.5
0.04	25	0.039	25.5
0.02	50	0.0198	50.5
0.01	100	0.00995	100.5

(after Langbein, 1949)

the year in which it occurs. The difference in recurrence interval between the two approaches is most pronounced for floods with small recurrence intervals (or large exceedance probabilities). The likelihood of a particular flood not being the annual maximum in a particular year is greatest for small floods. The likelihood that a relatively infrequent event, such as the 50-year flood, will be exceeded by an even larger, less frequent flood during the year when this event occurs is quite small. However, it is more likely that smaller, more frequent floods may occur during the year of the 50-year event. Thus, small floods of specific magnitudes will have larger exceedance probabilities or smaller recurrence intervals for flood-frequency analysis based on the partial duration series compared with the corresponding exceedance probabilities or recurrence intervals for flood-frequency analysis based on the annual maximum series (Table 6.1).

6.4.6 Why Is the Difference in Flood-Frequency Analysis Important Geomorphologically?

Geomorphologists interested in the influence of all channel-shaping floods on river morphodynamics desire information on the frequency and magnitude of all such floods, not just those that occur as annual maximums. Recurrence intervals based on the partial duration series tend to reveal more accurately how often on average particular channel-shaping flows occur compared with those based on the annual maximum series. Nevertheless, most flood-frequency analyses are based on annual maximum series, given the widespread availability of annual maximum data. The construction of partial duration series requires information on all peak flows that occur throughout time – information that generally is not readily available for the entire hydrological record of a gaging station. Differences in recurrence intervals (R.I.) for the two methods (Table 6.1) should be recognized and appreciated, particularly for the most frequent floods (R.I. < five years) (Keast and Ellison, 2013) (Table 6.1).

6.5 How Is the Frequency of River Flows Determined?

Flood-frequency analysis focuses on the largest flow events, characterized by the peak discharges of these events. From a geomorphological perspective, it ignores small or moderate flows, some of which may be capable of transporting sediment and shaping the form of rivers. From a hydrological perspective, the practical utility of flood-frequency analysis is to characterize large floods that may inundate land and potentially damage human infrastructure along rivers. Other hydrological concerns, such as the availability of water, require comprehensive information on the frequency of a broad range of flows in a river, not just the frequency of floods. This information may also be of value for water-quality studies examining pollution concentration levels or ecological analyses of habitat availability for aquatic organisms.

6.5.1 What Is Flow Duration Analysis?

Flow duration analysis characterizes the cumulative frequency distribution of nearly the entire spectrum of flows in a river. The qualifier "nearly" is necessary because it does not capture the largest, or peak, flows that are captured by flood-frequency analysis.

The traditional approach to flow duration analysis is based on mean daily discharge data reported by the United States Geological Survey for thousands of gaging stations throughout the United States (Searcy, 1959). Mean daily discharges represent the mean of individual stage-generated measurements of discharge at a gaging station over a 24-hour day (Rantz et al., 1982). Averaging over the day filters peak discharges, with the effect being most pronounced on small streams with hydrographs that vary substantially over a 24-hour period. Whereas the peak discharge and mean daily

discharge on the day of the peak discharge may not differ substantially for a major flood event on a large river, the peak discharge and mean daily discharge of a major event will differ greatly on a small stream, where the flow peaks and subsides over the duration of a few hours. Flow duration analysis is not restricted to the use of mean daily discharges, and the increasing availability of instantaneous discharge data at rates on the order of 15 minutes allows duration analysis to be conducted for small time increments of hours or even minutes for fast-responding, or flashy, stream systems (Rhoads, 1990a).

In the traditional approach, daily discharge data are classified into intervals based on magnitude. The number of class intervals is subjective, but it should be at least 20 to 30 and encompass the entire range of discharge values (Searcy, 1959). Because the resulting diagrams produced from the data involve plotting discharge values on logarithmic axes, the ranges of the intervals typically are not equal, but vary in size logarithmically. For a log base 10 cycle, or an increase by a factor of 10, the following base increments for class intervals are often appropriate: 1, 1.2, 1.4, 1.7, 2, 2.5, 3, 3.5, 4, 4.5, 5, 6, 7, 8, 10, where the base increments are multiplied by the discharge value at the lower bound of the cycle. This scheme may need to be adjusted to obtain an appropriate number of class intervals if the range of data encompasses only a single log cycle. Data are assigned to the classes based on whether discharge values equal or exceed the lower bound of the interval. Thus, a value of 1 would be assigned to the interval 1 to 1.2, whereas a value of 1.2 would be assigned to the interval 1.2 to 1.4. Once data are assigned to classes, frequencies are determined by dividing the number of days in each class by the total number of days in the period of record.

The class interval method of determining frequencies has been used largely because of the need to efficiently process large records of daily discharge data. The widespread availability of spreadsheet software for processing data has reduced the need for classification. Instead, the entire record of daily discharges can be sorted, ranked from largest to smallest by magnitude, and probabilities (p) estimated from Eq. (6.2) with $a = 0$, $b = 1$, and n_y replaced by n_d – the number of days in the record. Once probabilities are determined, the data are plotted to determine the relationship between discharge and probability. A line connecting the points on the plot establishes the flow–duration curve (Figure 6.8).

The duration curve provides the basis for determining the probability that a certain discharge will be equaled or exceeded as the mean daily discharge on a particular day. Because probabilities are determined empirically based on the amount of existing discharge data for a particular gaging station, these probabilities will change as new data become available for the

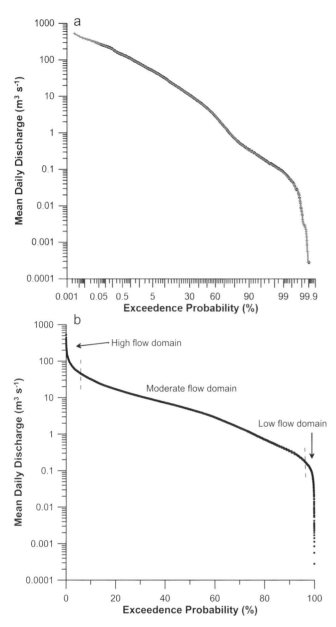

Figure 6.8. Flow–duration curve for Sangamon River at Monticello for 1915 to 2016. (a) Normal probability on *x* axis, flow–duration curve shown as gray line; (b) linear *x* axis and domains of high, medium, and low flow.

a mean daily discharge with an exceedance probability of 0.01 would be equaled or exceeded, on average, 3.6525 days per year.

6.5.2 How Are Flow–Duration Curves related to Runoff-Generating Mechanisms?

Many flow–duration curves, when plotted on arithmetic axes for probability, exhibit a characteristic S-shape with steep portions at the extremes of low and high flow connected to a relatively flat portion over a broad range of moderate flows (Figure 6.8). The three portions of the curve seem to represent distinct domains of hydrological response within which different factors govern the nature of the response (Yokoo and Sivapalan, 2011; Cheng et al., 2012). In general, the lower tail of the curve illustrates depletion of baseflow, whereas the high-flow domain reflects saturation-excess runoff. The flat middle part of the curve reveals a highly filtered response governed by combined contributions from subsurface and surface runoff processes. Abrupt changes in the discharge–probability relation between domains represent sharp transitions in runoff-generating mechanisms. At regional scales, this trifold structure of the flow duration curve can be related to climate and watershed characteristics (Coopersmith et al., 2012; Yaeger et al., 2012). Although flow duration curves and the three components of these curves often conform to three-parameter mixed gamma distributions (Cheng et al., 2012), the upper part of the flow–duration curve has been shown in some cases to exhibit power-law scaling (Molnar et al., 2006; Phillips and Jerolmack, 2016). In other words, the pattern of discharge data for the highest flows forms a straight line when both discharge and probability are plotted on logarithmic axes. Such scaling implies that these distributions may be heavy-tailed and that large flows of a specific magnitude occur more frequently than implied by thin-tailed distributions commonly used to characterize flow probabilities (Malamud and Turcotte, 2006).

6.5.3 How Are the Slopes of Flood-Frequency and Flow–Duration Curves Related to Hydrologic Variability?

The slopes of flood-frequency and flow–duration curves are direct indicators of flood and flow variability. As the slopes of these curves increase, the ratio of large to small discharges increases for a given pair of probabilities corresponding to the discharges. In other words, a river with a steeply sloping duration curve has a more variable hydrological regime than a river with a relatively flat curve. Although many factors can affect flow variability, the slopes of flood-frequency curves and of flow–duration curves both generally increase with increasing aridity (Molnar et al., 2006; Cheng et al., 2012).

station. Thus, the flow–duration curve provides a snapshot in time of the probability distribution of mean daily flows for a stream. Assuming stationarity in the hydrologic regime, an assumption challenged by global, regional, or local changes in climate, this distribution should over time converge on the distribution of the underlying population. Another common way of expressing flow duration is by multiplying the probability by 365.25, the average number of days in a year, to determine the number of days, on average, a flow of a specific probability is equaled or exceeded in a year. For example,

6.6 What Is a Channel-Formative Event?

Recognizing that the hydrology of river systems is characterized by flows of varying magnitudes and frequencies, a primary aim of fluvial geomorphology is to determine the extent to which various flows influence the form of river channels. This aim is rooted mainly in hydrology rather than hydraulics. It seeks to understand the role of particular hydrological events in changing or maintaining channel form. Those that are capable of influencing channel morphology, whether through change or maintenance, represent channel-formative or channel-shaping events. In alluvial rivers, the capacity to mobilize bed-material load often is considered a necessary requirement for a flow to be considered channel formative. Thus, channel-formative flows should transport at least some bed material. This criterion, however, is not necessarily a useful one for identifying differences in morphologic influence among varying flows. Most large floods mobilize more bed material than small floods, but as noted in Chapter 5, changes in channel form are governed by spatial variation in sediment flux, not by the magnitude of flux. Moreover, an emphasis on magnitude alone fails to consider the role of frequency. Small floods tend to occur more frequently than large floods, and this factor also plays an important role in the shaping of river channels.

Two major perspectives on the relation between channel form and channel-formative flows exist within fluvial geomorphology. One perspective tends to emphasize the persistence of channel form through time, whereas the other focuses mainly on changes in form through time. Both examine relations between channel form and the magnitude and frequency of different flows rather than relying on detailed mechanistic considerations of interrelations among flow hydraulics, sediment transport, and channel morphology.

6.7 What Is the Concept of Dominant Discharge and How Is It Related to River Equilibrium?

The first perspective on channel-formative flows invokes the concept of a dominant discharge (Q_{dd}). This concept can be thought of in at least two different ways, both of which are founded on the underlying assumption that channel forms are relatively stable through time and that this stability represents an equilibrium adjustment between process and form in natural rivers. One possible way to conceive of the dominant discharge is that it is a single value of discharge that represents the net morphological outcome of the entire distribution of channel-formative flows. Over time, many different flows affect channel form, but the net morphological result is the same as that produced by steady flow corresponding to the dominant discharge. This conception of the dominant discharge is essentially statistical in nature. The dominant discharge may not inordinately influence channel geometry, i.e., it may not be truly dominant, but it has an important representational function, just as the mean of a sample is an important representative property of the sample. This conception also implies that channel form can, or even will, change to some extent in response to all channel-formative flows, but that morphological fluctuations from the net, or average, form are constrained over time. Substantial changes in form either do not occur or are short lived; otherwise, the channel form cannot be considered stable.

The second conception of the dominant discharge is that a particular discharge within the probability distribution of discharges plays a truly dominant role in shaping channel form. Although other flows may influence channel morphology to a minor extent, one particular discharge predominantly shapes the form of the channel. This conception implies that channel form will not change systematically as long as the magnitude and frequency of the dominant discharge remain constant. Minor changes in channel morphology may occur in response to the influence of other channel-formative events, but the strong influence of the dominant discharge maintains a stable configuration.

It is important to recognize that the concept of a dominant discharge is a theoretical construct that consists of a hypothesis resting on an underlying assumption. The hypothesis is that channel form can be linked to a specific formative discharge to which that form is adjusted. The assumption is that equilibrium adjustments between form and flow occur within natural rivers to produce morphologically stable channels. This assumption has a long history in the science of rivers, dating at least to the work of Domenico Guglielmini in the late seventeenth century (Orme, 2013).

The notion of river morphological equilibrium has roots in regime theory within hydraulic engineering. This theory emerged in the late nineteenth and early twentieth century through attempts by engineers to characterize quantitatively morphological attributes of stable irrigation canals on the Indian subcontinent (Singh, 2003). A regime channel excavated in alluvium achieves physical stability, or a balance between erosion and deposition, through a dynamic equilibrium of forces, so that the channel cross-sectional geometry and gradient are maintained over time (Wolman, 1955).

Within geomorphology, the equilibrium assumption serves as the context for the concept of a graded river, alluded to by Gilbert (1877) and subsequently elaborated upon and expanded over the years (Davis, 1899, 1902; Rubey, 1933b; Kesseli, 1941; Mackin, 1948; Lane, 1954; Dury, 1966; Knox, 1976). According to a comprehensive definition:

A graded stream is one in which, over a period of years, slope, velocity, depth, width, roughness, pattern and channel morphology delicately and mutually adjust to provide the power and efficiency necessary to transport the load supplied from the drainage basin without aggradation or degradation of the channels.

(Leopold and Bull, 1979, p. 195)

The concept of a graded river explicitly frames stability and equilibrium within a mass balance context, a framing consistent with regime theory. Channel form can be considered stable or in equilibrium when sediment inputs equal sediment outputs and no net sediment accumulation or removal occurs within the stream reach of interest. Note that stability does not necessitate stasis. Mutual adjustment involves interaction; changes in one aspect of the system can trigger accommodating adjustments in other aspects through negative feedback to preserve mass balance and prevent systematic changes in form by net aggradation or degradation. The system is delicately adjusted, however, in the sense that fine-tuning amongst key variables prevents wholesale morphological restructuring in the face of inherent variability in external inputs and internal interactions.

The assumption of river equilibrium is closely aligned to the systems concept of steady state (Thorn and Welford, 1994; Church and Ferguson, 2015), whereby the mean state of a system remains constant over time. Fluctuations occur about the mean state, but the mean state does not change. It also links to the notion of characteristic forms (Brunsden and Thornes, 1979), or the development of morphological configurations in geomorphic systems that persist long enough under variable environmental conditions for the configurations to be viewed as characteristic of the system within that set of conditions.

Even though it provides a foundation for a wide range of research on river morphodynamics (see Chapter 7) and for practical applications of geomorphological principles to river management (see Chapter 16), dominant discharge is a useful conceptual construct rather than an objective fact. In this sense, it represents a regulative principle (Rhoads and Thorn, 1993) that facilitates scientific analysis of problems in river

science. Attempts to show conclusively that rivers attain an equilibrium state through adjustments related to a dominant discharge lie somewhat beyond the scope of scientific inquiry.

6.7.1 What Is the Bankfull Discharge and How Is It Related to the Dominant Discharge?

A stable alluvial river system consists of two fundamental components: the channel and the floodplain flanking the channel. The transition between the two occurs at the top of the banks of the channel where the floodplain surface begins. The elevation of this transition is referred to as the bankfull stage (Figure 6.9). If the water level rises to this stage it fills the channel completely. A further increase in stage will result in water spilling out of the channel onto the floodplain. The volumetric rate of flow associated with the bankfull stage is referred to as the bankfull discharge (Q_{bk}).

Assuming that the river channel form is stable over time, and that flow and form are mutually adjusted to one another in an equilibrium relationship, the bankfull discharge is presumed to equal the dominant discharge. The reasoning behind this conclusion is as follows:

Premise: If a specific dominant discharge is responsible for the stable form of a river channel, then the river channel form will conform to the magnitude of this discharge.

Supporting relation based on empirical evidence: The form of stable river channels conforms to the magnitude of the bankfull discharge. In particular, the cross-sectional area, top width, and mean depth of the flow at bankfull discharge are equal to the cross-sectional area, top width, and mean depth of the channel.

Conclusion: The bankfull discharge equals the dominant discharge.

Note that the supporting relation substantiates the effect in the premise (the "then" part of the premise), not the causal hypothesis (the "if" part of the premise). This logical form of reasoning from effect to cause is an example of abduction rather than deduction. The conclusion is not logically

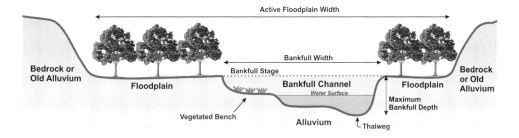

Figure 6.9. Diagram illustrating the relation of a bankfull channel to bankfull stage and the active floodplain in an alluvial river. (A black and white version of this figure will appear in some formats. For the color version, please refer to the plate section.)

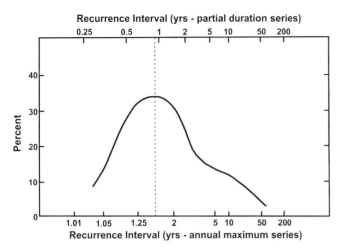

Figure 6.10. Probability distribution of recurrence intervals of bankfull discharge on annual maximum series and partial duration series for 36 stream sites in the United States. Dashed line shows the modal value of recurrence interval (adapted from Williams, 1978a).

necessary, given the supporting relation. For logical necessity to occur, supporting field evidence would need to confirm that a specific discharge is responsible for the channel form. In the absence of this evidence, other hypothetical causes could account for the strong relation between the bankfull discharge and channel form besides a specific discharge being responsible for this form. Nevertheless, this type of reasoning is common in the earth sciences and provides the basis for a wide range of inquiry (Rhoads and Thorn, 1993). In this particular instance, it maintains that because the channel form has certain stable dimensions, the discharge that fills the channel to capacity must be responsible for these dimensions. Presumably, if a discharge other than the discharge that fills the channel to the top of its banks were dominant, the size of the channel would be different. If a discharge larger than the bankfull Q were dominant, the channel would be larger, and if a discharge smaller than the bankfull Q were dominant, the channel would be smaller. Thus, bankfull stage corresponds to the discharge of importance for maintaining the average morphological characteristics of the channel.

Early work on bankfull discharge suggested that the recurrence interval of this flow might be similar for many rivers with an average of about 1.5 years on the annual series (Leopold et al., 1964, p. 319; Dunne and Leopold, 1978, pp. 309–311). This value corresponds to a recurrence interval of about 0.9 years on the partial duration series, indicating that the bankfull discharge will occur about every year on average. Subsequent work has shown that the recurrence interval of bankfull discharge is quite variable, ranging on the annual maximum series from one year to many decades, with a central tendency in the range of 1.2 to 1.5 years (Williams, 1978a; Petit and Pauquet, 1997; Castro and Jackson, 2001;

Ahilan et al., 2013) (Figure 6.10). Because the bankfull discharge on some rivers recurs on average once a year or more (e.g., Petit and Pauquet, 1997; Sweet and Geratz, 2003), frequency analysis based on the annual maximum series is not a useful method for estimating its recurrence interval; instead, the partial duration series should be used (Navratil et al., 2006). A unique flood-frequency analysis using instantaneous peak daily discharges derived from real-time gaging data indicates that the average recurrence interval of bankfull discharge for streams in Maryland, North Carolina, and New York is 0.68 years on the partial duration series or a frequency of about 1.5 events per year on average (Endreny, 2007).

6.7.2 How Is the Bankfull Channel Determined?

The bankfull channel is a basic morphological element of stable, alluvial rivers. Defining this morphological feature is fundamental to attempts to determine the formative flows that shape it. The bankfull channel must be accurately and consistently defined before its width, depth, or slope can be related to flow conditions that influence these morphological characteristics.

Determining the bankfull stage can be fairly straightforward when a well-defined channel with a simple, uniform geometry lies within a well-defined floodplain (Figure 6.11). However, in many circumstances, identification of a bankfull channel may be difficult, involving considerable subjective judgment. To avoid subjectivity, several objective methods have been proposed for defining the bankfull channel.

Simple methods rely on survey data of channel cross sections to identify either the top of bank (TOP), the beginning of the horizontal floodplain, or the inflection point (IP) in channel slope near the top of the banks, which marks the transition to the floodplain (Figure 6.12). Although either of these criteria can be used to define the bankfull channel, small differences in elevation between the TOP and the IP can lead to large differences in estimates of the magnitude and frequency of bankfull discharge (Navratil et al., 2006). Meandering rivers often exhibit differences in height between inner and outer banks (Lauer and Parker, 2008). In such cases, the elevation of the top of the lowest bank is considered the bankfull stage (Figure 6.12). In a channel with an approximately rectangular or trapezoidal shape, a plot of the change in the ratio of flow width to mean depth, W/D, versus stage will generally result in a minimum value of W/D at the top of the banks where the flow width increases dramatically for a slight increase in stage (Wolman, 1955) (Figure 6.13). This minimum can be associated with the bankfull stage. A related method involves plotting the flow area versus the flow width at different stages. For channels of uniform geometry, the bankfull stage occurs at an abrupt change in the relation between area and width (Figure 6.14).

Figure 6.11. Bankfull stage (black line) along the Embarras River, Illinois.

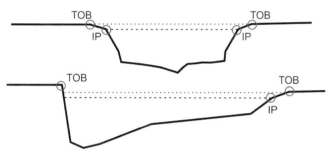

Figure 6.12. Diagram illustrating bankfull stages defined by the top of bank (TOB) and inflection point (IP) for channel cross sections with banks at the same elevation (a) and at different elevations (b).

Sedimentological and biological criteria have also been used to identify bankfull stage, including the upper limit of sand-sized particles in the channel perimeter and the stage corresponding to the lower limit of perennial vegetation, usually trees, growing along the channel (Williams, 1978a). These criteria generally complement, rather than supplant, morphological criteria. Vegetation characteristics generally are specific to certain environments, can be time dependent, and in many cases have not been strongly linked to specific channel-forming flow conditions. Interpretations of bankfull conditions based on changes in vegetation characteristics should therefore be employed carefully.

6.7.3 What Factors Complicate Determinations of the Bankfull Channel?

Several factors complicate the consistent identification of a bankfull channel based on morphological evidence, leading to uncertainty in estimating bankfull conditions (Johnson and Heil, 1996).

6.7.3.1 How Do Benches Complicate Bankfull Identification?

Benches are relatively narrow, flat depositional surfaces, often vegetated bank-attached bars, that form along the channel margins (Figures 6.9 and 6.15). The formation of benches may be related to the development of a floodplain within the bottom of a channel that has been enlarged through erosion or human excavation, to a change in hydrological regime that has decreased the magnitude of the dominant discharge, or to natural variability in depositional processes within a bankfull channel adjusted to the contemporary hydrological regime. In the first two cases, the formation of benches involves the reestablishment of mutual adjustment between channel form and the dominant discharge within an evolving river system. The benches represent the development of a new, smaller bankfull channel within a larger channel (Figure 6.15). The bankfull stage now corresponds to the elevation of the bench tops, not the top of the channel within which the benches sit. In the third case, the benches are depositional features that have formed within a contemporary bankfull channel adjusted to the prevailing hydrological regime (Figure 6.15). Evaluation of which case prevails in specific circumstances can often be difficult to determine without detailed historical information on changes in channel form in relation to the temporal sequence of flow events (see Chapter 15).

The development of depositional benches within a bankfull channel can complicate determinations of bankfull stage. Benches produce a distinct break in slope of the channel bank, a flat platform corresponding to the bench surface, and a channel within a channel (Figure 6.9). The minimum width–depth ratio and flow area–flow width methods do not work well when distinct benches develop within a bankfull channel. When benches are present, the minimum width–depth ratio

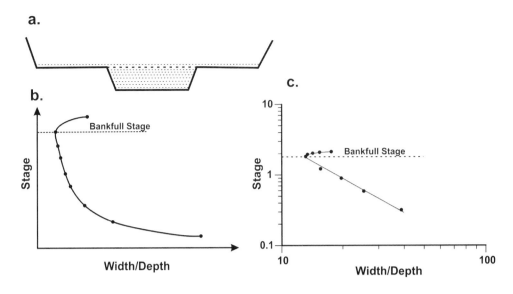

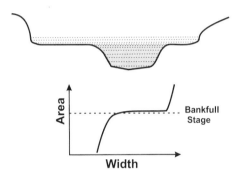

Figure 6.13. (a) Variation in water-surface elevation (dashed lines) within a channel cross section. Heavy dashed line indicates bankfull stage. (b) Variation in flow width–depth ratio with stage. Minimum W/D indicates bankfull stage. (c) Plot of width–depth ratio versus stage for Brandywine Creek (from Wolman, 1955).

Figure 6.14. (a) Change in water-surface elevation within a channel cross section. Shaded area indicates bankfull channel. (b) Variation in flow area with top width showing abrupt change in the relationship at bankfull stage.

or an abrupt change in the relation between flow area and flow width may occur in conjunction with a bench rather than near the top of the channel banks. To address this problem, the bench-index can be computed as (Riley, 1972)

$$BI = \frac{W_i - W_{i+1}}{h_{max(i)} - h_{max(i+1)}} \qquad (6.4)$$

where W is the width of the water surface, h_{max} is the maximum flow depth, and i is the flow stage with i increasing toward the channel bed. The bench index provides a measure of the steepness of the channel banks with changes in stage. Large values of BI indicate relatively flat slopes, or benches, whereas small values denote steep banks. Analysis of surveyed cross-channel profiles for rivers in Australia shows that bankfull stage can be associated with the peak in the BI farthest from the bed when BI is plotted against maximum depth (Figure 6.16) (Riley, 1972) – a recommendation consistent with other work on benches in Australian streams (Woodyer, 1968). This peak commonly

Figure 6.15. (a) Vegetated floodplain benches flanking a bankfull channel within a large, excavated drainage ditch, Illinois, USA. (b) Vegetated exposed benches contained within a bankfull channel in Illinois, USA.

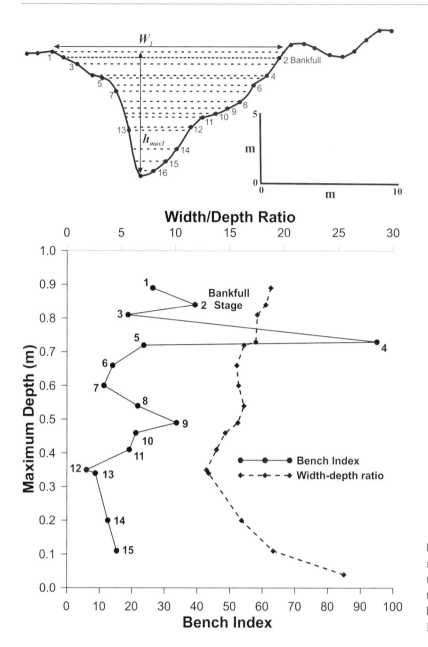

Figure 6.16. Change in bench index with changes in maximum depth for different stages (dashed lines, top). Numbers indicate index i in Eq. (6.4). Note that the minimum width–depth ratio occurs well below bankfull stage as indicated by the bench index (after Riley, 1972).

occurs below the elevation of the top of the banks. Peaks in the BI below the bankfull level define benches within the bankfull channel.

6.7.3.2 How Does Channel Instability Complicate Bankfull Identification?

One of the most important factors complicating identification of a bankfull channel is channel instability. If a channel is actively aggrading or degrading, channel form will vary over time rather than exhibit stability. Under such conditions, a channel may be identifiable, but this channel will not be the product of a prevailing formative discharge. In other words, form and flow are not mutually adjusted to one another. Progressive erosion of the channel bed can lead to incision and the development of a large trench that has little or no relation to a channel-forming flow of a specific recurrence interval. An example is the development of arroyos in the southwestern United States; the top of bank of an arroyo represents the edge of a terrace or abandoned floodplain, not the edge of the active floodplain (Figure 6.17a). A similar condition can occur in channels excavated by humans for the purpose of flood control or land drainage. If no within-channel benches have developed in response to excavation, mutual adjustment between the prevailing hydrological regime and channel form has not occurred (Figure 6.17b). Not all instability involves erosion. Progressive aggradation may infill a channel, producing an elevated streambed without well-defined banks (Figure 6.17d).

Figure 6.17. (a) Arroyo of Chaco Wash in Chaco Canyon, New Mexico. Height of arroyo banks is about 5 meters. A floodplain and well-defined channel have developed in the bottom of the arroyo. (b) Excavated stream channel in East Central Illinois where benches have not developed within the bottom of the enlarged channel. (c) Ephemeral bedrock stream in the McDowell Mountains, Arizona with steep valley walls that extend to the streambed (where person is standing). (d) Aggrading stream in Providence Canyon, Georgia. Note buried tree trunk at left and lack of well-defined channel banks.

6.7.3.3 What Other Factors Complicate Bankfull Identification?

Streams or rivers flowing through bedrock that lack alluvial deposits may not have a well-defined bankfull morphology (Figure 6.17c). Bankfull stage can be difficult to identify in braided rivers that consist of multiple channels arrayed across a floodplain with complex topography (Xia et al., 2010). A possible bankfull criterion for braided systems that lack well-defined channel banks (see Chapter 10) is the elevation

of bar tops (Wolman and Leopold, 1957). Along many alluvial rivers, bank heights differ on each side of the channel. This characteristic is quite common at bends in meandering rivers, where the inner bank is often lower than the outer bank. In these cases, the height of the lowest bank is considered the bankfull stage.

Finally, bankfull stage typically is treated in a highly two-dimensional manner, even when multiple cross sections are used to determine bankfull channel characteristics within a reach. The cross-sectional perspective of bankfull form leads to the impression that channel morphology is uniform and consistent along a reach. In reality, bank heights can vary spatially over short distances, and locally perched secondary channels can extend from the main channel onto the floodplain (Dzubakova et al., 2015). Under these conditions, bankfull "stage" does not correspond to an abrupt transition from in-channel flow to out-of-bank flow, as implied by two-dimensional representations, but instead occurs over a range of water-surface elevations that progressively lead to increasing amounts of flow on the floodplain (Czuba et al., 2019).

6.7.4 How Is the Bankfull Discharge Determined?

Once bankfull stage has been defined for a reach of interest, the bankfull discharge can be estimated by several methods (Williams, 1978a). The most direct method is to actually measure the discharge when the water stage equals the bankfull stage. Such measurements are rare because of the difficulties involved in accessing a site of interest at the exact time of bankfull flow. Nevertheless, with the availability of real-time gaging information for many streams, such measurements are now possible, especially on large rivers where bankfull conditions may persist for many hours or days. More commonly, bankfull discharge at a stream gaging site is estimated by using the rating curve for the site to determine the discharge when the water stage equals the bankfull stage. At ungaged sites, the bankfull discharge can be estimated using a reach-averaged flow resistance equation to compute mean velocity and then multiplying this velocity by the cross-sectional area of the bankfull channel (see Chapter 4). Such an approach involves selection of an appropriate roughness coefficient as well as information on the channel slope and is at best an approximate method.

In some cases, bankfull discharge is estimated from flood-frequency analysis by equating a discharge of a specific recurrence interval, usually $Q_{1.5}$, with bankfull flow (Q_{bk}). Such an approach is especially common in the practice of stream restoration, where $Q_{1.5}$ is sometimes equated with Q_{bk} for the purpose of channel design. Given the wide range of variability in the recurrence interval of bankfull discharge, this method has a high level of uncertainty. Another approach to estimating bankfull discharge at ungaged sites is to use regional power-function equations that have been developed through statistical analysis for different parts of the United States. These equations relate Q_{bk} to one or more predictor variables that are easily measured for a site of interest (Bent and Waite, 2013, table 8). A common predictor variable is the drainage area upstream of the site (A_d):

$$Q_{bk} = aA_d^{\ b} \tag{6.5}$$

where the exponent b usually has a value between 0.8 and 1.0. Measures of discharge, such as the two-year flood, also have been used to predict bankfull flow (He and Wilkerson, 2011). Because predictive relations typically are in the form of power functions, uncertainty associated with site-specific estimates of bankfull discharge can be quite large.

6.7.5 What Are Geomorphic Work and the Effective Discharge?

The attempt to associate the dominant discharge for stable rivers with the bankfull discharge, while intuitively appealing, is unsubstantiated without independent support. From a geomorphological perspective, a key role of rivers is to transport sediment supplied by the drainage basin. The movement of sediment by various flows can be viewed as the geomorphic work performed by the river (Wolman and Miller, 1960). Establishing a link between fluxes of sediment associated with various flows and the magnitude and frequency of these flows can provide insight into the relative importance of different flows in performing geomorphic work.

The information needed to evaluate geomorphic work include 1) the probability distribution of discharges of various magnitudes derived from a flow–duration analysis of mean daily discharges and 2) a function relating sediment load (Q_s) and discharge (Q) that holds for all discharges exceeding a critical threshold for sediment transport (Q_c) (Figure 6.18). The first component can be determined for any stream location where a gaging station has produced a record of mean

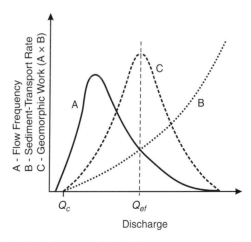

Figure 6.18. Flow frequency (A), sediment-transport rate (B), and geomorphic work (C) – the product of discharge frequency and the magnitude of sediment transport. The peak of the geomorphic work curve is the effective discharge (adapted from Wolman and Miller, 1960).

daily discharge over time. Ideally, the second component should be developed from actual measurements of discharge and sediment load used to produce a sediment rating curve. Because a rating curve typically defines the relation between suspended sediment load and discharge, and does not consider bedload transport, the analysis will underestimate sediment loads for streams with substantial bedload contributions to the total load. A predictive relation for sediment transport can be substituted for a rating curve, but such an approach is likely to include substantial error unless the relation can be calibrated to field measurements.

Multiplying the frequency of each discharge in the flow-frequency distribution by the magnitude of the sediment load corresponding to that discharge, the product of frequency and magnitude, yields a curve defining the geomorphic work associated with each discharge, where geomorphic work represents the frequency-weighted total amount of sediment transported by each discharge over time (Figure 6.18). The peak of this curve, which represents the discharge that performs the greatest amount of geomorphic work, is known as the effective discharge (Q_{ef}). In practice, the effective discharge is determined by dividing the range of measured discharges for a gaging station into intervals and then computing the total amount of transported sediment for each discharge interval to produce a histogram of geomorphic work, expressed as the percentage of the total sediment load transported by each discharge interval. Computations of total load for each interval can be based on a rating curve or, alternatively, on estimates of the average sediment load transported by flows within each interval derived from sediment load data for discharges within the interval. The method is sensitive to the selection of discharge intervals (Sichingabula, 1999), and estimates of Q_{ef} can vary substantially depending on the size of the intervals (Ma et al., 2010). Practical guidance for interval selection has been developed to help alleviate this problem (Biedenharn et al., 2000).

6.7.6 What Is the Relationship among Dominant Discharge, Bankfull Discharge, and Effective Discharge?

Initial empirical evaluation of geomorphic work in rivers revealed that a large proportion of the total sediment load is transported by flows that recur at least once each year and in some cases multiple times per year (Wolman and Miller, 1960). The effective discharge often has a frequency similar to that of the bankfull discharge. This analysis suggests that events of moderate magnitude and frequency have the strongest influence on the form of stable channels. The most frequent flows typically have limited transport capacity and do

not transport sufficient amounts of sediment to contribute greatly to the maintenance of channel shape. On the other hand, large floods, which transport enormous amounts of sediment, occur too infrequently to strongly influence channel form. Moderate flows approximately equal to the bankfull discharge have the appropriate combination of magnitude of transport and frequency of occurrence to transport the greatest amount of sediment over time. Analytical modeling of equilibrium adjustment of channel form to time-varying flows supports the contention that extreme events are less important than moderate magnitude–frequency events in determining the equilibrium geometry of river channels (Blom et al., 2017a).

Subsequent empirical research on geomorphic work has produced varied results (Table 6.2). In some cases, the effective discharge is similar to the bankfull discharge (Andrews, 1980; Carling, 1988; Rhoads, 1990a; Emmett and Wolman, 2001; Torizzo and Pitlick, 2004; Lenzi et al., 2006), providing support for the conclusion that the dominant discharge (Q_{dd}) conforms to the bankfull discharge and the discharge that performs the most geomorphic work. Moreover, when combined with the view that the recurrence interval of the bankfull discharge can be treated as a constant, usually 1.5 years, the tentative conclusion is that $Q_{dd} = Q_{bk} = Q_{1.5} = Q_{ef}$ (e.g., Simon et al., 2004).

The agreement between bankfull discharge and effective discharge tends to be strongest when considering bedload or total load estimates of Q_{ef}. Even in these cases, however, the effective and bankfull discharges may be dissimilar (Bunte et al., 2014; Hassan et al., 2014). Improvements in prediction and measurement of bedload transport are leading to advances in understanding the role of this type of transport in geomorphic work (Barry et al., 2008; Downs et al., 2016), but further research is needed to clarify the role of bedload transport in channel-shaping flows. For the most part, effective discharges estimated from suspended-load data are smaller and occur more frequently than bankfull discharges or an approximate equivalent, such as $Q_{1.5}$ (Table 6.2). A fairly comprehensive study found that median recurrence intervals of Q_{ef} are similar to $Q_{1.5}$ (Simon et al., 2004) but did not provide details on variability in estimated recurrence intervals of Q_{ef} about median values, which may have been considerable. Based on available evidence, the extent to which the bankfull discharge and effective discharge can be equated remains uncertain.

Several studies have explored geomorphic work in rivers from a theoretical perspective (Nash, 1994; Goodwin, 2004; Sholtes et al., 2014; Basso et al., 2015). If the flow-frequency distribution is log-normally distributed and the sediment load is a simple power function of discharge, it

Table 6.2. Examples of effective discharge determinations

Location	RI$_{ef}$ (yr)	RI$_{bk}$ (yr)	f$_{ef}$ (duration – day yr^{-1})	f$_{bk}$ (duration – day yr^{-1})	Notes	Source
Nine stream locations, United States			28 to 154; all but one site < 68	$Q_{ef} < Q_{bk}$	Effective discharge estimated from data on suspended-sediment load	Benson and Thomas (1966)
Four streams in Cumberland Drainage Basin, Australia	1.15 to 1.45, AMS 0.2 to 0.4, PDS $Q_{ef} < Q_{bk}$	1.92 to > 50, AMS			Effective discharge estimated using bedload-transport equations. Flows larger than Q_{ef} appear to shape channel capacity	Pickup and Warner (1976)
15 stream locations in Wyoming and Colorado, USA	$Q_{ef} \approx Q_{bk}$	1.01 to 4, AMS, modal value of 1.18 to 1.40	1.5 to 11, mean of 5.9		Effective discharge determined from total load using rating curves for measured suspended load and transport functions to compute bedload. Absolute percentage difference between Q_{ef} and Q_{bk} = 8%	Andrews (1980)
Two streams in England	$Q_{ef} \approx Q_{bk}$ for alluvial channel $Q_{ef} > Q_{bk}$ for constrained channel in till	0.9, PDS			Effective discharge determined using bedload rating curves derived from field measurements	Carling (1988)
21 stream locations in Canada			< 0.36 to 146; most between 3.65 and 36.5		Effective discharge determined using data on suspended-sediment load. Analysis of Q_{ef} complicated by a wide array of forms of geomorphic work curves	Ashmore and Day (1988)
One ephemeral stream, Arizona	2.38 to 2.53, AMS	$Q_{ef} = Q_{bk}$	0.03 to 0.04	$Q_{ef} = Q_{bk}$	Flow frequency determined from real-time hydrograph data at intervals of minutes rather than days. Effective discharge computed using a bedload-transport function	Rhoads (1990a)
43 stream locations, USA			0.09 to 50; mean of 8.7		Effective discharge computed from data on suspended-sediment load	Nash (1994)
Eight stream locations in Frasier River basin, Canada			0.09 to 59; mean of 30.7	Q_{ef}/Q_{bk} = 0.55 to 1.90	Effective discharge computed from data on suspended-sediment load; Q_{bk} approximated by $Q_{1.58}$	Sichingabula (1999)
Five stream locations in Wyoming and Idaho, USA	Q_{ef}/Q_{bk} = 0.98 to 1.30	1.5 to 1.7, AMS			Effective discharge computed using bedload rating curves derived from field measurements	Emmett and Wolman (2001)

Table 6.2. (cont.)

Location	RI$_{ef}$ (yr)	RI$_{bk}$ (yr)	f$_{ef}$ (duration – day yr^{-1})	f$_{bk}$ (duration – day yr^{-1})	Notes	Source
10 sites in Mississippi and 475 sites elsewhere across the United States	$Q_{ef}/Q_{1.5}$ = 0.56 to 2.72, mean of 1.04 for Mississippi sites. Median RI = 1.1 to 2.3 for ecoregion groupings of 475 sites	$Q_{1.5}$, AMS, assumed to equal bankfull			Effective discharge computed from data on suspended-sediment load	Simon et al. (2004)
12 gravel-bed streams in Colorado, USA			1.5 to 11.3, mean of 4.2	Q_{ef}/Q_{bk} = 0.7 to 1.5, mean of 1.0	Effective discharge estimated using a bedload transport equation	Torizzo and Pitlick (2004)
23 stream locations, Illinois, USA	< 1.01 to 1.23, AMS		7 to 194, mean of 67, PCM; 6 to 194, mean of 52, MIM		Effective discharge computed from data on suspended-sediment load using two methods: power curve method (PCM) and mean interval method (MIM)	Crowder and Knapp (2005)
One stream in the Italian Alps	Q_{efs} < 1, AMS Q_{efs}/Q_{bk} = 0.37 Q_{efb} = 1.58 to 2.31, AMS Q_{efb}/Q_{bk} = 0.98 to 1.4	1.6, AMS	Q_{efs} = 11 to 49 Q_{efb} = 0.10 to 0.42	0.38	Separate effective discharges computed for bedload (Q_{efb}) and suspended load (Q_{efs}) based on sediment rating curves for both types of load	Lenzi et al. (2006)
Several locations along the lower Waipoa River, New Zealand	Q_{ef} equal to ¼ of the mean Q_{bk}	< 2 to > 10, mean of 4, AMS			Effective discharge estimated from data on suspended-sediment load	Gomez et al. (2007)
10 stream locations, Wuding River Basin, China			0.095 to 11.5 Loess streams 68.5 to 334 Eolian-sand streams	Q_{ef} < Q_{bk}	Effective discharge estimated from data on suspended-sediment load. Q_{ef} changes, in some cases dramatically, with changes in discharge bins used to compute it. Many histograms exhibit two peaks, one at low Q and one at high Q	Ma et al. (2010)
44 stations in Yunnan and Tibet			≈120		Effective discharge estimated from data on suspended-sediment load. Q_{ef} is equivalent to the mean monsoon discharge, which persists seasonally for several months per year	Henck et al. (2010)

Table 6.2. (cont.)

Location	RI$_{ef}$ (yr)	RI$_{bk}$ (yr)	f$_{ef}$ (duration – day yr^{-1})	f$_{bk}$ (duration – day yr^{-1})	Notes	Source
36 stream locations, British Columbia, Canada	$Q_{ef} > Q_{bk}$ for seven streams, $Q_{ef} < Q_{bk}$ for 29 streams	1.01 to 500, AMS	0.036 to 173, mean of 26		Effective discharge estimated using a bedload transport equation	Hassan et al. (2014)
Six locations in the Ganges River system	1.14 to 5.12, AMS $Q_{ef} < Q_{bk}$	1.18 to 41, AMS	0.036 to 9, mean of 3.8		Effective discharge estimated from data on suspended-sediment load	Roy and Sinha (2014)
41 Rocky Mountain streams, USA	$Q_{bk} < Q_{ef} = Q_{max}$ for all but a few streams, where Q_{max} is largest observed Q for each stream	$Q_{1.5}$, AMS, assumed to equal bankfull			Effective discharge estimated from data on bedload transport used to produce bedload rating curves	Bunte et al. (2014)

AMS: annual maximum series; PDS, partial duration series.

can be shown analytically that the effective discharge equals (Goodwin, 2004; Sholtes et al., 2014)

$$Q_{ef} = \exp[(b-1)\sigma_{\ln Q}{}^2 + \overline{\ln Q}] \qquad (6.6)$$

where b is the exponent in the sediment load-discharge power function, $\sigma_{\ln Q}$ is the standard deviation of the natural logarithms of discharge, and $\overline{\ln Q}$ is the mean of the natural logarithms of discharge. This relation shows that the magnitude of Q_{ef} increases as the steepness of the sediment load curve increases, as the spread of the flow-frequency distribution increases, and with an increase in the mean size of flows (Figure 6.19). These three factors should be influenced by sediment-size characteristics of bed material, by variations in climate and land cover, and by watershed properties, such as drainage area, relief, and soil conditions. In mountain headwater streams, the magnitude of the effective discharge related to bedload transport is strongly influenced by the magnitude of b (Bunte et al., 2014). When this exponent exceeds a value of 4, the effective discharge shifts to a discharge with a recurrence interval that equals or exceeds the maximum discharge associated with a particular stream site. The squared term on the right side of Eq. (6.6) shows that the effective discharge is especially sensitive to variations in the standard deviation of the natural logarithms of discharge, indicating that rivers with high-variability hydrological regimes typically have larger effective discharges than rivers with low-variability hydrological regimes (Basso et al., 2015). In river systems dominated by seasonal variations in flow regime, the effective discharge may strongly reflect seasonality. In southeast Asia, the effective discharge of rivers equals the mean discharge during the monsoon season (Henck et al., 2010). The effective discharge will also increase as the threshold for sediment mobilization increases (Q_c), which will shift the sediment transport curve to the right (Baker, 1977) (Figure 6.19e). Similar evaluations of factors affecting Q_{ef} can be performed for flow-frequency distributions other than the log-normal distribution (Goodwin, 2004).

Alternatives to the effective discharge have been suggested for determining the importance of geomorphic work in fluvial systems. The half-load discharge (Q_{hl}) is the value of discharge above which and below which half of the total sediment load is transported (Vogel et al., 2003; Ferro and Porto, 2012). In most cases, $Q_{hl} > Q_{ef}$, unless Q_{ef} is a relatively rare event (Klonsky and Vogel, 2011). The half-load discharge also serves as a good predictor of bankfull discharge, generally outperforming other metrics such as Q_{ef}, $Q_{1.5}$, and Q_2 (Sholtes and Bledsoe, 2016). The functional-equivalent discharge (Q_{fe}) truly represents the entire probability distribution of geomorphic work (Doyle and Shields, 2008). It is a single value of discharge and associated sediment load, which, when integrated over time, reproduces the actual total sediment load generated over the entire range of

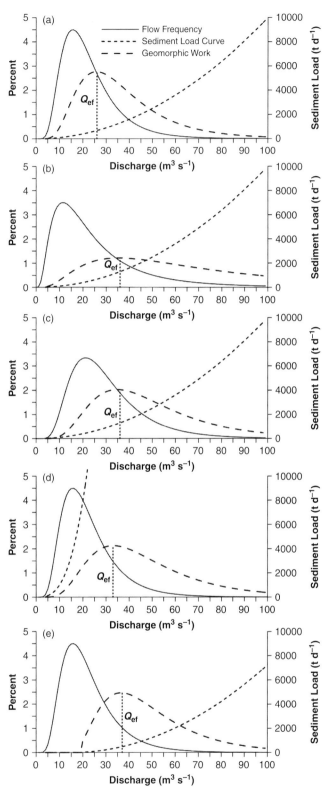

Figure 6.19. Change in effective discharge with changes in controlling factors for a log-normal flow-frequency distribution and power function of sediment-transport rate: (a) $\overline{Q} = 20$ m^3 s^{-1}, $\sigma_{\ln Q} = 0.5$, $b = 2$, $Q_c = 5$ m^3 s^{-1}, (b) $\overline{Q} = 20$ m^3 s^{-1}, $\sigma_{\ln Q} = 0.75$, $b = 2$, $Q_c = 5$ m^3 s^{-1}, (c) $\overline{Q} = 27$ m^3 s^{-1}, $\sigma_{\ln Q} = 0.5$, $b = 2$, $Q_c = 5$ m^3 s^{-1}, (d) $\overline{Q} = 20$ m^3 s^{-1}, $\sigma_{\ln Q} = 0.5$, $b = 3$, $Q_c = 5$ m^3 s^{-1}, (e) $\overline{Q} = 20$ m^3 s^{-1}, $\sigma_{\ln Q} = 0.5$, $b = 3$, $Q_c = 20$ m^3 s^{-1}.

discharge variability. Calculation of this metric is less straightforward than determinations of Q_{ef} and Q_{hb}, and the utility of Q_{fe} for characterizing geomorphic work has yet to be fully explored.

6.7.7 How Is Bed-material Transport Related to Equilibrium Adjustment in Gravel-Bed Rivers?

In gravel-bed rivers, which transport abundant bedload, the concept of geomorphic work based on suspended-sediment transport has been complemented by a view of equilibrium adjustment that considers bed-material mobilization in relation to bankfull channel form. This idea holds that gravel-bed rivers develop a morphology whereby the onset of mobilization of bed material occurs at or near bankfull conditions. The theory proposes that at bankfull discharge, the shear stress of the flow equals or slightly exceeds the critical shear stress for motion of the bed material. Empirical analysis of detailed event-based bedload-transport data for a stream in Puerto Rico indicates that at bankfull discharge, the ratio of the flow shear velocity ($u_{\star bk}$) to the critical shear velocity of bed material ($u_{\star c}$) is 1.1 – a value consistent with weak mobilization of bedload (Phillips and Jerolmack, 2016). Data for 186 stream locations across the United States reveal that $u_{\star bk}/u_{\star c} = 1.3$, despite dramatic differences in characteristics of flow-frequency distributions related to climate. Moreover, the magnitude–frequency distributions of potential bedload transport for these sites indicate that the largest potential amount of bedload transport occurs for events of moderate magnitude and frequency, even for sites with a preponderance of extreme floods. These results support the contention that the channel form of gravel-bed rivers is adjusted to the contemporary hydrological regime, particularly to events of moderate magnitude and frequency, so that rivers maintain near-threshold sediment-transport conditions at bankfull flow. Further analysis shows that consideration of sediment supply is important (Pfeiffer et al., 2017). The median value of Θ_{bk}/Θ_c is significantly higher for gravel-bed streams along the west coast of the United States ($\Theta_{bk}/\Theta_c = 2.35$), where sediment supply is relatively high, than the median value for gravel-bed streams elsewhere in the United States, where sediment supply is relatively low ($\Theta_{bk}/\Theta_c = 1.03$). The bankfull morphology of streams along the West Coast apparently has adjusted to accommodate the relatively large amounts of sediment supplied to these streams. This finding implies that adjustment of channel morphology to produce near-threshold conditions for bed-material mobilization at bankfull flow occurs mainly in gravel-bed rivers with low sediment supply. While the premise of near-threshold equilibrium adjustment seems to be relevant for at least some gravel-bed rivers, it is unlikely to hold in sand-bed rivers, where bed material is mobilized readily by flows smaller than the bankfull discharge.

6.8 What Is the Concept of Geomorphic Effectiveness and How Does It Relate to Channel-Formative Events?

The second major perspective on the relationship between channel-forming events and river morphodynamics, geomorphic effectiveness, adopts a time-dependent perspective on this relationship. This perspective focuses on the extent to which individual formative flows produce change in channel form over time. It contrasts with the equilibrium perspective based on the concept of a dominant discharge, which emphasizes the importance of a particular discharge, the bankfull flow, in shaping channel form and the mutual adjustment of process and form that results in a stable, time-independent channel configuration.

6.8.1 What Are the Limitations of the Equilibrium Viewpoint?

Although equilibrium thinking continues to play an important role in contemporary fluvial geomorphology, limitations of the equilibrium concept are now widely recognized (Phillips, 1992). Some of these limitations are apparent through empirical studies of river morphodynamics, whereby change in channel form through time is considerable and does not conform to any reasonable standard of stable channel forms or the lack of aggradation and degradation. Other objections are pitched at a conceptual level based on considerations of temporal and spatial scale and problems inherent in determining whether equilibrium actually exists in any real sense or whether it is just a convenient regulative principle or metaphor for referring to certain types of morphodynamic conditions (Rhoads and Thorn, 1993; Bracken and Wainwright, 2006).

Regulative principles and metaphors are useful for guiding the development of testable theories but cannot be directly tested empirically. In particular, any attempt to define the behavior of a geomorphic system as corresponding to a steady state is time dependent, a widely acknowledged proviso (Zhou et al., 2017). Note the inclusion of the phrase "over a period of years" in the definition of a graded stream (Leopold and Bull, 1979), which implies that beyond a "period of years" the stream may no longer be graded. Often the graded river is associated with the modern timescale of several years to centuries (Figure 1.9) (Schumm and Lichty, 1965). Testing can be conducted to determine whether a time series of data on channel form is stationary over a certain interval, but this test only confirms stationarity, not the actual existence of an equilibrium condition or steady state. In other words, the concept that rivers attain an equilibrium state is a philosophical stance as much as, if not more than, a scientifically testable hypothesis (Rhoads and Thorn, 1993).

6.8.2 What Is a Time-Dependent Perspective on River Dynamics and How Is It Related to Nonlinear Dynamics?

An alternative perspective on river dynamics is to view rivers both as persistent and also as continuously evolving over time and space. This perspective draws upon a philosophical position in which the fundamental nature of all reality is processual and components of this reality may endure for certain periods of time but also are subject to transformation, change, and even eradication (Rhoads, 2006a). As noted by the Greek philosopher Heraclitus, one cannot step into the same river twice because both the person and the river are ever changing. Nevertheless, an essential tension exists between change and endurance. To paraphrase philosopher Alfred North Whitehead's (1925, pp. 86–87) statement about change and endurance in relation to mountains: the river endures, but when after the ages it ceases to exist, it has gone.

An evolutionary, time-dependent perspective on river dynamics is consistent with recent developments that treat geomorphic systems, including rivers, as nonlinear dynamical systems (Phillips, 2006a). Evolutionary trajectories of change in system characteristics, including morphological characteristics, are important in such systems because patterns of change over time and space are diagnostic of underlying nonlinear dynamics (Phillips, 2006b). Nonlinear geomorphic systems are often characterized by thresholds that, once surpassed, lead to dramatic, abrupt, nonlinear change in system form and dynamics (Schumm, 1979). Thresholds can be either extrinsic or intrinsic. Extrinsic thresholds are related to changes in conditions external to the system that produce change within the system. The entrainment threshold for sediment mobilization is an example of an extrinsic threshold, where changes in hydraulic conditions associated with the delivery of water to a river from runoff lead to forces acting on bed material overcoming the forces tending to keep the bed material at rest. Intrinsic thresholds exist within nonlinear systems and can be transcended without a change in conditions external to the system. An example is progressive aggradation on valley floors until the gradient of the valley floor is locally increased beyond a threshold for incision of gullies into the accumulating alluvial deposits (Schumm, 1979).

6.8.3 What Are the Major Tenets of Geomorphic Effectiveness and Landscape Sensitivity?

The concept of geomorphic effectiveness has arisen through recognition of the need to understand the time-dependent behavior of geomorphic systems, particularly changes in morphological characteristics of these systems over time (morphodynamics) (Wolman and Gerson, 1978). It is closely related to the concept of landscape sensitivity (Brunsden and Thornes, 1979), which also adopts a time-dependent view of landscape dynamics. These related concepts include several different components.

Geomorphic effectiveness focuses on the occurrence of discrete formative events over time and space, where formative events are those that produce substantial morphological change. Formative events may be high-magnitude, low-frequency occurrences, but any event that produces substantial, lasting change is considered a formative event. Nonformative events may produce some change in river form, but generally this change occurs within the domain of natural variability of form about a characteristic mean state – a type of behavior consistent with notions of steady state, equilibrium, and stability.

In contrast to geomorphic work, which links the importance of a particular flow to the amount of sediment it transports, geomorphic effectiveness emphasizes the capacity of formative events to substantially alter the shape or form of the physical landscape. A formative event may not necessary transport the most sediment over some interval of time, but it does produce substantial change in river form. The more an event produces lasting change in form, the more effective it is. Thus, the notion of an effective discharge, as defined in geomorphic work, and the concept of geomorphic effectiveness refer to different aspects of geomorphic activity. Although this difference is somewhat confusing given the similarity in terminology (Lisenby et al., 2018), it is nevertheless an important distinction.

The concept of sensitivity complements geomorphic effectiveness and refers to the propensity for a river system to change morphologically or, alternatively, the capacity to resist morphological change (Downs and Gregory, 1995; Brunsden, 2001; Fryirs, 2017; Lisenby et al., 2018). Highly sensitive fluvial systems change readily, whereas systems with low sensitivity do not change easily. Rivers with large, frequent changes in morphological characteristics often are referred to as unstable or as exhibiting instability. In many respects, determinations of stability versus instability involve subjective judgment, because rivers fundamentally are dynamic landforms. From a scientific perspective, mutual adjustment between channel form and the prevailing hydrological-sediment regime is often held to be the normal state of a river system, whereas from a practical perspective, some types and rates of change in river form are viewed by humans as acceptable, whereas others are not. Stability exemplifies mutual adjustment and acceptable types and rates of change, whereas instability typifies lack of mutual adjustment or undesirable types or rates of change.

The concepts of effectiveness and sensitivity recognize that the persistence of change produced by a relatively rare formative event depends on how often the event recurs in relation to the time required for more frequently occurring processes to undo the morphological effects of the formative event (Wolman and Gerson, 1978; Brunsden and Thornes, 1979). If the time between formative events, the inter-event interval (t_f), is greater than the recovery time from the disturbance associated with intervening processes (t_r), the system can effectively recover from the disturbance (Figure 6.20). Such a system exhibits a high degree of resilience to disturbance (Brunsden, 2001; Phillips and van Dyke, 2016). Under these conditions, the development of characteristic forms, or morphological configurations that fluctuate to a minor extent around a mean morphological state, is possible. During the period when the system is recovering its characteristic form,

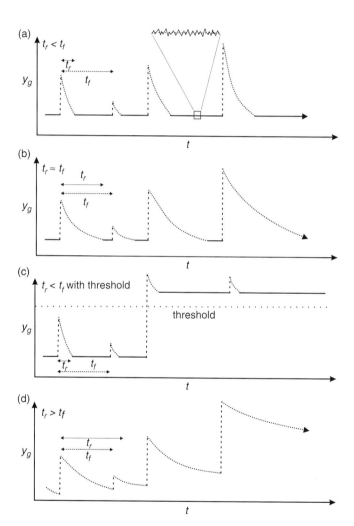

Figure 6.20. (a) Change in channel form (y_g) over time (t) for $t_r < t_f$. Horizontal solid lines indicate times of characteristic form (inset shows steady-state fluctuations around the characteristic form). Vertical dashed lines indicate change in form produced by formative events. Dotted lines indicate periods of recovery of form. (b) Pattern of morphodynamics for $t_f \approx t_r$. (c) Pattern of morphodynamics for $t_r < t_f$ with threshold that leads to the development of a new characteristic form. (d) Pattern of morphodynamics for $t_r > t_f$.

the system is said to be in a state of disequilibrium (Renwick, 1992). If $t_f \gg t_r$, characteristic forms will persist for long stretches of time over the history of the river system. As the difference between t_f and t_r decreases, the amount of time when the mean form of the river channel remains unchanged will decrease. For $t_r > t_f$, the morphodynamics of the system will be characterized by transient conditions in which form is either changing as the result of a disturbance or changing as the result of recovery from a disturbance. Such river systems do not exhibit characteristic forms and are in a state of nonequilibrium (Stevens et al., 1975; Renwick, 1992). Past forms produced by formative events persist in the landscape but are dynamically modified by processes occurring during intervals between major formative events (Calver and Anderson, 2004). Moreover, in contrast to systems experiencing disequilibrium adjustments that tend to move the system toward a stable configuration, the overall evolutionary trajectory of nonequilibrium river systems is difficult to predict.

Thresholds of change may exist within the river system that lead to a new characteristic form once the threshold is transgressed, either by the occurrence of a major formative event or by gradual change in external environmental conditions (Figure 6.20). The proximity of the river system to possible thresholds of change, both intrinsic and extrinsic, can strongly influence whether or not an event is formative and the morphological response of the system to a formative event (Downs and Gregory, 1995; Brunsden, 2001). A river system in close proximity to a threshold of channel change, for example the narrow band of bankfull discharge and channel slope separating meandering rivers from braided rivers (see Chapter 8), is more sensitive to major morphological change than a river system that is not close to the threshold. Events that otherwise would not be formative may become formative if these events drive the system over the threshold. The occurrence of thresholds can also contribute to the development of transient forms in nonequilibrium systems.

The effectiveness/sensitivity perspective calls attention to the importance of the sequencing of events of different magnitudes (Beven, 1981). The effectiveness of a particular magnitude of flood in altering the form of a river channel or floodplain may depend not only on its magnitude but on when it occurs in relation to other formative events. If two floods of equal size occur in close sequence, the first one may mobilize much of the sediment readily available for transport throughout the watershed, exhausting the sediment supply. As a result, the second flood will mobilize sediment from the channel boundary, producing major change in channel form. Conversely, if a flood with a specific peak magnitude produces major changes in channel form, and a second flood of the same magnitude occurs soon after the first flood, the second flood may have little influence on channel form,

because the form is already modified by an event of this magnitude. The two floods are equal in magnitude, but the ordering of the events plays a role in determining their effectiveness.

Geomorphic effectiveness stresses the relevance of geographic context in river morphodynamics (Wolman and Gerson, 1978). The extent to which a system will be modified by particular hydrological events and the extent to which it will recover from these events are dependent on climate, geology, vegetation characteristics, and watershed properties, such as soil thickness, drainage area, and relief. A runoff event of a particular magnitude that produces little or no morphological change in a low-relief, humid-temperate river system flowing through a forested floodplain may result in major, persistent morphological change in a high-relief, arid-region river system with little or no floodplain vegetation.

In general, ephemeral rivers in arid or semiarid regions tend to be characterized by transient forms and nonequilibrium morphodynamics (Graf, 1988a), whereas perennial rivers in humid-temperate environments commonly display characteristic forms and equilibrium morphodynamics (Leopold et al., 1964). Arid-region rivers often exhibit transient morphodynamic behavior in which channel form is not adjusted to prevailing hydrological and sediment-transport regimes. In such cases, river morphology reflects the nonsystematic influence of individual formative flows – large and small, recent and past – on the size and shape of the channel (Hooke, 2016a). Important factors influencing morphodynamics in dryland rivers include the extent to which connectivity of coarse-sediment transfer among adjacent reaches occurs during formative events (Hooke, 2004a) and the role of vegetation in modifying sediment fluxes and stabilizing channel form (Sandercock and Hooke, 2011; Hooke and Mant, 2015). In some cases, equilibrium and nonequilibrium forms may coexist in dryland rivers (Tooth and Nanson, 2000a). Small desert-mountain fluvial systems in Arizona exhibit both characteristic-form and transient-form components (Rhoads, 1990a) (Figure 6.21). A well-defined low-flow channel has bankfull dimensions adjusted to the effective discharge of the bedload-transport regime over a 24-year period of hydrological record. On the other hand, this channel is situated within a large incised trench that formed through a climate-related increase in the frequency of large floods, the occurrence of several high-magnitude events over a short time, the occurrence of an exceptionally large flood, or threshold-related incision associated with headward migration of a knickpoint. Thus, an equilibrium characteristic form is set within a nonequilibrium palimpsest morphology.

Spatial scale is an important consideration of geomorphic effectiveness (Fryirs, 2017). The effectiveness of events in changing fluvial forms can vary from the grain scale to the drainage-basin scale (Figure 1.7). Thus, events that are

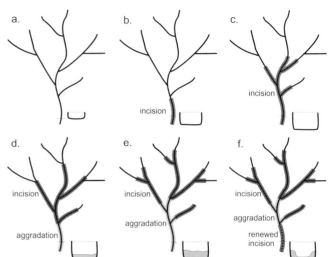

Figure 6.22. Complex response caused by a change in base level. (a) Fall in base level at network mouth. (b) Headward migration of channel incision. (c) Continued migration of incision. (d) Migration of channel incision; downstream flux of sediment from headward migration of incision leads to aggradation, (e) continued headward migration of incision and expansion of downstream aggradation, (f) renewed incision at network mouth. Changes in a channel cross section at a location near the basin mouth shown by inset figures. Shading in cross section indicates accumulated sediment.

Figure 6.21. (a) Ephemeral low-flow channel in South Mountain Arizona. (b) Low-flow channel within incised trench (trench wall on left).

formative at one scale may not be formative at another. A flood that changes a morphological component of the river at the bar-unit scale may not produce much change at the planform scale. Adjustments at the scale of the river network also can be distributed unevenly over space as the effects of a localized disturbance unfold over time.

In response to baselevel change, streams in small drainage basins can undergo cycles of erosion and sedimentation that vary spatially throughout the drainage network (Schumm, 1979). Baselevel lowering at the mouth of a basin can trigger channel incision that extends upstream through the network by migration of knickpoints. The spatial rate of migration is commonly an exponential function of distance, with rapid headward migration downstream and progressively diminishing rates of migration upstream (Graf, 1977a). The high flux of sediment downstream as the wave of erosion moves upstream through the tributaries can lead to aggradation within the incised main channel. Renewed incision downstream may occur in response to a local increase in channel gradient induced by aggradation and to exhaustion of available sediment in the tributaries. This spatially uneven pattern and

asynchronous timing of erosion and aggradation episodes throughout the network has been referred to as complex response (Schumm, 1979) (Figure 6.22).

Complex response occurs most commonly and most rapidly in small, high-energy stream networks. In intermediate-scale fluvial systems, complex response to baselevel fall may proceed slowly, continuing for millennia (Faulkner et al., 2016). In large lowland river systems, the response to changes in baselevel typically is spatially restricted and can be accommodated by changes in channel pattern, width, depth, and roughness (Schumm, 1993). For example, morphological effects of changes in sea level associated with cycles of Pleistocene glaciation in the Mississippi River valley extend only a few hundreds of kilometers upstream of the river mouth (Rittenour et al., 2007).

6.8.4 What Factors Influence the Geomorphic Effectiveness of Large Floods?

The event-based focus of geomorphic effectiveness, along with the emphasis on morphological change, has connected this perspective closely to interest in the role of extreme events, particularly large floods, on river morphodynamics. The definition of a large flood is subjective, but in general it is an event with a magnitude at least twice as large as the mean annual

High Potential	Low Potential
High Stream Power	Low Stream Power
High Hydrologic Variability	Low Hydrologic Variability
Coarse Bed Material	Fine Bed Material
Low Bank Resistance	High Bank Resistance

Figure 6.23. Factors influencing the potential for major morphological response of a river to a large flood.

flood. Geomorphic work indicates that flows of moderate magnitude and frequency are important in shaping channel characteristics, given the close association between the bankfull and effective discharges. Under certain conditions, however, large floods can substantially alter channel form.

The factors that influence the susceptibility of rivers to major morphological change during extreme floods are generally the same as those that influence river sensitivity and resilience. Important factors include the interplay between the energy of the flow and the resistance of the channel perimeter to erosion, especially when sediment supply is limited and excess capacity for transport of sediment comprising the channel boundary exists (Figure 6.23). A high degree of hydrological variability is important, because it tends to produce large floods that greatly exceed the magnitudes of moderate floods. Under these conditions, the morphological characteristics of the river system shaped by frequent floods will be increasingly unadjusted to the flow and sediment-transport conditions that occur during large, infrequent floods. Although rivers that transport coarse bed material may inherently have greater resistance to erosion than those that transport fine material (Wohl, 2008), once coarse material is mobilized during an extreme event it can act as an abrasional agent that contributes to morphological change (Kochel, 1988). Moreover, morphological recovery is difficult when large quantities of coarse material that only an extreme event can mobilize are redistributed throughout a river by such an event, resulting in persistent depositional and erosional forms.

6.8.4.1 How Is Flood Power Related to Morphological Change?

The energy expenditure of extreme events is referred to as flood power – the stream power per unit area (ω) of a flood (Baker and Costa, 1987). Most extreme floods in the United States for drainage basins with drainage areas between 0.39 and 3,100 km^2 have values of flood power in excess of 1,000 W m^{-2}, and a few have values in excess of 10,000 W m^{-2} (Figure 6.24).

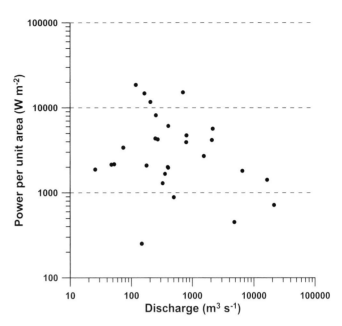

Figure 6.24. Relation between stream power per unit area and discharge of extreme floods (data from Baker and Costa, 1987).

Virtually all these floods occurred in steep, confined bedrock rivers. Confined rivers have a large percentage of the total stream length in contact with the valley margin (Fryirs et al., 2016) or a small ratio of bankfull channel width to floodplain width (Rinaldi et al., 2013). Confinement is effective at concentrating floodwaters within narrow valley bottoms, thereby generating relatively large flow depths and velocities, both of which contribute to high flood power (Fryirs et al., 2016).

Velocities for extreme floods range from 2.5 to 13 m s^{-1} (Baker and Costa, 1987). Because velocity and channel gradient have a strong influence on values of flood power, no strong relationship exists between flood discharge and flood power (Figure 6.24). Values of flood power for extreme flash floods in Europe for basins with drainage areas between 0.5 and 1,981 km^2 range from about 100 to 7,000 W m^{-2} (Marchi et al., 2016). Mean values of ω in bedrock channel reaches (4,195 W m^{-2}) are two to three times greater than mean values in alluvial and semialluvial reaches. Values of flood power within individual watersheds generally attain maximum values at drainage areas of about 10 to 100 km^2 (Marchi et al., 2016). This peak can be explained by interplay between the nonlinear increase in discharge with drainage area and the nonlinear decrease in channel slope with drainage area, which, in turn, produces a humped pattern in the relation between ω and drainage area with a peak at intermediate values of drainage area (Lecce, 1997a).

Values of flood power exceeding 300 W m^{-2} often result in substantial morphological change, especially in alluvial or semialluvial rivers (Miller, 1990; Magilligan, 1992a). This threshold is not exact, and reaches with values of flood power far in excess of 300 W m^{-2} may not exhibit signs of morphological change (Buraas et al., 2014; Yochum et al., 2017). In channels with resistant boundaries, the threshold for morphological response may be well above the threshold for alluvial channels, exceeding 1,000 W m^{-2} (Wohl et al., 2001; Yochum et al., 2017).

6.8.4.2 What Are the Common Geomorphic Responses of Rivers to Floods?

Common erosional responses to large floods include channel widening, channel incision, excavation of channel bars, bank retreat, channel avulsion, and stripping of soil and vegetation from floodplain surfaces (Osterkamp and Costa, 1987; Kochel, 1988; Dean and Schmidt, 2013; Magilligan et al., 2015; Righini et al., 2017). Depositional responses include the formation of within-channel bars, channel infilling, and vertical accretion on floodplain surfaces, including the formation of sand or gravel splays, levees, and sand ridges (Kochel, 1988; Dean and Schmidt, 2013; Magilligan et al., 2015). Erosional and depositional responses typically vary spatially. Two rivers affected by Tropical Storm Irene in New England widened mainly at channel bends and where the upstream ends of mid-channel islands were stripped of vegetation (Buraas et al., 2014). Estimates of the total length of the study reaches of the two rivers, both 16 km long, that widened vary from 5% to 35%, depending on what range of natural variability in channel width is used to determine widening.

6.8.4.3 What Is the Importance of Valley Confinement?

Confined reaches of rivers often generate high stream power and are susceptible to major erosional changes, whereas adjacent unconfined reaches may experience substantial depositional changes as sediment flushed from reaches within narrow valleys is deposited in reaches with relatively low stream power within broad valleys (Thompson and Croke, 2013). However, channel width freely adjusts in unconfined sections of valleys, and erosional increases in channel width tend to be larger in unconfined sections than in confined reaches (Righini et al., 2017). Thus, along the same river, erosion in confined reaches may be juxtaposed with deposition and erosion in unconfined reaches (Yochum et al., 2017).

6.8.4.4 What Is the Importance of the Duration of Flood Power?

The magnitude of peak flood power provides one possible index of potential geomorphic effectiveness, but the duration of flood power above a threshold for channel change (e.g., 300 W m^{-2}) may also be relevant to understanding the impact of large floods on river morphodynamics (Costa and O'Connor, 1995). For two events with the same maximum flood power, a flash flood that exceeds the threshold for a short time may not be as effective as a sustained flood that exceeds the threshold for a long time (Figure 6.25). Integrating power per unit area over either the entire duration of an event or the duration above a threshold of channel change (e.g., 300 W m^{-2}) yields the total energy per unit area of the event (E_T):

$$E_T = \int \omega dt = \int \frac{\rho g Q S}{W} dt \qquad (6.7)$$

The extent to which flood-power duration determines the geomorphic effectiveness of floods has yet to be thoroughly evaluated. During Tropical Storm Irene, the amount of total energy per unit area above the threshold for channel change was relatively small in relation to the total energy of the event because the power per unit area exceeded 300 W m^{-2} only for a few hours (Magilligan et al., 2015). Although flooding did not produce major erosive changes in channel morphology, it did mobilize coarse sediment, leading to the formation of gravel deposits within channels and on floodplains and low terraces. Thus, this short-duration, high–stream power event had profound sedimentological effects but limited erosive morphological impacts.

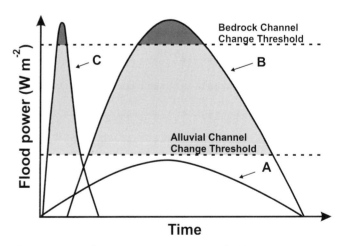

Figure 6.25. Total energy per unit area expended by time-varying flood power in relation to morphological impacts. Light and dark shading together represent total energy per unit area for alluvial channel change, and dark shading represents total energy per unit area for bedrock channel change. Event A is below threshold for channel change. Events B and C have equal peak flood power, but event B has much greater total energy for both alluvial and bedrock channel change than event C (adapted from Costa and O'Connor, 1995).

6.8.4.5 Why Are Antecedent Conditions Important in Morphological Response?

Antecedent conditions, including the sequencing of flood events and the state of channel form, can also influence the morphological response of a river to a major flood (Kochel, 1988). If a river channel susceptible to widening has recovered its form via narrowing, it will be susceptible to widening during a recurring extreme flood. On the other hand, the occurrence of a second major flood before the system has fully recovered from the occurrence of a preceding major flood that widened the channel likely will not result in additional widening. The relative increase in channel width produced by a major flood has been shown to be inversely related to the width of the channel prior to the occurrence of the flood, suggesting that antecedent conditions have an influence on channel erosion (Righini et al., 2017).

Within individual drainage basins, particularly those in mountainous environments, geomorphic effectiveness is related to topographic setting (Wohl, 2008). In headwaters, high channel gradients, the delivery of coarse clasts to channels from proximal hillslopes, valley confinement, and the potential for extreme precipitation and runoff associated with orographic effects promote high values of stream power, high hydrologic variability, and the transport of coarse bed material. Outcrops of bedrock enhance resistance of the channel perimeter to erosion, especially given that even the highest velocities in natural rivers are unlikely to produce cavitation of strong rock (Carling et al., 2017). Erosion of bedrock channels during extreme floods occurs mainly through the transport of large sediment particles that abrade bedrock surfaces. All these conditions favor the development of flood-dominated channel morphodynamics. As distance downstream from the most headwater locations increases, flood power initially increases before decreasing (Knighton, 1999a), bed material size generally decreases, and hydrological variability decreases with increasing drainage area. The potential for increased flood storage within the valley bottom also increases with drainage area. Collectively, these factors result in persistence of flood-dominated conditions within the zone of increasing flood power. Downstream of the peak in flood power, the geomorphic impact of large floods decreases.

6.8.4.6 How Do Responses to Floods Vary with Climatic Setting?

Climate plays an important role in distinguishing flood effectiveness in lowland environments. Studies of major floods along large rivers in humid-temperate or subtropical environments indicate either that such events have only minor sedimentological and morphological impacts on channels and floodplains (Moss and Kochel, 1978; Gomez et al., 1997; Magilligan et al., 1998; Fryirs et al., 2015) or that these types of fluvial systems recover rapidly from localized morphological effects of floods (Costa, 1974; Patton, 1988). Detailed analysis of changes in channel width and depth before and after a 100-year flood along the Des Plaines River in Illinois showed that this event produced less than 3% change in mean bankfull channel depth and no detectable change in channel width (Rhoads and Miller, 1991). Subsequent bankfull flows occurring a few months after the flood led to decreases in depth of comparable magnitude to those produced by the 100-year flood. The bankfull flows, which were of much longer duration than the flood, also produced small, but statistically significant, increases in channel width. The lack of morphological impact along this river, which is characteristic of many lowland meandering rivers, can be explained by low stream power, low hydrological variability, fine bed material, and well-vegetated channel banks containing abundant silt and clay that are highly resistant to erosion. The peak flood power for the 100-year flood within the main channel of the river was only 2 to 6 W m^{-2} because most floodwater flowed over the adjacent floodplain. Thus, broad floodplains along lowland rivers limit concentration of flow within the main channel, thereby preventing high values of flood power.

Dryland fluvial systems, including rivers in arid and semiarid environments, are often viewed as flood-dominated systems (Graf, 1988a). Hydrological variability tends to increase with increasing dryness, and sparse riparian or within-channel vegetation decreases resistance to erosion. Increasing aridity also inhibits weathering and soil development, so that alluvial rivers often lack abundant fines, which, when present, promote cohesion of channel banks. Morphological responses of arid-region rivers to floods are highly complex, exhibiting substantial spatial and temporal variation (Graf, 1988a). Large floods can lead to dramatic widening (Schumm and Lichty, 1963; Stevens et al., 1975; Huckleberry, 1994), to major aggradation (Hooke, 2016b), to complete eradication of channel planform and floodplain vegetation (Burkham, 1972), to channel incision through headcut migration (Graf, 1983a), and to major changes in channel arrangements and positions with respect to the floodplain (Graf, 1983b). Recovery from flood impacts in dryland rivers is also spatially and temporally uneven. Ephemeral or intermittent streams lack persistent flows, which tend to promote recovery of the system from disturbance. Post-flood morphological adjustments in these types of fluvial systems depend greatly on the timing and sequencing of subsequent flow events, both small and large, on sediment supply and connectivity, and on vegetation stabilization of areas of erosion and deposition (Hooke and Mant, 2015).

CHAPTER

7 The Shaping of Channel Geometry

7.1 How Is Channel Geometry Related to the Three-Dimensionality of River Form?

The form of river channels traditionally is treated as consisting of three distinct components: the geometry of the bankfull channel, the planform characteristics of the river, and the change in elevation of the river bed over distance, or longitudinal profile. Although these distinct components can be examined separately, together the components constitute the three-dimensional form of natural rivers. The study of these components, both separately and in relation to one another, lies at the heart of fluvial geomorphology. This chapter focuses specifically on the factors that influence the geometry, particularly the width and depth, of river channels. Chapters that follow examine other aspects of channel form.

7.2 How Is Channel Form Related to the Geometry of River Flow?

The amount of water within a river varies over time at a particular location as hydrologic variability produces increases and decreases in discharge. It can also vary over space along a river from upstream to downstream as increases in drainage area lead to increases in discharge. These variations in discharge locally and along a river provide the basis for examining changes in flow geometry, or hydraulic geometry, and for relating these changes to channel form. At a particular location, the geometry of the flow largely reflects constraints imposed upon it by the geometry of the channel boundary, unless the channel boundary is readily deformable and can change as the amount of flow

within it varies. Changes in the geometry of a flow of a particular frequency over space largely reflect changes in the dimensions of a river channel related to increases in discharge in the downstream direction. In this case, channel size increases to accommodate increasing amounts of flow. Change in the geometry of river flows over time and space has served as an important framework for exploring the connection between discharge and channel geometry in rivers. It has also provided the basis for extending the hydraulic geometry approach to consider channel geometry directly, rather than the geometry of the flow, and the factors that influence channel geometry.

7.2.1 What Is Hydraulic Geometry?

Hydraulic geometry relates hydraulic characteristics of the flow, namely flow width, mean depth, and mean velocity, to discharge as power functions (Leopold and Maddock, 1953):

$$W = aQ^b \tag{7.1}$$

$$D = cQ^f \tag{7.2}$$

$$U = kQ^m \tag{7.3}$$

Width and depth explicitly constitute geometric aspects of the body of water within a channel, whereas velocity is considered a third aspect of the geometry defining the rate of movement of the body of water. The choice of power functions is not based on theoretical considerations, but emerges from empirical analysis of field data. Substituting the terms on the right-hand sides of the three expressions for W, D, and U in the equation of continuity (Eq. (4.1)) yields

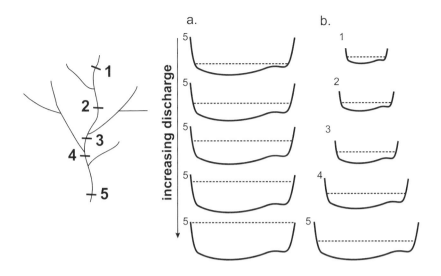

Figure 7.1. Definition of at-a-station and downstream hydraulic geometry for five locations (numbered) within a stream network (a). At-a-station hydraulic geometry examines changes in width, depth, and velocity as discharge varies at a particular location, in this case location 5 near the outlet of the network (b). Downstream hydraulic geometry examines how width, depth, and velocity change for a given flow frequency at multiple locations (1–5) throughout the network (c).

$$Q = aQ^b cQ^f kQ^m = ackQ^{b+f+m} \tag{7.4}$$

where

$$ack = 1 \text{ and } b + f + m = 1 \tag{7.5}$$

In other words, continuity requires that the product of the coefficients in the three power functions must equal 1 and the sum of the exponents in the power functions must equal 1.

Estimation of coefficients and exponents in Eq. (7.1)–(7.3) is accomplished statistically based on hydrological records for river gaging stations. Using Eq. (7.1)–(7.3) to examine how width, depth, and velocity change as discharge varies over time at a particular location along a river, either at a cross section or within short uniform reach, is known as at-a-station hydraulic geometry (Figure 7.1). Discharges determined from field data are based on measurements of flow width (W), mean depth (D), and mean velocity (U) (see Appendix D). Thus, the information needed to conduct an at-a-station hydraulic-geometry analysis is available from discharge measurements used to develop stage–discharge rating curves for gaging stations.

Eq. (7.1)–(7.3) can also be used to determine how width, depth, and velocity vary over space for locations along a river, throughout a drainage basin, or even throughout a set of drainage basins, for a discharge of a constant frequency or particular statistical value, such as the mean annual discharge. This type of analysis is known as downstream hydraulic geometry (Figure 7.1). Acquiring appropriate data for downstream hydraulic-geometry analysis is somewhat more complicated than it is for at-a-station analysis. Discharges used in the analysis may not have been measured directly at each location. If a discharge of a particular frequency is selected, the value of this discharge at each location can be determined from flow-duration or flood-frequency curves. If

the selected discharge is the mean annual discharge, this value can be computed from daily hydrological records for each location. Once a representative discharge is selected, corresponding data on flow width, mean depth, and mean velocity for this discharge must be acquired. If direct measurements of this discharge were obtained in the field at each location, data on W, D, and U are readily available. Commonly, direct measurements of the selected discharge will not be available at all locations. In this case, the stage corresponding to the discharge can be established from the rating curve for a site and the width and mean depth of the discharge estimated by relating the flow stage to survey data on the channel cross-sectional geometry at the location. Mean velocity is then determined as

$$U = \frac{Q}{WD} \tag{7.6}$$

This method is subject to uncertainty because it assumes that the channel cross-sectional geometry is fixed and does not change over time.

In early studies, values of the three dependent variables were plotted versus discharge on graphs with logarithmic axes and lines fitted by eye through the cluster of data (e.g., Leopold and Maddock, 1953). The widespread availability of ordinary least squares (OLS) regression analysis now allows values of the coefficients and exponents in hydraulic-geometry relations to be derived statistically (Figure 7.2). An advantage of OLS regression analysis is that it produces estimates of coefficients and exponents that satisfy continuity constraints for both the coefficients and the exponents (Rhoads, 1992).

The exponents in the hydraulic-geometry relations define relative rates of change in W, D, and U as discharge changes through time at a particular location (at-a-station hydraulic geometry) or as discharge of a constant frequency changes in the downstream direction (downstream hydraulic geometry). The coefficients are scale factors that equal values of W, D, and

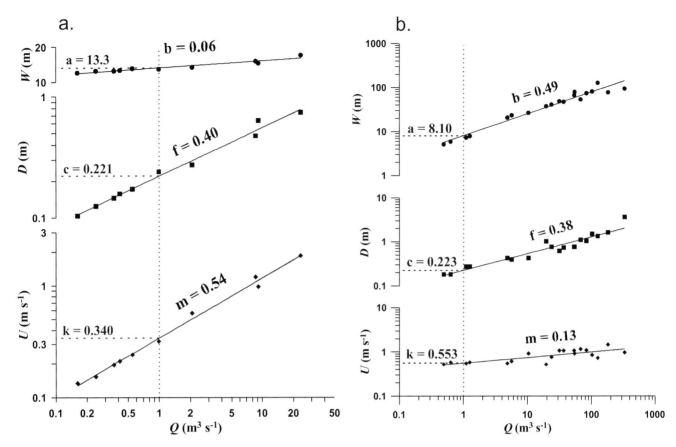

Figure 7.2. (a) At-a-station hydraulic-geometry relations for Brandywine Creek, Pennsylvania. (data from Wolman, 1955) (b) Downstream hydraulic geometry relations for Little Bighorn and Yellowstone River basins (data from Leopold and Maddock, 1953).

U when $Q = 1$ (Figure 7.2). If two rivers have the same exponent for W, D, or U, but different values of the corresponding coefficient, the river with the larger coefficient will have consistently larger values of W, D, or U than the river with the smaller coefficient.

7.2.2 How Do Flow Width, Depth, and Velocity Covary with Changing Discharge at a Particular Location along a River?

Studies of at-a-station hydraulic geometry have found that values of the exponents vary substantially not only among streams (Table 7.1) but also along individual streams (Knighton, 1975). Although the exponents are typically viewed as independent of scale, some evidence suggests that the magnitude of the exponents of at-a-station hydraulic geometry depends on drainage basin area (Dodov and Foufoula-Georgiou, 2004). Typically, values of b are less than values of f and m, but values for all three exponents have large ranges (Figure 7.3) (Table 7.1). Some summary studies suggest that mean or modal values of f are slightly greater than m (Leopold and Maddock, 1953; Williams,

1978b), whereas others indicate that m is slightly greater than f (Park, 1977; Rhodes, 1977; David et al., 2010).

For channels with relatively inerodible boundaries, at-a-station hydraulic geometry defines the varying influence of the channel geometry and flow resistance on W, D, and U as discharge changes (Ferguson, 1986b; Dingman, 2007). Channel shape is often reflected strongly in the value of b. Rectangular channels with vertical or near-vertical banks will exhibit little change in flow width with increasing discharge (Figure 7.3). In such channels $b \approx 0$ and $f + m \approx 1$ (Ferguson, 1986b). If flow resistance and water surface slope remain constant as discharge varies, and the channel is wide relative to depth, so that the hydraulic radius is nearly equal to the mean flow depth (D), the Manning equation (Eq. (4.35)) indicates that the mean velocity is proportional to $D^{0.67}$; therefore, $m = 0.67f$, $m/f = 0.67$ (Rhodes, 1977), and, for a wide rectangular channel with $b = 0$, $f = 0.60$, and $m = 0.40$. The relation between f and m largely reflects how flow resistance changes with increasing discharge. If resistance increases with increasing discharge, m/f is less than 0.67, whereas if resistance decreases with increasing discharge, m/f is greater than 0.67.

Table 7.1. At-a-station hydraulic-geometry exponents.

	Averages			Range containing modal value			Ranges			
	b	f	m	b	f	m	b	f	m	
Leopold and Maddock (1953)	0.26	0.40	0.34				0.06–0.59	0.13–0.63	0.07–0.55	20 sets of exponents for rivers in Great Plains and Southwest, USA
Park (1977)				0.0–0.1	0.3–0.4	0.4–0.5	0.00–0.59	0.06–0.73	0.07–0.71	139 sets of exponents for rivers in proglacial, humid-temperate, semiarid, arid, and tropical environments
Rhodes (1977, 1978)				0.0–0.1	0.4–0.5	0.4–0.5	0.00–0.84	0.01–0.84	0.03–0.99	315 sets of exponents for rivers in a variety of different environments
Williams, 1978b	0.22	0.42	0.36	0.0–0.1	0.5–0.6	0.3–0.4	0.00–0.82	0.10–0.78	0.01–0.81	165 sets of exponents for streams throughout the United States

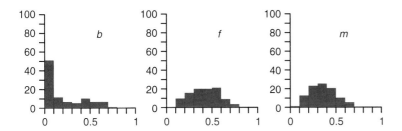

Figure 7.3. Histograms of values of b, f, and m for 165 streams throughout the United States (data from Williams, 1978b).

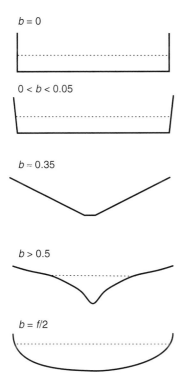

Figure 7.4. Variation in the exponent b in at-a-station hydraulic geometry with changes in channel form.

Flattening and lengthening of the banks in a trapezoidal channel will progressively increase b, but the relation between f and m still depends on change in resistance with increasing discharge (Figure 7.4). In parabolic channels, $W \propto D^{0.5}$ and therefore $b = 0.5f$ (Ferguson, 1986b). Convexity of the cross-channel profile of the channel bed, which can occur when bars are present in the river or when a narrow slot channel forms in bedrock, contributes to rapid increases in flow width with increasing discharge, enhancing values of b (Ferguson, 1986b; Turowski et al., 2008) (Figure 7.4). In such channels, b can exceed 0.5. Nevertheless, for the vast majority of natural channels, on the order of 70% to 90% (Rhodes, 1977; Williams, 1978b), $b < f$, a relation consistent with a trapezoidal or parabolic channel cross-sectional shape (Ferguson, 1986b; Dingman, 2007). Changes in water-surface slope with changes in discharge are not uncommon and complicate simple explanations of at-a-station hydraulic geometry based on the assumption of constant slope. As indicated by the Manning equation, such changes can strongly influence the magnitude of m and, because of interdependence among the exponents, the magnitudes of b and f.

Empirical analysis of streams throughout the United States supports general conclusions based on channel shape

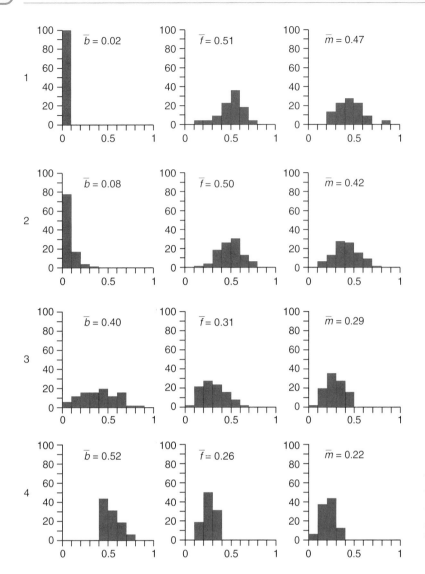

Figure 7.5. Histograms of exponent values and mean values of exponents for Case 1 (top row): channels with vertical, inerodible banks, Case 2 (second row): channels with nearly vertical inerodible banks, Case 3: channels with one erodible bank and one inerodible bank, and Case 4: channels with two erodible banks (data from Williams, 1978b).

considerations (Williams, 1978b). Channels with nearly vertical, inerodible banks have mean values of the width exponent ($\overline{b}$) nearly equal to zero, whereas those with steep, but not vertical, banks generally have small values of $\overline{b}$ (Figure 7.5, Cases 1 and 2). Channels with one erodible bank and one inerodible bank exhibit rather large mean values of the width exponent ($\overline{b} = 0.40$), but variability is large, as shown by the flat, broad frequency distribution of b values for this type of channel (Figure 7.5, Case 3). When both banks are erodible, the width is free to adjust to changing discharge and $\overline{b}$ is greater than 0.5 (Figure 7.5, Case 4). For all four cases, $f > m$, indicating that mean depth increases more rapidly than mean velocity with increasing discharge, but $m > 0.67f$, suggesting that flow resistance tends to decrease with increasing discharge. In steep mountain streams, m is often greater than f, a condition that reflects rapid decreases in resistance with increasing discharge (David et al., 2010; Reid et al., 2010). Explicit empirical analysis of changes in resistance do indeed show that resistance often declines as a power function of increasing discharge (Knighton, 1975). Changes in water-surface slope are more complex, with some locations exhibiting increases in slope and others exhibiting decreases in slope (Knighton, 1975).

In channels where changes in discharge result in erosion or deposition, changes in channel form contribute to scatter in at-a-station hydraulic-geometry relations. Erosion of channel banks that produces widening of the channel will increase b and decrease values of f and m. Deposition of bars with convex-upward surfaces within the channel will have the same effect. Scour and fill on the channel bed can produce hysteresis in changes of depth and velocity with discharge. Deposition and increases in bed elevation on the rising limb of a hydrograph followed by erosion and decreases in bed elevation near the peak discharge and on the falling limb generate a counterclockwise pattern of hysteresis in mean depth and a clockwise pattern of hysteresis in mean velocity (Leopold

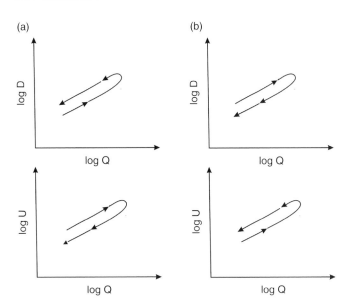

(a)   (b)

Figure 7.6. Patterns of hysteresis in change of depth and velocity with discharge for (a) deposition on the rising limb of a hydrograph and scour on the falling limb and (b) scour on the rising limb and deposition on the falling limb.

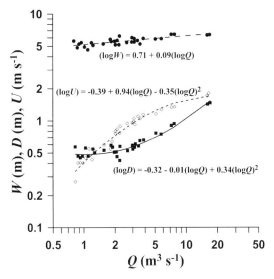

Figure 7.7. Hydraulic-geometry relations for River Ryton, Nottinghamshire, England showing log-quadratic relations for depth and velocity (data from Richards, 1973).

and Maddock, 1953) (Figure 7.6). Depths for a given discharge are smaller on the rising limb than on the falling limb, whereas the reverse pattern is true for mean velocity. Scour on the rising limb and fill on the falling limb will reverse the patterns of hysteresis for mean depth (clockwise) and velocity (counterclockwise). Changes in channel form caused by floods can substantially alter at-a-station hydraulic relations at a particular location along a river, resulting in pronounced differences in values of b, f, and m before and after the flood event (Knighton, 1975).

Despite the widespread use of power functions in at-a-station hydraulic geometry, no strong theoretical arguments exist to justify the use of a log-linear model to quantitatively characterize relations between W, D, or U and Q. Other models that have been used to examine these relations include log-quadratic models (Richards, 1973) and log piecewise-linear models (Bates, 1990). The log-quadratic model, a second-order polynomial relation, is capable of capturing nonlinear relations between $\log W$, $\log D$, or $\log U$ and $\log Q$ (Figure 7.7). Such nonlinearity can arise when bedforms, such as dunes, develop in sand-bed rivers as discharge increases, producing an increase in flow resistance. As resistance increases, the rate of increase in velocity diminishes, resulting in a compensatory increase in the rate of change in flow depth to maintain continuity. As a consequence, the relations between D and Q and between U and Q exhibit log-curvilinear patterns.

Testing of significant differences among exponents for at-a-station hydraulic-geometry relations at different locations is complicated by the fact that results are unit sum constrained; i.e., the sum of the exponents must equal 1. This constraint requires that compositional data analysis be used to examine potential differences among exponents or the relation of exponent values to external controls (Ridenour and Giardino, 1991). This type of analysis has shown that the ratio f/m depends on the rate of increase in roughness with increasing discharge (Ridenour and Giardino, 1995):

$$\frac{f}{m} = 1.48e^{2.29b_q} \tag{7.7}$$

where b_q is the exponent in a power function relating Manning's n to discharge:

$$n = a_q Q^{b_q} \tag{7.8}$$

Rapid increases in roughness with increasing discharge, or large values of b_q, increase f/m, which indicates that the rate of increase in depth increases relative to the rate of increase in velocity. Conversely, rapid decreases in roughness with discharge, or negative values of b_q, decrease f/m, which indicates that the rate of increase in velocity increases relative to the rate of increase in depth.

A variant on at-a-station hydraulic geometry is to use reach-averaged data for determining W, D, and U rather than hydraulic information obtained only at individual cross sections. In this approach, referred to as reach hydraulic geometry (Stewardson, 2005; Harman et al., 2008), the width, mean depth, and mean velocity are determined at numerous cross sections within a reach and the data averaged. Relations are then determined between the reach-averaged values of the three variables and discharge. An advantage of this approach is that it filters out local variability in channel form at the bar-element, bar-unit, and, depending on the length of the reach, planform scales. Nevertheless, variability among reach hydraulic-geometry relations is similar to that among traditional at-a-station relations, indicating that hydraulic characteristics vary substantially at a scale that encompasses many channel widths of length along rivers (Navratil and Albert, 2010). The increasing availability of high-resolution channel-geometry information has led to new approaches to examining variations in hydraulic characteristics of rivers over large lengths of rivers at multiple scales (Gonzalez and Pasternack, 2015). Such approaches should shed light on how scale-related variations in channel geometry influence the characterization of hydraulic relations at different scales. From a practical standpoint, predictive models of reach-averaged at-a-station hydraulic geometry can serve as flexible tools for evaluating how changes in channel geometry, such as those associated with river restoration schemes (see Chapter 16), will influence hydraulic conditions for aquatic organisms (McParland et al., 2016).

Although initial work on at-a-station hydraulic geometry held that the coefficients and exponents in the three basic relations (Eq. (7.1)–(7.3)) are independent, in at least some cases the exponents and coefficients of individual equations are highly correlated (Gleason and Smith, 2014). Plots of each exponent versus the logarithm of its corresponding coefficient for stations along individual river systems reveal strong negative linear relations with only minor amounts of scatter (Figure 7.8). In other words, large values of exponents are associated with small values of logarithms of coefficients. This type of relationship has been referred to as at-many-stations hydraulic geometry (AMHG). The principles of AMHG provide the basis for estimating the discharge of rivers from repeated satellite remote sensing observations of instantaneous river surface width (Gleason et al., 2014).

Subsequent work has explored the theoretical basis for AMHG, showing that it arises when independent at-a-station width–discharge relations within a reach intersect near the same values of width and discharge (Gleason and Wang, 2015). Although this condition serves as a mathematical constraint, possible geomorphological explanations for AMHG have been explored (Shen et al., 2016). The codependence of coefficients and exponents in AMHG has not been fully reconciled with theoretical analysis indicating that exponents and coefficients in at-a-station hydraulic-geometry relations do not necessarily depend to an equal extent on the same underlying influences of channel hydraulics and channel form (Dingman, 2007). This theoretical analysis implies that the exponents and coefficients should not necessarily covary systematically.

In erodible natural rivers and streams, where channel width, channel depth, bed and bank roughness, and water-surface gradient are changeable, mutual adjustability of flow width, mean depth, and mean velocity complicates attempts to predict at-a-station hydraulic geometry for specific stream locations. The complexity of mutual adjustments is reflected by the high degree of variability in values of b, f, and m (Table 7.1, Figure 7.3). Qualitative stability analysis suggests that the at-a-station relations are unstable in response to perturbations in hydraulic conditions (Phillips, 1990) – a finding consistent with large variability in exponent values.

A traditional approach to predicting average values of hydraulic-geometry exponents in erodible channels is based on the theory of minimum variance (Langbein, 1964; Williams, 1978b). This theory adopts a probabilistic view of hydraulic adjustment in which site-specific values of exponents cannot be precisely determined, but group averages for similar rivers conform to a minimization principle in which width, depth, and velocity mutually adjust to minimize the collective amount of change among the components of discharge. This type of principle, common to scientific and engineering problems dealing with adjustments in indeterminate systems, is an example of an extremal principle – a principle that invokes the minimization or maximization of some property of the system as a condition governing change in the state of the system. In hydraulic geometry the problem is to find exponent values that, subject to physical constraints, minimize the sum of the squares of the exponents. The variance of each component of discharge (W, D, or U) is proportional to the squared value of its corresponding exponent (b, f, or m). For the situation where all three hydraulic variables are completely unrestricted and free to adjust to a change in discharge, the minimum-variance condition corresponds to $b = f = m = 1/3$. Any other combination of exponent values will generate a sum of squared values $b^2 + f^2 + m^2$ that exceeds 0.3267, the sum of squared exponents for $b = f = m = 1/3$. Of course, in many cases physical constraints on one or more of the exponent values must be considered, where a constraint is any restriction on adjustment of width, depth, or velocity. For example, resistant banks will limit the adjustment of width and therefore the value of b. Although formal procedures have been developed for predicting average values of hydraulic-geometry exponents based on various constraints (Williams, 1978b), the imposition of constraints can be analyzed directly using hydraulic principles rather than through minimization of variance (Ferguson, 1986b; Dingman, 2007). Moreover, the extent to which river systems conform to the hypothesis of minimum variance remains speculative and is difficult to test directly. The same is true of efforts to predict hydraulic-geometry relations based on other extremal principles, such as minimum rate of energy dissipation and maximum entropy (Yang et al., 1981; Singh and Zhang, 2008a).

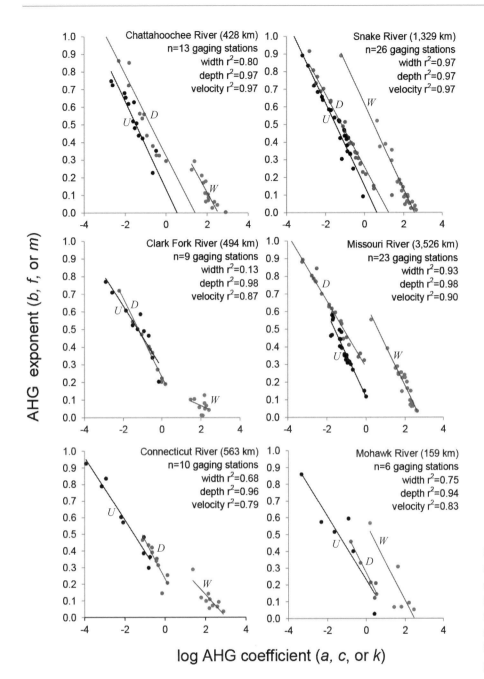

Figure 7.8. Inverse relations among at-a-station hydraulic-geometry exponents and logarithms of coefficients for width (blue), depth (red), and velocity (black) (from Gleason and Smith, 2014). (A black and white version of this figure will appear in some formats. For the color version, please refer to the plate section.)

Agreement between predicted and observed values of hydraulic-geometry relations (e.g., Singh and Zhang, 2008b) does not necessarily confirm that the agreement has occurred because river systems conform to the extremal principle on which the predicted values are based.

7.2.3 How Do Flow Width, Depth, and Velocity Covary as Discharge Changes over Distance along a River?

Downstream hydraulic-geometry analysis initially focused on changes in flow width, mean depth, and mean velocity in relation to changes in mean annual discharge (Q_m) in the downstream direction (Leopold and Maddock, 1953). Although downstream hydraulic geometry proposes to analyze downstream changes for a discharge of fixed frequency, the frequency of mean annual discharge (Q_m) usually varies somewhat among a set of locations within a river system. Nevertheless, variation in the frequency of mean annual flow is presumed to be minor (Leopold and Maddock, 1953, p. 3). The mean annual discharge generally is equaled or exceeded on almost half the number of days in a year.

Downstream hydraulic-geometry analysis based on Q_m indicates that $b > f > m$ (Table 7.2). Thus, in contrast to the

Table 7.2. Downstream hydraulic-geometry exponents.

	Averages			Range containing modal value			Ranges			
	b	f	m	b	f	m	b	f	m	
Leopold and Maddock (1953)	0.50	0.40	0.10							116 rivers throughout the United States; trend lines fitted by eye; based on mean annual discharge
Carlston (1969)	0.46	0.38	0.16							Reanalysis of Leopold and Maddock (1953) data using OLS regression
Rhodes (1987)				0.4–0.5	0.3–0.4	0.1–0.2	0.0–0.9	0.0–0.8	0.0–0.8	110 sets of exponents for rivers in a variety of different environments; based on a variety of flow frequencies
Singh et al. (2003)				0.4–0.5	0.3–0.4	0.1–0.2	0.0–0.8	0.1–0.5	0.0–0.6	456 data sets for rivers in a variety of different environments; based on a variety of flow frequencies

at-a-station case, increases in flow width typically accommodate downstream increases in discharge to a greater extent than increases in mean depth or velocity. Leopold and Maddock (1953) proposed that $W \propto Q^{0.5}$ based on fitting a line through the scatter of data by eye (Table 7.2). Regression analysis of the same data produced slightly different exponent values but did not change the general relation among these values (Carlston, 1969) (Table 7.2). Subsequent work summarizing results of a large number of studies showed that b, f, and m vary over broad ranges but have well-defined modal values (Rhodes, 1987; Singh et al., 2003) (Table 7.2).

At the time, the finding that velocity increases in the downstream direction was somewhat of a revelation (Leopold, 1953) and even controversial (Mackin, 1963), given that decreases in channel slope were presumed to produce decreases in velocity. Visual evidence often conveys the impression that mountain rivers flow faster than lowland rivers. Consideration of the Manning equation (Eq. (4.35)) indicates that, in addition to channel slope, mean velocity is influenced by flow depth and channel roughness. Downstream hydraulic geometry confirms that downstream increases in discharge of a given frequency are accommodated by relative increases in mean velocity (Rhodes, 1987), suggesting that downstream increases in flow depth and/or decreases in channel roughness appear to more than compensate for downstream decreases in slope.

7.2.3.1 How Do Bankfull Flow Width, Depth, and Velocity Covary as Bankfull Discharge Changes over Distance?

The downstream increase in the width, mean depth, and mean velocity of the mean annual discharge reflects adjustments of the three hydraulic components to downstream changes in channel geometry. This geometry, in turn, should reflect downstream changes in discharge. Informal physical reasoning suggests that river channels increase in size to accommodate an increasing amount of flow in the downstream direction; flows of increasing size should in most cases be capable of carving increasingly large channels. In contrast to at-a-station hydraulic geometry, which can capture the response of the river system to scour and fill over event timescales, downstream trends in channel form are typically viewed as conforming to modern timescales – the timescale of equilibrium adjustment (Figure 1.9).

One way to directly address the relation between discharge as a channel-shaping factor and the resulting channel form, particularly under the assumption that channel form is in a state of equilibrium, is to invoke the concept of dominant discharge. The mean annual discharge, a relatively frequent flow, does not correspond to the dominant discharge. Instead, a more meaningful approach to the relation between channel form and discharge focuses on the downstream hydraulic geometry of the bankfull discharge. Not only is the bankfull discharge typically viewed as the dominant discharge, but also, its width and mean depth correspond to the width and mean

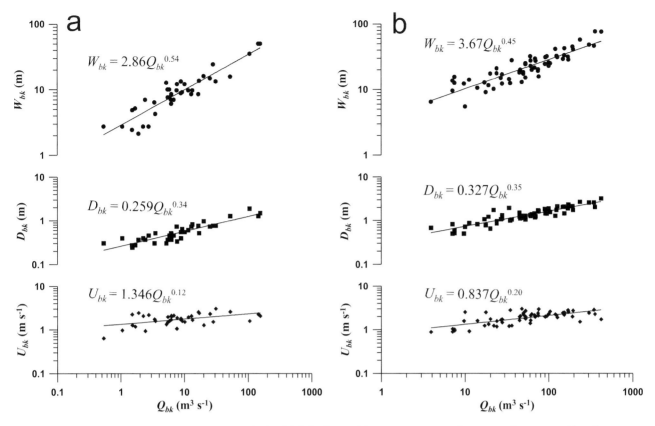

Figure 7.9. Downstream hydraulic-geometry relations for bankfull discharge: (a) 38 sites in upper Salmon River basin, USA (data from Emmett, 1975) and (b) 62 sites in the United Kingdom (data from Hey and Thorne, 1986).

depth of the river channel. Hydraulic-geometry relations for width and depth at bankfull flow are equivalent to bankfull channel-geometry relations; these relations quantitatively describe how bankfull channel-geometry changes as bankfull discharge changes. Thus, analysis of bankfull hydraulic geometry provides a link between hydraulic geometry and channel geometry.

Analyses of downstream hydraulic geometry for the bankfull discharge confirm that bankfull flow width and thus, bankfull channel width commonly increase downstream more rapidly than bankfull flow (channel) depth, so that the width–depth ratio, W_{bk}/D_{bk}, increases downstream (Figure 7.9). This result indicates that channels of headwater streams tend to be relatively narrow and deep compared with the channels of large rivers. The analysis also confirms that the mean velocity of the bankfull discharge increases downstream ($m > 0$).

An important point of emphasis in downstream hydraulic geometry is that the resulting statistical relations between hydraulic conditions and discharge represent general trends that emerge at a relatively coarse level of spatial sampling along a river system. This coarse level corresponds to a spacing of many channel widths between sampled locations.

At spatial intervals smaller than this level of sampling, considerable local variability in hydraulic conditions emerges (Fonstad and Marcus, 2010). Thus, the power functions of hydraulic geometry, which define continuous curvilinear relations between downstream increases in discharge and corresponding increases in flow width, depth, and velocity, do not hold at small spatial scales, where local effects produce variability that overwhelms general trends.

7.2.4 How Does Downstream Channel Geometry Differ from Downstream Hydraulic Geometry?

Given difficulties obtaining information on the width, mean depth, and mean velocity of bankfull discharge of many rivers, especially at locations not close to gaging stations, over time the focus of attention has shifted away from downstream hydraulic-geometry analysis to downstream channel-geometry analysis. The latter differs from the former in that it abandons the continuity requirement imposed on hydraulic geometry and considers only changes in channel width and depth in the downstream direction in relation to downstream changes in a flow of a specific frequency considered equivalent to the dominant discharge (Rhoads, 1991a). Downstream

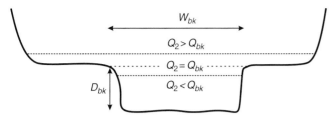

Figure 7.10. Relation of a flow of a particular frequency, in this case Q_2, to bankfull channel dimensions W_{bk} and D_{bk}. Only in the case where $Q_2 = Q_{bk}$ will the width and mean depth of the flow equal the bankfull channel width and mean depth. Under these conditions, $U_{bk} = Q_2/W_{bk} D_{bk}$ and continuity is satisfied. For $Q_2 > Q_{bk}$ and $Q_2 < Q_{bk}$, continuity based on the assumption that W_2 and D_2 correspond to the width (W_{bk}) and depth (D_{bk}) of bankfull flow (Q_{bk}) is not satisfied, and computation of U_{bk} as $Q_2/(W_{bk}D_{bk})$ will lead to erroneous estimates of U_{bk}.

variations in mean velocity are ignored. The advantage of this approach is that it can proceed using field measurements of bankfull channel width and mean depth combined with estimates of discharges of a specific frequency ($Q_{i,}$), for example, $Q_{1.5}$ or Q_2, which are often assumed to be similar in magnitude to the bankfull discharge:

$$W_{bk} = aQ_i^b \tag{7.9}$$

$$D_{bk} = cQ_i^f \tag{7.10}$$

Statistical analysis based on these equations can provide insight into how channel dimensions change in the downstream direction in relation to increases in a channel-forming flow. Values of the coefficients and exponents will vary depending on the specific Q_i used in the analysis (e.g. Wolman, 1955).

Note that because the relation of Q_i to channel form is not explicitly defined, imposing the continuity requirement (Eq. (7.4)), i.e., using a hydraulic-geometry approach to the analysis, will in many cases produce erroneous results for the estimate of bankfull velocity. Consider the case where Q_2 is used in hydraulic-geometry analysis as a surrogate for bankfull flow (Figure 7.10). If Q_2 does not equal the bankfull discharge, imposing continuity to determine the velocity of bankfull flow based on the dimensions of the bankfull channel (e.g., Brush, 1961), as is often done in hydraulic-geometry analysis when the bankfull discharge is known, will produce errors in the estimated bankfull velocity U_{bk} because all the difference between Q_2 and Q_{bk} will be assigned to U_{bk}. If $Q_2 > Q_{bk}$, U_{bk} will be overestimated, whereas if $Q_2 < Q_{bk}$, the bankfull velocity will be underestimated. Only in cases where $Q_2 = Q_{bk}$ will estimates of U_{bk} be accurate. Thus, this approach to hydraulic-geometry analysis should be avoided. A related technique, which uses values of W, D, and U corresponding to

Q_i selected from at-a-station hydraulic-geometry relations, satisfies continuity (Bomhof et al., 2015), but unless $Q_i = Q_{bk}$, the values of W and D obtained by this method do not represent the bankfull channel width and mean depth.

Empirical analysis of channel-geometry relations suggests that average values of exponents for bankfull channel width and depth for a discharge with a recurrence interval similar to that for bankfull flow, such as Q_2, are $b = 0.50$ and $f = 0.36$ (Knighton, 1987). Coefficients for the width relation (a) are generally an order of magnitude greater than coefficients for the depth relation (c). Values of a and c are also typically greater for sand-bed channels than for gravel-bed channels, indicating that for a discharge of 1 cubic meter per second, sand-bed channels are wider and deeper than gravel-bed channels.

Although the original goal of channel-geometry analysis was to explore quantitatively the relationship between a dominant channel-forming flow and stable channel form, the emergence of strong quantitative relationships has led to efforts to use these relations to predict the dimensions of stable channels. This approach is particularly useful in the practice of stream restoration, where channels that are disturbed and unstable are reconfigured in an attempt to achieve stability (see Chapter 16). To expand the domain of downstream channel-geometry relations, regional equations have been developed relating W_{bk} and D_{bk} to drainage area (A_d) in the form of power functions, avoiding the need for information on discharge (Bieger et al., 2015). The use of these equations to define stable channel dimensions inherently assumes that trends captured by channel-geometry equations reflect spatial patterns associated only with stable channel forms. Presumably, unstable channels would be associated with different spatial patterns of downstream variation in channel width and depth. The method also must be applied cautiously, given that scatter around power-function relations is associated with logarithmic scaling, which can lead to high levels of predictive uncertainty.

Bedrock channels generally exhibit downstream channel-geometry relationships similar to those for alluvial channels (Montgomery and Gran, 2001; Wohl and David, 2008). Field data representing a diverse range of environments suggests that for bedrock channels, $W_{bk} \propto Q^{0.5}$ and $D_{bk} \propto Q^{0.3}$, where Q is an infrequent peak flow that can mobilize coarse clasts (Wohl and David, 2008). Scatter around general trends typically is high, given the strong influence of local variability in bedrock erodibility and the varied effects of extreme events on channel form. Moreover, in mountainous environments, the development of downstream channel-geometry trends appears to be limited by the ratio of stream power per unit length to sediment size Ω/d_{84} (Wohl, 2004). This ratio represents a power-resistance metric. Mountain rivers with $\Omega/d_{84} < 10,000$ kg s^{-3}

tend not to have well-developed downstream hydraulic-geometry relations, suggesting that in these systems erosional forces are not sufficiently large in relation to substrate resistance to produce well-defined downstream trends in channel form. In clay-dominated nonalluvial channels, W_{bk} and D_{bk} can increase downstream at combined rates that exceed the rate of increase in Q_{bk} ($b + f > 1$), leading to downstream decreases in bankfull velocity ($m < 0$) (Fola and Rennie, 2010).

7.2.5 What Other Factors Are Related to Downstream Changes in Channel Geometry?

A major shortcoming of bivariate channel geometry models is that these models include only discharge as an independent variable influencing channel width or depth. Actually, many other factors can influence channel dimensions. Two main empirical approaches have been adopted to examine how factors other than discharge affect the width and mean depth of river channels. The first involves classifying or stratifying the data into distinct groups according to differences in a particular factor thought to have an important influence on channel form, whereas the second employs multivariate statistical models, particularly multivariate power functions, to explore the relative effects of different factors on channel morphology.

The first approach has emphasized the role of bank vegetation as an important control on channel form. Several studies have suggested that differences in bank vegetation can influence the scaling of channel geometry relations, resulting in systematic variation in the values of the coefficients in these relations (Table 7.3). For a given value of bankfull discharge, channels with abundant vegetation cover on the channel banks tend to be narrower and slightly deeper and have higher mean velocities than channels with sparse vegetation cover on the banks. The effect of vegetation seems to be especially pronounced in the case of the channel width–discharge relation. These studies imply that exponents, or rates of change, remain relatively uninfluenced by differences in bank vegetation.

Multivariate analyses have focused predominantly on the influence of resistive properties of boundary materials on channel geometry. The percentage of silt and clay in the channel perimeter has been viewed as an important factor influencing channel dimensions because cohesion associated with these materials increases resistance to erosion. Initial work related channel width and depth both to discharge and to the weighted percentage of silt and clay in the channel bed and banks (Schumm, 1960, 1968, 1969). This approach has been criticized as statistically problematic because the weighting procedure includes channel width and depth in ratio form, which can introduce spurious correlation (Melton, 1961). Also, small amounts of silt and clay in the beds of rivers transporting predominantly noncohesive sand and fine gravel are unlikely to substantially influence channel dimensions. To avoid these potential shortcomings, information on bank silt and clay content (SC_{bk}) alone should be considered. Multivariate analysis of data for 41 sites along rivers in the central United States (Schumm, 1960) yields

$$W_{bk} = 30.7 Q_{maf}^{0.57} SC_{bk}^{-0.64} \qquad (7.11)$$

$$D_{bk} = 0.055 Q_{maf}^{0.16} SC_{bk}^{0.62} \qquad (7.12)$$

where the independent variables are statistically significant at $p = 0.01$. These equations show that, when accounting for the

Table 7.3. Variation in bankfull downstream hydraulic-geometry relations for variations in bank vegetation by category.

Based on data from Andrews (1984)[a]			
$W_{bk} = 3.38 Q_{bk}^{0.54}$	$D_{bk} = 0.25 Q_{bk}^{0.34}$	$U_{bk} = 1.20 Q_{bk}^{0.13}$	Trees and thick brush
$W_{bk} = 4.17 Q_{bk}^{0.50}$	$D_{bk} = 0.26 Q_{bk}^{0.36}$	$U_{bk} = 0.92 Q_{bk}^{0.15}$	Sparse grass and brush
Hey and Thorne (1986)[b]			
$W_{bk} = 2.34 Q_{bk}^{0.50}$			> 50% tree and shrub cover
$W_{bk} = 2.73 Q_{bk}^{0.50}$			5–50% tree and shrub cover
$W_{bk} = 3.33 Q_{bk}^{0.50}$			1–5% tree and shrub cover
$W_{bk} = 4.33 Q_{bk}^{0.50}$			Grass, no trees and shrubs
Huang and Nanson (1997)[b]			
$W_{bk} = 1.80 Q_{bk}^{0.50}$	$D_{bk} = 0.64 Q_{bk}^{0.30}$	$U_{bk} = 0.95 Q_{bk}^{0.20}$	Dense trees for width relation, any trees for depth relation
$W_{bk} = 2.90 Q_{bk}^{0.50}$	$D_{bk} = 0.34 Q_{bk}^{0.30}$	$U_{bk} = 0.95 Q_{bk}^{0.20}$	Sparse or no trees for width relation, no trees for depth relation

[a] Functions determined using least squares regression;

[b] constrained relations with fixed exponents.

effect of silt and clay in the channel banks, width and depth change downstream in relation to discharge as expected, with width increasing faster than depth. On the other hand, for a constant mean annual flood (Q_{maf}), channel width decreases and channel depth increases with increasing percentage of silt and clay in the channel banks. High silt-clay content enhances bank strength and resistance to bank erosion, limiting channel widening. It also allows high banks to remain stable, resulting in relatively large channel depths. As a result, channels with high bank silt-clay content will be narrower and deeper than those with low bank silt-clay content. Multivariate analysis of data on bankfull width–depth ratio (Schumm, 1960) produces the relation

$$W_{bk}/D_{bk} = 564 Q_{maf}{}^{0.41} SC_{bk}{}^{-1.26} \qquad (7.13)$$

Eq. (7.13) confirms that downstream increases in discharge are associated with increases in channel width–depth ratio and that increases in bank silt-clay content correspond to reductions in width–depth ratio.

A variety of multivariate models of channel geometry have been developed for specific data sets (Table 7.4). Similarly to Eq. (7.11)–(7.13), the basic approach involves linear regression of log-transformed values of W_{bk} and D_{bk} against log-transformed values of multiple independent variables to generate multivariate power functions (Rhoads, 1992; Huang and Warner, 1995; Huang and Nanson, 1998; Lee and Julien, 2006). Channel geometry can also be explored within the context of continuously varying parameter models, which show how multiple independent variables influence values of the coefficients and exponents in basic bivariate models of channel width and depth (Rhoads, 1991a, 1992). Multivariate analyses confirm that the silt-clay content of material comprising the channel boundary has a significant influence on channel dimensions, with increases in silt and clay content resulting in decreases in channel width and increases in channel depth (Rhoads, 1991a, 1992). Increases in silt-clay content not only affect the scale of channel dimensions, by reducing a and increasing c, but also influence the downstream rate of change in these dimensions by decreasing b and increasing f (Rhoads, 1991a). The results of some multivariate analyses indicate that increases in bed-material size (d_{50}) result in reductions of channel width and depth (Parker, 1979; Bray, 1982; Lee and Julien, 2006; Thayer, 2017), whereas others indicate that increases in d_{50} are associated with increases in channel width or with the downstream rate of increase in channel width (Miller and Onesti, 1979; Rhoads, 1991a; Thayer, 2017). For the same discharge, size of bed material, and channel slope, sand-bed channels tend to be wider and shallower than

gravel-/cobble-bed channels. These channels also exhibit faster rates of change in channel width and slower rates of change in channel depth with increasing discharge than gravel-/cobble-bed channels (Thayer, 2017). Channel roughness in the form of Manning's n is positively related to channel width and channel depth (Huang and Warner, 1995) – a finding consistent with the argument that increases in roughness for a given magnitude of formative discharge reduce the mean velocity, requiring increases in channel width and depth to accommodate this discharge. Increases in channel slope clearly are associated with decreases in channel width and depth (Huang and Warner, 1995; Huang and Nanson, 1998, Lee and Julien, 2006), whereas increases in valley slope are associated with increases in width and depth (Rhoads, 1992). The choice of channel slope as an independent variable is somewhat questionable, however, given that local channel slope may change simultaneously with changes in channel width and depth to accommodate equilibrium adjustments over modern timescales.

Coefficients in some models (Huang and Nanson, 1998) serve as indices of bank strength, which vary according to the material and vegetation characteristics of the channel banks (Table 7.4). Multivariate analysis of vegetation effects shows that, in contrast to studies based on simple classification (Table 7.3), this factor can influence both the coefficients and the exponents of the basic channel-geometry relations (Rhoads, 1992). Discharge variability (Q_v), defined as the ratio of the five-year flood to the mean annual discharge (Q_5/Q_m), also has a significant influence on channel form. Rivers with highly variable discharge regimes have narrower and shallower channels than rivers with less variable regimes (Rhoads, 1992).

7.2.5.1 How Have Empirical Models Been Used to Explore the Role of Feedback in Downstream Channel-Geometry Relations?

Attainment of an equilibrium, graded, or regime state of river channels is commonly regarded as involving mutual adjustments between fluvial processes and channel form. In other words, many variables within the fluvial system can be viewed as interdependent. In the case of channel form, adjustments of width may influence channel depth, and adjustments of depth may influence channel width. Also, channel slope is not always independent of channel width and depth, but often mutually adjusts with width and depth, as implied by the concept of a graded river. Of course, the exact mechanisms of adjustment involve intermediary processes related to flow hydraulics and sediment transport, but the key point is that morphological components of the system do not necessarily adjust independently of one another. Traditional univariate or

Table 7.4. Examples of multivariate models of channel form.

Huang and Warner (1995); Huang and Nanson (1998): 140 stream locations in United States, United Kingdom, and Australia

$$W_{bk} = b_{1W}Q_i^{0.50}n^{0.36}S^{-0.16}$$

$$D_{bk} = b_{1D}Q_i^{0.30}n^{0.38}S^{-0.21}$$

where b_{1W} depends on type of channel banks: 6.00–6.25 – noncohesive sand, 4.5–6.0 – gravel, 3.5–5.0 – moderately cohesive sand, 2.5–3.5 – highly cohesive sand, 3.0–4.0 – moderately vegetated moderately cohesive sand, 2.1–3.0 – heavily vegetated and highly cohesive sand, $b_{1D} = b_{1W}^{-0.6}$, and Q_i includes a variety of different discharge frequencies as specified in various studies

Lee and Julien (2006): 1125 stream locations throughout the world

$$W_{bk} = 3.00Q_{bk}^{0.426}d_{50}^{-0.002}S^{-0.153}$$

$$D_{bk} = 0.20Q_{bk}^{0.336}d_{50}^{-0.025}S^{-0.060}$$

Rhoads (1991a): 252 sites in the midwestern United States

$$W_{bk} = aQ_m^{\,b}, D_{bk} = cQ_m^{\,f}$$

$$\ln(a) = 2.084 - 0.148X_d + 0.103\ln(Q_2) - 0.003SC_{bk} - 0.089\ln(SC_{bd})$$

$$b = 0.627 - 0.020\ln(Q_2) - 0.015\ln(d_{50}*) - 0.023\ln(SC_{bd})$$

$$\ln(c) = -0.675 + 0.080\ln(SC_{bd})$$

$$f = 0.177 + 0.029\ln(Q_2) + 0.001(SC_{bk})$$

SC_{bk} = percentage silt and clay in the channel banks; SC_{bd} = percentage silt and clay in the channel bed; $X_d = 1$ if $d_{50} > 2$ mm, otherwise $X_d = 0$; $(d_{50})* = d_{50}X_d$

Rhoads (1992): 62 sites in the UK

$$W_{bk} = aQ_{bk}^{\,b}, D_{bk} = cQ_{bk}^{\,f}$$

$$\ln(a) = -0.121 + 0.983(GR) + 0.355(LI) - 0.011\ln(Q_s) - 0.247\ln(d_{50})$$

$$b = 0.720 - 0.098(GR) + 0.027(ME) + 0.055\ln(d_{50})$$

$$\ln(c) = -1.889 - 0.508(GR) - 0.185(LI) + 0.940\ln(\sigma_s) - 0.393\ln(S_v)$$

$$f = 0.364 + 0.085(GR) - 0.019\ln(d_{50}) - 0.149\ln(\sigma_s) + 0.054\ln(S_v)$$

$GR = 1$ for grasslined banks, 0 otherwise; $LI = 1$ for 1% to 5% tree and shrub cover, 0 otherwise; $ME = 1$ for 5–50% tree and shrub cover, 0 otherwise; Q_s = sediment load (kg s^{-1}); σ_s = sediment sorting; S_v = valley slope

Rhoads (1992): 252 stream sites in the midwestern United States

$$W_{bk} = 15.80Q_2^{0.47}Q_v^{-0.17}e^{-0.0062SC_{bk}}e^{-0.0170SC_{bd}}S_v^{0.204}$$

$$D_{bk} = 1.06Q_2^{0.38}Q_v^{-0.22}e^{0.0018SC_{bk}}e^{0.0032SC_{bd}}S_v^{0.018}$$

$$Q_v = Q_5/Q_m$$

Davidson and Hey (2011): 48 gravel- and cobble-bed meandering rivers in the UK

$$W_{bk} = aQ_{bk}^{0.45}S_v^{-0.30}d_{84}^{0.09}d_{50}^{0.17}$$

$a = 2.55$, 0% tree and shrub cover; $a = 1.71$, 1–5% tree and shrub cover; $a = 1.53$, 5–50% tree and shrub cover; $a = 0.85$, > 50% tree and shrub cover

$$D_{bk} = bQ_{bk}^{0.37}S_v^{0.07}d_{84}^{0.07}d_{50}^{-0.16}$$

$b = 0.30$, 0% tree and shrub cover; $b = 0.31$, 1–5% tree and shrub cover; $b = 0.34$, 5–50% tree and shrub cover; $b = 0.41$, > 50% tree and shrub cover

Thayer (2017): 314 stream sites throughout the world

Gravel/cobble-bed channels ($n = 204$)

$$W_{bk} = 2.576Q_{bk}^{0.443}d_{50}^{0.043}S^{-0.109}$$

$$D_{bk} = 0.152Q_{bk}^{0.327}d_{50}^{-0.048}S^{-0.117}$$

Sand-bed channels ($n = 88$)

$$W_{bk} = 3.9726Q_{bk}^{0.571}d_{50}^{0.031}S^{0.007}$$

$$D_{bk} = 0.076Q_{bk}^{0.222}d_{50}^{-0.032}S^{-0.278}$$

All data ($n = 314$)

$$W_{bk} = 2.472Q_{bk}^{0.520}d_{50}^{-0.014}S^{-0.043}$$

$$D_{bk} = 0.130Q_{bk}^{0.255}d_{50}^{0.002}S^{-0.220}$$

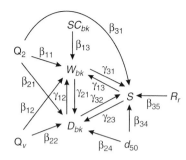

$$W_{bk} = \alpha_1 D_{bk}^{\gamma_{12}} S^{\gamma_{13}} Q_2^{\beta_{11}} Q_v^{\beta_{12}} e^{\beta_{13} SC_{bk}}$$

$$D_{bk} = \alpha_2 W_{bk}^{\gamma_{21}} S^{\gamma_{23}} Q_2^{\beta_{21}} Q_v^{\beta_{22}} d_{50}^{\beta_{24}}$$

$$S = \alpha_3 W_{bk}^{\gamma_{31}} D_{bk}^{\gamma_{32}} Q_2^{\beta_{31}} d_{50}^{\beta_{34}} R_r^{\beta_{35}}$$

Figure. 7.11. A simultaneous equation model of channel geometry showing mutual adjustment among channel width, depth, and slope. (Rhoads, 1991b) R_r is drainage-basin relief ratio.

multivariate approaches to analysis that involve specification of a dependent variable and one or more independent variables fail to capture this mutual adjustment.

Statistical models based on sets of simultaneous equations provide a statistical method for exploring mutual adjustment in fluvial systems (Miller, 1984, 1991a; Rhoads, 1988, 1992) (Figure 7.11). Although such models allow incorporation of feedback between variables, considerable statistical challenges arise in producing accurate estimates of parameters in the models (Rhoads, 1991b). In particular, statistical estimation procedures often enhance multicollinearity, which often is inherently high among fluvial variables, to problematic levels, resulting in imprecise and inaccurate estimates of model parameters. Such problems can be overcome by incorporating large amounts of data in the analysis. Simultaneous equation models may be useful for exploring feedback among fluvial variables as the amount of information on channel form, flow, and sediment properties grows.

7.3 How Has Physically Based Analysis Been Used to Examine Channel Geometry?

The empirical approach to downstream changes in the geometry of natural river channels has historical roots in the empirical regime methods of engineers interested in the characteristics of stable canals with mobile beds (Leopold and Maddock, 1953; Singh, 2003). Statistical analysis has provided general insight into the relationships between river-channel geometry and controlling factors (Rhoads, 1992), revealing regularities in these relationships, such as the consistent scaling of channel width and depth to discharge over a billionfold range in discharge (Ferguson, 1986b). Nevertheless, statistical models lack a strong physical basis, often do not balance

dimensionally, are restricted to the domain of data used to develop the relations, and may indicate different relationships between the dependent variable and an independent variable depending on other variables included in the analysis. As an alternative, river engineers and geomorphologists have pursued analytical methods relating river-channel geometry to flow, sediment transport, and properties of channel materials. These methods constitute what has become known as rational regime theory. All rational regime analyses invoke the assumption that river channels are fundamentally equilibrium systems and that the stable morphological configuration reflects the influence of process–form interactions at bankfull flow, which is equated to the dominant discharge. The timescale of concern reflects channel adjustment to prevailing environmental conditions and the development of a characteristic form over a period shorter than that associated with major changes in climate, topographic relief, or vegetation characteristics, but longer than the duration of individual events. Thus, the focus is on determining the channel form of a graded river over the modern timescale.

Whether or not channel geometry is indeed determinate has been a topic of debate (Maddock, 1970; Hey, 1978), but such debate has not impeded attempts to develop analytical solutions to the channel-geometry problem. Basically, the analytical approach involves specifying the mutually dependent geometrical variables of interest, usually channel width, depth, and gradient, and a set of governing equations specifying relations between these geometrical variables and independent variables, such as discharge, sediment load, and material properties of the channel boundary. If values of the independent variables are specified and the number of equations equals the number of unknowns, a solution for the system of equations can be found that provides values for the unknowns. Two fundamental approaches to channel geometry have been adopted: optimality analysis of models based on extremal hypotheses, and analytical or numerical analysis of models based solely on physical governing equations.

7.3.1 What Is the Optimality Approach to Rational Regime Theory?

The optimality approach to rational regime theory yields analytical solutions for a system of equations governing channel geometry by invoking an optimality criterion as a constraint in an otherwise underdetermined equation system, i.e., an equation system with more unknowns than equations. Under the assumption that a given water and sediment discharge will produce an alluvial channel with a characteristic bankfull width, depth, velocity, and slope, the system has four unknowns or degrees of freedom. In the most basic optimality models, these variables are related by three equations: the continuity equation for discharge, a bed-material or bedload transport function, and a flow resistance function (Chang, 1980; White et al., 1982). To achieve a solution, a fourth relation is required (Bettis and White, 1987). A standard approach has been to invoke an extremal hypothesis as the fourth relation. Examples of extremal hypotheses that have been used to close the equation system include minimum stream power per unit length of channel (Chang, 1980), minimum stream power per unit weight of water (Yang and Song, 1979), maximum friction factor (Davies and Sutherland, 1983), minimum energy dissipation rate over a reach of stream (Brebner and Wilson, 1967; Yang et al., 1981), maximum sediment transport rate (White et al., 1982; Millar and Quick, 1993), minimum Froude number (Jia, 1990; Yalin and da Silva, 2000), and maximum flow efficiency (Huang and Nanson, 2000, 2002). These hypotheses are proposed as required conditions for attainment of morphological equilibrium in natural river systems. For equilibrium to exist, the condition specified by the hypothesis must be satisfied.

The roots of extremal hypotheses derive from foundational work that viewed rivers as analogous to open thermodynamic systems (Leopold and Langbein, 1962). According to this perspective, river channels develop a quasi-equilibrium form that represents the most probable state between two opposing tendencies: minimum total rate of work in the whole fluvial system and uniform distribution of energy expenditure throughout the system (Langbein and Leopold, 1964). The notion of a most probable state of river channels also serves as the basis for the statistical optimization principle of minimum variance (Langbein, 1964). Over time, the sophistication of models based on extremal principles has increased; however, most models still consider channels of trapezoidal form to simplify the computations. Initial models specified fixed bank angles based on empirical relations between discharge and angle (Chang, 1980; White et al., 1982).

A major development in optimality modeling of channel geometry is the incorporation of a bank stability criterion into rational regime analysis (Figure 7.12) (Millar and Quick, 1993). The model as fully operationalized includes seven equations relating specified inputs to eight unknown dependent variables. Therefore, an extremal hypothesis is required to achieve a solution – in this case, the hypothesis of maximum sediment transport capacity. The model therefore predicts the configuration of a channel with a mobile bed and stable banks. An advantage of this model is that it allows the influence of variations in bank erodibility on channel geometry to be evaluated (Millar, 2005). Bank stability in the basic model assumes that bank material is noncohesive. Refined formulations include more realistic depictions of bank stability with separate stability criteria for mass failure and fluvial entrainment (Millar and Quick, 1998; Eaton, 2006).

Testing of rational regime models is challenging because few, if any, field studies report detailed information on all variables included in the models (Eaton et al., 2004), particularly data on bedload-transport rates and characteristics of bank material required to determine bank stability. A common approach to testing involves comparing model predictions of channel width, depth, and slope with observed values of these variables by using reasonable approximations of model inputs. Overall, optimality models perform reasonably well at reproducing general downstream trends in channel geometry derived from empirical analysis and at accounting for the influence of variations in bank strength on empirical relations (Millar and Quick, 1993; Huang and Nanson, 2002; Millar, 2005; Eaton and Church, 2007). Drawing upon this empirical success, the models can be used to explore computationally how equilibrium channel geometry changes in relation to variations in controlling factors, such as discharge, sediment load, and bank material properties. For example, for constant discharge and sediment load, increasing the strength of channel banks through an increase in the modified friction angle leads to dramatic reductions in channel width, increases in channel depth, and decreases in channel slope (Millar and Quick, 1993).

Although the capability to predict the geometric characteristics of stable alluvial channels is of practical value in efforts to manage unstable streams (Millar and Eaton, 2011), the scientific value of optimality methods remains somewhat controversial (Griffiths, 1984). A persistent criticism has been that extremal hypotheses lack a strong physical basis (Ferguson, 1986b) and that optimization based on these hypotheses essentially amounts to a teleological condition; i.e., that the geometric development of rivers exhibits goal-directed dynamics. Teleological conditions often are viewed suspiciously within science, particularly when applied to inanimate systems such as rivers. Moreover, if the empirical success of optimality analyses lies in mechanistic principles, then a mechanistic approach provides a more direct route to understanding the physical basis of channel-geometry relations. Despite these concerns, continued refinement combined with empirical success has fueled efforts to provide a physical justification for the optimality approach.

Maximization of flow resistance of the fluvial system in relation to the valley gradient may provide a physical basis for the success of analyses that employ extremal hypotheses (Eaton et al., 2004). Changes in channel length and gradient in relation to the valley gradient represent adjustments to maximize resistance of the entire fluvial system, thereby minimizing the potential for further adjustment. Extremal hypotheses

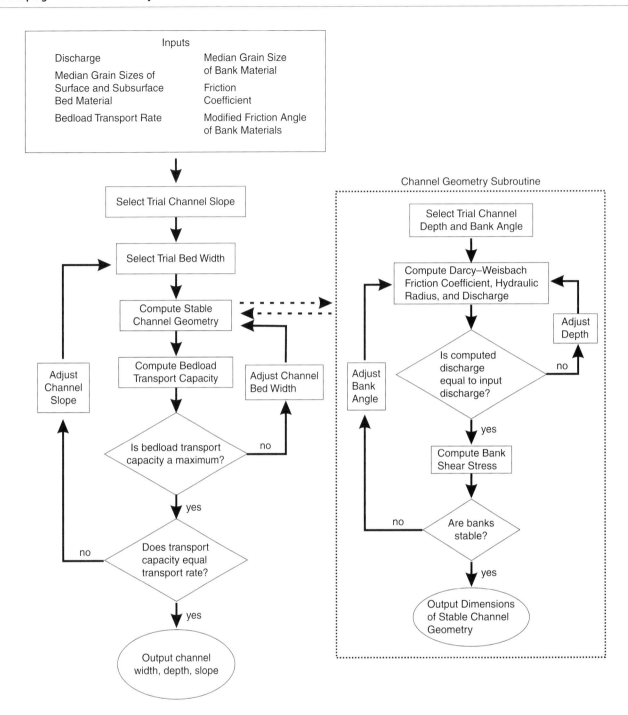

Figure 7.12. Flow chart for procedure to compute channel geometry using the rational regime model of Millar and Quick, 1993.

represent specific cases of the general tendency for the development of a stable state that maximizes flow resistance.

The most comprehensive development of an overarching theoretical framework for extremal hypotheses is based on the principle of least action as grounded in the physics of variational mechanics (Huang and Nanson, 2000; Nanson and Huang, 2008, 2017, 2018). This framework proposedly accounts for the utility of several different extremal

hypotheses as constraints for determining the geometry of equilibrium river channels. It also builds upon the application of variational mechanics in hydraulics (Yang, 1994) and of thermodynamic principles in the study of river morphology (da Silva, 2009).

Optimality models predict equilibrium states of channel geometry, not mechanisms of channel change and adjustment. Thus, the models are valuable for determining how changes in external controlling factors may result in new stable configurations (Eaton and Millar, 2017) but not necessarily for evaluating in detail the adjustments that occur in the transition from one stable state to another. Nevertheless, attempts have been made to employ optimality models to determine general evolutionary trajectories of channel form (Nanson and Huang, 2008, 2017). According to this evolutionary hypothesis of river-channel dynamics, adjustments of river-channel form to natural variability in discharge, sediment load, and bank erodibility lead to iterative changes whereby the outcome produced by a particular formative event becomes the initial condition to be reconfigured by the next formative event. Ultimately, however, the principle of least action implies that the maximally efficient, most stable state is also the most probable state, so that iterative changes that shift the river system toward stability are more likely than those that move the system away from stability. Thus, over time this state will act as an attractor and the river system will tend to converge on it through directional iterative change.

7.3.2 What Is the Analytical/Numerical Approach to Rational Regime Theory?

Designing straight, stable open channels capable of transporting flow without erosion of the channel bed or banks is a common task in hydraulic engineering. If the channel perimeter is composed of loose particles of uniform size, the goal is to determine the required dimensions and shape of the channel so that hydraulic conditions at bankfull stage are at or below the threshold for entrainment of the particles at all locations along the channel perimeter. Such a channel is known as a threshold channel. Although variational principles have been applied to the design of threshold channels (Cao and Knight, 1997), analysis typically involves application of physical principles based on vectorial mechanics to achieve analytical or numerical solutions for stable channel dimensions (Diplas and Vigilar, 1992). The vectorial mechanics approach to rational regime theory does not rely on extremal hypotheses to attain solutions for channel dimensions. Because the threshold approach to rational regime theory aims to determine the stable geometry of a channel with an immobile perimeter, it has limited relevance for natural rivers with mobile beds.

In contrast to threshold channels, alluvial channels are self-formed through interaction among flow, sediment transport, and the material in which the channels develop. In a river channel with a mobile perimeter, entrainment and transport of particles composing the channel banks involves net movement of eroded material away from the banks toward the center of the channel under the influence of gravity. This net inward flux of sediment, if not counteracted, should continuously erode the banks, resulting in an unstable, widening channel. This dilemma is known as the stable channel paradox: how can a self-formed river channel achieve stability under conditions of a mobile channel perimeter? Two fundamental solutions to this paradox have been proposed for alluvial rivers: one for rivers formed in gravel (Parker, 1978b) and the other for rivers in sand/silt (Parker, 1978a). Both solutions focus on stabilization of the channel banks.

The solution for rivers formed in gravel emphasizes the importance of the lateral turbulent transport of streamwise momentum over depth (M_{tl}) within a typical straight open channel:

$$M_{tl} = \int_{o}^{h_n} \rho \overline{u'v'} \tag{7.14}$$

where $h_n = h/\cos\theta_l$ is the local depth normal to the transverse slope (θ_l) of the channel perimeter. Because the streamwise velocity is greater in the center of a straight open channel than near the banks (Figure 4.23a), lateral turbulent diffusion of streamwise momentum occurs toward the banks. In other words, M_{tl} is positive toward the banks and becomes large in the near-bank region (Figure 7.13). Consider the relationship between particle mobility and bed shear stress, in this case expressed as a stress depth (δ_p):

$$\delta_p = \frac{\tau_b}{\rho g S} \tag{7.15}$$

where $\tau_b = \rho g h S$ is the local shear stress action on the channel perimeter and h is the local flow depth at a particular transverse position along the channel cross section. Particle mobility can be expressed as a critical stress depth (δ_c):

$$\delta_c = \frac{\tau_{bc}}{\rho g S} \tag{7.16}$$

where the critical shear stress (τ_{bc}) is for boundary material of uniform size. When δ_p in the center of the channel exceeds δ_c and M_{tl} is not considered, δ_p on part of the banks will also exceed δ_c, resulting in transport of bank sediment and instability of the banks (Figure 7.13b) (Parker, 1978b, 1979). Analysis using singular perturbation techniques shows that when the effects of M_{tl} are accounted for, $\delta_p < \delta_c$ over the entire length of the bank, whereas $\delta_p > \delta_c$ for the channel bed (Figure 7.13c) (Parker, 1978b). Thus, the banks do not erode, yet sediment is mobilized on the bed. The analysis predicts a stable gravel channel with a mobile bed and stable banks. It also provides a set of regime relations relating channel dimensions to discharge and sediment load. An important finding of the analysis is that under equilibrium conditions, the bankfull

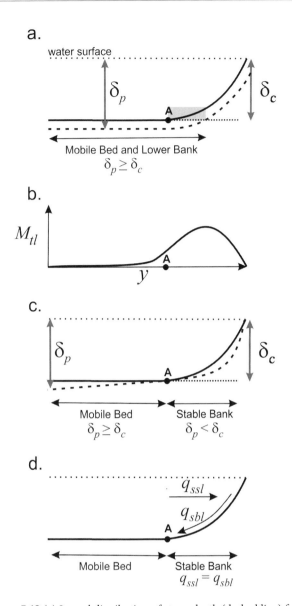

Figure 7.13 (a) Lateral distribution of stress depth (dashed line) for half of a symmetrical gravel river channel (solid line) without considering lateral transport of streamwise momentum. Shaded area indicates portion of the bank where stress depth exceeds critical stress depth for particle mobility. (b) Lateral distribution of lateral transport of streamwise momentum. (c) Lateral distribution of stress depth (dashed line) considering lateral transport of streamwise momentum. Stress depth equals critical stress depth at the base of the bank, resulting in a stable bank and a mobile bed. (d) Balance of lateral flux of suspended sediment and lateral flux of bedload in the bank region for a sand/silt river leads to a stable channel and mobile bed (adapted from Parker, 1978a,b).

bed shear stress at the center of the channel ($\tau_{b(c_l)}$) exceeds the critical bed shear stress (τ_{bc}) by about 20%, indicating that the geometry of gravel rivers adjusts to produce conditions that just barely exceed the threshold for sediment transport when

flow fills the channel (Parker, 1978b). Experimental studies designed to test the theory indicate that $\tau_{b(cl)}/\tau_c \leq 1.15$ (Diplas, 1990).

In an alluvial river with a bed and banks consisting of sand, the local boundary shear stresses are sufficient to entrain and transport bank material even when the effects of lateral turbulent transport of streamwise momentum are taken into account. Stabilization of the banks can be achieved in this case by a balance of sediment fluxes (Parker, 1978a). Lateral diffusion of sediment is related to the lateral gradient in sediment concentration as well as the lateral turbulent flux of streamwise momentum, which together produce a vertically integrated lateral volumetric flux of suspended sediment toward the bank within the center of the channel (q_{ssl}). Within the near-bank region, the progressive reduction in turbulence toward the water margin leads to diminished capacity for suspension of sediment. The diminished capacity for transport of suspended material results in deposition of suspended sediment on the bank face. Conversely, mobilization of bank sediment under the influence of τ_b and the transverse gravitational gradient of the channel bank results in a net lateral bedload flux (q_{sbl}) toward the center of the channel. When the fluxes are balanced, i.e., $q_{ssl} = q_{sbl}$, erosion keeps pace with deposition and the form of the banks remains constant (Parker, 1978a).

Refinements of the basic analysis have focused almost exclusively on gravel rivers. When graded, rather than uniform, sediment is included in the analysis, increasing particle-size heterogeneity increases channel depth and decreases channel width, producing relatively narrow deep channels (Ikeda et al., 1988). Similarly, the model predicts that increases in bank vegetation will decrease channel width and increase channel depth (Ikeda and Izumi, 1990), a finding consistent with empirical results (Table 7.3). Other refinements include formulation of the model in numerical form to allow predictions of stable geometry for the purpose of channel design (Vigilar and Diplas, 1997, 1998) and explicit consideration of the effects of lateral redistribution of streamwise momentum by secondary currents (Cao and Knight, 1998).

7.3.3 What Are the Limitations of Rational Regime Models?

Rational regime methods for determining stable channel geometry apply mainly to straight, single-thread, stable channels with noncohesive bed and bank materials. Although some optimality models consider effects of cohesive bank material, treatment of these effects on bank stability generally are incomplete for most rational regime models. Variation in channel planform is integrally connected to channel geometry, particularly channel shape and the number of channels in the river system (see Chapters 9–12). Rational regime models either do not deal at all with the effects of channel planform or include these effects to only a limited degree.

7.4 How Have Empirical and Physical Approaches Been Integrated in the Study of Channel Geometry?

A third type of analysis of channel geometry involves integration of empirical and physical methods. The standard channel-geometry relations are expressed in dimensionless form and include analysis of channel slope:

$$\widetilde{W}_{bk} = \alpha_a \hat{Q}_{bk}^{\eta_b}, \quad \widetilde{D}_{bk} = \alpha_c \hat{Q}_{bk}^{\eta_f}, \quad S = \alpha_g \hat{Q}_{bk}^{\eta_z} \qquad (7.17)$$

where

$$\widetilde{W}_{bk} = \frac{g^{1/5} W_{bk}}{Q_{bk}^{2/5}}, \widetilde{D}_{bk} = \frac{g^{1/5} D_{bk}}{Q_{bk}^{2/5}}, \quad \hat{Q}_{bk} = \frac{Q_{bk}}{\sqrt{g d_{50} d_{50}^{2}}}$$

$$(7.18)$$

Empirical analyses of these equations using data for 72 reaches along gravel-bed rivers yields (Parker et al., 2007)

$$\widetilde{W}_{bk} = 4.63 \hat{Q}_{bk}^{0.0667} \qquad (7.19)$$

$$\widetilde{D}_{bk} = 0.382 \hat{Q}_{bk}^{-0.0004} \qquad (7.20)$$

$$S = 0.101 \hat{Q}_{bk}^{-0.334} \qquad (7.21)$$

Given that $b = \eta_b + 0.4$ and $f = \eta_f + 0.4$, the results indicate that $b = 0.407$ and $f = 0.400$. Eq. (7.19)–(7.21) also provide accurate predictions of $\widetilde{W}_{bk}, \widetilde{D}_{bk}$, and S for three independent data sets consisting of a total of 97 stream reaches. Based on the small amount of scatter in the data, the estimated equations represent quasi-universal relations (Parker et al., 2007). Values of coefficients and exponents in the relations can be substituted into process-based expressions for flow resistance, the ratio of bankfull shear stress to critical shear stress, critical shear stress as a function of discharge, and gravel transport rate to estimate values of the parameters in these expressions. Moreover, relationships between values of parameters in the process-based expressions and the exponents and coefficients in the channel-geometry relations can be explored to determine how changes in underlying physical processes influence the relations (Parker et al., 2007). For example, increasing the ratio of bankfull shear stress to critical shear stress, which can be viewed as a measure of bank strength, increases bankfull channel width, decreases channel depth, and decreases slope. Alternatively, increasing the gravel transport rate results in decreased channel depth and increased width and slope. Increasing resistance increases depth, reduces slope, and leaves width unchanged.

A similar analysis for sand-bed streams yields different results (Wilkerson and Parker, 2011). A complicating factor is the strong relationship between the particle Reynolds number (R_p), a surrogate for the median size of the bed material, and the bankfull dimensionless shear stress (Θ_{bk}) in such streams, where $\Theta_{bk} \propto R_p^{-0.5}$ (Trampush et al., 2014). This relationship influences channel geometry, resulting in substantial variation in channel dimensions with R_p:

$$\widetilde{W}_{bk} = 0.00398 \hat{Q}_{bk}^{0.269} R_p^{0.494} \qquad (7.22)$$

$$\widetilde{D}_{bk} = 22.9 \hat{Q}_{bk}^{-0.124} R_p^{-0.310} \qquad (7.23)$$

$$S = 19.1 \hat{Q}_{bk}^{-0.394} R_p^{-0.196} \qquad (7.24)$$

For sand-bed streams, increases in particle size increase the channel width and decrease the depth. Although the exponent for R_p in the equation for slope is negative, when this equation is written in dimensional form, the relation between d_{50} and S is positive, confirming that increases in the grain size of bed material in sand-bed streams increase the channel slope. Subsequent work indicates that Θ_{bk} is related both to channel slope and to bed-material size (Li et al., 2015a). However, when considering the bankfull shear velocity u_{*bk} instead of Θ_{bk}, the dependency on grain size is eliminated, and instead u_{*bk} depends on S and the kinematic viscosity of water (v). This dependence of u_{*bk} on v may reflect increasing transport of fines, which is sensitive to variations in v, and the percentage of fines within channel banks. Empirical analysis confirms that the percentage of fines in the channel perimeter has a strong influence on channel shape and thus on the bankfull shear velocity (Schumm, 1971; Rhoads, 1991a).

The integrated empirical-theoretical approach has been proposed as an alternative to optimality models based on extremal hypotheses for explaining channel geometry (Parker et al., 2007; Wilkerson and Parker, 2011). By being expressed in dimensionless form, the resulting empirical relations are dimensionally balanced, and the coefficients are independent of the system of units. The attractiveness of the method is that it connects empirical outcomes to underlying process-based relations. Initial analyses exclude important factors, such as a direct expression of bank resistance to erosion, but the combined physical-empirical approach has provided insight into the physical factors that control channel geometry in specific contexts.

7.5 How Does Channel Geometry Change through Time?

At-a-station hydraulic geometry examines changes in the width and depth of flow with changing discharge, and although this change may include changes in channel geometry, it does not explicitly identify changes in channel dimensions. On the other hand, downstream channel-geometry analysis focuses on changes in bankfull dimensions of channels as discharge varies over space. Neither type of analysis explores how bankfull channel dimensions change over time in relation to variability in flow.

One way of conceptualizing river-channel dynamics is to view the current state of the channel as the product of a series of flows that have occurred over time. A quantitative framework for considering such changes can be expressed as (Pickup and Rieger, 1979; Yu and Wolman, 1987)

$$y_g(t) = \int_0^\infty w_g(j)Q(t-j)dj \tag{7.25}$$

where $y_g(t)$ is a geometric property of the bankfull channel (W_{bk}, D_{bk}) at time t, w_g is a weighting function, and j is the time step. The model indicates that the current form of the system is a convolution function of the weighted importance of the current and antecedent flows. Assuming that channel form and discharge are related as power functions, the model represented by Eq. (7.25) can be formulated as a linear statistical model (Rhoads, 1992):

$$\ln[y_g(t)] = \ln(\alpha) + \beta_0\ln(Q_t) + \beta_1\ln(Q_{t-1}) + \beta_2\ln(Q_{t-2}) + \ldots$$
$$= \ln(\alpha) + \sum_{j=0}^\infty \beta_j Q_{t-j} + e_t \tag{7.26}$$

where e_t is an error term assumed to have a mean of zero. The model contains an infinite number of lags but can be simplified by assuming that each value of discharge is associated with an equilibrium value of $y_g(t)$; e.g., for channel width,

$$W_{bk(eq)} = aQ_t^b \tag{7.27}$$

This relation proposes that each discharge, if sustained for a long enough period of time, produces an equilibrium width. Adjustment of the width to each discharge is presumed to be partial, with the amount of adjustment dependent on the extent to which the current state of the system differs from the equilibrium state:

$$\frac{W_{bk(t)}}{W_{bk(t-1)}} = \left(\frac{W_{bk(eq)}}{W_{bk(t-1)}}\right)^{(1-\varsigma)} \qquad 0 \leq \varsigma \leq 1 \tag{7.28}$$

Substituting Eq. (7.27) into Eq. (7.28) yields

$$W_{bk(t)} = a^{(1-\varsigma)}W_{bk(t-1)}^\varsigma Q_{(t)}^{b(1-\varsigma)} \tag{7.29}$$

The value of ς determines how sensitive the system is to antecedent flows, whereas the exponent $b(1-\varsigma)$ specifies

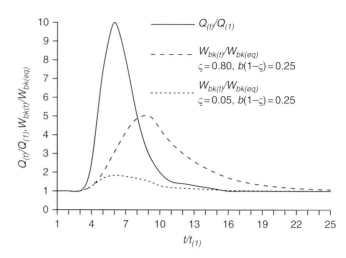

Figure 7.14. Response of dimensionless bankfull channel width to change in dimensionless discharge over time (t) as predicted by Eq. (7.29) for two different values of ς and the same $b(1-\varsigma)$. Values of discharge are expressed as ratios relative to values at $t = 1$.

the influence of the current flow on the morphological response of the system. If $\varsigma = 0$, the river system adjusts its width to the equilibrium state instantaneously in response to changes in discharge, and antecedent flows have no influence on width. Such a system is insensitive to antecedent conditions but highly sensitive to extant conditions. As the value of ς increases, the influence of previous flows increases, as does the lag time in system response. A system with $\varsigma \approx 1$ will strongly reflect the influence of antecedent flows on the current channel form. The morphology of such a system at any given time will largely be a product of past conditions (Figure 7.14). The scheme can be readily adjusted to predict channel width in relation to a threshold discharge for channel change (Q_{th}) by specifying $W_{bk(t)} = W_{bk(t-1)}$ for $Q_t < Q_{th}$ and applying Eq. (7.29) for $Q_t \geq Q$.

Although general convolution models in the form of Eq. (7.25) can be used to explore the possible dynamics of channel response to formative events when considering different types of weighting functions governing the impulse-response of the system (Howard, 1982; Wu et al., 2012), a shortcoming of convolution models is the lack of specification of physical equations governing channel dynamics. These models for the most part simply represent distributed-lag versions of the basic channel-geometry equations relating channel dimensions to discharge. In particular, channel form is likely to respond to variations in sediment load and rates of sediment transport as much as, if not more than, variations in discharge. Other major limitations include the assumptions that channel change occurs as a continuous function of change in flow and that increases and decreases in channel dimensions with changes in flow are smoothly reversible. The latter condition implies that processes of fluvial deposition and erosion essentially respond interchangeably to variations in flow, which almost certainly is not the case.

Although these simple models have serious shortcomings, the state of the science is still not advanced enough to allow deterministic predictions of channel adjustment based solely on mechanical principles. Reduced-complexity models are being developed in an attempt to link channel-geometry equations to process-based relations for hydraulics and sediment supply (Call et al., 2017; Davidson and Eaton, 2018). Such models allow iterative computation of changes in channel form and provide tools for exploring the role of antecedent morphological configurations in channel morphodynamics. Stochastic modeling of channel dynamics that includes mechanisms for bedload sediment transport, bank erosion, and deposition associated with vegetation colonization indicates that the response to specific discharges is historically contingent, with the probability of channel erosion or deposition depending on both flood magnitude and extant channel geometry (Davidson and Eaton, 2018). Channel widening and narrowing tend to occur cyclically. For a flood of a particular magnitude, recent narrowing increases the likelihood of high-boundary shear stresses that promote erosion, whereas recent widening increases the potential for low-boundary shear stresses that promote deposition. This modeling also shows that flood variability plays an important role in channel geometry, with the recurrence interval of the formative flow required to match the geometry generated by a traditional rational regime modeling increasing from two years in humid regions to nearly eight years in arid regions. Because it adopts an event-based approach to simulation of changes in channel geometry, this type of modeling provides a basis for exploring channel dynamics within a framework that does not depend on the equilibrium assumption underlying rational regime approaches. Of particular value is the capacity to evaluate the relation between recurrence intervals of formative events and recovery times, which lies at the heart of the geomorphic effectiveness concept (see Chapter 6).

Unfortunately, the data needed to estimate statistical models of channel dynamics (e.g., Eq. (7.29)) or to evaluate the predictions of predictive models of channel change over time are profoundly lacking. Although capabilities exist for continuously monitoring rates of flow and suspended-sediment transport in rivers, automated systems for continuously measuring changes in bankfull channel form have yet to be developed. A critical need exists for high-resolution information on changes in channel geometry in response to specific hydrological events.

CHAPTER

8 Channel Planform – Controls on Development and Change

8.1 What Is Channel Planform and Why Is It Important?

The pattern of a river when viewed from directly above is one of the most distinctive characteristics of river morphology. This characteristic defines the planform of the river, or the shape of the river as projected onto a horizontal two-dimensional surface, such as a map or an image. Before the development of modern remote sensing technology, the only way in which the planform of a river could be observed was by looking down on it from an elevated position on the landscape. Even from this vantage point, the pattern is seen obliquely rather than orthogonally. Today, the widespread availability of vertical aerial and satellite imagery of Earth's surface provides abundant visual information on river planform and on changes in planform over time. With the advent of new technologies, such as small unmanned aerial systems or drones, viewing rivers from above and extracting quantitative data about river form from aerial imagery is becoming commonplace (Woodget et al., 2017).

Recognition of different channel patterns lies at the core of attempts to classify rivers into different types. Over the past six decades, classification of rivers on the basis of differences in planform has been the focus of considerable attention in fluvial geomorphology, but as of yet no settled classification scheme or even settled terminology exists (Carling et al., 2014). Despite this caveat, differences in planform in many cases reflect differences in underlying fluvial processes that generate planform characteristics, indicating that planform variation provides a starting point for classification.

This chapter introduces the major planform types and the environmental conditions that influence the development of different planform characteristics. Subsequent chapters explore in detail the fluvial processes and forms for rivers with different planforms. Because the planform of a river is intimately connected to channel geometry, planform distinctions provide the basis not just for examining river morphology in two dimensions, but also for investigating the processes that influence the three-dimensionality of river form.

8.2 What Are the Major Types of Channel Planform?

The number of channels is an important criterion for classification of rivers based on planform. A fundamental distinction can be made between single-channel rivers and rivers with multiple channels (Figure 8.1). For single-channel rivers, sinuosity (S_I) has been used to further divide these rivers into straight versus meandering forms, where sinuosity is the ratio of river length (L_R) to valley length (L_V) or, alternatively, the ratio of river slope (S) to valley gradient (S_v):

$$S_I = \frac{L_R}{L_V} = \frac{S}{S_v} \tag{8.1}$$

The value of sinuosity used to distinguish straight from meandering rivers is rather arbitrary, but typical values range from 1.3 (Van den Berg, 1995) to 1.5 (Leopold and Wolman, 1957). Rivers with tortuous meanders often have $S_I \geq 2.5$ (Figure 8.2). Single-thread alluvial rivers are carved into a floodplain, which forms the banks of the river. For multichannel rivers, a distinction can

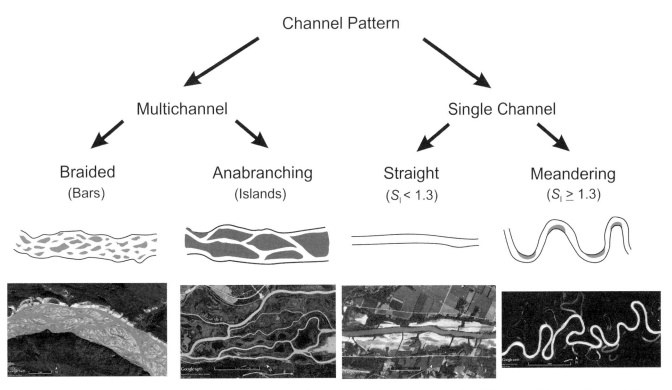

Figure 8.1. Major categories of channel patterns. Images from left to right are: Chitina River, Alaska, USA; upper Columbia River, British Columbia, Canada; Wabash River, Indiana, USA; Jurua River, Brazil.

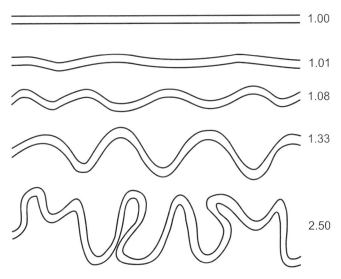

Figure 8.2. Values of sinuosity (right) in relation to variation in river planform for single-channel rivers.

be made between braided rivers, where the channels are divided by temporary, often unvegetated deposits of sediment known as bars, and anabranching rivers, where the channels are divided by permanent islands covered by vegetation. In braided rivers, active bars and intervening channels develop within a broad zone known as the braidplain (Sambrook Smith et al., 2006) or braid-train (Haschenburger and Cowie, 2009). The positions of bars

and channels can shift during events capable of transporting bed material, thereby rearranging the pattern of braiding. As flow stage rises, bars become progressively submerged. As a result, both the areal extent and the number of bars and channels changes. At the highest stages, all bars may be submerged and flow occupies the entire braidplain (Sambrook Smith et al., 2006). Thus, the character of braided rivers is highly stage dependent. In anabranching rivers, multiple channels are separated by vegetated or otherwise stable alluvial islands that divide flows up to bankfull discharge (Nanson and Knighton, 1996). Individual channels typically convey different amounts of flow, and the relative distribution of the total flow among the channels can change with changes in flow stage.

Among the four basic channel types, meandering and braided forms have been viewed as most common. Straight reaches of natural rivers are rather rare, especially over lengths exceeding 10 times the channel width (Leopold and Wolman, 1957). Until about 60 years ago, braided and anabranching rivers were seen as similar, but they are now viewed as distinct forms (Nanson and Knighton, 1996). The four basic types also do not completely encompass the rich variety of river patterns; some rivers have planforms that do not fit neatly within the fourfold classification. Like any classification scheme, categorization of rivers based on channel planform represents an attempt to impose order on river systems through a simple conceptual model defined on the basis of perceived similarities and differences in morphological configuration. Other schemes exist that include more than the four basic categories of pattern types (Schumm, 1981, 1985; Nanson and Knighton, 1996), but most alternative frameworks involve elaboration of subtypes within the four basic categories.

Assignment of a river to a planform category is based primarily on visual inspection of a river in the field, or, more commonly, on an aerial photograph or satellite image. Typically, no rigorous quantitative criteria are used in classifications, other than perhaps values of S_I for distinguishing between straight and meandering channels. Distinctions among rivers with different planforms may be clear for some cases (Figure 8.1), but transitional forms can exist that have characteristics of multiple planform types. Classification of rivers based on planform is to some extent subjective in the sense that it depends on expert judgment.

8.3 What Environmental Factors Are Associated with Differences in Channel Planform?

The characteristics of the environment within which a river is located strongly influence channel planform. Important environmental factors include discharge, valley slope, bed-material caliber, sediment supply, and bank erodibility. These factors represent either inputs to the river system (discharge and sediment supply) or constraints on the capacity of the river to adjust to these inputs (bed-material caliber, bank erodibility, and valley slope). Variation in planform typically is described as a continuum based on the reasoning that changes in planform characteristics occur continuously over the range of variation in controlling factors (Ferguson, 1987b). The extent to which the notion of a continuum adequately encompasses the actual planform variability of natural rivers has yet to be fully resolved.

8.4 How Do Environmental Conditions Differ for Meandering and Braided Rivers?

A considerable amount of effort has been devoted to exploring the difference between meandering and braided rivers, the two most common pattern types. Initial empirical work indicated that meandering rivers generally have lower channel slopes and bankfull discharges than braided rivers (Figure 8.3) (Leopold and Wolman, 1957). A threshold relation separating most of the rivers of each type can be defined as

$$S = 0.0125 Q_{bk}^{-0.44} \qquad (8.2)$$

Inclusion of channel slope and bankfull discharge in analysis of channel planform, while useful for characterizing

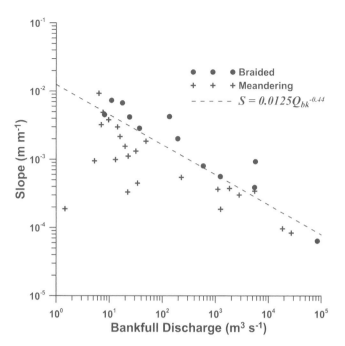

Figure 8.3. Relation of braided and meandering planforms to channel slope and bankfull discharge (data from Leopold and Wolman, 1957).

differences among rivers of different planform types, does not provide insight into the response of river morphology to external factors, because both bankfull discharge and channel slope are at least in part dependent on channel planform (Carson, 1984a). One way to address this problem is through experimental work where discharge and slope are imposed conditions that influence channel-planform development. An initially straight channel carved into sand in a small flume evolves into different planforms as the slope of the flume is increased for a constant discharge (Schumm and Khan, 1972). At low values of slope the channel remains straight. Although channels in sand cannot fully meander, as slope increases, flow within the relatively straight channel begins to meander, resulting in an increase in thalweg sinuosity (Figure 8.4). The sinuosity increases nonlinearly with increases in slope until, over a narrow range of slope, meandering abruptly diminishes and the stream transforms into a braided planform. This type of behavior is consistent with the empirical results showing that, for a constant discharge, braided rivers develop when slope is relatively high compared with conditions for the development of meandering rivers (Leopold and Wolman, 1957). Moreover, the transition from meandering to braiding occurs over a narrow range of slopes. To maintain equilibrium conditions within the experiments as the slope of the flume increases, sediment supply must be increased to offset increased rates of sediment transport. Thus, the supplied sediment load increases

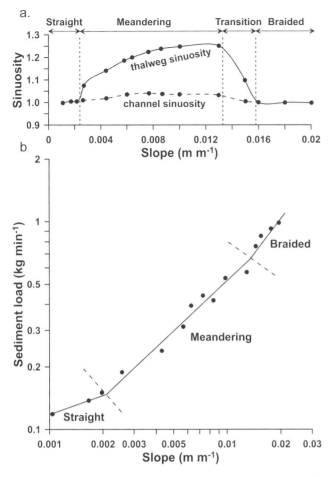

Figure 8.4. (a) Variation of the sinuosity and channel planform of a small experimental stream with changes in flume slope (dashed vertical lines indicate approximate boundaries of planform domains). (b) Changes in sediment load and channel planform with changes in flume slope (dashed lines indicate changes in planform and sediment load–slope relation) (from Schumm and Khan, 1972).

as new channel patterns emerge. Also, changes in planform type from straight to a meandering thalweg and from a meandering thalweg to braided result in increases in the rate of change of sediment load per unit increase in slope following each planform transition (Figure 8.4). These changes in the load–slope relation with changes in planform type suggest that pattern changes progressively increase the efficiency of sediment transport.

Recognition that discharge and slope are key variables influencing channel pattern has led to a focus on stream power, of which slope and discharge are components (see Chapter 4). The inverse relation between discharge and slope in Eq. (8.2) can be interpreted as a threshold of stream power per unit area (ω). If bankfull channel width is characterized by the downstream channel-geometry relation,

$$W_{bk} = aQ_{bk}{}^{b} \tag{8.3}$$

and Eq. (8.3) is substituted into an expression for bankfull stream power per unit area based on flow depth rather than hydraulic radius (see Eq. (4.23)),

$$\omega = \frac{\rho g Q_{bk} S}{W_{bk}} \tag{8.4}$$

the resulting stream-power threshold relation is

$$\omega_{th} = \frac{\rho g}{a} Q_{bk}{}^{1-b} S \tag{8.5}$$

where ω_{th} is the threshold stream power. Furthermore, if $1 - b \approx 0.44$, a reasonable assumption given that b typically has a value of about 0.5 (see Chapter 7), substituting Eq. (8.2) into Eq. (8.5) produces

$$\omega_{th} = \frac{123}{a} \tag{8.6}$$

for $\rho = 1000$ kg m^{-3}. Eq. (8.6) defines a threshold stream power per unit area (ω_{th}) for meandering versus braiding based on the value of the coefficient in the downstream hydraulic-geometry relation. For typical values of a, $\omega_{th} \approx 30 - 50$ W m^{-2} (Ferguson, 1981; Carson, 1984a).

A problem with the focus solely on discharge, slope, and stream power is that it emphasizes only the role of hydraulic action in planform change and ignores the erosional resistance of the channel boundary. Empirical analysis of data for braided streams shows that sand-bed braided streams differ from gravel-bed braided streams in terms of the bankfull discharge–channel slope conditions required for braiding (Figure 8.5). Gravel-bed streams plot above sand-bed streams, indicating that greater stream power is needed to produce braiding in gravel-bed rivers than in sand-bed rivers. Moreover, without consideration of grain-size effects, many sandy braided streams lie within the domain of meandering streams as delineated by Eq. (8.2). In fact, gravel meandering rivers generally plot above sandy braided rivers (Xu, 2004). This type of analysis suggests that grain size of bed material is an important factor determining the environmental domains of meandering versus braided rivers.

The primary approach to considering the role of bed material in channel-pattern development has been to relate a metric of potential stream power per unit area of a formative discharge to the median grain size of the bed material. Potential stream power per unit area is determined by substituting Eq. (8.3) into Eq. (8.4) with the bankfull discharge and channel slope replaced by the mean annual flood (Q_{maf}) and valley gradient (S_v). The use of mean annual flood and valley gradient avoids the problem of dependence of bankfull discharge on channel

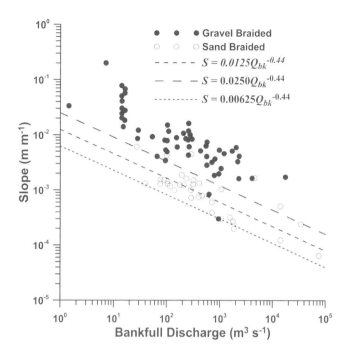

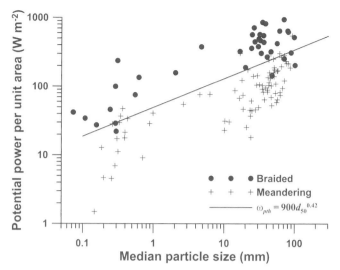

Figure 8.6. Braided versus meandering rivers in relation to potential stream power per unit area and median particle size (data from van den Berg, 1995).

Figure 8.5. Sand braided versus gravel braided rivers in relation to bankfull discharge and channel slope. c_t refers to the value of the constant in Eq. (8.2) (adapted from Ferguson, 1987b).

geometry and channel slope on sinuosity, where channel geometry and sinuosity are components of the channel planform. It results in a planform-independent estimate of the maximum stream power for a formative discharge in a particular environmental setting, i.e., the magnitude of formative stream power if the river flowed straight down the valley. For values of $a = 4.75$ for sandy rivers, $a = 3.0$ for gravel-bed rivers, and $b = 0.5$ for both types of rivers, substitution of Eq. (8.3) into Eq. (8.4) yields (van den Berg, 1995)

$$\omega_p = 2065Q_{maf}^{0.5}S_v \quad \text{sand-bed } (d_{50} \leq 2\,\text{mm}) \tag{8.7}$$

$$\omega_p = 3270Q_{maf}^{0.5}S_v \text{ gravel-bed } (d_{50} > 2\,\text{mm}) \tag{8.8}$$

where ω_p is the potential stream power per unit area (W m^{-2}). When plotted versus median grain size, the data for braided and meandering streams form two groups separated for the most part by the function (Figure 8.6)

$$\omega_{pth} = 900d_{50}^{0.42} \tag{8.9}$$

where ω_{pth} is the threshold potential stream power. This relation indicates that as the median grain size of bed material increases, the potential stream power needed to produce braiding increases. It also implies that braided rivers transport coarser bed material than meandering rivers, so that braided rivers have higher ratios of bedload to total load than meandering rivers. Meandering rivers typically transport a mixed load of suspended sediment and bedload, whereas braided

rivers typically transport large amounts of bedload. The use of potential stream power in the analysis of meandering-braided planform domains has been criticized on the grounds that incorporation of regime widths based on hydraulic-geometry relations (e.g., Eq. (8.3)), rather than actual channel widths, inflates separation between the two types of channels (Lewin and Brewer, 2001, 2003); however, others have defended the approach by arguing that the regime width provides an independent metric for the purpose of discriminating planform types (van den Berg and Bledsoe, 2003; Kleinhans and van den Berg, 2011).

Defining domains of braided and meandering rivers based on a distinct discriminant relation, without specification of the uncertainty of this relation, conveys the impression that changes from a meandering to a braided pattern will occur abruptly if stream power increases beyond the threshold relation for a river with a certain size of bed material. Instead, the boundary between meandering and braided regimes can be viewed as transitional, whereby the probability of a meandering stream becoming braided progressively increases with increasing potential stream power (Bledsoe and Watson, 2001a). Transitional zones defined by logistic regression analysis differ for sand-bed versus gravel-bed rivers, indicating that merely accounting for median particle size of river bed material is not sufficient to characterize planform differences over the full size range of this material (Figure 8.7). The domain of transitional probabilities is narrower for sand-bed channels than for gravel-bed channels, indicating that channels with sand beds are more susceptible to planform change than channels with gravel beds. Further analysis suggests that stable

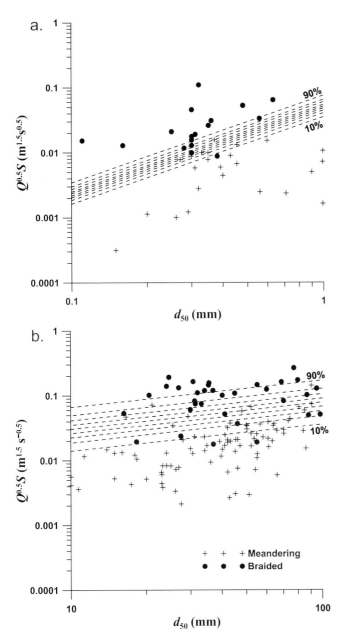

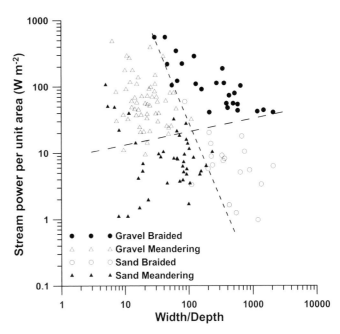

Figure 8.8. Domains of braided versus meandering rivers by sediment type (sand versus gravel) defined by stream power per unit area and channel width/depth ratio. Coarse dashed line separates gravel versus sand rivers, and fine dashed line divides meandering versus braided streams (from Xu, 2008).

Figure 8.7. Transitional probabilities (dashed lines) ranging from 10% to 90% for meandering versus braided rivers in relation to discharge–channel slope product and median grain size: (a) sand-bed streams; (b) gravel-bed streams. Discharge is represented by mean annual flood, with bankfull discharge substituted for streams without mean annual flood data (from Bledsoe and Watson, 2001a).

meandering channels without well-developed point bars tend to have less potential stream power per unit area for a given grain size than actively migrating meandering channels with well-developed point bars (Kleinhans and van den Berg, 2011).

Bank resistance is also an important factor determining whether a river has a meandering or braided pattern. When formative flows can readily overcome bank resistance,

resulting in bank erosion and channel widening, bars tend to form within the channel, leading to braiding. Thus, for a given magnitude of discharge, critical dimensionless shear stress, or stream power per unit area, braided channels typically have greater widths and larger width–depth ratios than meandering rivers (Xu, 2004, 2008) (Figure 8.8).

Theoretical analysis of characteristic scales of meandering versus braiding indicates that braided rivers have a larger width to depth ratio (W/D) for a given magnitude of the ratio of channel slope to Froude number ($S/\mathbf{F}$) (Parker, 1976) (Figure 8.9). Conversely, for a given W/D, braided rivers have higher values of $S/\mathbf{F}$ than meandering rivers. This analysis supports empirical observations that meandering rivers are narrower and have lower slopes than braided rivers. It also specifies that the transition between meandering and braiding can be approximated by the criterion $S/\mathbf{F} \approx D/W$. Combining this criterion with a rational regime model that solves for optimal channel geometry associated with maximum bedload transport capacity yields a multivariate meandering-braided threshold relation (Millar, 2000):

$$S = 0.0002 d_{50}^{0.61} \phi_b^{1.75} Q_{bk}^{-0.25} \tag{8.10}$$

where d_{50} refers to the median grain size of the bed and the bank materials (assumed to be uniform), and ϕ_b is a modified bank friction angle that accounts for cohesive effects of silt and

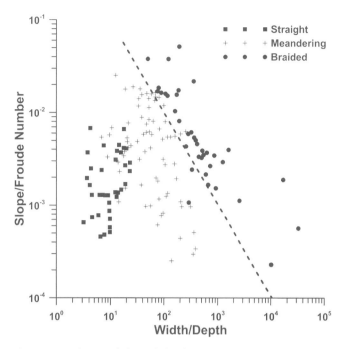

Figure 8.9. Relation of channel planform type to the channel slope/Froude number ratio and to width/depth ratio (from Parker, 1976).

clay as well as increases in bank strength associated with vegetation. Testing of this relation against empirical data using average values of ϕ_b shows that it discriminates well between meandering and braided streams (Millar, 2000). According to Eq. (8.10), the threshold between meandering and braided streams is highly sensitive to the resistance of bank materials to erosion, both through the median size of bank materials and through the modified bank friction coefficient. Values of ϕ_b can vary from 28° to 90° depending on the extent to which vegetation and cohesiveness strengthen the banks (Millar and Quick, 1993), with high friction angles corresponding to increasingly stable banks. The large value of the exponent for ϕ_b in Eq. (8.10) indicates that the threshold channel slope for the onset of braiding increases dramatically as ϕ_b increases. Steep channel slopes are required to produce a braided channel pattern in streams lined by thick, dense vegetation growing in banks containing abundant silt and clay.

The important influence of vegetation in delaying the onset of braiding has been confirmed through numerical simulations that predict bar growth and development on the channel bed in relation to plant stabilization of channel banks (Murray and Paola, 2003). Fast-growing plants that enhance bank strength inhibit channel widening, thereby suppressing bar development within the channel and the initiation of braiding. The strong influence of vegetation on channel planform is consistent with observations that braided rivers are most common in arid, alpine, and arctic environments where vegetation cover is limited.

Sediment supply and the transport of bed material is another important factor influencing whether a river is braided or single-threaded. High bedload concentrations are fundamental to the braided channel planform. In the Rocky Mountains, USA, braided streams occur within contributing watersheds underlain by erodible lithologies, whereas such streams are uncommon in watersheds underlain by resistant crystalline rocks (Mueller and Pitlick, 2013). The relation

$$C_{sb} = 0.006\hat{Q}^{-0.5} \tag{8.11}$$

where C_{sb} is the dimensionless volumetric bedload concentration, discriminates well between single-thread and braided streams (Mueller and Pitlick, 2014). For a given dimensionless discharge ($\hat{Q}$), braided streams have higher bedload concentrations than single-thread channels.

8.5 Under What Environmental Conditions Do Straight Channels Occur?

The conditions that promote the development of straight channels are less clear than those for meandering and braided channels. This state of affairs is not entirely surprising, given that the identification of straight channels is subjective and based on an arbitrary value of sinuosity. Many straight channels have meandering thalwegs, suggesting that the processes that promote meandering are operative within straight channels – an issue examined in detail in Chapter 9.

When data for straight rivers are plotted on bankfull discharge – channel slope or potential stream power per unit area – median grain size diagrams, these rivers fall within the domains of both meandering and braided rivers (Figure 8.10). When data for straight channels (Leopold and Wolman, 1957) are plotted in $\hat{Q} - S$ space, where $\hat{Q}$ is the dimensionless discharge that accounts for the confounding effect of grain size (Eq. 7.18), these data lie within the domain of meandering rivers (Eaton et al., 2010). However, the scatter of straight channels across a broad range of potential stream power values for narrow ranges of grain size (Figure 8.10) suggests that the domain of straight channels varies over the meandering and braided domains. For a given value of grain size, straight channels can have high or low values of potential stream power.

For low slopes and sediment loads, a straight experimental channel remains straight even when flow is introduced to the channel at an angle in an attempt to force it to meander (Schumm and Khan, 1972) (Figure 8.4). Also, linear stability analysis predicts that straight channels with gentle slopes and small width–depth ratios ($W/D < 10$) will remain straight (Parker, 1976; Fredsoe, 1978). Laboratory and field data used

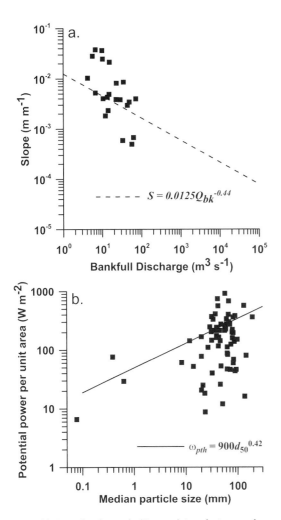

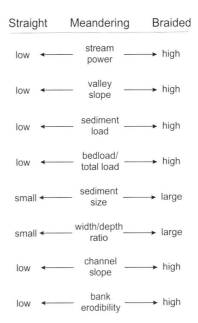

Figure 8.11. Environmental conditions for stable straight channels in relation to meandering and braided rivers (adapted from Schumm, 1981).

Figure 8.10. (a) Straight channels (S_I < 1.5) in relation to channel slope and bankfull discharge (from Leopold and Wolman, 1957). (b) Straight channels (S_I < 1.3) in relation to potential stream power and median particle size (data from van den Berg, 1995).

to evaluate theoretical predictions confirm that stable straight channels generally have relatively small width–depth ratios (Figure 8.9).

In straight rivers, the channel slope equals the valley slope; therefore, for a given formative discharge, such as Q_{bk}, the rate of energy expenditure, or stream power, is maximized. For such a river to remain straight, the valley slope (S_v) must equal the minimum channel slope required for transporting water and sediment loads without net erosion or deposition (S_{min}) (Huang et al., 2004). Thus, the character of a straight river will depend on the relation among the imposed valley gradient, sediment supply, and channel geometry required to transport the supplied sediment. The results of empirical and theoretical analyses suggest that at least two types of straight rivers exist – those with low stream power, low sediment supply, and small width–depth ratios that transport predominantly suspended load and those with high stream power, large sediment supply,

and large width–depth ratios that transport predominantly bedload (Chang, 1979; Schumm, 1985; van den Berg, 1995).

Relative to meandering and braided rivers, straight stable channels have low stream power and resistant banks with a high silt-clay content (Schumm, 1981) (Figure 8.11). Bedforms may migrate through these channels, but the combination of low stream power and resistant banks greatly limits changes in channel alignment over time. These types of channels are presumed to be rare (Leopold and Wolman, 1957; Schumm, 1985); however, the extent to which such channels occur has not been thoroughly documented. In at least some cases, straight low-energy channels occur within multichannel systems as individual anabranches within these systems (van den Berg, 1995). High-energy, bedload-dominated straight channels that are not braided have been recognized in detailed pattern classification schemes (e.g. Schumm, 1981), but examples of these channels are not abundant. In fact, some high-energy channels that have been identified as straight may actually be braided rivers (van den Berg, 1995). The stage dependence of bar exposure within braided rivers could influence visual classification depending on the amount of flow within the river at the time an image used to classify it was obtained. This possibility is consistent with experimental results showing that the overall sinuosity of braided channels is low if the path of the large channel containing the multichannel system interspersed with bars is considered (Figure 8.4a). Straight rivers characterized by well-developed alternate bars, a meandering thalweg, and a mixed sediment load have also been proposed as a distinct category of channel pattern (Schumm, 1981, 1985); however, as discussed in Chapter 9, such rivers usually are transient forms of meandering rivers.

8.6 What Is the Environmental Domain of Anabranching of Rivers?

Research on anabranching rivers has increased substantially over the past several decades, but understanding of these rivers is less well developed than for meandering and braided systems. Therefore, classification of these multichannel systems is somewhat poorly defined (Carling et al., 2014). When data for anabranching rivers are plotted in S-Q_{bk} space, scatter occurs over the domains of braided and meandering rivers, similarly to the case for straight rivers (Figure 8.12). Thus, anabranching can occur in both high–stream power and low–stream power rivers. Evaluation of case studies of anabranching river systems indicates that distinct subtypes can be identified, and as many as six different subtypes have been proposed (Nanson and Knighton, 1996). The major subtypes, for which at least a handful of cases have been studied in detail, include anastomosing and wandering gravel-bed rivers.

Anastomosing rivers plot within the meandering domain and therefore are relatively low-energy fluvial systems compared with braided rivers, even though both have multichannel patterns. The bank erodibility of anastomosing channels is relatively low, similar to that of straight or slowly migrating meandering rivers, whereas sediment supply is moderate to high, similar to that of active meandering rivers or even braided rivers (Knighton and Nanson, 1993). Low stream power, which restricts the transport capacity, combined with an ample sediment supply in many cases produces a net aggradational condition, resulting in avulsive behavior and the cutting of new channels through the floodplain. The upper Colombia River in British Columbia is a classic example of an anastomosing river (Figure 8.1).

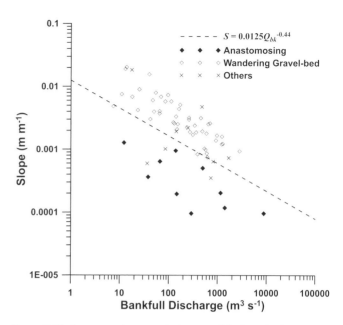

Figure 8.12. Anastomosing, wandering gravel-bed, and other anabranching rivers in relation to bankfull discharge and channel slope (data from Ferguson and Werritty, 1983; Carson, 1984b; Desloges and Church, 1989; Nanson and Knighton, 1996; Payne and Lapointe, 1997; Burge, 2005).

Wandering gravel-bed rivers have relatively high stream power per unit area, plotting above the threshold for braiding as defined by Eq. (8.2) (Figure 8.12). Traditionally, this subtype of anabranching has been viewed as transitional between meandering and fully braided conditions (Desloges and Church, 1989), conforming in experimental studies to the rapid decrease in thalweg sinuosity as meandering yields to braiding (Figure 8.4a). Wandering gravel-bed rivers typically consist of a main channel of irregular sinuosity or even a braided character flanked by secondary channels that constitute anabranches (Figure 8.13). Individual channels are separated by prominent gravel bars or vegetated islands. In some studies, the term anabranching has been used to refer

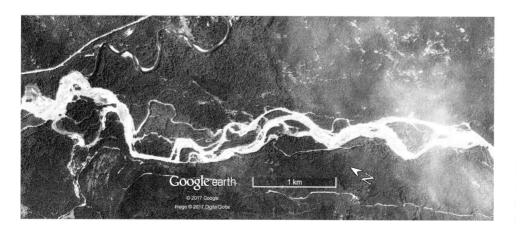

Figure 8.13. The Bella Coola River in British Columbia, Canada – an example of a wandering gravel-bed river.

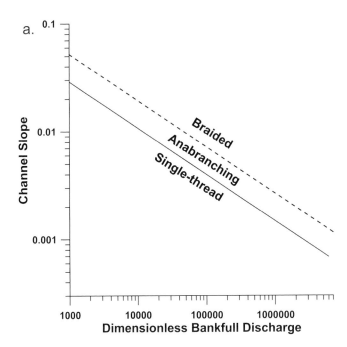

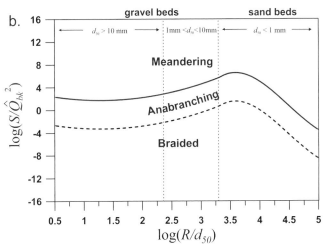

Figure 8.14. Thresholds for single-thread, anabranching, and braided rivers: (a) from Eaton et al. (2010) and (b) from Song and Bai (2015).

1996). To some extent, the rich array of channel patterns and the possibility of unique patterns are reflected in cases that do not neatly fit into traditional categories. For example, the largest rivers in the world by volumetric flow rate, those with mean annual discharges greater than 17,000 m³ s⁻¹, all exhibit anabranching patterns (Latrubesse, 2008); however, this simple designation fails to describe the complexity of planform characteristics of these rivers, details of which have yet to be fully identified. Individual threads of large ana-branching rivers may be activated at different flow stages, transport different amounts of sediment, and have different types of bank materials or vegetation. As a result, individual threads can exhibit straight, meandering, or braided plan-forms (Lewin and Ashworth, 2014a).

8.7 What Are the Implications of Changes in Environmental Conditions for Channel Planform Change?

The rather abrupt transitions among channel planform types documented in empirical and theoretical analyses suggest that the pattern of a river can transform from one type to another over time or space if changes in environmental conditions are great enough to drive the system over a threshold of pattern change. Over timescales of millions of years, tectonism, along with changes in climate, weathering rates, and vegetation characteristics, can alter hydrological conditions, valley gradients, sediment supply, sediment caliber, and bank strength, resulting in transformation of the planform of rivers. Over the past 10,000 years, the transition from the Pleistocene to the Holocene, along with fluctuations in climate throughout the Holocene, has produced changes in channel planform at many locations in mid- to high latitudes through modification of flood frequency and sediment supply (Leigh, 2008; Lewin and Macklin, 2010). Over shorter timescales, increases in flood frequency or even the influence of a single large flood event can transform meandering rivers into braided ones (Brewer and Lewin, 1998; Erskine, 2011), but the sensitivity of individual rivers to planform change is a complex problem that has yet to be fully resolved. Diagrams that relate planform change to factors such as bankfull discharge, valley slope, and median particle size cannot be used to predict changes in channel planform caused by erosional and depositional processes during an individual event or sequences of events. Improved understanding of the mechanisms by which changes in planform occur is important for attempts to predict how rivers may respond to human-induced changes in climate (Anisimov et al., 2008).

Changes in channel planform over space have been related to changes in valley slopes and sediment supply. For a constant discharge, downstream decreases in valley slope

either to wandering gravel-bed rivers, sometimes designated as island braided rivers (Beechie et al., 2006), or to anabranching rivers that differ from anastomosing rivers (Eaton et al., 2010; Beechie and Imaki, 2014; Song and Bai, 2015). These investigations confirm that the domain of relatively high-energy anabranching systems lies between those for meandering and braided rivers (Table 8.1; Figure 8.14).

Other types of anabranching rivers, which plot over a variety of slope–discharge conditions, represent either individual case studies or a small number of cases. Limited examples from geographically restricted locations call into question efforts to formulate new subtypes (Nanson and Knighton,

Table 8.1. Threshold relations for single-thread, anabranching, and braided rivers.

Eaton et al. **(2010)**

Single-thread–anabranching threshold $S = 0.40\mu' 1.41 \hat{Q}_{bk}^{-0.43}$

Anabranching–braided threshold $S = 0.72\mu' 1.41 \hat{Q}_{bk}^{-0.43}$

where μ' is the ratio of the critical bank shear stress to the critical bed shear stress. When bed and banks are equally erodible, $\mu' = 1$, and when banks are less erodible than the bed, $\mu' > 1$

Song and Bai **(2015)**

Sand-bed river thresholds

Meandering–anabranching threshold

$$\log\left(\frac{S}{\hat{Q}_{bk}^{2}}\right) = 231.03\log\left(\frac{R_{bk}}{\hat{\mu}d_{50}}\right) - 53.69\left(\log\left(\frac{R_{bk}}{\hat{\mu}d_{50}}\right)\right)^{2} + 3.994\left(\log\left(\frac{R_{bk}}{\hat{\mu}d_{50}}\right)\right)^{3} - 315.7$$

Anabranching–braided threshold

$$\log\left(\frac{S}{\hat{Q}_{bk}^{2}}\right) = 231.03\log\left(\frac{R_{bk}}{\hat{\mu}d_{50}}\right) - 53.69\left(\log\left(\frac{R_{bk}}{\hat{\mu}d_{50}}\right)\right)^{2} + 3.994\left(\log\left(\frac{R_{bk}}{\hat{\mu}d_{50}}\right)\right)^{3} - 320.7$$

Gravel-bed river thresholds

Meandering–anabranching threshold

$$\log\left(\frac{S}{\hat{Q}_{bk}^{2}}\right) = 0.98\left(\log\left(\frac{R_{bk}}{\hat{\mu}d_{50}}\right)\right)^{2} - 2.56\log\left(\frac{R_{bk}}{\hat{\mu}d_{50}}\right) + 3.4$$

Anabranching–braided threshold

$$\log\left(\frac{S}{\hat{Q}_{bk}^{2}}\right) = 0.98\left(\log\left(\frac{R_{bk}}{\hat{\mu}d_{50}}\right)\right)^{2} - 2.56\log\left(\frac{R_{bk}}{\hat{\mu}d_{50}}\right) - 1.6$$

where $\hat{Q}_{bk} = \frac{Q_{bk}}{A_{bk}\sqrt{8gR_{bk}}}$, R_{bk} = bankfull hydraulic radius, A_{bk} = bankfull flow area, $\hat{\mu}$ = bank strength impact factor, where $\hat{\mu} = 1$ is assumed to be the base case and decreases in $\hat{\mu}$ reflect increasing bank strength.

---

can result in the transformation of a braided river into a meandering channel, but the transition occurs gradually through the development of intermediate forms that do not fit neatly into traditional categories (Ferguson and Ashworth, 1991). Transformations in channel planform along the length of a river are not always indicative of changes in valley slope. Large influxes of sediment locally can transform river planform. A well-known example is the William River in northeastern Saskatchewan, Canada, which changes from a straight or slightly sinuous single-thread channel into a braided planform as it flows through a large dune field that supplies large amounts of eolian sand to the river (Smith and Smith, 1984). Other factors that can influence channel planform spatially include downstream changes in sediment size, bank strength, and the degree of valley confinement of the channel. For rivers close to the transition between braided and single-thread conditions as defined by Eq. (8.11), those in unconfined reaches with relatively fine bed material and low bank strength generally are braided, whereas those in confined reaches with coarse bed material and high bank strength exhibit single-thread channels (Mueller and Pitlick, 2014).

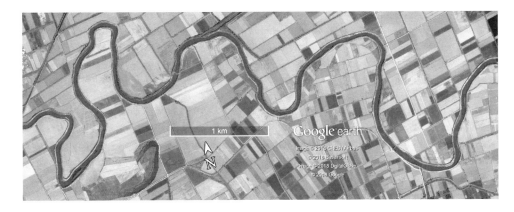

CHAPTER

9 The Dynamics of Meandering Rivers

9.1 Why Is the Meandering of Rivers Important?

The pattern of single-thread channels winding across landscapes as a series of reversing curves known as meanders is a well-known characteristic of many rivers. The origin of the term meandering is associated with the Büyük Menderes River in Turkey, which historically has been referred to as the Meander (from the Latin *Maeander* or Greek *Maiandros* [Μαιανδρος]) (Thonemann, 2011) (Figure 9.1). Since the dawn of civilization, the meandering of rivers has been of interest to humans. Those who live along meandering rivers have learned that such rivers are not fixed in position but migrate over time across valley bottoms, producing renewal of fertile floodplain soils. At the same time, the changing location of the river channel can lead to loss of land and represent a hazard to infrastructure located near the river.

From a geomorphological perspective, the prevalence of meandering rivers focuses attention on this type of river as a fundamental form in need of an explanation. The processes by which meandering rivers develop and evolve are central to understanding river dynamics. The dynamics of meandering rivers also are intimately connected to processes of floodplain development and change (see Chapter 14). Moreover, the long-term dynamics of meandering rivers are of basic importance for understanding how rivers shape landscapes and the role that rivers play in transporting sediment from continents to ocean basins. Ecologically, the physical forms and processes associated with meandering rivers are critical in the structuring of physical habitat for aquatic organisms. Attempts to manage meandering rivers to protect human infrastructure or to restore impaired ecological functions require sound knowledge of the dynamics of these rivers (see Chapter 16).

9.2 Why Do Rivers Meander?

The cause of river meandering has long been of interest to river scientists. In seeking an explanation for meandering, it

Figure 9.1. The Büyük Menderes River in Turkey.

typically is contrasted against the alternative of a straight channel. Explanations attempt to show how a straight configuration is not stable and deforms into a meandering pattern.

9.2.1 How Have Extremal Hypotheses Been Invoked to Explain Meandering?

The scaling of the width and wavelength of meandering streams over many orders of magnitude has fueled speculation that meandering must be the result of a general governing principle (Davy and Davies, 1979). As a result, various extremal hypotheses have been invoked as possible causes of river meandering, including minimization of the variance of directional change in the channel path (Langbein and Leopold, 1966), minimization of energy dissipation rate as water flows downhill (Yang, 1971), and minimization of stream power (Chang, 1988b). The latter two hypotheses can be viewed as examples of the more general principles of minimum energy of open channel flows (Huang et al., 2004) and least action (Nanson and Huang, 2008, 2017).

Similarly to regime analysis in hydraulic geometry (see Chapter 7), extremal principles can be invoked to constrain quantitative models of adjustments among river variables so that optimal solutions for channel geometry (width, depth, and slope) are obtained for a given set of inputs (bankfull discharge, grain-size characteristics of the bed and banks, sediment load, and bank resistance). Solutions for channel slope can be compared with valley gradients to determine channel sinuosity, which, if large, indicates that the channel will meander (Chang, 1979; Huang et al., 2004).

Extremal explanations of meandering rely on the premise that fluvial adjustments to imposed external conditions occur so that some underlying constraint is optimized. Even if the properties of meandering rivers conform to predictions based on extremal principles, the agreement may be the result of meandering rather than evidence that the extremal principle is a cause of meandering. Moreover, extremal accounts fail to specify how a channel that is initially straight evolves into a meandering river. The approach cannot explicitly describe the self-forming action of the fluvial system (Eaton et al., 2006).

9.2.2 What Mechanistic Arguments Have Been Proposed to Explain Meandering?

Mechanistic theories of river meandering try to determine the underlying processes that lead to the development of curves in a straight channel (Rhoads and Welford, 1991). In other words, these theories seek the mechanisms that initiate meandering. The development of curves in a straight channel requires erosion of the banks at regularly spaced intervals along alternate sides of the channel. Although bank erosion is a necessary condition for meandering (Friedkin, 1945), it results from flow and sediment-transport processes that trigger this erosion.

9.2.2.1 What Are Alternative Bars and Bar Units and How Are They Related to the Initiation of Meandering?

Field and experimental studies indicate that the development of meanders within an initially straight channel is preceded by the formation of alternate bars on the channel bed (Quraishy, 1944; Einstein and Shen, 1964; Keller, 1972; Lewin, 1976; Lanzoni, 2000a,b) (Figure 9.2). Alternative bars consist of bank-attached deposits of sediment that occur offset from one another on opposite sides of a straight channel. Further experimental work has shown that these bars actually are part of a large structural feature, referred to as a bar unit, that includes a zone of scour (pool) and a zone of deposition (alternate bar) (Chang et al., 1971; Ikeda, 1989). Material excavated from the zone of scour is deposited downstream as a fan-like wedge of sediment to form an alternate bar. Bar units alternate along opposing sides of the channel and overlap in a fish-scale pattern (Figures 9.3 and 9.4). Although the term "alternate bar" is frequently used to refer to bar units, the distinction between the two is important for understanding different morphological features of straight and meandering rivers at the bar-unit and bar-element scales (Figure 1.7). The distinction also differentiates between an alternate bar as a depositional feature, the common conception of a bar in a river, and the bar unit, which includes both erosional and depositional components.

Once bar units develop in a straight channel, flow at low stage is strongly influenced by the bar-unit topography, with flow converging in the pools and diverging downstream as the pool widens and bed elevation increases (Figure 9.3). Flow is forced laterally around the alternate bar over the side of the bar unit into the adjacent pool of the next successive bar unit. The zone of shallow water where flow moves laterally over the side of the bar unit constitutes a riffle. Thus, pool-riffle units, a common feature at the bar-element scale in meandering rivers, also can be identified in straight channels with bar units (Keller, 1972). The thread of maximum velocity strongly turns from side to side in a meandering pattern as the flow moves through the succession of bar units. At bankfull stage, the steering effect of the bar units on the path of the flow diminishes. Although this path still meanders, the degree of meandering is less pronounced and slightly out of phase with the variation in bed topography of the bar unit (Figure 9.3). Nevertheless, substantial spatial variation in flow depth both longitudinally and laterally over the bar unit results in pronounced convective (spatial) accelerations of downstream and cross-stream components of the flow and strong spatial

Figure 9.2. Alternate bars in a recently channelized section of the Embarras River, Illinois.

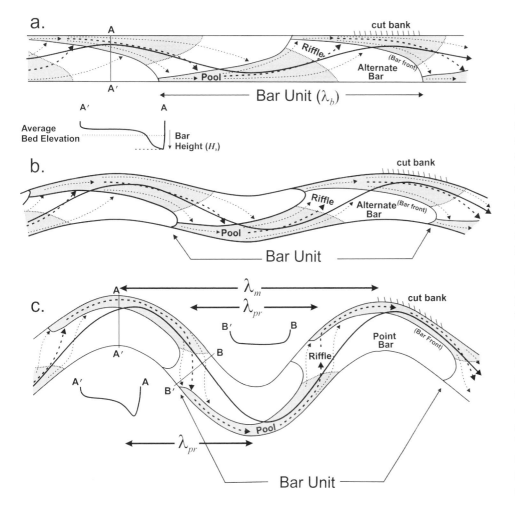

Figure 9.3. Conceptual model of evolution of bar units as a straight channel evolves into a meandering channel (a–c). Shaded areas are below the average bed elevation and unshaded (white) areas are above the average bed elevation. Light and heavy dashed lines with arrows show patterns of flow at low stage when alternate bars and point bars are exposed. Heavy dashed lines represent the main path of flow (thread of maximum velocity). Solid lines represent the main path of the flow at near bankfull stage. Insets for the straight (a) and meandering (c) channels illustrate typical cross sections. Height (H_b) and wavelength (λ_b) of a bar unit is shown for the straight channel. Wavelengths of meandering (λ_m) and pool-riffle sequences (λ_{pr}) are shown for the meandering channel.

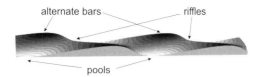

Figure 9.4. Three-dimensional view of alternating bar units (from Federici and Seminara, 2003).

variation in the pattern of bed shear stress and sediment sorting (Nelson, 1990; Whiting and Dietrich, 1991). Where the meandering thalweg approaches the sides of the channel, the boundary shear stress increases, enhancing the potential for bank erosion. Thus, bar units systematically deflect the flow laterally, which can produce erosion of the banks at regularly spaced intervals along a straight channel – a necessary condition for the initiation of meandering.

The development of bar units in straight channels has been related to the channel-forming index SW/D, where S is the channel slope, W is the flow width, and D is the flow depth (Ikeda, 1973, 1989; Welford, 1993). These units develop during transport-effective flows with a high channel-forming index, indicating that steep channels conveying flows with large width to depth ratios promote the formation of bar units. Subsequent flows with low values of SW/D may lead to eradication of bar units (Welford, 1994). Similarly, reductions in sediment supply can eradicate bar units (Venditti et al., 2012) or cause incision of the thalweg, or deepest part of the channel bed, thereby increasing subaerial exposure of bar tops (Lisle et al., 1993). Bar units also tend to evolve over time, initially having a wavelength (λ_b) of about three to six times the channel width, but elongating to as much as seven to 17 times the channel width (Lewin, 1976; Ashmore, 2009; Eekhout et al., 2013; Adami et al., 2016). The height, or amplitude, of the unit (H_b) varies irregularly over time around a mean height (Eekhout et al., 2013). The wavelength of well-developed bar units is positively related to the width and depth of the flow and inversely related to frictional resistance (Ikeda, 1984a), whereas the height is positively related to the width or width/depth ratio of the flow and to the size of bed material (Ikeda, 1984b; Jaeggi, 1984). However, the extent to which bar-unit morphology attains a stable state remains uncertain; long-term experiments indicate that the form of bar units varies cyclically over time and that units can even completely disappear and reform under constant flow conditions (Crosato et al., 2012).

In laboratory experiments with steady flow, bar units in straight channels typically migrate progressively downstream over time as sediment is transferred through the pools, across the riffle and bar top, and accreted onto the advancing bar front (Lanzoni, 2000a,b). Therefore, the rate of propagation of the bar units depends on the sediment-

transport rate. In the Alpine Rhine River, bar units migrate at a long-term average rate of about 1/10th of the mean bar-unit wavelength per year, with migration occurring mainly during flood events (Adami et al., 2016). When bed material consists of mixed sand and gravel, freely migrating bar units can stabilize through accumulation of immobile coarse material at the transition between the pool and the alternate bar, or the bar head (Lisle et al., 1991). Coarsening of the surface of alternate bars also prevents erosion of the bar top, which would otherwise occur due to enhanced shear stresses on this part of the bar unit (Whiting and Dietrich, 1991).

Forced bar units that are steady and do not migrate can develop downstream of obstacles to flow, such as local areas of bank or bed erosion, or where channel curvature fixes the bars in place. In experimental studies, steady bar units develop intrinsically in the absence of perturbations and coexist with migrating units (Lanzoni, 2000b; Crosato et al., 2011). Field evidence indicates that steady, forced bar units and migrating bar units occur together in conjunction with variations in channel morphology and hydrological regime (Rodrigues et al., 2015).

Two types of theory have arisen to try to explain both the development of bar units and the pattern of systematic bank erosion that leads to meandering of straight channels. The first type invokes oscillation of flow in an open channel as the mechanism that deforms the bed and also erodes the banks. The second type posits that interaction of the flow with a mobile bed and erodible banks is fundamentally unstable, leading to the development of bar units and systematic curves within a straight channel.

9.2.2.2 What Are the Major Flow Oscillation Theories of Meandering Initiation?

Flow oscillation theories are based on the idea that channelized flow moving along a straight path inherently oscillates, resulting in a meandering thalweg, the production of bar units, and subsequent systematic erosion of the channel banks. In part, the motivation for investigating flow oscillation as the cause of bar units and river meandering derives from the development of surface-tension meanders in sediment-free water rivulets flowing down smooth inclined surfaces (Tanner, 1960; Gorycki, 1973; Davies and Tinker, 1984). Meandering streams also form on the surface of glaciers, where the channels are carved into ice rather than rock or sediment (Ferguson, 1973; Marston, 1983; Karlstrom et al., 2013). These cases suggest that streams of water flowing down inclined surfaces have an inherent tendency to meander.

A conceptual model of the initiation of meandering links oscillatory behavior of the flow to helical motion, or spiral patterns of fluid rotation, a type of fluid motion that occurs in meandering rivers (Thompson, 1986) (Figure 9.5). Although channel curvature causes helical motion in meandering rivers, such motion can develop in straight channels through anisotropic boundary-generated turbulence (Einstein and Li, 1958), particularly near the corners of rectangular channels (see review in Rhoads and Welford, 1991). The turbulent production of rotational fluid motion, or vorticity, is related to differences in gradients of convective accelerations of turbulence associated with the cross-stream and vertical Reynolds normal stresses (Perkins, 1970):

$$\left[\frac{\partial}{\partial y} \left(\frac{\partial \overline{w'^2}}{\partial z} \right) - \frac{\partial}{\partial z} \left(\frac{\partial \overline{v'^2}}{\partial y} \right) \right] \neq 0 \tag{9.1}$$

Eq. (9.1) typically is positive on one side of the channel and negative on the other, giving rise to streamwise vortices, or cells of spiraling flow, with opposing senses of rotation (clockwise versus counterclockwise) on each side of the channel. If these dual surface-convergent helical cells compete with one another, producing alternating expansion and contraction of the cells along the length of the channel, the result will be a meandering thalweg (Einstein and Shen, 1964) (Figure 9.5a). The alternating pattern of expanding and contracting helical flow provides a mechanism for generating bar units by promoting scour along the path of the thalweg and deposition within regions of expanding helical flow adjacent to the thalweg. Through this process, pools, riffles, and alternate bars supposedly develop in straight channels (Figure 9.5b).

A shortcoming of this flow oscillation theory is that competing dual surface-convergent helical cells have yet to be documented in straight experimental or natural channels. Experimental studies and numerical simulations indicate that corner-induced cells do not seem to expand and contract across the channel through competition with one another (Naot and Rodi, 1982; Nezu et al., 1989). Moreover, the

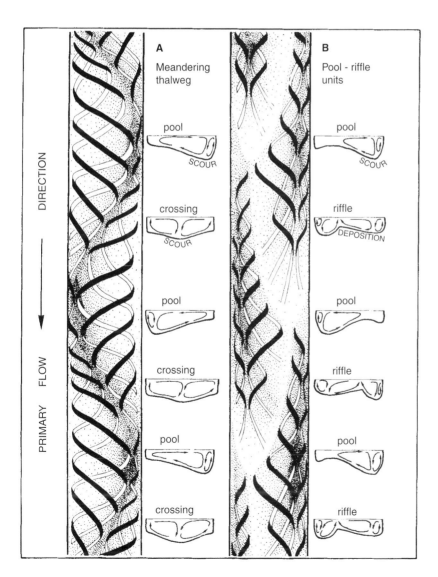

Figure 9.5. (a) Meandering thalweg (shaded) produced by competition between wall-generated surface-convergent helical cells in a straight channel as proposed by Einstein and Shen (1964). (b) Proposed modification of the basic pattern in (a) through the emergence of bar units that include pools (shaded zones), riffles (zones of crossover of the flow from one bar unit to the next), and alternate bars (white areas). Black ribbons are near-surface flow and white ribbons are near-bed flow (from Thompson, 1986).

same investigations show that corner-induced vortices typically do not extend from the walls more than two times the depth of flow. In natural rivers, which typically have bankfull width–depth ratios greater than 8, this restriction would limit helical motion to the near-bank region. Studies of flow structure in wide straight rivers and laboratory flumes have found that multiple cells develop across the width of the channel (Nezu et al., 1993; Rodriguez and Garcia, 2008; Blanckaert et al., 2010). The development of these cells has been attributed to lateral variation in the mean streamwise velocity ($\partial u/\partial y$) in the near-bed region of the flow (Yang et al., 2012). The development of multicellular helical vortices may produce small, parallel, longitudinal sand ridges in straight channels with erodible beds (Karcz, 1966; Colombini, 1993; Colombini and Parker, 1995) but has not been viewed as a mechanism of meander initiation.

An alternative flow oscillation theory proposes that meandering is initiated by the systematic production of macroturbulent eddies that scale with the flow width (Yalin, 1971). According to this theory, horizontal coherent structures (HCS) generated at the channel banks expand into the flow through a process referred to as horizontal bursting (da Silva, 2006; Yalin, 2006). The repeated development and expansion of HCS at periodic intervals along opposing sides of the channel results in deflection of the mean flow into a meandering pattern and the development of bar units. Evidence to support this theory is rather weak, and the size of macroturbulent eddies in rivers generally scales with flow depth rather than flow width. Nevertheless, experimental results indicate that HCS with a length scale of five to seven times the flow width exist in straight open channels (da Silva and Ahmari, 2009). Superimposition of the horizontal bursting on the mean flow produces slight meandering of flow streamlines, thereby providing a mechanism for flow oscillation (Figure 9.6). Although these experiments were conducted for an immobile bed, the length scale of the HCS and thalweg meandering is similar to the spacing of bar units in experimental studies with mobile beds. The extent to which HCS occur in natural rivers has yet to be ascertained.

9.2.2.3 How Has Meander Initiation Been Related to Bed and Bank Instability?

Since the 1970s, the prevailing general theory of meander initiation is that meandering occurs because small perturbations of the channel bed and channel banks in a straight channel with a mobile bed and erodible banks are, under certain conditions, fundamentally unstable. Analysis undertaken to test this general theory involves first formulating physically based mathematical models of flow and sediment transport in straight channels. Linear or nonlinear stability analyses are then used to determine whether analytical or

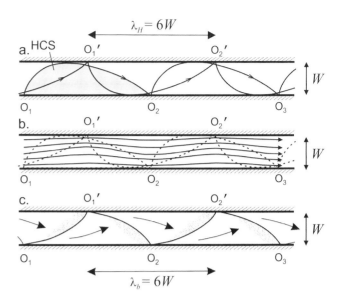

Figure. 9.6. (a) Planview schematic of horizontal coherent structures (HCS – shaded) and pattern of flow with water first moving toward the opposite bank as the HCS expands and then back toward the bank of origin as the HCS dissipates and contracts. λ_H is the HCS wavelength. Symbols O and O' indicate origins of successive structures (numbered subscripts) along opposite banks. (b) Slight meandering of the mean flow (solid lines with arrows) resulting from the influence of HCS generation on alternating sides of the channel. (c) Bar-unit development in response to the pattern of flow (from da Silva and Ahmari, 2009).

numerical solutions to the perturbed form of the system of equations are stable or unstable. Initial analysis of this type focused on the development of bar units in straight channels (bar theory). Subsequent analyses examined the stability of model solutions to perturbations of the channel banks (bend theory). Finally, models have been developed to consider interaction between perturbations of the bed and banks (bar-bend theory).

Bar theory seeks to determine the conditions under which doubly harmonic infinitesimal perturbations of the channel bed become unstable and grow in amplitude (Figure 9.7a). Basic equations in a two-dimensional approach to the problem consist of mathematical functions for momentum–force balance in the streamwise and lateral directions, continuity of flow, and continuity of sediment transport. Linearization of these functions yields expressions for the perturbations of the initial undisturbed flow, which is assumed to be steady, uniform, and unidirectional (Parker, 1976; Fredsoe, 1978). Solution of linearized equations for different wavelengths of perturbations yields amplification factors that determine whether perturbations grow or diminish. Positive values of the amplification factor indicate growth, whereas negative values indicate damping (Figure 9.7b). The wavelength of the perturbation with the maximum rate of amplification is

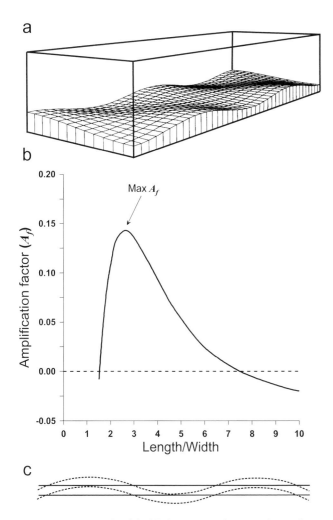

Figure 9.7. (a) Example of doubly harmonic infinitesimal perturbation of the channel bed used in linear bar theory (from Nelson, 1990). (b) Plot of bar amplification factor versus bar length/channel width ratio for a width–depth ratio of 20, a dimensionless bed shear stress of 0.2, and a grain resistance coefficient of 7. Maximum amplification rate corresponds to the peak of the curve and represents the assumed wavelength of alternate bars (from Rhoads and Welford, 1991 based on Fredsoe, 1978). (c) Perturbation (dashed lines) of straight channel planform (solid lines) in bend theory.

units into the characteristic fish-scale morphology (Columbini et al., 1987; Schielen et al., 1993) and bar-unit development under unsteady flow (Tubino, 1991; Hall, 2004). These theories also predict more accurately the dimensions of well-developed bar units (Knaapen et al., 2001). Numerical models of bar units have also been developed that are capable of accommodating strongly nonlinear interaction among flow, sediment transport, and the evolving bed morphology, thereby simulating the time-dependent evolution of bar units from an initial perturbation to a fully developed state (Nelson, 1990; Defina, 2003; Federici and Seminara, 2003; Francalanci et al., 2012; Qian et al., 2017).

A major limitation of bar theory is that it does not explicitly include mechanisms that lead to channel curvature, but simply assumes that bar units control the initial wavelength of meandering by deflecting flow laterally into the banks, which presumably leads to erosion. Bend theory focuses on the stability of a straight channel to infinitesimal periodic perturbations of the channel centerline (Figure 9.7c). Such models do not consider the development of migrating bar units, but instead include a function to allow the channel to migrate through erosion of the channel banks (Ikeda et al., 1981; Kitanidis and Kennedy, 1984). Predictions of selected wavelengths for bend theory are similar to those for bar theory, indicating that bar and bend instabilities operate at similar scales and that the formation of bar units might trigger a bend instability that leads to the development of a meandering channel.

A limitation of bend theory is that it does not explicitly account for the development of bar units in straight channels, which field and experimental evidence indicate, precede meandering. Bar-bend theory combines elements of bar theory and bend theory to provide a unified account of the development of bar units and erosion of the channel banks that leads to meander initiation (Blondeaux and Seminara, 1985; Johannesson and Parker, 1989; Parker and Johannesson, 1989; Seminara and Tubino, 1989). It includes interaction among bar units and channel curvature. A particular problem with bar theory is that it consistently underpredicts the wavelengths of well-developed bar units (Nelson and Smith, 1989; Garcia and Nino, 1993). Moreover, the rather large predicted migration rates of bar units deter erosion at fixed locations along a straight channel; instead, continual spatial changes in the locus of bank erosion associated with migrating bar units should result in uniform erosion of the banks along the length of the channel. Bar-bend theory addresses these problems by predicting that interaction between bar and bend instabilities results in a resonance phenomenon whereby incipient channel curvature forces the emergence of quasi-steady bed perturbations with wavelengths that exceed those of the fastest-growing perturbations in bar theory (Figure 9.8). Thus, the forcing effect of curvature suppresses free bar units and stimulates the development of forced bar units (Tubino and Seminara, 1990). Bar units initially present in the straight channel are deformed by curvature

assumed to be the one that emerges on the channel bed. In other words, this wavelength corresponds to the wavelength of bar-unit development.

A key finding of bar theory is that sediment transport is necessary for instability – a finding at odds with flow oscillation theories (Parker, 1976). Moreover, the theory also indicates that bar units generally will not develop when the width–depth ratio of the flow is less than 10–20 (Fredsoe, 1978; Seminara and Tubino, 1989). Linear bar theory applies only to the domain of infinitesimal perturbations; it does not account for interaction of evolving bar units with patterns of flow and sediment transport. Weakly nonlinear bar theories address the finite-amplitude effects that result in transformation of growing bar

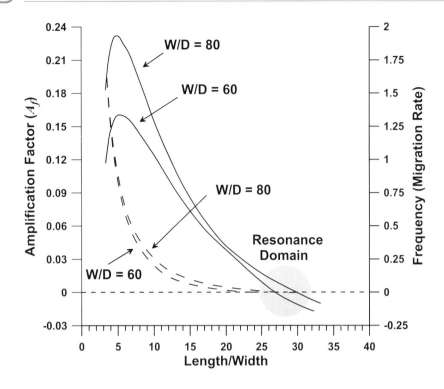

Figure 9.8. Nondimensional bar amplification factor (solid curves) and migration rate (dashed curves) for different values of width–depth ratio (W/D) and $\Theta = 0.25$, $D/d = 200$. Resonance domain (shaded circle) shows dimensionless bar unit wavelengths corresponding to excitation by bend instability (from Rhoads and Welford, 1991 based on Blondeaux and Seminara, 1985).

Figure 9.9. Stable vegetated alternate bars in a straight channel deflecting flow into channel banks and initiating meandering.

as meandering develops (Seminara and Tubino, 1992) (Figure 9.3). The result is a theory that predicts the evolution of a straight channel with migrating free bar units into a meandering channel with stable forced bar units through erosional and depositional processes affecting both the bed and the banks of the river. Experimental work in straight and mildly sinuous channels confirms that meandering inhibits the migration of free bar units and leads to the development of steady bar units within meander bends (Garcia and Nino, 1993). It also shows that free and forced bar units can coexist at low values of sinuosity – a finding consistent with bar-bend theory (Tubino and Seminara, 1990).

Neither bar nor bend theory accounts for the role of vegetation in alternate-bar development and the initiation of meandering. Vegetation on bar surfaces strongly influences bar stability and growth (Landwehr and Rhoads, 2003). Plants become established during periods of prolonged low flow when bar surfaces are exposed and may survive periodic inundation. The growth of vegetation stabilizes bar units by inhibiting migration. It also impedes flow over the bars, promoting deposition of fines and increases in bar size. Stable, vegetation-covered alternate bars can deflect flow laterally into the channel banks, initiating meandering of straight channels (Figure 9.9).

Application of bar-bend theory to river meandering requires translation of the concept of instability to experimental or field contexts. The mathematical constructs of spatially extended infinitesimal periodic perturbations used in stability analyses have little direct relevance for understanding perturbations in natural rivers or experimental channels. Although tests of the theory can be most rigorously controlled in experimental settings, producing true meandering in laboratory streams requires specific conditions that promote bank cohesion and floodplain development (Braudrick et al., 2009; Kleinhans et al., 2014). Experimental work and numerical simulations indicate that an initial upstream bend or oscillation of the incoming flow is needed to generate meandering in straight laboratory channels (Braudrick et al., 2009; van Dijk et al., 2012; Schuurman et al., 2016). Without an inflow perturbation, the straight channel remains straight despite the presence of migrating bars. The initiation of meandering upstream involves the development of a steady bar that initiates or reinforces bank erosion, generating an initial upstream bend, which in turn triggers the development of downstream bends. Free migrating bars, although present, do not deflect flow strongly enough toward the banks to produce erosion. With a static upstream perturbation, such as curvature of the flow at the entrance to the straight channel, sinuosity increases as initial bends develop, but over time the bends elongate and decrease in amplitude, reducing sinuosity. In other words, the effect of a static perturbation dampens over time and space. On the other hand, a dynamic inflow perturbation whereby water entering the straight channel shifts laterally over time produces more highly sinuous channels than does a static perturbation (van Dijk et al., 2012; Schuurman et al., 2016) (Figure 9.10). The lateral shifting of incoming flow corresponds in nature to an effect similar to temporal changes in the direction of flow exiting a migrating bend upstream of a straight reach that has yet to develop curvature. Both the numerical and experimental results indicate that a dynamic inflow condition is necessary to sustain bend dynamics after initiation of meandering. Dynamic inflow perturbations apparently generate convective bend instability, which stimulates downstream planform development (Lanzoni and Seminara, 2006).

Few, if any, field studies have examined the initiation of meandering in natural rivers at anywhere near the level of detail seen in experimental settings. Considerable opportunity exists for field investigations of this problem, particularly in artificially straightened rivers. Such work has direct relevance for efforts to restore channelized rivers. Although the exact cause of meandering remains the focus of ongoing research, the family of analytical, numerical, and physical models constituting bar-bend theory provides a solid foundation for the development of a general explanation of meandering grounded in basic principles of fluid mechanics and sediment transport.

9.3 What Are the Major Planform Characteristics of Meandering Rivers?

The most distinguishing attribute of meandering rivers is the development of systematic channel curvature at the planform scale (Figure 1.7). Series of reversing curves in the channel path

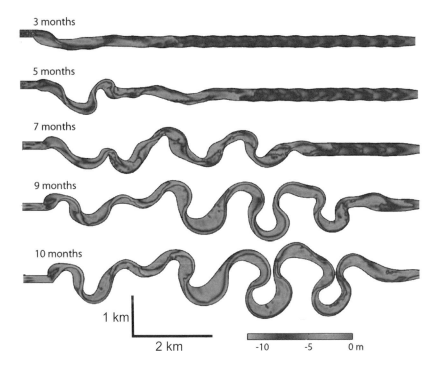

Figure 9.10. Numerical simulation of meander initiation in a straight channel with a dynamic inflow perturbation (from Schuurman et al., 2016). (A black and white version of this figure will appear in some formats. For the color version, please refer to the plate section.)

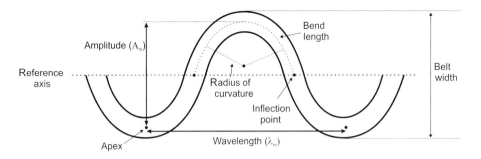

Figure 9.11. Planform geometry of a meandering river.

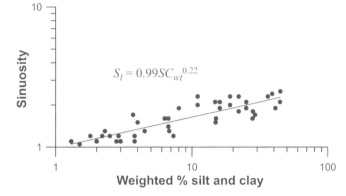

Figure 9.12. Relation between channel sinuosity and weighted percentage of silt and clay in the channel boundary (SC_{wt}) (data from Schumm, 1968).

are known as meander trains, and individual curves within a train are referred to as meander bends. The width of the zone of active bends along a meander train defines the meander belt (Figure 9.11). Inflection points occur at transitions between successive bends with opposing senses of curvature, and the reference line or down-valley axis connects a sequence of inflection points. The point where a meander bend reaches its greatest lateral distance from the reference line demarcates the bend apex. Meander trains constitute a series of spatial waves characterized by wavelengths and amplitudes. The wavelength of meandering is the distance between bend apexes on the same side of the valley in which the meandering river flows (Figure 9.11). The amplitude of meandering is the lateral distance between apexes of successive bends.

The wavelengths of meandering channels have been related to bankfull channel widths to explore the scaling relation between these two attributes of channel form. Typically, such relations have the form of power functions with exponents close to 1 (Leopold and Wolman, 1960; Williams, 1988), indicating that the relation is essentially linear. Moreover, it is reasonable to constrain the linear relation between the variables to pass through zero so that meander wavelength becomes zero for a channel width of zero. This type of analysis shows that the average meander wavelength is about 12 times the channel width (Richards, 1976a; Carling and Orr, 2000):

$$\lambda_m \approx 12 W_{bk} \qquad (9.2)$$

Values of the wavelength–width relation for specific meanders may differ substantially from this average relation. Similarly, although empirical power functions have been developed relating bend amplitudes and meander belt width to bankfull channel width (Williams, 1988), these functions are highly generalized. Bend amplitudes and meander belt width often vary over time as meander trains and individual bends evolve though meander migration.

Surprisingly, little work has examined the general factors controlling river sinuosity, the ratio of the river path length to the meander-belt path length or the channel slope to the valley slope. The sinuosity of meandering rivers is scale independent and does not vary systematically with increasing discharge or channel width. However, it does depend on the cohesiveness

of the channel boundary; rivers with high percentages of silt and clay in the bed and banks have higher sinuosities than rivers with low percentages of silt and clay (Figure 9.12).

Channel curvature (C_c), the rate of change in direction of the channel per unit length along the centerline path of the channel, is an important attribute of meandering rivers. The traditional method of characterizing bend curvature involves visually comparing arcs of circles of various sizes with bend curvature until an arc is identified that closely fits the path of the bend centerline. The radius of the circle corresponding to the fitted arc defines the radius of curvature (r_c) of the channel centerline over the entire length of the bend, where $r_c = 1/C_c$ (Figure 9.11). Dividing the radius of curvature by the channel width provides a scaled metric of relative curvature. Gently curving bends have large values of r_c/W_{bk}, whereas sharply curving bends have small values of r_c/W_{bk}. Typical values of r_c/W_{bk} range from about 1 to 7 with a geometric mean of 2.4 (Williams, 1986).

Local channel curvature can be determined based on cubic spline functions fitted to digitized points along the channel centerline (Guneralp and Rhoads, 2008). Using the coefficients of the cubic spline function, first-order (x', y') and second-order (x'', y'') derivatives of change in spatial coordinates (x, y) of channel position can be computed at any arc-length distance (s) along the composite fitted curve defining the channel centerline path. The curvature at any point s (C_{cs}) is

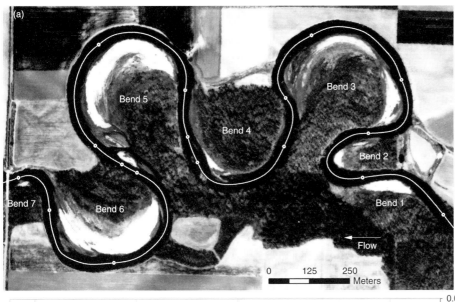

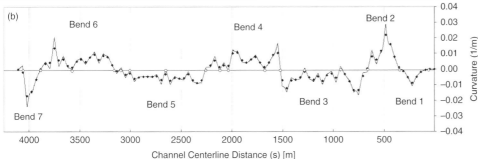

Figure 9.13. Plot of curvature (b) along the centerline of a meandering section of the Mackinaw River, Illinois (a). Values of zero curvature define the boundaries of discrete bends or loops within the reach (after Guneralp and Rhoads, 2008).

$$C_{cs} = \frac{x'y'' - y'x''}{\left[(x')^2 - (y')^2 \right]^{3/2}} \qquad (9.3)$$

This method allows curvature to be characterized continuously along the path of a meandering river (Figure 9.13).

Bend planform can be characterized by the relation between the chord length (C_{LC}), or distance between points of inflection bounding the bend, and the perpendicular distance from the center of the chord to the bend apex (P_L) (Figure 9.14). In simple bends, P_L is less than C_{LC}, whereas in elongate bends, P_L is greater than C_{LC}. Bend symmetry is determined by the planform shapes of the bend on each side of P_L: if the shapes are mirror images of one another, the bend is symmetrical, whereas if the shapes are not mirror images of one another, the bend is asymmetrical. Meander loops develop when the two sides of the bend curve back upon one another toward a point of intersection (Frothingham and Rhoads, 2003). The narrow area containing the point of intersection is referred to as the neck of the loop. In loops, the sum of absolute values of angles of the channel path relative to the downvalley direction at points of inflection near the neck of a bend, $|\alpha_1| + |\alpha_2|$, exceeds 180 degrees. The development of separate lobes along

an elongated bend, characterized by distinct arcs of curvature, produces forms referred to as multilobes or compound loops.

9.4 What Are the Major Morphological Components of Meandering Rivers at the Bar-Unit Scale?

In-channel morphological components of meandering rivers include the three major bar elements of steady, forced bar units – pools, riffles, and point bars; in addition, cut banks are a common feature of the channel boundary in these rivers (Figure 9.15). According to the bar-unit model of bed morphology in meandering rivers, pools, riffles, and point bars represent bar-element scale features of the bar units (Figure 9.3). Pools are topographic lows that extend from the upstream end of a bar unit around the outside portion of a bend. The pool usually reaches its maximum depth slightly downstream of the apex. Cut banks with vertical or near-vertical faces flank the pools along the outside of the bend. On the inner bank, the point bar represents the depositional downstream part of a separate, upstream bar unit. Riffles are local topographic highs within the center of the meandering

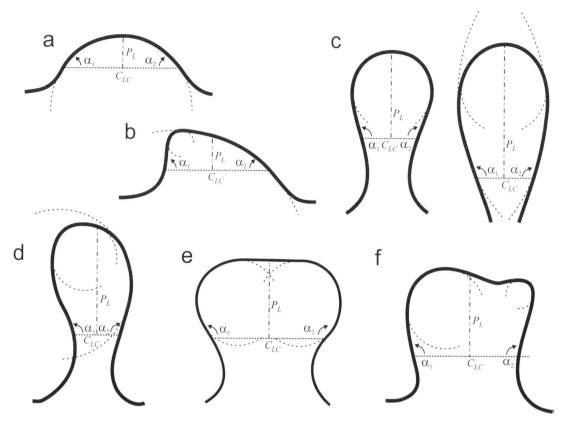

Figure 9.14. (a) Simple symmetrical bend. (b) Simple asymmetrical bend. (c) Elongate symmetrical loops. (d) Asymmetrical elongate loop. (e) Symmetrical compound loop. (f) Asymmetrical compound loop.

Figure 9.15. (a) Pool-riffle sequence at low flow in the East River near Mt. Crested Butte, CO. (b) Point bar and cut bank on Glenburn Creek, IL.

stream where flow moves over the upstream bar unit into the head of the pool in the next successive bar unit downstream. Riffles typically are located near or slightly downstream of bend inflections, where channel curvature is nearly zero.

The bar-unit model of in-channel morphology has its most complete expression in freely meandering gravel-bed rivers, where these units are most fully developed and where pronounced sorting of sediment by size occurs in relation to spatial variations in bed shear stress across the bar units. Cut banks and point bars are typical of sand-bed meandering rivers, and pool-riffle sequences have been identified in such rivers (Hudson, 2002), but the lack of coarse material inhibits the formation of prominent riffles. Despite the inclusion of pools and riffles as bar elements in the bar-unit model of meandering and straight alluvial rivers, research on pool-riffle sequences in such rivers has not been strongly tied to this model. As a result, the extent to which the bar unit model adequately represents pool-riffle morphology remains uncertain. Moreover, not all pool-riffle sequences are associated with the development of stationary bar units in straight or meandering alluvial rivers. Forced pools and riffles can develop when obstructions such as bedrock outcrops, large boulders, trees, or other in-channel features constrict the flow, promoting scour and associated deposition (Lisle, 1986;

Montgomery et al., 1995; Thompson and Fixler, 2017). External forcing plays a major role in the formation of these features, which differ from unforced pools and riffles that develop autogenically in unconstrained alluvial rivers.

9.4.1 How Are Pools and Riffles Identified?

Even though the bar-unit model indicates that the structure of bed morphology in meandering rivers is three-dimensional, most work on pools and riffles has identified these features based on one-dimensional analysis of longitudinal variations in the elevation of the channel bed. Common methods for identifying pools and riffles from bed-elevation survey data include the regression technique (Richards, 1976a; Carling and Orr, 2000), the differencing technique (O'Neill and Abrahams, 1984), and spectral analysis (Richards, 1976a; Carling and Orr, 2000) (Figure 9.16). The first two methods identify individual pools and riffles, whereas spectral analysis only yields information on groups of pools and riffles. Regression analysis involves fitting a linear regression function to bed-elevation data and then identifying pools as negative residuals and riffles as positive residuals about the trend line. To avoid including local variations in bed elevation that do not represent pools or riffles, a tolerance level around the zero line can be specified. The differencing technique is based on cumulative change in bed elevation (O'Neill and Abrahams, 1984). When cumulative differences in bed elevation exceed a specified tolerance level, a bedform is delineated, and the absolute maximum or minimum of a cumulative difference sequence following exceedance of the tolerance threshold defines the riffle crest or pool trough. Once an absolute maximum or minimum of a series is determined, the cumulated sum is zeroed at this location and the summation of differences resumes until a new absolute maximum or minimum is reached. For both the regression and differencing techniques, the choice of a tolerance level, defined in terms of the standard deviation of residuals or differences, is somewhat subjective and can influence the number of identified pools or riffles. Three-dimensional characterizations of bed topography in gravel-bed rivers indicate that the bed includes a variety of morphological units and that pools and riffles are not necessarily coupled, but may be separated by distinct intervening units (Wyrick and Pasternack, 2014).

9.4.2 What Are the Morphological and Sedimentological Characteristics of Pool-Riffle Sequences?

Pool-to-pool or riffle-to-riffle spacing along straight and meandering rivers averages five to seven channel widths

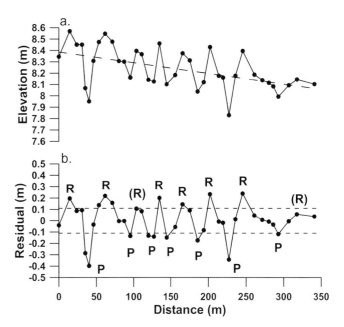

Figure 9.16. (a) Variation in bed elevation along a meandering section of the Embarras River, Illinois. (b) Detrended variation in bed elevation showing pools and riffles determined by the regression (P and R) and differencing (P, R, and (R)) methods with a tolerance level equal to 0.75 times the standard deviation of the detrended data (dashed lines).

(Richards, 1976a; Keller and Melhorn, 1978), but variability in the spacing of individual sequences is quite large, ranging from fewer than three to more than 11 channel widths (Hudson, 2002). The average spacing is about half the average wavelength of meandering. This relationship is consistent with the bar-unit model in relatively simple, nonelongated meander bends where each bend contains a bar unit and the wavelength of pool–pool or riffle–riffle spacing (λ_{pr}) is half the meander wavelength (Figure 9.3). In meandering rivers, the spacing of riffle-riffle sequences increases with increasing channel curvature, primarily because pool lengths increase within increasing curvature (Lofthouse and Robert, 2008). Some evidence suggests that pool spacing increases with decreasing gradient (Wohl et al., 1993), but the extent to which this relation can be generalized to all pools remains uncertain (Chartrand et al., 2018).

Data for pools and riffles in three reaches of the Severn River, UK show that riffle height and pool depth are positive power functions of the length of these bar elements (Carling and Orr, 2000). Pools and riffles have similar lengths, with pools only slightly longer on average than riffles. The longitudinal symmetry of pools and riffles also is similar. In both bar elements, the lengths of exit and entrance slopes are, on average, nearly identical. Mean angles of the bed from horizontal for entrance and exit slopes range from about 2 to 5 degrees. Riffle heights (H_r)

are a relatively constant proportion of bankfull channel depths with $H_r/D_{bk} \approx 0.16$. This general information on pool-riffle dimensions is constrained by the caveat that these dimensions depend strongly on the method employed to identify these features. If the regression method is used without a tolerance level, pools and riffles occupy the entire length of a reach with no portions lacking bar elements or consisting of transitional forms. If a tolerance level is introduced, the lengths, the heights, and possibly the spacing and symmetry of pools and riffles change, and some areas of the channel do not consist of pools or riffles.

The geometry of pools and riffles in meandering rivers differs rather markedly. Channel width, cross-sectional area, and gradient are greater at riffles than at pools (Richards, 1976b, 1978; Keller and Florsheim, 1993), but channel depth is greater at pools than at riffles. Widening at riffles reflects deflection of diverging flow against channel banks (Richards, 1976b), whereas narrowing at pools is related to point-bar deposition along the inner bank. The outer cut bank and the point bar along the inner bank produce a distinctly asymmetrical shape to the channel cross-section at pools compared with the relatively symmetrical shape of the channel at riffles (Knighton, 1982a) (Figure 9.3).

The standard perspective on the sedimentology of pools and riffles is that bed material on riffles is coarser than the material on the bed of pools. Several studies support this perspective, either qualitatively (Keller, 1971) or quantitatively (Hirsch and Abrahams, 1981; Bhowmik and Demissie, 1982). However, other work suggests that bed-material characteristics do not differ substantially between pools and riffles (Milne, 1982; Clifford, 1993; Milan, 2013b). In part, the issue may center around whether sampling in the pool includes the point bar. Inclusion of point-bar sediments as part of the pool enhances the distinction of pools as more fine-grained than riffles (Milne, 1982). Differences in sorting and structure (e.g.,

armoring, imbrication, and embeddedness) of sediment also may occur between pools and riffles (Hirsch and Abrahams, 1981; Clifford, 1993), but such differences have not been examined extensively. Pools can trap fine sediment, particularly when sediment supply is enhanced through land disturbance, resulting in decreases in pool volume and fining of pool-bed material (Lisle and Hilton, 1992; Goode and Wohl, 2007). Pool characteristics, including properties of bed material, can also be affected by collapse of eroded outer-bank sediment into the pool (Hudson, 2002).

9.4.2.1 What Factors Influence the Morphology of Forced Pool-Riffle Sequences?

The development of forced pool-riffle couplets requires sufficient constriction of the flow to produce the hydraulic conditions necessary to scour a pool and deposit the scoured material downstream in the form of a riffle (Figure 9.17). Thus, forced pools are associated with zones of channel narrowing, and intervening riffles develop where channels are locally wide (Chartrand et al., 2018). Field evidence indicates that the requisite degree of constriction is about one-third of the bed width (Lisle, 1986). Constrictions smaller than this width tend to produce local scour holes. The spacing of forced pools and riffles depends strongly on the abundance and spatial distribution of pool-forming elements (PFEs) (Buffington et al., 2002). In forested mountain drainage basins that introduce abundant pool-forming large woody debris (LWD) into streams, pool spacing decreases from greater than 13 channel widths to one channel width with increasing LWD loading (Montgomery et al., 1995). However, the influence of forcing on spacing is constrained somewhat by the minimum length needed for a pool-forcing element to create a pool-riffle couplet (Thompson, 2001). Interference between PFEs spaced more closely than the minimum length inhibits development of a pool-riffle couplet for each PFE, which may account for an

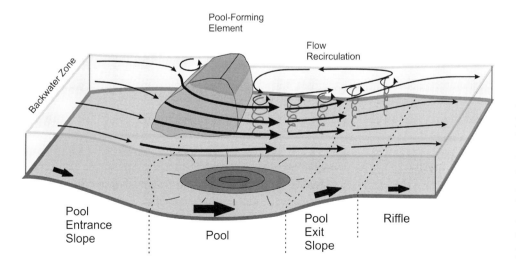

Figure 9.17. A forced pool-riffle sequence. Weight of surface streamlines corresponds to relative velocity magnitudes. Blue swirled lines represent vertical vortices bounding the zone of flow recirculation. Heavy arrows represent relative magnitudes of near-bed velocity. (A black and white version of this figure will appear in some formats. For the color version, please refer to the plate section.)

average pool-to-pool spacing of four to eight channel widths in many streams with forced pools and riffles. Experimental evidence for fixed two-dimensional pools and riffles indicates that the recovery length, the length over which the characteristics of the mean flow and turbulence recover from effects of a change in channel-bed geometry, is about three to four channel widths from a riffle to a pool (MacVicar and Best, 2013). If a similar recovery length exists for the transition from the pool to the riffle, this result implies an optimal spacing for forced pool-riffle sequences of six to eight channel widths.

The morphology of forced pool-riffle units depends on the characteristics of the obstacle producing the forcing, the characteristics of the material in which the units form, and the characteristics of the channel in which the units form (Buffington et al., 2002; Wohl and Legleiter, 2003; Thompson and McCarrick, 2010). As the degree of lateral constriction increases, pool depth and length increase, whereas the pool entrance slope decreases (Wohl and Legleiter, 2003). In pools formed in bedrock, pool depth increases with joint density, suggesting that rock erodibility is an important factor governing pool characteristics.

9.4.3 How Are Pools and Riffles Maintained?

In both straight and meandering rivers, the position and morphologic expression of pools and riffles often remain fairly stable, despite varying conditions of flow and sediment transport. Pools and riffles may erode or aggrade somewhat in response to flow variability, but tend to persist at specific locations over time (Gregory et al., 1994; Vetter, 2011a). This stability has fueled inquiry into the fluvial processes responsible for pool-riffle maintenance. At low flow, when morphological differences between pools and riffles and the influence of these differences on flow hydraulics are most visible, flow in pools has a relatively low cross-sectional mean velocity (U), large water mean depth (D), and low water surface slope (S_{ws}). In comparison, flow at riffles has a comparatively high U, small D, and high S_{ws}. These hydraulic conditions are not consistent with the hypothesis that pools, as topographic lows, should be sustained by scour and that riffles, as topographic highs, should be preserved by deposition. Consequently, changes in hydraulic conditions with changes in stage and discharge have been a major focus of research to determine how pools and riffles are maintained.

9.4.3.1 What Is the Velocity Reversal Hypothesis?
The velocity reversal hypothesis holds that as discharge through pools and riffles increases, near-bed velocity (u_b) and mean velocity (U) increase at a greater rate in pools than in riffles (Keller, 1971; Keller and Florsheim, 1993; Milan et al., 2001) (Figure 9.18). At a stage at or near bankfull,

the hydraulic relationship between pools and riffles actually reverses, so that the u_b and U of the pool exceed the u_b and U of the riffle. Moreover, the effect on the flow of differences in slopes of the channel bed of pools versus riffles diminishes within increasing stage, resulting in a decrease in water-surface slope over riffles and an increase in water-surface slope over pools. Because the mean flow depth of pools exceeds that of riffles at all flow stages, mean bed shear stress (Eq. (4.18)) may also reverse, so that near bankfull stage, the bed shear stress for pools becomes greater than the bed shear stress for riffles (Lisle, 1979; Milan et al., 2001).

The reversal hypothesis explains the maintenance of pools and riffles as erosional and deposition features, respectively, as well as the persistence of differences in grain size between these bar elements. At low flow, high near-bed velocity, or bed shear stress over the riffle winnows fine sediment from the riffles, depositing it in the pools (Figure 9.18). Coarse bed material on the riffles remains immobile at low flow stage. As flow stage rises, bed shear stress in the pool increases rapidly, leading to flushing of fine bed material out of the pool and mobilization of this fine material throughout the stream (Jackson and Beschta, 1982; Campbell and Sidle, 1985). When the reversal bed shear stress is exceeded, coarse particles mobilized on riffles, which serve as sources of these particles, are transported through the pools, where shear stress is greatest, resulting in riffle–riffle transport of coarse sediment (Jackson and Beschta, 1982). As discharge begins to decrease, competence in the pools initially exceeds competence on the riffles. In other words, the bed shear stress in pools remains above the critical bed shear stress for the coarsest mobile particles as stage decreases, but the bed shear stress on riffles falls below the critical shear stress for these particles. As a result, the coarsest particles can move through the pools but are deposited on the riffles. As coarse particles accumulate on riffles, progressive armoring, along with the development of imbrication, packing, and interlocking of individual grains (Clifford, 1993; Sear, 1996), impedes further mobilization of bed material, limiting transport of riffle sediment into the pool. As stage continues to fall, mobilization of relatively coarse sediment on the riffles ceases, and coarse particles in the pool continue to move onto the riffles. Conditions change when the bed shear stress in the pool falls below the bed shear stress on the riffle, which causes the coarsest mobile materials to be preferentially deposited within the pool. Thus, the coarseness of the pool substrate should depend directly on the magnitude of the reversal bed shear stress: the higher the magnitude of the reversal shear stress, the larger the particles in the pool substrate. As the bed shear stress in the pool continues to fall below that of the riffle, transport shifts to movement of fine bed material (Jackson and Beschta, 1982). The return to low-flow conditions once again results in

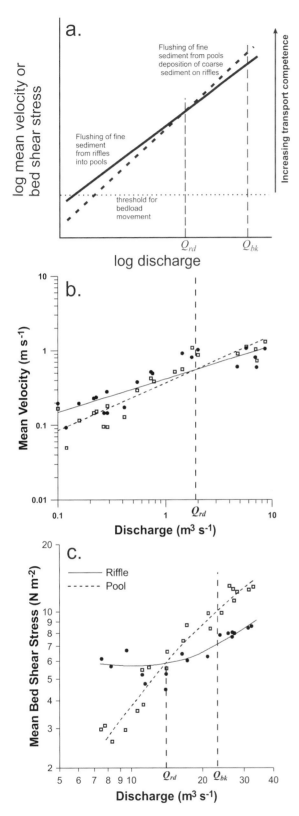

Figure 9.18. (a) Diagram depicting the concept of velocity/shear stress reversal for increasing discharge in pools and riffles. (b) Reversal of mean velocity between pools and riffles (from Milan et al., 2001). (c) Reversal of bed shear stress pools and riffles (from Lisle, 1979). Q_{rd} = reversal discharge.

flushing of mobile fine material, if available, into the pools, where it is deposited as a veneer over the relatively coarse substrate. This cycle of changing sediment-transport dynamics with varying discharge maintains pools as topographic lows and riffles as topographic highs on the channel bed.

The reversal hypothesis has been criticized on grounds that evidence to support the hypothesis is rather limited (MacWilliams et al., 2006). Clearly, reversal of mean velocity does not occur in some cases (Carling, 1991; Clifford and Richards, 1992; Vetter, 2011b), and effort has been devoted to determining the characteristics of pools and riffles required to produce velocity reversal (Carling and Wood, 1994; Caamano et al., 2009). An alternative hypothesis proposes that rather than reversal, a phase shift in the maxima and minima of bed shear stress in relation to the pool-riffle bed profile accounts for riffle deposition and pool scour at high flow (Wilkinson et al., 2004).

Analyses based on cross-sectional mean velocity or reach-averaged bed shear stress rely on a one-dimensional hydraulic representation of pool-riffle maintenance and neglect the influence of bed-material characteristics on maintenance. More sophisticated approaches have attempted to link pool-riffle dynamics to patterns of two- or three-dimensional fluid motion and the influence of this motion on boundary shear stresses and pathways of sediment routing (Booker et al., 2001; MacWilliams et al., 2006; Sawyer et al., 2010; Caamano et al., 2012; Milan, 2013b). Other work has examined the influence of turbulence generation and of spatial variation in local bed stresses on pool-riffle maintenance (Clifford and Richards, 1992; Carling and Orr, 2000). The characteristics of bed material and the influence of these characteristics on differential entrainment and transport of sediment within pools and riffles may also contribute to maintenance of these features (de Almeida et al., 2011, 2012; Hodge et al., 2013; Bayat et al., 2017). A change in cross-section shape from symmetrical at riffles to asymmetrical at pools produces flow convergence, acceleration, and the development of secondary currents within the pool that route coarse sediment along a narrow corridor of high bed shear stress adjacent to the deepest part of the pool, thereby preserving pool depth (Booker et al., 2001; MacWilliams et al., 2006; Milan, 2013b). In cases where external controls, such as valley constrictions, influence routing within pools for flows that exceed bankfull, the primary routing pathway may occur in the center of the pool, promoting scour and increases in pool depth (Sawyer et al., 2010). Differences in particle mobility on riffles versus pools can lead to higher rates of bed-material transport in pools at near-bankfull stage even when the bed shear stress in the pools at this stage is less than the bed shear stress in the riffle (de Almeida et al., 2011). Thus, sediment-transport reversal between pools and riffles is possible without a reversal in mean velocity or bed shear stress, indicating that pool-riffle maintenance does not necessarily depend entirely on a reversal of hydraulic conditions.

Despite these advances, a comprehensive theory of unforced pool-riffle maintenance has yet to emerge. Whatever details a particular explanation includes, preservation of pool and riffle dimensions requires net continuity of sediment transport through pool-riffle sequences. Imbalances in sediment transport between pools and riffles that result in net erosion of riffles and/or deposition in pools will result in elimination of pool-riffle topography.

9.4.3.2 How Are Forced Pools and Riffles Maintained?

Flow convergence/divergence and acceleration/deceleration play major roles in the maintenance of forced pool-riffle units (Harrison and Keller, 2007). Strong jet flow typically develops in forced pools, with pronounced vertically oriented rotating eddies forming on the margins of the jet along a region of flow separation in the lee of the constricting obstacle (Figure 9.17) (Thompson, 2004). At high stage, constriction of flow associated with the pool leads to clear velocity reversal between the pool and downstream riffle (Thompson et al., 1999), promoting pool maintenance. Moreover, forced pools are regions of high turbulence intensity, turbulent kinetic energy, instantaneous velocities, and mean near-bed velocities (MacVicar and Roy, 2007a,b; Thompson, 2006, 2007; Thompson and Wohl, 2009), all of which contribute to enhanced scour. Streamlines diverge over the pool exit slope as the effective flow area widens beyond the lateral flow separation zone, leading to flow deceleration downstream over the riffle (Figure 9.17). Flow acceleration and enhanced turbulence transport large particles through the pools, but particle mobility decreases as moving clasts encounter the pool exit slope where flow begins to decelerate (Thompson et al., 1996; Thompson and Wohl, 2009; McVicar and Roy,

2011). As a result, a characteristic pattern of sorting develops, with the largest clasts in the pool and progressive fining of coarse sediment toward the riffle (Thompson and Hoffman, 2001) (Figure 9.19).

9.5 What Are the Patterns of Flow in a Meandering River?

Fluid motion in meandering rivers is highly three dimensional, with components of fluid velocity directed downstream, laterally, and vertically. Two major factors generate this three-dimensional pattern of fluid motion: channel curvature and spatial variations in bed topography. The development of three-dimensional fluid motion in meandering rivers strongly influences patterns of bed-material transport as well as erosion and deposition.

9.5.1 How Does Channel Curvature Influence Fluid Motion?

The effect of channel curvature on flow structure in channel bends was first recognized in the nineteenth century (Thomson, 1876). Flow through a bend generates an inertial centrifugal force directed perpendicular to the curved path of the flow (Figure 9.20). In response to this inertial force, the water-surface elevation of the river increases slightly along the outer bank and decreases slightly along the inner bank, much as the surface of water contained within a rotating cylinder becomes depressed in the center of the cylinder and elevated along the walls of the cylinder. The difference in water-surface elevation between the inner and outer banks produces tilt of the water surface toward the inner bank, resulting in a cross-stream pressure-gradient force that counterbalances the centrifugal force. Expressed as forces per unit mass, or accelerations, the force-balance relation for the centrifugal and pressure-gradient forces over the entire cross section of the flow is

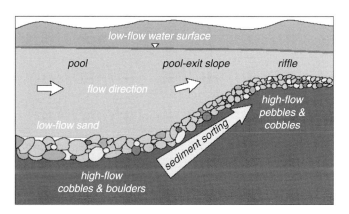

Figure 9.19. Typical pattern of sediment sorting in a forced pool (from Thompson and Wohl, 2009).

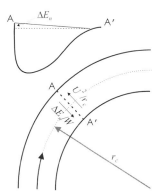

Figure 9.20. Balance of centrifugal and pressure-gradient forces in a curved channel and corresponding lateral tilting of the water surface.

$$\frac{U^2}{r_c} = g\frac{\Delta E_n}{W} \qquad (9.4)$$

where ΔE_n is the cross-stream change in water-surface elevation.

The balance of forces defined by Eq. (9.4) holds for the cross-sectionally averaged velocity of the flow (U). Near the surface, local velocities (u) are greater than U, whereas near the bed $u < U$. Therefore, the two forces are imbalanced locally within the flow. Near the surface, centrifugal forces exceed the pressure-gradient force, whereas near the bed, the pressure-gradient force exceeds the centrifugal force. As a result, flow near the surface is directed toward the outer bank and flow near the bed is directed inward. Helical motion develops as outward-moving near-surface flow descends in proximity to the outer bank and inward-moving near-bed flow moves upward close to the inner bank. The resulting pattern of fluid motion can be envisioned as a slight corkscrewing of flow streamlines – lines denoting the time-averaged pathways of water movement – through the curved channel (Figure 9.21). Water is mainly moving downstream but is spiraling somewhat as it does so. Although the strength of helical motion tends to increase with increasing curvature, experimental work suggests that the strength of this motion may reach a limit, or saturate, in sharp bends with $r_c/W_{bk} < 2$ (Blanckaert, 2009).

9.5.1.1 How Is Helical Motion Related to Secondary Flow?

Helical motion results in the development of secondary flow – components of flow oriented in directions other than that of the path of the flow or channel. Within a meandering river, the downstream direction (s) is defined by the tangent of local channel centerline curvature at each point along the channel, and downstream velocities (u) are oriented parallel to the tangential direction (Figure 9.22). The cross-stream (n) direction is oriented orthogonal to the local curved path of the channel centerline. In this frame of reference, cross-stream (v) and vertical (w) velocities in the cross-sectional (n-z) plane define the secondary flow. When information on w is not available or is not considered, variations in v over depth represent the secondary flow (Figure 9.22). Helical motion depicted by v only forms a pattern where values of v are directed toward the outer bank at the surface and toward the inner bank near the bed. This pattern corresponds to a form of secondary flow known as secondary circulation. If information on w is available, two-dimensional vectors of v and w exhibit an obvious rotational pattern depicting the secondary circulation (Figure 9.23).

An important consequence of helical motion is that the secondary circulation associated with this motion advects high-momentum flow in the center of the channel at the surface toward the outer bank. In a straight open channel, the maximum velocity typically is located in the center of the flow near the surface – the location farthest from the frictional effects of the channel boundary. Within a bend, outward-moving flow at the surface transports downstream momentum laterally, and vertical fluid motion near the outer bank transports downstream momentum downward. In many cases, the zone of highest downstream velocity within the channel in the downstream part of bends is submerged beneath the water surface near the outer bank (Figure 9.23). This phenomenon has important consequences for sediment transport and bank erosion. The location of the high-velocity core near the bed and face of the

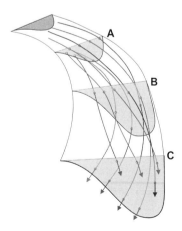

Figure 9.21. Idealized pattern of flow streamlines depicting helical motion in a curved channel moving through three successive cross sections (A, B, and C) (A black and white version of this figure will appear in some formats. For the color version, please refer to the plate section.) (from Frothingham and Rhoads, 2003).

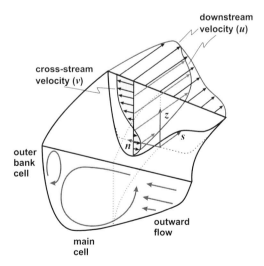

Figure 9.22. Flow pattern in a meander bend based on channel coordinate system (s,n,z) (A black and white version of this figure will appear in some formats. For the color version, please refer to the plate section.) (adapted from Ottevanger et al., 2012).

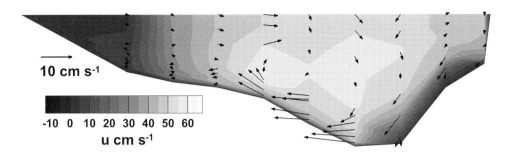

Figure 9.23 Filled contours of downstream velocity (u) with superimposed vectors (arrows) of vertical/cross-stream (w/v) velocities in a meander bend on the Embarras River, Illinois. Vector pattern of secondary flow exhibits clear evidence of fluid rotation related to helical motion. Note submerged position of maximum downstream velocity. Inner bank is to left, outer bank is to the right, view is upstream. (A black and white version of this figure will appear in some formats. For the color version, please refer to the plate section.)

BOX 9.1. SECONDARY FLOW AND DIFFERENCES BETWEEN THE FLOW PATH AND CHANNEL PATH.

In meandering rivers, the path of the flow may not always coincide with the path of the channel. The path of the flow can be defined either by the local orientation of the resultant discharge vector (Markham and Thorne, 1992) or by orientations of depth-averaged velocity vectors at different positions (n) across the flow (Rozovskii, 1957) (see Chapter 12). Components of individual three-dimensional velocity vectors parallel to the path of the flow are primary velocities (u_p) and components perpendicular to the path of the flow represent secondary velocities (v_s). Because any difference between the two flow frames of reference and the channel frame of reference involves rotation about a vertical axis, the vertical velocities (w) are the same in all three frames of reference. Both the flow frames of reference result in zero net lateral discharge in planes oriented perpendicular to the primary flow directions. In the frame of reference of the resultant discharge vector (RQ_{FOR}), secondary flow is defined within a single plane oriented perpendicular to the resultant discharge vector. In the frame of reference of the depth-averaged velocity vectors (DV_{FOR}), secondary flow is defined by a set of planes oriented perpendicular to depth-averaged vectors at different cross-stream positions. All three frames of reference produce identical depictions of secondary flow if the orientations of depth-averaged velocity vectors at every position across the flow are parallel to the local tangent to channel centerline. Under such conditions, $v_s = v$. Deviations of the flow path from the channel path will produce differences in patterns of secondary flow among the methods. The DV_{FOR} is useful for documenting patterns of secondary circulation produced by helical motion within converging or diverging flows (Rhoads and Kenworthy, 1999) (see Chapter 12), whereas the RQ_{FOR} reveals patterns of secondary circulation associated with helical motion when the meandering path of the flow is out of phase with the meandering path of the channel.

---

outer bank produces high boundary shear stresses that can mobilize bed or bank material, leading to bed or bank erosion.

In some (Bathurst et al., 1977, 1979; Thorne et al., 1985; Markham and Thorne, 1992; Engel and Rhoads, 2016) but not all (Frothingham and Rhoads, 2003; Nanson, 2010; Engel and Rhoads, 2012; Engel and Rhoads, 2017) field settings, a small cell of secondary circulation may develop near the outer bank with

a sense of rotation opposite that of the main cell associated with large-scale helical motion (Figure 9.22). The development of this cell has been attributed to feedback between turbulence-induced and centrifugal-induced vorticity near the outer bank. Momentum advection by weak turbulence-induced cells near the outer bank produces a negative gradient of u over depth in the upper part of the flow column, allowing centrifugal effects acting on this negative gradient to reinforce the pattern of turbulence-induced rotation as flow moves through a bend (Blanckaert and de Vriend, 2004). The outer bank cell, if present, can form a buffer that protects the outer bank from advection of high-momentum flow toward it by the main helical cell (Blanckaert and Graf, 2004); however, the outer bank cell also advects high momentum toward the bed, augmenting the effect of the main cell to enhance bed shear stresses near the base of the outer bank. The size of the outer bank cell increases with increasing steepness and roughness of the outer bank, shifting both the extent of the main cell and the zone of high bed shear stress toward the inner bank (Blanckaert, 2011; Blanckaert et al., 2012).

In sharp bends with high curvature, superelevation of the water surface along the outer bank may be pronounced enough to produce an adverse pressure gradient sufficient to cause flow stagnation along the outer bank (Blanckaert, 2010). This condition generally is met when

$$\frac{r_c}{W} < \left(0.5 C_{dm}^{-1} \frac{D}{W}\right)^{0.5} \tag{9.5}$$

where $C_{dm} = gRS_e/U^2$ is the dimensionless Chezy coefficient. Flow stagnation results in a protective zone of recirculating flow along at least part of the outer bank. The occurrence of flow stagnation along the outer bank is enhanced if this bank veers away from the inner bank, locally widening the channel and increasing the cross-sectional area of the flow (Blanckaert et al., 2013; Vermeulen et al., 2015). Local widening and the development of recirculating flow along the outer bank are common in sharp bends ($r_c/W < 2$–3) immediately upstream of the point of maximum curvature. Deposition of fine-grained sediment and organic material within the zone of recirculation forms a bank-attached bar known as a concave-

bank bench (Hickin, 1979, 1986; Page and Nanson, 1982; Vietz et al., 2012) (Figure 9.24). Acceleration of the flow within sharp bends may occur where both inner-bank separation and outer-bank stagnation reduce the cross-sectional area of the flow, producing a jet of high-velocity flow that descends into a zone of scour within the bend (Rhoads and Massey, 2012; Vermeulen et al., 2015).

Tilting of the water surface at bends in response to centrifugal acceleration is not only important in the generation of helical motion; it also influences cross-stream variation in bed shear stress and downstream velocity. In particular, water-surface gradients along the path of the channel are steeper between the outside bank of a bend and the inside bank of the next successive bend downstream than between the inner bank of a bend and the outside bank of the next successive bend (Figure 9.25). Consequently, downstream velocity and bed shear stress are greater from outer bank to inner bank than from inner bank to outer bank. Because successive bends have opposing senses of curvature, the thread of high velocity and bed shear stress shifts back and forth between alternating sides of the channels from one bend to the next.

Many, if not most, natural meander bends exhibit spatial variations in curvature ($\partial C_{cs}/\partial s$). Such variations play a major role in redistribution of momentum in sharp bends ($r_c/W < 3$) (Ottevanger et al., 2012) and must be considered in hydrodynamic models of flow through these types of bends (Blanckaert and de Vriend, 2010). In a sharp bend where curvature increases to a maximum at the apex and then decreases downstream of the apex, the spatial change in curvature increases the tilt of the water surface over distance upstream of the apex and decreases the tilt of the water surface over distance downstream of the apex. This effect, which contributes to the spatial redistribution of downstream

Figure 9.24. Concave bank bench (on left) deposited along the upstream end of a cut bank in a meander bend, Embarras River, Illinois (looking upstream).

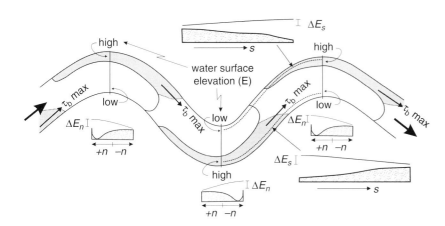

Figure 9.25. Diagram showing influence of cross-stream (n) tilting of the water surface in bends on downstream (s) gradients of the water surface (dashed lines) on each side of a meandering channel and the influence of alternating patterns of downstream water-surface gradients on the path of maximum boundary shear stress (τ_b max) (adapted from Dietrich, 1987).

momentum (Ottevanger et al., 2012), enhances the increase in bed shear stress along the inner bank upstream of the apex and the increase in the bed shear stress along the outer bank downstream of the apex (Figure 9.25).

9.5.2 How Does Spatial Variation in Bed Topography Influence Patterns of Flow in Curved Channels?

In a natural meandering channel that varies in curvature and that transports bed material, the combined effects of helical motion and changing curvature promote sediment flux convergence along the inner bank and sediment flux divergence along the outer bank, leading to the formation of forced bar-unit topography (Nelson and Smith, 1989) (Figure 9.3). The spatially nonuniform bed topography associated with bar units has a strong influence on patterns of flow within meandering rivers. Of particular importance is the development of the point bar along the inner bank of the channel. Point-bar development is most pronounced downstream of the bend apex (point of maximum curvature), where the inward flux of bed material associated with helical motion is reinforced by the decrease in downstream boundary shear stress along the inner bank (Nelson and Smith, 1989). The effect of the point bar on flow through the bend can be evaluated using equations defining the two-dimensional depth-averaged force balance for flow in a curved channel with nonuniform bed topography (Dietrich and Smith, 1983):

$$\tau_{bs} = \underset{(1)}{-\frac{\rho g h}{1-N}\frac{\partial E}{\partial s}} \underset{(2)}{- \rho\frac{1}{1-N}\frac{\partial \overline{u^2}h}{\partial s}} \underset{(3)}{- \rho\frac{\partial \overline{uv}h}{\partial n}} \underset{(4)}{+ 2\rho\frac{\overline{uv}h}{(1-N)r_c}}$$

$$(9.6)$$

$$\tau_{bn} = \underset{(1)}{-\rho g h\frac{\partial E}{\partial n}} \underset{(2)}{- \rho\frac{\overline{u^2}h}{(1-N)r_c}} \underset{(3)}{- \rho\frac{1}{(1-N)}\frac{\partial \overline{uv}h}{\partial s}} \underset{(4)}{- \rho\frac{\partial \overline{v^2}h}{\partial n}} \underset{(5)}{+ \rho\frac{\overline{v^2}h}{(1-N)r_c}}$$

$$(9.7)$$

where τ_{bs} is the streamwise (s) bed shear stress, τ_{bn} is the cross-stream (n) shear stress, h is the local flow depth, u is the downstream velocity, v is the cross-stream velocity, $(1-N) = (1-n/r_c)$ is a scaling factor that compares an arc length measured along the centerline with an arc length measured along any other line of constant n, and an overbar indicates that the quantity under the overbar is averaged over depth (Table 9.1). The main effect of the point bar is that it produces changes in local depth in the downstream direction. At high flows that submerge the point bar, the rising surface of the bar along the inner bank results in decreases in flow depth in the downstream direction ($\partial h/\partial s$) along lines of equal distance from the inner bank (n). Moreover, local depth also varies strongly in the cross-stream direction ($\partial h/\partial n$) as the bar surface slopes into the adjacent pool. If bedforms, such as ripples or dunes, or the size of particles on the bar surface vary spatially over the point bar, corresponding changes in bed friction will contribute to spatial changes in the magnitude of downstream velocity ($\partial u/\partial s$, $\partial u/\partial n$) and cross-stream velocity ($\partial v/\partial s$, $\partial v/\partial n$). These spatial changes in velocity magnitude contribute to convective accelerations or decelerations of the flow. Terms 2 and 3 in Eq. (9.6) and terms 3 and 4 in Eq. (9.7) account for changes in depth and in convective accelerations/decelerations of the flow. Whereas these terms can be neglected in curving experimental or artificial channels with uniform rectangular cross sections, they contribute substantially to the total force balance in natural meandering rivers with bar-unit topography (Legleiter et al., 2011).

The point bar represents an obstacle that steers the flow around it – a process referred to as topographic steering (Nelson and Smith, 1989). The downstream increase in bed elevation and decrease in flow depth along the inner bank increases the water-surface elevation relative to the elevation in the absence of the point bar. Similarly, the downstream decrease in bed elevation and increase in flow depth within the pool near the outer bank decreases the water-surface elevation relative to the elevation in the absence of a pool. The diminished cross-stream slope of the water surface allows the centrifugal force to exceed the pressure-gradient force over the point bar, leading to outward fluid motion throughout the water column (Figure 9.26). Helical motion still occurs over the point bar, but the motion is embedded in flow that is deflected toward the adjacent pool. Thus, the plane of secondary circulation, defined by zero net secondary

Table 9.1. Terms in Eq. (9.6) and (9.7) (based on Dietrich and Smith, 1983).

Eq. (9.6)	
Term (1)	Stress associated with downstream pressure gradient
Term (2)	Stress associated with downstream change in downstream momentum flux
Term (3)	Stress associated with cross-stream change in cross-stream flux of downstream momentum
Term (4)	Centrifugal stress associated with the cross-stream flux of downstream momentum
Eq. (9.7)	
Term (1)	Stress associated with the cross-stream pressure gradient
Term (2)	Centrifugal stress associated with the downstream momentum flux
Term (3)	Stress associated with the downstream change in downstream flux of cross-stream momentum
Term (4)	Stress associated with the cross-stream change in cross-stream momentum flux
Term (5)	Centrifugal stress associated with the cross-stream momentum flux

discharge normal to depth-averaged velocity vectors (Rozovskii, 1957), is curved and not oriented normal to the local channel direction over the point bar (Dietrich, 1987) (Figure 9.26). To maintain flow continuity, the net cross-stream discharge over the point bar is accommodated by convective acceleration of downstream flow in the adjacent pool, resulting in an abrupt crossover of the zone of maximum velocity from the inner to the outer bank immediately downstream of the bend apex (Figure 9.27). In the absence of this strong topographic effect, the maximum velocity and bed shear stress generally do not shift to the opposite bank until flow begins to enter the next successive bend downstream (Figure 9.27).

In sharp bends, flow may separate from the inner bank, producing a region of recirculating flow (Leeder and Bridges, 1975; Ferguson et al., 2003). Flow separation along the inner bank of rectangular curved channels is most common in situations where curvature of the bank increases abruptly and the inertia (velocity) of the flow is large (Blanckaert, 2015). In natural meandering rivers, pronounced change in point-bar topography, which influences topographic steering, as well as local changes in channel width within the bend may be contributing factors (Constantinescu et al., 2011a; Engel and Rhoads, 2012). Flow separation develops because turbulent and advective mechanisms cannot adequately transport momentum laterally toward the inner bank to maintain downstream flow.

9.6 How Is Bed Material Transported in a Meandering River?

The transport of bed material in bends has mainly focused on the situation where equilibrium adjustments exist between bed topography and sediment flux. In other words, the flux of sediment is balanced spatially and adjusted to spatial

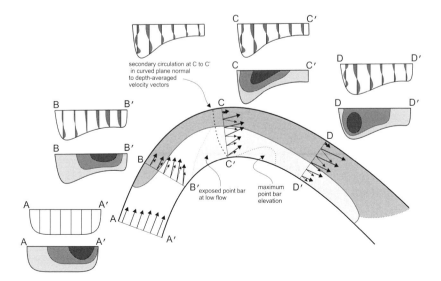

Figure 9.26. Idealized pattern of downstream and cross-stream flow through a bend for near-bankfull flow. Heavy arrows indicate near-surface velocity vectors and light arrows indicate near-bed vectors. Darkness of filled contours in cross sections corresponds directly to the relative magnitude of downstream velocity. Patterns of v are shown as deviations from zero over depth at verticals within the cross sections. At cross section C, the curved plane of secondary circulation and corresponding values of v_s are shown in addition to the patterns of cross-stream and downstream velocities.

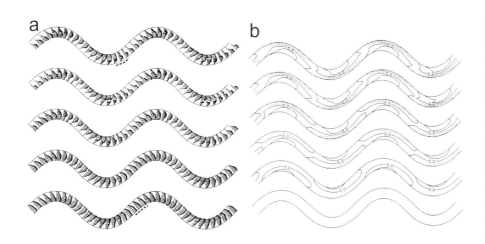

Figure 9.27. Simulated evolution of bed shear stress (a) and bed topography (b) in an idealized rectangular meandering channel with a mobile bed, inerodible banks, and an initially flat bottom. Dashed lines on top and bottom figures on the left show the location of the crossover of maximum bed shear stress from left side to right side of the channel. Note upstream shift of the crossover location as bar units develop through time (see b). On the right, dashed contours represent areas of scour and solid contours represent areas of deposition. Sequence from bottom to top shows emergence of equilibrium bar unit topography over time (adapted from Nelson and Smith, 1989).

variations in bed topography, so that transport of sediment over this spatially varying form does not lead to changes in topography. In rivers with mixed sand and fine gravel loads, this equilibrium appears to be achieved when the outward shifting zone of maximum boundary shear stress within the bend is balanced by convergent sediment transport caused by net outward flux of sediment toward the pool (Dietrich and Smith, 1984; Dietrich, 1987).

Upstream of the bend apex, or location of maximum curvature, the zone of maximum bedload transport is located along the inner bank in the region of high bed shear stress (Figure 9.28). The zone of maximum transport shifts outward through the bend in conjunction with the outward shift of the region of high bed shear stress. Downstream of the apex, the zone of maximum bed shear stress is located near the outer bank (Hooke, 1975; Dietrich and Smith, 1984). If sediment size is relatively uniform, the locus of maximum bedload transport rate will coincide with the zone of maximum bed shear stress (Hooke, 1975). However, in a mixture of bed material, the two zones may be displaced, with the zone of maximum bedload transport rate located closer to the channel centerline than the zone of maximum bed shear stress (Dietrich and Smith, 1984). This discrepancy reflects a grain-size effect, where the high bed shear stress along the outer bank is offset by the increase in grain size of bed material toward the outer bank, which decreases particle mobility.

The net cross-stream sediment flux associated with the outward shift of the zone of maximum sediment transport is caused by the cross-stream discharge over the bar top associated with shoaling and produces a sediment flux convergence that accommodates the net outward shift of the zone of maximum bed shear stress (Dietrich and Smith, 1984). The equilibrium slope of the point bar adjusts to provide just enough cross-stream flux of sediment to satisfy

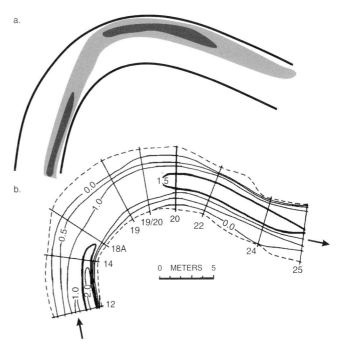

Figure 9.28. Spatial pattern of bedload transport rate in meander bends. (a) Corridor of maximum transport rate (shaded with darkest shading indicating highest rates of transport) as indicated by experimental studies of sand transport (e.g., Hooke, 1975). (b) Muddy Creek, Wyoming with fine to coarse sand (from Dietrich and Smith, 1984). Contour lines show ratio of actual rate to mean rate. Arrows indicate flow direction. For Muddy Creek, numbers indicate measurement cross sections and dashed lines are channel banks.

the change in downstream sediment-transport capacity associated with the outward shift in downstream boundary shear stress. Without this flux, the bed would scour, thereby increasing the lateral slope of the bed and the corresponding cross-stream flux until the supply of sediment equaled the transport capacity. The outward shift in

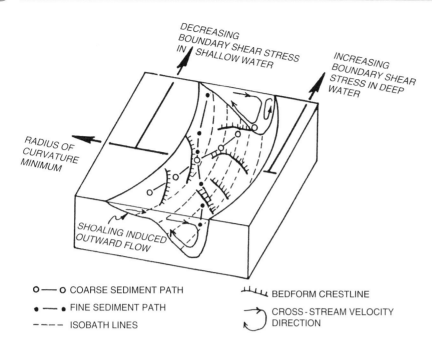

Figure 9.29. Pattern of bedload transport in a sand-bed meandering river showing pathways of coarse and fine sediment, changes in bedform crestlines, and secondary flow (from Dietrich and Smith, 1984).

bed shear stress can also be accommodated by coarsening of sediment within the pool, which reduces the dimensionless shear stress, thereby satisfying the increased transport capacity. This mechanism is important in gravel-bed meandering streams, where armoring of the pool limits scour and bedload transported through the pool is finer than the relatively coarse immobile substrate (Dietrich and Whiting, 1989).

The transport of sediment in meandering rivers also varies spatially according to grain size (Dietrich and Smith, 1984) (Figure 9.29). Coarse sediment moves along the inner bank upstream of the bend apex and travels outward over the bar platform and down the point-bar face into the pool downstream of the bend apex. On the other hand, fine sediment moves through the pool upstream of the apex and is progressively swept inward by near-bed secondary currents throughout the bend. As a result, the pathways of coarse and fine sediment cross within the bend, usually near the bend apex.

Dunes that develop inward of the channel centerline along the point-bar face have crests oriented obliquely across the channel, reflecting the cross-channel variation in downstream bed shear stress and its influence on crest migration speed. Upstream of the apex, where bed shear stress is greatest near the inner bank, the most downstream portion of the skewed crest lines is near the inner bank (Dietrich et al., 1979). Conversely, downstream of the apex, where the maximum bed shear stress has shifted into the pool, the most downstream portion of the skewed crest lines is toward the outer bank. The dune crest lines therefore rotate as these bedforms migrate through the bend.

Dune development and migration in sand-bedded meander bends generally occurs over a narrow zone along the point-bar face but can account for a substantial proportion of the bed material transported through bends (Kisling-Moller, 1992). Moreover, flow over the crests of these dunes can produce complex patterns of secondary currents between successive dunes, resulting in net inward transport of fine sediment upstream of the bed apex and net outward transport of fine sediment downstream of the apex – a flux that counters the tendency for inward transport of fine sediment by secondary circulation (Dietrich, 1987).

In bends in gravel-bed meanders, differential routing of fine and coarse gravel fractions promotes equal mobility by sweeping fine sediment inward, where movement of this material is less restricted by surrounding large grains, and by steering coarse grains outward, where they are exposed to high bed shear stress (Clayton and Pitlick, 2007). Despite cross-stream sorting of material by size, downstream fluxes of bedload remain equivalent across the channel because of an interplay between the size and abundance of material in transport. Large amounts of sand and fine gravel are transported along the inner parts of the channel, whereas a relatively small number of large clasts are transported within the pool, resulting in equal bedload transport rates across the channel. This spatial balance of transport rates contributes to the geomorphic stability of the bed morphology.

Few field studies have examined sediment transport at the scale of entire bar units, but experimental work indicates that the path length of particle transport in mildly sinuous channels corresponds to pool–point bar spacing (Pyrce and Ashmore, 2003b, 2005). For channel-forming flows, sediment mobilized within upstream pools tends to be deposited on the first point-bar surface downstream. The pathway of transport changes with bar growth, shifting from transport across the bar head and deposition on distal parts of the bar during initial stages of bar development to deposition on the bar margin near the head of the bar during late stages of bar formation.

The extent to which bed morphology in meandering rivers attains and maintains equilibrium with bedload transport fluxes remains incompletely understood. Changes in point-bar morphology do occur with changes in discharge and sediment supply (Anthony and Harvey, 1991). Increases in stage over the point bar reduce topographic steering, which should diminish the cross-stream transport of bedload toward the pool, resulting in bar aggradation and scour of the pool. High-resolution repeat surveys during floods indicate that point bars aggrade during prolonged periods of inundation but that patterns of erosion and deposition are complex and not necessarily consistently colocated with areas of high or low sediment-transport rates (Lotsari et al., 2014; Kasvi et al., 2015). Further work is needed to link changes in bed morphology within meander bends to variations in discharge and to modifications of flow structure and sediment transport patterns that accompany these variations in discharge (Kasvi et al., 2017).

9.7 How Do Riverbanks Erode?

Bank erosion is an inherent part of the dynamics of freely meandering alluvial rivers. Unless constrained by bedrock or other resistant material, meandering rivers migrate over valley bottoms, thereby contributing to floodplain development and to long-term storage and episodic remobilization of floodplain sediment (see Chapter 14). This migration occurs through erosion of outer cut banks or valley walls and corresponding deposition on point bars along inner banks (Figure 9.30). Although erosion of outer banks is characteristic of meandering rivers (Figure 9.3), the general processes that contribute to bank erosion are not specific to meandering rivers. These processes also occur in other types of rivers. Thus, an understanding of mechanisms of bank erosion is important for evaluating channel change in all rivers, but it is reviewed in this chapter on meandering rivers, given its central importance in lateral channel migration.

9.7.1 What Are the Major Processes Involved in Bank Erosion?

The major categories of processes affecting bank erosion include preparatory processes that weaken or strengthen the

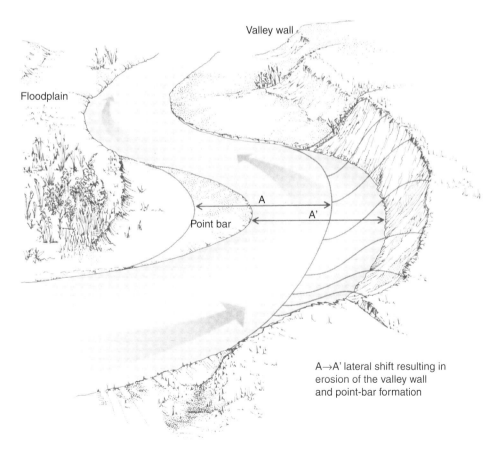

A→A' lateral shift resulting in erosion of the valley wall and point-bar formation

Figure 9.30. Lateral migration of a meandering river through erosion of the valley wall along the outer bank and deposition on the point bar along the inner bank (from Marsh and Dozier, 1981).

Table 9.2. Processes related to bank erosion in rivers.

Bank weakening and strengthening (preparatory) processes

Wetting – change in soil moisture content, pore pressure

Desiccation – shrinkage-related cracking and spalling

Freeze-thaw – disaggregation of sediment through frost action and needle ice growth

Vegetation growth – penetration of roots into the bank, surcharge by trees

Cattle trampling – mechanical weakening of the bank

Hydraulic action

Direct erosion of sediment from the bank face

Steepening or undercutting of the bank by erosion at the bank toe

Mass movement

Collapse of coherent, cohesive blocks of bank material into a river when gravitational forces acting on the bank exceed resisting forces

bank, hydraulic action, and mass movement (Table 9.2). Preparatory processes and mass failure are most important in banks that contain cohesive material, whereas hydraulic action influences all types of channel banks. Rooted vegetation adds strength to both noncohesive and cohesive banks, but typically occurs in cohesive banks.

9.7.2 How Do Noncohesive Banks Erode?

In extremely arid or cold environments, the banks of alluvial rivers may form entirely in noncohesive, unvegetated sand and gravel. Noncohesive banks consist of aggregates of individual particles, and erosion of the banks involves determining the fluid forces that result in particle entrainment, much like the problem of bed-material transport. Because particles are situated on a sloping bank, gravitational forces acting on the grains must be considered (Thorne, 1982). Force-balance analysis yields a relation specifying the bank shear stress (τ_{bnk}) required to mobilize unconsolidated, noncohesive bank material of the same composition as the bed material (Millar and Quick, 1993):

$$\frac{\tau_{bnk}}{(\rho_s - \rho)gd_{50}} > \Theta_c\sqrt{1 - \frac{\sin^2\theta'}{\sin^2\phi'}} \qquad (9.8)$$

where ϕ' is the angle of repose of the sediment and θ' is the angle of the bank. Approaches based on force-balance relations do not consider fluxes of sediment entering and leaving the bank region, which ultimately are responsible for changes in bank location. Morphodynamic models that include sediment-transport algorithms are required to evaluate how erosion of noncohesive banks leads to migration of the bank position (Stecca et al., 2017). Nevertheless, hydraulic action and the influence of this action on entrainment and transport of sediment are the main considerations for determining the erodibility of noncohesive banks.

9.7.3 What Are the Main Types of Cohesive Bank Failure?

Many alluvial rivers have composite banks where at least part of the bank is cohesive. For streams and small rivers, the upper part of the bank often contains abundant cohesive silt and clay deposited on the floodplain surface during overbank flows, whereas the base of the bank consists of noncohesive sand and gravel deposited by lateral accretion of sediment within the channel, including sediment on point bars and within pools. When banks contain upper cohesive layers, erosion typically occurs in part in the form of large coherent blocks or slabs of bank material falling into the stream as mass failures (Thorne and Tovey, 1981; Pizzuto, 1984b). Erosion of noncohesive material at the base of the bank through hydraulic action results in undercutting of the upper cohesive bank, producing overhanging blocks of material, or cantilevers. Failure of these cantilevers occurs in three forms: shear, beam, and tensile (Figure 9.31). Shear failures are common in weakly cohesive material and occur when the shear stress due to the weight of the block exceeds the shear strength of the bank material (Figure 9.32). In beam failures, the block rotates forward about a horizontal axis aligned perpendicular to the bank face with tensile forces acting above the axis and compressive forces below the axis (Figure 9.32). The tensile and compressive forces associated with the weight of the block about the axis exceed the compressive and tensile strengths of the block. Tensile failures develop when the weight of part of the block exceeds the tensile strength within the block. Rotational failures can also occur in high composite banks where a thick layer of cohesive material overlies noncohesive material at the base of the bank (Figure 9.31) (Thorne, 1982). In large rivers with complex histories of floodplain deposition, channel banks may exhibit stratigraphic profiles with multiple interbedded cohesive and noncohesive layers. In such cases, perched failures can occur when seepage erosion by outflow of water moving through pervious, noncohesive layers undermines overlying cohesive layers (Hagerty, 1991; Lindow et al., 2009).

9.7.4 What Is the Concept of Basal Endpoint Control?

The concept of basal endpoint control highlights the importance of processes at the base of riverbanks for bank stability (Thorne, 1982). This concept relates mass fluxes at the base of the bank, or bank toe, to different stability conditions. If the material supplied to the bank toe, usually by mass failures, is balanced

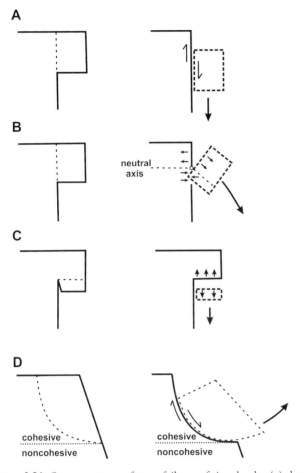

Figure 9.31. Common types of mass failures of river banks: (a) shear failure, (b) beam failure, (c) tensile failure, (d) rotational failure (slump) (after Thorne and Tovey, 1981).

by the bed-material transport capacity at this location, the height of the bank and the bank angle remain constant. For this balanced condition, the rate of bank erosion will be determined by the intensity of hydraulic action at the base of the bank, which governs bed-material transport capacity, and by geotechnical characteristics of the bank. Impeded removal occurs when material is supplied to the base of the bank at higher rates than the supplied material can be removed by hydraulic action. Under these conditions, sediment supply exceeds bed-material transport capacity. As material accumulates at the bank toe, bank height and angle decrease, increasing bank stability. As the bank stabilizes, the supply of material to the base of the bank decreases, shifting the bank-toe sediment flux toward a balanced state. Excess basal capacity develops when bed-material transport capacity at the bank toe exceeds the sediment supply at this location. Scour at the bank toe increases the bank height and bank angle, decreasing bank stability and increasing the supply of material to the base of the bank through failures, which also shifts the bank-toe sediment flux toward a balanced state.

9.7.5 How Has Slope Stability Analysis Provided Insight into Bank Erosion?

The problem of mass failure of river banks has been the focus of considerable quantitative analysis, particularly in the realm of river engineering, where it is of central concern to efforts to stabilize eroding banks or to protect critical locations along rivers from erosion. Slope stability analyses have been applied to this problem in an effort to predict conditions that lead to mass failure (Thorne and Tovey, 1981; Osman and Thorne, 1988; Thorne and Abt, 1993; Darby and Thorne, 1996; Darby et al., 2000; Klavon et al., 2017). These analyses evaluate streambank susceptibility to mass failure based on various geotechnical characteristics of the bank material in relation to gravitational forces acting on the bank. The stability of cohesive banks typically is expressed as a factor of safety (F_s):

Figure 9.32. (a) Shear failures, Wabash River, Indiana. (b) Beam failures, Embarras River, Illinois.

$$F_s = \frac{\text{resisting stress}}{\text{driving stress}} = \frac{\text{shear strength}}{\text{shear stress}} \tag{9.9}$$

where the driving or shear stress is the weight component of the soil material directed parallel to the plane of failure. The shear strength of the bank is governed by cohesion, by the effective normal stress, which equals the total normal stress minus pore water pressure, by the angle of internal friction of the bank material, and by matric suction – the difference between pore air pressure and pore water pressure. A value of $F_s > 1$ indicates stability, whereas $F_s < 1$ suggests that the bank is prone to mass failure. Bank stability decreases with increases in bank angle, bank height, and positive pore water pressure, whereas bank shear strength increases with increasing cohesion, angle of internal friction, and, for unsaturated soils, matric suction (Vanapalli et al., 1996). In banks prone to beam and shear failures, stability increases with increasing soil tensile strength and decreases with the length of tension cracks, which often develop at the bank surface (Thorne and Tovey, 1981). The preparatory process of prewetting is important, especially when prewetting leads to saturation of the bank material. Positive pore pressures develop under saturated conditions, reducing the effective normal stress and thus the bank shear strength. Erosion of composite banks generally tends to be associated with periods of high soil moisture (Hooke, 1979), when tension failures occur because the groundwater seepage force exceeds the soil shear strength, and when undercutting by seepage erosion results in mass failure (Fox and Felice, 2014). Other preparatory factors that enhance bank erosion include frost action (Lawler, 1993) and cattle trampling (Trimble, 1994) (Table 9.2).

Vegetation contributes substantially to spatial variability in the resistance of channel banks to erosion (Konsoer et al., 2016a). Dense growth of vegetation along inundated banks can impede flow, reducing near-bank velocities and boundary shear stresses (Kean and Smith, 2004). Plant roots also contribute to the tensile strength and cohesion of banks, thereby increasing bank stability. The roots of trees are particularly effective at increasing bank strength (Polvi et al., 2014). In some cases, large trees may contribute to bank instability by adding weight to the top of the bank, a process referred to as surcharge (Thorne, 1990). Refined analyses of bank stability consider the effects of vegetation through soil reinforcement by plant roots (Abernethy and Rutherfurd, 2000; Pollen and Simon, 2005; Pollen, 2007; van de Wiel and Darby, 2007) and surcharge (Simon and Collison, 2002).

9.7.6 How Has Bank Erosion Been Treated in Simple Models of Lateral Migration?

From a hydraulic perspective, treatments of bank erosion have ranged from simple to sophisticated. If the near-bank sediment

flux over the long term is assumed to achieve a balance between sediment supply and transport capacity, the rate of erosion should primarily reflect the intensity of hydraulic action near the bank. The rate of bank erosion $\xi_m (\text{m s}^{-1})$, or lateral migration of the channel centerline, in meandering rivers where erosion occurs along the outer bank has been expressed as (Ikeda et al., 1981; Pizzuto and Meckelnburg, 1989)

$$\xi_m = c_e(\bar{u}_{nb} - U) = c_e \bar{u}_{nb}' \tag{9.10}$$

where c_e is a dimensionless bank erodibility coefficient, $\bar{u}_{nb}$ is the depth-averaged velocity near the outer bank, and $\bar{u}_{nb}'$ is the excess depth-averaged near-bank velocity. This basic linear model does not explicitly describe the mechanics of bank erosion and simply assumes that the higher the depth-averaged velocity near the bank in relation to the mean velocity, the greater the rate of bank erosion. The value of c_e depends on bank strength, particularly the resistance of unvegetated bank material to fluvial shear (Constantine et al., 2009): the higher the value of c_e, the more erodible the bank material. A related linear model treats the rate of bank erosion as a function of the excess boundary shear stress (Darby et al., 2007; Motta et al., 2012a), similarly to the treatment of bed-material mobilization:

$$\xi_m = k_e(\tau_{bnk} - \tau_{cbnk}) \tag{9.11}$$

where $k_e (\text{m}^3 \text{ N}^{-1} \text{ s}^{-1})$ is a bank erosion coefficient and τ_{cbnk} is the critical shear stress for erosion of the bank material, i.e., the bank shear strength. An alternative formulation avoids the dimensionality of the erosion coefficient:

$$\xi_m = \hat{k}_e(u_{*bnk} - u_{*cbnk}) \tag{9.12}$$

where $\hat{k}_e$ is dimensionless, u_{*bnk} is the bank shear velocity, and u_{*cbnk} is the critical bank shear velocity.

Although Eq. (9.11) and (9.12) are simple linear relations, operationalizing these equations remains challenging. In particular, estimation of the total boundary shear stress acting on the channel banks in natural rivers is difficult. The ray-isovel method (see Chapter 4) has been used for this purpose (Kean et al., 2009), but the extent to which this method accurately represents the effects of turbulent stresses acting on banks has yet to be confirmed. Evaluations of near-bank Reynolds stresses require measurements of turbulence in the near-bank region, and relatively few studies have obtained these types of measurements (Hopkinson and Wynn-Thompson, 2016; Engel and Rhoads, 2017). The roughness of natural channel banks can be complex, complicating efforts to partition the total bank shear stress into components related to form drag and skin friction (Kean and Smith, 2006a,b). Estimation of the shear stress associated with skin friction is important because ultimately this component of the total stress drives bank erosion. Channel banks often exhibit irregular geometries that vary both vertically and horizontally (Konsoer et al., 2017). Vegetation growing on the banks and woody debris from trees that have fallen into the river along eroding banks also contribute to drag resistance (Thorne and Furbish, 1995; Gorrick and Rodriguez, 2014; Konsoer et al., 2016a), reducing velocities near the bank (Daniels and Rhoads, 2004). Blocks of failed cohesive bank material represent another component of bank roughness. Mass failures often occur during falling limbs of hydrographs when the flow is low (Rinaldi et al., 2008), so that failed material is not immediately removed by erosion. Because bank blocks both protect the bank toe and increase flow resistance, the endurance of these features is an important factor governing bank erosion (Wood et al., 2001; Parker et al., 2011) and rates of lateral migration in meandering rivers (Motta et al., 2014). Downstream of the apex of meander bends, where the high-velocity core is submerged beneath the surface near the outer bank, high levels of turbulent kinetic energy occur around blocks of failed material, which probably enhances erosion and rapid removal of this material (Engel and Rhoads, 2012). However, deflection of flow up and over persistent blocks can also promote erosion of the overlying bank (Hackney et al., 2015). Further research is needed to characterize the complexity of flow near the banks of rivers, especially the outer banks of meandering rivers, so that the forces acting directly on banks can be accurately determined.

As with Eq. (9.8) for noncohesive banks, Eq. (9.11) and (9.12) assume that if the critical shear stress for erosion of the bank is surpassed, the bank will retreat. These functions imply that threshold exceedance equates to a mass flux deficit, which in turn leads to net removal of sediment from the bank, thereby resulting in a change in the position of the bank. However, the two conditions, threshold exceedance and mass-flux deficit, are not necessarily equivalent. The shear stress at the bank can locally exceed the critical shear stress, but if spatial variations in downstream and cross-stream bed shear stresses produce convergence of the bed-material sediment flux near the bank, local sediment supply may equal or exceed local transport capacity, resulting in no erosion or even deposition. Thus, the spatial continuity of sediment flux associated with spatial variation in transport capacity ultimately governs the dynamics of bank erosion (Duan and Julien, 2010).

9.8 How Do Meandering Rivers Migrate over Time and Space?

Meandering rivers, unless incised into bedrock (Harden, 1990), migrate laterally through erosion at cut banks and deposition on point bars. The process of lateral migration in alluvial meandering rivers gives rise over time to the evolution of the planform characteristics of meander trains. Efforts to understand the dynamics of planform evolution through meander migration have involved both empirical and modeling approaches. More recently, the capacity to generate meandering rivers in the laboratory (Smith, 1998; Kleinhans et al., 2014) has led to contributions from experimental work.

Bedrock-incised meandering rivers can migrate laterally, but the timescale of migration usually is much greater than the timescale of migration of purely alluvial rivers. The inner banks of meandering rivers that are actively incising into bedrock are characterized by a sloping bedrock surface covered by a thin veneer of alluvium rather than by a point bar consisting entirely of alluvial deposits (Inoue et al., 2017). The equilibrium slope of this bedrock surface decreases with increases in outer-bank erosion rate and sediment supply. For large sediment supply, maximum incision occurs along the outer bank, but for restricted sediment supply, maximum incision occurs in the center of the channel and a bedrock bench develops along the outer bank. At broad spatial scales, the sinuosity and rate of migration of bedrock-incised meandering rivers have been positively related both to the frequency of extreme rainfall-runoff events and to the weakness of bedrock (Stark et al., 2010).

9.8.1 What Process Interactions Govern the Lateral Migration of Meandering Rivers?

Planform evolution in alluvial meandering rivers occurs through the influence of three-dimensional flow and turbulence on spatial patterns of bed-material transport, which, in turn, governs the spatial patterns of erosion and deposition that lead to the development of bar units and to erosion of outer banks along bends (Figure 9.33). Both the development of bar units and the erosion of outer banks result in changes in channel form that modify the three-dimensional structure of the mean flow and turbulence. Through this feedback, interaction occurs between point-bar deposition and outer-bank erosion, the processes that drive lateral migration of meandering rivers (Parker et al., 2011). Although erosion of outer banks may seem primary, deposition at inner banks is necessary for the channel as a whole to move laterally. Hydrological variability directly influences bank erosion through prewetting of the bank and changes in bank seepage associated with variations in stage, both of which affect the timing of bank failures (Fox et al., 2007). As the river planform evolves, changes in curvature and bend length produce changes in three-dimensional flow structure and turbulence. Through this feedback, planform-scale changes in the morphology of meandering rivers influence bar-unit development and outer-bank erosion.

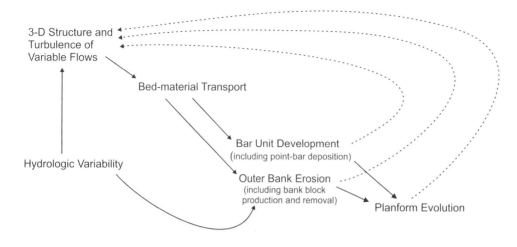

Figure 9.33. Conceptual model of planform evolution in alluvial meandering rivers. Dotted lines indicate feedbacks (from Engel and Rhoads, 2012).

9.8.2 How Do Straight Sections of Meandering Rivers Evolve over Time?

Empirical analysis of changes in the planform of meandering rivers over timescales from a year up to about 200 years can be assessed using evidence from maps, aerial photographs, and, for large rivers, satellite imagery (Legg et al., 2014; Schwenk et al., 2017). Observations of the response of sections of meandering rivers that have become straight, either because of natural processes or because of artificial straightening by humans, indicate that recovery of sinuosity is a common mechanism of adjustment. The development of meandering over time has been characterized as sequential, involving (Hooke, 1995a) 1) transformation of a straight channel into a series of simple bends (e.g., Figure 9.3), 2) growth of bends into symmetrical and asymmetrical elongate loops (Figure 9.14), and 3) cutoff of loops that reinitiates the development of simple bends or evolution of elongate loops first into compound forms (Figure 9.14) followed by cutoff. The extent to which meander bends conform to the sequential development scheme varies; the development of individual bends within a meander train can be complex and includes effects generated by changes in adjacent bends (Furbish, 1991). Modes of growth and adjustment include translation, or downvalley movement of bends, skewing, or rotation of the bend upstream or downstream, and extension, or increases in amplitude (Figure 9.34) (Hooke, 1977). These basic modes typically occur in combination to produce complicated patterns of planform evolution.

9.8.2.1 What Are Cutoffs and How Do They Occur?

Cutoffs, which involve the removal of bends from the path of meandering rivers, are an important part of the planform dynamics of meandering rivers. These events regulate channel sinuosity by shortening channel length and produce abrupt lateral change in the position of meandering rivers (Micheli and Larsen, 2011). Neck cutoffs develop when the limbs of a meander loop migrate toward one another until the limbs

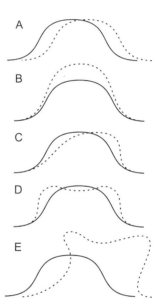

Figure 9.34. Styles of change (dashed lines) in an initial meander bend (solid line) through channel migration: (a) translation, (b) extension, (c) skewing/rotation, (d) lobing, (e) combination of translation, extension, skewing, and lobing.

intersect (Figure 9.35). The formation of this type of cutoff does not depend on overbank flow. Chute cutoffs form when overbank flow carves a channel across the interior of the bend, creating a new pathway for flow that is shorter than the path around the bend (Figure 9.35).

Three mechanisms have been identified that create chute cutoffs: slough enlargement, headward erosion, and embayment extension (Figure 9.36) (Viero et al., 2018). Sloughs are curvilinear topographic lows that develop in conjunction with lateral accretion of scroll bars on point bars along the inner banks of meander bends (see Chapter 14). Partial channelization of a slough at high or overbank stages can produce high bed shear stresses within the evolving slough that further erode and enlarge

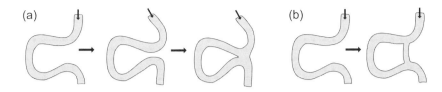

Figure 9.35. Illustration of (a) neck cutoff and (b) chute cutoff.

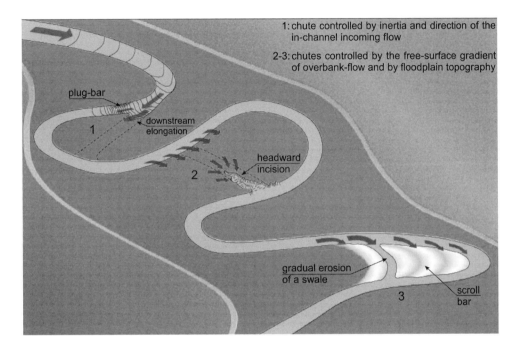

1: chute controlled by inertia and direction of the in-channel incoming flow

2-3: chutes controlled by the free-surface gradient of overbank-flow and by floodplain topography

plug-bar

downstream elongation

1

headward incision

2

gradual erosion of a swale

scroll bar

3

Figure 9.36. Mechanisms of chute cutoff: (1) downstream elongation of an embayment, (2) headward migration of a gully, (3) erosion of a swale or slough in scroll-bar topography (A black and white version of this figure will appear in some formats. For the color version, please refer to the plate section.) (from Viero et al., 2018).

it (Harrison et al., 2015), creating a chute channel that subsequently captures flow at subbankfull stages (Grenfell et al., 2012; Van Dijk et al., 2012, 2014). This process is common in rapidly extending meander bends, produces chute channels downstream of the bend inflection point, and may generate stable bifurcate channels within a meander bend (Grenfell et al., 2012, 2014). Headward erosion typically occurs when overbank flow moving across the interior of a meander loop initiates a knickpoint or headcut where it enters the downstream limb of the bend. Knickpoint initiation may also be promoted by seepage erosion from a strong gradient in groundwater flow across the neck of the bend (Han and Endreny, 2014). As the knickpoint migrates across the neck of the bend, it produces a gully that eventually intersects the upstream limb of the bend to create a chute channel that begins to capture the flow of the river (Gay et al., 1998; Zinger et al., 2011, 2013). In large rivers, the rapid enlargement of the chute channel following capture of the main flow can introduce enormous amounts of sediment into river downstream of the cutoff (Zinger et al., 2011). Embayments may develop downstream of the bend apex along the outer bank of a bend, where the path of overbank flow deviates most from the channel path, leading to erosion of the floodplain surface. During a series of overbank flows, the embayment extends across the neck of the

bend, eventually resulting in a chute cutoff (Constantine et al., 2010a). In both neck cutoffs and chute cutoffs, deposition at the upstream and downstream ends of the bend eventually prevents flow from entering the bend, resulting in the formation of an oxbow lake – a key process in floodplain dynamics (see Chapter 14).

9.8.2.2 What Bar-Unit Forms Develop in Elongated Meander Loops?

Recognition of the development of meander bends into characteristic elongate forms has led to experimental and numerical investigations of flow and bed morphology in meander loops with symmetrical and asymmetrical planforms. In laboratory studies of symmetrical loops, a series of overlapping bars, known as shingle bars, develop along the inner bank, and multiple pools form along the outer bank (Whiting and Dietrich, 1993a,b) (Figure 9.37). Maximum bed scour and near-bed velocities occur upstream of the bend apex along the outer bank, suggesting that maximum bank erosion should occur at this location (Termini, 2009). Changing curvature throughout the loop leads to damped oscillation of the water-surface elevation and near-bed velocity with pools developing in loci of accelerated near-bed flow.

Asymmetrical bends skewed upstream (classic Kinoshita form) and downstream also exhibit multiple pools along the outer bank (Abad and Garcia, 2009). In both skewed configurations, pools develop upstream and downstream of the bend

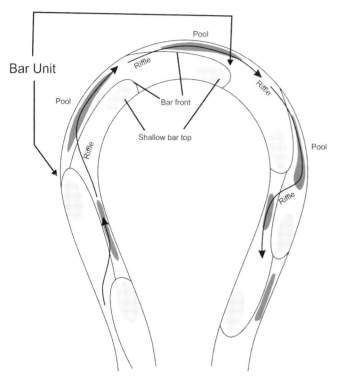

Figure 9.37. Shingle bars (bar units) around an elongate loop based on results of experimental studies by Whiting and Dietrich, 1993b (from Frothingham and Rhoads, 2003).

apex, with the deepest pools located downstream of the apex. The downstream-skewed configuration has the deepest pools of the two types of skewed loops, indicating that this type of loop should exhibit high downstream migration rates. Field studies confirm that multiple pools occur in elongate meander loops, but overlapping shingle bars observed in laboratory channels are not evident in natural loops (Hooke and Harvey, 1983; Frothingham and Rhoads, 2003; Engel and Rhoads, 2016; Konsoer et al., 2016b).

9.8.2.3 How Do Compound Loops Develop?

The development of compound, multilobe forms of river meandering involves upstream migration of the upstream limb of an elongate loop and downstream migration of the downstream limb of the loop (Engel and Rhoads, 2012) (Figure 9.38). This pattern of channel migration is consistent with experimental evidence indicating that pools develop along the outer bank upstream and downstream of the apex of elongate loops (Abad and Garcia, 2009). Presumably, bank erosion occurs in conjunction with this pool development in natural channels. As the two limbs of the bend migrate away from one another, lobes form on each side of an intervening region of low curvature. Decay of helical motion within the connecting low-curvature channel segment promotes riffle formation (Hooke and Harvey, 1983; Engel and Rhoads, 2012). Moreover, remnant material from bank failures preserved along the outer bank may deflect flow laterally toward the inner bank, initiating a reversal of curvature in the channel between the lobes (Figure 9.38).

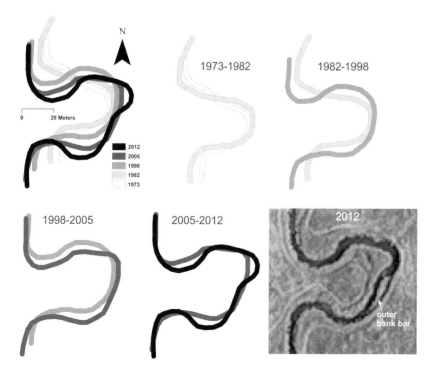

Figure 9.38. Evolution of a simple bend into a compound loop along the Embarras River, IL. Upstream limb of the bend in 1973 initially migrates upstream as downstream limb migrates downstream, elongating the head of the loop and forming upstream and downstream lobes. Extension of the upstream lobe and translation of the downstream lobe accentuate the compound form of the loop. Formation of a persistent bar from failed bank material along the outer bank between the lobes produces a reversal in curvature between the lobes in 2012 (photo).

9.8.3 What Factors Influence the Rate of Lateral Migration of a Meandering River?

The complexity of planform evolution has frustrated efforts to find simple explanatory relations between migration rates (ξ_m) and possible controlling factors. Generally, migration rates increase with increasing discharge, stream power per unit area, and channel slope, but statistical relationships have substantial unexplained variance (Hooke, 1980; Nanson and Hickin, 1986). For rivers in the Amazon basin, migration rate per unit channel width (ξ_m/W_{bk}) increases as a power function of total suspended-sediment flux, but the degree of scatter in the relationship is large (Constantine et al., 2014). As migration rate increases, the number of cutoffs per unit length of channels also increases. Meandering rivers with high sediment fluxes also exhibit greater fractional increases in sinuosity over time resulting from channel migration (Ahmed et al., 2019). These rivers are dominated by downstream-rotating meanders that increase sinuosity faster than extensional or upstream-rotating meanders.

Some evidence exists to suggest that maximal migration rates per unit channel width occur for bends with $2 \le r_c/W_{bk} \le 3$, with rates decreasing abruptly for $r_c/W_{bk} < 2$ and tailing off gradually for $r_c/W_{bk} > 3$ (Nanson and Hickin, 1983; Hickin and Nanson, 1984; Hooke, 2003b, 2007) (Figure 9.39). The evidence again is weak, given that the trends represent envelope curves bounding clusters of data, and many individual bends do not conform to these trends. The decrease in migration rate per unit width for $r_c/W_{bk} < 2$ is consistent with experimental work in sharp bends documenting the development of outer-bank flow separation that protects the outer bank from high fluid stresses and the saturation of secondary flow at high values of curvature (Blanckaert, 2011).

Attempts have been made to plot the evolutionary trajectories of ξ_m/W_{bk} versus r_c/W_{bk} of individual bends as these bends evolve over time to determine whether systematic nonlinear behavior organized around an attractor state can be identified (Hooke, 2003b, 2007). As straight reaches of meandering rivers begin to develop sinuosity, ξ_m/W_{bk} of emerging bends will increase as the r_c/W_{bk} of these bends decreases. Eventually, if a cutoff occurs, r_c/W_{bk} will abruptly increase and migration rate will suddenly decrease, resetting the system for renewed evolution toward cutoff. In meander bends where a cutoff does not occur, the planform of elongate bends will still evolve, leading to minor covariations in ξ_m/W_{bk} and average bend curvature that produce local orbital patterns on plots of migration rate versus r_c/W_{bk} (Figure 9.40). This empirical analysis of planform evolution differentiates various types of meandering dynamics in a visual descriptive sense, but it does not provide deep insight into the mechanisms that produce differences in meander dynamics.

9.8.4 How Have Models Been Used to Explore the Dynamics of Meandering?

Given that many meandering rivers may not exhibit major changes in planform over periods of years to decades, the timeframe over which direct observational evidence on planform change is available, the leading approach to exploring the dynamics of planform evolution of meandering rivers over timescales of decades to millennia has involved modeling. A wide array of models has been developed since the 1970s. These models fall into two main categories: 1) physics-based models, formulated and solved analytically, that may subsequently be operationalized numerically to simulate planform evolution over geologic timescales and 2) physics-based numerical models in which the fundamental equations are solved computationally to simulate lateral movement of individual bends of a meandering river over timescales of an event or several events. Although purely numerical models can solve highly nonlinear equation systems for which analytical solutions do not exist, the high computational demands of these models and the need to upgrade computational meshes to reflect progressive patterns of erosion and deposition over time limit the time and space domains over which these models can be applied.

In the development of models of the planform dynamics of meandering rivers, channel curvature has been viewed as a critical factor influencing lateral migration. Curvature,

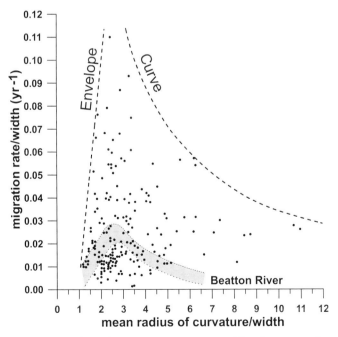

Figure 9.39. Relation between migration rate and bend curvature for meandering rivers. Shaded area indicates range of values for the Beatton River, British Columbia (from Nanson and Hickin, 1986).

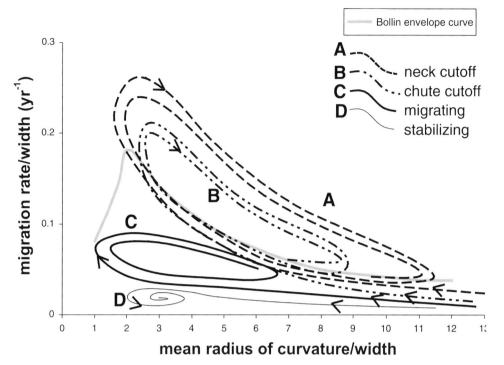

Figure 9.40. Hypothetical trajectories of relation between rates of lateral movement and bend curvature for (A) neck cutoff, (B) chute cutoff, (C) migrating bends without cutoff, and (D) bends that stabilize at low migration rates and constant mean radius of curvature. Bollin envelope curve demarcates the upper limit of curvature–migration relations for individual bends in River Bollin, UK (from Hooke, 2007).

through its effects on redistribution of momentum of the flow toward the outer bank, should lead to high near-bank velocities, which, if Eq. (9.10) is valid, causes bank erosion. Thus, shifting of the high-velocity core and zone of maximum bed shear stress from bank to bank through a sequence of bends leads to channel migration. If the flow adjusted immediately to changes in curvature, maximum migration would occur at the bend apex, or location of maximum curvature. As a result, many bends would increase in amplitude without translating or skewing downstream. However, momentum redistribution, as influenced by water-surface superelevation, helical flow, and topographic steering, occurs over a finite channel distance, resulting in spatial lag between channel-bend geometry and the path of the shifting thread of maximum velocity (Figure 9.3). As a result, the migration rate of the channel, as determined by the near-bank velocity, is not only a function of local curvature but depends also on upstream curvature (Howard and Knutson, 1984). This relationship can be expressed as (Parker and Andrews, 1986)

$$\hat{\xi}_m(\hat{s}) = \int_{-\infty}^{\hat{s}} f(\hat{s} - \hat{s}') C^*(\hat{s}') d\hat{s}' \tag{9.13}$$

where $\hat{\xi}_m(\hat{s})$ is the dimensionless migration rate of the channel centerline (ξ_m/U) at dimensionless position $\hat{s}$ along the river (s/h_{cl}), with h_{cl} designating the centerline depth, $\hat{s}'$ denotes a variable of integration that may be interpreted as the dimensionless distance upstream from position $\hat{s}$, C^* is the

dimensionless channel curvature ($W_{bk}C_{cs}$), and $f(\hat{s})$ represents a weighting function that depends on details of the underlying model but has an exponential form (Howard and Knutson, 1984; Furbish, 1991; Sun et al.,1996). This basic model characterizes lateral migration as a function not only of local curvature but also of curvature upstream of a particular location along the river. The specifics of the weighting function determine the extent to which upstream conditions influence downstream migration. If the exponential weighting decays rapidly, curvature upstream of a location only weakly influences migration at that location. On the other hand, if the weighting decreases slowly, upstream conditions contribute strongly to downstream migration. The net effect of strong upstream effects is to shift the locus of maximum migration downstream of the locus of peak curvature (Sylvester et al., 2019), enhancing translation and diminishing extension.

A simple physics-based linear model uses a first-order ordinary differential equation to characterize the planform dynamics of meandering rivers (Ikeda et al., 1981):

$$U\frac{\partial \overline{u}_{nb}'}{\partial s} + 2\frac{U}{D}c_f\overline{u}_{nb}' = \frac{W}{2}\left[-U^2\frac{\partial C_{cs}}{\partial s} + c_f C_{cs}\left(\frac{U^4}{gD^2} + A'\frac{U^2}{D}\right)\right] \tag{9.14}$$

where c_f is the friction coefficient (Eq. (4.42)) and A' is a scour factor that defines the role of secondary currents in determining the bed topography (Pizzuto and Meckelnburg, 1989; Sun et al., 1996). Integration of this equation yields a convolution

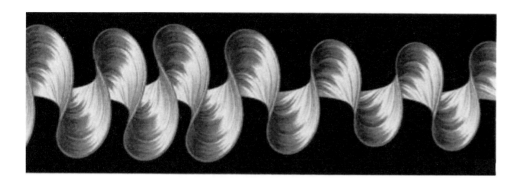

Figure 9.41. Simulation of bend evolution with a linear meander model showing development of upstream-skewed asymmetrical loops (Kinoshita curves) prior to cutoff (flow left to right) (from Sun et al., 2001).

integral relating the excess near-bank velocity to channel curvature in the form of Eq. (9.13) (Sun et al., 1996; Guneralp and Marston, 2012):

$$\bar{u}_{nb}'(s) = \frac{-WUC_{cs}}{2} + \frac{UWc_f}{2D}\left[F^2 + A' + 2\right]\int_0^\infty e^{\frac{-2c_f}{D}(s')}C_{cs}(s - s')ds'$$

$$(9.15)$$

where F is the Froude number. Note that the weighting, or impulse-response function, of curvature within the integral term has a negative exponential form. In other words, the influence of upstream curvatures on the local excess near-bank velocity diminishes exponentially as the distance upstream (s') increases. The rate of decay is directly related to the magnitude of the friction coefficient and inversely related to the flow depth. Spatial variation in $\bar{u}_{nb}'$ directly influences spatial variation in migration rates because these rates at various positions s along a meandering river are obtained from Eq. (9.10) using excess near-bank velocities generated by Eq. (9.15). Exploration of the model shows that it predicts growth of simple bends into upstream-skewed asymmetrical elongate loops known as Kinoshita curves (Parker et al., 1982c, 1983) that correspond with the general shape of many elongate natural meander bends (Carson and Lapointe, 1983) (Figure 9.41). It also predicts that downstream and lateral migration rates decrease as bends evolve into elongate loops.

Analytical models of meandering river dynamics have been used to numerically simulate planform development of meandering rivers, including the influence of cutoffs on these dynamics over geologic timescales (Howard, 1992; Sun et al., 1996; Stolum, 1996). Simulations suggest that the increase of sinuosity through bend elongation and the tightening of loop necks can cause the river system to transcend a critical state in which clusters of cutoffs occur to produce disordered, chaotic patterns of planform evolution (Stolum, 1996). This mode of planform evolution is supported to some extent by empirical evidence on the timing of cutoffs and planform response to cutoffs (Hooke, 2004b; Schwenk et al., 2015; Schwenk and

Foufoula-Georgiou, 2016). It also has underpinned efforts to frame the dynamics of meandering within the context of theory related to deterministic chaos, nonlinear dynamics, and autogenic self-organization (Hooke, 2007). The extent to which the planform dynamics exhibit chaotic or self-organizing behavior remains uncertain. Rigorous, formal analysis of spatial curvature series for natural meandering rivers has found no evidence of fundamental nonlinearity (Perucca et al., 2005), nor do simulations of long-term meander migration using numerical models support the hypothesis that planform dynamics are characterized by chaotic behavior or self-organized criticality (Frascati and Lanzoni, 2010). Instead, cutoffs serve as a mechanism through which the long-term planform characteristics of meandering rivers, including mean sinuosity and curvature, stabilize around a statistical steady state (Stolum, 1996; Camporeale et al., 2008).

Over the past several decades, analytical models of river meandering have been increasingly refined to include additional details on interactions among flow, sediment transport, and bed morphology (Johannesson and Parker, 1989; Seminara et al., 2001; Zolezzi and Seminara, 2001; Camporeale et al., 2007; Pittaluga et al., 2009) and to accommodate the effects of high bend curvature (Blanckaert and de Vriend, 2010; Ottevanger et al., 2013). A limitation of basic linear models is the failure to predict downstream-skewed asymmetrical meander loops and the development of compound loops prior to cutoff (Figure 9.34d,e), both of which occur as natural meandering rivers evolve over time. On the other hand, advanced models that consider resonance between bed topography and channel curvature (Seminara and Zolezzi, 2001; Zolezzi and Seminara, 2001) predict downstream skewing of bends and the development of compound loops prior to cutoff under certain conditions. These effects result from an impulse-response function for the excess near-bank velocity that includes terms defining oscillatory spatial perturbations to $\bar{u}_{nb}'$ (Frascati and Lanzoni, 2010). Empirical analysis of the curvature–migration relation for natural meandering rivers confirms that the impulse-response or weighting

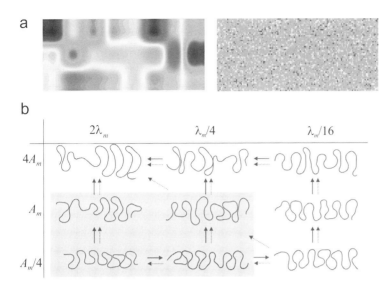

Figure 9.42. Influence of floodplain heterogeneity of meander evolution. (a) Coarse-grained (left) versus fine-grained (right) erosional heterogeneity of a floodplain (hot colors represent high bank erodibility and cool colors represent low bank erodibility). (b) Influence of the spatial scale of floodplain heterogeneity in erosional resistance on planform evolution of simulated meandering rivers. Dimensions of heterogeneous patches of erosional resistance are defined in terms of the amplitude (A_m) and wavelength (λ_m) of initial meander bends used in the simulations. Solid and dashed arrows represent the direction of increasing maximum cross-valley extent of the meander belt and increasing variability in bend amplitude, respectively. Shaded area corresponds to the spatial scales of erosional heterogeneity for which cutoffs commonly occur over shorter timescales compared with other scales. Complex bend geometries, such as compound loops, develop at all scales of heterogeneity prior to the development of cutoffs (A black and white version of this figure will appear in some formats. For the color version, please refer to the plate section.) (from Guneralp and Rhoads, 2011).

function ($f(s)$ in Eq. (9.13)) is more complex than simple exponential decay (Guneralp and Rhoads, 2010) and, for compound loops, includes oscillatory components (Guneralp and Rhoads, 2009).

Compound bend forms may develop not only through the internal dynamics of meandering but also through the influence of environmental heterogeneity on these dynamics. Simulations using the basic model (Eq. (9.15)) indicate that heterogeneity in the erosional characteristics of the floodplain across which a meandering river is migrating has a strong influence on bend evolution and can lead to the development of compound loops prior to cutoff (Guneralp and Rhoads, 2011; Motta et al., 2012b) (Figure 9.42). Field observations, geographical information system (GIS)-based analysis of satellite imagery, and numerical simulations demonstrate that migration of a meandering river into clay-rich floodplain deposits significantly reduces migration rates and channel width, and increases the angularity of bend geometry (Schwendel et al., 2015). Clay bodies also limit the propagation of planform disturbances, such as bend translation and response to cutoff, in the upstream and downstream directions. Besides spatial variation in floodplain sedimentology, patterns of riparian vegetation represent another factor contributing to the heterogeneity of erosional resistance to lateral migration (Perucca et al., 2007). Flood-induced erosion of outer banks has been documented to be 30 times more prevalent for unforested bends versus forested bends (Beeson and Doyle, 1995), and long-term rates of lateral migration are as much as three times greater for unforested versus forested bends (Burckhardt and Todd, 1998).

Most models of river planform dynamics assume that meander migration results from outer-bank erosion and that point-bar deposition tracks outer-bank movement. The models do not treat processes occurring at the inner bank independently. In other words, outer-bank erosion is the driver of migration and the inner bank simply follows along as the outer bank moves. This representation of meander migration constitutes bank "pull": the outer bank pulls the inner bank along with it. The possibility that changes in point-bar morphology, particularly growth of the point bar, might modify flow and sediment transport so that outer-bank erosion increases represents an alternative explanation for meander migration. This alternative scenario constitutes point-bar "push": changes at the inner bank of a meandering river lead to outer-bank erosion. To independently consider the effects of push-pull processes on lateral migration, inner bank and outer bank changes must be modeled separately within the context of physically based numerical models (Darby et al., 2002; Ruther and Olsen, 2007; Parker et al., 2011; Asahi et al., 2013; Eke et al., 2014). The relative importance of push versus pull effects in meander migration has yet to be fully assessed. Experimental studies indicate that point-bar deposition in the form of scroll bars laterally accreted onto the point bar is caused by channel widening induced by outer-bank erosion (van de Lageweg et al., 2014). These findings support the hypothesis that meander migration is driven largely by outer-bank pull, where erosion at the outer bank locally widens the channel and point-bar deposition occurs subsequently to reestablish the channel width that existed prior to outer-bank erosion. Field evidence suggests that outer-bank erosion can outpace inner-bank deposition, resulting in local overwidening of the channel and the development of regions of flow separation along the inner bank in the downstream part of bends (Engel and Rhoads, 2012). Deposition along the inner bank occurs slowly in regions of separated flow, resulting in asynchronous migration of the inner bank relative to the outer bank. Successional changes in vegetation occur on point bars as these features migrate over time (Robertson and Augspurger, 1999), and the influence of vegetation dynamics on point-bar deposition is likely an important factor in push-pull mechanisms of meander migration (Zen et al., 2016). In other words, vegetation dynamics seem to have a strong influence on meandering river morphodynamics (van Oorschot et al., 2016). That said, observations of paleo-meandering channels on Mars show that vegetation is not a necessary condition for meandering – a conclusion confirmed by numerical modeling of the planform dynamics of mud-dominated, unvegetated meandering rivers on Earth (Matsubara and Howard, 2014; Matsubara et al., 2015). The roles of mud and vegetation in the dynamics of meandering rivers are active topics of exploration (Kleinhans et al., 2018).

CHAPTER

10 The Dynamics of Braided Rivers

10.1 Why Are the Dynamics of Braided Rivers Important?

The distinctive morphological structure of braided rivers is characterized by an intricate, dynamically changing pattern of multiple interconnected channels and intervening ephemeral bars (Ashmore, 2013). Conditions of high stream power, erodible banks, and mobile bed material result in complex patterns of scour and fill, frequent changes in channel positions and connectivity, and migration and reshaping of bars. These conditions occur most frequently in environments with high sediment supply, sparse vegetation cover, and relatively steep slopes, such as proglacial mountain valleys and arid or semiarid regions. Braided rivers often are viewed as unstable because change rather than persistence typifies the morphodynamics of these rivers. Nevertheless, some attributes of braided rivers remain relatively constant over time, and the extent to which these rivers achieve equilibrium, or a graded condition, in which morphology adjusts so that neither net aggradation nor degradation occurs, remains somewhat controversial. The high mobility of sediment within braided rivers leads to change over a variety of flow conditions; however, formative conditions generally correspond to relatively high stages when flow and bed-material transport occur in several branches of the interconnected channel system. Large floods that completely inundate the braided network of channels and bars can produce substantial reconfiguration of river morphology (Bertoldi et al., 2010).

Because the form and dynamics of braided rivers seem to differ fundamentally from those of meandering rivers, these types of rivers have been investigated extensively to understand processes and forms associated with braiding. Moreover, knowledge of braided rivers is of increasing importance in river management. Although many braided rivers occur in relatively unpopulated environments, concern about the influence of human activities on braided rivers has grown in recent decades (Murphy et al., 2004; Surian, 2006; Gurnell et al., 2009). Braided rivers are valued ecological resources (Tockner et al., 2006), and efforts to manage these resources depend on fundamental knowledge of the processes that govern braiding (Piegay et al., 2006).

10.2 How Does Braiding Occur?

As with meandering rivers, explanations of how braiding occurs have mainly emphasized the importance of initial bar development, which subsequently triggers other adjustments that lead to the development of a braided channel. From a theoretical perspective, bar theory provides a physics-based explanation not only for the initiation of meandering (see Chapter 9) but also for the initiation of braiding. It provides a theoretical link between the meandering and braiding of rivers.

10.2.1 How Does Bar Theory Account for Braiding?

The growth of bar units within an initially straight channel can be assessed by mathematical stability analysis of the two-dimensional governing equations for flow and sediment transport to determine whether infinitesimal doubly harmonic perturbations of the bed grow into bars or are damped out (Engelund and Skovgaard, 1973; Parker, 1976; Fredsoe, 1978; Seminara and Tubino, 1989; Crosato and Mosselman, 2009). Results of such analysis for given values of dimensionless bed shear stress (Θ), width–depth ratio, and grain resistance coefficient yield plots of amplification factor (A_f), the rate at which the perturbations grow, versus dimensionless wavelength of the perturbations (see Figure 9.7b). The peak or maximum value of A_f corresponds to the maximum perturbation growth rate, which is assumed to represent the wavelength of bars that emerge on the bed of the initially straight channel. Plots of maximum values of A_f (A_{fmax}) versus width–depth ratio form a humped curve that can be used to determine whether bar

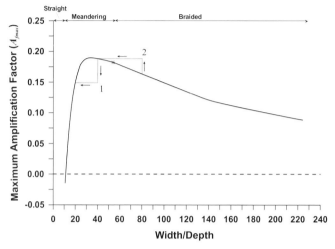

Figure 10.1. Maximum amplification factor versus width/depth ratio for $\Theta = 0.2$ and grain resistance coefficient of 7. Asterisk on humped curve indicates the critical width/depth ratio separating meandering from braiding. Dashed lines show examples of how (1) the value of A_{fmax} for $W/D = 40$, the actual W/D, is greater than the value of A_{fmax} for $W/D = 0.5(40) = 20$; therefore, bar units for $W/D = 40$ are selected and meandering occurs; 2) the value of A_{fmax} for $W/D = 80$, the actual W/D, is less than the value of A_{fmax} for $W/D = 40$, so bar units for $W/D = 40$ are selected, and double-row bar units develop within the channel with $W/D = 80$ to produce braiding (from Rhoads and Welford, 1991 based on Fredsoe, 1978).

development will produce a straight, meandering, or braided stream (Figure 10.1). If A_{fmax} is negative, no bars will develop and flow within the channel will remain straight and undivided. If A_{fmax} is positive, either a meandering or a braided stream will develop. On the plot, a critical width–depth ratio exists $(W/D)_c$ where A_{fmax} for $(W/D)_c$ equals A_{fmax} for $0.5(W/D)_c$ (Figure 10.1). For $W/D < (W/D)_c$, positive values of A_{fmax} for a given W/D always exceed A_{fmax} for $0.5(W/D)$, and only a single set of alternating bar units develops within the channel, leading to meandering (e.g., Figure 9.3). For $W/D \geq (W/D)_c$, A_{fmax} for a given W/D will be less than A_{fmax} for $0.5(W/D)$. Bar units with one-half the width–depth ratio of the flow will grow at a faster rate than bar units corresponding to the actual width–depth ratio. The result will be the development of two rows of alternate bars placed side by side in mirror image fashion – an arrangement known as double-row bars. In this configuration, flow will divide around a central bar, resulting in braiding (Figure 10.2). Moreover, given the humped shape of the curve, as the flow width–depth ratio increases further, additional critical thresholds will be crossed whereby A_{fmax} for $0.33W/D$, $0.25W/D$, and so on will exceed A_{fmax} for the actual value of W/D. Increases in the width–depth ratio can occur if the width of the initial straight channel is progressively increased. As additional thresholds are crossed, triple-row, quadruple-row, and higher-order sets of multiple-row bar units will form within the initial straight channel.

Bar theory suggests that meandering and braided rivers, despite appearing considerably different, develop from the same fundamental bar-unit structures. The main factor controlling differences between the two types of rivers for a given set of flow and sediment-transport conditions is the extent to which

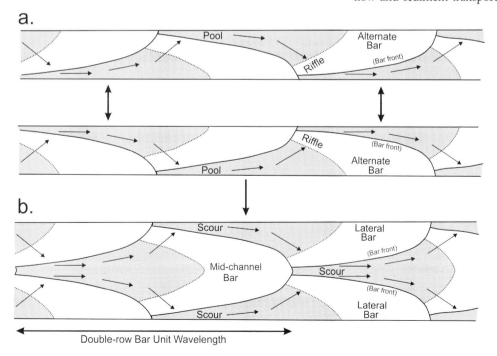

Figure 10.2. (a) Mirror images of alternate bar units combine in wide channels to produce (b) double-row bars and the development of a braided stream flowing around mid-channel bars and past lateral bars. Shaded areas are below the average bed elevation and unshaded areas are above the average bed elevation. Arrows indicate flow paths when bar tops are exposed.

the river system can widen in relation to these conditions. Channels with large width–depth ratios will braid through the development of multiple bar-unit structures across the width of the channel, whereas constraints on adjustment in width–depth ratio result in meandering or even straight channels.

10.2.2 How Have Experimental Studies Provided Insight into Mechanisms of Braiding?

In contrast to river meandering, which has proven difficult to study in experimental settings, braiding is readily produced in laboratory streams. Experimental studies indicate that braiding in an initially straight channel with flat beds occurs by four fundamental mechanisms (Ashmore, 1991a): chute cutoff, central bar initiation, transverse bar conversion, and multiple bar dissection (Figure 10.3). The importance of these four mechanisms in initiating braiding relates to initial conditions.

For relatively large dimensionless bed shear stresses ($\Theta \approx 0.08$–0.13) and moderate width–depth ratios (28–34), an initial straight erodible channel widens and bar units develop (Ashmore, 1982). Meandering flow through the bar units erodes the channel banks, leading to mild channel sinuosity that tends to

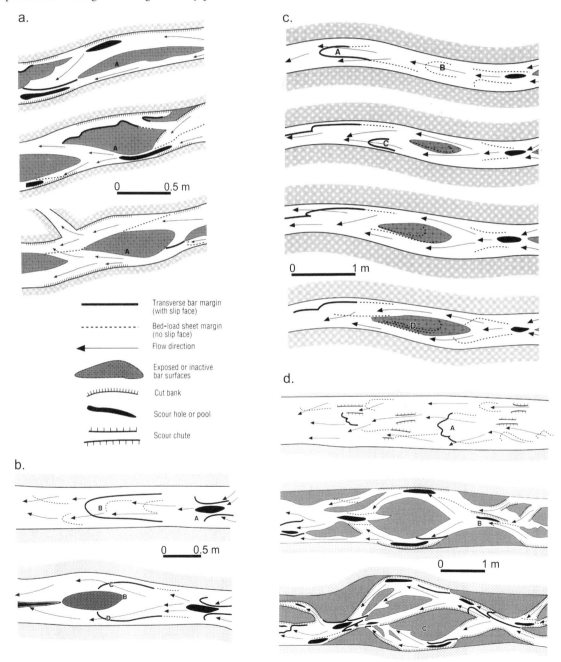

Figure 10.3. Mechanisms for initiation of braiding (from Ashmore, 1991a). (a) Chute cutoff – symbol A refers to initial exposed bar. (b) Traverse bar conversion – B refers to location of coarse bed material, C and D refer to submerged diagonal bars. (c) Central bar initiation – A and C refer to submerged transverse bars, B refers to bedload sheet, and D refers to lateral accretion zone on exposed bar. (d) Multiple bar braiding. Figures at different scales.

inhibit bar migration, which, in turn, enhances erosion of the banks (Hong and Davies, 1979; Federici and Paola, 2003; Bertoldi and Tubino, 2005). Flow diversion around the highest parts of alternate bars forms sloughs along the channel banks. Cutoff occurs when bank erosion at the upstream end of the slough produces a bedload pulse and local aggradation that enhance diversion of flow into the slough, causing headward erosion and scour that carve a new channel (Figure 10.3a) (Ashmore, 1991a). Bifurcation of flow at the head of the bar limits bedload-transport capacity along the former main path of the flow, leading to bar growth through lateral accretion as sediment is deposited on the margin of the bar along this flow pathway. In channels that evolve from a narrow initial condition into braided forms, chute cutoff has been proposed as the fundamental mechanism leading to bifurcation (Bertoldi and Tubino, 2005).

Transverse bar conversion occurs under similar hydraulic conditions to those for chute cutoff, but is less common. In this braiding mechanism, migrating transverse unit bars, lobate submerged bars with steep downstream slip faces (Smith, 1974) that typically occur downstream of zones of scour, gradually accrete vertically as progressive sequences of migrating bedload sheets stall within the center of the bar (Figure 10.3b). Through this process, the central portion of the submerged bar increases in height, coarse material accumulates on the head of the bar, and flow is diverted around it. The diverted flow erodes the channel banks, resulting in exposure of the bar (Figure 10.3b). The exposed bar enlarges through accretion of bedload sheets on its lateral and downstream margins. Submerged diagonal bars may develop upstream where flow is directed around the exposed bar.

Central bar initiation develops at relatively low values of dimensionless bed shear stress ($\Theta \approx 0.06$) in channels with moderate width–depth ratios. Stalling of migrating bedload sheets with coarse grains near the downstream margin of these sheets initiates the development of a central bar that subsequently builds by trapping additional fine and coarse particles on its surface (Leopold and Wolman, 1957; Ashmore, 1991a; Federici and Paola, 2003). Divergence of flow streamlines around the incipient bar promotes further deposition. Lateral deflection of the flow by the growing bar causes bank erosion, widening the channel locally and reducing flow depth. Continued bar growth and the reduction in flow depth eventually lead to protrusion of the bar top out of the water. The exposed bar divides the flow into two pathways around it and continues to enlarge through accretion of migrating bedload sheets onto the bar head and the lateral margins of the bar (Figure 10.3c). Unit bars evolve into compound bars through amalgamation of individual bars and through dissection of amalgamated features by chute channels.

Although braiding through widening of initially narrow channels often involves finite amplitude effects of single-row bar units on the flow that lead to channel bifurcation by chute cutoff (Bertoldi and Tubino, 2005), braiding can also occur through the spontaneous development of multiple bar units across the width of experimental channels with large initial width–depth ratios (> 100) (Fujita, 1989; Ashmore, 1991a). This finding is consistent with bar theory, which predicts that increases in width–depth ratio should produce an increase in the number of rows of bar units across the channel. In multiple bar braiding, discontinuous erosional chutes and corresponding lobate submerged bars form initially throughout the channel (Figure 10.3d). The bars grow through stalling of bedload sheets, vertical accretion on bar surfaces, and lateral accretion on bar margins, thereby concentrating flow into pathways adjacent to the bars. Erosion of channel banks by flow diverted around the growing bars also enhances bar exposure. Over time, the planform evolves into multiple exposed bars and an intervening network of interconnected channels. In natural braided rivers, multiple bar braiding can also occur through dissection of bars caused by variations in flow stage (Ferguson, 1993). A coherent submerged bar that forms during high stages of a flood event may subsequently be divided into multiple exposed bars when flow concentrates into narrow pathways that carve channels into the initial bar as water levels decrease. This mechanism can also contribute to the development of compound bars.

The process of bar development for all four cases involves local aggradation through spatial variation in flow competence or sediment-transport capacity. Moreover, strong feedbacks exist among evolving bar forms, zones of bed scour, regions of bank erosion, and spatial variability of bed-material transport. Loss of competence and transport capacity often are associated with the development of local regions of flow expansion associated with bank erosion. These regions of flow expansion occur in conjunction with bar growth and accompanying lateral deflection of flow into the banks. As long as bed material is mobile, the four processes can all occur under conditions of steady discharge; therefore, a highly variable discharge, sometimes thought to be an important factor that contributes to braiding in natural rivers, is not a necessary condition for braiding (Ashmore, 2013). Moreover, experimental studies show that net aggradation of the river system as a whole is not a necessary condition for braiding, even though some natural braided rivers are aggrading.

10.3 What Are the Basic Morphological Attributes of Braided Rivers?

Bar theory indicates that the basic morphological features of braided rivers consist of zones of scour, or pools, and bars, with flow tending to converge within the pools and diverge around the bars (Figure 10.2). This structure has been documented both in the laboratory and in the field and has been referred to variously as bar-pool, confluence-diffluence, and confluence-bifurcation units (Ferguson, 1993; Ashworth, 1996; Ashmore, 2013) (Figure 10.4). Material exhumed from the zone of scour within confluences is deposited downstream on bars, where competence and sediment-transport capacity decrease as flow exiting the zone of scour diverges laterally. As flow moves around the bar it converges downstream, initiating scour and the formation of another confluence-diffluence unit. From the standpoint of bar theory, this configuration of confluences with associated zones of scour and bifurcations at the head of bars conforms to the fundamental structure of double-row bar units (Figure 10.2).

10.3.1 What Is the Scaling of Bar Dimensions in Braided Rivers?

Bars in gravel-bed braided rivers often have a diamond or rhomboid shape that corresponds to the expected form based on the structure of bar units (Figures 10.4 and 10.5). To some extent, this form also corresponds to the streamlined shapes expected on the basis of minimum drag or hydraulic resistance (Komar, 1983). A variety of studies have examined the morphological scaling of bars in braided rivers (Sapozhnikov and Foufoula-Georgiou, 1996; Nykanen et al., 1998; Walsh and Hicks, 2002; Sambrook Smith et al., 2005; Kelly, 2006; An et al., 2013). Scaling has been quantified by relating the length (L_{br}) and width (W_{br}) of braid bars to one another in the form of a power function. Data for 22 braided systems obtained through field and laboratory studies yield the relation (Kelly, 2006)

$$L_{br} = 4.95 W_{br}^{0.97} \tag{10.1}$$

The exponent in this equation is nearly 1, indicating that, on average, bar dimensions increase proportionally for lengths ranging over six orders of magnitude (0.01 to 10,000 m). The geometry of bars tends to be self-similar or isometric in that bars maintain a similar average shape as size varies (Walsh and Hicks, 2002; An et al., 2013). The coefficient in Eq. (10.1) shows that bar lengths are generally about five times greater than bar widths – a value slightly greater than the optimal value of 3–4 for minimum drag (Komar, 1983). The perimeter length of bars (P_{br}) relates to bar area (A_{br}) as (Kelly, 2006)

$$P_{br} = 5.69 A_{br}^{0.50} \tag{10.2}$$

The exponent of 0.50 in this relationship between a length attribute of bars and the area of bars provides further confirmation of the self-similarity of bar shape. Sandy braided

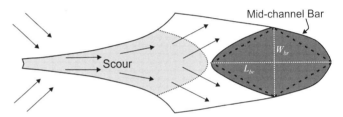

Figure 10.4. Basic morphological structure of a confluence-diffluence or pool-bar unit in a gravel-bed braided river. Light shading is below mean bed elevation, white area is above mean bed elevation, and dark shading corresponds to portion of the unit exposed subaerially at low to moderate flow stages. Ideal diamond or rhomboid shape is indicated by heavy black dashed lines. White lines show bar length (L_{br}) and width (W_{br}). Arrows indicate path of flow through the morphological structure.

Figure 10.5. Rhomboid-shaped bars in the Chitina River, Alaska (September 8, 2004; 61°22′18.82″ N, 143°48′18.82″W). Flow is from right to left.

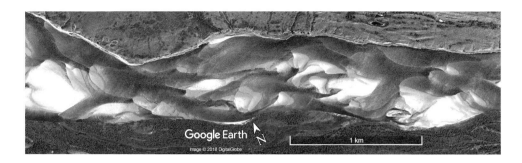

Figure 10.6. Linguoid bars in the South Saskatchewan River, Canada. (August 27, 2012; 51°25′06.67″N, 107°03′05.45″W). Flow is from right to left.

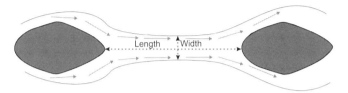

Figure 10.7. Length and width of a confluence-diffluence unit.

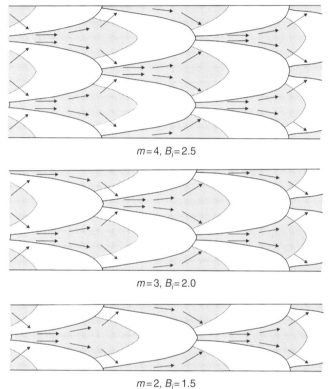

Figure 10.8. Structure of multiple-row bar units and flow pathways through units for modes (m_{br}) 2 to 4 and corresponding braid indices (B_i).

rivers typically include multiple overlapping transverse bars with linguoid, or tongue-shaped, forms (Figure 10.6). The scaling characteristics of linguoid bars are less well defined than those of rhomboid shapes in gravel braided rivers, but the available evidence indicates that bars in sandy and gravel braided rivers exhibit similar scaling (Sambrook Smith et al., 2005).

Although bar units in the form of multiple-row structures are basic building blocks of braided rivers, the wavelengths of these structures (Figure 10.2) have been difficult to define, in part because bar units often evolve and amalgamate over time (Fujita, 1989; Ashmore, 2009). An alternative length scale to bar wavelength is the distance between a confluence and a diffluence, or location where flow bifurcates, divided by the wetted width of flow in the channel linking the confluence and the diffluence (Figure 10.7). The confluence-diffluence length scale is about four to five times the channel width for both experimental and natural braided rivers (Hundey and Ashmore, 2009). The extent to which this length scale varies with changes in flow stage has yet to be examined in detail.

10.3.2 How Is the Intensity of Braiding Characterized?

A variety of indices have been used to characterize the degree of braiding, or the intricacy of interwoven channels and bars within the braided system (Bridge, 1993). The mode of bar development (m_{br}) refers to the number of rows of bar units that exist at given cross sections within a river channel (Fujita, 1989). A bar mode of one corresponds to the alternate-bar arrangement of bar units within a straight channel (Figure 10.2). Compounding this arrangement laterally across a wide channel (Figure 10.2) leads to higher modes that result in

braiding (Figure 10.8). In bar theory, the mode of bar development corresponds to the extent to which instability to perturbation promotes the development of a certain number of laterally adjacent single-bar units within channels with specific width–depth ratios (Figure 10.1).

The complexity of interconnected channels and intervening bars within a braided river system is referred to generically as braiding intensity (Ashmore, 2009). Indices of braiding intensity are based on one of three characteristics: the dimensions and frequency of channel bars, of which bar mode is a representative example; the number of channels in the network; and the total length of channels along a given length of the river (Egozi and Ashmore, 2008). A commonly used

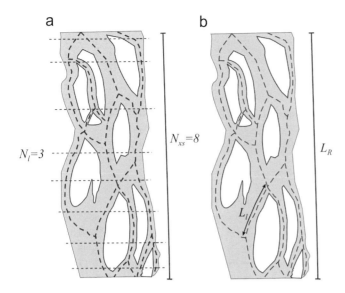

Figure 10.9. Diagram illustrating definitions of braiding index (B_i) (a) and total sinuosity (S_{IT}) (b).

braiding index (B_i) is the average number of channels along several transects oriented perpendicular to the river direction within a reach of a given length (Ashmore, 1991a; Chew and Ashmore, 2001) (Figure 10.9):

$$B_i = \frac{\sum N_l}{N_{xs}} \tag{10.3}$$

where N_l is a channel location crossed by a transect and N_{xs} is the number of transects. For a uniform braided river system of mode m_{br}, the relation between B_i and m_{br} is (Kleinhans and van den Berg, 2011)

$$B_i = \frac{m_{br} - 1}{2} + 1 \tag{10.4}$$

Another useful metric is the total sinuosity (S_{IT}), or total length of individual braid channels per unit length of river:

$$S_{IT} = \frac{\sum L_l}{L_R} \tag{10.5}$$

where L_l is the length of an individual braid channel and L_R is the reach length (Figure 10.9). The braiding index and total sinuosity are closely related to one another (Robertson-Rintoul and Richards, 1993). Moreover, B_i and S_{IT} are sensitive to variations in flow stage, so that no single measurement of these metrics at a specific value of discharge adequately characterizes braiding complexity (Egozi and Ashmore, 2008). Generally, B_i increases to a peak as discharge increases and then declines as high flows progressively inundate bar surfaces (Egozi and Ashmore, 2008). The relation of the braiding index to flow stage has yet to be resolved for natural braided rivers.

Total sinuosity of braiding increases with increasing potential bankfull stream power per unit length of channel ($Q_{bk}S_v$) and with decreasing grain size (d_{84}) of bed material (Robertson-Rintoul and Richards, 1993). The grain-size effect is particularly pronounced in sandy braided rivers. The braiding index also increases with increasing stream power (Ashmore, 1991b). In the downstream direction, B_i increases with increasing discharge (Ashmore, 2009; Bertoldi et al., 2009a; Egozi and Ashmore, 2009). For a constant discharge, the braiding index increases with decreasing grain size (Chew and Ashmore, 2001). Net aggradation increases the braiding index through the addition of bars, whereas degradation decreases the braiding index through a reduction in the number of bars through bar coalescence (Germanoski and Schumm, 1993).

10.3.3 What Are the Characteristics of Hydraulic and Channel Geometry in Braided Rivers?

Characterization of the average morphological attributes of channels within braided rivers is difficult given the unstable and highly changeable nature of these channels. Nevertheless, analysis of at-a-station hydraulic-geometry relations suggests that, despite considerable temporal variability in morphological attributes (Ashmore, 2013), systematic morphological adjustments occur in response to prevailing hydrologic conditions. Because channel banks are erodible or merge gradually with adjacent bars, and new channels may form through incision as discharge increases, exponents in at-a-station hydraulic-geometry relations for total wetted width of braided rivers exceed 0.3 (Fahnestock, 1963; Mosley, 1982) and often are greater than 0.5 (Smith et al., 1996; Bertoldi et al., 2009a). Exponents equal to or slightly exceeding 1.0 have been reported, indicating that increases in discharge in some cases are entirely accommodated by increases in total flow width (Ashmore and Sauks, 2006).

Analysis of channel morphology has focused on the total channel width, defined as the sum of bankfull widths of individual channels at transects across the river, and the average bankfull depth of these channels. It has also explored the widths and depths of individual channels within the braided system. For the same discharge, the total channel width of a braided river is about two to three times greater than the width of a single-thread channel, and average channel depth is two to three times less than the average depth of a single-thread channel (Ashmore, 2013). In the downstream direction, total channel width generally increases more rapidly in relation to channel depth than in single-thread channels (Ashmore, 1991b; Merritt and Wohl, 2003; Bertoldi et al., 2009a; Costigan et al., 2014). Apparently, braided rivers accommodate downstream increases in discharge primarily through adjustments in total channel width – a result consistent with the increase in B_i with increases in discharge. Interestingly, the individual channels of braided rivers and single channels of meandering rivers both exhibit the same dimensionless channel width–discharge relation with an exponent (b) of approximately 0.5, suggesting that channel morphology in braided and meandering rivers is the product of common underlying processes (Ashmore, 2013; Gaurav et al., 2017).

10.4 What Process–Form Interactions Occur at the Bar-Unit and Bar-Element Scales in Braided Rivers?

The fast morphodynamics of braided rivers, whereby the configuration of bars and channels changes during individual hydrologic events, complicates efforts to generalize interactions among flow, sediment transport, and channel form. In many rivers, adjustments of bed morphology and channel planform occur over longer timescales than adjustments of flow and sediment transport to changing inputs of water and sediment. This relatively slow morphodynamics facilitates efforts to link patterns of flow and sediment transport to channel changes. Flow and sediment transport can be determined for a stable morphological configuration, and then changes in flow and sediment transport can be assessed to evaluate the effects of changes on channel form. By contrast, change within braided rivers often occurs over timescales similar to adjustment timescales of flow and sediment transport. In other words, braided rivers are characterized by fundamental instability of the morphological configuration of individual channels and bars over event timescales (Figure 1.9). As a result, the dynamics of flow and sediment transport are inherently dependent on changes in channel form.

The arrangement of individual morphological features changes in response to the continuous or frequent movement of bed material within active parts of the braided river system, resulting in changes in local foci of erosion and deposition. In particular, morphological changes involve bed scour or aggradation, bank erosion or deposition, and the eradication or formation of new channels. Such changes entail the development or transformation of bar units and the confluences, bifurcations, and channel bars associated with these units. New channels may form through dissection of bars, preexisting channels may change location through avulsion, and channels may be eradicated through infilling. Studies of interactions between process and form in braided rivers have examined these interactions both when little morphological change occurs,

a situation typical of field studies, and when morphological change actively occurs, a situation common in laboratory studies.

10.4.1 What Are Confluence-Diffluence Units and How Are These Features Formed and Maintained?

Recognition of bar units as fundamental morphological building blocks of braided rivers has been emphasized in analyses that apply bar theory to the study of these rivers. The term confluence-diffluence unit has been used in field studies to refer to essentially the same morphological feature. A confluence-diffluence unit can be associated with a bar unit in the sense that a bar unit consists of a confluence, or zone of converging flow, at its upstream end and a diffluence, or zone of bifurcating flow, at its downstream end (Figures 10.4 and 10.10). The confluence-diffluence conception highlights the pattern of flow through a typical bar unit. The formation of confluence-diffluence units and change in these units have been recognized as fundamental to the dynamics of braided rivers.

In gravel-bed braided rivers, deposition of coarse material within the center of a channel, often downstream from a zone of scour associated with converging flow, acts as a nucleus for initiating bar development through subsequent trapping of bedload (Ashworth, 1996) (Figure 10.3c). The zone of scour, or confluence, may itself be a source of sediment for downstream deposition (Davoren and Mosley, 1986), but in some cases pulses of sediment may originate farther upstream and

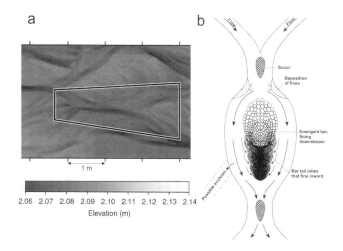

Figure 10.10. (a) High-resolution digital elevation model of a confluence-diffluence unit in an experimental braided river (based on data and image from J.T. Gardner published in Ashmore, 2013). (b) Planform morphology, patterns of flow, bar sedimentation, and relative particle size on the bar in a confluence-diffluence unit (modified from Ashworth, 1996).

simply translate through the confluence (Ferguson et al., 1992). Diverging flow around the growing submerged bar causes bank erosion, which, in turn, decreases flow depths, flow velocities, and the competence of flow over the bar, trapping coarse material on the bar head and depositing fine material on the bar margins and tail. Convergence of flow in the lee of the bar produces scour, initiating a downstream confluence-diffluence unit. Continued bank erosion and vertical growth of the bar eventually lead to emergence of the bar. Bed shear stresses are higher at the head of the exposed bar than at lateral positions adjacent to the bar head (Bridge and Gabel, 1992), promoting further deposition of relatively coarse material on the bar head. Where flow diverges around the bar, secondary currents directed toward the center of the channel may develop near the bed, promoting accumulation of bedload on the bar head (Richardson and Thorne, 1998). Deposition of fine sediment within a zone of weak or recirculating flow in the lee of the exposed bar often leads to downstream extension of the bar tail (Figure 10.10).

In sandy braided rivers, mid-channel bar growth is promoted by divergence of flow downstream from a zone of flow convergence along with pulses of sediment from bank erosion caused by diverging flow (Ashworth et al., 2000). Amalgamation of dunes form a bar nucleus, and subsequent bar growth occurs through dune superimposition and amalgamation on the nucleus. At low flows, the bar widens through lateral accretion of dunes onto the bar margins.

Field measurements within confluence-diffluence units that are not rapidly evolving confirm that velocities, bed shear stress, and rates of bedload transport decrease from the confluence to the head of the downstream bar (Ferguson et al., 1992). Flow accelerates through the zone of scour within the confluence where flow converges, resulting in high velocities, bed shear stresses, and rates of bedload transport (see Chapter 12). As flow diverges downstream of the confluence it decelerates, producing relatively low, laterally uniform rates of bedload transport across the channel. Generally, the size of bed material and the maximum size of bedload decrease downstream from the zone of scour to the bar head, but this spatial pattern can vary depending on the grain-size characteristics of pulses of sediment moving through the confluence and the extent to which the flow exceeds the threshold for sediment entrainment (Ashworth et al., 1992). Particle-tracing experiments in experimental braided channels highlight the importance of bars as depositional sinks for capturing mobile particles, with 81% of recovered tracers positioned on bar heads or bar margins (Kasprak et al., 2015). The average path length of particle movement corresponds closely to the average confluence-diffluence spacing, indicating that bar units exert a strong influence on the distance of particle travel.

10.4.2 What Is the Role of Confluences in the Dynamics of Bar Units?

The development of bars within braided rivers occurs through deposition associated with decreases over space in the competence or capacity of the flow to transport bed-material load. This general condition is characteristic of the four basic mechanisms involved in the initiation of braiding (Figure 10.3), all of which also operate in rivers that already are braided. Deposition in braided rivers is often coincident with scour, particularly scour that occurs at confluences of individual channels within the braidtrain (Ashmore, 1991a). Mechanisms of scour are linked to complex patterns of three-dimensional flow, turbulence, and patterns of bed-material transport at confluences, all of which are treated in detail in Chapter 12. Confluence scour provides a source of sediment for downstream deposition, with the depth and area of scour changing dynamically in relation to changes in discharge (Ashmore and Gardner, 2008). Under constant discharge, flow typically accelerates through confluences, transferring bedload from upstream to downstream.

As local sources and conveyors of sediment, confluences influence patterns of braiding, and confluence kinetics are often involved, either directly or indirectly, in changes in the spatial structure of confluence-diffluence units in braided rivers (Ashmore, 1993, 2013). Confluences in braided rivers are often highly dynamic (Dixon et al., 2018) and, in contrast to most confluences of rivers and streams in drainage networks (see Chapter 12), can change rapidly in planform in a virtually unlimited variety of ways. Some common types of change include lateral migration, downstream translation, expansion, rotation, and obliteration (Figure 10.11). Lateral migration leads to the development of point bars similar to those in meandering rivers, whereas downstream translation results in erosion of bar heads and deposition on bar tails. Moreover, the dynamics of confluences in braided rivers are sensitive to changes in the amount of flow in each tributary anabranch. As flow in one anabranch increases relative to flow in the other, the downstream channel tends to align with the orientation of the dominant anabranch, rotating the confluence and changing the planform symmetry of the tributaries in relation to the direction of the downstream channel (Mosley, 1976) (Figure 10.12). The resulting changes in scour and bed-material transfer are translated through the bar unit, producing changes in the mid-channel bar and bifurcation downstream of the confluence.

10.4.3 What Is the Role of Diffluences, or Bifurcations, in the Dynamics of Bar Units?

The development of channel bifurcations, or the splitting of individual channels into two or more channels, is a fundamental

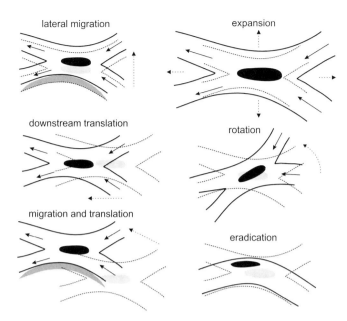

Figure 10.11. Kinetics of confluences in braided rivers. Dashed lines indicate initial position; solid lines indicate position after change. Dashed arrows indicate direction of change (modified from Ashmore, 1993).

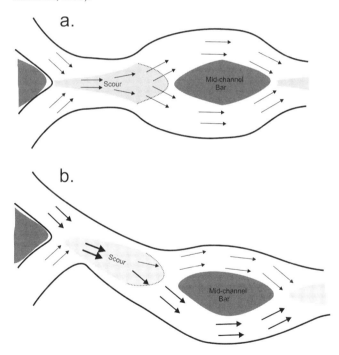

Figure 10.12. Rotation of a confluence with a symmetrical planform (a) in a braided river to align with the dominant upstream tributary and develop an asymmetrical planform (b).

process in braided rivers (Kleinhans et al., 2013). The onset of braiding inherently depends upon bifurcating behavior; without it, multiple channels would not occur. Bifurcations can form through both depositional and erosional processes. Depositional processes involve the formation and growth of bars. The braid initiation mechanisms of transverse bar conversion and central

bar initiation (Figure 10.3) are examples of bifurcation through deposition. Based on empirical evidence from the braided Brahmaputra-Jamuna River in Bangladesh, the division of the downstream velocity field within a single channel into multiple threads of high velocity has been proposed as a prerequisite for the development of depositional bifurcations (Richardson and Thorne, 2001). Multiple high-velocity threads develop in a single channel above a threshold of specific energy for a given width–depth ratio of the flow. According to this hypothesis, bar formation results from lateral nonuniformity of the downstream velocity distribution within a channel, with bars forming in zones of relatively low velocities between the threads of high velocity. Thus, flow division is the driver of depositional bifurcation, rather than bar formation being the driver of flow bifurcation. This hypothesis requires further testing to determine its general validity. Bifurcations can also develop erosionally through avulsions that create new channels branching away from a preexisting channel. The braid initiation mechanisms of chute cutoff and dissection of bars, a component of multiple-bar braiding, represent examples of bifurcation through erosion (Figure 10.3).

10.4.3.1 What Determines Whether Bifurcations Remain Stable?

Once a bifurcation develops, the extent to which it remains stable depends on the partitioning of water and sediment loads within each downstream channel. Insufficient capacity relative to the incoming load will lead to filling of the channels, whereas excessive capacity will lead to net erosion. Stability exists when the transport capacity of each channel matches its load. Determinations of the stability of bifurcations have involved one-dimensional or quasi-two-dimensional analysis of nodal conditions, particularly flow and sediment transport, at the bifurcation (Bolla Pittaluga et al., 2003, 2015; Miori et al., 2006). These simple theoretical analyses predict that symmetrical bifurcations, or those in which flow and sediment transport are divided equally between the two distributary channels, generally are not stable for dimensionless bed shear stresses and channel width–depth ratios normally found in natural braided rivers with erodible banks. A bifurcation in such a state will evolve into a stable, asymmetrical configuration where the two downstream branches convey different water and sediment discharges. Such analysis, although based on simplifying assumptions, suggests that bifurcations have an inherent tendency to develop asymmetrical conditions due mainly to local effects within the vicinity of bifurcations. Stability analysis also indicates that asymmetry of bifurcations is enhanced by nonuniformity of flow in the upstream channel, whereby flow on one side of the channel is greater than flow on the other (Bolla Pittaluga et al., 2003), a common situation in many natural channels.

Experimental and field studies generally confirm that bifurcations develop asymmetrical forms, supporting at

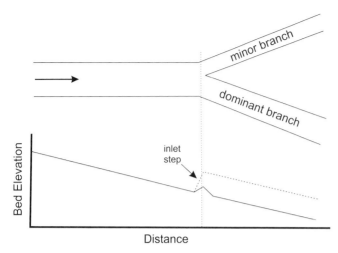

Figure 10.13. Longitudinal profile of the channel bed upstream and downstream of a bifurcation, illustrating the development of step at the inlet to the minor branch (dashed line) of a bifurcating channel. Local aggradation often occurs at the bifurcation, as indicated by the local increase in bed elevation of the main channel (solid line). Vertical dashed line indicates relation between planform and longitudinal profile of the bifurcation.

least qualitatively the findings of nodal analyses (Zolezzi et al., 2006; Bertoldi and Tubino, 2007). A key finding of quasi-two-dimensional nodal analysis is that the lateral flux of sediment at the bifurcation is an important factor determining bifurcation stability (Bolla Pittaluga et al., 2003, 2015). This lateral flux is driven by a difference in bed elevations at the inlets to the two distributary channels. Indeed, experimental and field studies confirm that an inlet step exists at the entrance to the minor branch channel (Bolla Pittaluga et al., 2003; Zolezzi et al., 2006; Bertoldi and Tubino, 2007), which determines the traverse bed slope that directs sediment preferentially toward the inlet to the dominant branch channel (Figure 10.13). Often, both branches have the same slopes downstream of the bifurcation, but at the bifurcation the bed tends to aggrade, possibly due to backwater effects, and the water-surface slope into the dominant branch increases (Zolezzi et al., 2006). The increase in water-surface slope promotes erosion in the main branch. Other experimental work has suggested that some bifurcations in braided systems remain stable, in that divided flow around a bar persists over time, whereas other bifurcations exhibit unstable switching behavior, whereby divided flow transitions to flow entirely on one side of the bar, followed by recurrence of divided flow (Federici and Paola, 2003). Unstable bifurcations develop at low values of dimensionless bed shear stress ($\Theta < 0.15$) and when flow in the incoming channel is not uniformly distributed across the channel.

10.4.3.2 What Is the Role of Flow Three-Dimensionality in Bifurcation Dynamics?

Nodal point analyses treat the hydraulics of bifurcation rather simplistically and do not fully consider the complex, three-dimensional structure of flow at these locations. The pattern of flow at bifurcations involves divergence of depth-averaged streamlines from the upstream channel into the two downstream branches (Figure 10.14). Streamlines of flow must curve to enter the downstream branches, and the senses of curvature within the two branches oppose one another. The opposing patterns of curvature produce a configuration similar to two meander bends placed back to back. Steering of the upstream flow into the downstream channels is governed by the balance between the depth-averaged centrifugal forces of curvature and the pressure-gradient effect associated with superelevation of the water surface at the apex of the bifurcation. Local imbalances between these forces over depth result in helical motion of the flow with secondary circulation directed toward the apex at the surface and away from the apex at the bed, especially immediately downstream from the apex within the two branching channels (Figure 10.14). This pattern of fluid motion preferentially directs near-bed fluid into the branch channels (Marra et al., 2014), which has important implications for the delivery of bedload to these channels.

Numerical modeling of flow at bifurcations confirms that superelevation of the water surface occurs in the vicinity of the bifurcation apex and that secondary currents near the apex are characterized by flow convergence near the surface and flow divergence near the bed (Miori et al., 2012). The

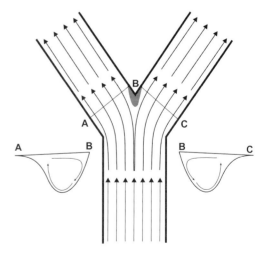

Figure 10.14. Depth-averaged flow streamlines at a bifurcation showing zone of superelevated water surface at bifurcation apex (shaded) and pattern of secondary flow in each branch channel associated with streamline curvature of diverging flow (letters indicate cross-section endpoints).

surface-convergent/bed-divergent pattern of secondary flow has also been observed in experimental studies of flow at bifurcations (Thomas et al., 2011; Marra et al., 2014), but field studies indicate that flow at bifurcations of large braided rivers may be characterized by multiple counterrotating helical cells, with the cells flanking the bifurcation apex exhibiting near-surface divergence and near-bed convergence (Richardson and Thorne, 1998), or by no helical cells (Parsons et al., 2007).

Disparities in the amount of flow moving through the channels downstream of the bifurcation influence the comparative strength of helical motion in each channel, with the strength of secondary circulation increasing at the inlet to the minor branch relative to the strength of this circulation in the dominant branch (Thomas et al., 2011). Enhanced secondary circulation should preferentially steer bed-material load into the minor branch, enhancing the sediment supply to this channel. Such an effect will moderate the tendency for less sediment to enter the minor branch due to the smaller discharge in this channel and the possible presence of a step in bed topography at its inlet. Although attempts have been made to incorporate effects of helical motion into nodal point analysis (van der Mark and Mosselman, 2013), two- or three-dimensional hydrodynamic models are required to adequately account for the effect of this motion on sediment division and morphological change at bifurcations.

10.4.3.3 How Are the Dynamics of Bifurcation Influenced by Bar Migration?

Because bifurcations in braided rivers involve the division of flow paths around bars, the dynamics of bars have a strong influence on the process of bifurcation. Theoretical and experimental results show that increases in the width/depth

ratio and decreases of the dimensionless critical shear stress of flow in the upstream channel lead to the development of large migrating bars that, upon reaching the bifurcation, periodically affect the symmetry of flow and bedload transport in the downstream branch channels (Bertoldi et al., 2009c). Exceptionally large migrating bars may completely close one of the branches.

In natural braided rivers, bifurcations evolve over the timescale of individual flood events, with increases in flood discharge producing increasing symmetry in the width of the two distributary channels and lateral migration of the upstream channel causing lateral movement of the bifurcation (Bertoldi, 2012). Numerical modeling of bar dynamics in braided rivers reveals several different modes by which symmetrical bifurcations evolve into asymmetrical configurations, including 1) deflection of flow by an upstream migrating bar, 2) development of an inlet step at the mouth of one of the branches, 3) reorientation of the upstream channel to align with the direction of one of the downstream branches, and 4) diversion of flow into one of the downstream branches due to a backwater effect produced by the development of a bar in the other branch (Figure 10.15) (Schuurman and Kleinhans, 2015). At the scale of the river planform, the evolution of any particular bifurcation is affected by the dynamics of adjacent bifurcations and bars. In other words, nonlocal factors within the interconnected network of confluences, bars, bifurcations, and channels are important in determining the evolution of particular bar-unit elements. Moreover, the simulations indicate that bifurcations in fully developed braided river systems form mainly by the carving of new channels into bars by cross-bar flow rather than by amalgamation of transverse unit bars into compound exposed bars. This bar-dissection mode of bifurcation is related mainly to headward erosion and avulsion

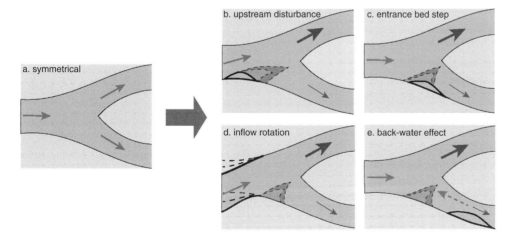

Figure 10.15. Types of evolution of a symmetrical bifurcation (a) to asymmetrical bifurcations in a braided river system: (b) deflection of flow by a bar, (c) restriction of flow and bed-material load by an inlet step, (d) rotation of the upstream channel, and (e) backwater effect from bar formation in one of the downstream channels (from Schuurman and Kleinhans, 2015). (A black and white version of this figure will appear in some formats. For the color version, please refer to the plate section.)

dynamics and is not captured well by depositional models of bifurcation development based on the growth of unit bars into compound bars.

10.4.4 What Is the Role of Avulsions in Bar-Unit Dynamics?

Avulsions within braided rivers are common, particularly as stage increases, creating opportunities for rising water levels to direct flow into new pathways defined by variations in bar morphology. Elongation of bar tails during low flows can result in amalgamation of the tail of an upstream bar onto the head of a downstream bar. As stage increases, this connected area may be progressively inundated and flow directed laterally across the relatively low area of the bar surface representing the former bar tail, leading to the development of a cross-bar channel (Schuurman and Kleinhans, 2015). Avulsions may also be produced by either complete or partial blockage of a channel through local stalling of migrating unit bars. Backwater effects associated with this blockage raise water levels upstream, producing overbank flow across bar surfaces (Leddy et al., 1993). If the path of overbank flow has a gradient advantage, headward erosion and incision will occur, leading to the formation of a new channel that captures the flow. Such avulsions are particularly common at channel bends, where superelevation of the water surface along the outside bank enhances backwater effects from local aggradation (Ferguson, 1993; Leddy et al., 1993). Finally, avulsion can result from lateral migration of individual channels within braided networks into one another, including migration of active channels into inactive remnants of formerly active channels, much like cutoffs in meandering rivers (see Chapter 9).

10.4.5 How Do Bar Unit–Scale Processes Produce Morphological Change?

The interaction among bedload transport, erosion and deposition, and changes in channel morphology at the bar-unit scale in braided rivers is exemplified by the response of the Rees River in New Zealand to a series of flow events (Williams et al., 2015) (Figure 10.16). Bedload moves in narrow bands within the braided channels that link zones of erosion and deposition. This linkage leads to coevolution of zones of erosion and deposition, such as areas of bed scour, bank retreat, bar dissection, channel infill, and central bar formation. The coevolution of erosional and depositional processes in turn drive morphodynamic changes that reconfigure the pattern of braiding at the bar-unit scale, including channel avulsion and bifurcation as well as channel abandonment and reactivation.

10.5 What Are the Dynamics of Braided Rivers at the Planform Scale?

10.5.1 How Are Bedload Fluxes Related to Channel Morphodynamics?

At the scale of entire reaches of braided rivers that encompass many confluence-diffluence units both laterally across the river and longitudinally along the river, bedload transport and channel morphodynamics are strongly interconnected. Frequent changes in channel and bar morphology related to bank erosion, channel incision and avulsion, scour hole and bar migration, bar formation or erosion, and channel infilling produce temporal and spatial variations in sediment supply that are independent of variability in discharge or external inputs of sediment to the river system (Goff and Ashmore, 1994). Even under completely steady flow conditions, sediment fluxes in laboratory models typically exhibit pulse-like variability (Ashmore, 1991b; Hoey and Sutherland, 1991; Young and Davies, 1991; Bertoldi et al., 2006).

Studies of changes in erosion and deposition patterns in the field at the scale of multiple confluence-diffluence units over event timescales imply that sediment fluxes associated with these patterns are highly variable, even under relatively constant hydrological conditions (Brasington et al., 2000, 2003; Lane et al., 2003). Fluctuations in discharge introduce additional complexity to understanding sediment pulses. Because the potential for erosion and deposition within braided rivers is high, local bedload fluxes are strongly related to pathways linking sources and sinks of sediment rather than to patterns of hydraulic forcing, such as pathways of high bed shear stress (Williams et al., 2015).

10.5.1.1 How Is the Active Braiding Index Related to Bed-Material Transport?

At the planform scale, the number of channel pathways through which flow travels typically differs from the number of pathways in which bed-material transport occurs. Thus, at any given time, the active braiding index (B_{ia}), defined on the basis of the number of channels transporting bed-material load, differs from the total braiding index (B_i), defined on the basis of wetted channel perimeter. In other words, at any given time, only part of the wetted width of the braided river system is actively transporting bed material and undergoing morphological change (Ashmore et al., 2011). The ratio B_{ia}/B_i varies between 0.2 and 0.6, with the ratio increasing with increases in stream power (Egozi and Ashmore, 2009; Bertoldi et al., 2009a). The development of a consistent relationship between total and active braiding index suggests that these two indices represent regime attributes of braided rivers (Ashmore, 2013). Moreover, the increase in the proportion of the bed conveying

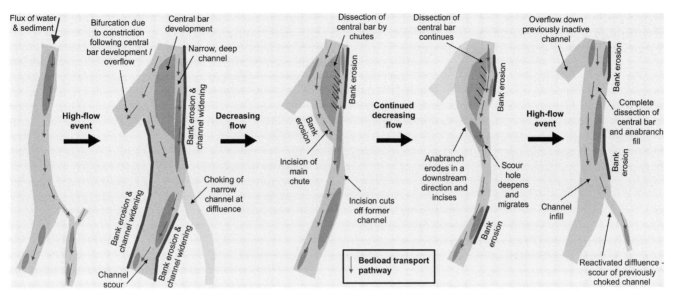

Figure 10.16. Evolution of a braided reach of the Rees River, New Zealand between January and March 2011 (from Williams et al., 2015). (A black and white version of this figure will appear in some formats. For the color version, please refer to the plate section.)

bedload with increasing stream power indicates that the dimensionless planform-scale bedload flux ($q_{sb}{}^*$) of experimental braided streams is strongly related to dimensionless stream power (ω^*) in the form of a power function:

$$q_{sbv}{}^* = 0.412\omega^{*\,2.27} \tag{10.6}$$

where

$$q_{sbv}{}^* = \frac{Q_{sbv}}{W\sqrt{g\left(\frac{\rho_s - \rho}{\rho}\right)d_m{}^3}}, \quad \omega^* = \frac{\rho g Q S}{\rho g W\sqrt{g\left(\frac{\rho_s - \rho}{\rho}\right)d_m{}^3}} \tag{10.7}$$

and Q_{sbv} is the total volumetric bedload transport rate ($m^3\ s^{-1}$) (Bertoldi et al., 2009b). As stream power increases, active width increases, resulting in increases in total bedload flux (Ashmore et al., 2011).

The dependence of active channel width on stream power emphasizes the importance of lateral variability in hydraulic conditions across the entire width of a braided river in determining the magnitude of bedload flux. If cross-sectionally averaged hydraulic conditions are used in bedload-transport functions to predict transport rates, the standard approach to predicting bedload transport in single-thread channels (see Chapter 5), the functions will underpredict bedload fluxes in braided rivers. Instead, local variations in hydraulic conditions within cross sections must be considered to produce accurate estimates of bedload fluxes in braided rivers using these transport functions (Bertoldi et al., 2009b).

Bed-material size could influence the active braiding index and widths of active transport in braided rivers. Sandy braided

rivers may have higher braiding indexes and active widths than gravel-bed braided rivers. Studies of active braiding index have been conducted mainly in the laboratory. Field investigations are needed to evaluate possible influences of differences in bed material on the spatial extent of active transport within braided rivers.

Field studies have shown that spatial variation in the structure of bed material influences bedload transport in gravel-bed braided rivers. The frequency of bedload transport in some cases can be over three times greater in zones of loose, open-framework gravels than where bed material is tightly interlocked and well imbricated (Powell and Ashworth, 1995). Despite these differences in the frequency of transport, total bedload yield may be similar for reaches with different bed-material structures, highlighting the lack of correspondence between the amount of bedload transported and the frequency at which flow is competent to transport bed material.

10.5.2 Do Braided Rivers Exhibit Statistical Uniformity at the Planform Scale?

Despite the complexity of change at the bar-unit and bar-element scales in braided rivers, analysis at the planform scale has focused mainly on the emergence of consistent statistical characteristics that seem to signal the development of morphological and dynamic uniformity at this scale (Ashmore, 2013). Even though individual channels, bars, confluences, and bifurcations change frequently within braided rivers, certain average properties remain fairly uniform or exhibit systematic relationships with controlling factors. This regularity has been interpreted variously as evidence that braided rivers, despite being highly changeable internally, achieve a steady state or statistical equilibrium at the planform scale.

From a morphological perspective, examples of statistical regularity at the planform scale in fully developed braided rivers include relationships between the braiding index and stream power (Ashmore, 1991b) or discharge (Egozi and Ashmore, 2009); between mean active channel width and stream power (Ashmore et al., 2011); between node density, including both confluences and bifurcations, and discharge (Bertoldi et al., 2009a); and between the number of confluences and the number of bifurcations, which are roughly equal (Kleinhans et al., 2013). The spatial scaling of exposed bars and submerged flow paths is another example (Sapozhnikov and Foufoula-Georgiou, 1996; Nykanen et al., 1998; Walsh and Hicks, 2002). This scaling implies that small parts of a braided river are similar to large parts, so that these rivers exhibit a fractal spatial structure.

Dynamic scaling applies to a system if small parts of it change identically to large parts when change over time for the two parts is standardized by the difference in spatial scale between the small and the large parts. During the evolution of an experimental laboratory river from an initial unbraided configuration to a braided form, dynamic scaling is not evident, indicating that mechanisms involved in the development of braiding are not similar at different scales (Sapozhnikov and Foufoula-Georgiou, 1999). As the system evolves, it eventually achieves statistical equilibrium by attaining a fully braided pattern that remains statistically similar over time and a longitudinal profile that has a constant gradient. Once the river system achieves this state of statistical equilibrium, change over time in the arrangement of channels and exposed bars exhibits dynamic scaling (Foufoula-Georgiou and Sapozhnikov, 1998; Sapozhnikov and Foufoula-Georgiou, 1999). Taken together, spatial scaling and dynamic scaling indicate that the morphology and dynamics of small parts and large parts of braided rivers are statistically indistinguishable from one another (Foufoula-Georgiou and Sapozhnikov, 2001). This statistical scale invariance implies universality of the underlying mechanisms responsible for the maintenance of braiding. Moreover, the lack of dynamic scaling during the development of braiding and the occurrence of dynamic scaling once the system is fully braided are characteristic of self-organized critical systems that, despite a large number of degrees of freedom for adjustment and highly nonlinear internal dynamics, self-organize into a critical statistically stable state. Thus, not only do braided rivers display a fractal spatial structure, but they also behave as self-organized critical systems.

10.5.3 How Do the Basic Mechanisms of Braiding Influence Planform Dynamics?

At the planform scale, the mechanisms that initiate braiding also are the fundamental mechanisms of braiding dynamics once a river has become braided. For a braided river in the Scottish Highlands, UK, the four classic mechanisms of braiding identified as important in the development of braiding (Figure 10.3) account for 61% of the total volumetric change in sediment storage over a four-year period (Wheaton et al., 2013). Chute cutoffs, where bank-attached bars become detached from the bank by flow over the bar initiating channel incision, contribute most to total volumetric change, accounting for 23%. Overall, total volumetric change for each of the four years increases exponentially with increasing mean annual discharge (Figure 10.17). During the year with the largest mean discharge, total volume of flow above bankfull stage, and peak instantaneous discharge, transverse bar conversion, i.e., diversion of flow around accumulating transverse unit bars in the middle of a channel, was the most important of the four mechanisms, accounting for 27% of the total volumetric change. The heightened importance of this mechanism is related to large amounts of bank erosion that widened channels, enhancing the development of large migrating bars.

Overall, bank erosion is widely recognized as a major factor contributing to volumetric changes in sediment storage in braided rivers and plays a critical role in supplying bed-material

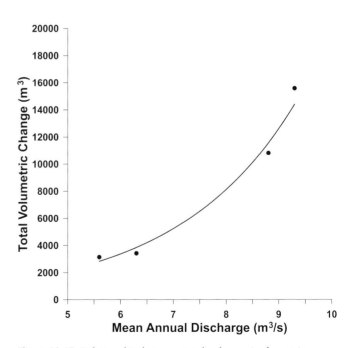

Figure 10.17. Relationship between total volumetric change in sediment storage associated with erosion and deposition versus mean annual discharge for a braided river in the Scottish Highlands, UK (data from Wheaton, 2013).

load to feed braiding mechanisms at the planform scale (Carson and Griffiths, 1989). Bank erosion is particularly active during events that fill individual channels within the braided system, whereas avulsive changes tend to occur during events that submerge these channels and flow across intervening bar surfaces. In the braided Tagliamento River in Italy (Bertoldi et al., 2010), flow pulses that recur only a few times per year concentrate morphological activity in a few active branches of the braided network that change through lateral migration of individual branches. This migration occurs through bank erosion at apexes of bends along these branches and is similar to lateral channel migration in weakly meandering single-thread channels. During flood pulses, or events with recurrence intervals greater than two years, morphological change is characterized by avulsive shifts in channel locations, which also influence confluence and bifurcation dynamics. These flood pulses rework and rearrange the braided network of channels at the planform scale. Increases in the stage of floods produce systematic increases in active-channel width and changes in mean bed elevation.

10.5.4 What Is the Influence of Vegetation on Planform Dynamics?

The development of the planform characteristics of braided rivers is strongly influenced by vegetation. Laboratory studies in which initial braided channels evolve on surfaces subsequently covered by differing densities of alfalfa show that increases in vegetation density decrease braiding intensity, the number of active channels, the width–depth ratio of individual channels, and lateral mobility of channels, confluences, and bifurcations (Gran and Paola, 2001). On the other hand, channel depths and bank stability increase. These general conclusions have subsequently been corroborated by field evidence (Tal et al., 2004; Welber et al., 2012). Other experimental evidence indicates that increasing the density of scattered shrub- or tree-like vegetation within the channels of ephemeral braided rivers enhances braiding intensity by promoting the development of bars in the lee of the plants (Coulthard, 2005). Moreover, bars formed downstream from the plants are stable and do not migrate like bars in unvegetated braided rivers. Accumulation of sediment within and around patches of vegetation can produce a distinctive topographic signature in natural braided rivers (Bertoldi et al., 2011a). These contrasting results suggest that the response of braided rivers to vegetation cover varies depending on plant characteristics, plant spacing, and whether plants cover only the floodplain or invade channels.

Rates of vegetation growth and establishment are an important consideration in determining the response of braided rivers to both human-induced and natural disturbances. These dynamics can be captured broadly by the relation between the time required for vegetation colonization and the time needed to rework the channel bed through bed-material transport and exchange (Paola, 2001). Reductions in flow and flood pulses in a braided river through, for example, the construction of dams or extraction of water upstream reduce bed-material transport while also exposing more bar surfaces for longer periods of time within the braided system. The resulting colonization of exposed surfaces by vegetation diminishes braiding intensity (Hicks et al., 2008). Once vegetation is established on exposed surfaces during periods of low flow, it is often resistant to removal by high flows (Bertoldi et al., 2011b). If vegetation growth becomes sufficiently dense, the braided river may be converted into an anabranching river or single-thread channel (Tal and Paola, 2007, 2010; Horn et al., 2012). In aggradational braided systems, such as those influenced by extreme sediment loading from volcanic eruptions, the establishment of vegetation increases as aggradation rates decline over time (Gran et al., 2015). Vegetation concentrates aggradation in narrow corridors of the braided corridor that are unvegetated, promoting avulsive dynamics of channels and inhibiting lateral migration. Vegetation dynamics and the corresponding effect of vegetation on morphodynamics of braided rivers can also vary spatially depending on reach-scale variations in groundwater availability, local inputs of water from tributaries, and surface-water retention characteristics of bed material, such as infiltration capacity (Bertoldi et al., 2011b). Spatial variations in inundation frequency and moisture conditions not only influence the density of vegetation but can produce differences in types of vegetation cover as well as differences in species within a cover type.

10.5.5 How Have Numerical Models Been Used to Simulate the Planform Dynamics of Braided Rivers?

Numerical modeling serves as a valuable tool for exploring the morphodynamics of braided rivers at the planform scale. This modeling involves the use of both cellular automata models and physics-based models (Williams et al., 2016a). Cellular automata models represent a braided river as a set of two-dimensional grid cells governed by rules that define fluxes of water and sediment between adjacent cells along with corresponding changes in bed elevation within cells. Simulations with this type of model confirm that braiding arises from local scour in zones of flow convergence and deposition in areas of flow divergence. These patterns develop when the flow can change its width freely in the absence of strong lateral constraints and the power-function relation between bed-load-transport rate and discharge or stream power is non-linear with an exponent greater than 1 (Murray and Paola, 1994). Other factors that contribute to braiding include redeposition of eroded sediment within channels and lateral fluxes of sediment (Murray and Paola, 1997). Cellular automata models reproduce many of the features observed in natural braided rivers, including bar formation, confluence scour, channel bifurcation, channel migration, and channel avulsion. These models continue to be refined in an effort to represent the actual mechanisms of water flow and sediment flux in real braided rivers (Thomas and Nicholas, 2002). The effects of vegetation on morphodynamics have also been incorporated (Murray and Paola, 2003; Nicholas et al., 2006a; Thomas et al., 2007; Ziliani et al., 2013). Whereas cellular automata models produce simulated braided rivers that look and behave like real braided rivers, detailed empirical evaluation of simulations indicates that the models best represent planform-scale processes but do not capture braided river dynamics at the bar-unit and bar-element scales (Doeschl-Wilson and Ashmore, 2005; Doeschl et al., 2006). This outcome is not unexpected given that details of flow and sediment-transport mechanisms incorporated into these reduced-complexity models are not sufficient to adequately represent interactions among the evolving bed morphology,

flow mechanics, and bedload flux at the bar-unit and bar-element scales (Lane, 2006).

To capture more adequately the influence of process mechanisms at bar-unit and bar-element scales on planform dynamics, physics-based numerical models have been applied to simulate the morphodynamics of braided rivers. Although three-dimensional numerical models have been used to simulate flow in braided rivers (Nicholas and Sambrook Smith, 1999), most morphodynamic models linking hydraulics to bed-material transport to determine the evolution of bed morphology are two-dimensional models based on depth-averaged shallow flow equations (Williams et al., 2016b), including models that fully couple flow and sediment transport (Siviglia et al., 2013). Such models now produce remarkably complete simulations of the development of braided river morphology from initial conditions, including the shapes, sizes, and dynamics of unit bars, compound bars, bifurcations, and confluences that compare closely to those in natural braided rivers (Figure 10.18). Over simulation periods of a few years, however, both the dynamics and the predicted characteristics of features at the bar-unit scale begin to deviate from natural conditions, indicating that further refinement of the models is necessary (Schuurman et al., 2013).

Physics-based models have been used to explore the influence of sediment heterogeneity on braiding, with results indicating that increases in heterogeneity increase braiding

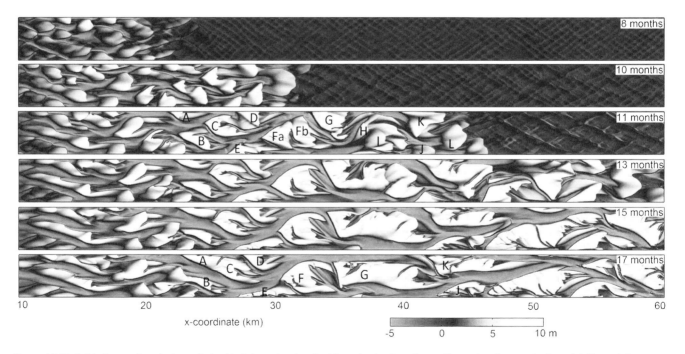

Figure 10.18. Initiation and evolution of a braided river simulated with a physics-based two-dimensional numerical model (from Schuurman and Kleinhans, 2015). Labels refer to bar elements discussed in original source. (A black and white version of this figure will appear in some formats. For the color version, please refer to the plate section.)

intensity and reduce the length and height of bar units (Singh et al., 2017). These models also have been used to simulate the effects of individual flow events on erosion, deposition, and channel morphology (Williams et al., 2016b). The capacity of these models to simulate erosion and deposition matches that which can be determined in the field based on differencing of sequential high-resolution digital elevation models of braided river morphology (Milan et al., 2007; Wheaton et al., 2010). This marriage between high-resolution field information and high-resolution simulations of channel change now provides the basis for rigorous evaluations of the predictive capabilities of physics-based models of braided river morphodynamics.

11 The Dynamics of Anabranching Rivers

11.1 Why Are Anabranching Channels Important?

Anabranching rivers exhibit complex planform configurations in which multiple channels are separated by stable islands (Nanson and Knighton, 1996). Individual channels within the network of interconnected channels may be meandering, straight, or braided. Anabranching, although not as common as meandering and braiding, occurs frequently enough that it is now regarded as a distinct, fundamental type of river planform. Anabranching rivers develop in a wide range of geographic contexts, ranging from subarctic to tropical and humid alpine to arid environments (Nanson, 2013), and the world's largest rivers exhibit anabranching planforms (Latrubesse, 2008). Anabranching rivers have received less scientific attention than meandering or braided rivers and thus are less well understood. The need to manage rivers with anabranching planforms necessitates understanding of the morphodynamics of these rivers and how humans can disrupt or transform these morphodynamics (Best, 2019).

Achievement of general understanding is complicated by recognition that anabranching rivers cannot be readily classified into a single distinct category based on plots of bank discharge versus channel slope, as is the case for meandering and braided rivers. Considered as a whole, anabranching rivers occur over a wide range of slopes for a given discharge (Figure 8.12). Attempts have been made to classify anabranching rivers into several different types based on observed characteristics (Nanson and Knighton, 1996), but some classes include only a few examples from specific geographic contexts. Visual similarity in the planform of many anabranching rivers suggests that these rivers are governed by common underlying

mechanisms of development and maintenance, but looks sometimes can be deceiving. Typologies based on visual similarities may not capture fundamental differences in morphodynamics among rivers, especially differences in underlying processes (Carling et al., 2014). Some basic distinctions among anabranching rivers can be made based on differences in bankfull discharge–slope regimes – for example, the distinction between wandering gravel-bed rivers and anastomosing rivers (Figure 8.12). Other anabranching rivers seem to defy classification, but investigations of individual cases are necessary to determine the appropriateness and scope of process-based categories.

11.2 Why and How Do Rivers Anabranch?

The lack of a clear relationship between slope and discharge, the components of stream power, and anabranching suggests that this pattern does not reflect a systematic response of rivers to changes in energy regime, as appears to be the case with the transition from meandering to braiding. Either some other factor or set of factors governs anabranching or it reflects a similar morphological adjustment to different sets of controlling conditions. Attempts to explain why rivers anabranch have drawn upon optimization principles, whereas explanations of how rivers anabranch have focused on differences in processes that produce multiple channels divided by islands.

11.2.1 How Have Optimization Principles Been Used to Explain Why Rivers Anabranch?

Theoretical analysis based on considerations of maximum flow efficiency (MFE) and the least action principle (LAP) indicates that, by developing anabranching planforms, rivers adjust so that maximum sediment-transport efficiency matches the prevailing supply of sediment without the need for changes in channel slope (Huang and Nanson, 2007). Through such adjustments, anabranching rivers achieve a graded or equilibrium state. Anabranching produces several small channels with relatively small width–depth ratios that enhance hydraulic efficiency and sediment-transport capacity compared with the efficiency and capacity of a single channel with a large width–depth ratio. On the other hand, increasing the number of channels also increases the number of islands, which enhances hydraulic resistance and reduces maximum sediment-transport capacity. Thus, a trade-off exists between the addition of channels and the decrease in channel width–depth ratios resulting from anabranching. For a particular formative discharge, channel slope, total channel width, and sediment supply, equilibrium states exist where sediment transport capacity is maximized and equals the sediment supply (Figure 11.1). Although a single-channel configuration may be able to transport the supplied sediment, this configuration may not maximize sediment transport efficiency, the most stable state according to the optimization principle of MFE. A multichannel state with a maximum transport efficiency that corresponds to the amount of supplied sediment will be the most efficient state for transporting the

supplied sediment; it will be more stable than a single-channel configuration that is not in the optimal state of maximum sediment transport efficiency.

According to MFE-LAP theory, in at least some anabranching rivers, channel morphology, flow, and sediment transport are mutually adjusted to produce stable, equilibrium conditions. Field and laboratory evidence indicates that in anabranching rivers where the total width–depth ratio of multiple anabranches at bankfull discharge is less than the width–depth ratio of a single channel, the anabranches transport sediment more efficiently, or a greater rate per unit of stream power, than the single channel (Jansen and Nanson, 2004). Experimental results also show that a single channel with a smaller width–depth ratio than the total width–depth ratio of several anabranches has a greater bed-material transport efficiency than multiple channels.

Changes in sediment supply can be accommodated by adjustments of both width–depth ratio and the number of channels to change the maximum transport efficiency of the anabranching system to match the new amount of delivered sediment. In overloaded systems, efficiency can be increased by decreasing width–depth ratios or the number of channels to increase maximum transport capacity, whereas in underloaded systems, channels and islands can be added to increase flow resistance and decrease maximum transport capacity. The anabranching adjustment mechanism may be especially important where low valley gradients restrict adjustments of channel gradients that would otherwise alter transport capacity (Nanson and Huang, 2008, 2017).

A shortcoming of MFE-LAP theory is that it does not provide detailed insight into the fluvial processes by which anabranching occurs. Moreover, many anabranching rivers develop where sediment supply overwhelms the capacity for adjustments by the river system to produce a matching transport capacity, resulting in systemwide aggradation. In such river systems, anabranching may actually reduce sediment transport efficiency (Tabata and Hickin, 2003). Although the theory argues that anabranching efficiently redistributes surplus sediment across aggrading floodplains (Jansen and Nanson, 2004; Nanson, 2013), it does not explicitly account for the influence of aggradation on anabranching.

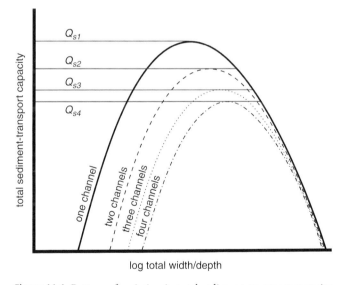

Figure 11.1. Pattern of variation in total sediment-transport capacity (Q_s) with the total width–depth ratio of a river system for different numbers of channels and a given discharge, channel gradient, and grain size of bed material. Maximum efficiencies correspond to peaks of curves. Note that for Q_{s2}, Q_{s3}, and Q_{s4}, maximum efficiency corresponds to river systems with two, three, and four channels, respectively. If the amount of sediment supplied to the river system equals Q_{s4}, only an anabranching system with four channels will transport this supplied sediment at maximum transport efficiency (based on Huang and Nanson, 2007).

11.2.2 What Are the Fundamental Processes of Anabranching?

Two fundamental processes have been recognized as important in anabranching (Nanson, 2013; Carling et al., 2014) (Figure 11.2). First, this pattern can develop through the bottom-up construction of bars by within-channel deposition and the conversion of these bars into stable islands around which flow divides. This process is similar to the development

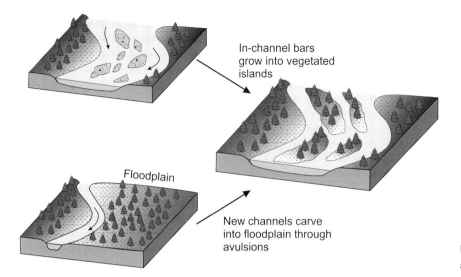

In-channel bars
grow into vegetated
islands

Floodplain

New channels carve
into floodplain through
avulsions

Figure 11.2. Two major processes of anabranching.

of bar units in braided rivers and depends on transport of sediment as bedload. The conversion of bars into islands results from lower mobility of the bed than in braided rivers; from stabilizing factors, such as vegetation growth, that inhibit bar dynamics; or from both of these factors. Numerical modeling suggests that anabranching may also arise from deposition of bars on floodplains, rather than within channels, and that wide sections of floodplains promote this type of deposition, a result consistent with the correspondence between some anabranched reaches and wide sections of floodplains (Moron et al., 2017). For this hypothesis to hold, considerable bedload transport would have to occur on floodplains of anabranching rivers, a process that seems unlikely in all cases, but that may occur in relatively high-energy situations.

The second major process involved in anabranching is top-down: water escaping from existing channels and flowing across the adjacent floodplain incises into the floodplain surface, producing new channels branching from and reconnecting with existing channels (Figure 11.2). This avulsive process carves sections of the floodplain into islands. Pathways of flow over the floodplain surface and the erosional energy of flow along these pathways that drives incision depend on elevation gradients related to spatial variations in channel and floodplain deposition (Tornqvist and Bridge, 2002). In particular, aggradation within a primary channel and on the adjacent floodplain can contribute to avulsive behavior by raising the channel bed above the elevation of distal portions of the floodplain and secondary channels that flow parallel to the primary channel (Brizga and Finlayson, 1990) (Figure 11.3). If the primary channel becomes perched on an aggradational alluvial ridge within the floodplain, a strong cross-valley slope develops between the channel and the floodplain margins. A breach in levees along the primary channel allows overbank

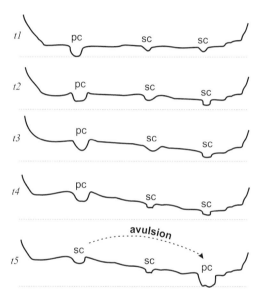

Figure 11.3. Cross sections showing aggradation of a floodplain and primary channel (pc) over time *(t1–t5)* relative to a base level (dashed line) and development of a lateral gradient to the floodplain. Avulsion changes a secondary channel (sc) to a primary channel.

flow to reach distal, low portions of the floodplain (Slingerland and Smith, 1998). Secondary channels form through down-valley extension from scour originating at levee breaches and/or up-valley extension from incision where flow reenters the primary channel (Schumm et al., 1996) (Figure 11.4). Eventually, concentration of floodwaters within a secondary channel on a low part of the floodplain enlarges this channel, transforming it into the new primary channel (Figures 11.3 and 11.4). The former primary channel becomes a secondary channel. The aggradation-avulsion process may then repeat itself, resulting in shifting loci of primary and secondary channels over time.

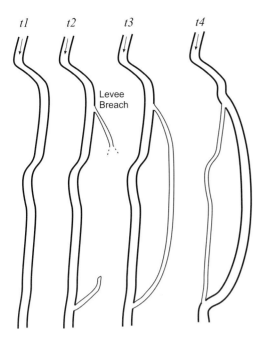

Figure 11.4. Development of an avulsion over time (*t1–t4*). Initial unbranched channel (*t1*) experiences a levee breach that results in erosion upstream and downstream, initiating the formation of a secondary channel (*t2*). Once the secondary channel forms (*t3*) it may capture most of the flow, transforming it into the new primary channel (*t4*).

11.2.2.1 What Factors Influence the Development of Avulsions in Anabranching Rivers?

The relative importance of aggradation in the development of anabranching patterns through avulsion can be expressed as the ratio of the timescale of avulsion (t_a) to the timescale of lateral migration (t_m) (Jerolmack and Mohrig, 2007):

$$\frac{t_a}{t_m} = \frac{D_{bk}/\zeta_a}{W_{bk}/\xi_m} = \frac{D_{bk}\xi_m}{W_{bk}\zeta_a} \tag{11.1}$$

where ζ_a is the vertical rate of aggradation, and ξ_m is the lateral migration rate. Anabranching patterns develop when $t_a/t_m < 1$, or when aggradation rates are large relative to lateral migration rates. Under these conditions, channel change is dominated by avulsions and the development of multiple channels.

The potential for levee breaches and the occurrence of avulsions is enhanced if channel aggradation rates in a primary channel exceed levee aggradation rates, resulting in progressive infilling of this channel, loss of channel conveyance capacity, and an increase in the frequency of overbank flows that can create new secondary channels or enlarge existing secondary channels (Makaske et al., 2009). The fate of secondary channels depends on the potential of a particular channel to capture flow from the primary channel and the capacity of flows within these channels to transport the supplied sediment. Many secondary channels persist until isolated by avulsions of the primary channel. Besides

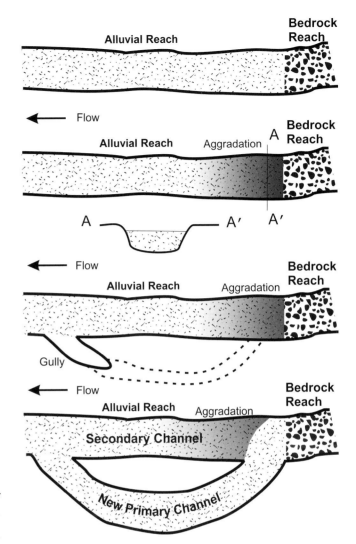

Figure 11.5. Sequence of development (top to bottom) of anabranching through avulsion in small, steep mixed bedrock-alluvial channels. Shading indicates aggradation (modified from Miller, 1991b).

aggradation, logjams and ice jams can reduce channel conveyance and produce backwater effects that divert water onto the floodplain or into incipient secondary channels, leading to avulsions (Jones and Schumm, 1999; Phillips, 2012).

A major factor influencing anabranching in mixed alluvial-bedrock channels is spatial variations in channel slope. Decreases in bed-material transport capacity at slope transitions from steep bedrock reaches to less steep alluvial reaches lead to local aggradation of the channel bed and local reductions in bankfull channel capacity (Miller, 1991b). Frequent overbank flows carve new anabranching channels into the adjacent floodplain through headward erosion of gullies that form where overbank flow reenters the channel downstream (Figure 11.5).

Longitudinal change in channel slopes that produces spatial variability in transport capacity relative to sediment supply is also an important factor governing the dynamics of the ana-branching mixed alluvial-bedrock Orange River in South Africa (Tooth and McCarthy, 2004a). In low-gradient, pre-dominantly alluvial reaches, sediment supply exceeds trans-port capacity, and channels divide around alluvial islands, formed by within-channel accretion of sediment that is stabi-lized by vegetation. A few anabranches also develop through dissection of the floodplain during floods. In high-gradient, predominantly bedrock reaches, transport capacity exceeds sediment supply. Here, multiple channels flow over an irre-gular bedrock surface and divide around rocky islands.

11.2.2.2 What Is the Importance of Bank Resistance and Vegetation in Anabranching?

Strong erosion-resistant banks supported by vegetation or cohesive sediment have been identified as essential for the development of anabranching in alluvial rivers (Nanson, 2013). Strong banks can attain the heights necessary so that individual channels are clearly separated at bankfull condi-tions by vegetated islands (Nanson and Knighton, 1996). Both vegetation and cohesiveness increase bank strength, thereby constraining lateral migration of channels that would erode islands and widening of channels that would reduce transport efficiency. The roots of vegetation can increase the erosional resistance of channel banks by as much as five orders of magnitude (Smith, 1976) – an effect that is vital where the banks of anabranching rivers consist of easily mobilized sand (Tooth and Nanson, 2004).

Vegetation contributes to anabranching by stabilizing islands, which without vegetation cover would be susceptible to erosion, as is the case for unvegetated bars in braided rivers. Although aimed mainly at distinguishing the effect of vegeta-tion on braided rivers, experimental work confirms that increases in vegetation within multichannel river systems confine flow into fewer channels, strengthen banks and bars against erosion, promote progressive deposition on bars that can convert these features into islands, reduce channel migra-tion rates, and limit bed-material exchange between channels and vegetated islands (Tal et al., 2004; Bertoldi et al., 2015). Dense root mats from trees and other plants on vegetated islands inhibit widespread erosion of the floodplain surface, while also focusing any erosive action that may occur on areas between plants (Sear et al., 2010).

In bedload-dominated anabranching systems, large stable wood jams within rivers can be focal points for deposition upstream or downstream, creating alluvial surfaces elevated above the surrounding floodplain (Collins et al., 2012). Protection from erosion and frequent flooding transforms these nuclei of deposition into stable, vegetated islands around which flow branches. The stable islands can resist erosion for hundreds of years.

In low-energy anabranching rivers, the accumulation of channel-spanning woody-debris dams can divert flow laterally into floodplain secondary channels or produce a local back-water effect that leads to overbank flow across the floodplain surface, enhancing the potential for the erosional develop-ment of new anabranches (Harwood and Brown, 1993; Sear et al., 2010). Vegetation encroachment within channels in some cases contributes to the loss of anabranches by increas-ing flow resistance and promoting deposition (Ellery et al., 1995; Marcinkowski et al., 2017). At the scale of entire water-sheds, vegetation intensifies rates of chemical weathering, soil formation, and the production of fine-grained cohesive parti-cles that contribute to the stability of islands and channel banks. Given the importance of vegetation in anabranching, the occurrence of anabranching rivers prior to the evolution of land plants seems unlikely (Nanson, 2013).

11.3 What Are the Dynamics of Wandering Gravel-Bed Rivers?

Wandering gravel-bed rivers occur at values of stream power per unit area (ω) of 30 to 100 W m^{-2} (Burge, 2005) and generally plot between sandy braided and gravel-bed braided rivers (Desloges and Church, 1989) or between gravel-bed meandering and gravel-bed braided rivers (Eaton et al., 2010) on diagrams of bankfull discharge versus channel gra-dient. This positioning of wandering gravel-bed rivers on discharge–slope diagrams is consistent with experimental work showing that as valley slope increases, channels with meandering thalwegs transition to braided patterns over a narrow range of slopes (Schumm and Khan, 1972) (Figure 8.4). For this reason, wandering gravel-bed rivers have been viewed by some not as a distinct planform type but as transi-tional, intermediate forms (Schumm, 1985; Carling et al., 2014). However, persistence of planform indicates adjustment to the prevailing hydrological regime and sediment supply (Desloges and Church, 1989). Although values of ω for wan-dering gravel-bed rivers overlap substantially with those for stream-power ranges for gravel-bed braided rivers (50–300 W m^{-2}) and for gravel-bed meandering rivers (20–80 W m^{-2}), these values differ markedly from those for low-energy anastomosing rivers (2–40 W m^{-2}) (Burge, 2005) (Figure 8.12). Thus, wandering gravel-bed rivers seem to represent a distinct type of anabranching state.

Classic work that led to the designation of wandering gravel-bed rivers focused on rivers draining mountain land-scapes in western Canada (Church, 1983; Desloges and Church, 1989). Examples include the Bella Coola River (Figure 8.13) and the lower Fraser River (Figure 11.6) in British Columbia. Subsequent research has identified rivers with similar planform characteristics and dynamics in

Figure 11.6. The lower Fraser River in British Columbia, Canada – a wandering gravel-bed river.

a variety of geographic settings (Ferguson and Werritty, 1983; Carson, 1986; Payne and Lapointe, 1997; Burge, 2005).

Wandering gravel-bed rivers often consist of unstable reaches, characterized by highly irregular sinuous channels that divide around medial or bank-attached vegetated islands and unvegetated migratory bars (Figure 11.6), interspersed with relatively stable straight or meandering sections of river. Unstable reaches are wider and steeper than intervening stable sections (Figure 11.7). These reaches constitute sedimentation zones that contain more abundant volumes of sediment and have greater mean grain size and frequency of large clasts than stable sections (Desloges and Church, 1989). Instability within sedimentation zones may be produced by tributaries that deliver large amounts of sediment to the river, sometimes in the form of tributary alluvial fans (e.g. Church, 1983), by upstream avulsions that flush large amounts of eroded floodplain sediment downstream (Payne and Lapointe, 1997; Fuller et al., 2003), or by local sources of readily erodible alluvium (Burge and Lapointe, 2005). Although multiple channels exist in unstable reaches, a single dominant channel, or primary channel, usually is evident (Wooldridge and Hickin, 2005). The sinuosity of the primary channel is generally less than that of freely meandering rivers, whereas the existence of multiple channels is less spatially extensive and less intense than in braided rivers. The primary channel migrates laterally and may occasionally change its path suddenly through avulsion. The development of anabranching within unstable reaches occurs both through flow splitting associated with the formation of new bars or islands and through carving of new channels into islands as the result of avulsions (Burge and Lapointe, 2005).

Along forest-lined unstable reaches, bank erosion introduces large woody debris into the primary channel that accumulates on bar heads and contributes to sedimentation and flow diversion. As flow is diverted into secondary channels, it tends to accelerate, producing a relatively wide, shallow cross-

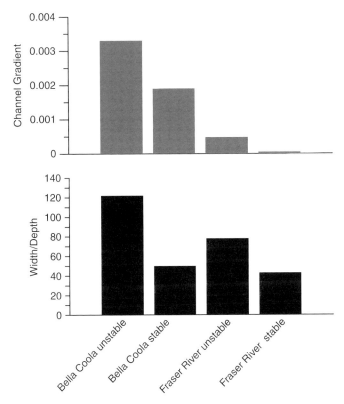

Figure 11.7. Morphological characteristics of unstable and stable reaches along the Bella Coola and Fraser Rivers (data from Desloges and Church, 1989).

sectional form with coarse bed material (Ellis and Church, 2005). Toward the downstream end of the secondary channel, where it reconnects with and is influenced by backwater in the primary channel, channel width–depth ratios increase and the bed is covered by fine material.

Stable reaches consist of single-thread channels that change at much lower rates than the multiple channels, bars, and islands in unstable sections. Sediment mobilized within an upstream unstable section is transferred through a stable reach and deposited downstream in another sedimentation zone. Stable reaches are mildly to moderately sinuous ($S_I \approx 1.2$ to 1.5) with prominent point bars and lateral bars covered by boulders, cobbles, and pebbles. Most bars are incised by chutes or sloughs active at high flows. Relatively high rates of bank erosion at some bends may lead to local overwidening of the channel relative to intervening inflection points between bends and the development of exceptionally wide point bars (Carson, 1986). Chute cutoffs commonly develop across these point bars, producing short-term anabranching (Fuller et al., 2003). Eventually the cutoff bend is abandoned and the sinuosity of the channel is reduced as it adopts a relatively straight course. Through this process, relatively low levels of sinuosity are maintained (Carson, 1986).

Only a few studies have examined in detail sediment transport in wandering gravel-bed rivers. Data collected in wandering gravel-bed reaches of the lower Fraser River indicate that sand constitutes about one-third of the total load and is transported as wash load (McLean et al., 1999). Gravel transported as bedload constitutes only 1% of the total load, but the transfer of gravel along the reach strongly influences channel change and morphology. Bank erosion associated with active channel shifting produces migrating waves of gravel that accrete onto bars. The growth of bars deflects flow into channel banks, causing erosion and further release of gravel into the river. Over the long term, the reach is undergoing net aggradation as bed-material fluxes generally decline downstream, but these fluxes vary substantially throughout the reach (McLean and Church, 1999). Most bed material moves from sites of persistent erosion to distinct sites of deposition, with the depositional sites being the primary locations of channel instability. Erosional and depositional sites shift spatially over decadal timescales in conjunction with changes in the planform of primary and secondary channels within the anabranching system.

Bars are a prominent component of wandering gravel-bed rivers, including discrete unit bars in primary or secondary channels and large compound bars that contain islands. As in braided rivers, unit bars are built through the vertical stacking of stalled migratory bedload sheets one to three grain diameters in thickness (Rice et al., 2009). Accretion of bedload sheets onto the slip face of unit bars contributes to translation of these bars within primary and secondary channels. Compound bars grow by the accretion of unit bars onto the head or margins of these complex forms, or within secondary channels that dissect these forms. The growth rate of compound bars decreases exponentially with increasing age and bar volume (Church and Rice, 2009). The length, width, and thickness of the bars increase as power functions of bar age, with the greatest rate of growth occurring in bar width.

Eventually, the bars attain equilibrium dimensions that scale with channel size, especially channel width. The texture of surface and subsurface sediments of the bars tends to fine from the bar head to the bar tail, but variations in the elevation of the bar surface above the bed of the primary channel are a source of variability in this relationship (Rice and Church, 2010).

Compound bars can persist for a century or more and are destroyed mainly by shifts in the position of the primary channel, which may be triggered by the development of nearby compound bars. The long-term development and persistence of compound bars are consistent with the relatively modest rates of bed-material transfer within these rivers (Church and Rice, 2009). As expected, bed morphology in wandering gravel-bed rivers is a hybrid of meandering and braided forms. The primary channel and compound bars form distinct bar units similar to those in meandering rivers, whereas the dissection of compound bars by secondary channels produces a pattern similar to that of braided rivers (Figure 11.8).

The dynamics of anabranching channels in wandering gravel-bed rivers include the development of new anabranches, the persistence of bifurcations with stable anabranching channels, and the elimination of anabranches (Burge and Lapointe, 2005). An analysis of channel change for the Renous River in New Brunswick, Canada indicates that over a 54-year period the frequencies of channel formation and abandonment are similar, sustaining the wandering pattern. Stable anabranching channels have similar values of stream power per unit area, channel geometries, and lengths, whereas unstable anabranching channels do not. Stability implies long-term persistence of the anabranching pattern, whereas instability signifies that one of the two channels will be abandoned. Abandonment occurs when the delivery of sediment to one of the anabranches exceeds its transport capacity because of a change in sediment supply or discharge. Increasing stability of newly formed bifurcations occurs when the incipient anabranching secondary channel evolves so that its stream power and dimensions become similar to those of the primary channel.

In contrast to results for braided rivers (Federici and Paola, 2003), stable bifurcations in wandering gravel-bed rivers have low values of dimensionless shear stress (Θ) compared with unstable bifurcations with values of Θ for stable bifurcations close to the threshold for particle mobility (Burge, 2006). Low sediment mobility apparently enhances the stability of bifurcations in wandering gravel-bed rivers. At stable bifurcations, accretion of bars onto island heads results in upstream extension of the island over time, and expansion of vegetation onto these bars contributes to bifurcation stability.

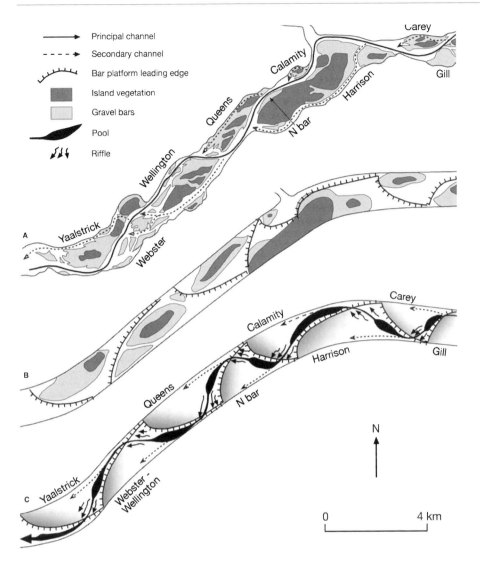

Figure 11.8. Bed morphology in the lower Fraser River. (a) Actual morphology showing compound bar dissection by secondary channels. (b) Representation of actual morphology as a sequence of alternating bank-attached bar units. (c) Idealized representation as bar units (from Rice et al., 2009).

11.4 What Are the Dynamics of Anastomosing Rivers?

Anastomosing rivers consist of multiple interconnected channels that enclose intervening floodbasins or flat, poorly drained floodplain depressions (Makaske, 2001). The beds of active channels typically consist of sand (Smith, 1986), and these channels are flanked by prominent levees that are prone to breaches during floods (Cazanacli and Smith, 1998; Adams et al., 2004) (Figure 11.9). Examples of anastomosing rivers include the upper Columbia River in British Columbia Canada (Smith, 1983; Makaske et al., 2017) (Figure 8.1), the Saskatchewan River in the Cumberland Marshes, East Saskatchewan, Canada (Smith et al., 1998; Morozova and Smith, 1999) (Figure 11.10), the lower Rhine-Meuse system in the Netherlands (Stouthamer and Berendsen, 2001, 2007), and the lower Neches River in Texas, USA (Phillips, 2014).

Anastomosing rivers plot well below the meandering–braided threshold on plots of bankfull discharge versus channel slope (Figure 8.12). These rivers also generally plot below meandering rivers on diagrams of $Q_{bk}^{0.5}S_v$ versus d_{50} (Makaske et al., 2009). Thus, these rivers have distinctly lower stream power than wandering gravel-bed rivers and in most cases have lower stream power than meandering rivers. In contrast to meandering or wandering gravel-bed rivers, anastomosing river channels tend to be relatively straight and laterally stable, exhibiting limited capacity for lateral migration (Makaske et al., 2009). Thickly vegetated banks and cohesive bank materials also limit lateral movement.

The dynamics of anastomosing rivers are dominated by avulsion, whereby overbank flow across the floodplain carves new secondary channels or results in capture of the flow in the primary channel by a secondary channel, which in turn becomes the new primary channel. The frequency of major channel-forming avulsions is on the order of 1.5 to 3 avulsions per 1000 years – rates much less than channel-forming rates of change for braided or wandering gravel-bed rivers (Makaske et al., 2002). Avulsions in some cases involve reoccupation of

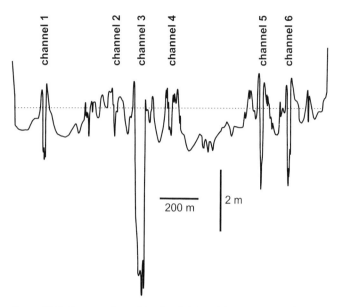

Figure 11.9. Cross section of the floodplain of the anastomosed upper Columbia River showing channels, levees, and intervening floodbasins. Dashed line is a datum of constant elevation. Levees adjacent to channels extend above this line; floodbasins occur between the channels and extend below the line (adapted from Makaske et al., 2002).

channels formerly abandoned by the active channel network (Morozova and Smith, 2000; Phillips, 2014). Not all levee breaches and corresponding overbank flows lead to avulsions; in many cases, the breaches produce crevasse splays that deposit channel sands on top of fine-grained overbank materials (Smith et al., 1989; Abbado et al., 2005). Prograding splay deposits often include incipient channels that are enlarged into well-defined secondary channels by subsequent levee-breaching flows (Makaske, 2001). The development of secondary channels by this process during periods of enhanced flooding and high sediment loads can transform meandering sand-bed rivers into anabranching systems (Slowik, 2018).

Islands form as secondary channels rejoin the primary channel or another secondary channel downstream, producing isolated sections of the floodplain. The development of natural levees along channels in the anastomosing network and the decrease in sedimentation rates away from these channels lead to the development of concave-upward profiles of the islands – a characteristic morphological signature of anastomosing rivers (Makaske, 2001). This morphology contrasts with surface profiles of bars in braided and wandering gravel-bed rivers, which generally have convex-upward shapes. The concave-upward morphology of the islands is effective at producing wetland environments characterized by persistent ponding of overbank flow and vertical accretion of fine organic-rich sediment. The organic content of the basins generally increases away from the channels toward the center of the floodbasins,

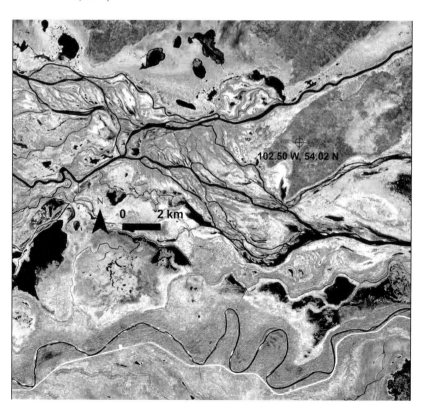

Figure 11.10. Anastomosing section of the lower Saskatchewan River near the Cumberland Marshes, Canada (Landsat image, NASA).

reflecting the diminishing delivery of fine sediment to these locations (Davies-Vollum and Smith, 2008).

Net aggradation of the floodplain and channels is a defining characteristic of anastomosing rivers. This factor drives avulsion by creating spatial gradients in floodplain topography and by reducing channel conveyance capacity. Factors that promote aggradation include rises in sea level (Stouthamer and Berendsen, 2007), increases in local base level from blocking action of cross-valley alluvial fans (Smith and Smith, 1980), basin subsidence (Smith, 1986), or overloading of a river reach by excessive delivery of sediment from upstream (Makaske et al., 2017). Net aggradation indicates that anastomosis is not a graded or equilibrium state in the sense that sediment supply and sediment transport by the river are balanced. Instead, the dynamics of anastomosing rivers involve interaction between channel formation and abandonment to accommodate a net surplus of sediment supply (Makaske, 2001; Abbado et al., 2005; Kleinhans et al., 2012). This process occurs much more slowly than in braided or wandering gravel-bed rivers, with individual channels persisting for hundreds to thousands of years (Makaske et al., 2002). Thus, anastomosis is not a short-lived or temporary condition but persists over geologic timescales.

Channel evolution within anastomosing rivers has been characterized as consisting of several stages (Makaske et al., 2002) (Figure 11.11). New channels form on crevasse splays and deepen through erosion into splay deposits accompanied by deposition on levees along the channels (Stage 1). Newly formed channels widen by undercutting of the bank toe, promoting bank failure, and continue to deepen through erosion of the bed. Trees begin to grow on aggrading levees (Stage 2). Following these developmental stages, channels start to infill through either lateral or vertical accretion, depending on the supply of bedload. Vertical accretion is favored by an abundant supply of bedload. Thick vegetation on levees promotes continued aggradation (Stage 3). Eventually, channels infill and are abandoned, with deposition of suspended sediment and organic material dominating the late stage of infilling (Stage 4). An abandoned channel completely infills and becomes covered by trees when it is close to an active channel that provides a supply of sediment for rapid sedimentation. On the other hand, a well-defined remnant channel without tree cover persists when the abandoned channel is far from an active sediment source and infilling occurs slowly.

11.5 What Are the Characteristics of Other Anabranching Rivers?

A variety of other anabranching river systems have been identified that do not fit neatly into distinct typologies. These systems seem to be adapted to specific conditions in distinct geographic contexts. The extent to which characteristics can be generalized to other fluvial systems beyond these contexts has yet to be ascertained.

11.5.1 What Are the Characteristics of Anabranching River Systems in Australia?

Several different types of anabranching river systems occur in Australia. These systems include rivers with strongly seasonal flow regimes in the wet-dry tropical environments of northern Australia (Wende and Nanson, 1998; Taylor, 1999; Tooth et al., 2008), ephemeral rivers in arid to semiarid central Australia (Tooth and Nanson, 1999, 2004; Fagan and Nanson, 2004), and perennial rivers in temperate southeastern Australia (Schumm et al., 1996; Kemp, 2010; Pietsch and Nanson, 2011). Anabranching rivers in southeastern Australia develop and change mainly through avulsions, but characteristics of these rivers differ from anastomosing rivers. In particular, these rivers have higher levels of stream power than anastomosing systems; as a result, primary channels often exhibit at least some tendency to migrate laterally and are characterized by sinuosities similar to those for meandering rivers (Schumm et al., 1996; Kemp, 2010). Moreover, net aggradation does not appear to be a major factor in avulsion dynamics.

Vegetation exerts a strong influence on the development of a distinctive channel pattern known as ridge-form anabranching in rivers of seasonal tropical northern Australia (Wende and Nanson, 1998; Tooth et al., 2008) and arid central Australia (Tooth and Nanson, 1999, 2000b). In both geographic settings, in-channel vegetation focuses patterns of erosion and deposition into distinct linear patterns, producing multiple subparallel channels separated by elongated vegetated ridges or islands (Figure 11.12). Ridges are initiated by sedimentation in the lee of trees and grow in size so that ridge tops are at the height of the floodplain. The development of this type of anabranching therefore is a bottom-up process involving the growth of bars into islands. Growth of vegetation on the ridges and channel margins stabilizes the morphological structure of these river systems, which transport sand and fine gravel as bedload. The stabilized ridges also contribute to scour in channels by concentrating flow along narrow linear corridors. Individual channels have small width to depth ratios – an adjustment that enhances sediment transport-efficiency (Tooth and Nanson, 2000b; Jansen and Nanson, 2010).

Cooper Creek is an extensive multichannel river system that drains into the Lake Eyre basin in arid central Australia. This system consists of relatively deep, narrow sand-bed anabranching channels interspersed with mud braids on the floodplain surface (Figure 11.13). Originally the braids were believed to be relict features (Rust, 1981), but subsequent work has shown that these features activate when large floods

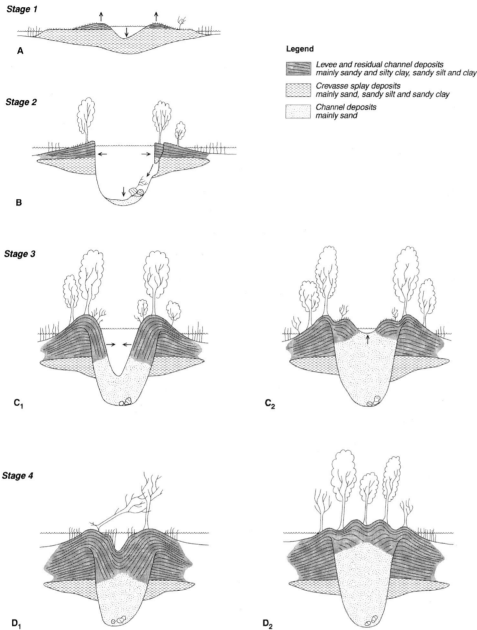

Figure 11.11. Stages of channel evolution in an anastomosing river. C_1 and C_2 correspond to lateral and vertical infilling, respectively. D_1 and D_2 correspond to channels far from and close to an active channel, respectively (from Makaske et al., 2002).

inundate the entire floodplain (Nanson et al., 1986). The braids occupy 44% of the floodplain area (Fagan and Nanson, 2004). Mud is transported as sand-sized aggregates in which silt (25–35%) and sand (10–20%) are bound by kaolinite clay (55–65%) (Maroulis and Nanson, 1996). The clay aggregates are mobilized as bedload on the floodplain surface during large floods and molded into braided forms, including bars and interwoven channels. The anabranching channels, which cut across the mud braids, vary in planform from straight to highly sinuous and have cohesive banks that are highly resistant to erosion. These channels are activated and maintained by moderate flows. The mechanism of ana-branch development has not been thoroughly documented,

but the wide spacing suggests that these channels develop through avulsions.

11.5.2 What Are the Anabranching Characteristics of Rivers in Arid and Semiarid Regions of the Americas?

Anabranching rivers also exist in other arid or semiarid environments. In the semiarid Altiplano Basin of Bolivia, anabranching channels develop mainly through the joining of upstream-migrating headcuts and downstream-extending crevasse channels at levee breaches (Figure 11.4) (Li et al., 2015b). On the other hand, some crevasse splays produced by levee breaches contribute to the infilling of abandoned

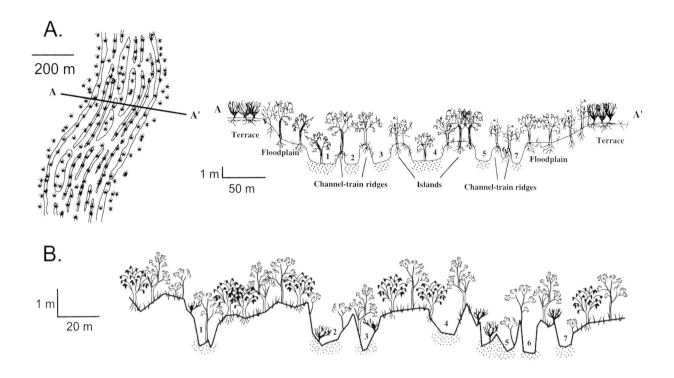

Figure 11.12. Characteristics of ridge-form anabranching rivers. (a) Planform view (left) and typical cross section of rivers in arid central Australia (from Tooth and Nanson, 1999). (b) Cross section of Magela Creek in seasonal tropical northern Australia (from Tooth et al., 2008). Numbers in cross sections correspond to channels.

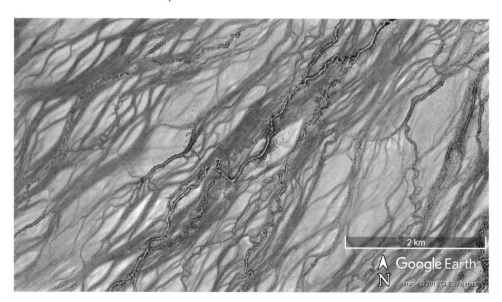

Figure 11.13. Mud braids (dark interwoven channels) and narrow sand-bed anabranching channels of Cooper Creek, Australia.

channels. Along Red Creek in the arid Red Desert of Wyoming, the high silt-clay content of the channel banks inhibits lateral migration, and avulsion is the dominant mechanism of channel dynamics. Net aggradation of the primary channel, primarily by lateral accretion, promotes frequent overbank flows and the formation of anabranches through avulsion (Schumann, 1989).

Anabranching ephemeral rivers are also evident in arid portions of southern Arizona in the Sonoran Desert, where numerous small interconnecting channels separated by portions of floodplain lie within broad, poorly defined flow zones (Rhoads, 1990b, 1991c). In some cases, a distinct primary channel flanked by numerous secondary channels is evident. The primary channel has a sandy bed and is

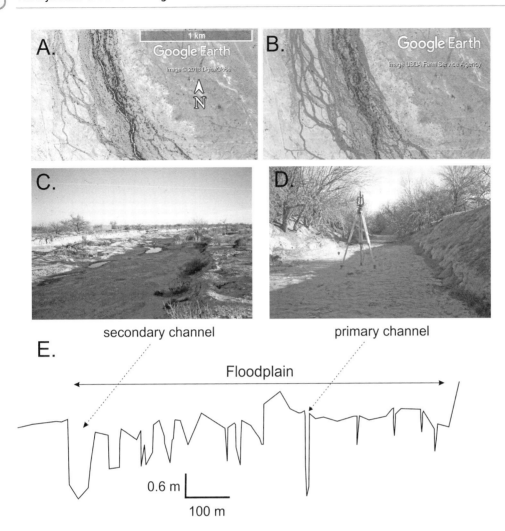

Figure 11.14. Santa Rosa Wash in southern Arizona. (a) Ephemeral anabranching channels within the floodplain. Primary channel is flanked by vegetation. (b) Floodwaters within the anabranching channels, June 2006. (c) Secondary anabranch channel. (d) Primary channel with sandy bed material. (e) Surveyed transect across the floodplain showing primary and secondary channels in photos. Other secondary channels correspond to local topographic lows across the floodplain.

relatively narrow and deep (Figure 11.14). Its well-defined banks are lined by small desert trees and shrubs. Secondary channels have beds consisting of mud, tend to be wide in relation to depth, and have steep, unvegetated banks (Figure 11.14). Historical accounts, hydraulic simulations, and evidence from aerial photography indicate that during floods secondary channels become activated as floodwaters spill into these channels.

11.5.3 How Do Biological and Physical Processes Interact to Produce Avulsive Change in Anabranching Rivers of the Okavango Delta, Africa?

In the Okavango Delta of Botswana, Africa, which, despite its common name, is actually an alluvial fan (Stanistreet and McCarthy, 1993), characteristics of the river systems traversing the fan change in the downstream direction. An anastomosing reach occurs in narrow, upper portions of the fan, known as the Panhandle Region, in relation to neotectonic activity that has produced increased aggradation and avulsion locally (Smith et al., 1997). For the most part, however,

channels in the Panhandle region are highly sinuous ($S_I > 2$), migrate laterally, and have width–depth ratios greater than 10 (Tooth and McCarthy, 2004b). Low-sinuosity ($S_I < 1.75$) anabranching channels develop in the middle part of the fan where channel dynamics are strongly influenced by both flora and fauna, resulting in what can be referred to as an organic river system. These channels have sandy beds and transport mainly bedload, but banks consist entirely of peat formed by the growth, death, and decomposition of lush vegetation in areas of water availability (McCarthy et al., 1991). Thick peat deposits comprising the banks are stabilized by dense growth of papyrus (*Cyperus papyrus*). As a result, the vegetation-confined channels tend to be narrow and deep, with width–depth ratios less than 10 (McCarthy et al., 1991).

The evolution of channels involves several stages (Figure 11.15). New channels formed by avulsion flow through old alluvium, transport sandy bedload, and are relatively wide and shallow. These initial channels often consist of a system of connected flooded pools, or malapos, within swamps. Although such channels lack substantial vegetation, the availability of water promotes abundant vegetation growth. Pioneer

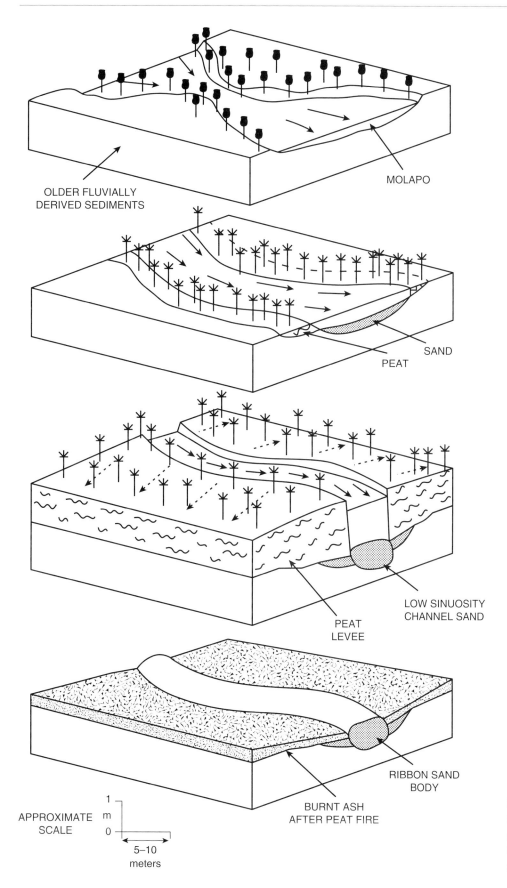

OLDER FLUVIALLY
DERIVED SEDIMENTS

MOLAPO

SAND
PEAT

LOW SINUOSITY
CHANNEL SAND

PEAT
LEVEE

RIBBON SAND
BODY

BURNT ASH
AFTER PEAT FIRE

APPROXIMATE
SCALE

1

m

0

5–10
meters

Figure 11.15. Sequential development of channels within anabranching rivers in the Okavango Delta (from Stanistreet et al., 1993).

species are replaced by papyrus, which, through continued growth and decay, initiates the development of peat deposits on channel margins. Over time, the formation of thick peat levees narrows the channels and confines the flow, promoting channel incision. Velocities of flow diminish downstream as water seeps through the peat levees into adjacent swamps. Blockage of flow by lodged floating rafts of debris, along with encroaching vegetation on the channel banks, contributes locally to reductions in velocity. Diminished bedload-transport capacity over distance downstream promotes progressive aggradation of the channels, which begin to infill with sand. Channel aggradation stimulates aggradation of the peat levees as accumulation of decomposing plant material from bank vegetation keeps pace with in-channel sedimentation. Over time, water levels in the channel may be several meters higher than the surrounding floodplain. Increases in water levels in aggrading channels result in substantial seepage through the levees into the adjacent swamps, further reducing flow in the channel, intensifying plant encroachment, and enhancing aggradation. The peat levees are highly resistant to erosion and breach only at locations of animal tracks, particularly at trampled trails used by hippopotami, who move from the river to backswamp areas daily for nocturnal grazing (McCarthy et al., 1998). Avulsive flows emanating from levee breaches typically carve small distributary channels along these trails that infill with bedload. Moreover, increased flow through the adjacent swamp as the primary channel loses water through seepage erodes preexisting hippopotamus trails within the swamp, producing a secondary channel system that flows adjacent to the primary channel but does not connect directly to it (McCarthy et al., 1992). The final phase of evolution of the primary channel involves complete abandonment and drying out of the adjacent peat levees, which eventually catch fire and are incinerated to a thin layer of ash. Left behind are sand ribbons elevated above the surrounding topography that mark the former aggraded bed of the channel. This final phase results in an inversion of topography. As the primary channel is completely abandoned, a secondary channel becomes the primary channel, and the cycle of evolution is reinitiated.

11.6 How Is Anabranching Related to the Dynamics of the World's Largest Rivers?

Anabranching is characteristic of the largest rivers of the world. Lists of the largest rivers in the world differ somewhat depending on which rivers are considered (Latrubesse, 2008; Ashworth and Lewin, 2012; Lewin and Ashworth, 2014a). In a study of 10 mega rivers with discharges of 17,000 m^3 s^{-1} or greater, only the lower Mississippi River, which has the smallest discharge of the 10 at 17,000 m^3 s^{-1} and is meandering, does not exhibit an anabranching pattern (Latrubesse, 2008). Anabranching mega rivers include the Amazon, Congo, Orinoco, Yangtze, Madeira, Negro, Brahmaputra, Japura, and Parana (Figure 11.16).

Anabranching mega rivers have combinations of channel slope and bankfull discharge that lie below the meandering-braided threshold (Figure 8.3), but otherwise cannot be discriminated from one another (Latrubesse, 2008). Because wandering gravel-bed rivers generally plot close to or above the meandering–braided threshold and most large rivers transport mainly sand rather than gravel, inclusion of these rivers with the wandering style seems inappropriate, although at least one large anabranching river, the Ganges, has been characterized as wandering (Carling et al., 2016). Comparison of anabranching mega rivers with known anastomosing rivers is complicated by the fact that mega rivers are all positioned in a distinct domain on the far right of the slope–discharge plot. Nevertheless, channels of many mega rivers migrate laterally and do not exhibit the stability and aggradational dynamics associated with anastomosing rivers. Separate anabranches of mega rivers are large enough to be considered individual large rivers and can exhibit straight, meandering, braided, anastomosing, or wandering forms (Ashworth and Lewin, 2012; Lewin and Ashworth, 2014a). Presumably the patterns of individual channels reflect general controls that govern planform characteristics in better-studied small rivers and streams, such as sediment load and stream-power relations, but such conditions have yet to be documented extensively.

Mega rivers commonly traverse multiple climatic and geologic settings, leading to greater intrinsic complexity and spatial differentiation of planform characteristics than are found in small rivers. The floodplains of many large rivers typically include a substantial array of inherited features, including terraces, paleochannels, and floodplain depressions and basins (Lewin and Ashworth, 2014b). These features reflect changes in base level and sediment supply produced by climatic, isostatic, or tectonic changes that have affected the river system over geologic timescales (Latrubesse and Franzinelli, 2005; Bettis et al., 2008). Modern human modifications of channel form and connectivity between channels and floodplains are superimposed on this natural legacy of change (Best, 2019). As a result, not all existing morphological and sedimentological components of large rivers may be fully integrated with one another. Instead, these systems exhibit what has been referred to as plurality, or a lack of strong spatial and temporal connectivity, resulting in fragmentation of hydro- and morphodynamics (Lewin and Ashworth, 2014a). From the standpoint of planform dynamics, some anabranches may not be well connected to main pathways of flow during most hydrological events and are effectively geomorphologically dormant. On the other hand, changes in the configuration of active channels through erosion and deposition can alter connectivity, reactivating formerly dormant channels.

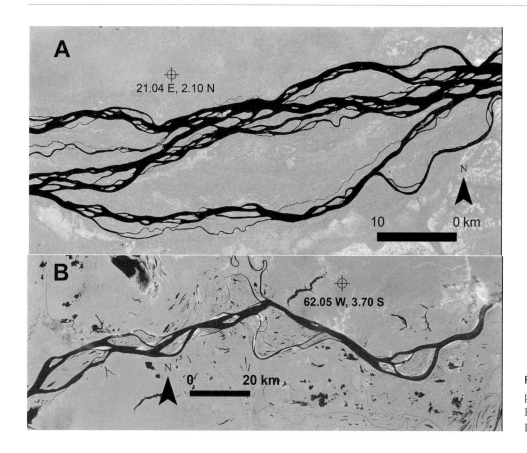

Figure 11.16 Anabranching patterns of large rivers. (a) Congo River; (b) Amazon River (NASA Landsat Imagery).

11.6.1 How Do Large Rivers Anabranch?

The development of anabranching in large rivers involves both within-channel island formation and excision of islands from floodplains through avulsions. Temporary mid-channel bars coexist with stable islands, much as in wandering gravel-bed rivers, so that divided flows with primary and secondary channels are commonplace (Ashworth and Lewin, 2012). As in braided or wandering gravel-bed rivers, local channel widening promotes the development of mid-channel bars, which then become vegetated to form islands (Frias et al., 2015). Because primary channels and large secondary channels often convey enough flow to migrate laterally, mechanisms of chute or neck cutoffs can lead to anabranching, particularly if bed-material transport is not sufficient to produce the deposition necessary to close the old channel path around a bend. Lateral movement can also cause amalgamation of secondary channels as one channel migrates into another (Frias et al., 2015).

Avulsion in large rivers is often related to reoccupation of relict or inactive secondary channels on the floodplain during floods. In contrast to avulsion dynamics in anastomosing rivers, where net aggradation along pathways of flow leads to perched channels and levee breaching, the complex floodplain topography of large rivers triggers avulsions at locations of exceptionally low banks where actively migrating channels intersect swales and paleochannels. Spillage of flow into these features produces erosion and the development of new active channels (Ashworth and Lewin, 2012).

11.6.2 What Factors Influence Anabranching in Large Rivers?

Tectonic effects can influence river pattern development in seismically active environments. In the Ganges River system, differential uplift or subsidence rates between competing anabranches may generate a gradient advantage for one of the channels, thereby producing an avulsion (Gupta et al., 2014). Nevertheless, modeling of the response to tectonic effects indicates that this river system is large enough to maintain high bed-material mobility in both channels following avulsion, thereby preventing substantial amounts of deposition in the secondary channel. Also, strong backwater effects upstream of bifurcations can maintain similar hydraulic conditions at the entry to each channel despite differences in discharge. These factors allow primary and secondary channels to remain open and morphologically active following avulsions, preserving an anabranching pattern.

Active meandering of individual channels can influence the dynamics of large anabranching rivers (Kleinhans et al., 2008). Modeling analysis indicates that over timescales of decades to

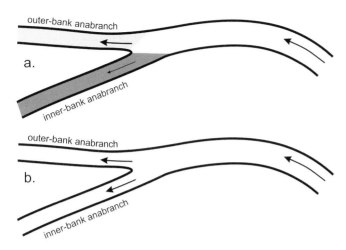

Figure 11.17. Influence of upstream curvature on bifurcation symmetry. (a) Dominance of outer-bank anabranch when anabranch gradients are equal – light shading indicates erosion, dark shading indicates deposition, (b) Symmetrical bifurcation develops when the gradient of the inner-bank anabranch is greater than the gradient of the outer-bank anabranch, which compensates for the curvature advantage of the outer-bank anabranch. Arrows indicate relative magnitudes of water and sediment loads (based on Kleinhans et al., 2008).

centuries, a bend in a river immediately upstream of a bifurcation of two anabranches with equal downstream gradients will favor the anabranch on the outside of the bend (Figure 11.17). The outer-bank anabranch will convey more discharge and sediment load than the inner-bank anabranch, resulting in an asymmetrical bifurcation. The curvature effect can counterbalance other factors that would favor the inner-bank anabranch, such as a gradient advantage, thereby producing a stable symmetrical bifurcation.

Field studies confirm that large anabranching rivers often exhibit, at least around islands that develop from bar growth, the confluence-diffluence structure typically found in braided rivers (Parsons et al., 2007; Szupiany et al., 2009, 2012; Hackney et al., 2018). Important differences between the two types of rivers, however, suggest that straightforward application of scaling relations or process mechanisms for braided rivers to anabranching large rivers is problematic. Although secondary flow typically is not included in basic one-dimensional assessments of bifurcation stability, it has been recognized as an important contributing factor to bifurcation dynamics (van der Mark and Mosselman, 2013). Investigations of flow structure in confluence-diffluence units of mega rivers, which typically have width–depth ratios in excess of 100, show that bed roughness associated with bedforms suppresses the development of helical motion, which often occurs in confluence-diffluence units of small rivers with small width–depth ratios (Parsons et al., 2007; Szupiany et al., 2009, 2012).

Discharge variability plays an important role in bifurcation dynamics in large rivers. Floods can strongly influence bed morphology at bifurcations, producing morphological changes that affect the division of flow in anabranches before and after these hydrological events (Szupiany et al., 2012). Moreover, asymmetry of flow and sediment transport through anabranches and the classification of the bifurcation as stable or unstable varies with changes in discharge (Hackney et al., 2018). Even though many bifurcations can be classified as unstable based on dimensionless stability criteria (see Chapter 10), wholesale changes in bifurcation planform characteristics occur slowly in large rivers, with many bifurcations remaining unchanged over many decades.

Island dynamics include both growth episodes, when active bars or migrating bedforms accrete onto islands, and degenerative episodes, when island margins are eroded, usually during large floods. Overall, however, many islands in large rivers are stable and persist for centuries (Ashworth and Lewin, 2012). This persistence contrasts with the transient nature of vegetated islands in braided rivers, where few islands endure for more than a quarter of a century (Zanoni et al., 2008). The fate of an island also depends on the fate of anabranches that divide around it. Infilling of an anabranch will incorporate an island into a larger island or into the floodplain.

Numerical modeling of the development of anabranching in river systems has not been pursued as extensively as modeling of meandering and braiding. Simulations of bar and island morphodynamics in mega rivers show that enhanced variability in flood magnitude promotes the development of emergent bars, which are converted into stable islands by colonizing vegetation (Nicholas et al., 2013). The predicted timescale of island development is about two to three decades. Islands, in turn, promote the development of new islands. Flow expansion in the lee of existing islands initiates bar development, growth, and vegetation colonization in the sheltered lee positions of islands, producing new islands downstream of existing ones. Through this process, large islands form through coalescence of small bars and islands (Nicholas, 2013). Modeling of the dynamics of large rivers has mainly explored the bottom-up formation of anabranching through bar formation but has not extensively examined the development of islands through avulsive processes that carve new channels into existing portions of the floodplain.

CHAPTER

12 The Dynamics of River Confluences

12.1 Why Are Confluences Important?

The arrangement of rivers in networks in which small streams combine to form large streams, large streams join to form small rivers, and waters from many tributaries unite to produce large rivers highlights the importance of confluences within fluvial systems. The perspective at the network and link scales emphasizes that the topological structure of river networks can be viewed as consisting of nodes, or confluences, and links, or the stream segments between nodes (Figure 1.7). In this sense, confluences are fundamental structural components of river networks. Within the context of these networks, confluences can be viewed as a type of planform element that occurs at nodes rather than within links, as is the case for meandering, braiding, and anabranching. Confluences within drainage networks develop over periods of drainage-basin evolution, which conform to geologic timescales.

Confluences are important components not only of river networks but also of channel networks within braided and anabranching rivers. In particular, confluences are elements of multiple-row bar units that serve as basic building blocks for the bottom-up construction of multichannel rivers (see Chapter 10). Erosion at confluences provides sediment that contributes to deposition associated with bar development and, when colonization by vegetation occurs, to island formation. Confluences at this scale are embedded within confluence-diffluence units that pair these elements with bifurcations. The evolution of confluences within multichannel rivers occurs over periods ranging from the event timescale in braided rivers to geologic timescales in large anabranching rivers. The material presented in this chapter focuses mainly on the dynamics of confluences in river networks but is relevant to all types of confluences.

From an ecological perspective, confluences have been recognized as having an important influence on ecological conditions in river systems. At the network scale, the network dynamics hypothesis maintains that habitat heterogeneity and biological diversity depend on the size and shape of a drainage basin, its drainage density, and the arrangement of streams of different size within the network (Benda et al., 2004a). Tributaries affect the ecology of a main river by producing changes in water temperature, suspended-sediment load, bed material, nutrient concentrations, water chemistry, and organic-matter content (Rice et al., 2008a; Czegledi et al., 2016). Such effects can both enhance (e.g., Kiffney et al., 2006) and degrade (e.g., Blettler et al., 2015) ecosystem quality. At the bar-element scale, local variability in bed material, bed morphology, and hydraulic conditions produces habitat heterogeneity within confluences. These locations often are viewed as biological hotspots in river systems (Benda, 2004a).

Confluences regulate the movement of water and sediment through drainage networks. Interaction between flood waves on a main stem and tributary can produce backwater effects that may extend many kilometers upstream on the tributary (Dyhouse, 1985). Also, confluences have been recognized as potential sites of long-term storage of sediment delivered to a main river from tributaries (Lancaster et al., 2010; Rice, 2017).

12.2 What Are the Planform Characteristics of Confluences?

The planform geometry of confluences is characterized by the angle of the confluence and by the orientation of the two tributaries upstream of the confluence in relation to the

A   SYMMETRICAL CONFLUENCE

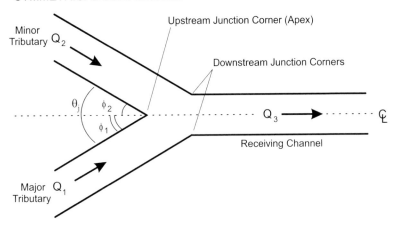

B   ASYMMETRICAL CONFLUENCE

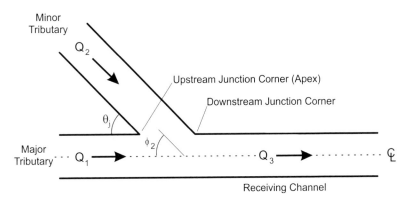

Figure 12.1. (a) Asymmetrical and (b) symmetrical confluence planforms.

downstream, or receiving, channel. The junction angle (θ_j) is the angle between the two upstream tributaries (Figure 12.1). Informally, the symmetry of a confluence planform can be characterized as Y-shaped, or symmetrical, and y-shaped, or asymmetrical. The degree of symmetry is defined more formally by angles of deviation about a centerline projected upstream along the center of the receiving channel (Figure 12.1). The sum of these two angles, ϕ_1 and ϕ_2, equals the junction angle. If angles from the centerline to each incoming channel are equal, i.e., $\phi_1 = \phi_2$, the confluence is symmetrical. If either ϕ_1 or $\phi_2 = \theta_j$, the confluence is asymmetrical. If the angles are unequal and neither equals θ_j, the ratio of the smallest to largest angle, which will have a value between 0 (asymmetrical) and 1 (symmetrical), provides a quantitative measure of the degree of symmetry. Note that symmetry as conventionally defined for confluences refers to planform geometry, whereas symmetry as normally defined for bifurcations (see Chapters 10 and 11) refers to the relative amounts of flow conveyed by the two bifurcating channels.

12.2.1 What Factors Influence Junction Angles at Confluences?

Empirical analysis of confluence planforms has shown that the mean junction angle $(\bar{\theta}_j)$ increases linearly as the difference in stream order between the two upstream tributaries increases (Lubowe, 1964). Thus, $\bar{\theta}_j$ for first-order streams joining other first-order streams is less than $\bar{\theta}_j$ for first-order streams joining fourth-order streams. Moreover, if the order relation is held constant, mean angles decrease with total basin relief. In other words, $\bar{\theta}_j$ for a first-order stream joining a first-order stream is greater in basins with low total relief than in a basin with high total relief.

Based on the assumption that flow should follow paths of steepest descent, the relation for junction angle of a tributary joining a main river is (Horton, 1945)

$$\cos \theta_j = S_3/S_2 \qquad (12.1)$$

where S_3 is the slope of the main river and S_2 is the slope of the tributary. According to Eq. (12.1), steep tributaries form large angles with low-gradient rivers. Evaluation of the relation indicates that it provides reasonable predictions of $\bar{\theta}_j$ using average values of S_3 and S_2 for streams of different order (Lubowe, 1964). A major shortcoming is that it predicts junction angles of zero for a tributary with a slope equal to that of the main river. The model also assumes that the main river does not change its slope at the confluence. To overcome these limitations, the model can be applied separately to each upstream channel (Howard, 1971c):

$$\cos\phi_1 = S_3/S_1 \tag{12.2}$$

$$\cos\phi_2 = S_3/S_2 \tag{12.3}$$

By convention, the relationship $S_3 \leq S_1 \leq S_2$ is assumed to hold. This convention is consistent with designation of the minor tributary, or the tributary with the smallest drainage area upstream of the confluence, by the subscript 2.

Eq. (12.1)–(12.3) represent optimal angular geometry models that minimize the sum of channel lengths weighted by channel slope (Roy, 1983). Various other optimization formulations are possible. It can be shown that, if channel slope and discharge are related by a power function (see Chapter 14), the angles ϕ_1 and ϕ_2 are functions of the mean discharges $\bar{Q}_1$ and $\bar{Q}_2$ of the upstream channels, where $\bar{Q}_2 \leq \bar{Q}_1$ (Roy, 1983). Alternatively, the angles can be expressed in terms of link magnitudes of the upstream channels (Pieri, 1984). In either case, the basic outcome is the same: as the difference in channel slopes, mean discharges, or link magnitudes of the tributary channels increases, the asymmetry of the junction increases as the ratio ϕ_1/ϕ_2 decreases toward zero. Conversely, when the mean discharges, channel slopes, or link magnitudes of the upstream channels are identical, the confluence is symmetrical ($\phi_1/\phi_2 = 1$). In other words, where small tributaries join large tributaries, confluences tend to be asymmetrical, and where two tributaries of the same size join, confluences tend to be symmetrical.

The capacity to extract data on stream-network geometry from digital elevation models now allows assessment of junction angles for large numbers of confluences. Analysis of nearly 1 million mapped river junctions throughout the United States suggests that $\bar{\theta}_j$ varies according to dominant mechanisms of network development in different climatic regimes (Seybold et al., 2017). In arid landscapes, presumably dominated by erosion from surface runoff, the mean junction angle is 45°. On the other hand, in humid environments junction angles average 72°, which corresponds to the theoretical value predicted by groundwater seepage models of network growth. A similar analysis of nearly 15,000 junctions shows that two distinct modes exist – one at 49.5° and the other at about 79° (Hooshyar et al., 2017). The mode of 49.5° occurs within the colluvial domain of the watershed slope-

area curve (see Chapter 2) and is attributed to channel formation by debris flows. The mode of 79° corresponds to the alluvial domain of the slope–area curve and is attributed to channel development by fluvial processes. Unlike other results (Seybold et al., 2017), the mean angles of junctions within the alluvial regime do not decrease with increasing aridity.

12.3 How Does Channel Form Change at Confluences?

Change in channel form at confluences has been addressed within the context of downstream channel geometry and the expected changes in geometry that should occur from upstream to downstream to accommodate an abrupt increase in the amount of flow (Richards, 1980; Roy and Woldenberg, 1986; Roy and Roy, 1988). In general, channel-geometry relations indicate that channel width and depth should increase and channel slope should decrease downstream of the confluence as discharge increases. Moreover, if subscripts 1 and 2 refer to the major and minor tributaries upstream of the confluence, differentiated on the basis of differences in discharge or its surrogate, drainage area, the basic channel-geometry relations (see Chapter 6) predict that:

$$W_{bk3} > W_{bk1} > W_{bk2} \tag{12.4}$$

$$D_{bk3} > D_{bk1} > D_{bk2} \tag{12.5}$$

$$S_3 < S_1 < S_2 \tag{12.6}$$

Empirical analysis of changes in channel form upstream and downstream of confluences reveals that changes in width, depth, and slope are complex and that the expected relations occur at only about 50% of the junctions (Rhoads, 1987b). This variability suggests that changes in channel dimensions at junctions are not controlled by discharge alone. When considering changes in channel dimensions along the main river, defined as the major tributary and receiving channel, the hydrological influence of the minor tributary becomes apparent for a drainage area ratio of the minor tributary to major tributary (A_{d2}/A_{d1}) greater than 0.7 (Figure 12.2). Below this value, width, depth, and slope both increase and decrease from upstream to downstream through the confluence, whereas above this value, width and depth only increase and slope only decreases. This threshold-like behavior suggests that asynchronous timing of flood waves, along with the influence of other factors, such as variations in bed material, bank strength, and vegetation characteristics, inhibit a systematic effect of the minor tributary on channel form when the minor tributary is small relative to the size of the major tributary. On the other hand, when the minor tributary is nearly as large as the major tributary, synchronous timing of flood waves, along

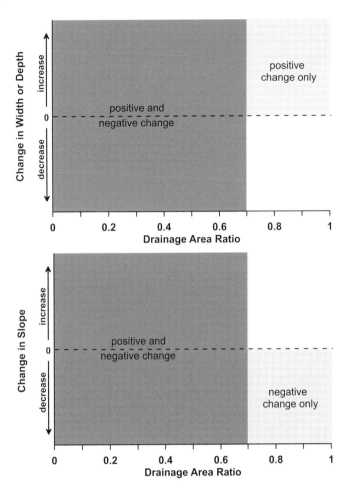

Figure 12.2. Change in channel form from upstream to downstream through confluences. Threshold of systematic change occurs at a drainage-area ratio of about 0.7 (based on Rhoads, 1987b).

with relatively equal discharges from the two tributaries, promotes a systematic change in channel form from upstream to downstream through the junction to accommodate the rather substantial addition of flow from the minor tributary.

Significant geomorphological effects of tributary influxes of sediment on main channel and valley characteristics at confluences in mountainous environments also depend on the drainage-area ratio (Benda et al., 2004b). Such effects can include the development of fans, bars, terraces, or other depositional features at or near a confluence. Generally, the probability of an effect increases as a logistic function of A_{d2}/A_{d1}, indicating that the proportion of geomorphically significant tributaries – those that can produce a confluence effect – increases as the size of the minor tributary increases relative to the size of the major tributary. The potential for a large tributary to join a main river is a function of basin shape, with round, compact basins more likely to result in such confluences than narrow, elongated watersheds. Besides basin shape, tributary gradient must also be considered when

examining the potential for geomorphologically significant tributary effects (Rice, 2017). Some small steep tributaries deliver relatively large amounts of coarse sediment to main rivers, thereby producing substantial aggradation at confluences.

12.4 What Are the Characteristics of Flow at Confluences?

12.4.1 What Is the Confluence Hydrodynamic Zone?

The angled convergence of flows from upstream tributaries within confluences results in some of the most complex hydrodynamic conditions found in natural rivers. Effects of this angled conjoining of flows do not only occur at the immediate location where the two channels meet, but extend upstream and downstream. The portion of the river system affected by merging of flows at a junction is known as the confluence hydrodynamic zone (CHZ) (Kenworthy and Rhoads, 1995). The extent of the CHZ depends on the morphological characteristics of the river channels forming the confluence and the hydrological characteristics of the confluent flows. Because river morphology and hydrology vary over time, the extent of the CHZ also changes dynamically.

12.4.2 How Do Flow Energy and Depth Change through Confluences?

From an engineering perspective, confluences have been recognized as loci of enhanced energy loss within river systems, resulting in changes in water depth as water moves through these locations in fluvial systems. Increased energy loss is related to internal resistance from high levels of turbulence, which can be of the same order of magnitude as energy loss associated with boundary friction (Lin and Soong, 1979). A variety of methods have been developed to predict changes in water depth from upstream to downstream through a confluence along a main channel (Taylor, 1944; Ramamurthy et al., 1988; Gurram et al., 1997; Hsu et al., 1998a; Shabayek et al., 2002; Creelle et al., 2017). Almost all of these methods adopt a control volume approach that uses one-dimensional hydraulic analysis to determine bulk changes in energy or momentum within a large volume representing the CHZ. Details of the flow within the CHZ, which is highly turbulent and three-dimensional, are not explicitly considered. In all cases, the confluences are asymmetrical, with angled downstream junction corners, and channel dimensions do not change from upstream to downstream along the main channel. Channels also have rectangular cross sections. Testing of the models draws upon experimental studies that conform to the assumptions underlying the

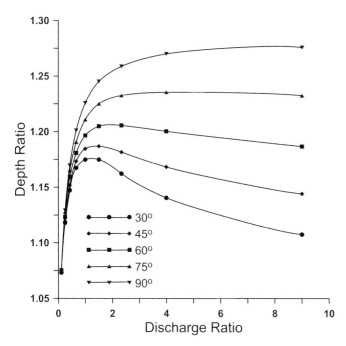

Figure 12.3. Variation in depth ratio (D_1/D_3) with discharge ratio (Q_2/Q_1) for different junction angles at an asymmetrical confluence for a downstream Froude number of 0.6 (adapted from Hsu et al., 1998a).

theoretical analyses (Webber and Greated, 1966; Lin and Soong, 1979; Hsu et al., 1998b; Weber et al., 2001; Lara Pinto Coelho, 2015).

Experimental results and model predictions indicate that the ratio of upstream to downstream depth (D_1/D_3) for a given downstream Froude number reaches a maximum at a discharge ratio (Q_2/Q_1) of about 1 for a junction angle of about 30 degrees (Hsu et al., 1998a; Shabayek et al., 2002) (Figure 12.3). The ratio increases for all values of Q_2/Q_1 as junction angle increases, and the maximum D_1/D_3 shifts progressively toward increasingly large values of Q_2/Q_1 as the angle increases. Given the simplicity of the predictive models, the extent to which they are useful for predicting changes in flow depth at natural confluences is uncertain. Few, if any, of the models have been tested using field data, and numerical simulations indicate that one-dimensional approximations do not fully capture the three-dimensional complexity of confluence hydraulics (Luo et al., 2018).

12.4.3 What Are the Main Hydrodynamic Features within the Confluence Hydrodynamic Zone?

Investigation of flow structure with the CHZ has been conducted mainly by earth scientists seeking to understand confluences as dynamic fluvial features within river systems. Based mainly on results of experimental work, general conceptual models of flow at symmetrical and asymmetrical confluences

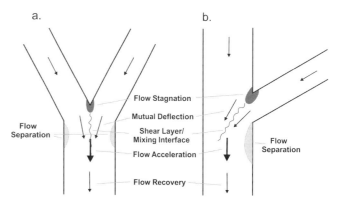

Figure 12.4. Characteristics of flow at symmetrical (a) and asymmetrical (b) stream confluences (adapted from Lewis and Rhoads, 2018).

have been developed (Mosley, 1976; Best, 1987) (Figure 12.4). These models define different types of hydrodynamic elements within the CHZ. The elements include 1) flow stagnation near the upstream junction corner, 2) mutual deflection of the two flows within the confluence, 3) flow separation near the downstream junction corner, 4) a shear layer/mixing interface between the two flows, and 5) flow acceleration within the receiving channel, and 6) flow recovery at the downstream end of the CHZ. Backwater effects can occur upstream of confluences, especially on low-gradient rivers (Dyhouse, 1985).

Stagnation, or at least deceleration of flow, occurs at the upstream junction corner because superelevation of the water surface develops where the combining streams meet (Biron et al., 2002). As flow encounters the adverse pressure gradient related to this increase in water-surface elevation, forward motion is impeded and the flow decelerates. If the adverse pressure gradient is strong, the flow will stall completely, producing a region of recirculating fluid. Often a tongue of reduced velocities separating the high-velocity cores of each incoming flow extends downstream from the stagnation zone, particularly when the two incoming flows have the same momentum flux (Lewis and Rhoads, 2018). Flow accelerates in the downstream direction within this region of reduced velocity (Konsoer and Rhoads, 2014).

Within the confluence the tributary flows mutually deflect one another. This deflection results from pressure gradients produced by the spatial pattern of water-surface elevations that steers the confluent flows into the receiving channel. Mixing is initiated along an interface that may be visible if the streams transport different amounts of suspended material (Figure 12.5). The mixing interface is also the locus of shearing related to differences in velocity between the confluent flows. This feature is also referred to as the shear layer when the two flows have dissimilar velocities.

Separation in the receiving channel near the downstream junction corner develops when the momentum of incoming

Figure 12.5. Mixing interface at the confluence of the Chilcotin and Fraser rivers, British Columbia (from Rice et al., 2008b).

flow from an angled tributary has sufficient inertia that it detaches from the channel wall (Best and Reid, 1984; Qing-Yuan et al., 2009). As with flow stagnation, separation is characterized by recirculating fluid. Separation is common in laboratory junctions with sharply angled junction corners, whereas rounding or flaring of junction corners – a common feature in many natural confluences – inhibits separation (Webber and Greated, 1966; Joy and Townsend, 1981). The characteristics of separation also vary with differences in the shape of the receiving channel (Schindfessel et al., 2017).

The extent to which flow entering the receiving channel accelerates depends on the geometry of the receiving channel and the degree of flow separation. Decreases in total cross-sectional area between the two upstream channels and the receiving channel cause acceleration, an effect augmented by separation, which constricts the portion of the receiving channel through which flow passes. As flow moves downstream, the effects of the confluence diminish, producing recovery as hydraulic conditions become governed by characteristics of the receiving channel rather than by convergence of flows within the confluence.

12.4.3.1 What Factors Influence the Characteristics of CHZ Hydrodynamic Features?

The development, spatial extent, and spatial position of different hydrodynamic features within the CHZ are governed by four primary factors: the momentum flux ratio of the incoming flows, the junction angle, the symmetry of the junction, and the morphology of the channel bed. The momentum flux ratio (M_r) is

$$M_r = \frac{\rho_2 Q_2 U_2}{\rho_1 Q_1 U_1} \tag{12.7}$$

As the momentum flux ratio increases, the region of stagnation shifts progressively around the upstream junction corner into the mouth of the major tributary, the mixing interface is deflected away from the mouth of the minor tributary, and the size of the region of flow separation increases at the downstream junction corner corresponding to the minor tributary, which in turn enhances flow acceleration as constriction of the downstream flow increases (Figure 12.6) (Schindfessel et al., 2015). An increase in junction angle at an asymmetrical confluence will have the same effect as an increase in momentum flux ratio (Best, 1987; Penna et al., 2018). At a symmetrical confluence, increasing the junction angle produces more angled flow from the tributaries, which enhances flow stagnation and the potential for flow separation at downstream junction corners. Decreases in symmetry cause the confluence to operate increasingly like an asymmetrical confluence, where flow separation is likely at only one of the downstream junction corners and increases in momentum flux ratio strongly deflect flow from the major tributary.

The relationship between the bed elevations of the two upstream channels is an important morphological attribute that can influence flow structure at confluences. If the channel beds have the same or nearly the same elevation, the bed

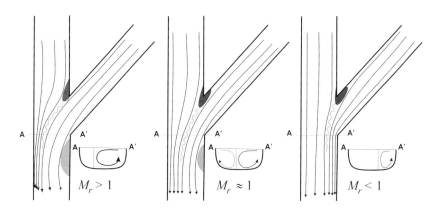

Figure 12.6. Changes in flow structure at an asymmetrical confluence with changes in momentum flux ratio. Solid lines with arrows are depth-average flow streamlines. Dotted lines show the position of the mixing interface. Patterns of secondary flow shown at cross section A–A'.

morphology is characterized as concordant. Discordance occurs when the elevation of the bed of the minor tributary where it joins the confluence and the bed elevation of the main channel, defined by the major tributary and receiving channel, differ (Kennedy, 1984). In discordance, the bed of the minor tributary is perched above the bed of the main river and is connected to the main-river bed by a steeply inclined ramp or step. Before discussing the effects of discordance on flow structure, the hydrodynamics of flow at concordant confluences will first be considered to provide a basis for comparison with discordant confluences.

12.4.4 What Are the Characteristics of Three-Dimensional Flow at Confluences?

The main hydrodynamic elements establish the spatial complexity of flow at confluences, and the influence of controlling factors shows that these elements can change dynamically at a single confluence as the momentum flux ratio varies and differ between confluences as junction angle and junction symmetry vary. However, the complexity of flow at confluences consists of more than just the development of these different hydrodynamic elements. It is also characterized by three-dimensional fluid motion. Two important mechanisms contribute to three-dimensional patterns of fluid motion: 1) the formation of coherent turbulent structures and associated high levels of turbulence within the mixing interface, or shear layer, between the two flows, and 2) the occurrence of large-scale coherent structures, particularly cells of helical motion, within the mean flow.

12.4.4.1 What Coherent Structures Are Generated within the Mixing Interface?

The development of a shear layer/mixing interface between confluent flows has been viewed as analogous to a plane mixing layer that forms when two parallel flows separated by a barrier, or splitter plate, begin to flow past one another downstream of the barrier (Figure 12.7). Embedded within such layers are turbulent vortices rotating around vertical axes extending through the flow depth (Browand, 1986; Rogers and Moser, 1992). The development of these vortices occurs as a result of fundamental instability of shearing across a lateral gradient in streamwise velocity between the two flows (Lam et al., 2016). Initial small vortices grow and pair over distance to form large vortices (Winant and Browand, 1974), with the strength of individual vortices and the growth rate of the mixing layer dependent directly on the lateral velocity gradient (Chu and Babarutsi, 1988; Lee and Kim, 2015). This mechanism of vortex development is known as the Kelvin–Helmholtz (KH) instability, and the resulting vortices, all of which tend to rotate in the same direction, are known as KH vortices. Similar quasi-two-dimensional vortices develop in shallow mixing layers (Uijttewaal and Tukker, 1998), such as those that occur in rivers, but bed friction reduces the velocity difference between the two flows and diminishes the coherence of the vortices, thereby inhibiting the growth of the mixing-layer width over distance (Chu and Babarutsi, 1988; Uijttewaal and Booij, 2000; Sukhodolov et al., 2010). A mixing layer is considered shallow when the largest turbulent vortices greatly exceed the flow depth but are smaller than the flow width.

The mixing-layer model of KH vortex generation within confluences has been challenged to some extent by field data showing a connection between turbulence in the stagnation zone and mixing interface (Rhoads and Sukhodolov, 2008) and by the results of numerical modeling capable of simulating vortex dynamics (Constantinescu et al., 2011b). When a well-developed region of flow stagnation exists near the upstream junction corner, this region acts like a wake behind a solid body in a shallow open-channel flow (Chen and Jirka, 1995; Herrero et al., 2016). Interaction between shear layers on the two sides of the wake leads to the shedding of successive vortices with opposite senses of rotation into the mixing interface as a von Karman vortex street (Figure 12.7).

Numerical simulations indicate that both KH and wake-like vortices can be generated at confluences (Constantinescu et al., 2014). The development of KH vortices occurs at extreme values of M_r, or, more specifically, velocity ratio ($U_R = U_2/U_1$), when

A.

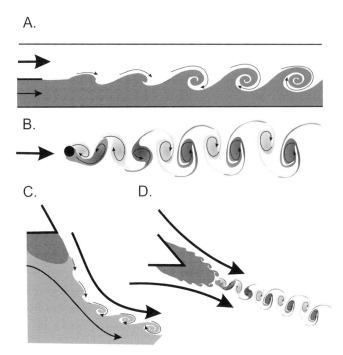

B.

C.   D.

Figure 12.7. (a) Planview of the development of KH vortices in a plane mixing layer downstream of a splitter plate of two parallel flows (b) Planview of the development of a vortex street of counterrotating vortices downstream of a cylinder (black circle). (c) Planview of KH mode near the upstream junction corner of an asymmetrical confluence at high M_r.(d) Planview of wake mode near the upstream junction corner for $M_r \approx 1$.

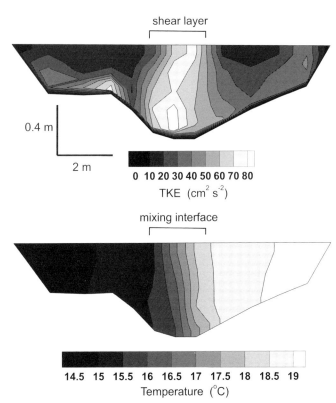

Figure 12.8. Coincidence of the shear layer, defined on the basis of elevated TKE, and of the mixing interface, defined on the basis of a strong lateral gradient in water temperature, at a cross section of the flow within a stream confluence (data from Rhoads and Sukhodolov, 2001 and Sukhodolov and Rhoads, 2001).

lateral shear dominates and the region of stagnation is located in the mouth of the tributary with low momentum. Under these conditions, the mixing interface is in KH mode (Figure 12.7). When $M_r \approx 1$, stagnation tends to be centered on the upstream junction corner, and shear along the margins of this region of stagnant fluid is strong. As a result, wake-like vortices are produced and the mixing interface is in wake mode (Constantinescu et al., 2011b) (Figure 12.7). At natural confluences where incoming flows converge at an angle, conditions can transition from wake mode to KH mode and even into jet-like conditions over distance from the upstream junction corner as streamwise momentum is transferred laterally into the mixing interface by the converging mean flows (Rhoads and Sukhodolov, 2008). Moreover, large-scale intrusions of one flow into the other that exceed the size of individual coherent structures by an order of magnitude sometimes occur along the mixing interface (Rhoads and Sukhodolov, 2004). The origin of these intrusions has yet to be ascertained but may reflect periodicity in hydrological interactions between the incoming flows.

The shear layer, a hydrodynamic feature, is characterized by a narrow, vertically aligned band of elevated turbulent kinetic energy (TKE) (De Serres et al., 1999; Sukhodolov and Rhoads, 2001). Similarly, the mixing interface is delineated by a narrow,

vertically aligned gradient in a property of the flowing water (e.g., temperature) or a transported constituent (e.g., fine sediment) (Rhoads and Sukhodolov, 2001). Typically, the shear layer and mixing interface coincide when the two flows first meet within the confluence (Figure 12.8). However, elevated turbulence within the shear layer dissipates substantially within the CHZ even though mixing may be limited. Persistence of the mixing interface well downstream of locations where elevated levels of turbulence are no longer detectable indicates that the shear layer and mixing interface are not identical but have different distinguishing attributes and can have different spatial extents (Sukhodolov and Rhoads, 2001).

12.4.4.2 How Does Large-Scale Helical Motion Develop at Confluences?

Large-scale helical motion is another common hydrodynamic feature at confluences. The occurrence of helical motion at concordant confluences, where spatial variations in bed morphology do not strongly influence flow, has been attributed to interaction between centrifugal and pressure-gradient forces, similar to the generation of helical motion in meander bends (Ashmore and Parker, 1983; Rhoads and Kenworthy, 1995;

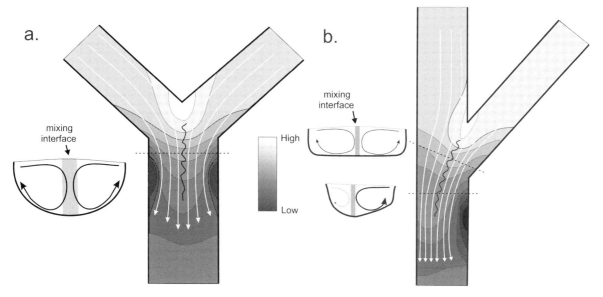

Figure 12.9. Patterns of depth-averaged streamlines (white curved arrows), relative water-surface elevations (shading), and secondary flow (cross sections) at symmetrical (a) and asymmetrical (b) confluences. Dashed lines indicate locations of cross sections for depiction of secondary flow.

Sukhodolov and Sukhodolova, 2019). Angled incoming tributary flows curve within the CHZ to become aligned with the direction of the receiving channel. Centrifugal forces associated with this curvature are counteracted by pressure-gradient forces related to spatial variations in water-surface topography. Because the forces are balanced only in relation to the depth-averaged velocity, near the surface the centrifugal force exceeds the pressure-gradient force, whereas near the bed the pressure-gradient force exceeds the centrifugal force. As a result, the orientations of flow streamlines differ over depth, leading to helical motion.

At symmetrical confluences, both tributaries are angled in relation to the receiving channel, and mutual deflection of the two flows within the CHZ results in a pattern of depth-averaged streamline curvature resembling two meander bends placed back to back (Mosley, 1976; Ashmore et al., 1992). The water-surface topography varies spatially to generate a pressure-gradient field required to counteract depth-averaged centrifugal forces associated with the pattern of streamline curvature. This topography is characterized by superelevation of the water surface near the upstream junction corner, a ridge of superelevated water extending downstream into the receiving channel, and regions of low elevation near the downstream junction corners (Figure 12.9). Similarly to a meander bend, the water surface in each bend is superelevated near the outer "bank," only in this case the "bank" consists of fluid from the other tributary rather than a solid boundary. Twin surface-convergent helical cells form within the CHZ in response to streamline curvature, with the mixing interface forming the boundary between the two cells. The

relative sizes of the cells will vary depending on M_r and the position of the mixing interface within the receiving channel (Baranya et al., 2015).

At asymmetrical confluences, only one of the tributaries is angled in relation to the receiving channel, but mutual deflection of the two incoming flows can still produce opposing patterns of streamline curvature where the two flows meet at the upstream end of the CHZ (Rhoads, 1996). However, the channel bank opposite the mouth of the lateral tributary, if resistant to erosion, constrains the orientation of the two flows and forces both to have the same sense of curvature upon entering the receiving channel. As a result, depth-averaged streamlines of flow from the minor (lateral) tributary curve uniformly into the receiving channel, whereas those from the major tributary reverse their sense of curvature from upstream to downstream within the CHZ (Rhoads, 1996). Superelevation of the water surface again occurs near the upstream junction corner and extends along the angled mixing interface toward the right bank of the receiving channel (Bradbrook et al., 2000; Weber et al., 2001; Huang et al., 2002; Yang et al., 2013) (Figure 12.9). A region of low water-surface elevation forms at the downstream junction corner, where flow may separate from the channel boundary. Twin surface-convergent helical cells begin to develop in the upstream part of the CHZ, where patterns of streamline curvature from the two incoming flows oppose one another (Rhoads and Kenworthy, 1998; Rhoads and Johnson, 2018). Numerical modeling suggests that paired corotating helical cells may develop on each side of the mixing interface (Constantinescu et al., 2011b, 2012), but these paired cells have yet to be confirmed on the basis of field evidence. In the receiving channel, the cell within flow from the minor tributary strengthens, whereas the cell within flow from the major tributary weakens as the sense of curvature of streamlines in this flow reverses (Figure 12.9). Here, helical motion typically consists of either a single large cell within flow from the minor tributary for $M_r > 1$ or a dominant cell on the minor tributary side of the mixing interface and a weak counterrotating cell on the major tributary side of the interface for $M_r < 1$ (Rhoads and Kenworthy, 1995; Rhoads, 1996; Shakibainia et al., 2010) (Figure 12.9). The lateral extent of the cell associated with the minor tributary across the width of the receiving channel will depend on the M_r of the incoming flows, with the lateral extent increasing with increasing M_r (Figure 12.8).

Some field studies have found that helical motion either does not occur at concordant confluences of large rivers (Parsons et al., 2007) or is restricted in spatial extent to portions of the flow near the mixing interface (Szupiany et al., 2009). The lack of helical motion may reflect the influence of dunes and other bedforms, which disrupt near-bed flow patterns typically associated with pressure-gradient effects (Parsons et al., 2008b). The development of helical motion only in the vicinity of the mixing interface also may be related to nonlinear lateral patterns of water-surface superelevation in large rivers, which commonly have large width–depth ratios ($W/D > 100$) (Rhoads, 2006b). Water-surface elevations may only be markedly superelevated in the vicinity of the mixing interface. Nevertheless, helical motion has been observed in some river confluences (Rhoads and Johnson, 2018), indicating that the development of this motion is not governed by scale alone. Moreover, at least some large river confluences exhibit the major hydrodynamic elements associated with small confluences (Gualtieri et al., 2018).

12.4.4.3 Why Is Frame of Reference Important for Documenting Secondary Flow at Confluences?

Documentation of helical motion at confluences based on patterns of secondary flow is sensitive to the frame of reference used to represent this flow (Rhoads and Kenworthy, 1998, 1999). Several different frames of reference (FOR) have been proposed to analyze secondary flow in rivers, including the cross-stream FOR (Dietrich and Smith, 1983), the resultant discharge vector FOR (RQ_{FOR}) (Markham and Thorne, 1992), and the depth-averaged velocity vector FOR (DV_{FOR}) (Rozovskii, 1957). The first two frames of reference involve establishing cross sections of uniform, straight-line orientations across the river. The cross-stream frame of reference is oriented perpendicular to the local path of the channel centerline. Rigorous application of this frame of reference to evaluate force-balance relations also requires that the cross-

stream discharge (Q_n), estimated from field measurements of cross-stream velocities, satisfies mass-balance considerations for a two-dimensional depth-averaged flow (Dietrich and Smith, 1983). The RQ_{FOR} involves determining from field measurements of downstream and cross-stream velocities the orientation of a straight-line cross section that aligns the cross section perpendicular to the resultant discharge vector.

Within confluences, orientations of depth-averaged two-dimensional velocity vectors often vary laterally across the channel, so that the alignment of any straight-line cross section will be at an angle to at least some of these vectors (Rhoads and Kenworthy, 1999). Consider converging flows in a symmetrical confluence with opposing patterns of helical motion of the two incoming flows generated by streamline curvature (Figure 12.10a). Measurements of cross-stream (v) and vertical velocities (w) along a cross section aligned perpendicular to the direction of the receiving channel will show a pattern of v-w vectors on one side of the channel oriented opposite to vectors on the other side of the channel (Figure 12.10b). This pattern arises because the paths of both flows are skewed in relation to the cross-section orientation. The cross-stream FOR will capture flow convergence but will mask any helical motion embedded in the converging flows. The same problem applies to the RQ_{FOR}.

The DV_{FOR} determines components of each local two-dimensional velocity vector ($\mathbf{u_{xy}}$) aligned parallel (u_p) and perpendicular (v_s) to the depth-averaged two-dimensional velocity vector ($\mathbf{U_{xy}}$). Vertical velocities (w) remain unaltered in this frame of reference. Values of u_p and v_s are computed as

$$u_p = \mathbf{u_{xy}} \cos(\theta_\mathbf{u} - \phi_\mathbf{U}) \tag{12.8}$$

$$v_s = \mathbf{u_{xy}} \sin(\theta_\mathbf{u} - \phi_\mathbf{U}) \tag{12.9}$$

where

$$\mathbf{u_{xy}} = \sqrt{u^2 + v^2} \tag{12.10}$$

$$\theta_\mathbf{u} = \tan^{-1}\left(\frac{v}{u}\right) \tag{12.11}$$

$$\phi_\mathbf{U} = \tan^{-1}\left[\frac{\frac{1}{h}\left(\int_0^h v\,dz\right)}{\frac{1}{h}\left(\int_0^h u\,dz\right)}\right] = \tan^{-1}\frac{\bar{v}}{\bar{u}} \tag{12.12}$$

In the example of converging flow at confluences, values of u_p and v_s represent components of the flow perpendicular to the local orientation of depth-averaged streamlines (Figure 12.11). The pattern of v_s-w vectors clearly defines a distinct pattern of secondary circulation, thereby revealing opposing patterns of helical motion within the two incoming flows (Figure 12.10c), each of which exhibits a distinct core of primary velocity (Figure 12.10d).

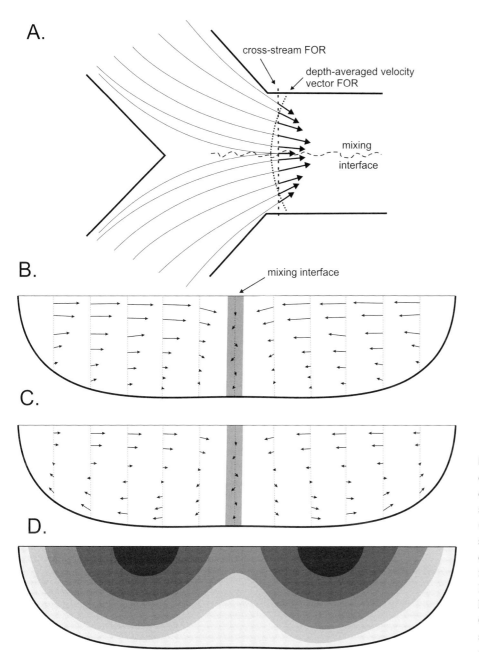

Figure 12.10. (a) Cross-section FOR and depth-averaged velocity vector FOR for converging depth-averaged flow streamlines at a symmetrical confluence. (b) Pattern of secondary flow in cross-section FOR (v-w vectors) showing converging flows. (c) Pattern of secondary flow in depth-averaged velocity vector FOR (v_s-w vectors) showing embedded helical motion in converging flows. (d) Contour pattern of u_p with darkness of shading corresponding to velocity magnitude.

12.4.5 How Does Bed Discordance Influence Confluence Hydrodynamics?

Although Playfair's Law (see Chapter 1) implies that tributaries join main rivers at the same elevation, hydraulic considerations indicate that the elevation constraint applies to the surfaces of confluent flows and not necessarily to the channel beds. Field investigation in mountain environments has shown that many confluences are discordant, with the beds of tributaries perched above the beds of main rivers and a step or steep slope connecting the two beds (Kennedy, 1984) (Figure 12.12). The degree of discordance has been characterized in terms of A_{ex}/A_T, where A_{ex} is the excess area of the deepest tributary channel below the bed level of the shallowest tributary channel and A_T is the total area of the deepest channel (Figure 12.12) (Kennedy, 1984). The value of $A_{ex}/A_T \geq 10$ has been arbitrarily designated as indicating discordance. This metric is at best approximate because it does not directly represent the slope of the channel bed of the minor tributary as it joins the confluence, which is the major factor affecting hydrodynamics. It also assumes that the channel banks, which serve as a reference level for cross-section definitions, are equal in elevation upstream of the confluence on each tributary.

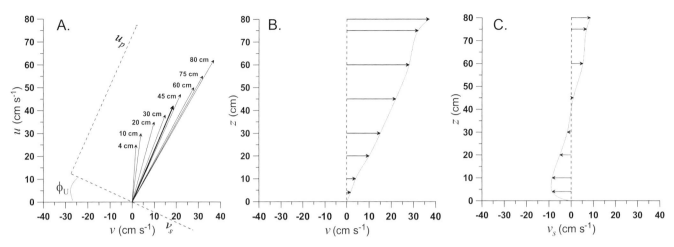

Figure 12.11. (a) Changes in orientation of two-dimensional (u-v) velocity vectors with distance above the bed (values in cm) for a strongly skewed flow where the v axis corresponds to the cross-stream (y) direction and the u axis corresponds to the streamwise (x) direction. Depth-averaged vector ($\mathbf{u_{xy}}$) is shown as unlabeled dark arrow near 45 cm depth. Axes corresponding to u_p and v_s are shown as dashed lines. (b) Vertical profile of v corresponding to vectors in (a). (c) Vertical profile of v_s corresponding to vectors in (a).

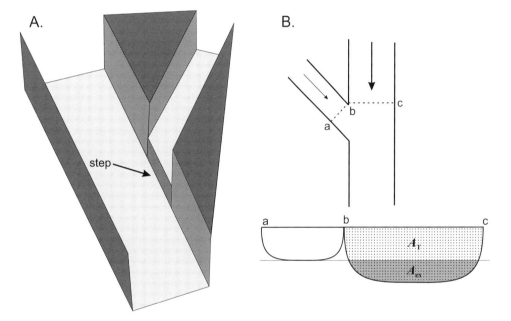

Figure 12.12. (a) Illustration of a discordant confluence. (b) Definition of discordance showing excess area of the deepest tributary (A_{ex} – shaded) total area of the deepest tributary (A_T – dotted) (based on Kennedy, 1984).

Discordance of the tributary channel bed has a strong influence on flow structure within the CHZ. Experimental and numerical studies show that if the step is vertical or nearly so, separation of flow from the bed occurs in the lee of this feature. Fluid motion associated with this flow separation displaces the lower part of the shear layer/mixing interface toward the mouth of the lateral tributary, leading to upwelling of fluid within the region of lateral flow separation near the downstream junction corner (Best and Roy, 1991; Biron et al., 1996a,b; Bradbrook et al., 2001). In extreme cases, upwelling can prevent lateral flow separation near the water surface (Dordevic, 2013). Although strong secondary circulation develops locally near the bed in the lee

of the step and contributes to upwelling downstream (McLelland et al., 1996; Bradbrook et al., 2001), the strongly three-dimensional flow induced by the step appears to disrupt the development of coherent helical cells related to curvature of flow streamlines. The height of the step also influences mutual deflection of the flows; deflection of the near-bed main-channel flow by the tributary flow diminishes with increasing height of the step (Dordevic, 2013).

The experimental and numerical studies of flow over vertical or high-angle steps represent idealized conditions. Tributaries with alluvial beds cannot maintain a vertical face, and the degree of discordance is unlikely to exceed the

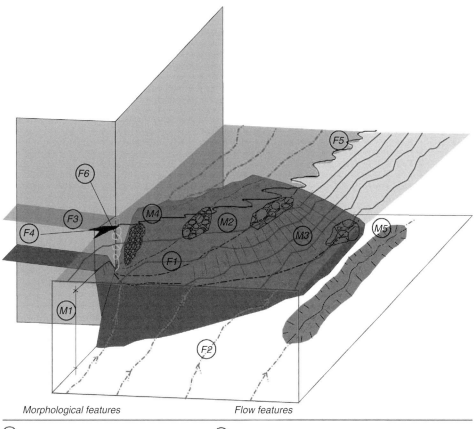

Morphological features   Flow features

(M1) bed discordance

(M2) sediment deposition bar

(M3) faces of the sediment deposition bar
     and corridors of coarse sediment transport

(M4) corridors of fine sediment transport

(M5) erosion zone

(F1) near-surface flow originating from the main channel

(F2) near-bed flow originating from the main channel

(F3) flow originating from the tributary

(F4) flow stagnation zone

(F5) shear layer zone

(F6) spiral vortices at the downstream corner of the tributary
     that lift fine material in suspension

Figure 12.13. Flow and sediment dynamics at an experimental discordant confluence of a steep tributary and a main channel (from Leite Ribeiro et al., 2012a). (A black and white version of this figure will appear in some formats. For the color version, please refer to the plate section.)

submerged static angle of repose of mixed sand and gravel (25° to 40°) (Kleinhans, 2004). Nevertheless, shear-layer distortion, strong upwelling, and the lack of curvature-related helical motion have been reported for a natural confluence with moderate discordance (De Serres et al., 1999).

In experiments with a mobile bed, discordant confluences readily form when a steep tributary with supercritical flow joins a relatively low-gradient main channel with subcritical flow (Leite Ribeiro et al., 2012a). The abrupt change in flow depth produces a bed gradient at the tributary mouth on the order of 25°. As the tributary flow enters the main-channel flow, it penetrates into the main-channel flow only near the surface, and main-channel flow moves beneath the penetrating tributary flow (Figure 12.13). As a result, much of the shear layer between the flows is aligned horizontally rather than vertically (Leite Ribeiro et al., 2012b).

Jet flow can occur at discordant confluences when the velocity ratio between the minor and major tributaries (U_2

$/U_1$) exceeds 2 (Sukhodolov et al., 2017). This type of flow is characterized by a V-shaped wedge of flow from the minor tributary penetrating into the confluence. The wedge is bounded by shear layers and becomes disconnected from the bed, giving rise to a two-layer vertical structure of the flow similar to that observed experimentally.

12.5 What Are the Dynamics of Bed Morphology at Confluences?

The structure of flow at confluences produces distinctive patterns of erosion and deposition that, in turn, shape the morphology of the channel bed at confluences. Mutual adjustment between flow and form results in characteristic patterns of bed-material transport, but changes in flow conditions can trigger net erosion and deposition that reconfigure the bed morphology. The spatial structure of bed morphology at confluences

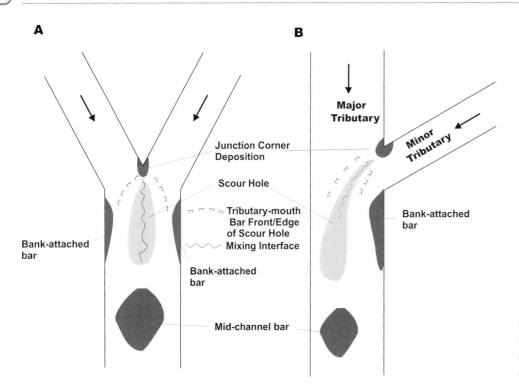

Figure 12.14. Basic elements of bed morphology at (a) a symmetrical confluence and (b) an asymmetrical confluence.

responds dynamically to changes in controlling factors (junction angle, junction symmetry, momentum flux ratio, and discordance/concordance) that influence the hydraulic characteristics and spatial extent of the basic hydrodynamic elements.

Major features of bed morphology include a scour hole within the confluence that extends into the receiving channel, tributary-mouth bars that may prograde into the confluence, bank-attached lateral bars in the receiving channel downstream of the mouths of angled tributaries, sediment accumulation near the upstream junction corner, and a mid-channel bar in the receiving channel toward the downstream end of the CHZ (Best and Rhoads, 2008) (Figure 12.14). Not all these features develop at every confluence. The presence of specific features depends on local hydrodynamic conditions and the extent to which these conditions produce spatial variation in bed-material transport.

12.5.1 How Does Scour of the Bed Occur at Confluences?

Scour holes are a prominent feature of bed morphology at many confluences. Several factors contribute to scour at confluences, including convective acceleration of flow, high levels of turbulence along the shear layer, and divergent patterns of near-bed flow related to helical motion. As a transport-effective flow accelerates from upstream to downstream through a confluence, bed shear stress and sediment transport capacity increase over distance, resulting in excavation of bed material. High levels of near-bed turbulence also enhance bed shear stresses, although levels of turbulence generally decline over distance within and downstream of confluences. At symmetrical confluences, the development of helical motion, particularly dual surface-convergent cells flanking the mixing interface (Figure 12.9), can position the high-velocity core below the water surface (Rhoads, 1996), thereby enhancing near-bed shear stresses (Constantinescu et al., 2012). These cells produce near-bed patterns of fluid motion that sweep bed material away from the center of the channel toward the banks. At asymmetrical confluences, dual cells within the confluence (Figure 12.9) operate similarly to those at symmetrical confluences. The development of a single prominent helical cell within the receiving channel (Figure 12.9) promotes scour along the bank across from the tributary mouth and transports bed material toward the opposite bank, much like flow in a meander bend (Yuan et al., 2017) (Figure 12.14).

Although depth of scour is ultimately a product of the spatial variation in transport capacity, it can be related generally to the controlling factors of confluence dynamics. Experimental studies indicate that scour increases as the junction angle increases and as the discharge of the minor tributary increases relative to the discharge of the major tributary (Mosley, 1976; Best, 1988; Liu et al., 2012) – a finding supported by numerical modeling of confluence scour (Ahadiyan et al., 2018). The increase in scour with increasing junction angle is nonlinear, with little or no scour occurring from 0° to 15°, a steep increase in the rate of scour between 15° and 80°,

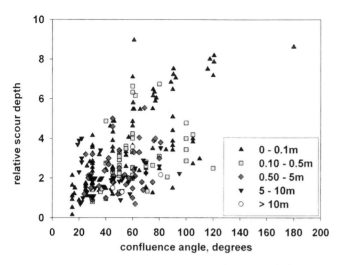

Figure 12.15. Relative scour depth versus confluence angle for confluences of different scour depths (from Best and Rhoads, 2008).

and leveling off of the rate of increase at about 80° (Mosley, 1976; Best, 1988). Scour also decreases with increasing total bed-material load (Mosley et al., 1976) and decreases with increases in width ratio, $W_2/W_{1,3}$ (Nazari-Giglou et al., 2016), where $W_{1,3}$ is the width of the main channel (major tributary and receiving channel).

A summary of field and laboratory data shows that depth of scour is positively related to junction angle, but that the degree of scatter is large (Figure 12.15). Also, large rivers generally have smaller depths of scour for a given junction angle than small streams and experimental channels (Best and Rhoads, 2008). Nevertheless, scour depths at junctions in large rivers can exceed 70 meters (Ianniruberto et al., 2018).

Only a few field studies have examined scour at discordant confluences, and the results of such studies are mixed: some note a general absence of scour (Biron et al., 1993; Boyer et al., 2006) and others report notable scour (Sukhodolov et al., 2017). Diminished scour may be attributable to the movement of the tributary flow over the main-channel flow, which reduces interaction of the two flows near the bed – a key factor in turbulence generation. Mutual deflection of flows that occurs only in the upper part of the water column will also prevent near-bed movement of tributary flow toward the channel banks by helical motion. Mobile-bed experimental results show that limited scour occurs at discordant confluences for modest flow rates and no sediment supply in the main channel (Leite Ribeiro et al., 2012a) (Figure 12.13). This lack of scour reflects armoring of the bed within the main channel under these conditions (Guillen-Ludena et al., 2016).

Armoring has been observed to limit depth of scour at confluences of natural streams transporting coarse bed material (Roy et al., 1988). Experiments with flow rates capable of

transporting sediment supplied to the main channel produce scour at discordant confluences, particularly along the bank of the main channel opposite the tributary mouth (Guillen-Ludena et al., 2015, 2016). This pattern of scour has been attributed to flow acceleration resulting from constriction of the cross-sectional area by the formation of a bank-attached bar near the downstream junction corner.

12.5.2 Does Deposition Occur in Association with the Flow Stagnation?

Flow stagnation characterized by recirculating or relatively low-velocity flow may lead to sediment accumulation near the upstream junction corner. This type of fluid motion promotes deposition of suspended sediment transferred into the stagnation zone by coherent turbulent structures along its margins. Typically, sedimentation of fines does not produce large bars, but it may create mud drapes or the formation of upstream-migrating ripples (Best and Rhoads, 2008). Bed material in this part of confluences is typically finer than the mean size of material throughout the confluence (Biron et al., 1993).

12.5.3 What Are Tributary-Mouth Bars?

Tributary-mouth bars consist of accumulations of bed material at the mouth of one, or both, of the tributaries. The fronts of these bars often merge with the side slopes of the scour hole and may form depositional wedges that migrate into the scour hole, partially filling it (Figure 12.16). The bar front in some cases can be quite steep, approaching the angle of repose of the bed material, in which case these fronts are referred to as avalanche faces (Best, 1988). Such conditions often occur at discordant confluences where a steep tributary delivers abundant coarse material at a high angle to a relatively fine-grained main channel, forming a steep bar front at the mouth of the tributary (Leite Ribeiro et al., 2012a; Guillen-Ludena et al., 2015). Migration of these wedges of coarse sediment into the confluence can infill initial scour in this region, restricting the formation of a scour hole under equilibrium conditions to a location along the bank opposite the tributary mouth. Under concordant conditions, the slope of the bar front is at an angle much less than the angle of repose, and the front merges gradually with the bed of the main channel or the flanks of the scour hole (Rhoads and Kenworthy, 1995).

12.5.4 How Do Bank-Attached Tributary Bars Form in the Receiving Channel?

The development of bank-attached tributary bars below the mouth of angled tributaries at the downstream junction

Figure 12.16. (a) Tributary-mouth bar from the Copper Slough (top) penetrating into the confluence of the Kaskaskia River (left) and Copper Slough. Note the abrupt change in slope delineating the bar front and that the bank-attached bar in the receiving channel of the Kaskaskia River (right) represents the exposed surface of a continuous wedge of sediment that also forms the tributary-mouth bar. This wedge of sediment wraps about the downstream junction corner and extends downstream along the left bank of the Kaskaskia River (photo by author, June 24, 1994). (b) Kaskaskia River–Copper Slough confluence looking upstream along Kaskaskia River. Prominent bank-attached bar is visible on the right along the left bank of the Kaskaskia River. Flow from the Copper Slough enters the confluence on the right immediately upstream of the bank-attached bar and traverses the tributary-mouth bar, which is part of a continuous wedge of sediment that includes the bank-attached bar (photo by author August 27, 1994).

corner has been attributed in mobile-bed experimental studies to deposition of sediment within the region of flow separation that typically forms at this location (Best, 1988). Although flow separation has been documented at natural confluences and can contribute to bar development (Rhoads and Kenworthy, 1995; Ianniruberto et al., 2018), it often develops at low flow in response to strong lateral deflection of the flow by an existing bar platform or by irregularities in the channel bank lines (Rhoads and Kenworthy, 1995; Lewis and Rhoads, 2018). On the other hand, bar-formative events often correspond to relatively high-stage, transport-effective flows that are not influenced strongly by bed or bank morphology. The rounded downstream corner of many natural confluences limits or prevents the development of large-scale flow separation during high-stage flows. Bar formation under these conditions reflects both the influence of helical motion within the tributary flow entering the confluence, which sweeps near-bed sediment toward the adjacent channel bank, as well as streamwise decreases in bed-material transport capacity around the junction corner, which promote progressive deposition of transported bed material over distance. These depositional processes result in relatively coarse-grained bars dominated by material transported as bedload (Petts and Thoms, 1987; Rhoads and Kenworthy, 1995; Mosher and Martini, 2002) (Figure 12.16). Tracer experiments indicate that gravel-sized particles placed on the surface of bank-attached bars are mobilized during high-flow events (Petts and Thoms, 1987). Moreover, measurements of bed-load transport show that

gravel-sized material is transported on the bar surface (Rhoads, 1996), leaving behind an armor layer of coarse material on the bar head (Rhoads and Kenworthy, 1995). The downstream fining of both transported bed material and sampled bed material over the surface of some bank-attached bars suggests that transport competence decreases downstream over these features (Rhoads, 1996; Rhoads et al., 2009).

The occurrence of bed discordance and its effects on the hydrodynamics of flow near the downstream junction corner also limit lateral flow separation. The development of large bank-attached bars at sandy discordant confluences appears to involve deposition of transported bed material, possibly in the form of migrating ripples (Biron et al., 1993). Material on the surface of these sandy bank-attached bars fines in the downstream direction. In laboratory experiments, bank-attached bars at discordant confluences with a steep lateral tributary consist of coarse bed material from the tributary (Guillen-Ludena et al., 2015). For this type of confluence, the size of bank-attached bars increases as the width ratio between the lateral tributary channel and the main channel (major tributary channel and receiving channel) decreases (Guillen-Ludena et al., 2017a).

12.5.5 What Is the Relationship between Tributary-Mouth and Bank-Attached Bars?

Mounting evidence suggests that under certain conditions, the bank-attached tributary bars and tributary-mouth bars become elements of a large bar unit that protrudes from the mouth of the tributary and wraps around the downstream junction corner. In other words, the tributary-mouth bar transitions without an obvious break in morphology into the bank-attached bar. At the asymmetrical confluence of the Kaskaskia River and Copper Slough in Illinois, USA, which has a junction angle of 70°, this unit forms when a wedge of sediment penetrates into the confluence from the lateral tributary and the penetrating wedge extends around the downstream junction corner as mobile bed material turns into the receiving channel (Rhoads and Kenworthy, 1995) (Figure 12.16). A similar morphology has been observed at an asymmetrical confluence with a junction angle of 70° in Hudson Bay Lowland of Canada (Mosher and Martini, 2002). Experimental studies of mobile-bed discordant confluences indicate that merging of the mouth bar and the bank-attached bar occurs for a junction angle of 70°, but that at 90° the mouth bar remains within the minor tributary and a distinct transition exists between the two bars (Guillen-Ludena et al., 2016).

12.5.6 How Do Mid-Channel Bars Develop Downstream of Confluences?

Mid-channel bars in the downstream portion of the CHZ within the region of flow recovery are most common at symmetrical confluences with relatively coarse bed material. These features are characteristic of confluence-diffluence units in braided or wandering gravel-bed rivers, but are not always present at symmetrical confluences in river networks. In braided and wandering gravel-bed rivers, mid-channel bars constitute downstream depositional elements of coherent bar units (see Chapters 10 and 11). At asymmetrical confluences, scoured bed material may be deposited as an alternate bank-attached bar downstream of and opposite the bank-attached bar near the downstream junction corner (Mosher and Martini, 2002). The source of material for bar formation is typically scour within the confluence. If flow within a confluence with coarse bed material is at or near the threshold for transport of this sediment during most transport-effective events, bed material mobilized by scour within the CHZ, where mean flow accelerates and levels of turbulence are high, likely will be deposited a short distance downstream in the region of flow recovery, where the flow no longer exceeds the threshold for transport. On the other hand, in a confluence with sandy bed material, where flow conditions during transport-effective events greatly exceed the transport threshold, scoured material is likely to be flushed far downstream, well beyond the CHZ. Thus, the occurrence of mid-channel or alternate bank-attached bars largely depends on local hydraulic conditions and the relation of these conditions to the mobility of scoured bed material.

12.5.7 How Are Morphological Features at Confluences Related to Controlling Factors?

The spatial arrangement of the different morphological elements at confluences is responsive to changes in junction angle, discharge or momentum-flux ratio, and the relative sediment loads of the tributaries. Experimental studies of asymmetrical confluences with mobile beds and fixed side walls show that the longitudinal axis of the scour hole becomes more aligned with the lateral tributary as junction angle and discharge ratio increase, reflecting the greater influence of flow from this tributary on the pattern of erosion within the confluence (Best, 1988). Correspondingly, penetration of sediment from the main channel into the confluence decreases with increases in junction angle and discharge ratio (Best, 1988). The mouth bar of the minor tributary does not penetrate into the confluence until $Q_r \gg 1$, where $Q_r = Q_2/Q_1$. For asymmetrical experimental confluences with a steep tributary transporting a large sediment load, penetration of the lateral tributary-mouth bar decreases with increasing discharge ratio (Guillen-Ludena et al., 2016), perhaps because of enhanced turbulence within the confluence that erodes the bar front. The size of bank-attached bars increases at asymmetrical confluences with increases in junction angle and discharge ratio due to enlargement of the region of separated or reduced-velocity flow near the downstream junction corner (Best, 1988). At a concordant natural confluence with an asymmetrical planform, high flows with $M_r < 1$ generally produce scour that extends from within the confluence into the receiving channel, penetration of the major tributary-mouth bar into the confluence, and erosion of the bank-attached bar, whereas a transition to flows with $M_r > 1$ results in penetration of the minor (lateral) tributary-mouth bar into the confluence, infilling of scour and confinement of the region of scour to the bank opposite the lateral tributary mouth, and enlargement of the bank-attached bar (Rhoads, 1996; Rhoads et al., 2009) (Figure 12.17). Thus, alternating transport-effective flows with high M_r and low M_r lead to episodic storage ($M_r > 1$) and flushing ($M_r < 1$) of sediment within the confluence. At a discordant asymmetrical confluence, the lateral tributary-mouth bar also advances with increases in M_r and retreats with decreases in M_r, but changes in scour are not pronounced (Boyer et al., 2006).

12.6 What Are the Dynamics of Sediment Transport at Confluences?

Experimental studies of sediment transport through symmetrical and asymmetrical confluences with concordant beds under conditions of equilibrium bed morphology indicate that bed material from each tributary remains segregated

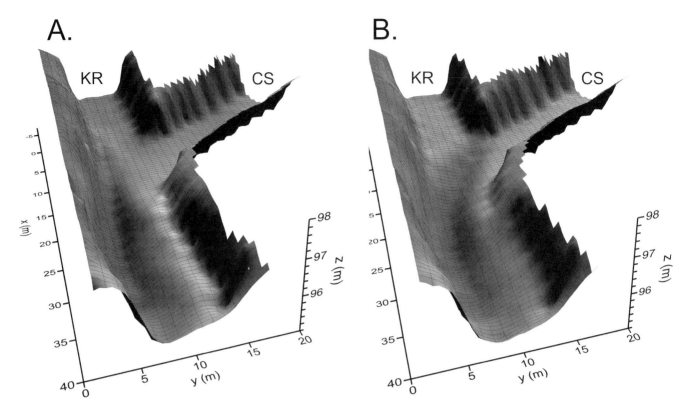

Figure 12.17. Bed morphology at Kaskaskia-Copper Slough confluence formed by (a) high–momentum flux ratio flows ($M_r > 1$) and (b) low–momentum flux ratio flows ($M_r < 1$). In (a), a wedge of sediment from the Copper Slough (CS) penetrates into the confluence and wraps around the downstream junction corner, confining the scour hole to the far bank within the confluence. In (b), the zone of scour extends upstream into the confluence, and the bank-attached bar downstream of the confluence is lower in height than in (a). (A black and white version of this figure will appear in some formats. For the color version, please refer to the plate section.)

and moves around, rather than through, the scour hole (Mosley, 1976; Best, 1988) (Figure 12.18). Segregation of material has been attributed to the development of twin surface-convergent helical cells within the scour hole that sweep incoming bed material from each tributary around the flanks of this morphological element (Mosley, 1976; Best, 1988). Segregation of sediment loads has also been observed in experimental asymmetrical discordant confluences, with tributary sediment moving along the bank-attached bar and main-channel sediment moving through the zone of scour (Leite Ribeiro et al., 2012a; Guillen-Ludena et al., 2015) (Figure 12.19). In this case, segregation seems to be related to the development of a coherent turbulent vortex at the lateral tributary mouth that steers tributary bed material toward the region along the bank where the bank-attached bar develops (Guillen-Ludena et al., 2017b).

At natural confluences, particle-tracing studies indicate that at low-angle ($\theta_j \approx 15°$ to $20°$) junctions with modest scour, coarse bed material from the two upstream channels moves directly through the scour region along intersecting trajectories (Roy and Bergeron, 1990), whereas at high-angle ($\theta_j \approx 80°$ to $85°$) junctions with deep scour, bed material from the upstream channels moves around the scour hole

and remains segregated (Imhoff and Wilcox, 2016). Direct measurements of bedload transport at an asymmetrical concordant confluence demonstrate that for events with $M_r < 1$, bed material from the upstream channels is transported through the scour hole but remains mostly segregated (Rhoads, 1996) (Figure 12.20). This pattern of bedload transport has been attributed to the development of dual helical cells that produce bed-divergent secondary flow, preventing the two loads from mixing within the upstream part of the scour hole. The transition to a single–helical cell flow structure toward the downstream end of the scour hole directs fine sediment from the main channel toward the bank-attached bar, where it mixes with coarse tributary sediment (Figure 12.20). The spatial pattern of bed material generated by transport-effective flows with $M_r > 1$ conforms to that produced by a single dominant helical cell associated with curving lateral tributary flow: fine main-channel sediment moves toward the bank-attached bar within the confluence and mixes with tributary sediment adjacent to a region of scour that is largely devoid of sediment and consists mainly of exposed glacial clay (Rhoads et al., 2009) (Figure 12.20). Daily measurements of bedload transport over a three-month period in a South American

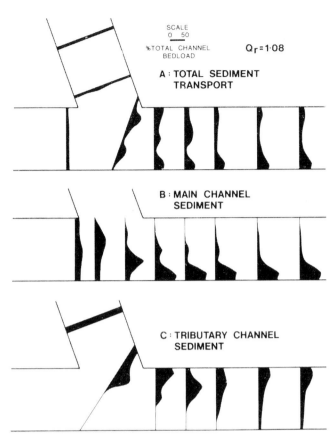

Figure 12.18. Patterns of bedload transport in an experimental confluence (from Best, 1988).

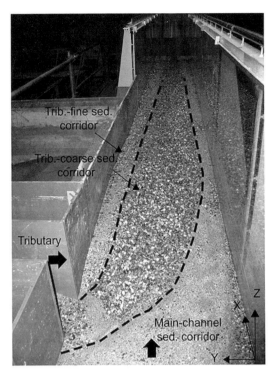

Figure 12.19. Segregation of fine bedload of the main channel (right) from the coarse bedload of the tributary (left) at a discordant confluence where a steep tributary joins a main channel (from Guillen-Ludena et al., 2015).

12.7 What Are the Dynamics of Confluent Meander Bends?

12.7.1 What Is a Confluent Meander Bend?

Physical models of confluences, as well as many field studies, are based on planform configurations where the tributary channel and receiving channels have straight or relatively straight alignments. Many natural rivers, especially those that meander, curve and bend. The potential therefore exists for tributaries to join the main river along bends. Moreover, in a freely meandering river, the migration of bends down valley will tend to capture flow from lateral tributaries, so that these tributaries will preferentially enter the river along the outer erosional bank of meander bends rather than along the inner depositional bank (Davis, 1903). The joining of a tributary with a main river along the outer bank of a meander bend is referred to as a confluent meander bend.

The dynamics of these features combine fluvial processes and forms for meander bends and confluences (Riley and Rhoads, 2012; Riley et al., 2015). The portion of the bend upstream of the confluence is largely unaffected by the confluence, and patterns of flow, bed morphology, and sediment transport in this region are similar to those in a meander bend. An exception can occur when the momentum flux of the tributary exceeds the momentum flux of flow in the bend at

confluence confirm that two-thirds of the total bedload volume is transported through the scour hole at the confluence (Martin-Vide et al., 2015). During this period, bedload transport varied 200-fold for a fivefold variation in discharge.

At a natural discordance confluence, where no obvious helical motion develops, the most active corridors of bedload transport correspond to margins of the shear layer. At this location, bedload-transport rates correlate most closely with the root mean square of $\rho u w'$ – the cross stress defined by the mean streamwise velocity (u) and the instantaneous vertical velocity fluctuation (w')(Boyer et al., 2006). As the shear layer fluctuates back and forth across the confluence due to changes in momentum flux ratio, corridors of bedload transport exhibit corresponding shifts. These fluctuations lead to changes in patterns of erosion and deposition within the confluence, thereby controlling the locations of morphological change. This dynamic mechanism also distributes incoming bed material from each tributary, so that this material is not segregated but instead becomes well mixed within the confluence.

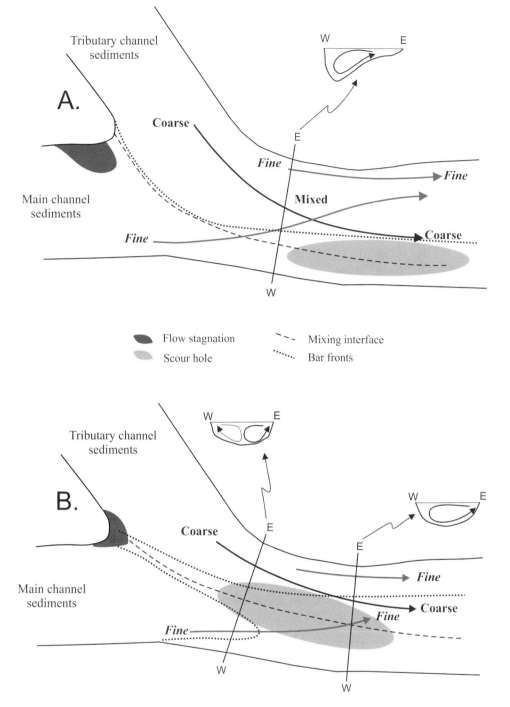

Figure 12.20. Conceptual model of spatial patterns of surficial bed material, bed-material transport pathways, and bed morphology at an asymmetrical confluence with coarse tributary sediment and fine main-channel sediment for (a) $M_r > 1$ and (b) $M_r \leq 1$ (from Rhoads et al., 2009).

a high-angle confluent meander bend, in which case the region of flow stagnation will be located along the outer bank immediately upstream of the tributary mouth. Upstream of the confluence, a point bar forms along the inner bank and a pool is present along the outer bank (Figure 12.21). The flow begins to develop helical motion as it moves through the bend. At the confluence, the incoming tributary flow disrupts the typical pattern of flow and bed morphology associated with meander bends. Hydrodynamic elements of confluences, such as flow stagnation, flow separation, flow acceleration, mutual

deflection, and a mixing interface/shear, and morphologic features, including a scour hole and bank-attached bar, develop where the tributary enters (Figure 12.21).

12.7.2 What Are the Dynamics of High-Angle Confluent Meander Bends?

At high-angle confluent meander beds, the influence of the tributary varies with momentum flux ratio (Figure 12.21a, b). For $M_r < 1$, the mixing interface is located between the center of

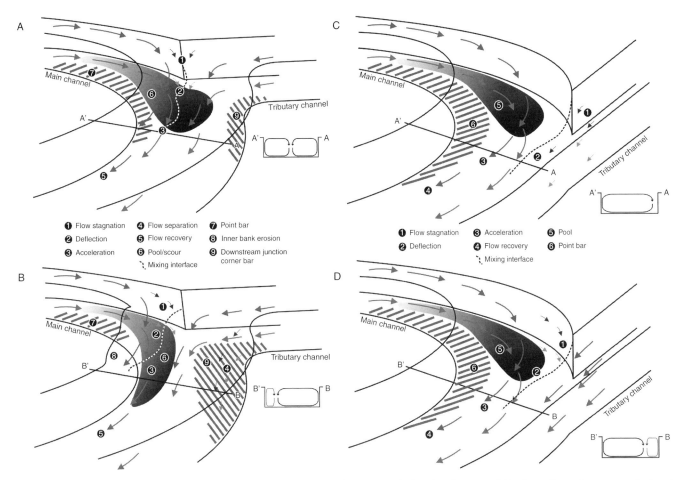

Figure 12.21. Flow structure and bed morphology at (a) high-angle confluent meander bend with $M_r < 1$, (b) high-angle confluent meander bend with $M_r > 1$, (c) low-angle confluent meander bend with $M_r < 1$, and (d) low-angle confluent meander bend with $M_r > 1$ (from Riley et al., 2015). (A black and white version of this figure will appear in some formats. For the color version, please refer to the plate section.)

the confluence and the tributary mouth. Helical motion develops in the tributary flow, which curves in the opposite direction to flow in the bend, resulting in counterrotating helical cells on each side of the mixing interface. Scour within and immediately downstream of the confluence extends the zone of erosion associated with the pool in the upstream part of the bend. A bank-attached bar may develop along the outer bank of the bend where sediment from the tributary accumulates in a region of reduced velocity or flow separation, thereby locally protecting this bank from erosion. The point bar in the bend, which normally becomes widest downstream of the bend apex, has a limited lateral extent downstream of the tributary mouth. Incoming flow from the lateral tributary deflects flow toward the inner bank, producing high velocities and turbulence in this region, which restrict deposition. For $M_r > 1$, the tributary flow extends far into the main channel, increasing flow separation along the outer bank (Sui and Huang, 2017), shifting the mixing interface toward the inner bank, confining flow in the bend to the inner-bank region, and enlarging the helical cell associated with the tributary flow. The bank-attached bar along the outer

bank increases in size and the scour hole shifts toward the inner bank, preventing the formation of a point bar in the downstream part of the bend and in some cases causing erosion of the inner bank.

12.7.3 What Are the Dynamics of Low-Angle Confluent Meander Bends?

Effects of the lateral tributary are less pronounced for low-angle confluent meander bends (Figure 12.21c,d). The trajectory of the tributary flow does not change markedly in this case, limiting flow separation and reductions in transport capacity that promote the formation of bank-attached bars. Because the tributary flow does not deflect the bend flow strongly toward the inner bank, the point bar along this bank persists downstream of where the tributary joins the bend. Bend flow produces a well-defined pool along the outer bank upstream of the tributary mouth, but scour at the mouth and downstream of this location diminishes. The tributary flow, which does not curve and therefore does not

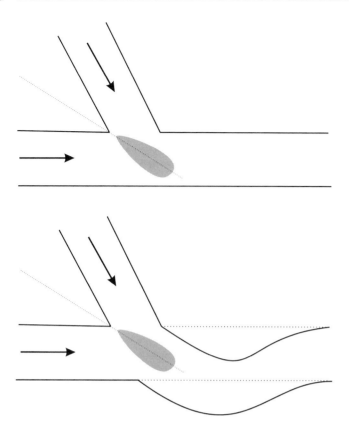

Figure 12.22. Planform adjustment (top to bottom) of an asymmetrical confluence with equal discharge and sediment load of the minor and major tributaries (adapted from Mosley, 1976).

develop strong helical motion that could enhance near-bed shear stresses, occupies the region along the outer bank. Also, the delivery of bed material to this region from the tributary limits erosion of the bed.

12.8 How Do Confluences Change over Time?

Few studies have examined the stability of confluence planform over time. Experimental work on initial asymmetrical confluences formed in erodible material shows that if incoming discharges and sediment loads of the tributaries are equal, the receiving channel shifts through erosion of one bank and deposition on the other to become aligned with the orientation of the scour hole, which bisects the upstream junction angle (Mosley, 1976) (Figure 12.22). In other words, the confluence develops a more symmetrical planform in response to the equal contributions of each tributary to flow and sediment load in the receiving channel. Unless the receiving channel is free to completely change its downstream path, the change in planform will likely be constrained to the immediate vicinity of the confluence, resulting in the development of curvature in the receiving channel (Figure 12.22). Such adjustments are common in braided rivers, where changes in contributions of

tributaries are common and material forming the channel perimeter is highly erodible (Figure 10.12).

Most confluences in river networks develop in conjunction with the structure of the stream network, with the planform of the confluence configured by adjustments of junction angles to tributary gradients and contributing areas over geologic timescales. Many confluences within river networks exhibit stability over modern timescales and evolve mainly over geologic timescales. Stability over the modern timescale depends on the extent to which adjustments to changes in hydrological conditions of the tributaries are constrained by resistance to erosion. At asymmetrical confluences, frequent and persistent $M_r > 1$ conditions due, for example, to changes in land use within the watershed of the minor tributary can lead to net erosion of the bank of the receiving channel opposite the minor tributary and formation of a bank-attached bar on the other bank, resulting in minor realignment of the receiving-channel planform over a timescale of a few decades (Rhoads et al., 2009). Satellite analysis of large rivers, many of which have anabranching or braided planforms, indicates that some confluences in these river systems exhibit considerable amounts of change over timescales of decades (Dixon et al., 2018). Those that do are associated with large rivers flowing across broad alluvial plains where changes in planform characteristics, including confluences, are relatively unconstrained. Many large-river confluences do not exhibit change over decadal timescales. Adjustment over modern timescales is constrained in rivers with relatively narrow floodplains or bedrock control.

12.9 What Are the Dynamics of Mixing at Confluences?

Confluences are important locations for mixing within drainage networks. Often the two incoming flows have different water properties, such as temperature, conductivity, pH, or isotopic characteristics, or are transporting different types of suspended material (organic particles or sediment). The flows meet at the upstream end of the confluence, form a distinct mixing interface (Figure 12.5), and progressively mix in the downstream direction. Mixing can occur both locally within the CHZ as well as downstream of the CHZ, where flow characteristics no longer are strongly influenced by hydrodynamic conditions within the confluence.

12.9.1 What Are the Fundamental Mechanisms of Mixing in Rivers?

Mixing in rivers involves three fundamental mechanisms: molecular diffusion, turbulent diffusion, and advection (Rutherford,

1994). When flow is turbulent, as it is in rivers, molecular diffusion is small in relation to the other two mechanisms and can be ignored. Turbulent diffusion occurs through turbulence, or variations in instantaneous velocities about the mean velocity, whereas advection involves secondary currents that transport fluid and materials dissolved or suspended in the fluid laterally and vertically. Buoyancy can also influence mixing when one fluid has a substantially different density than another. The property or characteristic used to examine the degree of mixing between the streams is referred to as the tracer. The rate of transverse mixing traditionally is assessed using the transverse mixing coefficient (k_y), defined by the time rate of change in the transverse spatial variance of the tracer (σ_y^2) as it is travels downstream (Rutherford, 1994):

$$k_y = \frac{1}{2}\frac{d\sigma_y^2}{dt} \tag{12.13}$$

Determining the time rate of travel can be complicated in a nonuniform flow where velocity varies both across and along the channel. An alternative approach to ascertaining lateral mixing is to examine how the standard deviation of tracer concentration $\sqrt{\sigma_y^2}$ varies over distance at different cross sections along a river channel downstream of a confluence (Biron et al., 2004b). Because transverse mixing at a confluence occurs in relation to the difference in tracer concentrations of the incoming flows, a standardized metric of mixing for cross sections within and downstream of confluences is

$$\psi_m = 1 - \frac{\sqrt{\sigma_y^2}}{\sqrt{\sigma_{yus}^2}} \tag{12.14}$$

where σ_{yus}^2 is the variance of the tracer at a composite cross section consisting of tracer data collected at two cross sections upstream of the confluence, one on each tributary (Lewis and Rhoads, 2015). For a conservative tracer, ψ_m will have a value close to 0 within the upstream part of the confluence where the two flows initially meet, and it will increase toward a value of 1 over distance downstream as the flows mix (Lewis and Rhoads, 2015). The standardized metric of a tracer usually does not attain a value of 1, which ideally would correspond to complete mixing, but becomes quite large (e.g., > 0.95) and steady over distance when the flows are fully mixed (Umar et al., 2018).

12.9.2 What Is the Influence of Lateral Turbulent Diffusion on Mixing at Confluences?

The main mechanism of turbulent diffusion within the CHZ is the development of the shear layer between the flows, which, because it is generated by differences in velocities of the flows,

typically coincides with the mixing interface. The shear layer is characterized by the development of coherent turbulent structures, either Kelvin–Helmholtz vortices if it is in KH mode or a vortex street if it is in wake mode (Constantinescu et al., 2011b). Pairing or amalgamation of vortices over distance increases the lateral extent of the vortices and thus of turbulent diffusion associated with these vortices. Increases in the lateral extent of vortices leads to an increase in the width of the mixing interface and a decrease in the transverse gradient, or variance, of the depth-averaged tracer concentration. In shallow-shear flows, such as rivers, where bed friction effects are important, the growth rate of the mixing interface is inhibited (Chu and Babarutsi, 1988; Uijttewaal and Booij, 2000).

In small stream confluences, enhanced levels of TKE associated with the shear layer can dissipate rapidly as the difference in velocity on each side of the interface diminishes over distance (Rhoads and Sukhodolov, 2008; Mignot et al., 2014). As a result, levels of TKE in the mixing interface approach background levels within the CHZ (Rhoads and Sukhodolov, 2001; Sukhodolov and Rhoads, 2001). If a narrow, vertically aligned mixing interface between the confluent flows still exists at the downstream end of the CHZ and turbulence at this location has reached background levels, further mixing must occur through turbulent or advective processes within the river channel downstream of the CHZ (Rathbun and Rostad, 2004). Under such conditions, the two flows exit the CHZ unmixed and may remain unmixed far downstream of the confluence. In large rivers, lack of substantial mixing has been reported tens or hundreds of kilometers downstream from confluences (Mackay, 1970; Bouchez et al., 2010; Campodonico et al., 2015; Umar et al., 2018). Over such distances, chemical transformations of tracers can become important and should be considered in determinations of mixing (Guinoiseau et al., 2016).

12.9.3 What Is the Influence of Lateral Advection on Mixing at Confluences?

Advection can substantially influence mixing at confluences. At discordant asymmetrical confluences, secondary flow in the lee of the bed step distorts the mixing interface by directing near-bed flow inward and upward into the lateral separation zone (Best and Roy, 1991; De Serres et al., 1999; Bradbrook et al., 2001). The step also enhances turbulent diffusion by increasing levels of turbulence within the confluence compared with levels in the absence of a step (Biron et al., 1996b). Together, these effects produce rapid mixing within the CHZ, so that within about 10 channel widths of the junction apex, mixing reaches about 80% to 90% (Gaudet and Roy, 1995). Mixing depends on flow stage, with more rapid mixing occurring at low stage, when the effects of discordance are most pronounced, than at high stage (Gaudet and Roy, 1995; Biron et al., 2004b). At concordant asymmetrical confluences, curvature-induced helical motion can enhance mixing, particularly when the helical cell associated with the minor (lateral) tributary is large and has strong circulation relative to the size and strength of the cell associated with the major tributary (Lewis and Rhoads, 2015). The lateral extent and strength of the lateral tributary cell increase with increasing momentum flux ratio.

Experimental and numerical studies indicate that the effect of dominant helical motion in the receiving channel associated with flow from the curving lateral tributary is to tilt the mixing interface and draw near-bed fluid from the major tributary inward and upward along the flank of the separation zone or the face of the bank-attached bar (Riviere et al., 2015; Yuan et al., 2016; Chen et al., 2017; Tang et al., 2018). This pattern of mixing has been observed in the field at a small asymmetrical confluence, although the same basic pattern occurs across a range of momentum flux ratios, mainly because helical motion within flow from the lateral tributary

is relatively strong at all values of M_r (Kenworthy and Rhoads, 1995; Rhoads and Sukhodolov, 2001; Lewis and Rhoads, 2015; Constantinescu et al., 2016) (Figure 12.23).

The rate of mixing over distance is positively related to M_r for $M_r \leq 5$ but seems to become relatively constant for $M_r > 5$ (Lewis and Rhoads, 2015). Discharge and stage are major factors influencing the length scale of mixing, with the most rapid mixing over distance occurring for the smallest, shallowest flows. For a set of 12 mixing events, values of ψ_m exceeds 0.80 (>80% mixed) for a distance of 3.5 channel widths downstream of the confluence apex for six flows and are greater than 0.50 (>50% mixed) for nine flows (Lewis and Rhoads, 2015). Thus, helical motion produces substantial mixing within the CHZ.

12.9.4 What Is the Influence of Density Differences on Mixing at Confluences?

Density differences between the confluent flows also affect mixing when differences are large enough to cause buoyancy. The magnitude of density difference required to produce

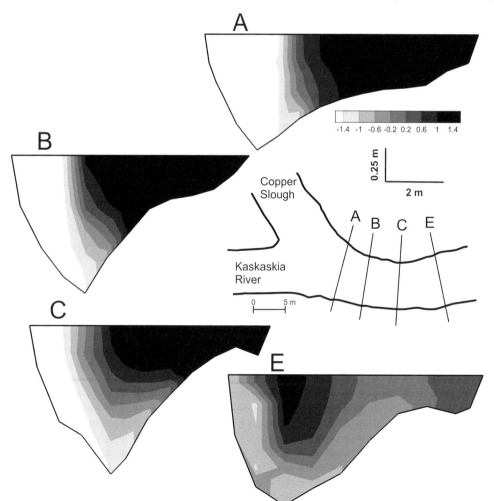

Figure 12.23. Pattern of thermal mixing for $M_r = 1.03$ at the asymmetrical confluence of the Kaskaskia River and Copper Slough, Illinois, USA. Contour values of temperature (T °C) are standardized relative to the mean at each cross section. View is looking upstream. Note inward and upward penetration of cool water from the Kaskaskia River (left) into warm water of Copper Slough (right) near the bed caused by counterclockwise helical motion of flow from the Copper Slough (from Lewis and Rhoads, 2015).

buoyancy effects on mixing has yet to be fully ascertained, but the potential for buoyant behavior depends on the relation between inertial and buoyant forces. For an inflow of less dense water from the minor tributary into water of greater density within a main river, this relation can be expressed in the form of the densimetric Froude number:

$$\mathbf{F}_D = U/\sqrt{g'D} \tag{12.15}$$

where

$$g' = \frac{(\rho_1 - \rho_2)}{\rho_1} g \tag{12.16}$$

is the reduced gravity, U is a representative cross-sectionally averaged velocity, and D is a representative mean flow depth. No consensus exists on values of U and D to use in Eq. (12.16) (Rutherford, 1994), but values for the lateral tributary are a reasonable choice (Fischer et al., 1979; Lewis and Rhoads, 2015). The issue at hand is whether the buoyant force associated with the reduced density of the inflow exceeds the inertial force associated with turbulent diffusion of momentum toward the channel bed. When $\mathbf{F}_D \ll 1$, buoyant forces dominate, and the tributary flow detaches from the bed and moves over the denser main-channel flow. In this situation, horizontal stratification of the flows, i.e., flows moving side by side separated by a vertical mixing interface, is replaced by vertical stratification of the flow, i.e., less dense fluid moving over fluid of greater density separated by a horizontal mixing interface. When $\mathbf{F}_D \gg 1$, inertial forces dominate, and the inflow remains attached to the bed. Under these conditions, horizontal stratification with a vertical mixing interface is preserved. For $\mathbf{F}_D \approx 1$, both forces may be important, resulting in combined effects on mixing.

The exact value of $\mathbf{F}_D$ required for density effects to become manifest in mixing is uncertain, but horizontal stratification of confluent flows is evident when $\mathbf{F}_D < 1$ (Ramon et al., 2013). The occurrence of underflows of denser water beneath less dense water has been documented at some large river confluences (Lane et al., 2008; Laraque et al., 2009; Park and Latrubesse, 2015; Herrero et al., 2018), but these studies do not report values of $\mathbf{F}_D$. A rather extensive analysis of mixing at small confluences reveals some evidence of horizontal stratification when $\mathbf{F}_D \approx 2$, but not when $\mathbf{F}_D > 5$ (Lewis and Rhoads, 2015). Values of densimetric Froude number less than 1 are common in river confluences associated with reservoirs, where depths are large and velocities are low. Under such conditions, pronounced vertical stratification of confluent flows can occur, with denser water plunging beneath less dense water upstream of the confluence (Ramon et al., 2013, 2014; Lyubimova et al., 2014).

The effect of vertical stratification on mixing rates remains unclear. In some cases, movement of one flow beneath another seems to enhance rates of mixing at confluences (Lane et al., 2008), perhaps by increasing the contact area between the two flows along an inclined or horizontal mixing interface (Ramon et al., 2016). In other cases, vertically stratified flows are relatively stable and completely mix over long distances (100 km or >40 channel widths) downstream from the confluence (Laraque et al., 2009).

Few, if any, studies have examined the influence of bed discordance on mixing of flows with density contrasts. Flow over a step at the entrance to a confluence should strongly affect the relation between inertial and buoyant forces near the bed in the lee of the step. The result may be highly complex, three-dimensional patterns of mixing.

CHAPTER

13 The Vertical Dimension of Rivers: Longitudinal Profiles, Profile Adjustments, and Step-Pool Morphology

13.1 Why Is the Longitudinal Profile of Rivers Important?

Rivers are the primary conveyors of water, sediment, and dissolved constituents from land to the oceans on Earth (see Chapter 3). This net flux of material moves downward under the influence of gravity. Put simply, rivers flow under the influence of gravitational gradients from high elevations to low elevations. Thus, the third component of channel morphology, the longitudinal profile, or change in elevation of the river bed over distance, is essential for maintaining the movement of water and transport of sediment within rivers. This dimension evolves over geologic timescales in relation to the interplay between fluvial erosion, which reduces the elevation of landscapes above sea level, and tectonic uplift, which increases the elevation of landscapes above this level. Early ideas about equilibrium and graded conditions in river systems underscored the importance of adjustments in channel slope for developing a balance between sediment supply and transport (Davis, 1899; Mackin, 1948). As understanding of the complexity of river systems advanced, the emphasis on slope adjustments as primary diminished. Nevertheless, such adjustments are still recognized as important in the short-term response of rivers to changes in inputs of water or sediment (Eaton and Church, 2004) (see Chapter 6), including changes caused by human activity (see Chapter 15), and in the long-term evolution of landscapes (see Chapter 2). Over the past few decades, concern about management of aquatic resources in upland or mountain landscapes has also led to intensive investigation of steep streams, where channel slope is a major factor influencing fluvial processes and forms.

13.2 What Are the Characteristics of the Longitudinal Profile of a River?

13.2.1 What Is the Shape of the Longitudinal Profile?

The longitudinal profile of a river depicts change in elevation of the channel bed over the entire length of the river from its origin to its mouth (Figure 13.1). The longitudinal profiles of rivers develop in conjunction with the evolution of drainage basins and stream networks over geologic timescales. Within large drainage basins (>50 km^2), the longitudinal profiles of rivers determine to a large extent the difference in elevation that drives erosional processes. More than 80% of total relief in such basins is related to the decrease in elevation along rivers (Whipple, 2004).

The form of the longitudinal profile in many rivers consists of relatively rapid change in elevation over distance (dZ/dx) in the headwaters and relatively slow change in elevation over distance toward the mouth. Because channel slope (S), or gradient, equals dZ/dx, the tangent to the longitudinal profile curve, slope also

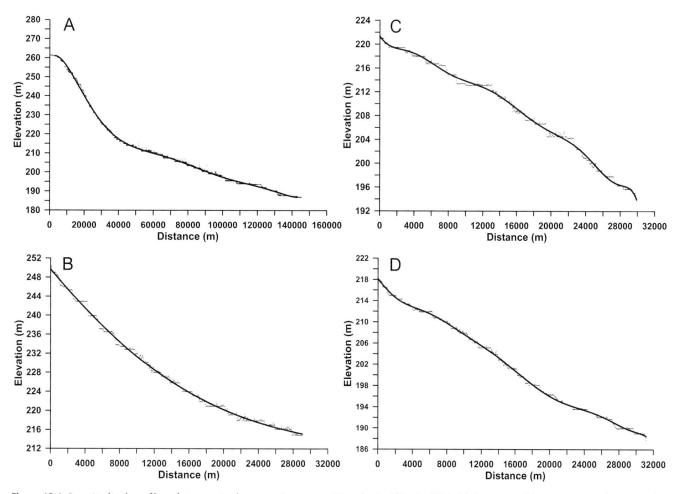

Figure 13.1. Longitudinal profiles of streams in the upper Sangamon River basin, Illinois, USA: (a) Sangamon River main stem (segmented concave-upward), (b) Drummond Creek (smooth concave-upward), (c) Goose Creek (convex-upward), and (d) Friends Creek (approximately linear). All profiles (dark lines) are high-order polynomials (fourth order or greater) fitted to 10 m digital elevation data (light irregular lines).

changes rapidly over distance in the headwaters and gradually over distance near the mouth. The result is a concave-upward longitudinal profile (Figure 13.1a,b). This shape has been viewed as the classic form of the profile. Within the Davisian perspective on river development it was associated with the concept of grade or equilibrium of river systems (e.g., Mackin, 1948). Although this form is common, exceptions are not uncommon and include convex-upward and relatively straight profiles (Figure 13.1c,d).

13.2.2 What Mathematical Relations Have Been Used to Represent the Longitudinal Profile?

A variety of mathematical relations have been used to represent longitudinal profiles of natural rivers, particularly profiles that are linear or concave-upward (Table 13.1). The functional forms of the mathematical relations are based on underlying assumptions about the rate of erosion with elevation (Strahler, 1952a; Tanner, 1971), variation in particle size of bed material over distance

(Shulits, 1941; Mackin, 1948; Tanner, 1971), or changes in sediment flux over distance (Rice and Church, 2001). Coefficients and exponents in the models are determined through statistical analysis of elevation, slope, and distance data derived from field, map, or digital elevation model (DEM) data (Hack, 1957; Ohmori, 1991; Rice and Church, 2001) (Figure 13.1). Some models treat the relation between elevation and distance directly, whereas others relate channel slope to distance. Exponential models of longitudinal profiles are common (Morris and Williams, 1999a), but linear, logarithmic, power, and polynomial models have also been proposed or fitted to data (Table 13.1).

13.2.3 How Does Representation of the Longitudinal Profile Change with Data Resolution?

The representation of the longitudinal profile depends on the resolution of elevation information. As resolution increases, fine-grained irregularity in the profile generally increases, which for field or high-resolution airborne LiDAR data (0.5 to 2 m) reflects

Table 13.1. Functional forms used to depict river longitudinal profiles.

Functional form	Elevation	Sources	Slope	Sources
Linear	$Z = b_0 + b_1 x$	Shepherd (1985); Ohmori (1991); Rice and Church (2001)		
Logarithmic	$Z = b_0 + b_1 \log(x)$	Jones (1924); Hack (1957); Shepherd (1985); Ohmori (1991)	$S = b_0 + b_1 \log(x)$	Jones (1924)
Exponential	$Z = ae^{b_1 x}$	Strahler (1952a); Tanner (1971); Shepherd (1985); Ohmori (1991); Rice and Church (2001)	$S = ae^{b_1 x}$	Shulits (1941)
Power	$Z = ax^{b_1}$	Ohmori (1991); Shepherd (1985); Rice and Church (2001)	$S = ax^{b_1}$	Hack (1957); Goldrick and Bishop (2007)
Quadratic	$Z = b_0 + b_1 x + b_2 x^2$	Church (1972); Rice and Church (2001)		
Three-parameter exponential	$Z = ae^{b_1 x} + b_0$	Rice and Church (2001)		
Three-parameter power	$Z = ax^{b_1} + b_0$	Hack (1957)		
Exponential plus linear	$Z = ae^{b_2 x} + b_1 x + b_0$	Rice and Church (2001)		
High-order polynomial	$Z = b_0 + b_1 x + b_2 x^2 + b_3 x^3 \dots b_n x^n$	Figure 13.1		

Z = bed elevation, x = distance downstream, S = channel slope.

local variability in river bed topography associated with bar unit–scale features. Local irregularity in profiles based on relatively coarse-grained (10 to 30 m) DEM data often arises because the process of converting discrete altitude measurements into gridded elevation information introduces errors and artifacts along valley bottoms (Schwanghart and Scherler, 2017). Successive zero-gradient sections with vertical steps at the downstream side are particularly common artifacts of longitudinal profiles derived from coarse-grained DEM data (Figure 13.1). Also, coarse-resolution DEMs commonly overestimate elevations along rivers in steep topography.

13.2.4 What Factors Influence Channel Slope?

In alluvial rivers transporting bed material, three major factors influence channel slope: 1) the discharge, 2) the sediment load, and 3) the size of the bed material. Increases in discharge reduce slopes, whereas increases in sediment load and grain size increase the equilibrium slope (Rubey, 1952; Blom et al., 2016). Generally, discharge and sediment load increase downstream as basin area increases, unless the river flows from a wet environment upstream to an arid environment downstream, in which case discharge and sediment load may decrease downstream. Given that many streams and rivers exhibit concave-upward profiles, the issue of concern is whether this profile shape, which corresponds to decreasing slope downstream, is sufficient to maintain

transport of an increasing sediment load as discharge increases downstream. If the volumetric sediment transport capacity of a river (Q_{svc}) is a function of stream power per unit length (e.g., Bagnold, 1966), $Q_{svc} \propto QS$, the downstream decrease in gradient reduces transport capacity in opposition to downstream increase in discharge, which increases transport capacity. Reasonable proportionalities for changes in gradient and discharge with increasing drainage area are $S \propto A_d^{-0.45}$ and $Q \propto A_d^{0.85}$. For $x \propto A_d^{0.6}$, where x is distance downstream, these proportionalities result in $QS \propto x^{0.67}$, showing that stream power and therefore sediment-transport capacity increase downstream for a concave-upward profile. A downstream increase in stream power will occur as long as the rate of increase in discharge exceeds the rate of decrease in slope.

Multivariate analysis has shown that valley slope (S_v) has a statistically significant positive relationship with channel slope independently of discharge, sediment load, grain size, bank or bed silt-clay percentage, and bank vegetation (Rhoads, 1992; Davidson and Hey, 2011). In other words, steep landscapes have higher-gradient channels than relatively flat landscapes, even when accounting for differences in other factors that influence slope. This relationship emphasizes the importance of geologic factors that influence landscape steepness, particularly rates of tectonic uplift and rock erodibility, in controlling the slope of channels.

13.2.5 How Are Adjustment of Grain Size and Channel Slope Interdependent?

The extent to which grain size influences the longitudinal profile of rivers through adjustments of channel slope is complicated by interdependency between grain size and slope. The accumulation of coarse material within a river can steepen the channel gradient to produce the bed shear stress required to transport this material; on the other hand, local increases in flow competence associated with increases in channel slope can coarsen the bed material through size-selective transport of fine fractions within the bed. Multivariate analysis that controls for the negative relation between channel gradient and discharge, or its surrogate drainage area, illustrates that gradient is positively related to the size of bed material, but does not resolve the causal relationship between these two variables (Table 13.2). To some extent this relationship may be scale dependent, with local variations in bed-material size influencing channel gradient at the bar-unit scale, but the relationship becoming more complex and interdependent at the scale of the longitudinal profile (network scale) (Penning-Rowsell and Townshend, 1978; Prestegaard, 1983). The downstream change in size of bed material, typically characterized by fining, has received a considerable amount of attention both as a possible controlling factor on the shape of the longitudinal profile and also as a product of aggradation associated with downstream decreases in channel slope. Exploration of this issue is necessary before examining the influence of discharge, sediment load, and sediment caliber on the shape of the longitudinal profile.

13.3 What Factors Influence Downstream Change in the Size of Bed Material?

Downstream fining of bed material in rivers is commonly characterized by what has become known as Sternberg's (1875) Law:

$$d_i = d_o e^{-\delta x} \tag{13.1}$$

where d_i is a characteristic particle size (mm) of the bed-material size distribution at x (km), d_o is a scaling factor corresponding to initial characteristic size of the bed material at $x = 0$ (mm), and δ is the fining rate (km^{-1}). The characteristic particle size often is represented by either the median (d_{50}) or the mean (d_m) grain size. This relation indicates that particle size declines exponentially over distance.

Numerous empirical studies have used Eq. (13.1) to characterize downstream changes in particle size in streams and rivers (Hoey and Bluck, 1999, table 4; Morris and Williams, 1999b, table 1). Values of δ vary over an enormous range ($\approx 5 \times 10^{-4}$ to $\approx 1 \times 10^{0}$) with the largest values associated with steep mountain torrents and glacial outwash streams. The lowest values occur in large lowland sand-bed rivers. In steep headwater tributaries with large cobbles and boulders (e.g., Rhoads, 1989) the distance required to halve the particle diameter of the reference grain size (x_h) may be less than half a kilometer (Hoey and Bluck, 1999), whereas in large sand-bed rivers x_h may be hundreds of kilometers (Frings, 2008). Assuming that channel slope is proportional to the size of bed material, an exponential decline in particle size should be associated with an exponential decline in channel slope over distance (Shulits, 1941). Alternatively, equating the exponential decline in grain size (Eq. (13.1)) to the exponential decline in bed elevation (Table 13.1) yields a power-function relation between elevation and grain size (Tanner, 1971). On the other hand, some empirical studies have found that the downstream trend in particle size is best described by power functions rather than exponential relations (Brush, 1961; Brierley and Hickin, 1985).

Two factors contribute to the downstream reduction in particle size in rivers: abrasion and selective sorting. Abrasion involves mechanical wear of bed material as it moves along the stream and collides with other particles or underlying bedrock. Selective sorting is related both to

Table 13.2. Multivariate analyses of channel gradient.

Source	Equation		Locale
Hack (1957)	$S = 0.006(d_{50}/A_d)^{0.6}$	Map-derived gradient	Appalachian Mts.
Schroder (1991) based on data from Hack (1957)	$S = 0.0076(d_{50}/A_d)^{0.4}$	Field-measured gradient	Appalachian Mts.
Schroder (1991)	$S = 0.0066(d_{50}/A_d)^{0.4}$	Field-measured gradient	The Eifel, West Germany
Schroder (1991)	$S = 0.002d_{50}^{1.23}A_d^{-0.42}$	Field-measured gradient	The Eifel, West Germany
Bray (1982)	$S = 0.059d_{50}^{0.59}Q_2^{-0.33}$	Field-measured gradient	Gravel-bed rivers, Alberta, Canada
Hey and Thorne (1986)	$S = 0.096d_{50}^{0.71}Q_{bk}^{-0.31}$	Field-measured gradient	Gravel-bed rivers, United Kingdom

differential entrainment and transport of bed material by particle size and to differences in the transport distance of mobilized particles according to size. Although the downstream fining relation (Eq. (13.1)) was originally developed to examine the effects of abrasion, it has been expanded to include the effects of selective sorting (Tanner, 1971):

$$\delta = \delta_a + \delta_s \tag{13.2}$$

where δ_a is the contribution of abrasion to fining and δ_s is the contribution of selective sorting to fining. Determining the separate contributions of abrasion and sorting to downstream fining has been a major focus of research over the past several decades.

13.3.1 What Is the Role of Abrasion in Downstream Fining?

Abrasion was long thought to be the major factor producing downstream reduction in grain size and provided the motivation behind the formulation of Sternberg's Law, which originally was expressed in terms of reduction in particle weight rather than particle size. It involves a variety of wearing mechanisms, including splitting, chipping, crushing, cracking, grinding, dissolution, and sandblasting (Kuenen, 1956; Frings, 2008) (Figure 13.2). Splitting and chipping caused by particle collisions are the most effective wearing mechanisms. Vibration of particles that are not moving can result in grinding (Schumm and Stevens, 1973), the least effective wearing mechanism.

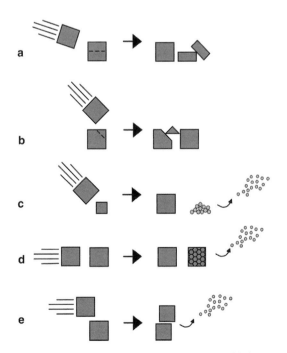

Figure 13.2. Forms of abrasion in rivers: (a) splitting, (b) chipping, (c) crushing, (d) cracking, and (e) grinding (from Frings, 2008).

Abrasion has been studied experimentally using tumbling mills, or rotating barrels containing particles (Wentworth, 1919; Krumbein, 1941; Kodama, 1994a), and abrasion tanks, or circular flumes in which mobile particles within a current are transported over an abraded substrate (Kuenen, 1956: Lewin and Brewer, 2002; Attal and Lave, 2009). Such experiments confirm that lithology has the strongest influence on abrasion rates, with differences in abrasion rates for weak versus strong rocks exceeding two orders of magnitude (Morris and Williams, 1999b; Attal and Lave, 2009). Presumably, abrasion rates should increase both with grain size and with the number of moving particles. For a given particle velocity, large particles have more kinetic energy than small particles, and particles moving as bedload are in contact with other particles nearly constantly. As more particles move, particle interactions should increase. Experimental results are somewhat mixed regarding an increase in abrasion rates with increasing particle size or with the amount of material in motion (Frings, 2008).

For hydrodynamic conditions approximating those in mountain rivers, mass attrition loss by grinding increases with particle velocity but is only weakly dependent on particle size (Attal and Lave, 2009). On the other hand, the production of fragments by chipping and splitting is enhanced by large particles, high impact velocities, and the presence of joints. Other experimental work has shown that abrasion rates of coarse particles generally decrease as fine particles are added, probably because these fine materials dampen impacts (Kuenen, 1956; Kodama, 1994a). Additional factors that affect abrasion include grain rounding (less abrasion) versus angularity (more abrasion) and chemical weathering, which increases the susceptibility of grains to mechanical wear (Jones and Humphrey, 1997; Frings, 2008). Besides affecting the size of particles, abrasion plays an important role in the rounding of river rocks.

Typically, laboratory experiments determine reductions in particle weights and yield abrasion coefficients for weight reduction over equivalent travel distance. The need to convert weight-loss abrasion coefficients to size-diminution abrasion coefficients for comparison with field studies has been problematic because the relation between weight loss and size diminution is complex and cannot be captured by a single universal conversion factor (Lewin and Brewer, 2002). For example, weight loss through abrasion for angular particles initially occurs through rounding without a substantial change in particle diameter; once the particle is well rounded, subsequent abrasion results in reduction of particle diameter (Domokos et al., 2014; Miller et al., 2014). Despite these complications, a variety of values for δ_a have been reported for abrasion experiments (Kodama, 1994a; Morris and Williams, 1999b). Initial estimates were on the order of 10^{-5}

to 10^{-2} km^{-1} with most values between 10^{-4} and 10^{-3} km^{-1} (Krumbein, 1941; Kuenen, 1956). These values are generally an order of magnitude less than many values of δ, suggesting that abrasion is a minor factor in particle-size diminution. More recent experimental work has reported values of δ_a on the order of 10^{-2} to 10^{-1} km^{-1} with some values exceeding 10^{-1} km^{-1} (Kodama, 1994a; Lewin and Brewer, 2002; Attal and Lave, 2009). These results have fueled renewed interest in the role of abrasion in downstream fining, particularly in high-energy mountain rivers transporting large clasts as bed-load. Also, the role of in situ chemical weathering in abrasion is worthy of further consideration. Abrasion rates in tumbling mills for natural particles sampled from fluvial environments are high initially but decline markedly over time. This decline in the rate of abrasion has been attributed at least in part to the removal of easily abraded weathered layers from particle surfaces (Jones and Humphrey, 1997). In natural river systems, individual particles are likely to experience brief episodes of mobility with intervening long periods of storage within the channel bed, bars, or the floodplain. During storage, weathered surfaces develop that are readily removed during a subsequent episode of transport. This weathering effect may explain differences between abrasion rates in tumbling barrels, which do not account for storage and weathering, and downstream rates of fining in natural rivers, which include effects of storage and weathering.

Isolating the effects of abrasion in the field is difficult, but it has been studied by fixing particles in place or tethering them to the bed and noting reduction in size by the abrasional effects of moving bedload (Brewer et al., 1992). Downstream fining related to abrasion can be assessed by determining change in the relative abundance of particles of different lithologies within particular grain-size classes along the length of a river that does not have additional inputs of bed material from hillslopes or tributaries (Werritty, 1992; Kodama, 1994b). If particle size is the major determinant of mobility and selective sorting through size-selective transport dominates downstream fining, the relative abundances of particles of different lithologies within a grain-size class should remain relatively constant in the downstream direction. Systematic changes in the relative abundance of particles of different lithologies within a size class over distance indicate that some particles of that size with specific lithologic characteristics are no longer as abundant as they were upstream, presumably because the size of these particles has been reduced through abrasion. For example, in the Watarase River in Japan, andesite, quartz-porphyry, and sandstone/hornfels particles initially are about three times more abundant by weight than chert clasts in the $d_\phi = -5.5$ to -6.0 size range, but about 20 km downstream, particles of that size for all three lithologies become much less abundant compared with chert

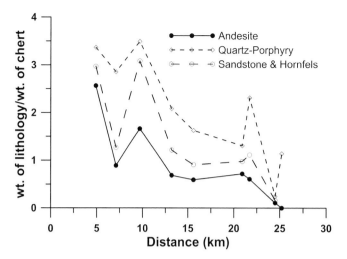

Figure 13.3. Downstream changes in relative weight fractions of particles for $d_\phi = -5.5$ to -6.0 for different rock types within the Watarase River, Japan (from Kodama, 1994b).

(Kodama, 1994b) (Figure 13.3). In fact, andesite and sandstone/hornfels clasts in the size range $d_\phi = -5.5$ to -6.0 no longer exist at that distance. The loss of clasts in this size range from the bed for the three lithologies is attributed to abrasion. Neither field studies of abrasion in place nor studies of changes in lithological composition have provided specific information on values of δ_a.

13.3.2 What Is the Role of Selective Sorting in Downstream Fining?

Selective sorting of transported particles has become a major focus of interest over the past several decades in relation to the problem of downstream fining. Sorting of particles can occur due to differences in grain shape and density, a somewhat neglected area of research, but the term as used here refers to sorting of material by size over distance under the assumption that variability in density and shape play a minor role in this process. A variety of factors contribute to sorting of transported material by size over distance.

13.3.2.1 How Does Selective Sorting Operate in Gravel-Bed Rivers?
Differences in entrainment thresholds for particles of different sizes represent a commonly identified mechanism for producing selective sorting in gravel-bed rivers. Entrainment of bed material in rivers was long thought to be strongly size dependent, with small grains being more easily mobilized than large particles (see Chapter 5). Based on consideration of total weight alone, in a mixture of bed material of different sizes and varying flows, small particles should be mobilized more frequently and deposited less frequently than large particles.

Conversely, coarse particles should be mobilized less frequently and deposited more frequently. If fine grains are entrained more easily, deposited less readily, and transported at higher rates than coarse grains, and also travel farther than coarse grains during episodes of transport, these fine grains will outrace the coarse grains, leading to separation of coarse and fine grains over distance. Over time, fine particles move downstream in greater quantities than coarse particles, producing downstream fining.

Differential mobility of particles alone is not sufficient to sustain downstream fining by selective sorting. In gravel-bed rivers it must be supported by a downstream decrease in transport competence (Frings, 2008). If a gravel-bed river is competent to transport coarse grains everywhere along its length, eventually these grains will be moved downstream, eliminating the longitudinal separation of sizes associated with selective transport. The fining pattern remains stable only if bed shear stress decreases downstream. Under this condition, the maximum size of grains that can remain mobile diminishes over distance, maintaining the downstream fining pattern. Fining occurs as coarse particles of steadily decreasing size are preferentially extracted from the transported load over distance. A downstream decrease in channel slope associated with a concave-upward profile is one way in which bed shear stress can decrease downstream. But here the interdependence between grain size and slope becomes apparent. Channel slope is directly related to grain size, implying that grain size influences slope, and that slope decreases as grain size decreases downstream (Table 13.2). However, the downstream decrease in grain size, if dominated by selective sorting, is only stable if a downstream decrease in bed shear stress exists, such as that produced by a concave-upward profile and corresponding downstream decrease in slope. Thus, grain size influences slope, and slope influences grain size.

Studies in small streams and laboratory flumes too short for substantial abrasion to occur highlight the importance of selective sorting in downstream fining (Ashworth and Ferguson, 1989; Paola et al., 1992; Ferguson et al., 1996b; Seal et al., 1997). Also, numerical modeling indicates that only a small degree of size-selective transport is required to produce downstream fining of gravel in streams with strongly concave-upward longitudinal profiles (Hoey and Ferguson, 1994). For such profiles, the reduction in channel slope and excess shear stress in the downstream direction for a constant discharge and channel width force size-selective deposition that leads to downstream fining in the absence of abrasion.

Fining through size-selective deposition, whereby coarse particles cease moving at steeper gradients than fine particles, indicates that fining is often coincident with bed aggradation (Cui et al., 1996). The downstream decline in slope associated with a concave-upward profile entails a decline not only in competence but also in sediment-transport capacity, given the dependence of both competence and capacity on bed shear stress. Profile concavity therefore has been related to long-term aggradational tendencies (Parker, 1991a). For a concave-upward longitudinal profile, the amount of fining is not highly sensitive to changes in controlling factors but tends to increase with decreasing discharge and increasing size-selectivity of transport, critical dimensionless shear stress, and sediment supply rate (Hoey and Ferguson, 1997). The rate of fining is more rapid than the rate of aggradation, suggesting that fining represents an important adjustment mechanism to disturbance of fluvial systems over modern timescales.

Where channel form is not constant, downstream fining can vary with changes in channel dimensions along a river, which influence the magnitude of bed shear stress. Along the Cosumnes River in California, upstream and downstream reaches with relatively large width–depth ratios have bankfull values of dimensionless bed shear stress (Θ) close to the threshold for transport, whereas an intermediate section with relatively small width–depth ratios has values of Θ about twice as large as those for upstream and downstream reaches (Constantine et al., 2003). Selective transport of sediment is most sensitive to differences in grain size for near-threshold conditions and less sensitive to grain size for high values of dimensionless shear stress. As a result, pronounced selective transport and associated downstream fining occur in upstream and downstream reaches, but not in the intermediate reach.

The influence of size-selective transport on downstream fining is somewhat at odds with the notion that some gravel-bed rivers exhibit equal mobility, in which particles of all sizes are entrained at the same critical dimensionless bed shear stress (see Chapter 5). Under these conditions, selective sorting and associated downstream fining should not occur. Nevertheless, even when equal mobility is satisfied locally within a river, discrete spatial variation in the median grain size of bed material, a phenomenon referred to as patchiness, can lead to selective sorting and downstream fining (Paola and Seal, 1995). Although equal mobility is satisfied locally within patches associated with a specific median grain size, section-averaged selective mobility arises because of spatial variability in local median grain size between patches. If the streambed consists of a coarse patch on one side of the channel and a fine patch on the other, the median grain sizes (d_{50}) will differ for the two patches. Particles within each patch may be equally mobile in relation to the d_{50} of the patch; i.e., all particles within each patch will be mobilized at the same critical bed shear stress, but the difference in d_{50} between patches results in differences in the critical bed shear stresses (τ_c) of the patches. Given that $\tau_c = \Theta_c(\rho_s - \rho)gd_{50}$ for each patch, for a constant Θ_c the value of τ_c for the fine-grained patch will be

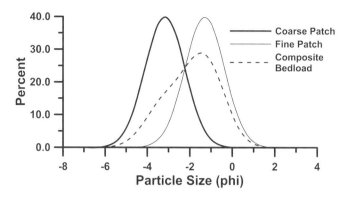

Figure 13.4. Hypothetical example of a river reach consisting of a coarse patch and fine patch of bed material with different median grain sizes (d_{50}). Each patch displays equal mobility of particles within the patch, but, given the differences in d_{50} between the patches, the transport rate of material in the fine patch exceeds the transport rate of material in the coarse patch, resulting in preferential mobility of fine sediment (dashed line) for the entire reach. The example assumes a twofold difference in transport rate between the patches (adapted from Paola and Seal, 1995).

less than τ_c for the coarse-grained patch. Moreover, because the bedload-transport rate is a function of $\tau - \tau_c$ (see Chapter 5), the transport rate for the fine-grained patch will be greater than the transport rate for the coarse-grained patch. As a result, more fine sediment than coarse sediment will be mobilized within the reach as a whole (Figure 13.4). This effect will be most pronounced at low excess shear stresses, when the difference in transport rate is greatest for coarse versus fine patches. As the magnitude of the shear stress increases, the effect will diminish. Moreover, selective sorting and downstream fining will be enhanced by well-sorted patches with distinctly different values of d_{50} and small variances in grain size within patches. The patchiness mechanism appears to account for downstream fining along the North Fork Toutle River near Mount St. Helens, Washington (Seal and Paola, 1995). The origin of patches remains a topic of inquiry, but it may be related to the breakdown of equal mobility in strongly bimodal sediments and sorting of bed material with unequal mobility into patches until equal mobility develops within patches (Paola and Seal, 1995). A possible unexplored consequence of this proposed mechanism is that over time it should lead to an overall reduction in patchiness within a patchy reach, which, in turn, should lead to a diminished influence on downstream fining.

Differences in distance of transport of grains of different sizes is another factor that can affect downstream fining. Even if particles of different sizes are mobilized at the same critical shear stress, if fine particles move downstream on average greater distances than coarse particles during discrete transport-effective events, a net increase in the flux of fine particles

relative to coarse particles will occur over distance. Particle-tracing studies indicate that travel distance is related to particle size (see Chapter 5), suggesting that size-selective distance of transport plays a role in downstream fining.

Local sorting of bed material at the bar-unit scale also can potentially influence downstream fining in gravel-bed rivers (Bluck, 1987; Clifford et al., 1993). According to this model, conversion of fine-grained unit bars to coarse-grained bars, which occurs by trapping of large clasts on bar heads, initiates a turbulence template that sweeps away fine sediments and promotes further trapping of clasts large enough to remain stable within this template. Through this process, stable gravel sheets are produced that resist erosion. The trapping process upstream affects the surface grain-size distributions of bars downstream by starving these bars of coarse material. The result is downstream fining of bar-head clast size and total-bar clast size over a sequence of successive bars (Bluck, 1987). Downstream fining over a sequence of bars has been observed in the lower Fraser River, British Columbia (Rice and Church, 2010). Along a 50 km gravel-bed reach that lacks major tributary inputs of sediment, the median grain size of surficial bed material decreases downstream at an average rate of 0.76 mm km^{-1}. The average downstream reduction in median grain size from head to tail over individual bars is 6.3 mm km^{-1}. Because the length of the bars is far too short for abrasion to occur, the high rate of fining demonstrates the substantial effectiveness of selective sorting at the bar-unit scale in gravel-bed rivers. Bar unit–scale sorting can exceed link-scale sorting by an order of magnitude.

13.3.2.2 How Does Selective Sorting Operate in Sand-Bed Rivers?

Low rates of downstream fining in large lowland sand-bed rivers reflect the limited effectiveness of both abrasion and sorting in these fluvial systems (Frings, 2008). Factors that restrict abrasion include substantial grain hardness, relatively small grain sizes, and high degrees of grain rounding. Abrasion rates for sandy rivers are usually too small to cause downstream fining. Selective sorting also is constrained because unimodal distributions of sand produce strong hiding-exposure effects that equalize entrainment thresholds among different grain sizes (see Chapter 5) and because a wide range of flows in large rivers generally are competent to fully mobilize sand grains of all sizes. As a result, suspended–bed material transport, rather than bedload transport, dominates size-selective sorting in sand-bed rivers (Wright and Parker, 2005; Frings et al., 2011).

Size-related sorting of suspended bed material is related to vertical gradients in concentrations of suspended particles, which are more pronounced for coarse grains than for fine grains (see Chapter 5). The concentration of coarse grains is highest near the bed, where flow velocities are lowest. By contrast, fine grains are distributed relatively uniformly over depth, including near the surface, where velocities are highest. Because of these differences, the downstream flux of coarse grains is less than the downstream flux of fine grains, where the fluxes are determined by integrating the product of flow velocity and sediment concentration over depth. Moreover, fine sands will be fully suspended during a wide array of flows, whereas coarse grains typically saltate – a relatively slow, episodic transport process. Downstream decreases in bed shear stress, which may develop in conjunction with a concave-upward longitudinal profile, lead to deposition of suspended coarse grains of progressively decreasing size over distance, which stabilizes the downstream fining pattern (Frings et al., 2011).

Other factors that contribute to downstream fining in sand-bed rivers include differential deposition related to bedform dynamics and overbank sedimentation during floods. Differential deposition involves coarse sands preferentially becoming buried in dunes compared with fine sands (Frings et al., 2011). Overbank deposition of fine suspended sand during floods may slow the rate of downstream fining by preferentially removing fine bed material from the channel, but the importance of this effect has yet to be fully determined (Wright and Parker, 2005; Frings et al., 2011).

13.3.3 What Causes Abrupt Transitions from Gravel to Sand in Downstream Fining?

As coarse gravel-size material introduced to rivers in headwaters fines in the downstream direction, the transition from bed material dominated by gravel to bed material dominated by sand can sometimes occur abruptly over short distances (Sambrook Smith and Ferguson, 1995; Radoane et al., 2008). Such abrupt transitions contribute to what has been referred to as the "grain-size gap," whereby rivers with bed material dominated by size fractions between 1 and 10 mm in size are uncommon. From an abrasional perspective, the grain-size gap suggests that pebble-sized materials do not always progressively diminish in size into sand-sized material (Yatsu, 1955). Chipping and grinding of boulders, cobbles, and large pebbles produce sand-size grains, but viscous damping of grain collisions as coarse material diminishes in size sets a lower limit on the size of gravel ($\approx$10 mm) (Jerolmack and Brzinski, 2010). According to this perspective, material in the size range of 1–10 mm in diameter truly is deficient in river systems.

An excess of sand input into a river system may result in selective deposition, particularly if the supply exceeds the transport capacity (Dade and Friend, 1998), resulting in clogging and even burying of gravel locally to produce an abrupt transition to a sand bed. Such a transition can occur where

a river flows through a highly erodible landscape that generates sand-size material (Sambrook Smith and Ferguson, 1995). It also can arise when human activity, such as mining, introduces large amounts of sand into a river (Knighton, 1999b).

A local reduction, or break, in channel slope can also produce an abrupt transition in bed material size by locally changing transport competence and capacity. Local reductions in slope are often associated with a rise in base level that initiates a depositional response in the lower reaches of a river (Sambrook Smith and Ferguson, 1995). The effectiveness of slope reductions in producing size-selective deposition, aggradation, and changes in bed-material texture from gravel dominated to sand dominated has been documented in flume experiments with a bimodal sediment feed of sand and gravel (Sambrook Smith and Ferguson, 1996).

Other work has proposed that the abrupt transition between sand and gravel occurs naturally through threshold-like changes in the mode of bedload transport in a river system with a concave-upward longitudinal profile transporting a mixed load of gravel and sand. The initiation of motion and transport of sand and gravel as bedload are highly sensitive to the ratio of sand versus gravel in the surface or subsurface bed-material size distribution (Wilcock, 1998; Wilcock and Kenworthy, 2002). An increase in sand content in the bed reduces the threshold of entrainment for both gravel and sand; in other words, sand and gravel are more easily mobilized as sand content increases. However, the ratio of the critical shear stress of sand (τ_{cs}) to the critical stress of gravel (τ_{cg}) varies in a highly nonlinear manner as the percentage of sand in the bed surface or subsurface increases (Figure 13.5). This ratio decreases rapidly for small percentages of sand and then stabilizes as sand content exceeds about 40% of the total weight of bed material. Thus, sand becomes more mobile relative to gravel as sand content increases, but this change in mobility occurs highly nonlinearly over a narrow range of sand content. Moreover, the greater the difference in the size of sand (d_s) versus gravel (d_g), the more mobile the sand becomes relative to the gravel as the percentage of sand increases (Figure 13.5).

In a gravel-bed river, sand can be introduced externally from hillslope erosion, from additions of sediment by tributaries, or through abrasional processes of chipping and grinding of large clasts. The sensitive dependence of transport conditions on sand content indicates that even small amounts of sand will promote differences in mobility of sand relative to gravel, resulting in preferential downstream transport of sand as bedload. Where shear stress decreases downstream, aggradation related to the decline in bed shear stress along with selective downstream transport of sand will produce an increase in sand content of bed material over distance. The interplay between declining bed shear stress and the rapid

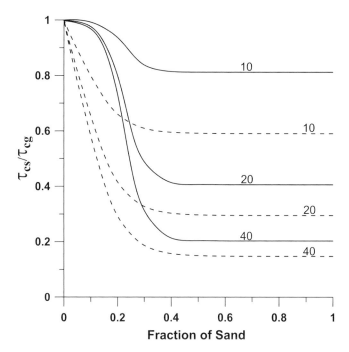

Figure 13.5. Variation in the ratio of the critical shear stress for sand (τ_{cs}) versus gravel (τ_{cg}) with the fraction of sand in the channel bed. Dashed lines refer to surface sand content with τ_{cs} and τ_{cg} based on surface grain size. Solid lines refer to subsurface sand content with τ_{cs} and τ_{cg} based on subsurface grain size. Numbers refer to ratio of gravel grain size to sand grain size (adapted from Wilcock and Kenworthy, 2002).

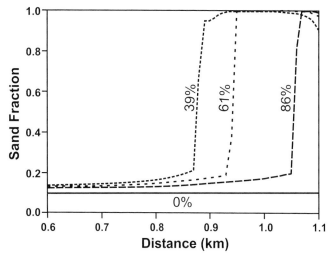

Figure 13.6. Simulation results using the two-fraction model of sand and gravel transport (Wilcock and Kenworthy, 2002) for an aggrading river system with a concave-upward longitudinal profile illustrating the development of an abrupt transition from gravel to sand at a particular distance downstream and migration of the transition downstream over time (% of total simulation time) (adapted from Ferguson, 2003).

increase in relative mobility of sand compared with gravel with increasing sand content, particularly in the domain of 10% to 40% sand content (Figure 13.5), results in an abrupt gravel–sand transition over a short distance along the river (Ferguson, 2003) (Figure 13.6). Upstream of the transition, the bed consists of gravel that is too immobile relative to sand to move farther downstream in great quantities, given the downstream decrease in bed shear stress. Alternatively, sand, which can move at much lower shear stresses, remains mobile and continues to move downstream as bedload. Therefore, downstream of the transition, the bed consists mainly of sand. A break in channel slope often occurs at the transition (Sambrook Smith and Ferguson, 1995; Ferguson et al., 2011; Venditti and Church, 2014), but given the interdependency between bed-material size and slope, it is unclear whether the break in slope is the cause or the effect of the transition. Although the transition has a tendency to migrate downstream over time (Ferguson, 2003; Ferguson et al., 2011) (Figure 13.6), this migration can be arrested by abrasion of gravel, basin subsidence, a rise in base level, and delta progradation (Cui and Parker, 1998; Parker and Cui, 1998; Blom et al., 2017b). In natural rivers, not all transitions from gravel to sand are abrupt; some occur over extended distances and

include discrete patches of fine and coarse bed material (Singer, 2008a). The exact mechanisms that promote abrupt versus extended transitions have yet to be conclusively determined.

Change in the dynamics of sediment suspension along rivers transporting mixed loads represents an alternative hypothesis for the development of gravel–sand transitions. Suspension of sand is related to the shear velocity (u^*) (see Chapter 5), and a downstream reduction in u^* can result in deposition of suspended sand on a gravel bed. Theoretical analysis indicates that the capacity for rivers to transport sand as wash load diminishes sharply over a fairly broad range of sand grain sizes if $u^* < 0.1$ m s^{-1}, resulting in rapid accumulation of sand on the bed (Lamb and Venditti, 2016). The abrupt loss in transport competence of sand as wash load over a broad range of grain sizes produces a step change in the median size of bed material as abundant amounts of suspended sand are deposited on the bed. The range in median grain size over which this step change occurs is between 1 and 5 mm, which encompasses the grain-size gap. Thus, median grain sizes of bed material within the range of 1 to 5 mm will only occur if $u^* \approx 1$ m s^{-1} – a situation that is probably rare in most rivers. Analysis of a gravel–sand transition within the Fraser River of British Columbia suggests that the transition between sand transported as wash load (fully suspended) and as bed-material load (intermittently suspended or bedload) occurs where channel gradient decreases at the downstream end of a gravel reach (Venditti et al., 2015). This reduction in gradient induces sudden deposition of coarse fractions of the sand wash load onto the bed during low flows. Subsequent high flows redistribute this deposited sand downstream over an extended reach containing predominantly sand with local patches of gravel. The presence of some coarse material within the sandy reach is probably related to enhancement of gravel mobility by the sandy bed.

13.3.4 How Do Tributary Inputs Affect Downstream Fining?

Downstream fining in some cases can be influenced by additions of sediment along a river by tributaries. Rivers are arranged in networks and do not occur in isolation; therefore, inputs of material from tributaries may disrupt patterns of fining. The extent to which tributaries influence the bed-material characteristics of a river depends both on the size of delivered sediment and on the quantity of delivered sediment in relation to the amount of flow added to the main river (Rice, 1998; Ferguson et al., 2006). Generally, the size of the delivered sediment is contingent on the particle-size distribution of source material, the distance this source material travels along the tributary, and the extent to which tributaries to the tributary alter trends in downstream fining along the tributary. The supply of material is directly related to the erodibility of rocks or soil, the relief, and the drainage area of the tributary watershed.

A tributary can modify bed-material characteristics in a river when it delivers large amounts of sediment coarser than the bed material in the river. Such inputs increase the median grain size of the bed material locally. Small tributaries with large fluxes of relatively coarse bedload, a situation typical of steep mountain tributaries, have a strong effect on the main river's bed-material characteristics by adding large amounts of sediment that the main river is not adjusted to transport, yet not contributing much additional flow to enhance the transport competence and capacity of the main river (Ferguson et al., 2006). If a tributary contributes relatively fine sediment, changes in grain size are less likely, because the river usually has the capacity to transport this sediment in large quantities, given that it is adjusted to transport coarse bed material. In such cases, inputs large enough to overwhelm the transport capacity of a river are required to produce deposition and a decrease in grain size (Knighton, 1991).

In drainage basins with spatial variation in lithology, downstream fining can be strongly influenced by differences in the size distributions of sediment supplied to the heads of tributaries by various rock types. In the drainage network of Standing Stone Creek of Pennsylvania, USA, tributaries contribute large amounts of coarse quartzite to the creek in the headwaters, but farther downstream, tributaries contribute relatively fine material associated with

other rock types (Pizzuto, 1995). A simple routing model indicates that over 80% of the observed downstream fining in Standing Stone Creek results from this spatial variation in sediment supply, with abrasion and selective sorting accounting for the remaining 20%.

Field studies of downstream change in grain size have noted that delivery of coarse material to a main river by tributaries can disrupt fining trends locally by increasing mean grain size (Church and Kellerhals, 1978; Knighton, 1980, 1982b; Swanson and Meyer, 2014). Such inputs locally reset the fining process. Recognition that some tributaries disrupt fining along rivers and others do not provides the basis for the concept of sedimentary links (Rice and Church, 1998; Rice, 1999) (Figure 13.7). In contrast to a topological link, or a stream segment bounded by confluences (see Chapter 2), a sedimentary link is a river segment bounded by confluences where tributary inputs substantially influence the bed-material characteristics of a river. Thus, if both tributaries that form confluences bounding a topological link along a river change the grain size of bed material at the upstream and downstream ends of this link, a sedimentary link is the same as a topological link. In many cases, however, a sedimentary link will extend over many topological links, as not all tributaries will influence bed-material characteristics of a river. Other lateral sources of sediment, such as inputs from bedrocks, outcrops, or highly erodible alluvial deposits, can also produce changes in bed-material characteristics that

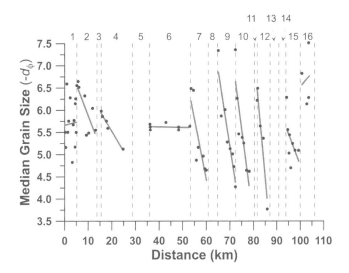

Figure 13.7. Demarcation of 16 sedimentary links (dashed vertical lines) along the Pine River, British Columbia, Canada. Solid lines indicate trends in median particle size (in $-d_\phi$ units). Exponential declines in size occur in 8 of the 16 links. Boundaries of most links are tributaries, with inputs of coarse material at many tributaries producing local increases in median grain size (from Rice and Church, 1998).

constitute sedimentary links (Figure 13.7). Generally, the decrease in particle size between links is defined by negative exponential relations that seem to represent the influence of selective sorting on downstream fining, given that abrasion is limited by relatively short sedimentary-link lengths (Rice, 1999).

Simulations based on a landscape evolution model suggest that downstream fining patterns can develop at the scale of river networks as a dynamic natural adjustment among water and sediment fluxes under equilibrium conditions, whereby the rate of erosion balances the rate of uplift, and no net storage or deposition of sediment occurs within the river system (Gasparini et al., 1999). For a constant supply of sand and gravel everywhere throughout an equilibrium drainage basin, the rate of bedload transport must increase downstream to ensure that no sediment accumulates within the basin. Because slope is decreasing toward the mouth of the basin, transport efficiency must increase to accommodate the required increase in bedload-transport rate – an outcome achieved in simulations based on a two-fraction model of bedload transport by an increase in the proportion of the bed covered by sand in the downstream direction. The resulting downstream decrease in mean grain size of bed material enhances the mobility of gravel and sand, thereby contributing to the downstream increase in sediment flux required by equilibrium conditions.

13.4 What Factors Influence Equilibrium Longitudinal Profiles of Alluvial Rivers?

The analysis of downstream fining, a common phenomenon in natural rivers, indicates that a concave-upward profile is typically viewed as necessary to produce this fining. Correspondingly, given the relation between channel slope and grain size (Table 13.2), a decrease in grain size downstream should also reinforce the downstream decrease in slope associated with a concave-upward profile. This interdependency between grain size and slope provides some insight into why many rivers have concave-upward profiles. Nevertheless, it ignores the influence of other important factors, such as discharge and sediment supply, on the shape of the longitudinal profile.

Given the difficulties of conducting experimental or field studies of longitudinal profile development over geologic time-scales, analysis of the influence of different factors on equilibrium longitudinal profiles of alluvial rivers has relied on mathematical modeling. Simulations with a model based on conservation of momentum and mass for water and conservation of mass for sediment provide insight into the independent influences of downstream changes in sediment caliber, discharge, and sediment load on the shape of the graded river profile (Snow and Slingerland, 1987). In a graded state, the profile has developed so that conservation of mass (water and sediment fluxes) exists

Table 13.3. Adjustments of longitudinal profile shape to change in controlling variables over distance (based on Snow and Slingerland, 1987).

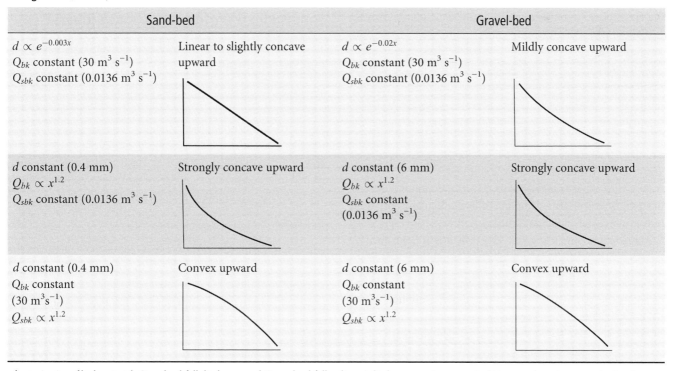

	Sand-bed		Gravel-bed
$d \propto e^{-0.003x}$ Q_{bk} constant (30 m³ s⁻¹) Q_{sbk} constant (0.0136 m³ s⁻¹)	Linear to slightly concave upward	$d \propto e^{-0.02x}$ Q_{bk} constant (30 m³ s⁻¹) Q_{sbk} constant (0.0136 m³ s⁻¹)	Mildly concave upward
d constant (0.4 mm) $Q_{bk} \propto x^{1.2}$ Q_{sbk} constant (0.0136 m³ s⁻¹)	Strongly concave upward	d constant (6 mm) $Q_{bk} \propto x^{1.2}$ Q_{sbk} constant (0.0136 m³ s⁻¹)	Strongly concave upward
d constant (0.4 mm) Q_{bk} constant (30 m³s⁻¹) $Q_{sbk} \propto x^{1.2}$	Convex upward	d constant (6 mm) Q_{bk} constant (30 m³s⁻¹) $Q_{sbk} \propto x^{1.2}$	Convex upward

d = grain size of bed material, Q_{bk} = bankfull discharge, and Q_{sbk} = bankfull sediment discharge. x axis represents distance and y axis represents elevation along the entire length of the river, and the form of the profile remains constant over time. For constant discharge and sediment load, and an exponential decrease in particle size over distance, the model predicts that sand-bed rivers will have straight or slightly concave-upward profiles, whereas gravel-bed rivers will have concave-upward profiles (Table 13.3). Thus, changes in the size of bed material have a stronger influence on profile concavity in gravel-bed rivers than in sand-bed rivers, assuming that abundant amounts of material are in transport. Discharge has a pronounced independent effect on concavity. As discharge increases as a power function of distance in the downstream direction for a constant grain size and sediment load, the gradient of the river decreases nonlinearly (Table 13.3). The increase in discharge promotes an increase in sediment transport capacity; however, because sediment load and grain size are constant, transport capacity must remain constant to maintain equilibrium. As a result, slope declines nonlinearly to offset the increase in discharge. The independent effect of a nonlinear downstream increase in sediment load is to produce profile convexity. Transport capacity must increase to accommodate this increase in load. For constant discharge and grain size, slope increases nonlinearly downstream to generate an increase in transport capacity.

Obviously, few, if any, natural rivers have discharges, particle sizes, or sediment loads that remain constant over their lengths. Discharge and sediment load usually increase downstream as power functions of distance, whereas particle size decreases exponentially. The influence of combinations of these downstream changes in discharge, sediment load, and particle size on the form of the longitudinal profile depends on the relative magnitudes of the different trends (Snow and Slingerland, 1987). Downstream increases in discharge that exceed downstream increases in sediment load will favor concave-upward profiles, and concavity will be reinforced by downstream reductions in particle size, especially in gravel-bed rivers. Downstream increases in sediment load that exceed downstream increases in discharge produce convex-upward profiles, although this convexity will be moderated by downstream decreases in particle size, particularly in gravel-bed rivers. The convex-upward form develops to maintain continuity of sediment transport under conditions of increasing concentrations of transported bed material in the downstream direction.

Mathematical modeling has also provided insight into the separate roles of abrasion and selective sorting in the development of concave-upward longitudinal profiles (Parker, 1991a, b). The relative importance of the two fining mechanisms depends mainly on the resistance of clasts to abrasion. High resistance results in lower rates of downstream fining than low resistance. Profile concavity increases with increased rates of downstream fining corresponding to low resistance to abrasion.

Modeling indicates that in the absence of uplift, subsidence, or base level change, downstream fining solely by selective sorting occurs only when a sediment flux imbalance exists over the length of the longitudinal profile, resulting in net aggradation (Blom et al., 2016). This scenario includes the development of sand–gravel transitions. Under these conditions, the longitudinal profile coevolves with the pattern of downstream fining from an ungraded state toward a graded state. For constant discharge and channel dimensions, the graded, equilibrium state corresponds to a uniform channel slope and constant mean grain size along the entire profile. Under the same conditions, downstream fining related to abrasion and selective transport of the products of abrasion results in the development of an equilibrium longitudinal profile characterized by mild concavity (Blom et al., 2016). A longitudinal profile in which vertical aggradation balances subsidence in the absence of selective sorting also has a concave-upward form (Sinha and Parker, 1996). The effect of tributary additions on profile form is difficult to discern through modeling (Sinha and Parker, 1996), but field evidence suggests that most sedimentary links exhibit pronounced concavity (Rice and Church, 2001). High supply rates of sediment by tributaries enhance concavity by increasing channel slope locally where tributary-supplied material accumulates in the main river (Blom et al., 2016). Few studies have considered the influence of spatial variations in channel dimensions on the form of the longitudinal profile, but theoretical considerations suggest that downstream increases in channel width at the network scale contribute to profile concavity (Ferrer-Boix et al., 2016).

13.5 What Factors Govern the Equilibrium Longitudinal Profiles of Bedrock Rivers?

13.5.1 How Are Uplift and Erosion Related to the Equilibrium Profile?

The problem of the development of equilibrium or steady-state profiles of bedrock rivers in areas of active tectonism has been examined by assuming that river incision into bedrock occurs through detachment-limited erosion related to stream power and that this erosional incision is balanced by uplift. Incision related to bed shear stress or stream power per unit area can be approximated as (Whipple and Tucker, 1999)

$$I_r = k_i A_d{}^m S^n \tag{13.3}$$

where I_r is the incision rate, k_i is an erosion coefficient, and the drainage area (A_d) is a substitute for discharge. This simple expression of detachment-limited erosion is referred to as the stream power incision model (SPIM) (Lague, 2014). The rate of change in the elevation of the river bed (dZ/dt) is the product of the difference between the rate of tectonic uplift (U_r) and the rate of incision:

$$\frac{dZ}{dt} = U_r - k_i A_d{}^m S^n \tag{13.4}$$

Assuming uplift rate and incision are balanced, the condition for steady state ($dZ/dt = 0$), and solving for S yields

$$S = k_{si}A_d{}^{\Phi} \tag{13.5}$$

where

$$k_{si} = \left(\frac{U_r}{k_i}\right)^{1/n} \tag{13.6}$$

$$\Phi = -m/n \tag{13.7}$$

Eq. (13.5) shows that the equilibrium channel slope is related to drainage area as a power function for spatially uniform uplift rate and bedrock erodibility. Values of k_{si} and Φ can be estimated based on regression analysis of data on S and A_d derived from digital elevation models (Sklar and Dietrich, 1998), but care must be exercised to ensure that the slope is determined along the path of the river rather than a path of steepest descent (Wobus et al., 2006a). The coefficient k_{si}, which represents a scaling metric that defines the channel gradient for $A_d = 1$ km^2, is referred to as the steepness index. It increases with rock uplift rate or erosion rate, but is unaffected by variations in lithology and climate (Whipple et al., 2013). The exponent Φ defines the rate of change in channel gradient as drainage area increases. Referred to as the concavity index, it varies over a narrow range of about -0.4 to -0.7 (Whipple, 2004; Lague, 2014) with a representative value of -0.45 (Wobus et al., 2006a). The negative values of the exponent confirm that most bedrock rivers have concave-upward longitudinal profiles. The lack of variation in Φ among drainage basins with different, but spatially uniform, tectonic, climatic, and lithologic conditions suggests that, in contrast to k_{si}, the concavity index is not sensitive to differences in these conditions. In other words, equilibrium profile concavity is similar for many rivers in different geographic settings.

13.5.2 How Has the Basic Model of Equilibrium Profiles Been Refined?

If U_r and k_i vary spatially within a drainage basin, which is often the case, this variation will influence concavity, and application of the model assuming constant values is not appropriate (Sklar and Dietrich, 1998). Consider a river system in which the rate of uplift and drainage area vary as power functions of distance along the river:

$$U_r = U_{ro}x^{\vartheta} \tag{13.8}$$

$$A_d = k_d x^{\gamma} \tag{13.9}$$

For these conditions, the steepness and concavity indices in Eq. (13.5) become (Kirby and Whipple, 2001)

$$k_{si} = \left[\left(\frac{U_{ro}}{k_i}\right)^{1/n} k_d{}^{-(\vartheta/\gamma n)}\right] \tag{13.10}$$

$$\Phi = -(m/n) + (\vartheta/\gamma n) \tag{13.11}$$

According to these relations, uplift rates that decrease downstream ($\vartheta < 0$) will enhance concavity and steepness, whereas rates that increase downstream ($\vartheta > 0$) will diminish concavity and steepness. Subsequent refinements have focused on incorporating an erosion-threshold function, stochastic distributions of erosive events, and effects of changing channel width into the SPIM (Snyder et al., 2003; Tucker, 2004; Finnegan et al., 2005; Yanites and Tucker, 2010; Lague, 2014; Yanites, 2018). All these factors have first-order nonlinear effects on the relation between channel slope and drainage area. For example, the stochastic-erosion effect can account for profile adjustments that otherwise would require a highly nonlinear form of the basic model (Eq. (13.3), $n > 3$) (Snyder et al., 2003), and channel width adjustments decrease the sensitivity of the steepness index to increases in rock uplift rate (Yanites, 2018).

13.5.3 How Does Bed-Material Transport Influence Bedrock Channel Erosion and Longitudinal Profiles?

An important issue related to the longitudinal profiles of bedrock rivers, or at least those in mountainous environments subject to tectonic uplift, is whether detachment-limited erosion provides an adequate representation of incision in these systems. Models based on detachment-limited erosion ignore effects related to sediment supply and the transport of sediment. By contrast, models based on transport-limited erosion attempt to account for these effects (Sklar and Dietrich, 2004; Turowski et al., 2007). The transport of bedload or saltating suspended load is viewed as an important abrasional agent promoting incision of bedrock channels.

Theoretical considerations suggest that at high rates of sediment supply, where coarse-sediment delivery exceeds bedload-transport capacity, bedrock becomes buried and transient alluvial cover protects the bed from erosion. If the supply of coarse sediment is zero, bedload transport rates are zero, and no abrasional tools exist to erode the bed (Sklar and Dietrich, 2001). Maximum rates of erosion occur at moderate rates of supply relative to bedload-transport capacity, when the bedrock is only partially exposed, but an abundance of coarse particles is transported to produce abrasion (Sklar and Dietrich, 2004) (Figure 13.8). Erosion rates by bedload abrasion also reach a maximum at intermediate levels of excess bed shear stress, or transport stage, because the impact frequency of saltating grains decreases as stress values approach the threshold for suspension.

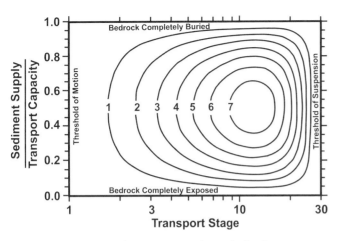

Figure 13.8. Variation of nondimensional river bedrock erosion rate (contour value $\times 10^{-15}$) as a function of transport stage (Θ/Θ_c) and relative sediment supply expressed as ratio of total sediment supply to total bedload transport capacity as predicted by an abrasion-saltation model of bedload transport (adapted from Sklar and Dietrich, 2004).

The saltation-abrasion bedload model does not consider the effects of suspended-sediment load on bedrock incision. Accounting for the effects of suspended sediment indicates that it can play an important role in bedrock erosion, particularly at high transport stages when suspended particles impact the bed through advection by turbulent eddies (Lamb et al., 2008b). Because impact velocities are enhanced by turbulent fluctuations at high transport stages, erosion rate increases with transport stage in the total load bedrock-erosion model and does not decrease to zero as predicted by the saltation-abrasion bedload model (Figure 13.9).

Application of the saltation-abrasion bedload model to the problem of longitudinal profile development indicates that under steady-state conditions rivers exhibit concave-upward profiles in which spatially uniform incision rates balance spatially uniform uplift rates (Sklar and Dietrich, 2008). This concave-upward form develops in response to the need to maintain the bed shear stress required to erode the channel bed at the rate of uplift as discharge increases and gradient decreases downstream. The transport rate of bedload progressively increases downstream, which increases abrasional effects, whereas the area of exposed bedrock decreases downstream, which reduces these effects. These trends balance one another to produce uniform rates of abrasional erosion along the length of the profile. The spatial rate of downstream fining has a strong influence on profile concavity, with increases in fining rate producing increases in concavity. Increases in uplift rate reduce concavity by increasing the supply of sediment to the river, which requires steeper gradients, particularly along downstream portions of the profile, to maintain erosion rates that balance uplift. Spatial gradients in runoff

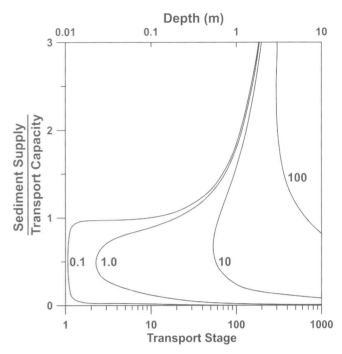

Figure 13.9. Variation of river bedrock-erosion rate (mm/yr) (values of contours) predicted for 1 mm sand by total load erosion model. Channel slope is constant at 0.053, so transport stage (Θ/Θ_c) is a function of flow depth. Relative sediment supply is expressed as a ratio of sediment supply per unit channel width versus bedload-transport capacity per unit width (adapted from Lamb et al., 2008b).

that produce faster rates of increase in discharge with drainage area enhance concavity, as do uplift rates that decrease from upstream to downstream.

13.6 How Do Longitudinal Profiles Adjust to Changes in External Forcing?

When influenced by external forcing, longitudinal profiles undergo a period of adjustment characterized by transient evolution (Whipple et al., 2013). External forcing is related to changes in relative base level or climate. Relative base level can be defined as sea level for a large river, a higher-order river for a tributary, or an active structural boundary that locally controls the elevation of the river. Changes in relative base level are caused by uplift, subsidence, incision of a main river, or changes in sea level. As base level changes, adjustments are transmitted headward along the river network. Changes in climate alter water and sediment fluxes throughout river networks, resulting in profile adjustments that are spatially extensive.

Except in some small alluvial systems, the period of transience associated with adjustment of longitudinal profiles occurs over geologic timescales (10^4 to 10^6 years) (Whipple, 2001). The temporal response of the river system depends on the duration of forcing (Whipple et al., 2013). An abrupt forcing, such as a rapid drop in sea level, produces an initial response following by progressive adjustments that restore the profile to its original steady-state form. Persistent forcing, such as continual uplift or a change in climate regime, produces a new steady-state condition with a different form. Responses to cyclic forcings, such as periodic changes in climate or waxing and waning of uplift, depend on the periodicity of forcing relative to the timescale of adjustment. Short periodicities of forcing in conjunction with long timescales of adjustment lead to a state of disequilibrium in which profile form is constantly evolving (see Chapter 6).

13.6.1 What Are Knickpoints and How Do They Develop?

The transient response of the longitudinal profile to external forcing often is characterized by the development of knickpoints – positions along the river where the channel gradient locally increases (Whipple et al., 2013). Knickpoints that develop in response to external forcing demarcate spatial transitions between zones of active adjustment to disturbance downstream versus zones that have yet to adjust to disturbance upstream. These types of knickpoints are migratory, progressively moving upstream to produce change in the elevation of the channel bed. Not all knickpoints are migratory; these features can develop in steady-state longitudinal profiles if the river locally encounters a change in lithology or input of different material from a tributary that requires a local adjustment in channel slope to maintain continuity of bed-material transport (Whipple

et al., 2013; Lague, 2014). Static knickpoints may also form through scouring of bedrock at confluences in steep mountain rivers (Hayakawa and Oguchi, 2014). Knickpoints can become anchored even in transient landscapes, at least temporarily, where they migrate into resistant strata or cross active structures such as faults, or where drainage area diminishes to the extent that headward migration can no longer be sustained by channel erosion, given the discharge and sediment supply at that location (Whipple et al., 2013).

Knickpoint morphology varies, but two basic forms can be identified: vertical-step knickpoints and slope-break knickpoints (Figure 13.10). Vertical-step knickpoints are short, steep reaches of a river that in some cases are vertical or nearly vertical. Examples in bedrock rivers include waterfalls and rapids; vertical knickpoints in alluvial channels are referred to as headcuts (Figure 13.11). These types of knickpoints typically develop in response to a rapid decrease in base level, resulting in the migration of the knickpoint upstream to adjust the spatial distribution of bed elevations along the longitudinal profile to the new elevation of the river mouth. Slope-break knickpoints define transitions between separate concave-upward segments of a longitudinal profile with different steepness indices. These knickpoints usually form in response to persistent changes in rates of external forcing rather than to abrupt episodes of forcing (Whipple et al., 2013). Examples include changes in the rate of rock uplift or changes in climate that alter the efficacy of river incision. Downstream of the slope break, the longitudinal profile has achieved a steepness adjusted to the new forcing conditions, whereas upstream of the knickpoint, channel gradients have

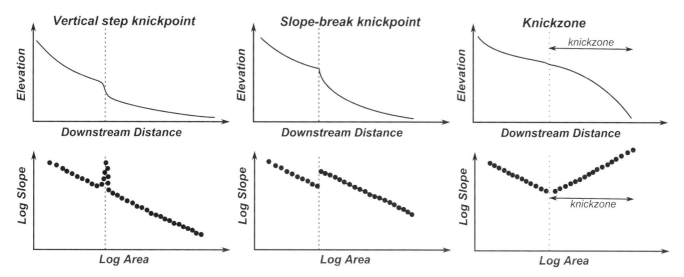

Figure 13.10. (top row) Expressions of vertical-step knickpoint, slope-break knickpoint, and knickzone in longitudinal profiles. (bottom row) Slope–area relations corresponding to the two types of knickpoints and the knickzone. Slope–area relations are for drainage areas greater than the critical drainage area below which nonfluvial processes dominate (e.g., mass movement). Dashed vertical lines in the first two columns indicate location of knickpoints, and dotted vertical line in the last column indicates a possible knickpoint location (adapted from Lague, 2014).

Figure 13.11. (a) Waimea Falls, Oahu, Hawaii. (b) Headcut in small tributary to the Sangamon River, Illinois. (c) Headcut in Santa Rosa Wash, Arizona.

not yet adjusted to achieve balance with the new conditions. In some cases, vertical-step knickpoints can occur in conjunction with slope-break knickpoints where the vertical-step knickpoint is located immediately downstream from the slope-break knickpoint (Berlin and Anderson, 2007; Whipple et al., 2013). Related features, knickzones, correspond to prolonged changes in profile shape, particularly convex or straight segments of the longitudinal profile (Lague, 2014). Whereas knickpoints are local features, knickzones typically extend over several kilometers to as much as hundreds of kilometers.

The identification of different types of knickpoints is often conducted using logarithmic plots of channel slope and drainage area derived from analysis of digital elevation data (Figure 13.10). Vertical-step knickpoints are defined by an abrupt spike in the gradient–area relationship as the channel slope increases greatly over a small increment in drainage area. Slope-break knickpoints are characterized by an abrupt offset in the channel slope as the river abruptly transitions between profile segments with different steepness indices over a small increase in drainage area. For knickzones where profile shape transitions from concave-upward to straight, the slope becomes constant with increasing drainage area, whereas for transitions from concave-upward to convex-upward profiles, slope increases with drainage area.

13.6.2 What Processes Are Involved in the Evolution of Vertical-Step Knickpoints?

The migration of vertical-step knickpoints involves both advective and diffusive processes. Advective processes are related to retreat of the knickpoint upstream, whereas diffusive processes tend to flatten the knickpoint by producing rotation of the knickpoint face. Headcuts in noncohesive alluvial materials are dominated by diffusive processes that

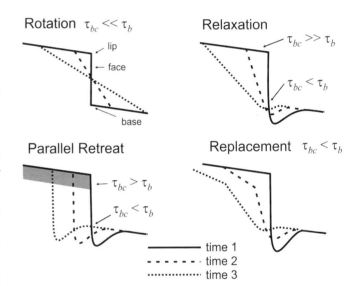

Figure 13.12. Longitudinal bed profiles showing different modes of evolution of vertical-step knickpoints through time (adapted from Gardner, 1983 and Frankel et al., 2007).

flatten the knickpoint to produce a sloping channel bed (Figure 13.12). Net erosion occurs upstream of the headcut lip, where the water surface steepens as flow moves over the lip and bed shear stress greatly exceeds the critical shear stress (Gardner, 1983). Deposition occurs downstream where the load arriving from upstream exceeds the transport capacity. Advective processes dominate when a resistant upper layer overlies an easily erodible layer at the base of the headcut face (Gardner, 1983; Dey et al., 2007) (Figure 13.12). Under these conditions, the bed shear stress at the top of the knickpoint is less than the critical shear stress, but the bed shear stress at the base of the knickpoint face exceeds the critical shear stress (Gardner, 1983). As a result, the headcut undergoes parallel retreat, with the rate of retreat proportional to the rate of flow and downstream sediment transport (Dey et al., 2007).

Some experiments in uniform, highly resistant cohesive sediments indicate that headcut migration is dominated by inclination without downstream deposition, resulting in flattening of the knickpoint as it migrates upstream (Begin et al., 1981; Gardner, 1983). This inclinational mode of knickpoint evolution has been referred to as relaxation (Frankel et al., 2007) and involves erosion of the knickpoint face that decreases its inclination without incision at or upstream of the knickpoint lip (Figure 13.12). Testing of a theoretical model relating the timescale of erosion above the headcut face to the timescale of erosion at the base of the face reveals that headcuts in resistant cohesive material can undergo parallel retreat rather than flattening during upstream migration (Stein and Julien, 1993). The mode of retreat, relaxation versus parallel, depends on the Froude number of the flow upstream of the headcut and the ratio of the headcut height to the flow depth upstream of the headcut. For Froude numbers less than 1, i.e., subcritical flow upstream of the headcut, parallel retreat will generally occur if the height of the headcut face exceeds four times the normal flow depth. This finding calls attention to the importance of a ventilated plunging nappe in producing scour at the base of the headcut face. Once a plunge pool has formed, entry of the nappe into the pool generates a turbulent impinging jet that promotes scour (Flores-Cervantes et al., 2006) and produces strongly recirculating flow at the base of the headcut face that enhances migration by eroding this face (Alonso and Bennett, 2002; Bennett and Alonso, 2005). Not all vertical knickpoints are steep enough or high enough to have ventilated nappes, and in some cases flow remains attached to the face of the knickpoint to produce an unventilated waterfall. In such cases, the erosive mechanism for knickpoint evolution becomes the shear-induced fluvial erosion along the face of the knickpoint. Such knickpoints generally are dominated by advective dynamics that produce parallel retreat, but they can exhibit decreases in the inclination of the knickpoint face during high flows, which have relatively strong diffusive effects compared with low flows (Bressan et al., 2014).

Knickpoint evolution in moderately resistant cohesive sediments involves knickpoint replacement, whereby incision at and upstream of the knickpoint lip is accompanied by erosion of the knickpoint face (Gardner, 1983) (Figure 13.12). This diffusive mode of knickpoint development does not involve deposition downstream of the knickpoint face. Knickpoint replacement is also characteristic of river incision into steeply dipping beds of resistant bedrock (Frankel et al., 2007).

Undercutting of the knickpoint face is often invoked to explain the retreat of vertical knickpoints, especially those with a resistant layer at the lip of the knickpoint (Stein and LaTray, 2002). In bedrock channels, the persistence of waterfalls during upstream migration typically has been attributed to undercutting of an overlying strong caprock layer by plunge-pool erosion of a weak rock layer at the waterfall base, leading to mass failure of the caprock (Gilbert, 1907). Many waterfalls do not exhibit basal undercutting, nor are all waterfalls associated with strong caprock overlying weak rock. Bedrock knickpoints commonly consist of a series of closely spaced waterfalls that vertically erode, or drill, into rock to create plunge pools (Scheingross et al., 2017a). Although clear, free-falling water can generate high bed shear stresses downstream of waterfalls (Pasternack et al., 2007), vertical drilling occurs mainly through particle impact abrasion with bedrock, which enlarges the plunge pool until deposition of sediment within the evolving pool protects the bottom, at which point deepening ceases (Scheingross et al., 2017b). Although waterfall retreat rate has been related to discharge over the waterfall and the area of the waterfall face (Hayakawa and Matsukura, 2003), vertical drilling highlights the importance of bedload transport and the availability of abrasional tools in controlling the rate of knickpoint retreat (Cook et al., 2013). Propagation of multi-waterfall knickpoints increases with an increase in the number of waterfalls and a decrease in the spacing of the waterfalls (Scheingross et al., 2017b). Toppling of rock columns in uniform bedrock with vertical joints can explain retreat in the absence of undercutting (Lamb and Dietrich, 2009).

The morphology and migration rate of vertical-step knickpoints in bedrock is strongly influenced by geology, particularly the strike and dip of rock layers (Miller, 1991c; Frankel et al., 2007) and the strength of bedrock (Baynes et al., 2018a). Resistant, upstream dipping rocks that become exposed during river incision are especially effective at producing persistent vertical-step knickpoints (Miller, 1991c). Fracture characteristics, weathering, and mass wasting of rock forming the knickpoint face are important factors governing migration rates of waterfalls (Haviv et al., 2010).

13.6.3 What Factors Influence Knickpoint Dynamics at the Drainage-Basin Scale?

At the drainage-basin scale, a sudden or progressive decrease in base level triggers a wave of incision that migrates headward, changing the longitudinal profiles of rivers within the drainage network (Whipple and Tucker, 1999). Incisional response may occur in the form of successive waves of migrating knickpoints (Loget and van den Driessche, 2009; Cantelli and Muto, 2014; Grimaud et al., 2016). The rate of knickpoint retreat is related to upstream discharge, sediment supply, and spatial variability in substrate erodibility. Migration rate also appears to increase with uplift rate (Whittaker and Boulton, 2012). Field evidence indicates that migration of knickpoints is strongly influenced by drainage area, which serves as a surrogate for discharge in the SPIM (Crosby and Whipple, 2006). Knickpoint propagation is more rapid and extends over greater distances where the drainage area of a river within the network is large than where it is small (Bishop et al., 2005; Castillo et al., 2013). Given the dependence of knickpoint retreat on drainage area, retreat of a knickpoint along a main river with a large drainage area can stall where the knickpoint propagates into a tributary with a small drainage area, resulting in the formation of hanging valleys, where the tributary joins the main river at a knickpoint or waterfall (Wobus et al., 2006b; Crosby et al., 2007).

Overall lowering of the longitudinal profile resulting from river incision has been described as a diffusive process:

$$\frac{\partial z}{\partial t} = k_{di} \frac{\partial^2 z}{\partial x^2} \qquad (13.12)$$

where k_{di} is a diffusion coefficient (Begin et al., 1981). This simple diffusion model does not explicitly account for knickpoint migration, an advective process. It does, however, call attention to the need to consider diffusive processes in transient adjustments to external forcing. Refined models based on the SPIM (Eq. (13.3)) now include a diffusive term on the right-hand side of Eq. (13.4) to predict profile adjustments to tectonic uplift (Pritchard et al., 2009).

13.6.4 How Can Transient Adjustments Be Identified from Analysis of Longitudinal Profiles?

Evaluating whether river profiles reflect steady-state or transient conditions has been performed using plots of slope versus drainage area (Figure 13.10), but noisy topographic data can complicate interpretations of these plots. In particular, step-like changes in channel elevation over distance associated with digital elevation models introduce imprecision into determinations of channel slopes. An alternative approach involves integrating Eq. (13.5) under the assumption of spatially invariant uplift and erodibility (Perron and Royden, 2013; Royden and Perron, 2013). This approach yields the model defined by Eq. (2.13) and (2.14) relating channel-bed elevations (Z) to a spatial integral of drainage area χ, which has units of distance (see Chapter 2). Values of χ are determined from Eq. (2.14) using assumed values of m/n between 0 and 1. A series of regression analyses are then performed for measured values of Z and the sets of values of χ computed for different assumed values of m/n. The regression analysis for the set of χ values with the highest coefficient of determination (R^2) is selected as the best fit. A plot of bed elevations versus χ (a chi plot) is produced using values of χ for the best-fit regression analysis. If a profile is in steady state, the plot should be linear. Moreover, a comparison of Eq.

(13.6) and (2.13) shows that, for $A_o = 1$ km^2, the slope of a linear chi-plot profile represents the steepness index (k_{si}).

An advantage of the chi-plot approach is that it removes the effect of drainage area, so that locations within a drainage network with similar elevations have similar values of χ, even if the drainage areas of those locations differ. Thus, all rivers in steady state within an area of spatially uniform uplift and resistance should exhibit collinear chi plots. Transient adjustments of longitudinal profiles, characterized by knickpoints and knickzones, are depicted on chi plots as local increases in change of elevation per unit change in χ, which produce positive deviations from a linear profile (Neely et al., 2017) (Figure 13.13). By removing the effect of drainage area, vertical-step knickpoints moving through a stream network from a common origin in response to an abrupt drop in base level should plot at the same χ coordinate on chi plots. In other words, the analysis collapses knickpoints to a common plotting location, providing insight into the systemwide response of stream profiles to disturbance (Figure 13.13). Differences in slopes of transformed profiles upstream and downstream of knickpoints also define differences between adjusted sections of profiles below knickpoints and unadjusted sections of profiles above knickpoints.

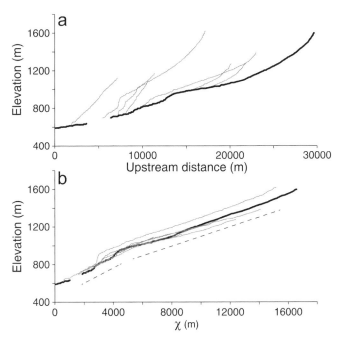

Figure 13.13. (a) Longitudinal profiles of the main stem (black) and tributaries (gray) in the Big Tujunga drainage basin, California, USA. (b) Chi plot showing collapse of knickpoints (local steep segments) on main stem and tributary to similar χ coordinates (3000–4000 m) and collinearity of channel steepness (indicated by dashed lines) for main stem and tributaries downstream and upstream of knickpoints. Gap in main stem in (a) and (b) is location of Big Tujunga Reservoir (from Perron and Royden, 2013).

13.7 How Does Vertical Adjustment Influence the Morphology of Steep Channels?

Differences in landscape characteristics, such as valley slope and the size of supplied sediment, can lead to differences in channel planform among rivers (see Chapter 8). Meandering versus braiding represents an example of planform adjustment to differences in potential stream power and grain size. In steep mountain environments, bedrock control and narrow valley bottoms limit horizontal morphological adjustments, including changes in planform. Under these conditions, accommodation of imposed flow and sediment regimes may be possible only through adjustments of the vertical dimension of channel morphology. These adjustments involve bar unit–scale modifications of the longitudinal profile of rivers. Settings in which such adjustments occur have been defined by classification schemes that relate differences in the morphology of mountain river channels to position along the longitudinal profile (Montgomery and Buffington, 1997) (Table 13.4; Figure 13.14).

13.7.1 How Is Channel Morphology Related to the Longitudinal Profile in Mountain Rivers?

In the most headward parts of mountain watersheds, where channels are carved into steep hillslopes, colluvial processes, such as debris flows, dominate transport, scour, and deposition (Figures 13.14 and 13.15). These small hillslope channels may not connect directly to the stream network. Downstream of the steepest channels, fluvial processes become prominent and channels sequentially transform from cascade, to step-pool, to plane bed, to pool-riffle, to dune-ripple morphologies as channel slope decreases (Table 13.4; Figure 13.14). Although sequential changes along individual streams are common, substantial overlap exists among channel-gradient ranges of the different types when groups of streams are considered (Montgomery and Buffington, 1997; Chartrand and Whiting, 2000; Wohl and Merritt, 2008).

Table 13.4. Channel types along the longitudinal profile.

Type	Bed characteristics
Cascade	Steep gradients (median $S \approx 0.12$ m/m, range 0.03 to 0.2), confined by narrow valley walls, randomly distributed boulders and cobbles, partial channel-spanning steps occur locally
Step-pool	Moderately steep gradients (median $S \approx 0.04$ m/m, range < 0.01 to 0.19), confined by valley walls, numerous channel-spanning steps and intervening pools
Plane bed	Moderate gradients (median $S \approx 0.025$ m/m, range < 0.01 to 0.075). Armored gravel beds that lack rhythmic bedforms and are characterized by even, featureless topography; moderate confinement
Pool-riffle	Moderately low gradients (median $S \approx 0.015$ m/m, range 0.001 to 0.045; gravel or mixed sand and gravel beds with pools, riffles, and point bars within an alluvial, often sinuous, channel; unconfined
Dune-ripple	Low gradients (median $S < 0.001$ m/m). Sand bed or sand bed with small amount of fine gravel; bedforms prominent, including dunes and ripples; unconfined

(after Montgomery and Buffington, 1997; Chartrand and Whiting, 2000; Wohl and Merritt, 2008)

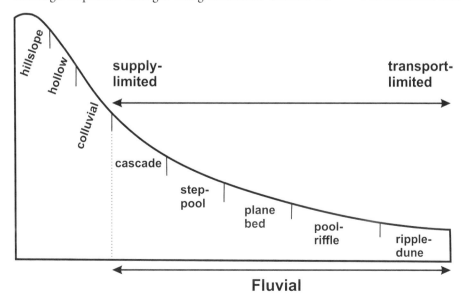

Figure 13.14. Changes in channel types along the longitudinal profile (after Montgomery and Buffington, 1997). Dashed vertical line indicates transition from hillslope to fluvial processes.

Figure 13.15. Colluvial channels on a steep hillslope near Mt. Crested Butte, Colorado, USA.

Pool-riffle and dune-ripple channel types are representative of meandering lowland rivers in which sediment supply is abundant and often exceeds transport capacity, resulting in net deposition that produces alluvial valley fills and broad floodplains. Here, river channels are carved into alluvium. Cascade and step-pool types develop in steep headwaters where transport capacity is relatively high compared with sediment supply, limiting deposition and the accumulation of alluvium. If transport capacity consistently exceeds supply, a bedrock channel will develop. Sub–reach scale variations in alluvial cover are common, producing mixed alluvial and bedrock channel segments. Plane-bed channels represent transitional forms between supply-limited and transport-limited types but typically have alluvial beds. These channels have relatively high width–depth ratios and relative roughness (d_{90}/D_{bk}) (Figure 13.16).

In cascade and step-pool channels, increases in roughness reduce the transport capacity to accommodate the limited sediment supply (Montgomery and Buffington, 1997). In cascade channels, this increase in roughness primarily occurs through the accumulation of coarse particles that only rarely are mobilized by the stream. These obstacles produce high flow resistance through tumbling flow over individual grain steps and jet-and-wake flow around individual grains (Figure 13.16). Relatively immobile large woody debris (LWD) may contribute to flow resistance where LWD becomes lodged behind large individual clasts.

Step-pool channels exhibit distinctive vertical adjustment of the longitudinal profile of the river locally. The beds of these channels are organized into a series of channel-spanning steps comprised of large clasts and intervening pools containing relatively fine sediment (Figure 13.16). Logs may partly or even wholly constitute steps (Marston, 1982; Wohl et al., 1997; Gomi et al., 2003), and in some cases steps may consist entirely of bedrock (Duckson and Duckson, 1995; Wohl and Grodek, 1994) (Figure 13.16). Steps lead to substantial loss of potential energy through the abrupt drop in bed elevation, and additional energy loss occurs as flow transitions from supercritical over the steps to subcritical within the pools.

13.7.2 What Are the Dynamics of Step-Pool Channels?

Over the past several decades, a considerable amount of research has examined the dynamics of step-pool channels, which now are recognized as a distinct channel type in high-gradient mountain streams (Zimmermann, 2013). The vertical dimension of channel form is of primary importance in step-pool systems, given that these systems are characterized by pronounced local variation in bed elevations. Investigations of step-pool channels have sought to characterize formative processes, morphological characteristics, morphological stability, hydraulics/flow resistance, and sediment transport.

13.7.2.1 How Do Step-Pool Structures Form?

In general, the formation of step-pool structures in high-gradient streams requires large clasts, usually boulders, that serve as keystones for the development of steps and relatively small width–depth ratios, whereby the largest clasts occupy a substantial proportion of the channel width (Grant et al., 1990). The formation of step-pool channels has focused on two primary hypotheses: 1) hydraulic origin and 2) the jammed state mechanism. Hydraulic theories include several variants relating steps to standing waves that develop at Froude numbers equal to or greater than 1 (Church and Zimmermann, 2007). The most prominent of these theories proposes that steps are analogous to antidunes (Whitaker and Jaeggi, 1982; Chin, 1989). During high discharges with supercritical flow, antidunes develop within steep channels. Mobile large clasts stall on the crests of the antidunes, trapping other large particles that armor these features to form steps. As stage

Figure 13.16. (a) Cascade, Little Cottonwood Creek, Utah, USA. (b) Boulder and bedrock steps, Cement Creek, Colorado, USA. (c) Boulder steps, Cement Creek, Colorado. (d) Bedrock steps, Fall Creek, Indiana. (e) Gravel plane bed, River Twiss, Yorkshire, United Kingdom.

declines, tumbling flow over the accumulated clasts produces scour downstream, initiating a pool.

Antidunes develop under specific hydraulic conditions, i.e., Froude numbers greater than 1, and exhibit a narrow range of wavelengths (see Chapter 4). Some field data indicate that estimated hydraulic conditions for step formation as well as step wavelengths conform to those for antidunes (Chin, 1999a; Chartrand and Whiting, 2000; Lenzi, 2001). Although this

correspondence has been viewed as support for the antidune theory, detailed flume experiments show that steps develop in the absence of antidunes, even though hydraulic conditions and wavelengths for some of the experimental steps are similar to those predicted for antidunes (Weichert et al., 2008). Instead, the agreement appears to represent a general adjustment of streams to maximize flow resistance and bed stability; in sand-bed streams, this adjustment involves the formation of antidunes, whereas in high-gradient mountain streams transporting boulders, it involves the formation of step-pools. This perspective is consistent with general extremal hypotheses proposing that step-pool structures develop through self-organization of bed structure to maximize flow resistance (Abrahams et al., 1995) and minimize stream power (Chin and Phillips, 2007).

According to the jammed state hypothesis, the development of steps and pools is not caused by active structuring of the bed by a transport-effective flow. Instead, it is caused by clogging, or jamming, of the channel either by large immobile keystones delivered to the channel by hillslope processes or by mutual grain-on-grain interference that inhibits downstream movement of large clasts during events otherwise capable of transporting these clasts (Church and Zimmermann, 2007). Drawing upon physical theory of a jammed state in mobile granular media, the hypothesis proposes that steps represent expressions of such a state in high-gradient streams. Once steps are formed by jamming, local scour by plunging jets downstream of the steps produces pools (Comiti et al., 2005). Key factors influencing the development of a jammed state are the jamming ratio, or ratio of channel width to coarse particle size (W_{bk}/d_{84}), the transport stage, or ratio of actual to critical bed shear stress (τ_b/τ_{bc}) where $\tau_b/\tau_{bc} > 1$, and the concentration of transported bed material. Jammed states tend to develop for low values of all three key factors. In particular, experimental results indicate that step-pool stability increases as jamming ratio decreases for $W_{bk}/d_{84} < 6$ (Zimmermann et al., 2010). Numerical simulations show that, in agreement with the jammed state hypothesis, jamming leads to the development of steps at high flows and that steps persist at low flow under sediment-starved conditions (Saletti et al., 2016). The exact role of sediment supply in step-pool stability remains somewhat uncertain; experimental work demonstrates that pulses of fine mobile gravel, which increase the concentration of transported bed material, can result in overall coarsening of the channel bed, which presumably enhances morphological stability (Johnson et al., 2015).

The jammed state hypothesis is consistent with field studies indicating that steps develop randomly in streams in conjunction with the occurrence of keystones that are large in relation to channel width and that trap other particles during transport-effective high flows, thereby forming an interlocking, imbricate arrangement of clasts (Zimmermann and Church, 2001; Billi et al., 2014). Many of these relatively immobile keystones are delivered to the streams by debris flows or other mass movement processes. In experiments in which step-initiating large clasts are mobilized, the vast majority of steps form in conjunction with deposition or exposure of a large keystone clast on the bed (Curran and Wilcock, 2005; Curran, 2007). In both the field and the laboratory, the location of keystone clasts is random, promoting irregularity in step spacing (Zimmermann and Church, 2001; Curran and Wilcock, 2005).

13.7.2.2 How Stable Are Step-Pool Structures?

The stability of step-pool units has been examined in relation to large floods that may be capable of changing or even eliminating these units. Experimental results indicate that instability of step-pool units is mainly related to movement of keystone clasts within steps (Zhang et al., 2018). Simple estimates of flow competence for the largest particles in step-pool structures suggest that these units are stable for flows with recurrences intervals of decades or even centuries (Grant et al., 1990; Chin, 1998). Such analyses do not consider the interlocking, imbricated structure of most steps, which has a strong influence on stability (Zimmermann and Church, 2001). Documentation of changes in step-pool morphology before and after floods provides insight into the stability of these features in response to potential channel-shaping flows (Lenzi, 2001; Turowski et al., 2009; Molnar et al., 2010). Floods with recurrence intervals of 20 to 50 years are capable of transforming reaches with step-pool units into riffle-pool sequences, forming new step-pool structures in reaches where none existed previously, and partly or completely rearranging existing step-pool morphology, including infilling pools with sediment. More frequent events, with recurrence intervals of one to five years, do not substantially modify step-pool morphology but can produce scour of infilled pools and armoring of pool beds, resulting in increases in hydraulic resistance (Zimmermann and Church, 2001; Lenzi, 2001).

13.7.2.3 What Are the Morphological Characteristics of Step-Pool Structures?

Identification of pool-riffle units for analysis of morphological characteristics involves both visual observation in the field (Wooldridge and Hickin, 2002) and analysis of longitudinal profile data (Zimmerman et al., 2008). Studies of the morphology of step-pool units have focused mainly on dimensions of length and height (Chin and Wohl, 2005). Several different metrics have been used, including step wavelength (λ_s), step spacing (L_s), step amplitude (A_{st}), step height (H_s), and step drop height (ΔE_{st}) (Figure 13.17). Values of ratios such as height to length can differ by as much as 30% because of differences in definitions of metrics, leading to calls for standardization (Nickolotsky and Pavlowsky, 2007).

Step height is primarily determined by the size of clasts comprising the step, whereas step wavelength is positively related to drainage area and negatively related to channel slope (Chin, 1999b). Wavelength-amplitude ratios (λ_s/A_{st}) have been explored to determine the extent to which step-pool morphology is similar to meandering in the vertical dimension (Chin, 1999b, 2002). Ratios range from 17:1 to 5:1, with some evidence that values increase in the downstream direction

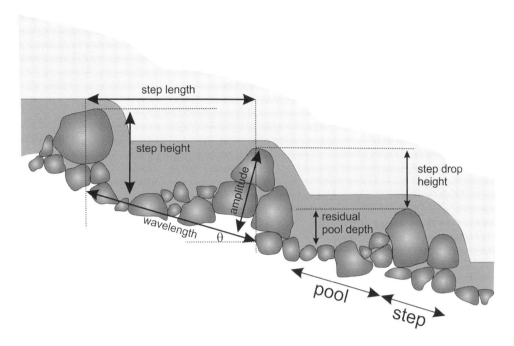

Figure 13.17. Morphological characteristics of step-pool sequences. Light shading corresponds to high flow stage and dark shading to low flow stage.

(Chin and Wohl, 2005), consistently with the downstream increase in step wavelength with increasing drainage area and decreasing slope (Chin, 1999b). Bed step wavelength or spacing generally ranges between one and four channel widths with a mode of about one to two channel widths (Chin and Wohl, 2005). Probability distributions of step length differ significantly from those for random series of spacings, suggesting that step spacing is not entirely random and that hydraulic control influences step-length distributions (Milzow et al., 2006). In particular, the scouring of pools should reflect the influence of the upstream step on flow dynamics within pools over a variety of flow stages (Comiti et al., 2005). Large step heights with relatively shallow flows over the step result in nappe flow characterized by a free-falling jet that promotes scour over a constrained distance, producing deep, short pools. As step height diminishes and flow depth increases, flow transitions from nappe flow, where flow free falls over the step, to impinging flow, where free fall no longer occurs, but the step directs a submerged jet downward toward the bed. Under these conditions, scour depth and length depend on the angle of the jet and the degree to which impinging flow produces erosion. Generally, this effect will produce longer, shallower pools than nappe flow. Drop height scaled by channel width ($\Delta E_{st}/W_{bk}$) exhibits a strong positive relation to channel slope (S), indicating that the heights of drops in relation to channel size increase as the steepness of the channel increases (Chartrand et al., 2011). This relation may reflect the increasing availability of large clasts to form steps in upstream locations. As channel width increases downstream and the size of clasts decreases with decreasing channel slope, relative step heights decrease, which should increase pool lengths and decrease pool depth.

13.7.2.4 How Do Step-Pool Structures Influence Flow and Sediment Transport?

For regularly spaced step-pools with pools characterized by reverse bed slopes, step steepness, H_s/λ_s, should be slightly greater than the channel slope (S) to maximize flow resistance, or $1 \leq (H_s/\lambda_s)/S \leq 2$ (Abrahams et al., 1995; Canovaro and Solari, 2007). Both flume and field data for step-pools exhibit considerable scatter that extends beyond this narrow range, especially for gradients less than 0.07 (Chartrand et al., 2011; Zimmermann, 2013). The available evidence thus does not strongly support the contention that the morphology of clast-structured step-pools conforms closely to an idealized form associated with maximum flow resistance. On the other hand, limited data for bedrock step-pools indicate that $1 < (H_s/\lambda_s)/S < 1.12$ for channels in igneous rock (Duckson and Duckson, 2001).

Although flow resistance in step-pools may not always be maximized, the morphological structure of step-pools does produce substantial flow resistance. The influence of step-pool morphology on flow resistance varies with flow stage. As stage increases, the effect of step-pools on flow diminishes, but in many cases these features still greatly influence hydraulics, except perhaps during rare floods with large flow depths. At low stages, when nappe flow occurs, spill resistance dominates, whereas at high stages, when steps are submerged, form and grain resistance become important (Chin, 2003). At low stages, flow over the crest of the step is supercritical and, commonly, flow in the pool is subcritical, particularly where the bed of the pool has a reverse slope. At high stages, reach-averaged Froude numbers generally are less than 1, i.e., flow is subcritical (Zimmerman, 2013).

Profiles of streamwise velocity differ markedly between pools and steps, often deviate locally from log-law shapes, and may include negative (upstream) velocities near the bed of the pool in the wake of the step (Wohl and Thompson, 2000). The complex influence of channel form on water movement through step-pool channels produces highly turbulent, three-dimensional flow at all flow stages, with vertical movement of fluid being especially pronounced (Wilcox and Wohl, 2007). Despite this local complexity, at the reach scale, flow resistance can be characterized reasonably accurately using relations derived from the assumptions that the velocity profile is logarithmic and that grain roughness, in this case the d_{84} of the step material, is the major component of resistance (Lee and Ferguson, 2002).

Trapped large wood associated with step-pool channels can enhance flow resistance, thereby promoting channel stability. Efforts to unravel this effect have attempted to partition the total resistance f_f, the Darcy–Weisbach resistance coefficient (see Chapter 4), into components associated with individual grains (f_{grain}), with spilling of flow over steps (f_{spill}), and with form roughness, including woody debris (f_{debris}). Assessing additively the contributions of individual components to total resistance is complicated by interacting, synergistic effects among these components, even when evaluated in an experimental context (Wilcox et al., 2006). The location of wood is important, because wood added to steps affects not only form resistance but also spill resistance by increasing the height of steps (Curran and Wohl, 2003; Wilcox and Wohl, 2006). Wood also can increase resistance by locally inducing scour or deposition that adds to the complexity of channel shape.

Overall, energy loss in step-pool systems is typically dominated by decreases in potential energy from flow over steps rather than by changes in kinetic energy associated with flow acceleration and deceleration (Wilcox et al., 2011). In reaches without large wood, flow resistance is positively correlated with step height–length ratio, demonstrating the increasing importance of spill resistance with the relative height of the step (MacFarlane and Wohl, 2003). Hydraulic jumps, which dissipate energy (see Chapter 4), commonly occur as flow moves from steps into pools. The geometry of steps and residual water depths in pools can strongly influence the characteristics of these jumps and the total energy loss associated with the step (Vallé and Pasternack, 2006; Pasternack et al., 2008).

The considerable flow resistance associated with step-pool morphology has a strong influence on sediment transport during events capable of mobilizing fine gravel but incapable of mobilizing clasts within steps. Sediment-transport equations that do not account for form roughness of the steps overpredict rates of transport of mobile gravel. Instead, partitioning of the total bed shear stress to account for stress on the immobile and mobile grains is required to produce accurate predictions of bedload transport in step-pool channels (Yager et al., 2007, 2012). Measurements of sediment transport in step-pool systems indicate that bankfull flows often are not competent to mobilize surface median grain sizes, and floods with as much as 10-year recurrence intervals are required to transport material of this size (Thompson and Croke, 2008). Particle-tracing studies for clasts of different sizes indicate that these clasts display equal mobility during flood events, but that mobility in step-pool units is an order of magnitude less than for pool-riffle systems (Lamarre and Roy, 2008). When examined at high frequency, bedload-transport rates in step-pool channels are highly stochastic, with coarse grains moving only intermittently in intense bursts of transport, some of which are associated with displacement of large step-forming clasts (Saletti et al., 2015). Despite being infrequent, these bursts account for the bulk of the total transport.

CHAPTER

14 The Dynamics of Floodplains

14.1 Why Are Floodplains Important?

Floodplains are integral components of alluvial river systems. The common conception of a river is water flowing within a well-defined channel. When water overtops the banks of the channel and begins to inundate land adjacent to the river, the floodplain, this type of event is often viewed as a natural hazard that can damage human infrastructure and even lead to loss of life. In this sense, flow is perceived to be "out of the river" when it overtops the channel banks. From a geomorphic perspective, the floodplain and channel, or network of channels in anabranching or braided rivers, together comprise the river system. When viewed from this perspective, a floodplain represents the part of the system that is activated during infrequent hydrological events. This component conveys water when flows exceed the capacity of the channel for a single-thread river or set of channels for multi-thread rivers. It is also a repository of sediment transported and deposited by flows contained within the channel banks and by floods that overtop the banks.

Hydrologically, floodplains slow down the flow of water during floods by allowing it to spread out over a broad expanse. Flow over a floodplain is shallow relative to flow within channels, and low depths inhibit high velocities. Moreover, many floodplains are vegetated, and the growth of trees, grass, and shrubs on the surface generates substantial flow resistance. This inhibiting effect limits the rate at which water from different sources accumulates over distance downstream within a drainage basin, thereby reducing peak discharges compared with those that would result if flow throughout the entire basin were channelized. Thus, floodplains help to attenuate the size of peak discharges within watersheds. Besides this hydrological role, floodplains have a wide variety of other important functions, including buffering adjacent areas from flooding, filtering sediment and pollutants from floodwaters, providing habitat for plants and animals, recharging shallow groundwater aquifers, generating fertile soils for farming, and sequestering carbon through burial within alluvial deposits.

14.2 What Is a Floodplain?

Floodplains can be defined on the basis of geomorphological, topographic, stratigraphic, hydrologic, engineering, and legal perspectives (Graf, 1988b). Hydrological and engineering perspectives relate the floodplain to the extent of inundation during flood events of different recurrence intervals. Within the framework of these perspectives, the floodplain is highly dynamic, and its extent varies depending on the area of inundation. As greater discharges are considered, the floodplain extent increases. In the United States, this approach serves as the foundation for flood hazard mapping by the Federal Emergency Management Agency in support of the National Flood Insurance Program. Flood insurance rate maps designate areas inundated by the 100-year and 500-year floods and are used to set rates of insurance against the risk of flooding. The legal perspective is often complex and involves adjudication by courts. It commonly focuses on areas prone to inundation. This perspective therefore corresponds closely to hydrological and engineering perspectives.

The geomorphological perspective integrates topographic and stratigraphic perspectives to identify floodplains as natural fluvial landforms. According to this perspective, a floodplain is a relatively flat topographic surface adjacent to a main channel or zone of active channels. It is separated from the main channel or zone of active channels by banks, and consists of sediment transported and deposited by the river in its current hydrologic

Figure 14.1. Floodplain of the Embarras River in Illinois, USA. Low terrace is visible as a slight rise on far left of image at the edge of floodplain.

Figure 14.2. Snake River, Jackson Hole, Wyoming showing floodplain (low tree-covered surfaces adjacent to the channel) and terraces (shrub-covered elevated surfaces).

regime (Nanson and Croke, 1992) (Figure 14.1). When a floodplain naturally becomes disconnected from contemporary hydrological and geomorphological processes, often because of uplift of the valley bottom or incision of the river channel, it is referred to as a terrace – a relict floodplain elevated above the surface of the modern floodplain (Figures 14.1 and 14.2).

Elevation differences between terraces and floodplains provide a topographic basis for differentiating between these two types of landforms (Stout and Belmont, 2014; Yan et al., 2018); however, caution must be exercised to ensure that such differences are not part of the natural dynamics of floodplain formation (Lauer and Parker, 2008). Even when terraces are not present, the modern floodplain may not correspond to the extent of the valley floor. In wide valleys carved by past runoff regimes, the modern river system may be underfit to the valley (Dury, 1964), so that distal parts of the valley floor are composed mainly of relict alluvium and are not strongly affected by contemporary fluvial processes. Distinguishing the modern floodplain from remnants of old floodplains in such cases requires detailed exploration of the characteristics of valley-floor deposits (Saucier, 1994a; Aslan

and Autin, 1999; Blum et al., 2000). The modern floodplain also evolves continuously as it is affected by formative processes. Not all these processes may be distributed equally over the entire extent of the floodplain, so that over timescales of decades to millennia, processes associated with different hydrological regimes may affect some parts of the floodplain more than others. This time-transgressive development can complicate conceptions of floodplains as genetically homogeneous landforms. For the most part, floodplains are considered as being formed and reformed predominantly by contemporary processes, while recognizing that these features may contain elements influenced by past hydrological conditions.

Although floodplains have traditionally been characterized as flat surfaces, high-resolution topographic data reveal that these surfaces display considerable morphological complexity (Scown et al., 2015, 2016). The standard deviation of elevations, the skewness of elevation distributions, the coefficient of variation of elevations, and the standard deviation of total surface curvature all exhibit considerable spatial variability both within and between floodplains. Moreover, the spatial structure of this variability varies with the spatial scale of analysis. Many flood-plains, even those flanking a single, well-defined main channel, contain a variety of secondary channels (David et al., 2017). These secondary channels contribute both to floodplain morphological complexity and to hydrological connectivity between the main channel and the floodplain – important aspects of habitat diversity and ecological potential of floodplains (Dzubakova et al., 2015).

14.3 Why Do Floodplains Develop along Rivers?

The development of floodplains along rivers reflects an imbalance between sediment supply and total transport capacity of the river system, including flow within the channel and flow that exceeds the capacity of the channel. Floodplains develop where sediment supply exceeds transport capacity, resulting in sediment storage. Many fluvial systems flowing from mountainous headwaters to lowlands transition from supply-limited to transport-limited in the downstream direction as channel slope decreases faster than depth increases, resulting in a decrease in total boundary shear stress (Montgomery and Buffington, 1997) (Figure 13.14). Floodplain initiation has been associated with positions along longitudinal profiles of rivers where channel slope decreases abruptly at the transition between steep headwaters and gentle lowlands (Jain et al., 2008). The downstream change in slope and total boundary shear stress is often accompanied by a decrease in valley confinement, which increases the accommodation space for sediment deposition on valley floors. The downstream pattern of increasing deposition and accommodation space favors the development of wide floodplains, which are most characteristic of lowland alluvial rivers.

14.4 What Are the Major Depositional Processes on Floodplains?

Floodplains are predominantly depositional fluvial landforms within which large amounts of alluvium are stored. Major depositional processes of floodplain formation include 1) lateral accretion of bars and 2) vertical accretion on floodplain surfaces, including overbank sedimentation, levee building, splay formation, and infilling of channels. The extent to which these different processes influence floodplain development varies both with geographical variability in the morphology of fluvial systems and with temporal variability in discharge and sediment supply.

14.4.1 What Are the Dynamics of Lateral Accretion?

Lateral accretion involves deposition of bedload or suspended load onto the margins of bars along the channel banks within rivers. This process is most commonly associated with meandering rivers where progressive migration of the outer bank downstream and laterally through erosion is balanced by advance of the inner bank through point-bar deposition (Wolman and Leopold, 1957). As the river migrates laterally, old floodplain is consumed along the outer bank and new floodplain is created along the inner bank. Lateral accretion also can occur in braided rivers as migrating unit bars stall along the margins of large stationary medial or alternate bars and become incorporated into these bars (Ashmore, 1982, 1991a). This type of deposition is common in early stages of the development of braiding (see Chapter 10). Because lateral accretion involves deposition on within-channel bars, it can occur at stages below bankfull.

In meandering rivers, the point bar consists of a transverse sloping face that extends into the channel thalweg and a relatively flat platform near the inner bank (Nanson, 1980). Typically, the bar platform lies slightly below the elevation of the floodplain along the inner bank. The initial development of the point bar involves aggradation of relatively coarse-grained sediment at the bar head and lateral accretion of migrating transverse unit bars (Lewin, 1978; Peakall et al., 2007). Experimental studies indicate that lateral accretion of the point bar occurs largely in response to erosion of the outer bank, which results in local widening of the channel and deposition of bedload on the point-bar face (bank-pull mechanism; see Chapter 9) (van de Lageweg et al., 2014). Episodic deposition of migrating sand waves, or transverse bars, on the point-bar surface during transport-effective events results in progressive lateral accretion (Figure 14.3) and leads to a distinctive internal stratigraphy characterized by large-scale inclined strata (Bridge et al., 1995; Leclerc and Hickin, 1997; Sambrook Smith et al., 2016) (Figure 14.4).

The bar platform is inundated at flows slightly below or near bankfull stage. During these events, vertical accretion by migrating bedforms (Jackson, 1976; Bridge et al., 1986) or by deposition of suspended sand (Nanson, 1980) builds a curvilinear ridge, or scroll bar, on the point-bar platform (Figure 14.4). Scroll bars tend to be best developed close to the inner bank along the downstream end of the point bar in the lee of locally emergent topographic highs (Jackson, 1976). The orientation of the scroll bar mirrors channel curvature to some extent, but may deviate from this curvature downstream of the bend apex (Jackson, 1976). A trough or chute often develops between the inner bank and the ridge (Bridge et al., 1986) with the trough becoming more pronounced downstream of the apex where the direction of the scroll bar deviates from the direction of the inner bank (Jackson, 1976). When inundated, flow tends to be directed inward over the scroll bar, particularly at the downstream end of the bar, which promotes migration of the scroll-bar crest toward the inner bank (Jackson, 1976).

Figure 14.3. Maier Bend along the Wabash River, USA, showing point bar along the inner bank and accreted stacked, transverse unit bars along the face of the point bar at the edge of the water (image from Google Earth, September 5, 2013). Low bedrock platform is exposed along the outer bank at the left side of the image. Flow is from right to left.

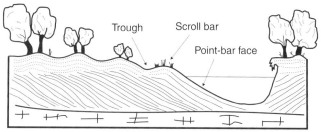

Figure 14.4. Conceptual model of floodplain deposition along a meandering river. Lateral accretion on point-bar face produces inclined strata (solid lines). Vertical accretion (dashed lines) occurs on top of the lateral accretion deposits (adapted from Nanson and Croke, 1992; Leclerc and Hickin, 1997).

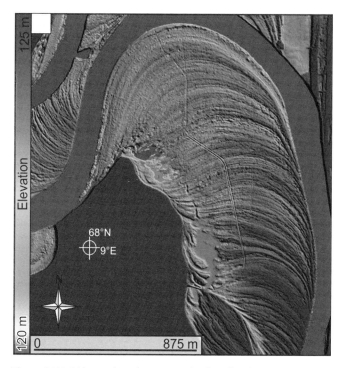

Figure 14.5. Ridge and swale topography (scrolling) along the Ivalojoki River, Finland (from Strick et al., 2018).

The scroll bar and intervening trough eventually become incorporated into the floodplain along the inner bank of the bend. Over time, vertical accretion of the bar diminishes as it grows in height to the elevation of the floodplain. As it becomes farther removed from the migrating channel, the bar is draped by fine sand and mud, and becomes colonized by vegetation. The intervening trough accumulates mud and organic debris (Bridge et al., 1986).

The development of successive scroll bars and troughs over time as the channel migrates laterally gives rise to distinctive ridge and swale topography, also referred to as scrolling, over the inner portion of meander bends (Figure 14.5). The scrolls have a convex configuration mirroring the curvature of the inner bank. Scroll spacing is often remarkably regular, occurring on average at about 50% of the channel width (Strick et al., 2018). Although floodplain swales separated by ridges are most commonly attributed to depositional processes, in coarse-grained meandering rivers, scrolled topography may also form though erosion of a chute channel between the inner bank and the floodplain during high flows (Nanson and Croke, 1992). The regular development of chute channels as the river migrates produces an undulating surface of ridges and swales.

Along actively migrating meandering rivers flowing through wide alluvial valley floors shaped by ancestral hydrological regimes of much greater volume than the current regime, such as the lower Mississippi River Valley, not all of the valley floor will be affected by processes of lateral accretion associated with the modern river. The portion of the valley floor reworked by contemporary lateral accretion is restricted to the meander belt (Saucier, 1994b). Over time, the position of the meander belt can change, resulting in new portions of a wide valley floor being reworked by remobilization of sediment at cut banks and deposition on point bars.

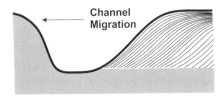

Figure 14.6. Oblique accretion by deposition of fine sand and mud (from Page et al., 2003).

Lateral accretion in slowly migrating meandering rivers transporting large amounts of clay and silt in suspension has been referred to as oblique accretion. This type of lateral accretion consists of "lateral accumulation of fine-grained floodplain sediment by progradation of a relatively steep convex bank in association with channel migration" (Page et al., 2003, p. 5). Deposition of a succession of alternating layers of sand and muddy fine sand on the steep inner bank as the channel migrates produces highly inclined (>20°) strata with concave-upward bounding surfaces (Figure 14.6). Oblique accretion can be quite effective at trapping suspended sediment prone to deposition near the channel margin, thereby limiting vertical accretion on portions of the floodplain surface distant from the river channel. This restriction of deposition to near the channel results in abrupt tapering of accretionary layers toward the floodplain surface (Figure 14.6).

Lateral accretion can also occur along the outer bank upstream of the apex of sharp bends in conjunction with the occurrence of flow deceleration or stagnation at this location. The accumulation of sediment within this region of relatively low-velocity flow has been referred to as counterpoint accretion (Nanson and Croke, 1992) or counter-point-bar deposition (Smith et al., 2009a). This depositional process produces concave-bank benches (Figure 9.24), which, once formed, become vegetated. The establishment of vegetation helps to trap suspended sediment and enhance deposition, thereby promoting bench growth (Page and Nanson, 1982). The benches, which consist mainly of silt or fine sand mixed with clay and organic material resulting from the deposition of suspended sediment within a slackwater environment, eventually increase in height to nearly the level of the floodplain (Nanson and Page, 1983). Bends that translate downstream preserve counter-point-bar deposits, and progressive lateral accretion along a downstream-migrating outer bank produces concave scroll-bar patterns (Smith et al., 2009a).

14.4.2 What Are the Dynamics of Overbank Sedimentation?

When flow exceeds the capacity of a river channel and begins to spill onto the floodplain, suspended sediment transported by the flow may be deposited on the floodplain surface. This depositional process is referred to as overbank sedimentation, and the increase in floodplain elevation produced by overbank sedimentation is called vertical accretion. Sediment deposited by overbank sedimentation will eventually be reworked by lateral accretion if the river channel is actively migrating across its floodplain. Unchecked by lateral accretion or floodplain erosion, vertical accretion of the floodplain is self-limiting in alluvial river systems because increases in the floodplain elevation correspond to increases in channel-bank heights, which will in turn decrease the frequency of floodplain inundation and overbank sedimentation (Wolman and Leopold, 1957; Moody and Troutman, 2000). Under such conditions, the rate of vertical growth of the floodplain progressively diminishes toward zero.

14.4.2.1 What Are the Major Mechanisms of Overbank Sedimentation?

Two major mechanisms are involved in moving suspended sediment onto floodplains: lateral turbulent diffusion and lateral advection. Lateral turbulent diffusion is related to pronounced differences in velocities and suspended sediment concentrations between flow in the main channel and flow on the floodplain. This difference results in the development of a shear layer near the top of the channel banks at the channel–floodplain interface. The shear layer is characterized by turbulent vortices that transfer momentum and sediment between the channel and the floodplain (Figure 14.7). Both laboratory and field measurements indicate that the channel–floodplain interface is a region of elevated turbulence (Knight and Shiono, 1990; Carling et al., 2002). Because the sediment-transport capacity, and hence the suspended-sediment concentration within the main channel, is greater than over the floodplain, a net transport of suspended sediment occurs toward the floodplain. The reduced transport capacity of flow on the floodplain relative to flow in the main channel, in turn, leads to deposition of suspended sediment transferred to the floodplain.

If flow above the level of the floodplain moves parallel to the channel and floodplain, as is the case for a straight channel flanked on each side by floodplain, lateral turbulent diffusion will be the dominant mechanism of sediment transfer. Where the direction of overbank flow deviates from the channel direction (Figure 14.8), components of overbank flow normal to the local orientation of the channel will develop, and sediment from the main channel will be laterally advected onto the floodplain (Bathurst et al., 2002). Advection reinforces diffusive transfer of sediment on the down-current side of the channel but counters diffusive transfer on the up-current side of the channel (James, 1985). Overbank deposition should therefore be asymmetrical along a meandering river,

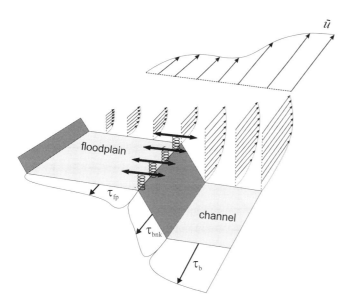

Figure 14.7. Flow structure in a compound channel–floodplain system (adapted from Shiono and Knight, 1991).

with more overbank deposition on the down-current sides of channels than on the up-current sides.

14.4.2.2　How Does Overbank Sedimentation Vary Laterally across the Floodplain?

The thickness of sediment deposited on the floodplain (h_{sd}) over time (t) by lateral turbulent diffusion is a function of the sediment fall velocity (v_f), the vertical sediment diffusivity (ε_{sv}), and the vertically integrated excess sediment concentration (ξ_{ex}), where ξ_{ex} is the difference between the vertically integrated sediment concentration of the main channel and the vertically integrated suspended sediment–transport capacity of the floodplain (Pizzuto, 1987):

$$h_{sd} = \int_0^t \frac{v_f^2}{\varepsilon_{sv}} \xi_{ex} dt \qquad (14.1)$$

For steady-state conditions in which the suspended-sediment concentration across the floodplain does not vary with time, the dimensionless thickness of deposition varies as

$$\frac{h_{sd}}{h_{sdo}} = \frac{\sinh(Gy_d)e^{-G}}{\cosh(G)} + e^{-Gy_d} \qquad (14.2)$$

where h_{sdo} is the thickness at the top of the channel bank, assumed to be the maximum thickness, y_d is the dimensionless lateral distance from the channel (y/W_{fp}), and G is a function of the distance from the channel margin to the valley wall (W_{fp}), lateral sediment diffusivity (ε_{sy}), fall velocity, and vertical sediment diffusivity:

$$G = \frac{W_{fp}v_f}{(\varepsilon_{sy}\varepsilon_{sv})^{0.5}} \qquad (14.3)$$

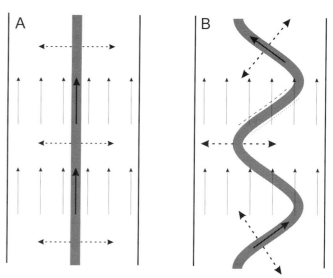

Figure 14.8. (a) Straight channel (shaded) flanked by a floodplain. Diffusive transport of suspended sediment (dashed arrows) occurs perpendicular to the channel, and advective transport of sediment (solid arrows) by overbank flow on floodplain is directed downstream, parallel to the channel. (b) Meandering channel (shaded) flanked by a floodplain. Lateral transport by turbulent diffusion is perpendicular to the channel. Path of overbank flow (light solid arrows) oblique to the channel path produces lateral advection of suspended sediment onto the floodplain, with advective transport reinforcing diffusive turbulent transport on the down-current side of the channel (dashed line along channel) and countering diffusive transport on the up-current side of the channel (dotted line along channel).

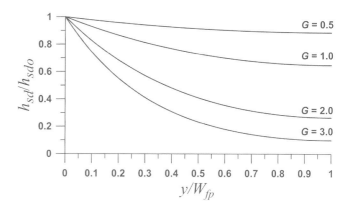

Figure 14.9. Variation in dimensionless thickness of sediment deposition with dimensionless lateral distance from the channel for four different values of G (Eq. (14.2)).

Eq. (14.2) predicts that the thickness of vertical-accretion deposits decreases exponentially with lateral distance from the channel (Figure 14.9). The same basic functional form describes the decrease in mean grain size of deposited sediment with distance from the channel (Pizzuto, 1987).

Field studies generally confirm that the thickness of sediment deposition and the size of deposited material decrease exponentially with distance from the main channel, suggesting that lateral turbulent diffusion influences the general spatial pattern of vertical accretion on floodplains (Pizzuto, 1987; Marriott, 1992; Asselman and Middelkoop, 1995; Walling and He, 1997, 1998). Nevertheless, besides lateral advection, a variety of other factors affect floodplain overbank sedimentation, including local floodplain morphology and microtopography (Walling and He, 1998; Cabezas et al., 2010); the distribution of floodplain vegetation, organic debris, and other roughness elements (Jeffries et al., 2003; Yang et al., 2007); spatial variations in channel capacity (Pierce and King, 2008); the depth and duration of inundation (Walling and He, 1998; Benedetti, 2003); total sediment concentration in the main channel during a flood; variations in channel bank height (Dunne et al., 1998); and spatial variations in floodplain width, valley confinement, or stream power (Lecce, 1997b; Knox, 2006; Nicholas et al., 2006b). Suspended-sediment concentrations are particularly important. The 1993 flood on the Mississippi River, which lasted for several months and inundated the floodplain to a depth of several meters, generally produced less than 4 mm of vertical accretion (Gomez et al., 1995, 1997), presumably because suspended-sediment concentrations within the river were mostly less than 100 mg l^{-1} (Magilligan et al., 1998). By contrast, the development of new floodplain along flood-widened sections of the Powder River in Wyoming, which has typical suspended-sediment concentrations of 1000 to 5000 mg l^{-1}, is dominated by vertical accretion, with rates around 2 to 8 cm yr^{-1} (Moody et al., 1999). A similar role for vertical accretion in channel narrowing has been observed on the Green River, which has suspended-sediment concentrations similar to those of the Powder River (Allred and Schmidt, 1999). Exceptionally high rates of overbank sedimentation, on the order of 1.4 to 1.8 cm hr^{-1}, have been estimated for floods along the Waipaoa River in New Zealand, where suspended-sediment concentrations can exceed 30,000 mg l^{-1} (Gomez et al., 1998).

During flood events, complex patterns of floodplain topography or interaction between flow within the channel and flow over the floodplain can strongly influence the hydraulics of overbank flows (Nicholas and Mitchell, 2003). Under these conditions, advective transport of suspended sediment must be considered. One approach to this problem is to use two-dimensional depth-averaged numerical models to predict spatial patterns of deposition (Nicholas and Walling, 1997, 1998; Hardy et al., 2000). Such models include components for simulating flow and suspended-sediment transport by lateral diffusion and advection. Comparison of predicted patterns of sedimentation with local samples of floodplain deposits on floodplains with complex topography suggests that advective effects can strongly influence both the amount and the texture of material deposited on the floodplain surface by floods (Nicholas and Walling, 1998). Measurements of flow and suspended-sediment transport on floodplains during floods are needed to improve the calibration of numerical floodplain-deposition models, and rigorous testing of the predictive accuracy of these models requires detailed mapping of patterns of deposition following floods.

14.4.2.3 What Are the Dynamics of Levee Development?

Overbank sedimentation plays an important role in constructing natural levees – low wedge-shaped aggradational ridges at the interface between the channel and floodplain (Brierley et al., 1997). These features may be located at the top of the channel banks or displaced slightly from the top of banks. In either case, natural levees generally are highest close to the channel and decline in elevation with lateral distance from the channel. Levees are most common along low-gradient alluvial channels that transport considerable amounts of suspended sediment. Lateral stability of a river contributes to levee formation by inhibiting reworking of these features by bank erosion. Thus, natural levees are particularly common along anastomosing rivers (Adams et al., 2004), but also occur along meandering rivers, particularly where migration rates are relatively low (Saucier, 1994b; Hudson and Heitmuller, 2003).

Levees form incrementally, increasing in height and width over time. The development of natural levees typically is linked to lateral turbulent diffusion of sediment during overbank flows, which, as indicated by Eq. (14.2), decreases exponentially with distance from the channel (Figure 14.9). According to this model, diffusion produces the greatest amount of deposition immediately adjacent to the channel, building a linear ridge that slopes laterally away from the channel. The diffusion model also indicates that high values of G result in more pronounced lateral variation in the thickness of sediment deposition (Figure 14.9). Because lateral sediment diffusivity is inversely related to G (Eq. (14.3)), high values of diffusivity will generate low, gently sloping levees, and low values of diffusivity will produce high, steep levees. Also, levees consisting of coarse sediment, which has higher fall velocities than fine sediment, will be high and steep compared with those consisting of fine sediment.

The extent to which lateral turbulent diffusion accounts for levee formation along natural rivers has yet to be thoroughly evaluated and may vary geographically. Contrasts in levee morphology for upper Columbia and lower Saskatchewan Rivers have been attributed to differences in the importance of diffusive versus advective processes for these two systems (Adams et al., 2004) (Figure 14.10). Floodwaters rise

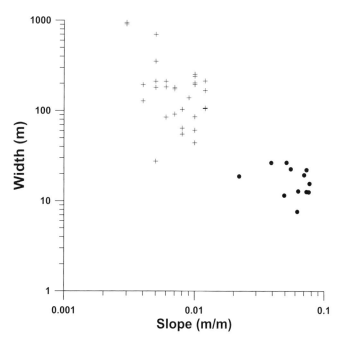

Figure 14.10. Contrasting width and slope of levees along the upper Columbia (dots) and lower Saskatchewan rivers (crosses) (data from Adams et al., 2004).

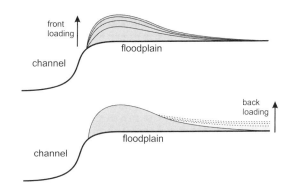

Figure 14.11. Front loading (a) and back loading (b) of levees.

synchronously in the channel and on the floodplain along the upper Columbia, but those in the main channel move faster than those on the floodplain. The narrow shear layer between main-channel and floodplain flows promotes low diffusivity and, presumably, the formation of relatively steep, narrow levees next to the channel. By contrast, along the lower Saskatchewan River, flow rises rapidly in the main channel and then spills laterally onto the floodplain at overbank stages. This spilling flow results in advective transport of suspended sediment onto the floodplain. As the velocities of the spilling flow decrease gradually over distance, extended deposition of suspended material produces relatively broad, gently sloping levees.

The regularity of flow spilling abruptly rather than gradually onto floodplains as it exceeds the capacity of the main channel has yet to be determined. Moreover, direct evidence of diffusive deposition of sediment on levees remains somewhat elusive. In fact, even on the upper Columbia River, documentation of levee building during moderate floods indicates that deposition occurs mainly through advective transport of sediment, rather than turbulent diffusion, although diffusion may play a role during large floods (Filgueira-Rivera et al., 2007). On the other hand, the initial stage of levee formation along newly formed channels often involves deposition of coarse sediment on banks close to the channels, which suggests that diffusion is important at this stage (Cazanacli and Smith, 1998). Once levees are in place, lateral spillage of overbank flow onto the floodplain will be promoted by the existence of elevated, laterally sloping

levees, unless secondary channels and sloughs provide pathways for floodwaters to rise on the floodplain at the same rate as in the main channel.

Levee morphology is strongly affected by diffusive or advective processes that produce deposition close to the channel near the crest of the levee, referred to as front loading, and those that produce deposition far from the channel on distal parts of the levee, referred to back loading (Filgueira-Rivera et al., 2007) (Figure 14.11). Front loading occurs mainly during floods that overtop the levee crest when relatively coarse sediment from the channel is deposited on top of the levee, increasing its height and steepness (Smith and Perez-Arlucea, 2008). Deposits produced by front loading generally consist of sand and decrease in grain size away from the channel. These deposits also generally fine upward but can coarsen upward if the energy of a levee-overtopping flood increases on rising stages of the hydrograph (Skolasinska, 2014). Decreasing deposition over distance from the channel produces strongly concave-upward transverse levee profiles (Hudson and Heitmuller, 2003).

Back loading is caused by sedimentation during floods when the floodplain is inundated but the levee crest is exposed (Filgueira-Rivera et al., 2007). These events result in deposition that decreases toward the levee crest on the floodplain side of the levee. Aggradation of fine material on the distal part of the levee reduces its relief and steepness. Deposits associated with back loading generally consist of relatively fine sand or mud. Draping of this fine material on the distal part of the levee contributes to the decrease in sediment texture over distance from the channel. As levee height and steepness increase initially through front loading, continuation of this growth necessitates increases in flood stages to inundate the levee crest. Increased exposure of levee crests during moderate floods favors back loading, which progressively decreases the relief and steepness of the levee (Cazanacli and Smith, 1998).

The height of levees varies spatially and can differ on opposite sides of the channel (Smith et al., 2009b). Deposition often is concentrated preferentially on relatively low portions of levees during floods, but spatial variability in deposition and erosion sustains spatial variation in levee morphology. Channel planform also influences levee characteristics. Along meandering rivers, levee width at outer banks exhibits a tendency to increase with decreasing radius of bend curvature (Hudson and Heitmuller, 2003). This relationship may reflect greater advective transport of sediment onto the floodplain along the outer banks of highly curved versus gently curved channels during floods. Both enhanced super-elevation of the water surface and outward-directed near-surface flow within highly curved channels contribute to this effect. Also, levees generally are lower on the inside of meander bends, where new levees are in the process of being built, compared with outer banks, where established levees have yet to be reworked by lateral migration.

14.4.2.4 What Are the Dynamics of Splays?

Splays are bodies of sand deposited on top of fine-grained sediment or organic matter comprising the presplay flood-plain surface. These deposits are associated with concentrated flow capable of transporting sand either as bedload or in suspension out of the channel and onto the floodplain. As this flow moves onto the floodplain, it spreads out and decelerates, resulting in deposition of the sandy bedload or suspended load. Thus, splays result from advective transport of sediment. Given the importance of advective transport, splay deposits are distinctly coarser than material deposited by diffusive transport on levees and weak advective transport on distal areas of the floodplain. Splays may form during individual flood events but often grow and develop over a series of floods that occur over many years.

Splays are common in many different types of fluvial systems, including meandering (Izenberg et al., 1996), braided (Bristow et al., 1999), anastomosing (Smith et al., 1989; Smith and Perez-Arlucea, 1994), and dryland ephemeral (Tooth, 2005; Li and Bristow, 2015) rivers. The development of splays is also fundamental to the dynamics of deltas and the delivery of sediment from distributary channels to delta surfaces (Cahoon et al., 2011; Shen et al., 2015b). The important role of splays in floodplain and delta sedimentation has been acknowledged through attempts to create intentional breaches in artificial levees along rivers to promote the development of splay deposits (Florsheim and Mount, 2002; Nichols and Viers, 2017; Nienhuis et al., 2018).

Two types of splays are recognized: bank-top splays and crevasse splays (Lewin et al., 2017). Bank-top splays involve ascent of bedload or elevation of suspended sediment onto the floodplain in the absence of major erosional failure of the channel banks. This type of deposition occurs where the gradually sloping surfaces of bars provide ramps for bedload to move upward onto the floodplain or where the channel path deviates

Figure 14.12. Bank-top splay extending downstream from outer bank of a meander bend along Big Creek, Illinois, April 1998 (Illinois State Geological Survey).

from the downvalley direction, allowing sand suspended in floodwaters to spread across the floodplain (Figure 14.12). Extreme floods along gravel-bed rivers may locally create splay-like deposits consisting of gravel that extend from channel banks onto the floodplain (Ritter, 1975; Magilligan et al., 2015). Crevasse splays develop in conjunction with local breaks in well-defined levees. Water from the channel rushing through the break decreases in velocity as it spreads across the floodplain, often producing lobate deposits with distinct margins that often terminate in depressional wetlands or zones of standing water (O'Brien and Wells, 1986; Bristow et al., 1999). Crevasse splays also typically are incised by a main crevasse channel that divides into distributary channels.

The spatial extent of crevasse splays is governed both by the amount of sediment supplied to the crevasse channel and by the lateral slope of the water surface across the floodplain (Millard et al., 2017). Large amounts of sediment and relatively steep lateral water-surface slopes promote the development of large crevasse splays. Sediment supply likely also governs the rate at which splays develop, which can vary from deposition of sediment more than a meter thick during individual floods (Joeckel et al., 2016) to progressive development at average rates of a few centimeters per year over several decades (Smith and Perez-Arlucea, 1994).

The persistence of the crevasse channel that feeds a crevasse splay depends on the extent to which this channel has incised into the levee along the main channel (van Toorenenburg et al., 2018). Minor incision into an elevated levee preserves a steep gradient toward the floodplain, promoting continued delivery of sediment to the splay during floods. However, incision that progresses further will decrease the slope of the crevasse channel, diminishing rates of outflow from the main channel and causing deposition within the crevasse channel during floods. This process can infill the crevasse channel and heal the levee breach. Accumulation of sediment within floodplain basins in which the splay terminates may increase the base level of the crevasse channel, resulting in backwater effects during floods that reduce sediment delivery (Yuill et al., 2016a). Incision that progresses rapidly into a levee during a flood can redirect substantial amounts of main-channel flow, leading to avulsion as the crevasse channel incises into the floodplain to become a secondary channel or the new path of the main channel.

14.4.2.5 What Are the Dynamics of Abandoned Channel Infills?

When avulsions occur and the flow of a river occupies a new channel, the abandoned channel gradually fills with sediment.

This process is common along all types of rivers but occurs at different timescales, depending on the rate of avulsion and the efficacy of postavulsion sedimentation processes to deposit material within the abandoned channel. Timescales generally are shortest for fluvial systems with relatively high stream power that transport predominantly coarse bed material and longest for those with relatively low stream power that transport mainly fine suspended sediment.

Along meandering rivers, cutoffs are the most common type of avulsion (see Chapter 9). These events shorten the length of the river by rerouting flow across the neck of a bend. Once flow into the bend diminishes, the entrance and exit of the abandoned bend become plugged with sediment to form a curvilinear oxbow lake. Eventually, sediment fills the lake, forming a meander scar (Gagliano and Howard, 1984; Saucier, 1994b) (Figure 14.13). The length of oxbow lakes is directly proportional to the sinuosity of a river – highly sinuous rivers with elongate bends have longer lakes than low-sinuosity rivers (Constantine and Dunne, 2008).

Abandonment of a meander bend involves three stages: 1) cutoff initiation, when some or all of the flow of the river is redirected across the neck of the bend; 2) plug-bar formation, in-channel sedimentation that progressively blocks the entrance and exit of the bend; and 3) disconnection, complete or nearly complete blockage of within-channel flow into the former bend and conversion of the bend into an oxbow lake to

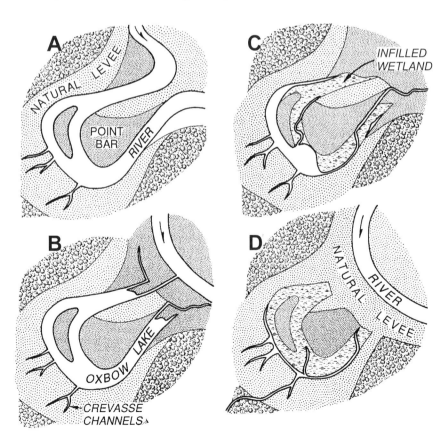

Figure 14.13. Stages in the evolution of an abandoned meander bend: (a) pre-cutoff, (b) incipient cutoff with plug-bar deposition and oxbow lake formation, (c) reduction in lake area through infilling, and (d) complete infilling of oxbow lake and formation of meander scar (adapted from Saucier, 1994b).

which suspended sediment is delivered only during floods (Toonen et al., 2012). The rate at which plug-bar formation occurs depends on the bed-material load of a river, the ratio of cutoff length to bend length, and the geometry of the cutoff (Gagliano and Howard, 1984; Shields and Abt, 1989). High bed-material loads and a small ratio of cutoff length to bend length, which diverts a large percentage of the total flow and bed-material transport into an incipient cutoff, favor rapid plug formation. Enhanced movement of bed-material load favors more rapid formation of plugs in sand-bed than in gravel-bed rivers (Schwendel et al., 2018). The diversion angle, or the angle between the entrance to a bend that has been cut off and the path of the flow in the cutoff, is an important characteristic of cutoff geometry (Fisk, 1947; Shields and Abt, 1989). High diversion angles promote rapid plug-bar formation through deposition of bed-material load by rapidly decelerating flow at the entrance to an incipient oxbow (Constantine et al., 2010b; Toonen et al., 2012). When the diversion angle is low, transport of bed-material load into a channel undergoing abandonment can continue, resulting in deposition of coarse sediment farther into the channel. Thus, abandoned channels with low diversion angles accumulate relatively large amounts of coarse bed material (Shields and Abt, 1989; Constantine et al., 2010b) and can be major storage areas for transported bed material (Dieras et al., 2013).

Chute cutoffs produce distinctive hydrodynamic conditions at the entrance and exit of a cutoff bend that influence patterns of sedimentation (Figure 14.14). Once a chute channel incises across the neck of a bend, flow bifurcates at the upstream end of this channel and converges at its downstream end. Because the chute channel is shorter and steeper than the bend, much of the flow of the river is diverted through the chute channel, and the diverted flow moves at high velocity relative to flow around the bend. Thus, flow in the chute channel has much greater momentum than flow entering and exiting the bend. As a result, zones of flow stagnation and separation form at the upstream bifurcation and downstream confluence (Zinger et al., 2013). These zones of reduced velocity or flow recirculation promote deposition, leading to the development of bars. In the upstream bifurcation, growth of a bar in the region of flow stagnation steers water into the chute channel, enhancing both the capture of flow by the chute channel and continued growth of this bar. Expansion of the bar across the entrance to the bend and vertical accretion on the bar surface eventually prevent low and moderate flows from entering the bend. The lack of flow through the upstream entrance to the abandoned bend results in recirculating flow at the bend exit, which enhances deposition on the downstream side of the channel at this location (Le Coz et al., 2010). During floods that overtop plug bars, water may enter an abandoned channel from downstream and

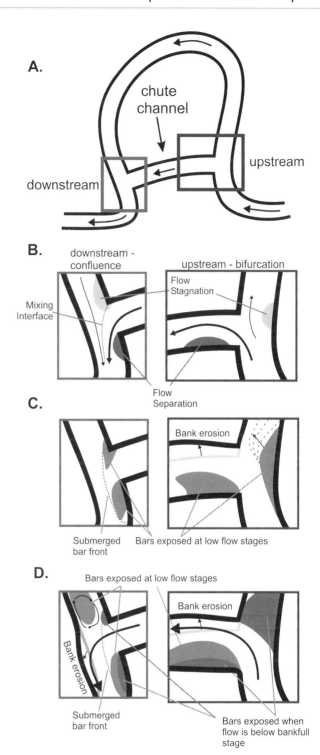

Figure 14.14. Conceptual model of flow and deposition at an incipient chute cutoff. (a) Chute cutoff. (b) Patterns of flow at bend entrance and exit following cutoff (arrows indicate relative magnitudes and direction of flow). (c) Initial patterns of deposition following cutoff. (d) Closure of bend entrance and late-stage deposition in bend exit with recirculating flow (modified from Zinger et al., 2013). (A black and white version of this figure will appear in some formats. For the color version, please refer to the plate section.)

upstream, resulting in complex patterns of flow within this channel (Costigan and Gerken, 2016).

In neck cutoffs, no chute channel exists; instead, the upstream limb of a bend migrates into the downstream limb, forming a breach that eliminates the neck and allows flow upstream to cascade directly into the downstream limb (Gagliano and Howard, 1984). The sudden increase in energy gradient produces deep scour at and downstream of the breach (Richards, 2018). As a result, the channel bed at the bend exit where the downstream limb of the bend joins the newly formed cutoff often becomes perched above the level of the bed at the cutoff (Figure 14.15). In essence, a neck cutoff produces a tight new bend along the river where the flow must change direction nearly 180 degrees over a short distance. Jetting of flow into the bank opposite the breach results in erosion and retreat of this bank, whereas the development of flow separation along the opposite bank produces deposition in the form of a cutoff bar. This process initiates translation and extension of the newly formed tight bend, generating longitudinal scroll bars in the upstream limb that mark the progress of migration. As the newly formed tight bend rotates in the downstream direction, flow within the upstream limb of this bend becomes reoriented so that it moves obliquely across the entrance to the cutoff bend, promoting deposition of a plug bar across this entrance (Figure 14.15). The downstream diversion angle is often greater than the upstream diversion angle in neck cutoffs, leading to more rapid plug-bar formation at the exit of the abandoned bend than at the bend entrance (Schwendel et al., 2018).

Once plug bars develop and an oxbow lake is established, rates of sedimentation within the lake depend on connectivity between the main river and the lake, the depth of water in the lake, overbank flow frequency, the sediment concentration of overbank flows, and the distance of different locations within the oxbow from the main river. Generally, sediments deposited within the lake are finer than those that accumulate from bed-material transport before plug bars are in place (Toonen et al., 2012). Sedimentation in oxbow lakes typically consists of muds (silts and clays) (Fisk, 1947; Saucier, 1994b; Ishii and Hori, 2016), but accumulations of sand are not uncommon (Erskine et al., 1992; Hooke, 1995b). Grain size of accumulating sediments can vary over time depending on the frequency of floods and on changes in the proximity of the main river to the oxbow lake as the river migrates across its floodplain. Often, small channels, known as tie channels or batture channels, extend through the plug bars, allowing some exchange of water and sediment between the main river and the oxbow lake (Gagliano and Howard, 1984; Rowland et al., 2005, 2009). Within the lake, spatial variation in sedimentation

Table 14.1. Rates of sedimentation in abandoned channels and oxbow lakes.

Source	Location	Rates (mm yr^{-1})
Lewis and Lewin, 1983	92 cutoffs in Wales, UK	3–71
Erskine et al., 1992	Four oxbow lake cutoffs along lower Hunter River in Australia	45–140
Davidson et al., 2004	Two sites within an oxbow lake along the Mississippi River	2 to 13
Piegay et al., 2008	14 cutoffs, Ain River, France	6.5–24
Citterio and Piegay, 2009	39 cutoffs in southeastern France	0–26
Ishii and Hori, 2016	Four cutoffs, Ishikari lowland, Japan	4–90

is linked to water depth, with greater amounts of sedimentation occurring in former pools within the bend where depths initially are largest (Citterio and Piegay, 2008). Rates of infilling are on the order of a few millimeters per year up to several centimeters per year (Table 14.1). Generally, rates decline over time as the lake fills with sediment (Hooke, 1995b), unless increases in the frequency of overbank flows counteract this tendency (Erskine et al., 1992). Migration of the river away from an oxbow lake can decrease rates of sedimentation, whereas changes in land use, such as clearing of forest and implementation of agriculture on floodplains, can increase rates of sedimentation by one to two orders of magnitude (Wren et al., 2008).

Channel infills in multichannel fluvial systems dominated by bed-material transport, such as braided and high-energy anabranching rivers, typically occur through stalling of transverse unit bars at entrances to and within anabranches (Ashworth et al., 2011). Stalling of the bars reduces flow within the anabranch, leading to bar growth through lateral and downstream accretion. Reductions in flow by the growing bars promote stalling and superimposition of migrating dunes within the anabranch, further contributing to closure and infilling. Once the anabranch is closed, further sedimentation occurs through deposition of suspended fine sand and mud within the abandoned channel by floods that exceed the level of the bar tops. Overall, however, the sedimentological characteristics of channel-fill deposits are quite similar to those of bar deposits. In low-energy anastomosing rivers, infilling of channels often occurs

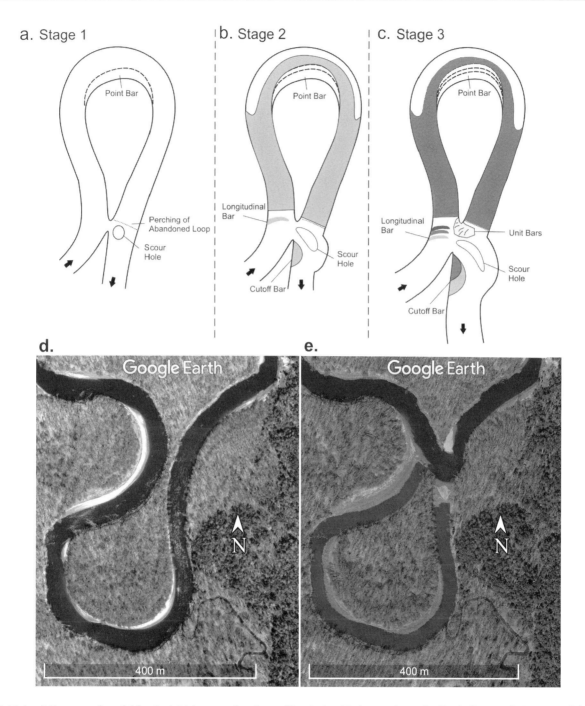

Figure 14.15. (top) Conceptual model for the initial stages of neck cutoff evolution (darkness of gray shading indicates relative age and thickness of deposits). (a) Stage 1: breach and formation of a scour hole. (b) Stage 2: scour hole elongation, erosion and deposition in downstream meander limb, and shallowing and narrowing of abandoned loop. (c) Stage 3: growth of cutoff bar, migration of longitudinal bar, unit-bar formation, and continued deposition in abandoned loop (from Richards, 2018). (bottom) Example of neck cutoff along the Bad River, Wisconsin. (d) Narrow neck, May 2011. (e) Neck cutoff, April 2015, showing sharp new bend, deposition across abandoned loop entrance and exit, cutoff bar, and downstream rotation of new bend by bank erosion in downstream limb. Flow is from left to right.

through lateral and vertical accretion of sandy bedload within the predominately muddy and highly organic avulsion-belt deposits (Makaske, 2001). Thus, channel-fill deposits are generally coarser than surrounding floodplain deposits.

14.5 What Are the Major Erosional Processes on Floodplains?

Although floodplains develop through accumulation of alluvium, once formed, these fluvial landforms can be modified by

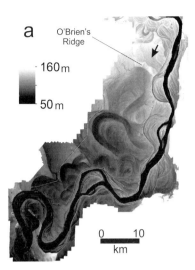

Figure 14.16. (a) LiDAR image of floodplain topography along the Mississippi River at O'Brien's Ridge. Arrow shows direction of floodwaters during 2011 flood, which flowed perpendicular to a meander scar (image courtesy of Alison Goodwell). (b) Scouring of large gullies on the floodplain of the Mississippi River at O'Brien's Ridge produced by the 2011 flood (photo by author).

erosional processes. Major erosional processes involve 1) stripping of floodplain surfaces, 2) lateral migration of cut banks and associated shaving of floodplains, 3) carving of secondary channels into floodplain surfaces, and 4) avulsive incision of a new main river channel. As with depositional processes, the efficacy of erosional processes varies geographically and with river type. In general, floodplains of rivers with high stream power that transport coarse bed material are more prone to erosional change than rivers with low stream power transporting predominately wash load.

14.5.1 What Is Floodplain Stripping?

Large floods capable of generating shear stresses on the floodplain surface that exceed the erosional resistance of floodplain materials can strip alluvium from floodplain surfaces. Because the erosive potential of flow typically decreases dramatically in lowland rivers once stage exceeds bankfull level and shallow, slow-moving flow covers the floodplain, erosional stripping of floodplains is most common within steep, narrow valleys that concentrate floodwaters so that overbank flows have high stream power (Nanson, 1986; Warner, 1997; Hauer and Habersack, 2009; Tranmer et al., 2015). In some cases, stripping can be catastrophic, removing virtually the entire floodplain. On floodplains of large, low-gradient meandering rivers, erosion can occur locally where overbank flow moves perpendicular to topographic gradients associated with relict planform features, such as meander scars. During the 2011 flood along the Mississippi River, widespread overbank flow generated by intentional levee breaches produced upstream-migrating gullies where the energy of this flow increased as it crossed the relict outer bank of a meander scar (Goodwell et al., 2014) (Figure 14.16). Erosion may also within secondary channels

on the floodplain that concentrate flow during floods, leading to scour of these channels (Riquier et al., 2017).

14.5.2 How Do Avulsions Contribute to Floodplain Erosion?

Flood-induced avulsions that reroute a main channel or carve secondary channels are common floodplain erosional processes. Avulsions occur regularly in braided rivers during floods and also represent a major long-term mechanism of planform change in anabranching rivers (see Chapters 10 and 11). Along meandering rivers, chute cutoffs can remove large quantities of stored sediment from floodplains (Zinger et al., 2011). Much of this mobilized material represents stored bed-material load, and over the long term, the remobilization of this coarse material by cutoffs generally exceeds the amount of bed-material load trapped within oxbow lakes (Dieras et al., 2013). Thus, chute cutoffs in meandering rivers generally constitute a net source of bed material to the main channel.

14.5.3 How Is Sediment Lost from Floodplains through Channel Migration?

Two processes associated with the dynamics of river meandering – floodplain shaving and bend extension – produce net transfers of sediment from floodplains to channels (Lauer and Parker, 2008). Shaving results from a difference between the heights of inner and outer banks along bends. This difference occurs because vertical accretion lags behind lateral accretion as the inner bank advances in conjunction with outer-bank retreat. Therefore, the inner bank generally is lower than the

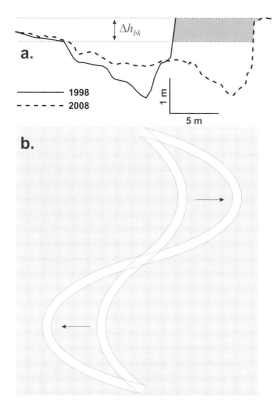

Figure 14.17. (a) Floodplain shaving produced by lateral channel migration along the Embarras River, Illinois (data from Engel and Rhoads, 2012). Shaded area represents material transferred to the channel from the floodplain. (b) Planview showing how extension of river meanders leads to a decrease in the area of floodplain (shaded box) as the channel lengthens. For a constant channel depth, the decrease in area will produce a loss in floodplain volume.

outer bank, which has been built upward by long-term over-bank deposition (Figure 14.4). As the higher, outer bank is eroded, a volume of floodplain material is delivered to the channel that is not immediately replaced on the inner bank by lateral and vertical accretion (Figure 14.17). Correspondingly, the upper part of the floodplain is removed or shaved off along the inner part of the bend.

Extension of a bend produces channel elongation, which, for constant mean channel width and depth, increases the volume of material excavated by the channel (Figure 14.17), reducing floodplain storage of sediment. The removal of sediment from floodplains by extension of bends is limited by cutoffs, which shorten the length of a meandering river. As the abandoned bend produced by a cutoff gradually fills with sediment, a net transfer of material occurs from the main channel to the floodplain. However, chute cutoffs often develop suddenly and excavate large amounts of sediment from the floodplain (Zinger et al., 2011), whereas infilling of oxbow lakes extends over centuries or millennia (Table 14.1). Thus, over the short

term along individual bends, chute cutoffs result in a net transfer of sediment from floodplains to river channels.

Estimates of sediment transfer rates from floodplains to channels by shaving and extension for four meandering rivers in the United States show that these two processes constitute on average about 22% of total outer bank erosion (Lauer and Parker, 2008). Each process contributes approximately equally to the total percentage of sediment transfer to channels. The importance of floodplain shaving is accentuated where meandering rivers migrate laterally across floodplains that have aggraded vertically at accelerated rates during a period of increased sediment supply from uplands (see Chapter 15). In these situations, the enhanced thickness of overbank deposits results in large volumes of contributed sediment from the high eroding outer bank relative to the low height of the laterally accreting inner bank.

14.6 How Are Types of Floodplains Related to Stream Power and Material Properties?

The morphology and sedimentology of floodplains can be related to bankfull stream power per unit area (ω) and material properties of river systems (Figure 14.18) (Nanson and Croke, 1992). Stream power per unit area (ω) is a useful metric because it is related to sediment-transport rates (Chapter 5), bedform type (Chapter 4), bank erosion rates (Chapter 9), and channel planform characteristics (Chapter 8). A general distinction among types of floodplains according to material properties recognizes those consisting of noncohesive (gravel to fine sand) alluvium versus those comprised of cohesive (silt and clay) alluvium. Because the size characteristics of material transported and deposited by rivers are related to fluvial energy expenditure, only rivers with low values of bankfull stream power per unit area ($\omega < 10$ W m^{-2}) produce predominantly cohesive floodplains. For noncohesive floodplains, the size of deposited sediment mediates the effect of stream power on floodplain dynamics.

The typology of floodplains can be divided into three classes: A – high-energy, noncohesive ($\omega > 300$ W m^{-2}), B – medium-energy, noncohesive ($\omega = 10$–300 W m^{-2}), and C – low-energy, cohesive ($\omega < 10$ W m^{-2}) (Figure 14.18). Within each of these classes, a variety of subclasses have been identified (Nanson and Croke, 1992). Not all of these types are as well defined and distinctive as others, but some subclasses correspond to rivers with distinctive planform types.

Class A floodplains typically are disequilibrium forms where the floodplain is episodically eroded, either partially or totally, by extreme flood events with exceptional levels of erosional capacity. Rebuilding of the floodplain during subsequent moderate floods or frequent low flows occurs predominantly through vertical accretion. Because the floodplain

Class A: High Energy - Noncohesive

A1: Confined coarse-textured floodplain
$\omega > 1000$ W m^{-2}

A2: Confined vertical-accretion sandy floodplain
$\omega = 300$ to 1000 W m^{-2}

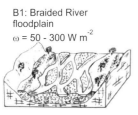

Class B: Medium Energy - Noncohesive

B1: Braided River floodplain
$\omega = 50 - 300$ W m^{-2}

B3: Meandering river floodplain
$\omega = 10 - 60$ W m^{-2}

Class C: Low Energy - Cohesive

C2a: Anastomosing river, organic-rich floodplain
$\omega < 10$ W m^{-2}

C2b: Anastomosing river, inorganic floodplain
$\omega < 10$ W m^{-2}

Figure 14.18. Examples of floodplain types (from Nanson and Croke, 1992).

Figure 14.19. (a) Cobble-boulder floodplain in a confined valley setting along Cherry Creek, Arizona. (b) Close-up of main channel and coarse-grained floodplain.

morphology evolves in relation to the current hydrological regime, it does not achieve a steady-state condition. Although valley confinement is common for Class A floodplains (Figure 14.18), unconfined forms have been recognized, including unconfined sandy vertical accretion floodplains (A3) and cut-and-fill floodplains (A4) (Nanson and Croke, 1992). Confinement limits lateral accretion, but even unconfined class A floodplains develop mainly through vertical accretion following erosion of the floodplain. In high-gradient confined gorges and canyons, transport capacity generally exceeds sediment supply, and deposition occurs only in protected areas such as local expansions in canyon width, the lee of bedrock obstacles, alcoves in canyon walls, and tributary mouths (Baker and Kochel, 1988). Fully formed, continuous floodplains are unlikely to exist in these environments, but local pockets of deposition may produce isolated coarse-grained floodplains (A1) (Tranmer et al., 2015) (Figure 14.19). Given the coarse-grained character of these floodplains, nearly all deposited material is noncohesive.

In confined or partially confined valley settings with an abundant supply of sand, lateral migration of river channels is restricted by the close proximity of valley walls, but locally deposition can occur on opposite sides of the river as it shifts from one side of the narrow valley floor to the other (Jain et al., 2008). Under such conditions, overbank deposition builds discontinuous vegetated floodplains consisting of sandy alluvium (A2) (Nanson, 1986). The timescale of floodplain construction may be on the order of hundreds or even thousands of years. Sustained vertical accretion produces prominent levees along the channel that confine flow and increase stream power within the main channel during floods (Ferguson and Brierley, 1999). Eventually, a single large flood or a sequence of moderate floods erodes the fine-grained sediment to underlying basal lag gravel – a process known as catastrophic stripping (Nanson, 1986) (Figure 14.20). Mechanisms of stripping include scouring of parallel or subparallel chutes on the floodplain surface, incision of chutes across the interior of bends, and erosion of convex channel banks (Warner, 1997; Hauer and Habersack, 2009). Less dramatic, more localized stripping may occur on

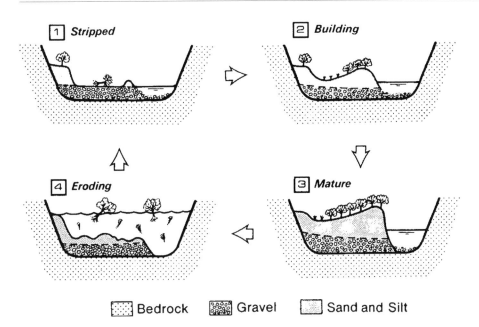

Figure 14.20. Cycle of building of a confined sandy floodplain (A2) through vertical accretion on a stripped surface, development of mature levees, and subsequent erosion resulting in stripping (from Nanson, 1986).

relatively unconfined floodplains during extraordinary floods with exceptionally high values of stream power per unit area (Magilligan et al., 2015).

Class B floodplains correspond to braided (B1), wandering gravel-bed (B2), and meandering rivers (B3). Generally, these rivers are viewed as equilibrium forms in which an average configuration of the fluvial system is maintained by the flow regime corresponding to the modern timescale (decades to centuries). Such a perspective is of course complicated by climate-related changes in flow regime that occur over this timescale. Major processes of floodplain development vary between braided and meandering rivers. Floodplains of braided rivers form mainly through channel infills associated with frequent avulsive changes in channel position and through vertical or lateral accretion of bars. Meandering rivers are shaped by lateral accretion of bed material on point bars associated with meander migration and by overbank sedimentation of fines during floods. Locally, channel infills are important in abandoned channels produced by cutoffs. Floodplains of wandering gravel-bed rivers develop through a mixture of processes related to braided and meandering rivers. Although major floods sometimes locally reconfigure morphological characteristics of Class B floodplains, erosional change is limited because floodwaters spread widely over these broad, unconfined floodplains. Alluvium comprising these floodplains consists of a mixture of noncohesive and cohesive materials, but noncohesive sediment is abundant enough to limit the effect of cohesion on floodplain resistance to lateral migration. The amount of cohesive material generally increases with decreasing stream power.

Class C floodplains correspond to low-gradient, laterally stable single-thread (C1) or anastomosing channels (C2).

These channels predominantly transport wash load in suspension, producing cohesive floodplains. Low stream power ($\omega <$ 10 W m^{-2}) and cohesive, resistant channel banks limit lateral migration. Flows frequently exceed the capacity of relatively small main channels and spread widely across broad floodplains. Through this process, Class C floodplains develop mainly through vertical accretion of fine sediment (C2b) or fine sediment and organic material (C2a) (Figure 14.18). Infrequent avulsive events produce abandoned channels that gradually fill with fine-grained sediment.

14.7 How Is Floodplain Sedimentology Related to River Planform Type?

Connecting fluvial processes of river systems to sedimentological properties of deposits comprising floodplains is a major area of research in the field of sedimentology. Information on the connection between fluvial processes and floodplain sedimentology is important not only for understanding the role of particular fluvial processes in floodplain development, but also for interpreting the genesis of sedimentary structures evident in the rock record. This type of information provides insight into the depositional environments of sedimentary rocks and contributes to geological interpretations of Earth history. In-depth treatments of fluvial sedimentology provide details on relations between floodplain deposits and river dynamics (e.g., Bridge, 2003; Miall, 2006). Here, a brief overview is provided to illustrate how floodplain sedimentology relates to fluvial processes associated with rivers of different planform types. The emphasis is on typical depositional assemblages and the characteristic three-dimensional structure of floodplain deposits, also known as alluvial architecture.

14.7.1 What Are the Sedimentological Characteristics of Floodplains of Meandering Rivers?

The floodplains of meandering rivers are characterized by a variety of distinctive morphological features, including levees flanking the main channel, oxbow lakes, meander scars, crevasse splays, and, in wide valleys, backswamp areas distal to the meander belt (Figure 14.21). The texture and sedimentary structure of the floodplain are typically dominated by lateral accretion deposits. Besides the large-scale inclined strata produced by discrete episodes of point-bar deposition (Figure 14.4), lateral accretion deposits display distinctive vertical sequences of texture and bedding. As the point bar migrates laterally, basal gravels on the deepest part

of the channel bed, or thalweg, are first covered by cross-bedded coarse sands produced by migrating dunes on the overriding lower part of the point bar. Continued lateral advance of the point bar results in ensuing burial of the dune deposits by progressively finer materials toward the top of the point bar, including cross-laminated medium to fine sands generated by migrating ripples. These cross-laminations may be interspersed with horizontally laminated fine sand and silt.

Sediment that accumulates on top of the lateral-accretion deposits via overbank deposition generally consists of fine sand, silt, and clay. Over time, this vertical-accretion layer also incorporates substantial amounts of organic matter. Splays can deposit sand locally on top of this fine-grained material (Figure 14.21). Channel infills develop following

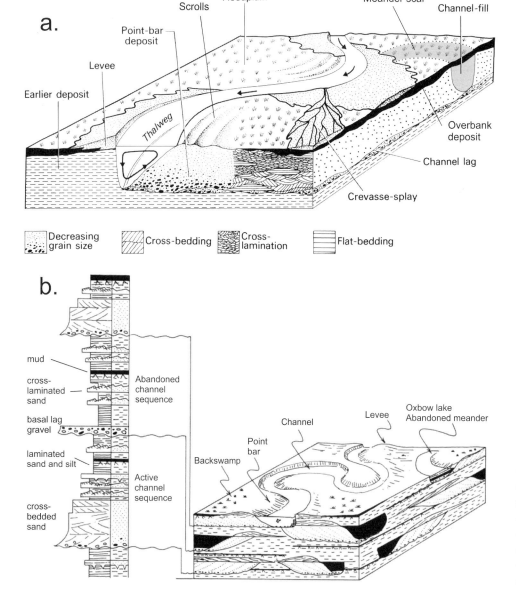

Figure 14.21. (a) Idealized depositional model of a meandering river floodplain (adapted from Collinson, 1996). (b) Typical types of vertical depositional sequences associated with meandering river floodplains (adapted from Selley, 2000).

cutoff and the formation of oxbow lakes. The resulting abandoned channel sequences typically consist of basal lag gravels overlaid by fine sand and mud.

The basic model of meandering-river floodplain alluvial architecture generated by meander migration and point-bar deposition draws mainly upon the expected vertical sequence of deposition along the inner bank downstream of the bend apex. Here, the bed shear stress decreases from the thalweg toward the top of the point bar, producing corresponding inward fining of bed material and transitions in bedform type. Upstream of the apex, the bed shear stress increases from the thalweg to the top of the point bar, resulting in an inward increase in bed-material size (Figure 9.29). Thus, characteristics of preserved sedimentary structures depend on position along the inner bank (Ghinassi et al., 2016). The model also assumes that meander bends grow primarily through extension, so that accretion is indeed in the lateral, or cross-valley, direction. The mode of bend migration – translation versus extension versus rotation (Figure 9.34) – has a strong influence on the preservation potential of point-bar deposits at different locations along the inner bank (Ghinassi et al., 2016; Durkin et al., 2018). Overall, aggradation of the entire floodplain system, which is implied in the basic model (Figure 14.21), tends to enhance preservation potential (Ghinassi et al., 2016; Durkin et al., 2018).

14.7.2 What Are the Sedimentological Characteristics of Floodplains of Braided Rivers?

Studies of the depositional environment of braided-river systems have focused mainly on sedimentary architecture within the active-channel environment, or the portion of the braided river system where flow actively reworks bars as it moves through intervening, interconnected channels (Lunt et al., 2004; Bridge and Lunt, 2006; Miall, 2006). This interest in the active-channel environment in part reflects the abundance of braided-river deposits in the sedimentary record prior to the evolution of rooted vascular plants in the late Silurian, approximately 425 mya (Long, 2011). The depositional environment of braided rivers is dominated by infilling of active and abandoned channels by migrating bars, bedforms, and gravel or sand sheets. Lag gravels representing coarse bed material moving in active corridors of bedload transport constitute basal units in active channel-fill sequences (Selley, 2000). Basal erosion surfaces typically have concave-upward configurations perpendicular to the channel direction, a form that reflects the arcuate shape of channel beds (Figure 14.22). Characteristic facies overlying the lag gravels include large-scale inclined strata formed by migration of the margins or fronts of unit or compound bars, medium- and small-scale cross-stratification formed

by migration of sinuous-crested dunes, and low-angle strata produced by migration of low-amplitude dunes or unit bars (Sambrook Smith et al., 2006). The migration of sinuous-crested dunes within infilling channels produces trough cross-bedding with concave-upward surfaces perpendicular to the migration direction. This trough cross-bedding, most prevalent above basal erosion surfaces (Sambrook Smith et al., 2006), is the most common internal structure of bars and channel infills (Lunt et al., 2004). The scale of trough cross-bedding decreases upward if dunes diminish in size and are replaced by ripples in shallowing flow within an infilling channel (Figure 14.22, sequence C). Sediment size generally fines upward but will coarsen upward if bar migration results in a bar head overriding a bar tail (Bridge and Lunt, 2006). Although high-angle cross strata can be generated by progressive migration of unit-bar fronts into aggrading channels (Cant and Walker, 1978; Lunt et al., 2004), in many cases migrating unit bars have relatively low bar fronts that do not produce well-defined cross strata (Sambrook Smith et al., 2006). Such strata are most likely to form where high compound bars with well-defined margins advance into adjacent channels. The sedimentological similarity between fills in channels undergoing progressive sedimentation and compound bars indicates that infilling of active channels does not differ markedly from bar development (Lunt et al., 2004; Ashworth et al., 2011). In channels abandoned suddenly through avulsions, fills beyond the upstream plug typically consist of horizontally bedded or cross-laminated fine sand transported into the channels from adjacent bar surfaces during floods. These strata grade upward into layers of mud representing suspended sediment transported by floodwaters.

Stacked sequences of scour and fill deposits are characteristic of aggrading braided rivers (Figure 14.22). New basal erosion surfaces develop through excavation of deposited sediment. The formation of these surfaces is caused mainly by confluence scour, by avulsions that result in incision of new channels into bars, and by local scour, rather than by extensive lateral channel migration (Best and Ashworth, 1997; Gardner et al., 2018). Excavation of sediment through scour eventually leads to infilling and subsequent bar development in response to frequent variations in spatial patterns of erosion and deposition within braided rivers (see Chapter 10). As aggradation continues, local scour and fill occurs repeatedly across the width of the river, resulting in successive vertical depositional sequences bounded by underlying basal erosion surfaces (Figure 14.22).

Although braided channels can cover large areas of valley floors, lateral instability of these channels does not preclude the existence of floodplains within braided river systems (Bridge, 1985). Braided-river floodplains generally consist of discontinuous vegetated landforms composed of an assemblage of

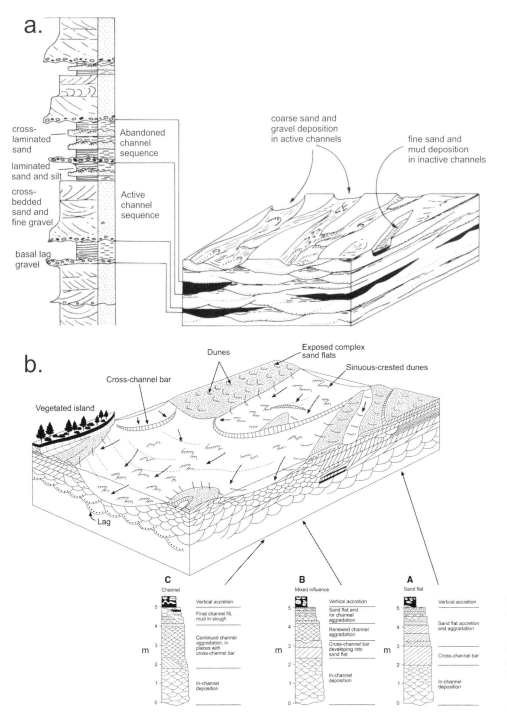

Figure 14.22. (a) Idealized depositional environment for an aggrading gravel-bed braided river (adapted from Selley, 2000). (b) Idealized depositional environment for a sand-bed braided river (redrawn by Sambrook Smith et al., 2006 from Cant and Walker, 1978).

components at various stages of development (Reinfelds and Nanson, 1993). Discontinuous floodplains occur within or adjacent to the active-channel environment, also known as the braidplain (Sambrook Smith et al., 2006) or braidtrain (Haschenburger and Cowie, 2009) (Figure 14.23). These landforms, like those along meandering rivers, are formed by the present hydrological regime of the river and periodically become inundated by overbank flow. The floodplains consist internally of the same assemblages of sand and gravel deposits associated with the active-channel environment (Figure 14.22), but typically are overlain by vertically accreted fines and localized splays.

The main mechanism contributing to the development of floodplains along braided rivers is shifting of the active-channel environment, or braidtrain, from one part of the valley floor to another (Reinfelds and Nanson, 1993). Other mechanisms include localized aggradation during floods that produces an elevated erosion-resistant platform and localized incision that isolates part of the valley floor from the active

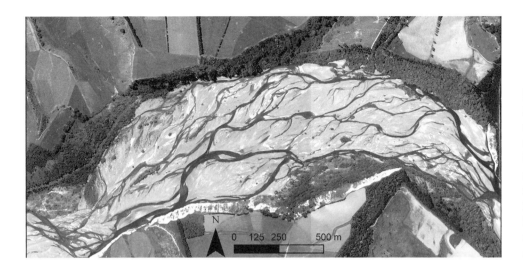

Figure 14.23. Ngaruroro River, New Zealand, showing the unvegetated active channel environment, or active braidplain, flanked by discontinuous segments of vegetated floodplain. (Sourced from LINZ Data Service and licensed by New Zealand Aerial Mapping Ltd. for reuse under the Creative Commons Attribution 4.0 International License.)

braidtrain. Similarly, floodplains can be reactivated by lateral shifts in the position of the braidtrain and reactivation of abandoned floodplain channels by floods.

The conversion of bars to floodplain, either in the form of islands within the braidtrain or as extensive surfaces adjacent to the braidtrain, involves stabilization of bar surfaces through growth of vegetation and accumulation of woody debris (Reinfelds and Nanson, 1993; Gurnell et al., 2001; Haschenburger and Cowie, 2009). Studies of braided rivers in New Zealand indicate that floodplains consist of an ensemble of discrete depositional units with different stages of development, frequency of inundation, amounts of accreted fines, vegetation coverage, and surface stability (Reinfelds and Nanson, 1993; Haschenburger and Cowie, 2009). Early stages include both erosional and depositional processes, but dominance of the latter leads to expansion of floodplain extent. Over time, as zones of deposition stabilize through increases in surface elevation and vegetation growth, the likelihood of erosion diminishes and the longevity of floodplains increases. The rather substantial extent and thicknesses of accreting fines on the surfaces of established and mature floodplains (Table 14.2) shows that these landforms store large amounts of fine sediment within gravel-bed braided river systems.

14.7.3 What Are the Sedimentological Characteristics of Floodplains of Wandering Gravel-Bed Rivers?

Wandering gravel-bed rivers have been viewed to some extent as hybrid forms of meandering and braided rivers (see Chapter 12). This perspective is consistent with the development of floodplains along these rivers by three primary processes: infilling of abandoned channels following avulsions, lateral accretion of channel bars, and overbank sedimentation (Desloges and Church, 1987). The first process is characteristic of braided rivers, the second process of meandering

rivers, and the third process is common to both meandering and braided rivers.

Following recognition of wandering gravel-bed rivers as a distinct planform type (Church, 1983; Desloges and Church, 1989), initial investigation of depositional characteristics of rivers with wandering gravel-bed reaches called into question the validity of distinguishing rivers sedimentologically based on differences in planform. Over a 20 km reach, the planform of the gravel-bed Squamish River in British Columbia, Canada changes from braided to wandering to meandering. Types of bars change downvalley, with midchannel compound bars common in the braided reach, bankattached compound bars predominant in the wandering reach, and point bars characteristic of the meandering reach (Brierley, 1991a). Despite distinct differences in bar form, bars in the three reaches cannot be readily distinguished from one another on the basis of distinctive sedimentological characteristics (Brierley, 1989; Brierley and Hickin, 1991). Depositional processes by which sediment becomes incorporated into bars and the floodplain in the three reaches are similar and not controlled by differences in planform. The extent to which this finding for a short reach of a single river can be generalized remains uncertain. It does seem to indicate that within specific geographic contexts, similarities in general environmental conditions (sediment caliber, sediment supply, and hydrological regime) may prevent differences in sedimentology that otherwise might emerge from differences in channel planform among rivers in widely different geographic contexts.

Continued work on the alluvial architecture of wandering gravel-bed rivers has led to the development of a three-dimensional depositional model that includes several major types of in-channel deposits: coarse-grained vertical-accretion deposits produced mainly by stacking of migrating bedload sheets, inclined slip-face deposits formed by the migration of distinct bar margins, lateral-accretion deposits associated with

Table 14.2. Characteristics of floodplain elements of different stages in braided rivers in New Zealand.

Stage	River	Inundating discharge (Q_{in})	Extent (E_{af}) and depth (D_{af}) of accreting fines	Vegetation and coverage (C_{vg})	Surface stability (years)
Stabilizing bar	Waimakariri	$Q_{in} \ll Q_{maf}$	Some infill by fines, but cobble exposed	Patchy lichen, mosses, and small plants	3–30
	Ngaruroro	$Q_{in}/Q_{maf} = 0.3$	$E_{af} = 10$–85%. Infill by fines, but gravel still exposed; $D_{af} = 1$–30 cm	Herbaceous with some woody plants, $C_{vg} = 10$–90%	3–10
Incipient floodplain	Waimakariri	$Q_{in} < Q_{maf}$	Some large clasts exposed, $D_{af} < 7$ cm	Grasses, tree seedlings, and small plants. $C_{vg} = 50$–100%	30–50
	Ngaruroro	$Q_{in}/Q_{maf} = 0.5$	$E_{af} \approx 100\%$, no large clasts exposed. $D_{af} = 10$–40 cm	Balanced mix of herbaceous and woody plants. $C_{vg} \approx 100\%$	10–35
Established floodplain	Waimakariri	$Q_{in} \approx Q_{maf}$	$E_{af} \approx 100\%$, $D_{af} = 10$–30 cm	Trees with grass understory; sedges. $C_{vg} \approx 100\%$; isolated bare areas	50–100
	Ngaruroro	$Q_{in}/Q_{maf} = 0.9$	$E_{af} \approx 100\%$, $D_{af} = 35$–110 cm	Increase in extent, age, and diversity of woody plants, $C_{vg} \approx 100\%$	20–60
Mature floodplain	Waimakariri	$Q_{in} > Q_{maf}$	$E_{af} \approx 100\%$, $D_{af} = 20$–200 cm	Mature trees with dense ground cover; $C_{vg} = 100\%$	100–250
	Ngaruroro	$Q_{in}/Q_{maf} = 1.2$	$E_{af} \approx 100\%$, $D_{af} = 15$ to greater than 160 cm	Increase in extent, age, and diversity of woody plants, $C_{vg} \approx 100\%$	> 60

(adapted from Reinfelds and Nanson, 1993; Haschenburger and Cowie, 2009)

deposition on point bars, downstream-accretion deposits formed at the distal ends of unit or compound bars, channel infills deposited in abandoned channels, upstream-accretion deposits generated by stalling of migrating bedload sheets on upstream-dipping bar surfaces, and scour-fill deposits that accumulate in local areas of bed scour (Figure 14.24) (Wooldridge and Hickin, 2005). Coarse-grained vertical-accretion deposits dominate, constituting about 40% of the alluvial architecture. These deposits consist of horizontal, parallel, continuous strata. Slip-face deposits, which produce cross beds, comprise about 20% of the architecture. The remaining types each account for 10% or less of the total amount of deposition. Interestingly, lateral accretion, a dominant depositional process in lowland meandering rivers, and cut and fill, an important morphodynamic process in braided rivers, are not major components of this depositional model. Lateral accretion is important on point bars, but these bars are not as pervasive as in meandering rivers. Similarly, scour-fill deposits are limited by less dynamic shifting of channels and confluences than in braided rivers. The limited role of lateral accretion requires further substantiation; other research has indicated that this process is the dominant mode

of deposition in wandering gravel-bed rivers (Desloges and Church, 1987).

Floodplain deposits in wandering gravel-bed rivers do not necessarily reflect the architecture of within-channel deposition. In relatively sandy wandering gravel-bed rivers, sedimentary sequences often consist of vertically accreted sands overlying bar-platform sediments consisting of coarse sand or gravel (Desloges and Church, 1987; Brierley, 1991b; Brierley and Hickin, 1992) (Figure 14.25). Bar platform sediments produced by vertical accretion of coarse-grained bedload sheets are typically stripped away by chute channels, which are subsequently infilled by low-energy fine-grained sediment. Channel infills on floodplains consist of cross-laminated medium-fine sands and laminated silty fine sand, deposited from suspension during floods, overlain by organic-rich, silty, very fine sand deposited in standing water following channel abandonment (Brierley, 1991b; Hickin et al., 2009). Locally, bar deposits may be preserved on the floodplain in the interiors of abandoned bends. In gravel-rich wandering gravel-bed rivers, these vertical sequences consist of basal cobbles and boulders overlain by horizontal to gently inclined strata of pebbles to cobbles representing stacked gravel bedload sheets (Hickin

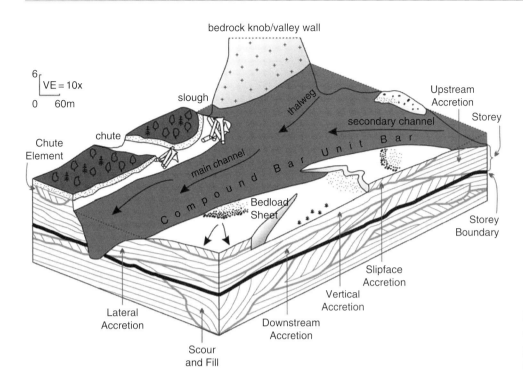

Figure 14.24. Depositional environment of a wandering gravel-bed river (from Wooldridge and Hickin, 2005).

et al., 2009). The sequence is capped by laminated sandy silt deposited from suspension by overbank flows. In all cases, the texture of floodplain sediments exhibits pronounced fining upward (Figure 14.25).

14.7.4 What Are the Sedimentological Characteristics of Floodplains of Anastomosing Rivers?

Information on the alluvial architecture of anastomosing rivers is rather limited, given the relatively small number of sedimentological studies of this type of river and the small number of rivers that have been investigated. Characteristic surface features of floodplains of anastomosing rivers in humid-temperate environments, such as those in western Canada, include active primary and secondary channels with sandy beds flanked by well-developed levees comprised of silty or clayey sand; abandoned secondary channels containing sand, mud, and organic matter; crevasse splays consisting of sand or muddy sand; and distal areas beyond the levees underlain by organic-rich mud and possibly peat (Figure 14.26) (Makaske, 2001; Stevaux and Souza, 2004). Distal areas include marshes, lakes, and mires, with marshes containing organic and clastic mud, lakes containing laminated clay and silt, and mires containing thick deposits of peat. Mires develop only during times of reduced sediment supply, when areas of slow-growing peat can become established on the floodplain (Makaske et al., 2017). Because

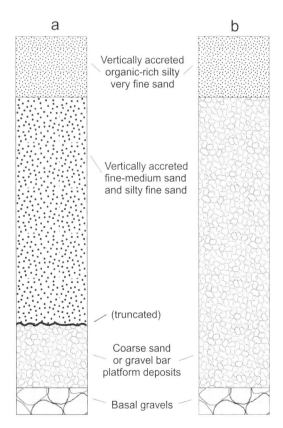

Figure 14.25. Vertical textural sequences of floodplain deposits in wandering gravel-bed rivers. (a) Typical sequence for most of the floodplain. (b) Sequence for preserved bars on abandoned-bend interiors (based on information in Brierley, 1991b; Hickin et al., 2009).

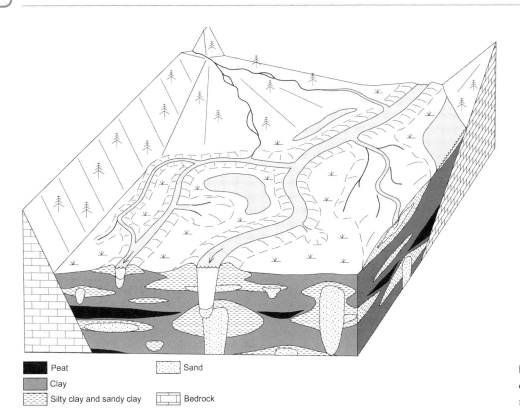

Peat
Clay
Silty clay and sandy clay
Sand
Bedrock

Figure 14.26. Depositional environment of an anastomosing river (from Makaske et al., 2017).

anastomosing river systems tend to be net aggradational, both the channels and the floodplains build upward over time. Moreover, occasional avulsion events shift the locations of channels within the aggrading valley floor. As the floodplain aggrades, thick sequences of clay are produced from extensive overbank deposition. The sequences of clay locally contain isolated bodies of sand representing channel infills and levees associated with former channel locations. Locally, crevasse splays can also produce bodies of sand within the aggrading floodplain.

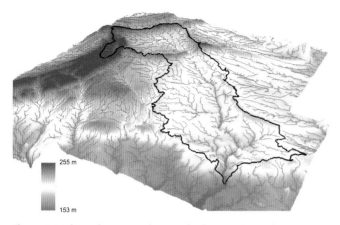

Figure 1.1. Three-dimensional view of a drainage basin for the upper part of the Sangamon River in Illinois, United States, showing drainage divide (black line), basin outlet (red dot), and stream network (blue lines).

Figure 2.33. Landsat 7 image of a distributary drainage pattern on the Lena River Delta, Siberia, 2–27-2000.

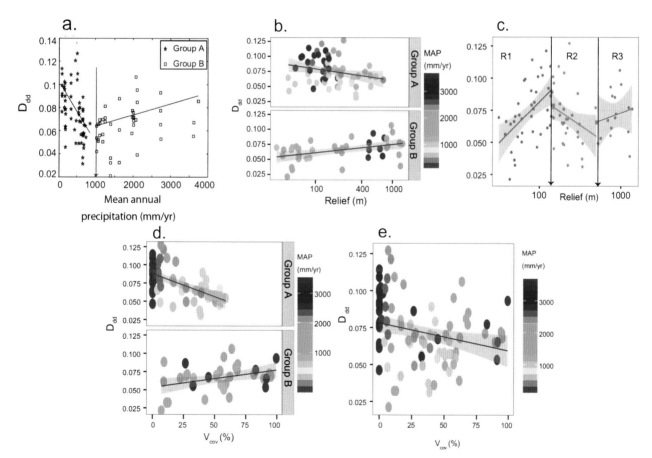

Figure 2.12. Dimensionless drainage density and controlling factors for 101 drainage basins in the United States. (a) Variation in drainage density with mean annual precipitation (MAP) for two groups: A: MAP < 1050 mm yr^{-1} and B: MAP ≥ 1050 mm yr^{-1}. (b) Variation in drainage density for groups A and B with relief. Colors of dots correspond to MAP. (c) Variation in drainage density for three classes of relief. (d) Variation in drainage density with percent vegetation cover (V_{cov}) for groups A and B and (e) for all data. Colors of dots correspond to MAP (from Sangireddy et al., 2016a).

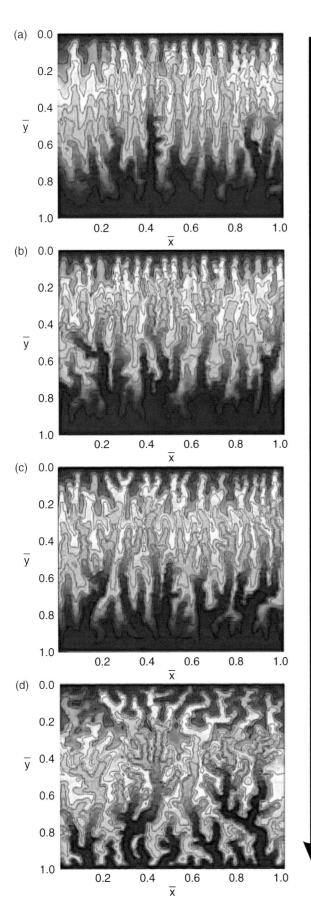

(a)

(b)

(c)

(d)

increasing roughness

Figure 2.30. Influence of roughness on stream dissection of an initial slope. As the roughness of the initial surface increases, the branching of streams and valleys dissecting the surface also increases (red corresponds to high elevations, blue to low elevations) (from Simpson and Schlunegger, 2003).

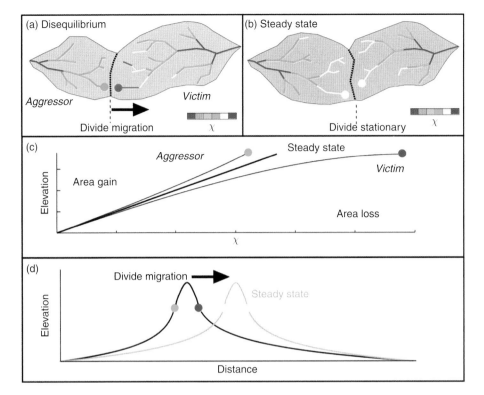

Figure 2.31. Mapping of χ across drainage divides indicates disequilibrium and a tendency for the divide migration (a) or steady state and a stationary divide (b). Plots of elevation versus χ show basin conditions at the divide relative to the steady-state condition (c). Channel locations (aggressor) that plot above (to the left, green dot) of the steady-state curve have high erosion rates relative to channel locations (victim) that plot below (to the right, red dot) of the curve, leading to divide migration (d) until steady state (yellow dots) is achieved (from Willett et al., 2014).

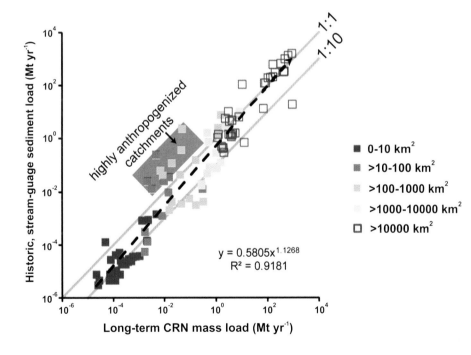

Figure 3.14. Stream sediment loads versus CRN mass loads for 103 matched locations. Sediment loads are colored according to drainage-basin area. Dashed line represents best-fit power function. Light lines show ratios of CRN to stream loads. Highly anthropogenized watersheds exhibit increase in stream loads relative to CRN loads (from Covault et al., 2013).

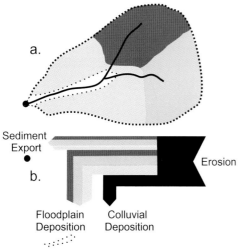

Figure 3.17. Schematic diagram illustrating different sediment sources within a watershed (a, colors) and apportionment of sediment fluxes by sources at watershed outlet (dot) and floodplain deposition (yellow area) based on fine-sediment tracing analysis (b).

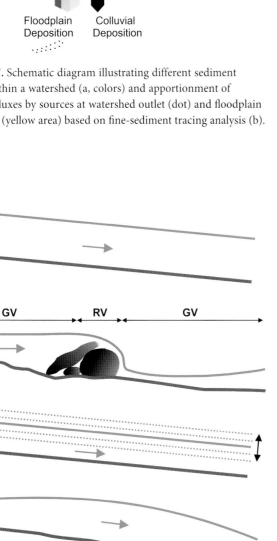

Figure 4.1. Idealized diagram of a channel reach illustrating water-surface width (W), mean flow depth (D), and mean flow velocity (U) at a cross section. Three-dimensional coordinate system is also shown.

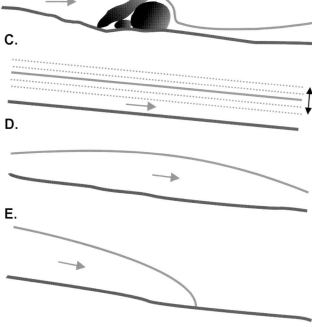

Figure 4.2. Classification of flow types: (a) steady, uniform flow, (b) gradually varied (GV) and rapidly varied (RV) flows, (c) unsteady, uniform flow (rare), (d) gradually varied, unsteady flow (flood wave), (e) rapidly varied unsteady flow (flood bore). Brown lines are channel bed and blue lines are water surface.

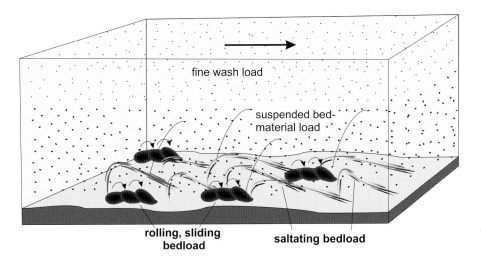

Figure 5.2. Diagram illustrating the different types of sediment transport in rivers.

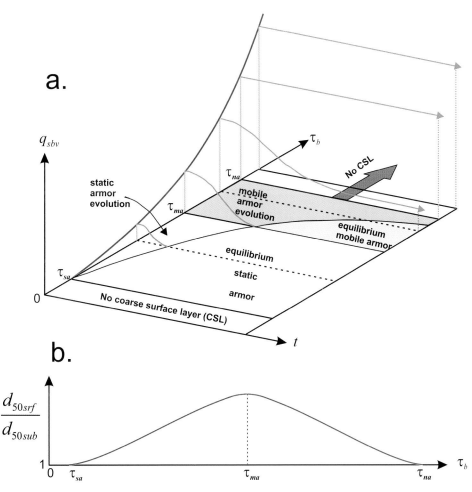

Figure 5.19. Diagram illustrating hypothesized, ideal model of armor layer development and coarseness for zero sediment influx but constant bed shear stress through time. (a) Evolution of equilibrium static armor and mobile armor over time (t) in relation to increasing bed shear stress (τ_b) and bedload transport rate (q_{sbv}). For $\tau_b < \tau_{sa}$, no bedload transport occurs and no coarse surface layer develops. For $\tau_{sa} < \tau_b < \tau_{ma}$, fines are progressively winnowed from the surface bed material until a static-armor layer develops and q_{sbv} goes to zero. For $\tau_{ma} < \tau_b < \tau_{na}$, the surface layer coarsens and transport rate decreases as fines in mixture are hidden beneath the mobile armor. All particles are mobile, and at equilibrium the grain-size distribution of the bedload approximates the grain-size distribution of the subsurface bed material. For $\tau_b > \tau_{na}$, mobility of coarse particles can be maintained without the regulating effect of mobile armor, and no coarse surface layer develops. Sediment influx will change the rate of armor evolution. (b) Coarseness of armor layer at equilibrium in relation to shear stress thresholds. Coarseness of surface layer reaches a maximum at the transition between static and mobile armor (τ_{ma}) and decreases with both increasing and decreasing τ_b (modified from Jain, 1990).

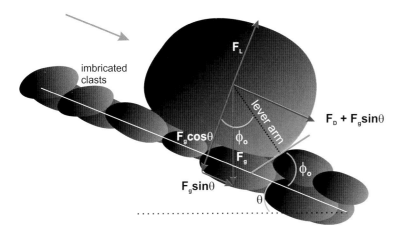

Figure 5.6. Components of the force balance acting on a grain on a sloping riverbed.

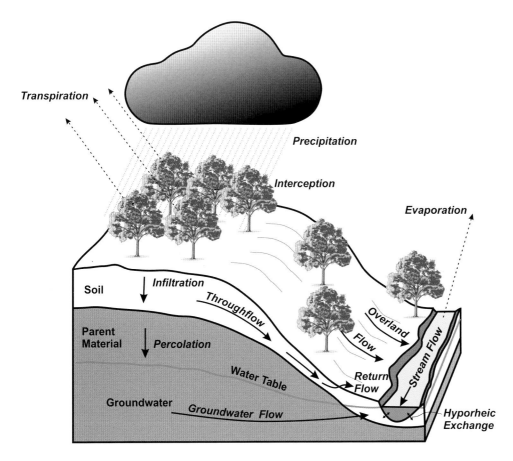

Figure 6.1. Pathways of the terrestrial water cycle.

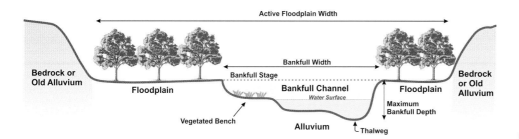

Figure 6.9. Diagram illustrating the relation of a bankfull channel to bankfull stage and the active floodplain in an alluvial river.

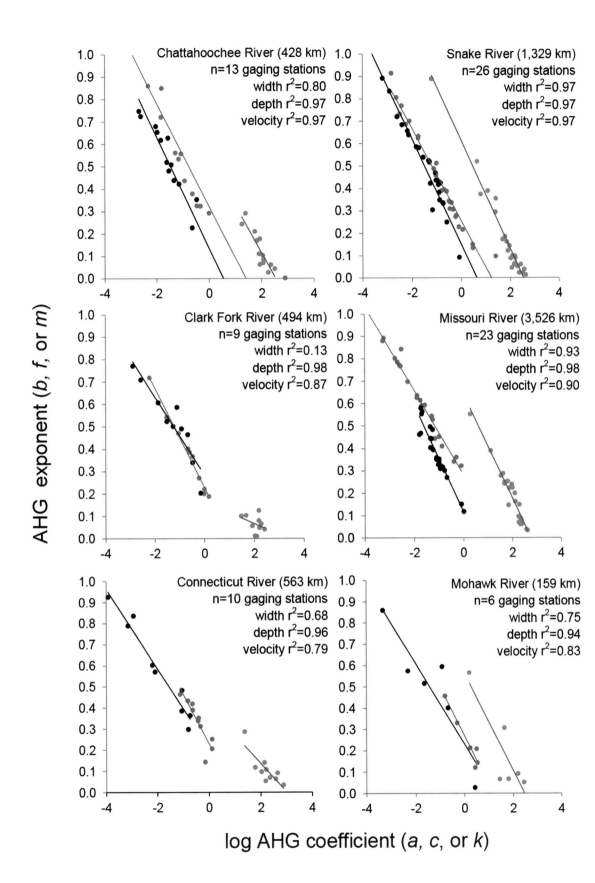

Figure 7.8. Inverse relations among at-a-station hydraulic-geometry exponents and logarithms of coefficients for width (blue), depth (red), and velocity (black) (from Gleason and Smith, 2014).

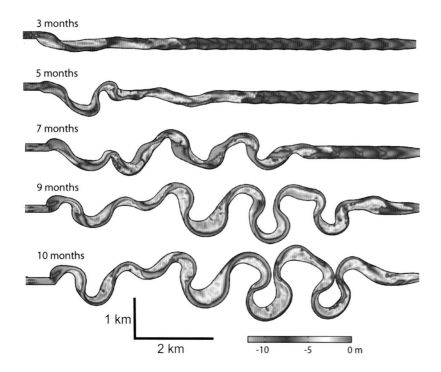

3 months

5 months

7 months

9 months

10 months

1 km

2 km

-10   -5   0 m

Figure 9.10. Numerical simulation of meander initiation in a straight channel with a dynamic inflow perturbation (from Schuurman et al., 2016).

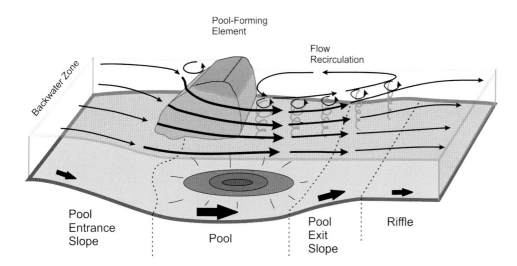

Pool-Forming Element

Flow Recirculation

Backwater Zone

Pool Entrance Slope

Pool

Pool Exit Slope

Riffle

Figure 9.17. A forced pool-riffle sequence. Weight of surface streamlines corresponds to relative velocity magnitudes. Blue swirled lines represent vertical vortices bounding the zone of flow recirculation. Heavy arrows represent relative magnitudes of near-bed velocity.

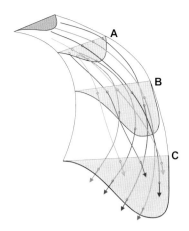

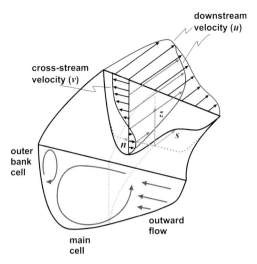

Figure 9.21. Idealized pattern of flow streamlines depicting helical motion in a curved channel moving through three successive cross sections (A, B, and C) (from Frothingham and Rhoads, 2003).

Figure 9.22. Flow pattern in a meander bend based on channel coordinate system (s,n,z) (adapted from Ottevanger et al., 2012).

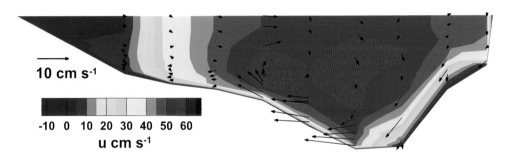

Figure 9.23. Filled contours of downstream velocity (u) with superimposed vectors (arrows) of vertical/cross-stream (w/v) velocities in a meander bend on the Embarras River, Illinois. Vector pattern of secondary flow exhibits clear evidence of fluid rotation related to helical motion. Note submerged position of maximum downstream velocity. Inner bank is to left, outer bank is to the right, view is upstream.

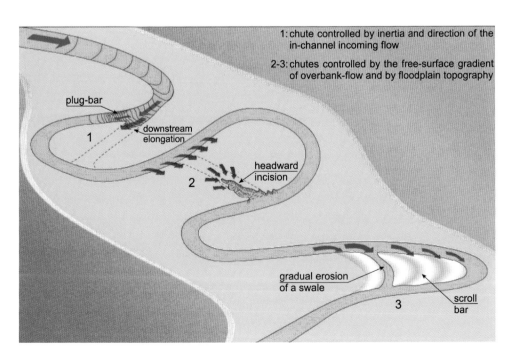

Figure 9.36. Mechanisms of chute cutoff: (1) downstream elongation of an embayment, (2) headward migration of a gully, (3) erosion of a swale or slough in scroll-bar topography (from Viero et al., 2018).

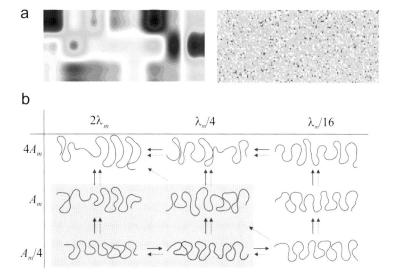

a

b

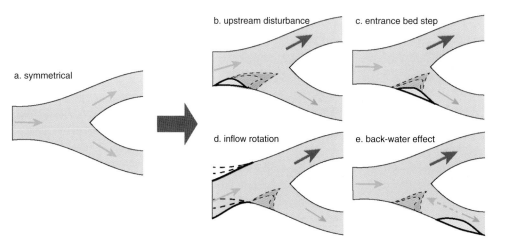

Figure 9.42. Influence of floodplain heterogeneity of meander evolution. (a) Coarse-grained (left) versus fine-grained (right) erosional heterogeneity of a floodplain (hot colors represent high bank erodibility and cool colors represent low bank erodibility). (b) Influence of the spatial scale of floodplain heterogeneity in erosional resistance on planform evolution of simulated meandering rivers. Dimensions of heterogeneous patches of erosional resistance are defined in terms of the amplitude (A_m) and wavelength (λ_m) of initial meander bends used in the simulations. Solid and dashed arrows represent the direction of increasing maximum cross-valley extent of the meander belt and increasing variability in bend amplitude, respectively. Shaded area corresponds to the spatial scales of erosional heterogeneity for which cutoffs commonly occur over shorter timescales compared with other scales. Complex bend geometries, such as compound loops, develop at all scales of heterogeneity prior to the development of cutoffs (from Guneralp and Rhoads, 2011).

Figure 10.15. Types of evolution of a symmetrical bifurcation (a) to asymmetrical bifurcations in a braided river system: (b) deflection of flow by a bar, (c) restriction of flow and bed-material load by an inlet step, (d) rotation of the upstream channel, and (e) backwater effect from bar formation in one of the downstream channels (from Schuurman and Kleinhans, 2015).

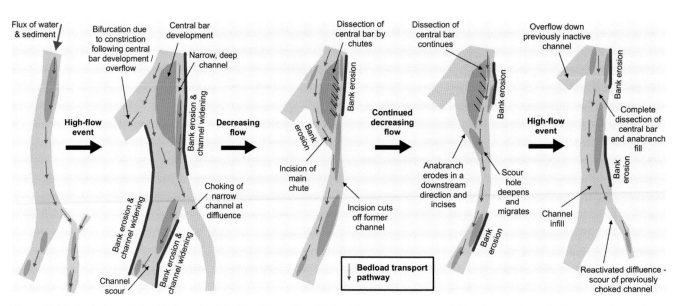

Figure 10.16. Evolution of a braided reach of the Rees River, New Zealand between January and March 2011 (from Williams et al., 2015).

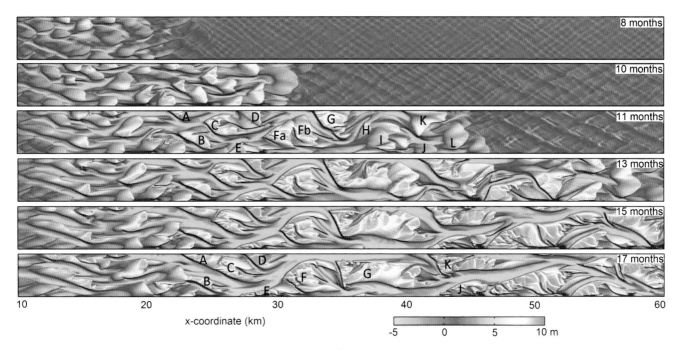

Figure 10.18. Initiation and evolution of a braided river simulated with a physics-based two-dimensional numerical model (from Schuurman and Kleinhans, 2015). Labels refer to bar elements discussed in original source.

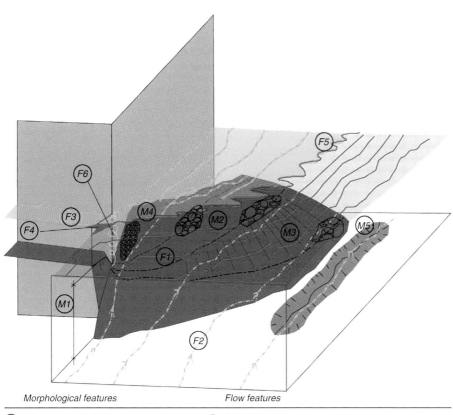

Morphological features

Flow features

(M1) bed discordance

(M2) sediment deposition bar

(M3) faces of the sediment deposition bar and corridors of coarse sediment transport

(M4) corridors of fine sediment transport

(M5) erosion zone

(F1) near-surface flow originating from the main channel

(F2) near-bed flow originating from the main channel

(F3) flow originating from the tributary

(F4) flow stagnation zone

(F5) shear layer zone

(F6) spiral vortices at the downstream corner of the tributary that lift fine material in suspension

Figure 12.13. Flow and sediment dynamics at an experimental discordant confluence of a steep tributary and a main channel (from Leite Ribeiro et al., 2012a).

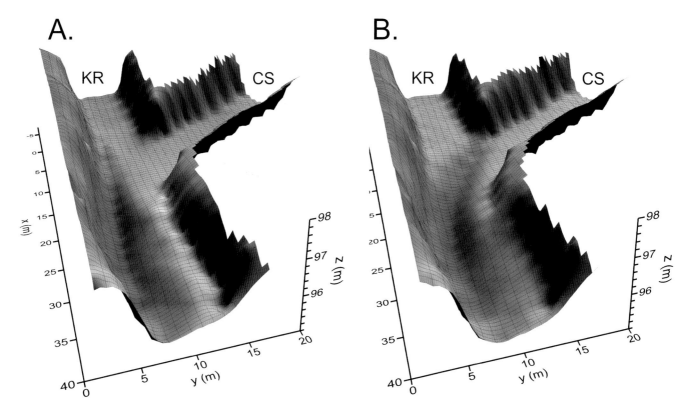

Figure 12.17. Bed morphology at Kaskaskia-Copper Slough confluence formed by (a) high–momentum flux ratio flows ($M_r > 1$) and (b) low–momentum flux ratio flows ($M_r < 1$). In (a), a wedge of sediment from the Copper Slough (CS) penetrates into the confluence and wraps around the downstream junction corner, confining the scour hole to the far bank within the confluence. In (b), the zone of scour extends upstream into the confluence, and the bank-attached bar downstream of the confluence is lower in height than in (a).

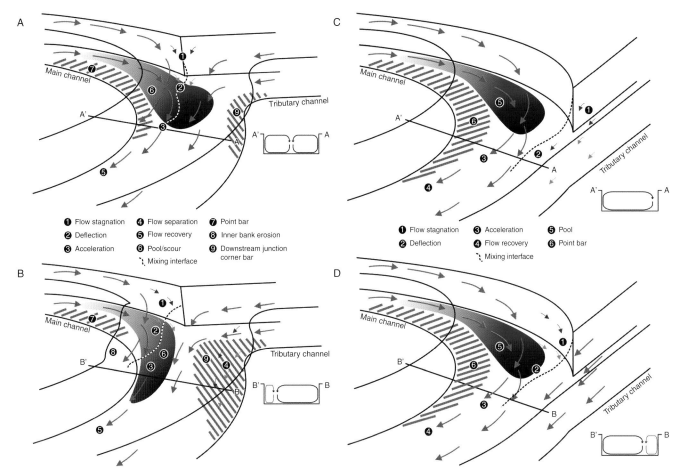

Figure 12.21. Flow structure and bed morphology at (a) high-angle confluent meander bend with $M_r < 1$, (b) high-angle confluent meander bend with $M_r > 1$, (c) low-angle confluent meander bend with $M_r < 1$, and (d) low-angle confluent meander bend with $M_r > 1$ (from Riley et al., 2015).

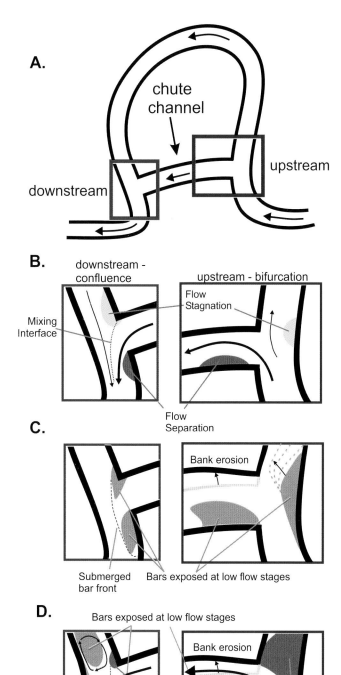

A.

chute
channel

downstream

upstream

B.

downstream -
confluence

upstream - bifurcation

Flow
Stagnation

Mixing
Interface

Flow
Separation

C.

Bank erosion

Submerged
bar front

Bars exposed at low flow stages

D.

Bars exposed at low flow stages

Bank erosion

Bank
erosion

Submerged
bar front

Bars exposed when
flow is below bankfull
stage

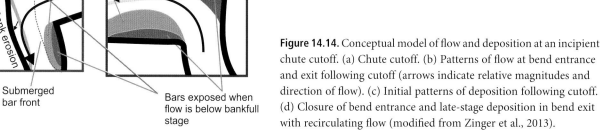

Figure 14.14. Conceptual model of flow and deposition at an incipient chute cutoff. (a) Chute cutoff. (b) Patterns of flow at bend entrance and exit following cutoff (arrows indicate relative magnitudes and direction of flow). (c) Initial patterns of deposition following cutoff. (d) Closure of bend entrance and late-stage deposition in bend exit with recirculating flow (modified from Zinger et al., 2013).

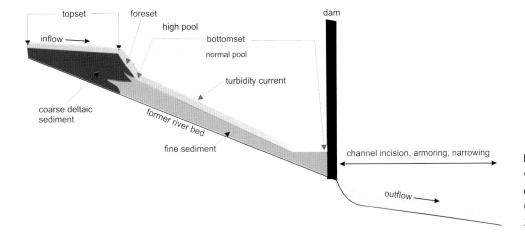

Figure 15.24. Pattern of deposition within a reservoir and channel erosion downstream of a high dam (adapted from Csiki and Rhoads, 2010).

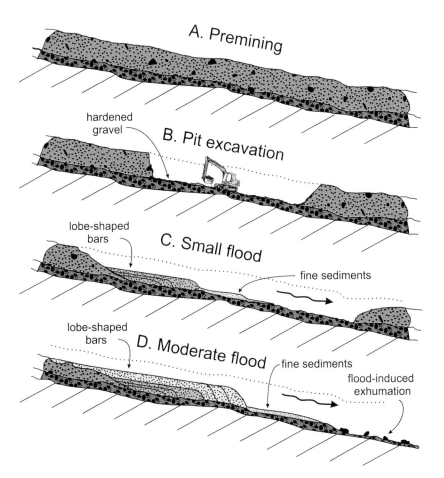

Figure 15.28. Pattern of deposition and erosion in a pit mine in an ephemeral river showing development of lobe-shaped bars at upstream lip of mining pit and erosion at downstream lip during small and moderate floods (modified from Calle et al., 2017).

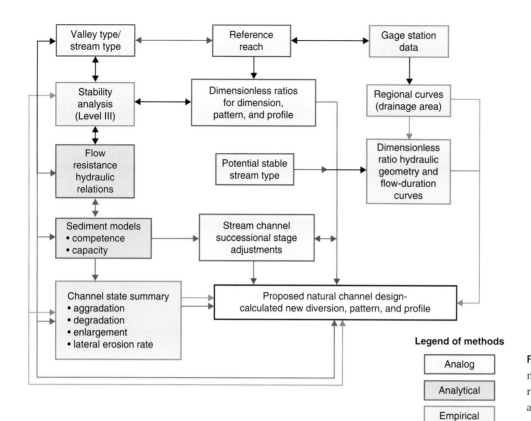

Figure 16.9. Methodology of natural channel design showing relations among analog, analytical, and empirical components (from Rosgen, 2007).

CHAPTER

15 Human Impacts on River Dynamics

15.1 Why Are Human Impacts on River Dynamics Important?

The environment of Earth has become strongly influenced by activities of humans. Effects of these activities occur over a variety of scales, ranging from the local to the global. Transformation of the environment by humans has been particularly effective since the beginning of the Industrial Revolution in the mid- to early 1800s, with a pronounced acceleration occurring after 1950 (Steffen et al., 2015). Dramatic population growth, along with the development and spread of technologies powered by fossil fuels, has led to fundamental, widespread changes in many aspects of the environment (Lewis and Maslin, 2015). These changes are so remarkable that a new label, the Anthropocene, or age of humans, has arisen to designate this time of human dominance (Crutzen, 2002). Although debate surrounds formal use of the term as an epoch of geologic time (Brown et al., 2013, 2017; Lewin and Macklin, 2014; Zalasiewicz et al., 2017), informal use has become embedded in vocabulary aimed at drawing attention to the impact of humans on Earth's environment (Steffen et al., 2011).

Of particular importance geomorphologically has been the increasing capacity of humans to influence global sediment fluxes (Hooke, 1994, 2000). Sediment fluxes associated with excavation and dredging activities, about 316 Gt yr^{-1}, are over 20 times greater than the amount of sediment supplied by major rivers to oceans (Cooper et al., 2018). This estimate does not include effects of humans on sediment fluxes associated with agriculture, which generate an additional flux of about 75 Gt yr^{-1} (Wilkinson and McElroy, 2007). Over 50% of Earth's ice-free land area has been directly modified by human action, either through moving of earth material or through changes in land cover conditions that alter sediment fluxes (Hooke, 2012). Without question, humans have become the most important geomorphological agent on the planet in terms of sediment redistribution.

Regarding rivers in particular, humans have affected virtually all aspects of fluvial systems, including hydrological, hydraulic, geomorphological, ecological, and biogeochemical characteristics (Meybeck, 2003). Human agency has become an important factor influencing river dynamics across all scales of time (Figures 1.6 and 1.9). This influence is comparable to, and may exceed, that of change in environmental conditions related to tectonics, climate, vegetation, and topography. As a result, exploration of the ways in which humans influence the dynamics of river systems has become a major focus of research in fluvial geomorphology (Hooke, 1986; Gregory, 2006).

15.2 What Are the Major Impacts of Humans on Rivers?

From a geomorphological perspective, humans alter river systems both directly and indirectly (Park, 1981) (Table 15.1). Indirect effects relate mainly to changes in watershed conditions that modify inputs of water and sediment to river systems. These effects extend over watershed scales and can have wide-ranging, spatially and temporally variable impacts on river dynamics. Direct effects involve physical manipulation of rivers, resulting in changes to the hydraulics of flow

Table 15.1. Human activities that affect river systems.

Indirect effects	Direct effects
Changes in land cover/ land use	Dam and reservoir construction
Agriculture	Channelization
Land clearing/ deforestation	Flood control
Crop cultivation	Navigation
Livestock grazing	Drainage
Irrigation	Mining of river sediment
Land drainage	
Forestry	
Timber harvesting	
Surface mining	
Introduction of mining waste	
Urbanization	
Human-induced changes in climate	
Interbasin water transfers and flow diversions	

and river morphology. These effects tend to be localized but can produce dramatic change in river morphodynamics. Direct effects often trigger adjustments to altered conditions that extend upstream or downstream from the site of manipulation. These adjustments change the way in which the river operates, often leading to unanticipated responses, some of which may threaten human welfare or environmental quality.

15.3 How Does Human-Induced Change in Land Cover/Land Use Affect River Systems?

Throughout the course of human civilization, humans have continuously been modifying the character of Earth's terrestrial surfaces to accommodate their needs. The most common modification has been to remove native vegetation by clearing the land to use it for agriculture, grazing, isolated settlements, or, more recently, intensive residential, commercial, or industrial purposes (Harden, 2013). In addition, much forest clearing has occurred historically to access timber for fuel or construction. Initial clearing of land and implementation of new surface-cover conditions typically trigger a cascade of hydrogeomorphic responses within affected watersheds.

These responses include changes in hillslope hydrology and sediment delivery as well as connectivity between hillslope runoff and erosion (Royall, 2013). Because inputs of water and sediment are fundamental fluxes governing river dynamics, changes in these inputs often result in changes in river systems. The extent to which river systems are affected by human-induced changes in land cover depends on the spatial extent of the changes as well the rapidity, or intensity, of the changes. Thus, scale plays a fundamental role in the response of fluvial systems to human alterations of land-cover conditions. Changes that are both extensive and intensive can greatly alter the form and function of rivers. System response also depends on the natural characteristics of the landscape other than vegetation cover, such as climate, relief, and the erodibility of soil or rock, which determine how well river systems are buffered against change in land cover. Rivers in steep, erodible landscapes with high-intensity precipitation will be more sensitive to changes in land cover than those in low-relief, erosion-resistant landscapes with low-intensity precipitation.

15.3.1 What Is the Impact of Agriculture on River Dynamics?

The history of systematic human-induced changes in land cover spans at least the last 10 millennia since the advent of agriculture in the early Neolithic (James, 2013; James and Lecce, 2013). Major agricultural uses of land include the production of crops as well as grazing by sheep, cattle, and other ruminants. The global impact of agriculture on geomorphic processes is highly uneven in space and time, and largely conforms to the patterns by which different societies around the world have historically advanced the extensiveness and intensity of agricultural production (Harden, 2013; James, 2013). Evidence of accelerated rates of soil erosion and sediment delivery to rivers between 10,000 before present (BP) and the end of the Middle Ages (500 BP) has been documented in China, the Middle East, Central America, and Europe (James, 2013; James and Lecce, 2013). Although the effects of these historical changes in some cases have been locally pronounced, the extent and intensity of agricultural impacts on rivers have increased dramatically since the early 1800s. The development and implementation of advanced farming technologies, including major land-clearing and soil-working equipment, has helped to accelerate the impacts of agriculture on river systems. To some extent, intensive impacts have been ameliorated by conservation practices, but the spread of agriculture across the globe continues (Lanz et al., 2018).

15.3.1.1 What Are the Impacts of Agriculture on Runoff and Flood Frequency?

The main impacts of agriculture on watershed hydrology are to decrease soil infiltration and reduce flow resistance over hillslopes, thereby amplifying runoff generation and flood magnitudes. Exposure of bare soil during initial clearing of vegetation and seasonally thereafter by tillage or grazing results in decreased soil infiltration rates, both during wetting and when the soil is saturated (Figure 15.1). The infiltration rate under saturated conditions is referred to as the saturated hydraulic conductivity. Globally, saturated hydraulic conductivities are two to three times smaller for arable land than for natural vegetation and forests (Jarvis et al., 2013). The removal of trees or grasses and replacement with seasonal crops decreases canopy interception, so that precipitation intensities exceed infiltration capacities more frequently when bare soils are exposed or when annual crops are small and have limited capacity to intercept rainfall. The result is an increase in the likelihood of infiltration excess overland flow. Without the resistance provided by stems, leaves, and trunks, rates of overland flow are also greater on bare soil than on grassed or forested slopes, increasing rates of runoff. In poorly drained, low-relief landscapes, extension of stream networks through the addition of drainage ditches and installation of subsurface tile-drain systems to facilitate soil drainage enhance the efficiency of stormflow runoff (Rhoads et al., 2016; Kelly et al., 2017). Together, these changes in watershed hydrology can increase flood magnitudes of rivers across a wide range of recurrence intervals (Knox, 2001).

The influence of agricultural land use on hydrologic response is complex and can vary geographically.

Conversion of land from agriculture to forest does not always produce consistent reductions in streamflow, suggesting that the effects of agricultural land use either are not substantial or cannot be readily changed (Cruise et al., 2010). On the other hand, conservation practices, such as tillage methods that leave substantial crop residue on the soil surface and/or produce a rough, porous surface (Mannering and Fenster, 1983), can reverse the trend toward increasing runoff, resulting in reductions in flood magnitude and streamflow in agricultural watersheds (Potter, 1991; Knox, 2001). In water-deficient environments where rivers serve as a source of water for agriculture, abstraction of water for irrigation reduces, rather than increases, flow, resulting in changes in the hydrology, ecology, and geomorphology of rivers (Johnson, 1994).

15.3.1.2 What Are the Impacts of Agriculture on Soil Erosion?

One of the most notable aspects of agricultural land use is enhanced exposure of bare soil, increasing the potential for soil erosion. Soils are exposed both through initial land clearing and through cultivation of annual crops that seasonally leave tilled soils without substantial vegetation cover. Intensive grazing by sheep and cattle also results in soil compaction and exposure of bare soil by trampling (Trimble and Mendel, 1995; Evans, 1998; Grudzinski and Daniels, 2018). Rainsplash and overland flow are more effective at mobilizing and transporting downslope bare soil compared with vegetated soil. Accelerated erosion that leads to the development of rills and gullies on bare, exposed hillslopes can mobilize enormous amounts of soil material. Simple estimates using the universal soil loss equation (see Chapter 3) indicate that erosion rates for bare soil can exceed those for natural woodland by a factor of 1000 when slope, rainfall intensity, and soil properties are held constant (Jacobson et al., 2001). This simple estimate is generally consistent with a global compilation of data from a wide range of environments showing that conventionally plowed agricultural fields have soil-erosion rates one to two orders of magnitude greater than rates of erosion under natural vegetation (Montgomery, 2007b). Estimates of soil erosion for 220 countries constituting 84% of Earth's land area show that cropland constitutes about 11% of this area but generates over 50% of the estimated soil erosion (Borrelli et al., 2017). Between 2001 and 2012, soil erosion increased by 2.5% as cropland expanded by 0.22 million km^2.

In the United States, comparison of erosion rates between noncropped land in conservation reserve and cropped land provides insight into the influence of growing crops on soil erosion (Nearing et al., 2017) (Figure 15.2). Average rates for cropland have declined since the 1980s as a result of conservation practices but still are around 7 Mg km^{-2} yr^{-1}. In contrast,

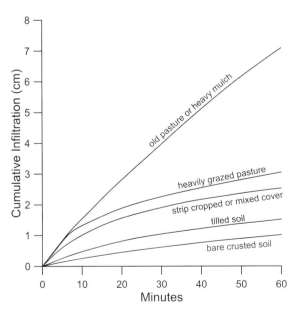

Figure 15.1. Variation in infiltration rates for different agricultural land cover types (Holtan and Kirkpatrick, 1950).

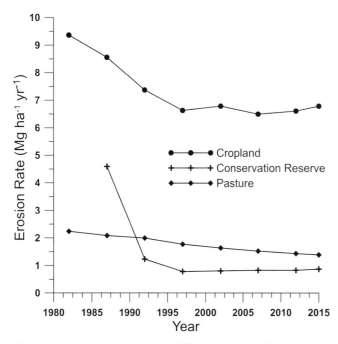

Figure 15.2. Soil erosion rates for different agricultural land uses (data from US Department of Agriculture, 2018).

pastureland has rates of about 2 Mg km^{-2} yr^{-1}, and rates for land in conservation reserve are less than 1 Mg km^{-2} yr^{-1}. A similar situation exists in northeastern China, where conversion of forest and grassland to cropland has occurred over the past three centuries with rapid acceleration during the twentieth century. There, soil erosion rates for cropland (12.2 to 23.8 Mg km^{-2} yr^{-1}) are an order of magnitude greater than rates for grassland and forest (1.3 to 5.2 Mg km^{-2} yr^{-1}) (Nearing et al., 2017).

15.3.1.3 What Factors Determine How Changes in Runoff and Soil Erosion Affect Rivers?

The extent to which river systems are influenced by increases in soil erosion related to land clearing and farming depends on the degree of connectivity of sediment fluxes between hillslopes and the stream network. In a general sense, this connectivity is governed by factors that control hillslope sediment fluxes. Conditions that promote connectivity include steep slopes, low surface roughness to impede overland flow, high rainfall amounts and intensities, and narrow valley bottoms. As connectivity increases, the sensitivity of the geomorphic landscape system increases, in the sense that human-induced changes to hillslopes are likely to be transmitted more readily to river systems.

Average transport distance of sediment mobilized by upland erosion is an important consideration in ascertaining how upland erosion alters sediment delivery to streams.

Evaluations of soil erosion rarely consider explicitly how far the eroded material is transported, and in many landscapes a considerable amount of eroded material is redeposited on hillslopes before reaching a stream channel. Typical transport distances during transport-effective events are < 1 m for rainsplash erosion, < 100 m for sheet, rill, and seepage erosion, and < 1 km for gully erosion (Poesen, 2018). The development of gullies therefore is particularly effective at transporting large amounts of hillslope material over substantial distances.

The flux of sediment from hillslopes generally has greater inertia than the flux of water, so that enhanced movement of sediment into stream networks may not coincide with increases in runoff and streamflow. Consequently, changes in discharge and sediment load within rivers may not occur together but may be asynchronous, leading to complex, nonlinear responses of river systems to anthropogenic impacts. Nonlinearity is also influenced by thresholds of water discharge and sediment flux that, once exceeded, lead to fundamental changes in the dynamics and morphology of fluvial systems. Whereas recovery from exceedance of thresholds is possible in some cases, in others human disturbances are great enough that recovery does not occur. Such thresholds have been identified as tipping points (Notebaert et al., 2018). Often these thresholds are related to the degree of sediment-flux connectivity between hillslopes and channels; however, the complex, nonlinear relation between the intensity of land-cover change and the magnitude of change in sediment flux complicates efforts to develop a general understanding of how river systems respond to human impacts (Verstraeten et al., 2017).

If accelerated rates of soil erosion involve only increases in rainsplash and sheet erosion, the translocation of soil material may be confined mainly to hillslopes, resulting primarily in the formation of new colluvial deposits (Figure 15.3). Under such conditions, sediment yields in rivers and streams will not increase dramatically. Sediment delivery ratios decrease as sediment production increases, but delivery of sediment to streams remains relatively constant. If changes in land-cover conditions enhance runoff generation to the extent that it exceeds the threshold for gully development, delivery of sediment to valley floors and streams may increase dramatically (Trimble, 2013; Wilkinson et al., 2018). Under these conditions, sediment yields increase, but substantial storage of eroded upland sediment also occurs within river systems, resulting in sediment delivery ratios much less than 1 (Trimble, 1977) (Figure 15.3). Changes in channel morphology may also occur as rivers adjust to increased sediment loads and flood magnitudes.

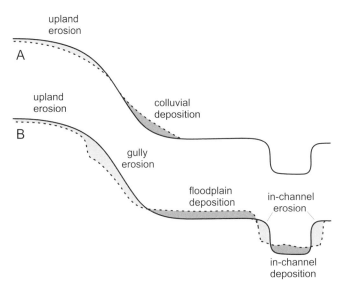

Figure 15.3. (a) Upland agricultural erosion resulting in colluvial deposition on lower slopes. (b) Upland and gully agricultural erosion resulting in deposition on floodplains and within stream channels. Changes in runoff and sediment load may also produce morphological adjustments of stream channels.

15.3.1.4 What Has Been the Impact of Agriculture on Floodplain Sediment Storage and Channel Morphology?

In central Europe and the United Kingdom, early Neolithic farming was limited in intensity and area, so that eroded soil was redeposited on hillslopes and did not reach stream channels; however, during the medieval period, farming intensified, gullies developed, and accelerated floodplain sedimentation occurred (Lang et al., 2003; Macklin et al., 2014). Following European colonization of North America, episodes of rapid land clearing and the use of technology-based farming practices without consideration of soil conservation led to unprecedented levels of soil erosion. By the end of the 1930s, decades of imprudent farming resulted in excessive soil loss, including widespread gully erosion (Happ et al., 1940; Trimble, 1974). Flushing of eroded sediment from hillslopes to valley bottoms produced substantial aggradation of floodplains in the eastern and midwestern United States (Costa, 1975; Knox, 1987; Miller et al., 1993; Jackson et al., 2005; Grimley et al., 2017). This accumulated sediment is referred to both as legacy sediment and as postsettlement alluvium (James and Lecce, 2013; Wohl, 2015a; James, 2018) (Figure 15.4).

In the hilly Driftless Area of Wisconsin, Illinois, Minnesota, and Iowa, a region of the Midwest that was not glaciated during the Pleistocene Epoch, clearing of forest and grassland land in the mid- to late 1800s for cropland and grazing led to widespread upland erosion that resulted in delivery of large amounts of sediment to the stream network. A wave of aggradation first affected headwater tributaries

Figure 15.4. Postsettlement alluvium (light) on top of presettlement soil (dark) in the Blue River watershed, Wisconsin (from Lecce, 1997b).

and progressed downstream into main valleys, producing thick deposits of postsettlement alluvium on floodplains. The impact on floodplain morphology and sedimentology has been greater than any natural environmental changes over the past 10,000 years (Knox, 2006). The thickness of legacy sediment on floodplains, which can be determined by identifying the vertical position of the presettlement floodplain soil within sediment cores or cut bank exposures (Figure 15.4), ranges from 0.3 to 0.5 m along tributaries to 3 to 4 m along main valleys (Knox, 1987; Magilligan, 1992b). The pattern of floodplain sedimentation is spatially variable, with the thickness of deposits directly related to watershed size and valley-bottom width, and inversely related to bankfull stream power per unit length (Magilligan, 1985; Lecce, 1997b; Faulkner, 1998). The implementation of soil conservation practices in the 1940s slowed the rate of sediment delivery and has even led to net sediment export within

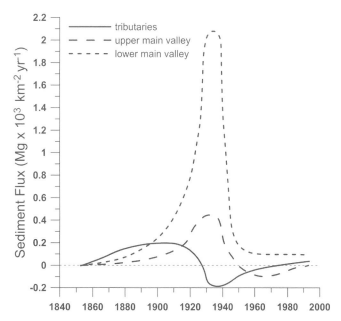

Figure 15.5. Changes in net sediment flux over time in headwater tributaries, upper main valley, and lower main valley of Coon Creek watershed, Wisconsin, USA. Positive flux indicates net storage; negative flux indicates net erosion (adapted from Trimble, 2009).

headwater portions of watersheds (Trimble, 1993) (Figure 15.5).

The impact of changes in runoff and sediment flux in the Driftless region on channel form has been complex, triggering sequences of aggradation and erosion that vary in time and space (Trimble, 1995). In some cases, changes in runoff associated with land clearing resulted in more frequent floods that enlarged the channels of headwater streams (Knox, 1977). In other cases, channel aggradation occurred as large amounts of sediment entered headwater streams (Happ, 1945; Trimble, 1983). Accelerated overbank sedimentation contributed to changes in channel form and morphodynamics by initially increasing the depth of channels, which confined high flows, thereby increasing the stream power and bed shear stress of these flows. Adjustments to the enhanced energy of flow involved channel widening and increased rates of lateral migration because the coarse substrate of the channels limited incision (Knox, 2001) (Figure 15.6). The erosional response of streams, particularly in upper parts of watersheds, has also been facilitated by reductions in sediment load associated with implementation of soil conservation practices (Figure 15.5). This response has produced a new floodplain inset between

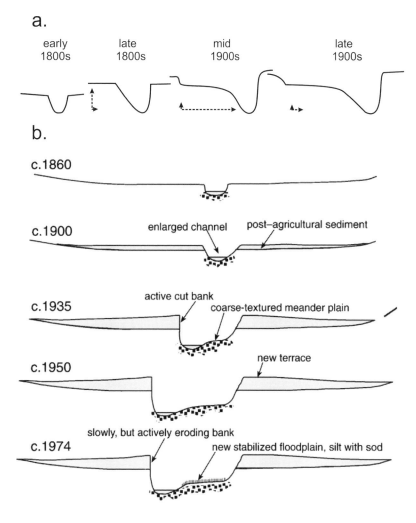

Figure 15.6. Change in form of headwater streams in Driftless region of Wisconsin. (a) Change in channel cross sections over time. Dashed arrows indicate relative magnitudes of vertical versus lateral adjustments at different times (adapted from Knox, 2001). (b) Change in channel and floodplain morphology related to floodplain deposition and channel erosion (from Trimble, 2009).

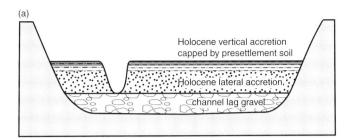

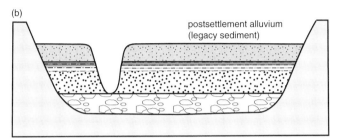

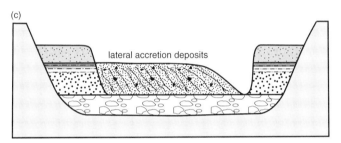

Figure 15.7. Evolution of valley floors and stream channels in Driftless Area of Wisconsin. (a) Presettlement configuration. (b) Accumulation of postsettlement alluvium and deepening of stream channels. (c) Entrenchment of the floodplain resulting in terrace formation and development of lateral accretion deposits on inset floodplain (adapted from Lecce, 1997b).

terraces representing the abandoned surface of the aggradational floodplain (Lecce, 1997b; Knox, 2006; Trimble, 2009) (Figure 15.7). Overbank deposition no longer occurs on the abandoned surfaces except during extreme floods. Accelerated lateral migration also mobilizes the thick vertical accretion sequences produced by enhanced overbank deposition, flushing this material downstream. The adjustment process has been transgressive over time and space, with the sequence of initial floodplain aggradation and inset floodplain development advancing downstream over a period of many decades (Knox, 2006; Trimble, 2009). Other tributary streams in the region have developed an entrenched form not through the aggradation–lateral migration mechanism, but through upstream migration of headcuts (Faulkner, 1998). In either case, the confinement of flow within an inset floodplain or incised channel by flanking terraces concentrates floodwaters, promoting high-energy flows that transport suspended sediment downstream, where it is deposited on floodplains and within channels along the lowest reaches of the river systems (Lecce, 1997b; Trimble, 2009). Because these systems drain

directly into the Mississippi River, the lowest reaches are frequently influenced by backwater conditions, which limits sediment-transport capacity (Knox, 2006). Since the 1940s, decreases in flood magnitudes and upland erosion rates following implementation of soil conservation practices have reduced rates of overbank sedimentation and lateral channel migration (Magilligan, 1985; Trimble, 1993; Knox, 2001).

In southeastern Australia, responses of rivers to agricultural land-use change related to Anglo-European colonization have exhibited considerable complexity, with responses varying both within watersheds and between watersheds. In the Bega River drainage basin, extensive clearing of native vegetation in the 1800s for grazing and growing crops subsequently altered the hydrologic and sediment regime of the river, increasing the geomorphic effectiveness of floods. As a result, the lower Bega River was transformed from a narrow, deep channel transporting a mixed load into a wide, shallow channel transporting sandy bedload (Brooks and Brierley, 1997). In some locations, the channel widened by as much as 340%. At the time of settlement, many watercourses in the upper and middle parts of the watershed were discontinuous, flowing through extensive swamps on thick valley fills. Increases in runoff led to incision of the fills and the formation of continuous channels. The flux of sediment from this incision served as a primary source of sediment downstream where channel widening occurred. Grazing and deforestation following the arrival of Europeans also produced gully erosion and the accumulation of thick deposits of postsettlement alluvium along rivers of the Goulburn Plains (Portenga et al., 2016). In the upper Hunter River watershed of Australia, river adjustments since European settlement have been localized and fragmented, with less than 20% of the total length of rivers exhibiting change. Longitudinal connectivity of sediment flux along this system is limited by bedrock control of river morphology within confined or partly confined settings (Fryirs et al., 2009). A variety of human impacts continue to affect rivers in this watershed, producing continued local fluvial adjustments nearly 200 years after European settlement.

Clearing of riparian vegetation for agriculture in some cases has had a major influence on channel and floodplain evolution of rivers. Removal of native floodplain vegetation and in-channel woody debris along the Cann River in Australia since the time of European settlement has resulted in a 360% increase in channel depth, a 240% increase in channel slope, a 700% increase in channel capacity, and a 150-fold increase in the rate of lateral migration (Brooks et al., 2003). Studies of deforestation in tropical environments have shown that clearing of riparian vegetation can lead to accelerated rates of channel migration along meandering rivers (Horton et al., 2017) and increased sedimentation within anabranching rivers (Latrubesse et al., 2009).

Besides clearing of vegetation, grazing of floodplains by cattle can also impact channel form and the delivery of sediment to streams. The main effect of cattle on channels and floodplains is to expose bare soil through trampling and to erode streambanks locally through the creation of paths or ramps leading from the floodplain into the channel. Rates of erosion for grazed streambanks three to six times greater than those for ungrazed streambanks have been reported (Trimble, 1994). Most erosion is related directly to hoof action, but ramps can also direct high flows onto bare cattle paths on the floodplain and be scoured by return flow from the floodplain reentering the channel. Exposure of trampled bare soil on floodplains, as well as decreased resistance of streambanks to erosion where grazing has reduced the biomass and rooting depths of bank vegetation, can result in enhanced floodplain erosion and delivery of this eroded material to streams (Grudzinski et al., 2016; Yu and Rhoads, 2018). The direct effect of grazing on stream morphology seems to be spatially limited, given the tendency for cattle to enter streams at ramps. Changes in channel geometry produced by riparian grazing tend to be either minor or restricted to local areas of intense bank trampling (Agouridis et al., 2005; Lucas et al., 2009; Grudzinski and Daniels, 2018).

The longevity of responses of fluvial systems to human alteration of land cover for agriculture at watershed scales highlights the important role of anthropogenic landscape memory in the contemporary dynamics of these systems (Brierley, 2010). The accumulation of legacy sediment on floodplains and in other storage locations within river systems provides a source of sediment that can, through remobilization of this stored material, maintain elevated sediment fluxes for decades, centuries, or possibly even millennia into the future (Jackson et al., 2005; Trimble, 2009). Also, the development of enlarged trenches within floodplains through incision or lateral migration can have long-lasting impacts on the energy regime of floods (Lecce, 1997b).

15.3.2　How Do Timber Harvesting and Mining Affect River Dynamics?

15.3.2.1　What Is the Effect of Timber Harvesting?

The effects of logging, including clear cutting and the construction of logging roads, on hydrological response varies. Some evidence suggests that logging increases peak discharges of all floods (Jones and Grant, 1996), whereas other evidence indicates that it only increases the peaks of small to moderate events (Wright et al., 1990; Thomas and Megahan, 1998; Beschta et al., 2000). The hydrological effect of logging generally decreases with drainage-basin size, so that increases in peak flow are more pronounced in small versus large watersheds.

Although the influence of logging operations on streamflow is not entirely clear, the influence of this activity on sediment production in some cases is pronounced, particularly in high-relief environments. Exposed soils and unpaved logging roads generate large amounts of mobile fine sediment, thereby increasing suspended sediment concentrations in streams (Beschta, 1978). Increased loads of sand can infill pools and gravel substrates in streams, degrading habitat for aquatic organisms (Lisle and Hilton, 1992; Wood and Armitage, 1997). Such effects are less likely to occur in low-relief environments where rates of delivery of sand to streams are low and ample stores exist to trap sand before it enters streams (Kasprak et al., 2013).

The removal of trees enhances root rot and increases the amount and rate of water infiltration into soils by eliminating canopy interception of precipitation. In steep mountain environments, these changes decrease the stability of hillslopes, resulting in the occurrence of landslides at lower hillslope gradients in logged areas than in unlogged areas (Wolter et al., 2010). Logging also leads to an overall increase in the frequency of landslides, which can rapidly deliver large amounts of sediment to streams.

The response of Redwood Creek, a gravel-bed stream in northern coastal California, to high sediment loading is one of the best-documented cases of the influence of logging-related landscape disturbance on a fluvial system (Nolan et al., 1995; Madej and Ozaki, 1996). Large-scale timber harvesting in the watershed of Redwood Creek began in the watershed in the early 1950s, and by 1992 80% of the basin was logged, mainly by clear cutting. To support timber harvesting, thousands of kilometers of unpaved logging roads were constructed. Between 1955 and 1975 a series of large floods resulted in basin-wide hillslope erosion and the delivery of large amounts of coarse-grained sediment to the creek, particularly in the upstream part of the drainage basin. Landslides, along with surface erosion and gullying associated with logging roads, were major contributors of sediment. The large influx of sediment induced widespread channel aggradation and widening, with bed elevations increasing by as much as 9 m in some locations. Extensive bank erosion accompanied the process of widening, and aggradation resulted in loss of pools through infilling. Since the mid-1970s a reduction in large floods has led to progressive channel incision with bed elevations decreasing exponentially over time (Madej and Ozaki, 2009). In the upper and middle parts of the basin, the incising creek attained stable bed elevations by the mid-1990s. The substantial excavation of channel-stored sediment upstream generated a downstream-translating wave of bed material that at first produced aggradation of the channel bed along lower reaches of the creek. This period of aggradation has been followed by net incision, and lower reaches are still in a state of recovery. Throughout the basin, channel incision has coarsened the bed material and increased the spatial frequency of pools.

The response of Redwood Creek to timber harvesting represents an example of an aggradational-degradational episode – a common type of response in fluvial systems affected by large influxes of sediment. In such episodes, net sedimentation initially occurs in response to high sediment loads, followed by net erosion when sediment loads decline (James, 2018). The response of rivers to agricultural land-use change represents another example of an aggradational-degradational episode (Figures 15.5 and 15.7). Large increases in sediment load from upland disturbance in agricultural watersheds led to sedimentation on floodplains, whereas subsequent declines in sediment production when land clearing ceased and conservation practices were implemented resulted in channel incision and terrace formation.

15.3.2.2 What Is the Effect of Mining Activities?

Mining activities generally produce large amounts of spoil or tailings. Gradual erosion of spoil heaps (Kincey et al., 2018) or catastrophic failure of tailings ponds (Bird et al., 2008) can deliver contaminated mining waste into rivers, severely impairing water and sediment quality. In particular, contamination of bed material or floodplain sediment by heavy metals, such as copper, zinc, lead, and mercury, commonly occurs in mining-affected watersheds (Leigh, 1994; Miller et al., 1998; Hudson-Edwards et al., 1999; Lecce et al., 2011). Radioactive waste has also entered rivers in some instances (Graf, 1990). Contamination may be widespread, extending tens or even hundreds of kilometers downstream from the source (Pavlowsky et al., 2017).

When relatively small amounts of mining waste are introduced to a river compared with its transport capacity, sediment-related contaminants are dispersed through processes of bed-material and suspended-load transport (Miller, 1997). Rivers able to transport the load supplied by the influx of mining waste will not exhibit pronounced instability or major

morphological change (Nelson and Church, 2012; Ferguson et al., 2015). On the other hand, when introduced volumes are large and exceed the capacity of the river to transport the material, major morphological adjustments occur (Miller, 1997).

Rivers respond to a major influx of mining waste in the form of an aggradational-degradational episode that involves movement of the slug of introduced sediment downstream as a bed wave (James, 2006). Major channel and floodplain aggradation occurs initially near the source of the mining waste and progresses downstream (Gilbert, 1917; Knighton, 1989, 1991). Because the wave of sediment both translates and disperses (Lisle et al., 2001), the amount and rate of aggradation diminish over distance (Knighton, 1989) (Figure 15.8). Zones of storage of material along the path of the bed-material wave contribute to downstream reductions in aggradation (Gran and Czuba, 2017). Both the rate of movement and the degree of storage depend on transport efficacy, with high rates of movement and low amounts of storage corresponding to portions of the river system with high efficacy (James, 2006). When the supply of material is depleted, a wave of degradation sweeps through the system, again progressing in the downstream direction and diminishing in intensity over distance. The timescale of morphological adjustment to the aggradation-degradational episode depends on the amount of sediment delivered to the river system and the capacity of the river to mobilize the introduced sediment.

Major inputs of sediment to rivers accompany hydraulic mining, a practice in which high-pressure jets of water are used to dislodge rock or soil material and direct the water-sediment slurry through sluices to extract ore, notably gold and tin. Hydraulic mining has occurred throughout the United States and in other countries, notably New Zealand and Australia. In Tasmania, hydraulic mining introduced an estimated 40 million m^3 of mining waste to the Ringarooma River between 1875 and 1982 (Knighton, 1989). By the late 1980s, lower reaches of the river were still aggrading, and the downstream-moving wave of degradation had yet to affect these reaches (Knighton, 1991).

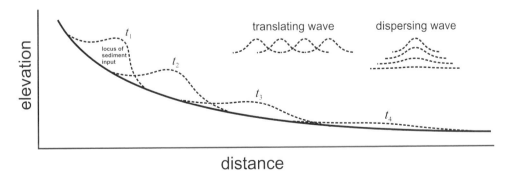

Figure 15.8. Downstream translation and dispersion of a bed wave caused by the introduction of a major influx of sediment in the headwaters of a river system. Solid line represents original longitudinal profile and dotted lines represent modifications of bed elevation at different times (t_1–t_4) caused by translation and dispersion of the sediment wave. Vertical dimension of the wave is exaggerated for enhanced visibility.

a

b

Figure 15.9. (a) Hydraulic-mining debris within the Yuba River in 1908 near the Smartsville Narrows, California. (b) North Fork of Shady Creek near Patterson, California in 1909. Valley was deeply filled by hydraulic-mining debris, which the creek has partially excavated. (Photographs taken by G.K. Gilbert, U.S. Geological Survey Denver Library Photographic Collection.)

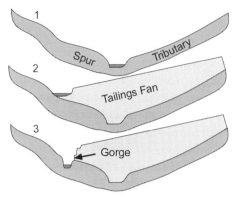

Figure 15.10. Influence of hydraulic-mining debris from tributaries on main channels. (1) Premining main channel in center of valley floor. (2) Burial of valley floor and deflection of main channel by tailings fan from tributary valley on right. (3) Incision of main channel through tailings into valley spur to form a bedrock gorge (from James, 2004).

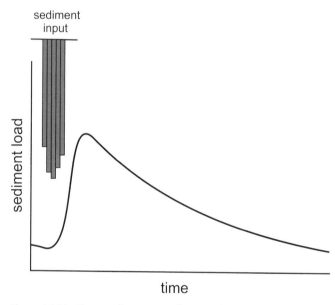

Figure 15.11. Temporal response of river sediment load to an input of mining waste (after James, 1989).

The most notable use of hydraulic mining was during the California Gold Rush in the mid-nineteenth century, and the most dramatic effects of mining waste on river morphology can be found in the Sacramento River system in California. Hydraulic mining in the 1800s introduced an estimated 1.1 billion m^3 of sediment to headwater tributaries of this river system, including the Yuba, Bear, American, and Feather rivers. The massive influx of sediment rapidly filled valley bottoms throughout the Sierra Nevada foothills and Sacramento Valley (Figure 15.9), extending into bays in the vicinity of San Francisco, destroying agricultural land, and impeding navigation (Gilbert, 1917; James et al., 2017). In some cases, sedimentation in the headwaters was so great that tailings fans issuing from tributaries completely filled main valleys, forcing main channels toward the side of the valley opposite the tributary mouth (James, 2004). When the supply of tailings ceased, the main channel incised through

the mining waste into valley spurs to form narrow bedrock gorges (Figure 15.10). The massive devastation caused by hydraulic mining led to a virtual moratorium on it by 1884 (James, 1989), the time when aggradation in the headwaters peaked (James et al., 2019). Since initial aggradation, tributaries in the region generally have been undergoing net incision in response to the diminished sediment supply (James, 1991, 1997; Ghoshal et al., 2010) (Figures 15.9 and 15.10). Nevertheless, substantial amounts of mining sediment are stored in the system, and tributary sediment loads remain high as remobilization of stored sediment produces a long-tailed temporal response in sediment-transport rate (Figure 15.11). At present,

Figure 15.12. (a) Rill erosion of exposed fine-grained soil during urban construction. (b) Exposure of bare soil along a small stream during clearing of land for residential construction.

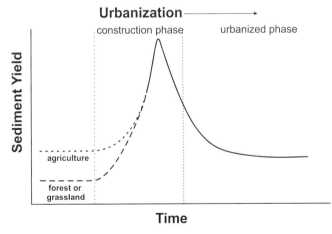

Figure 15.13. Change in sediment yield related to the construction and urbanized phases of urbanization in relation to sediment yields associated with native vegetation (forest and grassland) or agriculture (adapted from Wolman, 1967).

downstream flushing of hydraulic-mining material into lower portions of the Sacramento River system continues, nearly 150 years since hydraulic mining ceased.

15.3.3 How Does Urbanization Affect River Dynamics?

Urban areas constitute less than 3% of Earth's land surface (van Vliet et al., 2017) but contain more than 50% of the world population (United Nations, 2018). Despite occupying only a small percentage of the total land area, the extent of urbanized land is growing at a rapid rate (Seto et al., 2011). Urbanization, the process of an area becoming more urbanized, has a profound influence on all aspects of river systems, including hydrological and geomorphological conditions. This process contributes to degraded environmental quality of urban streams, which has been referred to as urban stream syndrome (Booth et al., 2016; Vietz et al., 2016).

15.3.3.1 How Does Urban Construction Affect Stream Dynamics?

The process of urbanization occurs in two major phases: the construction phase and the urbanized phase (Wolman, 1967). During construction, natural vegetation or agricultural land is cleared to prepare land for the building of roads, structures, and other infrastructure. The exposure of bare soil not only increases susceptibility to erosion by overland flow and rain-splash (Figure 15.12) but also enhances runoff by decreasing interception and infiltration. Thus, soil erosion rates can be several times higher in areas undergoing construction than in adjacent undeveloped areas (Jeong and Dorn, 2019). Construction activities near streams enhance the connectivity of sediment fluxes between streams and the adjacent landscape, greatly enhancing the delivery of sediment to streams (Figure 15.12 and Figure 15.13).

Recognition of the need to control erosion at construction sites has increased (Tyner et al., 2011), partly in response to early studies demonstrating that sediment concentrations of streams in watersheds undergoing construction can be one to two orders of magnitude higher than in adjacent natural or agricultural watersheds (Wolman and Schick, 1967; Walling and Gregory, 1970). Despite efforts to control erosion at construction sites, the effect of construction activities on suspended-sediment yield remains sizable (Chin, 2006); watersheds with active construction have yields 3 to 420 times greater than forested or agricultural watersheds, with median values of 60 times compared with forested watersheds and 11 times compared with agricultural watersheds (Russell et al., 2017). The impact of construction on sediment yield tends to decrease with increasing drainage area as the local increase in sediment delivery associated with construction becomes diluted by additions of flow from tributaries unaffected by construction (Wolman and Schick, 1967).

The influence of the influx of sediment during the construction phase on channel form has been hypothesized to

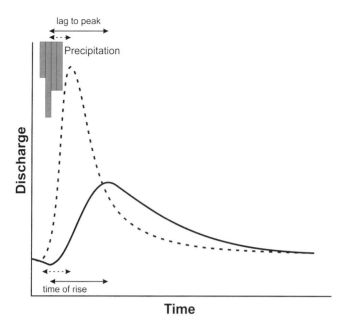

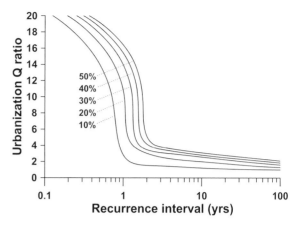

Figure 15.15. Increase in peak discharges of different recurrence intervals (urbanization Q ratio) for various percentages of impervious surface (derived from Hollis, 1975).

Figure 15.14. Preurbanization (solid line) and posturbanization (dashed line) hydrographs including lag to peak and time of rise for each hydrograph.

involve net aggradation, including clogging of small channels with sand (Wolman, 1967; Chin, 2006). The extent to which such aggradation occurs is somewhat uncertain. Few studies report major channel infilling resulting from urban construction. A long-term study of the effects of urbanization on Watts Branch in Maryland, USA, documented a reduction in channel capacity through accretion of sediment along channel margins and growth of point bars during the construction phase of residential development (Leopold, 1973; Leopold et al., 2005) but did not establish a direct connection between this change in channel form and an influx of construction sediment. Rather than being deposited within channels in large quantities, sediment derived from construction sites may in some instances contribute to floodplain development (Graf, 1975). Sediment eroded from construction sites often consists mainly of fine-grained particles transported in suspension (Figure 15.12). This type of material is unlikely to accumulate in large quantities in most streams or rivers, although it may be deposited on floodplains (Leopold et al., 2005).

In gravel-bed rivers, influxes of fine sediment from urbanization can infill voids separating coarse clasts, thereby degrading habitat for benthic organisms. Also, if heavy metals from industrial activities are adsorbed onto fine sediment, elevated levels of bed-material contamination can develop in urban rivers (Thoms, 1987; Rhoads and Cahill, 1999). Information on coarse bed material supplied to streams by construction generally is lacking but may be important in some locations (Russell et al., 2017). Countering the tendency for channel aggradation is an increase in amounts of runoff when land is

cleared for construction, which allows streams to transport augmented sediment loads.

15.3.3.2 How Does Urban Infrastructure Affect River Dynamics?

In the second, urbanized phase of urbanization, the landscape is progressively covered with a variety of impervious surfaces, such as roofs, paved streets, roads, or sidewalks, all of which decrease infiltration, thereby increasing the amount and rate of overland flow. Moreover, storm sewer systems artificially expand the drainage network and efficiently convey runoff into streams (Graf, 1977b). The result is fundamental change in hydrological characteristics of watersheds affected by urbanization.

Peak discharge and total volume of runoff for urbanized watersheds increase relative to preurbanized conditions, whereas time of rise, lag to peak, and recession time decrease (Anderson, 1970; Leopold, 1990; Hung et al., 2018) (Figure 15.14). This change in hydrograph characteristics represents an increase in the flashiness of hydrologic response (Rosburg et al., 2017). Increases in mean and high flows generally occur when the total watershed area covered by impervious surfaces exceeds 10% (Oudin et al., 2018); however, the spatial distribution of impervious cover is also important. Watersheds where the impervious cover is located near the basin outlet have greater increases in peak flow and decreases in time to peak than those where the impervious cover is positioned away from the outlet (Roodsari and Chandler, 2016).

Unraveling the effects of urbanization on flood frequency can be complicated by climate variability (Hejazi and Markus, 2009), but increases in impervious surface cover generally increase flood magnitudes, especially for events of high to moderate frequency, i.e., those with recurrence intervals of two to three years or less (Hollis, 1975; Huang et al., 2008; Hawley and Bledsoe, 2011) (Figure 15.15). The ratio of the urban two-year flood (Q_2) to rural Q_2 commonly exceeds 2 and generally increases with increasing percentage of impervious surface cover (Bledsoe and Watson, 2001b). The effect on infrequent events is less pronounced because such floods are generated by storms that would produce copious runoff in the absence of impervious cover. Even though peak discharges of high-frequency flows increase, the total duration for which streamflow exceeds these peaks may decrease because of the increased flashiness of the hydrologic regime (Konrad et al., 2005).

The traditional perspective on the urbanized phase of urbanization is that sediment supply to streams decreases because of the increase in impervious cover and revegetation of soils exposed during the construction phase (Wolman, 1967; Gurnell et al., 2007). Detailed studies of sediment dynamics in the urbanized phase are less abundant than those for the construction phase (Chin, 2006), resulting in some uncertainty about the extent to which the sediment reduction hypothesis holds. Some studies indicate that sediment yields decrease following the construction phase (Chin, 2006), including the supply of bed material (Annable et al., 2012). On the other hand, sediment yields can remain high if increases in runoff and the magnitude of floods produce channel erosion that contributes large amounts of sediment to streams (Trimble, 1997; Gellis et al., 2017). In addition, a wide variety of nonpoint sources of sediment exist in urbanized areas. Sediment buildup and wash off is common on impervious surfaces, especially streets and roads (Bai and Li, 2013). Disintegration of paved surfaces by mechanical wear as well as from freeze-thaw and other weathering processes produces particles that can be flushed into storm sewer systems and delivered to streams. Landscaping activities and infill construction are other sources of sediment. The production of sediment in urbanized watersheds not only contributes fine sediment to rivers (Thoms, 1987) but can also supply coarse material transported as bedload (Russell et al., 2018).

The changes in watershed hydrology and sediment delivery associated with the urbanized phase of urbanization have substantial consequences for the stability and dynamics of urban streams. Increases in peak discharges and water-surface gradients associated with flashy, unsteady flows produce more frequent events with high stream power (Rhoads, 1995). Consequently, the frequency of events capable of mobilizing bed material increases (Plumb et al., 2017) and the bed-material transport capacity for a flood event of a specific recurrence interval also increases (Whipple and DiLouie, 1981). Increases in stream power and sediment mobilization, along with reductions in sediment delivery related to increases in impervious surface cover, greatly enhance the potential for alluvial urban streams to erode. Two types of erosional responses are common: channel incision and channel

widening (Figure 15.16). The occurrence of one or both of these processes typically produces enlargement of channels (Hammer, 1972). Modes of enlargement include uniform, progressive increases in both width and depth (Leopold et al., 2005; Taniguchi and Biggs, 2015), initial rapid incision through knickpoint migration followed by widening as high, steep banks fail through mass movement (Hawley et al., 2012), and/or progressive widening (Pizzuto et al., 2000; Galster et al., 2008) (Figure 15.17). Bank erosion is characteristic of all these modes (Figure 15.17). Enlargement may also be accompanied by net aggradation of the channel and floodplain if the increase in overbank flows enhances floodplain aggradation and the channel bed also aggrades in response to input of sediment from channel widening (Leopold et al., 2005). The enlargement ratio, the ratio of the preurbanized bankfull channel area to the posturbanized area, typically ranges from 1 to 4, with a mean value of 2.5 (Chin, 2006). In highly erodible, semiarid and arid urban channels, enlargement ratios as great as 10 to 14 have been reported (Hawley and Bledsoe, 2013; Chin et al., 2017). Other fluvial responses to the urban phase can include coarsening of the bed substrate, which increases flow resistance, and shortening of riffles, which occurs as a consequence of decreases in channel slope caused by headcut migration (Hawley et al., 2013). Channels can also transition from single-channel to braided forms if flow–sediment load relations change to the extent that the system transcends a channel pattern threshold (Arnold et al., 1982; Hawley et al., 2012).

Although enlargement is the most common response in the urbanized phase, not all channels in urban areas exhibit substantial enlargement (Nelson et al., 2006; Jordan et al., 2010; Navratil et al., 2013; Phillips and Scatena, 2013). Moreover, the extent and mode of channel adjustment often are difficult to predict. A variety of factors can affect channel response, including spatial variation in the erodibility of the channel bed and banks; infrastructure constraints, including bridges, culverts, levees, bank-protection schemes, or grade control structures; vegetation management and in-stream wood loadings; and stormwater management programs. This complexity has led to the development of various conceptual schemes for characterizing the evolution of urban channels over time (Hawley et al., 2012; Booth and Fischenich, 2015; Bevan et al., 2018). A key concern is the timescale over which adjustment occurs, which is highly variable but seems to be on the order of decades (Chin, 2006). The response of urban channels in urbanized watersheds can also vary spatially. Enlargement generally decreases with drainage area, presumably because the density of impervious surfaces also decreases as drainage areas become large (Navratil et al., 2013; Bevan et al., 2018). Moreover, the progression of channel incision upstream through migrating headcuts can produce

Figure 15.16. (a) Actively incising urban stream in St. Louis, USA with exposed bridge footing along right bank and headcut eroding upstream into concrete protective apron (lower middle of the photo). (b) Actively widening urban stream in Chicago, USA. (c) Bank erosion along an incised urban stream in St. Louis.

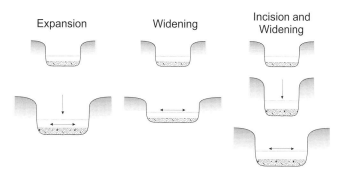

Figure 15.17. Modes of channel enlargement related to the urbanized phase of urbanization, including expansion, widening, and incision and widening.

aggradation downstream as channels become overloaded with incoming sediment (Colosimo and Wilcock, 2007; Hawley et al., 2012).

15.4 How Might Anthropogenic Climate Change Affect River Dynamics?

Assessment of the science related to climate change by the Intergovernmental Panel on Climate Change (IPCC) concludes with high confidence that global climate has changed relative to preindustrial conditions and that human-induced global warming has already caused multiple observed changes in the climate system (Hoegh-Guldberg et al., 2018). Although the primary focus has been on increasing temperatures, emerging evidence suggests that warming is affecting the global water cycle. Overall, however, changes in the water cycle are more uncertain than changes in temperature. Documented and projected changes include shifting regional patterns of mean precipitation; the frequency, intensity, and amount of heavy precipitation; and drought and dryness. Disentangling the effects of changes in precipitation on runoff and streamflow from those of other factors, such as land-cover change, irrigation, and flow regulation, is complicated. Streamflow trends since 1950 are not statistically significant for most of the world's largest rivers (Dai, 2016), but the frequency of floods has increased in some regions (e.g., Stevens et al., 2016). Overall, the signal of anthropogenic climate change remains difficult to discern within the hydrological record.

Climate is generally viewed as an independent variable in relation to river dynamics over both modern and geologic timescales (Figure 1.9). Changes in climate have occurred throughout Earth's history, but given the relatively fast dynamics of rivers over geologic timescales, most attention has focused on the effect of climate change on rivers during the past 10,000 years, i.e., the Holocene Epoch. Paleoflood records show that the frequency and magnitude of floods are highly sensitive to Holocene changes in climate, with changes in flood characteristics occurring at timescales ranging from decades to millennia (Ely et al., 1993; Knox, 1993). These changes in climate often result from hemispheric or global variations in atmospheric circulation patterns that influence storm tracks and air mass boundaries (Knox, 2000) – variations that are expected to contribute to extreme events within the context of global warming (Diffenbaugh et al., 2017). The effects of changes in climate on river systems over centuries to millennia during the Pleistocene and Holocene Epochs have been documented extensively (Macklin et al., 2012), but the extent to which channel form reflects spatial variations in climate remains controversial (Phillips and Jerolmack, 2016). In many cases, changes in river systems related to changes in climate that occur over centuries or millennia may have as much to do with changes in vegetation cover, which is influenced strongly by long-term changes in climate, as with changes in precipitation or temperature (Knox, 1972).

Less certain is the extent to which changes in climate affect river morphology over shorter periods, particularly decades – the timeframe of immediate concern in regard to global warming. Climate shifts at this timescale have the potential to change flood frequency and sediment flux, thereby impacting channel form (Slater and Singer, 2013). Nevertheless, isolating the specific effect of climate change from the effects of other factors can be problematic. The arroyo problem in the southwestern United States is a good example of the complexity of isolating the role of climate in channel change (Miller, 2017).

Arroyos are steep-walled, flat-bottomed trenches cut into alluvium found throughout the southwestern United States (see Figure 6.17a). Numerous cycles of arroyo cutting and infilling have occurred over the last 12,000 years, and changes in climate have been proposed to explain these cycles. A major, widespread episode of arroyo cutting that extended over a multidecadal period from the mid-1800s to the early 1900s, and which was documented historically, was possibly caused by changes in the frequency, intensity, or seasonality of precipitation (Graf, 1983c). However, other factors, including overgrazing by livestock, may have either contributed to or even wholly triggered this historical episode of arroyo cutting. The role of climate change in historical arroyo formation is, at best, ambiguous and difficult to ascertain.

Given the lack of strong evidence of hydrological changes associated with global warming, the extent to which anthropogenic climate change is influencing river dynamics remains unclear. In regard to this issue, the geomorphic concept of river sensitivity (Fryirs, 2017) becomes important, because the influence of climate change on river dynamics is likely to vary geographically depending on the extent to which this change alters drainage-basin hydrology and sediment delivery, and on how responsive river systems are to alterations of watershed-scale processes. Prevailing notions of the inherent morphological stability of river systems in relation to formative events (see Chapter 6) suggest that rivers affected strongly by individual formative events will exhibit the greatest sensitivity to climate change. Alluvial river systems in arid and semiarid environments often display this type of behavior. Of particular interest is an increase in storminess, which in turn increases the frequency of floods capable of altering channel form (Naylor et al., 2017). In systems where recovery of form occurs relatively slowly in relation to the frequency of formative events, an increase in the frequency of formative events may lead to high degrees of instability, characterized by frequent, unpredictable changes in channel form. The response to climate change is likely to occur more gradually in river systems that are well buffered against changes in inputs of water and sediment. Nevertheless, anthropogenic climate change that produces persistent, long-term change in environmental conditions, including the distribution of flows capable of transporting bed material as well as of events that supply bed material to the river, should result in change in the characteristic form of river channels.

As with attempts to determine how anthropogenic forcing is modifying climate, predictive numerical-simulation modeling provides a potential tool for exploring how future changes in climate might affect river dynamics (Lotsari et al., 2015; Praskievicz, 2015). Such approaches, if calibrated against data on channel change, serve as useful tools for managing rivers in the context of anthropogenic climate change. Advances in physical modeling also are being explored to represent experimentally the effects of decadal-scale changes in climate on river systems to help manage and mitigate these effects (Baynes et al., 2018b).

15.5 How Does Channelization Affect River Dynamics?

15.5.1 What Is River Channelization and Why Is It Undertaken?

Channelization is an engineering practice that involves direct modification of river channels. The purpose of channelization varies, but the most common applications are to alleviate flooding, improve land drainage, enhance navigability, control channel erosion, or relocate a stream to access land occupied by it (Brookes, 1988; Simon and Rinaldi, 2013). Methods of channelization include widening, deepening, and straightening channels, constructing artificial levees along river banks, covering the channel bed or banks with rock or concrete, and clearing channels of obstructions, particularly live trees growing along the channel banks or woody debris that has accumulated within the river (Brookes, 1985).

For flood protection, widening, deepening, straightening, levee construction, and clearing are typically undertaken to increase the bankfull conveyance capacity (Q_{bk}) of the channel, where Q_{bk} can be expressed in the form of the Manning equation (see Chapter 4) as

$$Q_{bk} = A_{bk}R_{bk}^{0.67}S^{0.5}/n \qquad (15.1)$$

Such modifications allow the channel to convey more flow than it did previously by increasing its cross-sectional area, hydraulic radius, and slope and by decreasing flow resistance. Increases in conveyance capacity reduce the occurrence of overbank flows, thereby diminishing the frequency of flooding. A consequence of channelization is that floodplains and channels often become disconnected. Channelized sections of rivers also move water downstream rapidly, increasing the flashiness of hydrographs and the magnitudes of peak discharges along reaches downstream from these sections (Wyzga, 1996).

In land-drainage schemes, channelization is implemented mainly to increase channel depth, thereby providing outlets for subsurface drainage tiles (Urban and Rhoads, 2003). These tiles, which consist of buried perforated pipes, allow water that has accumulated in the soil to drain rapidly into nearby channels, reducing the amount of time that the soil remains saturated. The need for such artificial drainage to sustain agricultural production of crops is often acute in low-relief landscapes.

The primary purpose of lining the banks or bed of a river with various types of hard materials is to protect alluvial channels from erosion. Often this type of work is performed after a river has already been widened, straightened, and deepened. Types of material used to line channels include loose rock, often referred to as rip rap; gabions, wire cages

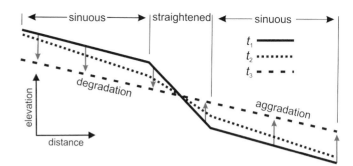

Figure 15.18. Local steepening of the longitudinal profile by channel straightening (solid line) and subsequent adjustment of the profile (dotted and dashed lines) through degradation upstream and aggradation downstream (after Parker and Andres, 1976).

filled with rock; and concrete. In extreme cases, channels may be completely lined with concrete (Vaughn, 1990).

Channelization has occurred for millennia, but it became more common following the Industrial Revolution as increasingly sophisticated mechanized machinery was developed to modify rivers and streams. Advances in the capacity of humans to reshape rivers over the past 150 years has led to widespread channelization. In some locations, such as Denmark and the headwaters of watersheds in Illinois, USA, nearly all streams are channelized (Brookes, 1987a; Mattingly et al., 1993). Adverse consequences of channelization began to attract attention in the 1960s and 1970s (Daniels, 1960; Emerson, 1971; Keller, 1975), particularly the tendency for this practice to produce erosional instability in streams and to eradicate geomorphic features, such as pools and riffles, that provide habitat for aquatic organisms (Frothingham et al., 2001; Rhoads et al., 2003; Lau et al., 2006). Subsequently, a considerable body of geomorphological research has developed focusing on the effects of channelization on river dynamics.

15.5.2 How Does Channelization Affect Channel Stability?

The main influence of channelization on rivers is to confine flows within a channel of increased depth and decreased roughness, both of which enhance flow velocities. Channelization often involves straightening of meandering channels, which increases channel slope by decreasing channel sinuosity and the length of the channel between any two points of elevation along the path of the channel (Figure 15.18). The resulting increases in discharge, velocity, flow depth, and channel slope enhance stream power per unit area and bed shear stress (see Chapter 4), both of which are related to the bed material–transport capacity of the river (see Chapter 5). Because direct modification of the river system occurs at the planform scale rather than the network or watershed scale, sediment supply to the modified channel from the watershed is not greatly affected by channelization. The increase in bed material–transport capacity triggered by channelization often leads to erosional instability that can threaten bridges, structures, and land along the river.

Empirical evidence indicates that erosional instability in response to channelization is a threshold phenomenon that occurs when the bankfull stream power per unit area (ω_{bk}) of the modified river exceeds a critical value. Channelized streams in England and Denmark do not exhibit instability, at least over a period of years to decades, when $\omega_{bk} < 35$ W m^{-2} (Brookes, 1987a,b). Similarly, many channelized agricultural streams in east-central Illinois, which have values of ω_{bk} less than 20 W m^{-2} (Rhoads and Herricks, 1996), remain unchanged decades after being modified (Urban and Rhoads, 2003). Other work has noted that domains of stability and instability can overlap and therefore has treated erosional instability probabilistically. Logistic regression analysis of data on stable meandering versus incising channels indicates that the transition between the two can be defined as (Bledsoe and Watson, 2001a)

$$p(instability) = \frac{\exp\left[4.63 + 14.39\log\left(S\sqrt{\frac{Q}{d_{50}}}\right)\right]}{1 + \exp\left[4.63 + 14.39\log\left(S\sqrt{\frac{Q}{d_{50}}}\right)\right]} \quad (15.2)$$

where $p(instability)$ is the probability the channel will incise. Discharge (Q) in this model is represented by both the mean annual flood and bankfull discharge, depending on data availability. The model shows that increases in discharge and channel slope, along with decreases in median grain size, increase the probability of instability.

15.5.3 How Do River Channels Respond to Channelization?

15.5.3.1 What Is the Channel Evolution Model of Response to Channelization?

Channel incision, or degradation of the channel bed, is a common response to channelization (Galay, 1983; Bledsoe et al., 2002). Straightening of a meandering channel often results in degradation of the bed within the oversteepened, straightened reach (Figure 15.18). The spatial extent of degradation progresses upstream, usually through migrating knickpoints and headcuts (Darby and Thorne, 1992; Kesel and Yodis, 1992). Downstream of the channelized reach, aggradation occurs as excessive amounts of sediment from excavation of bed material upstream are flushed downstream (Figure 15.18).

The frequent occurrence of incision in response to channelization, as well as to other natural and human disturbances that increase bed material–transport capacity without a corresponding increase in the supply of coarse material to the channel (Bledsoe et al., 2002), has led to the development of a channel evolution model (CEM) to characterize this type of response (Schumm et al., 1984; Simon, 1989a) (Figure 15.19). The CEM consists of six stages, progressing from the initial channel condition prior to channelization (stage I) to stabilization of the disturbed channel (stage VI). Following channelization in Stage II, which modifies an original sinuous stream into a straight, enlarged trapezoidal form, degradation occurs in Stage III in response to the increased erosional energy of the channelized stream. Where gravel exists in the bed material, degradation may be accompanied by the development of a coarse armor layer (Talbot and Lapointe, 2002a). During Stage III, bank heights (h_b) do not exceed a critical height (h_{bc}) for mechanical stability, and failure of the banks does not occur. Stage IV begins when degradation produces bank heights that exceed the critical height for stability. During this stage, the channel widens through mass failure of the banks.

Incision and widening during Stages III and IV are efficient at reducing bed shear stress and stream power per unit weight of water over time, thereby promoting stabilization of the evolving system by decreasing sediment-transport capacity (Simon, 1992). Degradation decreases bed shear stress and stream power by reducing the channel slope. As the channel widens, bed shear stress diminishes as flows of a specific frequency decrease in depth. Velocity decreases through reductions in depth and slope and through increases in frictional resistance. Within a reach that has incised and widened, sediment supply increases as the zone of incision and widening migrates headward and erosion upstream flushes large amounts of sediment downstream (Simon, 1989b; Landemaine et al., 2015). The result of these changes in bed shear stress, stream power, and sediment supply is a transition to Stage V, characterized by aggradation of the channel bed in conjunction with continued widening by failure of high channel banks. As aggradation progresses, bank heights are reduced below the threshold for instability, resulting in stabilization of the evolving channel system and formation of a bankfull channel within the bottom of an incised trench generated by degradation and widening (Stage VI). The growth of pioneering woody vegetation on the aggrading bed and accreting banks in the latter stages of adjustment also contributes to stabilization (Hupp and Simon, 1991; Hupp, 1992).

The longitudinal component of the model characterizes the upstream progression of degradation and the corresponding transition from degradation to aggradation to stabilization at any particular location as the locus of degradation migrates

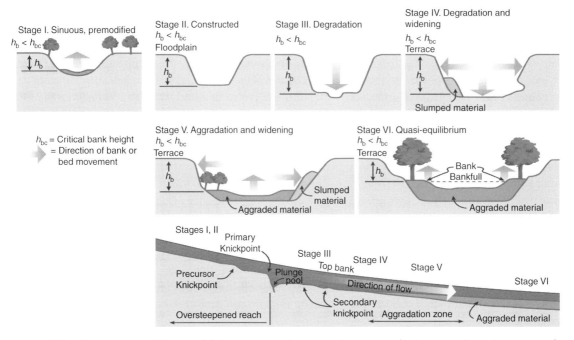

Figure 15.19. The channel evolution model characterizing the erosional response of a stream to channelization or other causes of stream incision (modified from Simon and Rinaldi, 2006).

upstream (Figure 15.19). At a location, rates of degradation or aggradation decrease exponentially over time (Simon, 1989a). Maximum amounts and rates of degradation occur closest to the locus of maximum increase in channel slope, with amounts and rates of degradation decreasing upstream (Simon, 1989a, 1992). Similarly, amount and rates of aggradation are greatest immediately downstream of the locus of maximum increase in channel slope and decrease downstream.

Channelized streams exhibiting responses consistent with the CEM have been identified throughout the midwestern United States, although the exact sequence of evolutionary development varies somewhat with differences in channel materials (Simon and Rinaldi, 2000). The timescale of recovery depends on the postchannelization hydrological and sediment regimes, which govern the rate at which the system moves through the stages of adjustment. For channelized streams in the Midwest, recovery times vary from one to two decades to more than 50 years, depending on the availability of sediment to initiate aggradation following initial degradation. In channelized ephemeral streams in arid regions, the timescale of the evolution process may be prolonged by the lack of persistent flow to maintain the adjustment process (Rhoads, 1990b). In these fluvial systems, large, active headcuts can persist decades after channelization (Figure 15.20).

15.5.3.2 What Other Types of Responses Can Occur?

Although incision is a common mode of adjustment, a variety of responses to channelization have been documented,

including armoring of the channel bed, armoring and the development of a sinuous thalweg, recovery of channel sinuosity through bank erosion, and the development of a sinuous thalweg within an aggrading bed (Figure 15.21). In some cases, multiple responses occur at the same time (Talbot and Lapointe, 2002b). Recovery of channel sinuosity has been documented in straightened gravel-bed rivers (Lewin, 1976) and low-gradient rivers in agricultural landscapes of the United Kingdom, Denmark, and the midwestern United States (Noble and Palmquist, 1968; Brookes, 1987a,b; Barnard and Melhorn, 1982). In these cases, erosion of the channel banks occurs more readily than erosion of the bed. The increase in sinuosity reduces the channel gradient, decreasing the sediment-transport capacity of the flow and the potential for continued channel erosion. The development of sinuosity may also occur within channels that have incised and widened if the banks are erodible locally (Grissinger and Murphey, 1984). On the other hand, the capacity for straightened rivers to widen or recover sinuosity is constrained if bank-protection schemes prevent erosion of the banks (Scorpio et al., 2018). The timescale for recovery of sinuosity has not been examined extensively but appears to vary with the overall energy of the fluvial system. In relatively high-energy gravel-bed rivers, recovery can happen over a few years (Lewin, 1976), whereas in low-gradient, channelized agricultural streams, complete recovery may require a century or more (Barnard and Melhorn, 1982).

Not all rivers exhibit aggradation downstream of zones of channelization. Enlargement of channel cross sections is

Figure 15.20. Channelized section of Santa Rosa Wash in southern Arizona. Arrow points to headcut on the bed of the channelized wash 24 years after initial channelization (shown from a vertical perspective in the inset in upper right corner).

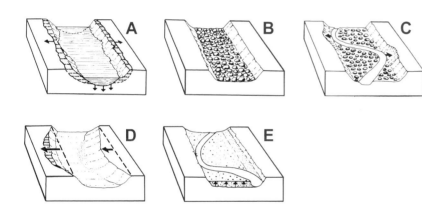

Figure 15.21. Responses to channelization: (a) degradation and widening, (b) armoring of the bed, (c) armoring and development of a sinuous thalweg, (d) increase in sinuosity, (e) deposition and the development of a sinuous thalweg (after Brookes, 1987a).

common downstream of channelization projects in England and Wales (Brookes, 1987b). In most cases, this enlargement is related to increases in channel width. The cause of enlargement has been attributed to enhanced velocities of flood flows within the channelized reach that produce erosion of channel banks when these flows enter the downstream unchannelized reaches.

Wholly aggradational responses can occur when channelization produces large increases in channel width, particularly in relatively low-energy streams. Under such conditions, the bed shear stress of a wide range of transport-effective flows is reduced, resulting in net decreases in sediment-transport capacity. In the midwestern United States, some small headwater streams have been modified into large drainage ditches with wide bottoms. Adjustment to this altered channel form involves the formation of depositional benches within the bottom of the ditch through vertical accretion (Landwehr and Rhoads, 2003). Bar platforms consisting of relatively coarse bed material are initially deposited on the bottom of wide, newly excavated ditches and stabilized by vegetation

that becomes established on the bar surfaces at low flow. Interaction between vegetation growth and flow resistance leads to further deposition and progressive increases in the elevation of the bar surface. The height of the vertically accreting bars eventually stabilizes to form vegetated benches within the ditch. Confinement of flow by alternating benches produces a distinctive morphology characterized by an inset meandering channel flanked by a discontinuous floodplain (Figure 15.22; see also Figure 6.15a).

15.6 How Do Dams Affect River Dynamics?

Countless numbers of dams have been constructed on rivers around the world for purposes of flood control, water supply, power generation, and recreation. At a global scale, only 37% of rivers longer than 1000 km remain free flowing over their entire length, and dams are the leading factor contributing to alteration of these rivers (Grill et al., 2019). In the United States alone, more than 90,000 dams have been inventoried (U.S. Army Corps of Engineers, 2018), but this inventory does

Figure 15.22. Meandering channel flanked by alternating depositional benches within a drainage ditch in Illinois, USA.

not include innumerable small structures (Graf, 1999). The total number of dams in the United States alone may be on the order of 2 million (Graf, 1993; Naiman and Turner, 2000). Although the peak of dam building globally occurred in the 1960s (Petts and Gurnell, 2013), dams continue to be built (Miao et al., 2015; Zarfl et al., 2015) and ambitious plans for future dam construction are being developed (Winemiller et al., 2016; Latrubesse et al., 2017) to meet growing global demands for water and energy.

15.6.1 How Do Dams Affect the Flow of Rivers?

Dams represent barriers that impede the downstream movement of water and sediment within rivers. In this sense, dams fragment rivers by disrupting longitudinal connectivity (Nilsson et al., 2005; Grill et al., 2019). The extent to which connectivity is disrupted depends mainly on the size of the dam and its associated reservoir in relation to the amount of flow in the river. Although dams are often classified by height or volume of reservoir storage capacity, a particularly useful index of the potential impact of a dam on a river system is the ratio of the volumetric reservoir storage capacity to the volume of mean annual runoff – the impoundment/runoff index (Graf, 1999). This index defines the number of years of mean annual runoff a reservoir can hold. It represents a temporal metric indicating how much the downstream movement of water is, on average, delayed by the dam if water is released through the dam at the mean annual rate of runoff.

Low run-of-river dams that do not impound water outside the river channel and allow all water to pass over the crest of the dam (Csiki and Rhoads, 2010) have impoundment/runoff indexes of nearly zero (Figure 15.23). On the other hand, large,

Figure 15.23. (a) Run-of-river dam in Illinois (courtesy of Shane Csiki). (b) Glen Canyon Dam and Lake Powell, Colorado River, Arizona (courtesy of U.S. Bureau of Reclamation).

high dams with vast reservoirs can impound substantial amounts of runoff (Figure 15.23). Lake Powell, impounded by the 216 m high Glen Canyon Dam, has a storage capacity of 32.3 km^3 and an impoundment/runoff index of about 2.25 yrs. By contrast, the Three Gorges Dam on the Yangtze River in China impounds 39.3 km^3 of water, yet this amount is less than 10% of the annual runoff volume of the river. Within the United States, the impoundment/runoff index varies regionally, with low values of approximately 0.25 yrs in New England and the mid-Atlantic states and high values in excess of 3 yrs in the upper Colorado and Rio Grande basins (Graf, 1999).

Dams, particularly large dams that regulate the release of impounded water, have profound effects on the timing, magnitude, and duration of river flows. These effects are greater than anticipated changes in flow related to anthropogenic climate change (Graf, 1999). Although effects vary depending on reservoir operation schemes, river regulation by dams typically reduces high flows and increases low flows (Petts and Gurnell, 2013), resulting in homogenization of hydrological regimes by decreasing flow variability (Poff et al., 2007). Hydrological analysis based on mean daily discharges confirms that dams generally increase minimum flows of various durations and decrease maximum flows of different durations (Magilligan and Nislow, 2005; Wang et al., 2016). Dams also decrease the magnitudes of large floods, in part because flood protection is a purpose of many dams. Of particular importance geomorphologically is the influence of dams on the magnitude of moderate-magnitude, moderate-frequency floods, which are viewed as important channel-forming flows (see Chapter 6). Throughout the United States, reductions in the magnitude of the two-year flood are common, averaging approximately 60% with a range of 20% to 90% (Magilligan et al., 2003). The biggest reductions in flood magnitude occur in areas of the western United States with large impoundment/runoff indexes (FitzHugh and Vogel, 2011). For example, below Hoover and

Parker Dams on the lower Colorado River, the magnitude of the two-year flood has been reduced by about 80% (Schmidt and Wilcock, 2008). The influence of dams on hydrological conditions decreases as the contributing area downstream of the dam increases and flow in a main river is augmented by additions from tributaries. Nevertheless, hydrological alterations by large dams may extend considerable distances downstream, particularly in arid or semiarid environments, where tributary inflows downstream of dams are limited.

15.6.2 How Do Dams Affect the Movement of Sediment in Rivers?

Reservoirs behind dams intercept and trap sediment transported by rivers. Trapping of sediment by reservoirs on a global scale has important implications for sediment fluxes reaching coastal oceans (see Chapter 3). The mechanism of deposition within reservoirs depends on the influence of the dam on upstream hydraulic conditions. Large reservoirs act as lakes where deceleration of incoming river flow leads to deposition of transported sediment. The trap efficiency of a reservoir is the percentage of the total incoming sediment load deposited within a reservoir over a specific period of time. Sand and gravel transported as bed-material load are typically trapped by large reservoirs with an efficiency of 100% (Toniolo et al., 2007). This coarse sediment is deposited near the entrance to the reservoir, forming a prograding delta characterized by vertically accreted topset deposits with a nearly horizontal surface and laterally accreted foreset deposits with a steeply dipping surface at the front of the delta (Figure 15.24). Deposition may extend upstream of the reservoir, leading to aggradation of the inflowing river (Morris et al., 2008; Petts and Gurnell, 2013). Beyond the delta, fine sediments transported as wash load settle out of the incoming flow to form bottomset deposits. Turbidity currents can also redistribute

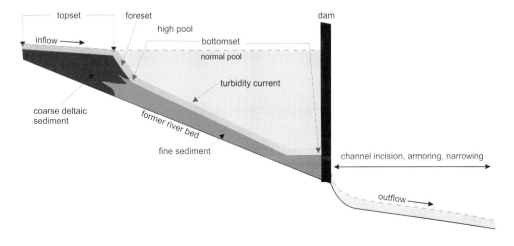

Figure 15.24. Pattern of deposition within a reservoir and channel erosion downstream of a high dam (adapted from Csiki and Rhoads, 2010). (A black and white version of this figure will appear in some formats. For the color version, please refer to the plate section.)

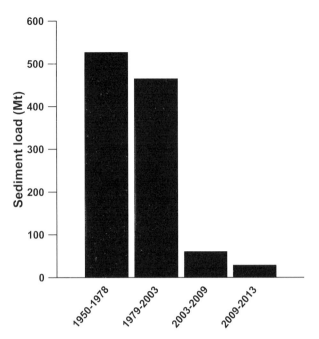

Figure 15.25. Change in the sediment load of the Yangtze River at Yichang Station 38 km downstream of the Three Gorges Dam (TGD). 1950–1978: no large dams upstream; 1979–2003: many large dams constructed throughout upstream basin; 2003–2009: TGD begins to impound water; 2009–2013: TGD in full operation (data from Wang et al., 2018).

fine sediments within reservoirs, especially when sediment-laden floodwater enters the reservoir (De Cesare et al., 2001). The trap efficiency of fine suspended sediment in reservoirs depends on many factors, including size distribution of sediment, the reservoir size and shape, the rate of water inflow to the reservoir, the location of the outlet, and the schedule of water discharges from the reservoir (Jothiprakash and Garg, 2008). Simple relations between trap efficiency and the impoundment/runoff index indicate that for ratios greater than 0.2 yrs, trap efficiencies generally exceed 90% (Brune, 1953).

The capacity of large reservoirs to efficiently trap sediment is illustrated by changes in sediment loads and concentrations downstream of dams. Along the Yangtze River, sediment loads have declined dramatically because of upstream dam construction, particularly following impoundment of water by the Three Gorges Dam and its massive reservoir (Figure 15.25). Although the impact of the Three Gorges Dam diminishes downstream, sediment loads remain about 25% of predam values hundreds of kilometers downstream (Wang et al., 2018). Not all the decline in sediment load far downstream is attributable to the Three Gorges Dam, but this dam alone accounts for about 65% of the total decrease (Yang et al., 2015a). On the Colorado River, the construction of the Glen Canyon Dam reduced the median sediment concentration

and annual sediment loads immediately downstream of the dam by more than 99% (Howard and Dolan, 1981; Topping et al., 2000). Reductions in total sediment loads exceeding 95% are not uncommon immediately downstream from many dams, and decreases greater than 75% often occur hundreds of kilometers downstream from major dams (Schmidt and Wilcock, 2008).

At the other end of the spectrum, run-of-river dams impound small amounts of water relative to the inflow and therefore trap little suspended sediment. Although bed material can be trapped behind such dams, in some cases little or no accumulation of sediment occurs (Csiki and Rhoads, 2014). The lack of sediment accumulation may be related to turbulent structures that develop behind the dam during large floods and that are capable of lifting coarse sand and fine gravel over the dam (Csiki and Rhoads, 2010). Also, ramps of bed material may form immediately behind run-of-river dams, allowing particles the size of pebbles and cobbles to move as bedload up the ramps and over the dams (Pearson and Pizzuto, 2015).

15.6.3 How Do Dams Influence River Morphology?

Modification of river flow and sediment regimes by dams can produce substantial change in channel morphology downstream from these structures (Brandt, 2000). Generalizing downstream adjustments of rivers to dams has proven elusive, mainly because of the myriad factors that influence these adjustments. At the most general level, change involves either aggradation or degradation of the downstream channel (Schmidt and Wilcock, 2008). These two types of responses depend on changes in the bed material–transport capacity of transport-effective flows in relation to changes in bed-material supply (Grant et al., 2003). Such changes, in turn, reflect the influence of the dam on discharge, bed-material load, and transport competence downstream compared with upstream of the dam (Schmidt and Wilcock, 2008). The effects of the dam on flow and sediment regimes are superimposed on environmental conditions, such as the size distribution of bed material, the material composition of channel banks, characteristics of riparian vegetation, valley floor width and topography, contributions of sediment from tributaries, outcropping bedrock, and the influence of geologic setting on the longitudinal profile (Grant et al., 2003; Phillips et al., 2005; Petts and Gurnell, 2013). Variations in the effects of dams on flow and sediment regimes combined with diversity in local environmental conditions lead to a wide variety of possible adjustments (Brandt, 2000; Petts and Gurnell, 2005).

A common effect of dams is to reduce the supply of bed material to the river immediately downstream of the dam, resulting in a deficit of supplied bed material relative to bed material–transport capacity (Grant et al., 2003). In the western United States, almost two-thirds of the rivers affected by

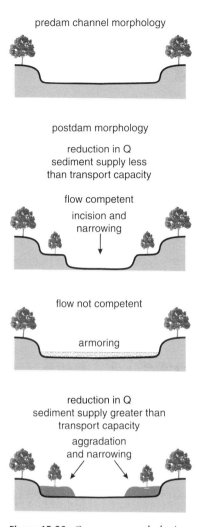

Figure 15.26. Common morphologic responses of river channels downstream of dams.

upstream dams have a sediment deficit (Schmidt and Wilcock, 2008). To fulfill the transport capacity for bed-material load, the "sediment-hungry" flow below the dam often excavates bed material (Kondolf, 1997), resulting in channel incision (Williams and Wolman, 1984) (Figure 15.26). Incision serves as a mechanism for decreasing the channel slope, thereby decreasing the sediment-transport capacity of the flow.

Whether or not incision occurs depends on the competence of the flow to transport bed material downstream of the dam (Wilcock and Schmidt, 2008). If the dam does not affect or enhances the magnitude of a transport-effective flow of a specific frequency, such as the two-year flood, the competence of this flow will remain the same or increase. Under these conditions, the flow should be competent to transport material on the bed of the river downstream of the dam, and incision may occur. Conversely, if the dam reduces the magnitudes of floods of particular frequencies, the competence and transport capacity of these floods will also be reduced.

Incision may still occur in rivers with relatively fine-grained sediment if the reduction in bed material–transport capacity associated with decreased magnitudes of flow is smaller than the reduction in sediment supply; in other words, if the transport capacity downstream of the dam exceeds the amount of sediment supplied downstream. Incision under these conditions generally results in narrowing of the channel relative to predam width (Grams et al., 2007) (Figure 15.26). Where bed material downstream consists of coarse gravel, the reduced competence of formative flows may not exceed the entrainment threshold of coarse material. Winnowing of fines by such regulated flows further coarsens the bed, leading to the development of a static armor layer. The immobility of the coarse channel bed produces a relatively passive response to regulation that prevents or limits incision (Church, 1995; Ayles and Church, 2015) (Figure 15.26). Excess transport capacity may also be accommodated by erosion of the channel banks (Brandt, 2000). The role of bank erosion will depend on the erosional resistance of the banks and the extent to which sediment-transport capacity is diminished by increasing erosional resistance of an armor layer or by reductions in channel slope associated with incision.

Adjustments downstream of dams can also influence the planform characteristics and dynamics of rivers. A well-known example is the conversion of the north and south branches of the Platte River from a wide, sandy, braided river several hundreds of meters wide to a single-thread meandering channel several tens of meters wide (Petts and Gurnell, 2013). This transformation corresponded to a reduction of the peak annual discharge by more than 75% following flow regulation. It also was accompanied by encroachment of vegetation into channels and conversion of braid bars into stable islands and floodplain through rapid colonization by riparian forest (Johnson, 1997). Scour and channel incision downstream of dams can convert braided rivers into single-thread rivers of high sinuosity, but if the banks of incised channels are highly erodible, influxes of sediment from bank erosion can subsequently reestablish braided patterns (Xu, 1996). In meandering rivers, regulation of flow downstream of dams commonly results in diminished rates of lateral channel migration (Wellmeyer et al., 2005).

Decreases in flood magnitudes and sediment loads below dams can substantially diminish the geomorphic complexity of rivers, where complexity is defined as the number of discrete islands, bars, floodplain elements, and terraces per unit length of river (Graf, 2006). Moreover, many surfaces of these morphologic units that previously were actively affected by hydrological and sedimentological processes become inactive after dam closure. Across the United States, complexity is 37% less and the area of active surfaces is 72% less in regulated reaches than in unregulated reaches. Thus, net erosion may not always involve incision; it can also include mobilization and eradication of depositional features within the river. The construction of Glen Canyon Dam resulted in channel incision immediately below the dam in Marble Canyon (Grams et al., 2007), but farther downstream in the Grand Canyon, sand bars that formed in protected bedrock alcoves have been eliminated or reduced in size (Kearsley et al., 1994). The loss of these sandbars has negatively affected the ecology of the river as well as the recreational boating industry, which uses the bars as campsites.

Although incision and armoring are common immediately below dams, decreases in bed material–transport capacity associated with reductions in discharge can result in net aggradation if abundant local sources of sediment are available downstream of a dam. Tributaries represent potential local sources of sediment, and the total supply of sediment by tributaries increases with increasing drainage area downstream of the dam (Curtis et al., 2010). Tributaries also can contribute to a river coarse bed material that exceeds the competence of regulated flows, resulting in accumulations of sediment at confluences (Howard and Dolan, 1981; Church, 1995; Curtis et al., 2010). Excess sediment supply leads to aggradation, which results in channel narrowing (Everitt, 1993; Church, 1995; Grams and Schmidt, 2002, 2005) (Figure 15.26). Vegetation dynamics typically play an important role in this process. Colonization of depositional surfaces by vegetation stabilizes these surfaces and, by increasing flow resistance, also promotes additional aggradation (Petts and Gurnell, 2013). The introduction of large amounts of fines into a regulated gravel-bed river with reduced competence and transport capacity can also increase the embeddedness, or amount of fines, within gravel matrices (Kondolf and Wilcock, 1996). This change in the structure of the bed material negatively affects habitat for aquatic organisms, particularly salmon and trout, which rely on matrix gravels for spawning (Kondolf, 2000).

Changes in river form and sedimentology that occur downstream of dams represent adjustments to an altered regime of flow and sediment load. Eventually, under the assumption that rivers attain an equilibrium state, characteristics of the river system should completely adjust to the altered regime and become stable. Estimates of the timescale for equilibrium adjustment range from decades to centuries (Petts and Gurnell, 2013). The length of time depends both on the extent to which the dam has disrupted the hydrological and sediment regimes of the river and on the capacity of fluvial processes downstream of the dam to produce evolutionary responses that transform the river system into a stable configuration.

Compared with large dams, less attention has been given to the impacts of run-of-river dams on river dynamics. Some small run-of-river dams in the United States, also known as mill dams, originally had crests that exceeded the bankfull elevations of streams that existed at the time of European settlement. In the eastern United States, tens of thousands of such mill dams were constructed between the late 1600s and the early twentieth century (Walter and Merritts, 2008). Ponding of water behind the dams led to aggradation, which buried presettlement floodplain wetlands beneath 1 to 5 meters of accumulated sediment. The depths of sedimentation roughly corresponded to the height of the dam crests. Subsequent breaching of many of these dams has produced incised channels flanked by valley-flat terraces consisting of reservoir sediments. The crests of dams that have not breached now correspond to the elevation of accumulated sediment upstream of the dams and these structures are now viewed as run-of-river dams.

Because run-of-river dams do not alter the hydrological regime of a river and may not substantially impede the transport of bed material, the effects of these structures on the channel generally are either undetectable or relatively minor (Skalak et al., 2009; Csiki and Rhoads, 2014; Fencl et al., 2015). Channels are not consistently wider or deeper upstream as compared with downstream of run-of-river dams. Channel substrates tend to be somewhat finer upstream as compared with downstream of these dams, but the difference in particle-size characteristics is not always statistically significant.

15.7 How Does Instream Mining Affect River Dynamics?

By transporting and sorting bed material, rivers produce deposits that serve as a valuable natural resource for a variety of commercial and industrial purposes. These deposits typically are well sorted and readily accessible, limiting processing and transportation costs. Sand and gravel mined from river beds and floodplains are used to make concrete, for road construction, as construction fill, as paving material, for landscaping, to enhance traction on icy roads, and in water filtration systems. Extraction of sand and gravel from rivers occurs around the world and has been a major factor affecting river dynamics since the beginning of the twentieth century (Rinaldi et al., 2005). Growth in extraction seems likely, given that depletion of sand as a natural resource is a growing global concern (Torres et al., 2017; Bendixen et al., 2019).

15.7.1 How Does Mining Trigger River Adjustments?

Instream mining can be classified into three major types: dry-pit mining, conducted on ephemeral streams or portions of perennial rivers that are removed from active areas of flow; wet-pit mining, where draglines and dredges are used to remove submerged bed material; and bar skimming, where gravel is scraped off the surface of exposed bars (Kondolf, 1994). The main control of channel response to these various types of mining is the ratio between sediment extraction rate and sediment replenishment rate (Rinaldi et al., 2005). If extraction greatly exceeds replenishment, bed-material availability becomes less than bed-material transport capacity, and the alluvial river system is sediment starved (Kondolf, 1997). Typically, such conditions develop during widespread pit mining. This type of mining produces large excavations in the channel bed that may extend many meters below the level of the bed upstream and downstream. It can result in millions of cubic meters of sand and gravel being removed from the river bed. Sediment deficits and changes in hydraulic conditions associated with the pit combine to produce a variety of adjustments, including upstream incision, downstream incision, changes in channel geometry and planform, lateral channel instability, and bed armoring (Rinaldi et al., 2005). All these adjustments represent responses to the new bed-material supply regime of the river. Isolating adjustments to instream mining is not always straightforward, because rivers affected by mining often are also influenced by channelization and dam construction (Rinaldi et al, 2005) as well as by watershed-scale factors, such as changes in land use or climate (Wishart et al., 2008).

15.7.2 What Morphological Adjustments Occur in Response to Instream Mining?

River incision upstream of mining pits has been extensively documented (Rinaldi et al., 2005). This response has been attributed to knickpoint or headcut migration initiated at the upstream lip of the mining pit (Figure 15.27). Trapping of incoming bed-material load within the pit reduces the sediment supply downstream, resulting in downstream erosion. Trapping of sediment is enhanced by the development of upstream flow near the bottom of the pit (Barman et al., 2018). Upstream migration and downstream propagation of incision can extend over many kilometers, flushing downstream millions of cubic meters of sediment (Martin-Vide et al., 2010; Arrospide et al., 2018). Incisional responses that transform channels into deep, incised trenches can confine floods and promote high bed shear stresses, producing prolonged erosional channel instability in which incision-related loss of sediment greatly exceeds losses associated with gravel mining (Martin-Vide et al., 2010).

Experimental studies of mining-pit evolution indicate that the upstream lip migrates downstream through deposition and the downstream lip migrates downstream through erosion (Lee et al., 1993; Barman et al., 2018). Through these processes, the pit initially translates downstream and over time begins to reduce its depth as diffusive processes

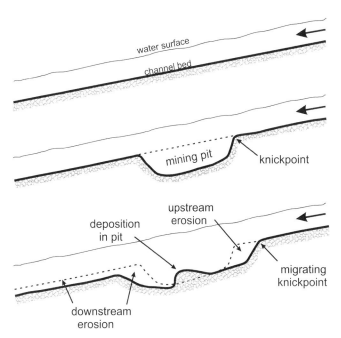

Figure 15.27. Upstream and downstream erosion of a river mining pit (modified from Kondolf, 1994).

gradually eradicate the pit (Lee et al., 1993). These experimental findings are at odds with numerous field studies that have documented upstream migration of incision above intensively mined areas (Rinaldi et al., 2005). Some field investigations have documented downstream-advancing lobes of deposition at the upstream end of mining excavations (Yuill et al., 2016b; Calle et al., 2017) (Figure 15.28). Further work is needed to determine conditions that promote upstream migration of incision versus downstream migration of deposition in rivers affected by mining excavations.

Changes in channel planform are commonly associated with mining impacts. Through mining-related incision, braided rivers often develop meandering or wandering gravel-bed forms (Kondolf and Swanson, 1993; Rinaldi et al., 2005). In meandering rivers, rates of lateral migration may be enhanced through disturbance of the floodplain by mining excavations (Mossa and McLean, 1997). Gravel pits in inactive parts of braided rivers or on floodplains of meandering rivers have the potential to capture flows of large floods, redirecting flooding and producing instability of newly activated incised channels (Bull and Scott, 1974).

Gravel harvesting through bar skimming is generally less disruptive of river systems than pit mining but can produce gradual incision if extraction rates consistently exceed replenishment rates. In gravel-bed streams of the Olympic Peninsula, Washington, USA, extraction rates that exceeded replenishment rates by a factor of 10 over a period of two to

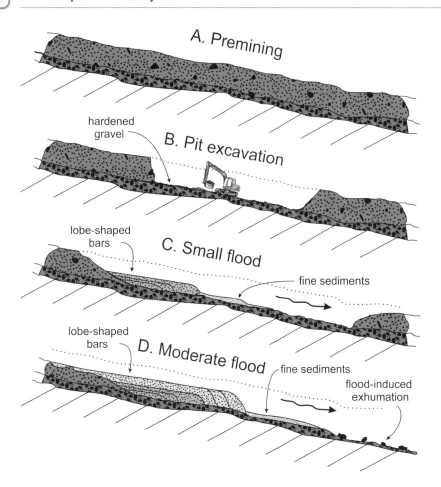

Figure 15.28. Pattern of deposition and erosion in a pit mine in an ephemeral river showing development of lobe-shaped bars at upstream lip of mining pit and erosion at downstream lip during small and moderate floods (modified from Calle et al., 2017). (A black and white version of this figure will appear in some formats. For the color version, please refer to the plate section.)

three decades led to rates of bed lowering of 0.03 m/yr (Collins and Dunne, 1989). Another consequence of bar skimming is that it removes natural armor layers, which regulate rates of bedload transport, from bar surfaces. Removal of these layers enhances the mobility of the underlying fine bed material (Kondolf, 1994).

16 River Dynamics and Management

16.1 Why Is an Understanding of River Dynamics Important in River Management?

A prominent development in fluvial geomorphology since the emergence of the process approach in the middle of the twentieth century (see Chapter 1) has been growing interest in connecting new knowledge of fluvial forms and processes to practical concerns regarding river management. This growing interest springs largely from the abundance of research conducted at the modern and event timescales, which correspond to lengths of time over which management typically is targeted. Given the fluid social, political, and economic contexts within which management policies are formulated and enacted, most policies have durations ranging at most from years to several decades. Management timelines rarely extend over more than a century. The focus on river geomorphology at modern and event timescales also dovetails with temporal outlooks of river engineering and applied river ecology – two fields that generate knowledge relevant to river management. Geomorphological insights can therefore be readily integrated with engineering and ecological understandings to broadly inform management decisions.

An important trend over the past several decades in regard to river management has been a shift from approaches to management that view rivers primarily as natural resources to be utilized primarily for economic benefit or as natural hazards from which humans must be protected to approaches that seek to conserve or improve ecological, recreational, and aesthetic qualities of rivers. This shift represents an explicit move toward environmental management of rivers. Because this type of management implicitly seeks to work within the framework of the myriad natural processes that constitute rivers, geomorphological information is foundational to such management.

Although geomorphological knowledge is fundamental to environmental management of rivers, it does not necessarily provide straightforward prescriptions about how to manage rivers effectively. As other chapters in this book have shown, rivers are complex, nonlinear systems, the dynamics of which can be unpredictable and highly uncertain. Attempts to manage rivers occur in the context of this unpredictability and uncertainty. What fluvial geomorphologists can do is provide sound knowledge about river dynamics to river managers so that this knowledge can be considered in the management process. Recognition that management constitutes a process is an important point of emphasis. Both management and the science informing it evolve, and, over time, refinement is continuous as lessons are learned both about how rivers work and about how management affects rivers. In this sense, the process of management adapts through discoveries of science and through efforts to manage rivers.

16.2 How Is Scientific Inquiry on River Dynamics Related to Management?

The pursuit of scientific research to help find solutions to practical problems of immediate concern to society has often been characterized as applied science. This type of inquiry contrasts with basic science, the primary aim of which is to contribute to knowledge of the world for its own sake without concern about any immediate practical benefits of the acquired knowledge. Although the distinction between applied and basic science has been subject to criticism (Roll-Hansen, 2017), it is widely recognized among working scientists (Bentley et al., 2015).

16.2.1 What Is the Distinction between Applied Science and Basic Science?

Formalizing the distinction between basic and applied science is difficult because it is based mainly on intuition (Rhoads and Thorn, 1996). Moreover, over the long run the distinction can change as knowledge once thought to be basic turns out to have clear practical value. Despite these complications, a useful way to establish the distinction is on the basis of the aims of inquiry and the nature of scientific statements (Table 16.1). Basic science seeks truthful (veridical) knowledge with high explanatory power. Powerful explanations are typically viewed as causal in nature, so basic research focuses on relating effects to underlying causes. Most of the information presented in this book falls into this category, but a specific example is basic inquiry to determine the cause of river meandering (see Chapter 9).

Table 16.1. Basic and applied science in fluvial geomorphology.

	Primary scientific aim/purpose	Structure of scientific statements	Example research question
Basic science	Truth, knowledge, explanatory power	Causal: Y is caused by X	What is the cause of river meandering?
Applied predictive science	Predictive accuracy, forecasting	Probabilistic: Y is highly likely to occur if X occurs	How does artificial straightening of a meandering river influence its dynamics?
Applied design science	Usefulness of knowledge for producing technological advances to achieve societally desired goals	If you want to achieve Y, you ought to do X (technical norm)	What is the best way to reestablish meandering in an artificially straightened meandering river?

(adapted from Rhoads and Thorn, 1996)

Applied science can be divided into two different types: applied predictive science and applied design science. The aim of applied predictive science is to develop relations that are grounded at least to some extent in truthful knowledge, but that also yield accurate predictions of outcomes in relation to problems of societal interest (Table 16.1). Although such relations may include specification of cause and effect, reliability of predictive accuracy is valued over explanatory power invoking causality. The development of quantitative relations or models to predict the response of rivers to human disturbance, such as channel incision and widening associated with channelization (Langendoen and Alonso, 2008), epitomizes this type of science.

16.2.2 What Is an Applied Design Science?

Whereas basic science and applied predictive science implicitly invoke values when deciding what questions to ask and how to address those questions (Blue and Brierley, 2016), applied design science explicitly considers societal values within the context of the research. It addresses concerns about how things ought to be to attain certain goals valued by society (Niiniluoto, 1993, 2014). Such concerns are known as technical norms (Table 16.1). The aim of applied design science is to provide knowledge for the production of technological products, or science-based tools, that help society achieve valued goals. Engineering and medicine represent applied design sciences – these fields provide knowledge for technological developments that allow buildings or roads to be constructed and that maintain human health. The environmental management of rivers now involves reconfiguring channels, floodplains, and vegetation to achieve ecological, water-quality, and geomorphological goals through activities variously referred to as restoration, naturalization, and mitigation (Rhoads et al., 1999; Spink et al., 2010; Wohl et al., 2015). These design-based activities rely on knowledge about fluvial forms and processes (Rhoads and Thorn, 1996). Thus, a branch of fluvial geomorphology has arisen that can be characterized as a design science (Gregory, 2006). The knowledge in this science is intended to directly support management by informing decisions about what ought to be done to achieve management goals.

16.2.3 How Are Basic Science, Applied Science, and Management Interrelated?

Environmental management of rivers occurs at the level of professional practice that utilizes knowledge generated by basic research, applied predictive research, and applied design research conducted within fluvial geomorphology

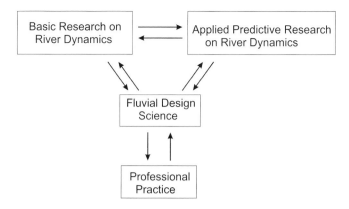

Figure 16.1. Conceptualization of a fluvial design science for geomorphology and its relation to basic research on river dynamics, applied predictive research on river dynamics, and professional practice to support environmental management of rivers.

(Figure 16.1). A wide variety of tools now exist to support river management based on knowledge emerging from fluvial geomorphology (Kondolf and Piegay, 2016). Although basic and applied research generate knowledge useful to professional practice, the flow of knowledge and information is not one way. Trial-and-error attempts to manage rivers to achieve certain goals, particularly in cases where existing knowledge may not provide firm guidance about what ought to be done, can lead to outcomes that inform design science. Moreover, lessons learned from trial-and-error attempts at management can in some cases constitute natural experiments that contribute to knowledge in basic science and applied predictive science. These lessons may improve basic understanding of mechanisms governing river-channel dynamics, thereby allowing more accurate predictions of river response to human disturbance. Over time, however, the goal is to increase the flow of information from science to practice as the capacity of science to generate and support effective technological products improves. It is also important to emphasize that not all technological products involve methods to actively manipulate or restructure river systems; a recommendation or set of guidelines that promotes doing nothing to a river and allowing it to adjust on its own to best achieve desired goals (e.g., Kondolf, 2011) can constitute a technological product.

16.3 What Are the General Goals of Environmental Management of Rivers?

In the most general sense, the goals of environmental management of rivers are to protect existing environmental benefits of rivers and to undo degradation of benefits that has occurred through direct and indirect human impacts on rivers. Many human activities have altered rivers (see Chapter 15), including agriculture, urbanization, mining, channelization, the construction of dams, and deforestation. Although the effects of some of these activities were recognized prior to the middle of the twentieth century (e.g., Happ et al., 1940), concern about human impacts on rivers only began to become widespread following the global environmental movement of the 1960s. Contemporary interest in sustainability is rooted in this movement, which involved a shift in the values of society toward enhanced recognition of the dependency of human life on characteristics of the environment, reassessment of the influence of modern technological developments on environmental conditions, and concern about trade-offs between technology-driven economic growth and the sustainability of environmental resources (McCormick, 1989). With regard to rivers, initial attention focused mainly on the effects of industrial pollution on water quality, which in the United States led to the establishment of the Clean Water Act in 1972. Around the same time, recognition of the adverse impacts of traditional river engineering schemes, such as channelization, on channel stability and ecological conditions led to calls for the development of alternative river-management practices based on geomorphological principles (Keller, 1975; Leopold, 1977; Nunnally, 1978). By the 1980s, the participation of fluvial geomorphologists in practices aimed at undoing the effects of river engineering, such as channel straightening, began to emerge (Brookes, 1987c). The linkage between the environment management of rivers and fluvial geomorphology was formalized in the 1990s (Sear, 1994; Brookes, 1995). Since that time, the development of both a geomorphological design science to support the environmental management of rivers and a professional practice to engage in this type of management has exploded.

16.3.1 What Role Do Social Processes Play in Goal Setting?

The process that translates societal values into goals of river management is complex and includes consideration of social, economic, political, cultural, and environmental issues. Given that the process involves interaction among people, it is inherently social in nature (Lave, 2016). Participants in decision-making are commonly referred to as stakeholders – individuals who have a stake in decisions. The political authority of particular institutions and stakeholders involved in decision-making often plays a key role in goal setting. When political authority is centralized, as is typically the case when government agencies lead management, goals commonly reflect the policies of these agencies. When political authority is decentralized, and decision-making occurs democratically, goals reflect negotiation among the competing values of stakeholders (Emery et al., 2013). In such cases, power relations still influence which stakeholders are included in decision-making, who makes decisions about whom to include in decision-making, and whose voices have the most authority in decision-making (Junker et al., 2007). Although scientific and technical information, including information on river dynamics, commonly is a component of centralized and decentralized deliberations about river management, decisions about how to manage a river may not predominantly reflect biophysical ideals in relation to goals. Instead, actual biophysical outcomes often are strongly influenced by social, political, cultural, and economic factors (Rhoads et al., 1999; Eden et al., 2000; McDonald et al., 2004).

16.3.2 What Are the Major Geomorphological and Ecological Goals of Management?

Societal concern about the environmental quality of rivers has led to broad environment-management goals of relevance to geomorphology and ecology: 1) stabilization of rivers undergoing net erosion or deposition, particularly erosion, and 2) enhancement of degraded ecological conditions, particularly physical habitat and characteristics of aquatic communities (Hawley, 2018). The goals are connected largely through the consideration of physical habitat. This connection has increasingly led to interaction between geomorphologists and ecologists to provide scientific input relevant to achieving the two goals.

When targeted toward river stabilization, management often inherently adopts the perspective that streams ought to be in a state of sediment mass balance. In other words, a river should transport the sediment supplied to it without net aggradation or degradation. This condition, which has been referred to as the equilibrium, graded, or stable state (see Chapter 6), is implicitly valued as the ideal configuration of a river that management should strive to attain; it provides a reference condition for the morphology and dynamics of a natural river in the absence of disturbance. The use of the terms "stable" and "unstable" to describe a state of mass balance versus the lack thereof is highly suggestive of the values underpinning these alternative descriptions of the condition of a river. Anything unstable generally is less desirable than something stable.

The preference for a stable condition is associated with myriad societal concerns associated with unstable rivers, such as channel enlargement or infilling, loss of land adjacent to rivers through bank erosion, excessive sedimentation, and increases in sediment loads that degrade water quality. In many cases, these problems are induced by direct and indirect impacts of humans on the dynamics of river systems (see Chapter 15). Such impacts trigger mass imbalances and generate rates of erosion or deposition that are accelerated relative to those that would otherwise occur.

Equilibrium theories of river adjustment provide a scientific underpinning for design science, both by informing how environmental management ought to proceed to produce stable states and by reinforcing the valuation that the inherent natural condition of a river, when not subjected to disturbance, especially disturbance by humans, is one of stability. Obviously, valuation of this type may be misplaced in cases where rivers exhibit nonequilibrium dynamics in undisturbed settings – a common condition in highly sensitive fluvial systems subjected to episodic morphological restructuring by natural events at a frequency higher than the timescale of recovery during intervening inter-event intervals (see Chapter 6). Geomorphological knowledge can help inform this issue, providing insight into whether or not equilibrium concepts are appropriate when considering how to manage a particular river in accordance with its undisturbed dynamics.

Management aimed at mass-balance stability is a human choice, guided by human values. The river, an inanimate feature of Earth's surface, does not have a preference for any particular configuration (Phillips, 2010). Disturbances, whether human induced or biophysical, trigger interactions among hydrological conditions, sediment supply, channel morphology, and transport capacity that result in change. But such change, even when it involves mass imbalances, is not abnormal in relation to the dynamics of the river. It simply reflects the response of the river to conditions imposed upon it. Fluvial geomorphology can play a role in river management by helping society understand how much rivers have been altered by human activity, the consequences of this alteration for river dynamics, and possible long-term implications for sustainability of river resources. Through such efforts, valuations that drive decision-making and management policies can be grounded on a sound foundation of geomorphological knowledge.

With regard to the ecological goal of enhancement of degraded conditions, the role of values in river management becomes more complex than it is for the goal related to channel stability and morphodynamics. Ethical issues related to the intrinsic value of living organisms as well as to the role of ecosystems in providing services that sustain human well-being enter into the debate. Proposed scientific concepts such as stream health (Meyer, 1997; Blue, 2018), which adapts the notion of human health to stream ecosystems, and ecosystem integrity (Karr, 1981; Maloney, 2019), which indicates that some aquatic ecosystems lack integrity, can complicate the relationship between scientific/technical input and societal preferences in policy formulation. The content of these value-laden concepts may be associated with certain norms, such as the inherently greater value of pristine ecosystems compared with human-altered ecosystems, thereby inserting personal values into the management process under the guise of scientific impartiality (Lackey, 2001). Scientists and technical experts must exercise care in, on the one hand, providing information to help society make informed choices about river management and, on the other hand, advocating as a concerned citizen for particular management outcomes. Maintaining separation between opinions that reflect personal values and those informed by objective scientific appraisal of information is often difficult (Rhoads et al., 1999), and mixing of the two types of opinions within the context of decision-making can be misleading to the public.

Concern about physical stability and physical habitat has led to integration of geomorphological and ecological research to support environmental management of rivers. The geomorphological characteristics of rivers are important components of physical habitat for fish, macroinvertebrates, and instream plants (Vaughan et al., 2009). The quality of physical habitat has been a main focus of environmental management of rivers, especially given that past management practices, such as stream channelization, have eliminated physical habitat, in some cases at the scale of entire river networks (Frothingham et al., 2002). The geomorphological characteristics of rivers therefore are vitally linked to ecological aspects of river management. Nevertheless, in managing rivers to improve ecological conditions, physical habitat is not the only consideration. Many other abiotic and biotic factors, including hydrologic variability, water quality, nutrient dynamics, thermal regime, sources of organisms, connectivity to these sources, food availability, and predator–prey interactions, affect the aquatic ecology of a river or stream. Holistic approaches to ecological management should extend beyond degraded physical habitat and consider the complex interactions among hydrologic, hydraulic, geomorphologic, biogeochemical, and biological processes that determine ecosystem function (Palmer et al., 2014; Hawley, 2018).

16.4 What Are the Main Environmentally Oriented River Management Strategies?

A variety of terms exist to distinguish different types of environmentally oriented river management strategies (Table 16.2). All these terms, except protection and possibly creation, refer to efforts to undo perceived adverse effects of human activities on rivers. The usage of some terms does not always conform to formal definitions, and meanings have evolved to some extent over time.

16.4.1 What Is River Protection?

Protection seeks to preserve the existing environmental quality of rivers. Some protection policies target rivers that

have yet to be dramatically impacted by human activity. In the United States, recognition of the need to protect rivers with outstanding natural, cultural, and recreational values in a free-flowing condition led to passage by Congress of the National Wild and Scenic Rivers (NWSR) Act in 1968. Parts of hundreds of rivers in 40 states and Puerto Rico are now included in the NWSR system. Either a state or a federal agency administers the Act for each river. The extent to which this system truly protects the wild and scenic nature of rivers is debatable, given that the Act is aimed primarily at dam construction. Although the intent of the Act and the goal of its administration are to safeguard or even enhance qualities that led to designation, development within watersheds is not prohibited. Protection of rivers depends on voluntary stewardship by landowners or river users and on regulatory programs of federal, state, local, or tribal governments.

At the federal level, protection policies have been developed to maintain existing environmental quality and prevent further degradation. At the watershed scale, the U.S. Environmental Protection Agency's (USEPA) Healthy Watersheds Protection (HWP) program focuses mainly on protecting water quality. The HWP program encourages nonregulatory collaborative approaches to maintaining clean water in rivers through partnerships among the USEPA, state and local agencies, and the private sector. The USEPA also has regulatory authority through the Clean Water Act to control point sources of pollution in rivers. Discharges of pollutants from a point source are unlawful unless a permit to do so is obtained from the USEPA. Sediment discharges are regulated under section 404 of the CWA, which requires a permit before dredged or fill material is discharged into a river, unless the activity is exempt from this requirement. Exempt activities include certain practices related to farming and forestry.

In Europe, protection of water quality is a priority of European Water Policy as specified by the European Union Water Directive Framework (WDF). Key objectives are general protection of aquatic ecology, specific protection of unique and valuable habitats, protection of drinking water resources, and protection of bathing water. The innovative aspects of this policy are its focus on management by drainage basins, rather than by political boundaries, and its emphasis on the ecological goals for assessing environmental quality (Smith et al., 2014). The focus on ecological quality includes evaluations of hydromorphological quality, which is viewed as having an important influence on ecological conditions. Concern about evaluations of dynamic connections among hydrological regime, channel morphology, substrate conditions, physical habitat, and the

Table 16.2. Terminology of some common environmental management practices.

Term	Definition
Protection	Creating laws or other mechanisms to protect rivers deemed to be of high environmental quality from degradation. Also referred to as preservation.
Restoration	Restoring the structure and function of a river system as closely as possible to predisturbance conditions. In the United States, "predisturbance" usually refers to pre–European settlement.
Rehabilitation	Partially restoring the structure and function of a river system to predisturbance conditions. Also known as partial restoration.
Enhancement/ improvement	Improving or enhancing the environmental quality of a river.
Naturalization	Increasing the hydraulic, geomorphologic, and ecological diversity of river systems in human-dominated environments where rivers have been highly modified.
Creation	Creating a new river system where none existed previously.
Mitigation	Compensating for damage to a river system by alleviating this damage either on site or at an alternative site. Strives to achieve no net loss of environmental quality.

Table 16.3. Common goals of stream restoration.

Enhance water quality
Improve stream habitat
Improve stream ecosystems
Enhance fish passage
Enhance riparian zones
Enhance resilience to disturbance
Stabilize eroding streambanks
Reduce channel instability
Reconfigure modified channels into more natural forms
Reconnect floodplains and channels
Enhance aesthetic, recreational, and educational value
Remove or modify dams
Acquire land to reduce hazards from erosion and flooding
Manage instream flow

composition of ecological communities provides challenging opportunities for fluvial geomorphology to contribute to the scientific basis of WDF implementation (Newson and Large, 2006).

16.4.2 What Is River Restoration?

Restoration is the most well-known and commonly used of the terms designating attempts to alleviate adverse effects of humans on the geomorphological stability and ecological quality of rivers. This term initially was used to label practices related to its definitional meaning of restoring a human-modified or human-disturbed freshwater system to its condition prior to modification or disturbance (National Research Council, 1992). In the United States, the reference state for the predisturbance condition typically corresponds to the state of the system prior to European settlement (Shields et al., 2008). This state represents a nearly pristine condition.

Over time, the meaning of the term "stream restoration" has gradually expanded with the growing popularity of environmental management of rivers. Also, in most cases, the extensiveness of human impacts, particularly at watershed scales, prevents true restoration that reestablishes the predisturbance form and function of a river system. Consequently, stream restoration now refers to a wide variety of activities that seek to undo perceived deleterious human impacts (Wohl et al., 2015). To a large extent, it has subsumed terms such as rehabilitation, improvement, and even creation (Wilcock, 2012).

Although the scope of the term has broadened to reflect the various practices designated as restoration, this trend also has the potential to weaken the meaning of the term if these practices become all things to all people. The notion of improvement can be highly subjective, and, within the context of common societal goals (Table 16.3), competing values about what a river should be can produce management outcomes that depart radically from the reestablishment of predisturbance water quality, ecology, geomorphology, hydrology, and hydraulics (Wohl et al., 2015). Considerable attention has focused on the social dimensions of restoration, including how competing interests lead to different types of restoration outcomes (Eden et al., 2000; McDonald et al., 2004; Emery et al., 2013; Ashmore, 2015). Given that restoration is a human activity, to a large extent it is first and foremost a social process.

As a professional practice, stream restoration has become a multibillion-dollar global industry. The rapid growth of this industry has fueled the development of an applied design science to support restoration practices but has also posed enormous challenges for linking science and practice to accommodate a diverse set of environmental goals (Wohl et al., 2005, 2015). The applied design science of stream restoration involves integration of geomorphology, ecology, hydrology, hydraulics, engineering, landscape architecture, chemistry, and even social science. This broad integration of scientific and technical knowledge has yet to be refined to the point that, based on a holistic understanding of river systems, achievement of goals is assured through the practice of restoration.

16.4.3 What Is Stream Naturalization?

In human-dominated environments, such as industrial agricultural landscapes and urban settings, streams typically have been extensively modified through channelization (Rhoads and Herricks, 1996; Frothingham et al., 2002). The net effect of widespread channelization is to homogenize the geomorphological characteristics of streams, producing straight, uniform, trapezoidal forms designed to convey flow efficiently downstream for the purpose of land drainage or flood control. Straightening also limits the amount of space on the landscape the stream channel occupies, an important consideration where the economic value of land is high. Modification can be so widespread that few, if any, streams remain untouched to serve as templates for unaltered conditions (Rhoads and Herricks, 1996; Frothingham et al., 2002). Moreover, the history of channelization may be so long that detailed scientific information on stream characteristics prior to modification does not exist. Because human activity has altered hydrological responses and sediment fluxes at the watershed scale, the predisturbance characteristics of streams, even if known, are no longer sustainable under contemporary conditions. Finally, in human-dominated environments, people are highly likely to continue to manipulate streams and strongly influence fluvial processes over the foreseeable future. Thus, humans have become part of the biophysical dynamics of these river systems.

Attempts to manage streams in human-dominated contexts commonly includes coalitions of stakeholders involved in community-based decision-making (Rhoads et al., 1999; Wade et al., 2002; Petts, 2007). Scientists and technical experts not only provide information to support this decision-making but also participate as stakeholders (Figure 16.2). The process of decision-making involves social negotiations among stakeholders holding shared and competing values. Through this process, which is fundamentally social in nature, outcomes are determined about how to naturalize streams in relation to existing conditions.

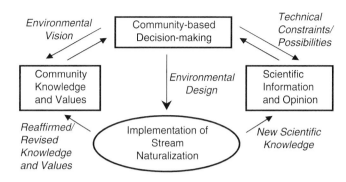

Figure 16.2. The adaptive process of stream naturalization.

Evaluations of community-based decision-making about stream management in both rural-agricultural and highly urbanized settings show that outcomes of decision-making lead to place-based conceptions of how to best transform channelized streams into features that are more natural as conceived by the community (Rhoads and Herricks, 1996; Rhoads et al., 1999; Wade et al. 2002). This decision-making process, along with implementation of the resulting conceptions, is referred to as stream naturalization. A key element of naturalization is recognition that the idea of what is "natural" is socially negotiated rather than some fixed condition that either can be rigorously defined (e.g., Newson and Large, 2006) or corresponds to pristine, undisturbed conditions (Wohl and Merritt, 2007). Outcomes of naturalization attempt to establish sustainable, morphologically and hydraulically varied, yet dynamically stable fluvial systems that are capable of supporting biologically diverse aquatic ecosystems. The goal is to move the stream system away from its highly modified state, in which form is often highly homogeneous and function is either static or limited, toward a condition that enhances the diversity of form and function (Figure 16.3). How this goal is accomplished depends on the place-specific desires of the local community, informed by scientific and technical input. Should the community desire to engage in restoration, i.e., restore a highly homogenized stream system to a predisturbance state, and this type of management is affordable and feasible, the implementation of restoration would represent a type of naturalization. In this sense, naturalization subsumes restoration. The social process of naturalization is also adaptive and evolves over time. Through postproject evaluations of the functional performance of naturalized streams, new knowledge and understandings emerge that inform community values related to environmental visioning as well as scientific/technical strategies to achieve specific goals of naturalization (Figure 16.2).

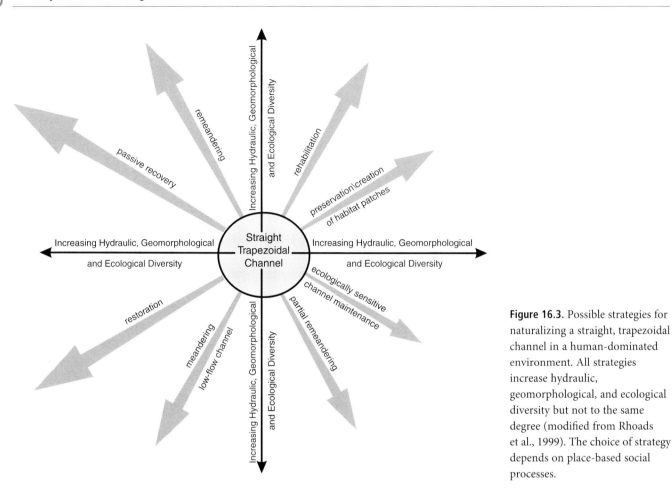

Figure 16.3. Possible strategies for naturalizing a straight, trapezoidal channel in a human-dominated environment. All strategies increase hydraulic, geomorphological, and ecological diversity but not to the same degree (modified from Rhoads et al., 1999). The choice of strategy depends on place-based social processes.

16.4.4 What Is Stream Mitigation?

The focus of stream mitigation is on proposed activities that, if undertaken, will degrade the environmental quality of a stream. The process of mitigation involves three sequential considerations: avoid, minimize, and compensate (Lave, 2018). If the proposed activities can be modified to avoid or minimize impact, these modifications should first be implemented. Any remaining impacts should be accommodated through compensation, which involves enhancing a degraded system elsewhere to balance, or offset, the impact at the site of interest.

Stream mitigation is closely tied to regulation of discharges of dredged and fill material into waters of the United States under Section 404 of the Clean Water Act (Doyle et al., 2013). Anyone proposing an activity that could lead to such discharges must first obtain a permit from the U.S. Army Corps of Engineers. Permits go through a review process that includes state agencies and other federal agencies, such as the U.S. Fish and Wildlife Service and

U.S. Environmental Protection Agency (Mattingly et al., 1993). In the vast majority of cases, permits are approved even if the proposed activities cannot be mitigated by avoiding or minimizing impacts (Lave, 2018). This situation has led to the need for compensatory mitigation, or offsetting adverse effects at the impact site by enhancing degraded stream conditions at another location, or compensation site. The goal of compensatory mitigation is no net loss of stream function (Doyle and Shields, 2012).

The need for compensatory mitigation is evaluated using a scoring system that assigns numerical values to adverse impacts based on the existing condition of the stream, the duration of the proposed activity, the type of proposed activity, and the length of the reach affected. The total score for adverse impacts indicates how many mitigation credits are required to compensate for the activity. These credits are determined by assigning numerical values to categories of benefits associated with compensatory-mitigation activities that enhance environmental conditions within a stream or its adjacent riparian corridor (Table 16.4). Collectively,

Table 16.4. Examples of compensatory-mitigation activities in different categories of benefits for determination of mitigation credits.

Excellent benefits

- Creating floodplains with native vegetation adjacent to a stream.
- Removing low head dams, levees, and dikes that fragment aquatic habitat and/or interfere with natural hydrologic functions (e.g., flooding, recharge, connectivity to floodplain).
- Restoring a stream channel to its former location or restoring sinuosity and channel dimensions of a degraded stream reach to appropriate design based on reference reach or other appropriate standards.
- Building a new, stable channel at a higher elevation and connecting it to its natural floodplain.
- Creating or reconnecting floodplains adjacent to streams artificially disconnected from their floodplain.
- Reconnecting artificially cut-off or abandoned oxbows, side channels, or meanders where functionally appropriate.

Good benefits

- Most streambank stabilization projects that employ bioengineering (i.e., vegetative) techniques to restore bank stability in actively eroding areas. Includes reshaping banks if native vegetation is successfully planted following construction.
- Stream restoration methods utilizing rock/riprap materials to modify flow characteristics and enhance channel stability/aquatic habitat. Includes bendway weirs, stream barbs, Newbury weirs, constructed riffles, but not rock armoring of streambanks alone.
- Replacement of inappropriately designed culverts (undersized or impassable by fish) with open span bridges or structural arch culverts.
- Fish ladders, baffles, or similar structures to allow passage of fish where previously difficult or impossible.
- "Daylighting" of piped or culverted stream segments into an appropriately designed open channel.
- Localized reshaping or terracing of streambanks to restore natural channel dimensions and induce native vegetative growth.
- Construction of pools, riffles, and runs in an existing channel.
- Routing a stream around an existing impoundment into a morphologically stable reach.
- Creation of instream fish habitat using Lunker structures, tree revetments, boulder placement, or root wads.
- Removal of culverts, weirs, pipes, and other minor instream structures.

Moderate benefits

- Linear applications of rock/riprap such as Longitudinal Peak Stone Toe Protection (LPSTP) on or along eroded streambanks.
- Replacement of undersized culverts with appropriately sized closed culverts.
- Removal of culverts, weirs, pipes, and other minor instream structures.
- Removal or realignment of natural debris or manmade instream structures to prevent scour and/or bank erosion.

(from U.S. Army Corps of Engineers, 2010)

activities that produce credit-generating benefits are referred to as stream restoration (Doyle and Shields, 2012). The numerical value of the benefits is weighted by the length of the stream over which compensatory-mitigation activities are implemented.

Compensatory mitigation involves three approaches: permittee-responsible mitigation, in-lieu fee mitigation, and mitigation banking (Womble and Doyle, 2012). In permittee-responsible mitigation, the permittee assumes responsibility for implementing restoration at a compensation site that produces enough credits to offset adverse impacts at the impact site. In-lieu fee mitigation programs involve payment of fees by permittees into a fund that is subsequently used to support restoration activities that offset impacts. Mitigation banking establishes large, allocated sites of stream restoration that generate surplus credits. Often these banks are developed by for-profit companies, which then sell the accumulated credits to permittees who need to offset adverse effects at impact sites (Doyle et al., 2015; Lave, 2018).

Mitigation not only consists of improvement of impaired streams to offset impacts elsewhere; it can also include the creation of entirely new streams. In the Appalachian Mountains, entire mountaintops have been removed to access coal beneath the surface, with the removed waste rock dumped within adjacent valleys, burying existing streams

(Palmer et al., 2010a). To mitigate this effect, new streams are constructed to compensate for the buried streams. Constructing a stream that functions like a natural stream is a complicated endeavor, and in most cases, lost ecosystem functions are not replaced (Palmer and Hondula, 2014).

16.5 How Can Geomorphological Assessments Contribute to River Management?

Ideally, the environmental management of rivers should occur within the context of watershed planning. Because individual rivers and reaches of rivers are affected by material fluxes from contributing watersheds, attempts to manage a portion of a river without considering its context within the watershed may fail to achieve management goals. Management strategies that ignore watershed-scale processes run the risk of being unsustainable. Perceived local problems in a stream, such as accelerated bank erosion or in-channel deposition, often are related to the supply of water and sediment to the stream by the watershed. Efforts to fix these perceived problems without accounting for the connection of the stream to its watershed are unlikely to be successful, because the root cause of the problem has not been addressed. Similarly, restoration or naturalization schemes that seek to enhance aquatic habitat by increasing the geomorphological heterogeneity of a degraded river may be ineffective if this heterogeneity cannot be sustained by watershed-scale processes.

Watershed planning typically involves several steps (Table 16.5). Although the planning process is primarily social in nature (Rhoads et al., 1999), in most cases participants seek to make informed choices based on reliable scientific and technical information. The assessment step provides an opportunity for input of this type of information into the planning process, including information on the geomorphological characteristics of watersheds. In response to the public desire for planning based on sound science, watershed-scale geomorphological assessment methodologies have been developed to inform watershed planning and the associated management of river systems. Geomorphological knowledge gained from assessment can also contribute to the development of management strategies by providing guidance on how desired outcomes of management can be best achieved. This contribution is consistent with the role of geomorphology as a design science. It involves participation of fluvial geomorphologists in the formulation of river management strategies, including approaches to river restoration, naturalization, and mitigation, to achieve societal goals.

16.5.1 What Are the General Goals of Geomorphological Assessments?

The general goals of geomorphological assessments are to understand 1) the current form and function of rivers within a watershed, 2) how current river form and function are related to watershed-scale water and sediment fluxes, 3) if possible, the condition of the river system prior to human disturbance, 4) the evolution of the river system following human disturbance, 5) whether portions of the river system are still evolving in response to human disturbance, 6) if evolving, what trajectories of evolution are likely in the future, 7) what human disturbances continue to affect or are anticipated to affect river form and function, and 8) how different management actions will influence the future form and function of the river system. The extent to which these goals can be achieved depends to a large extent on the availability of financial resources to acquire and analyze geomorphological information. Detailed geomorphological assessments may require considerable investments of time, money, and personnel. Such assessments help to establish connections between local river conditions deemed to be problematic, such as sites of excessive bank erosion or channel incision, and altered watershed-scale processes, including fluxes of water and sediment. Information on such connections is useful for recognizing that site-specific actions may not solve apparent problems if watershed-scale conditions responsible for the problems are not adequately addressed.

Comprehensive approaches to watershed assessment seek to integrate geomorphological analysis with other relevant evaluations targeted at hydrological, water-quality, and ecological conditions (Roni and Beechie, 2013). Integration of these various components within a comprehensive assessment scheme defines interconnections among the components and reveals how various potential management schemes may affect the interconnected components. Comprehensive approaches to assessment provide the basis for scientifically sound decision-making in the watershed planning process.

Geomorphological assessments can be conducted using either customized or standardized approaches. In customized approaches, individual practitioners with expertise in fluvial geomorphology develop tailored methods deemed suitable for specific situations. Assessments are grounded on basic geomorphological principles and the application of common investigative tools (e.g., Kondolf and Piegay, 2016), but details differ from case to case and rely on the expert judgment of practitioners. In standardized approaches, assessments are based on consistent methods involving prescribed steps and protocols. Assessments of this type are often favored by

Table 16.5. The watershed planning process.

Step 1	Organize the planning process	Identify participants Determine responsibilities of participants Inform the public
Step 2	Identify concerns and preferences	Solicit opinions of participants Receive intake from the public Set goals for assessment based on concerns and preferences
Step 3	Conduct watershed assessments	Gather and analyze biophysical information on the condition of the watershed, particularly in relation to concerns and preferences
Step 4	Determine problems, causes of problems, and viability of preferences	Use the assessment results to evaluate whether initial concerns constitute actual problems and, if so, specify what is causing the problems. Also use the results to assess whether preferences can be realistically accommodated within the framework of existing biophysical processes
Step 5	Establish specific objectives	State what management seeks to accomplish to solve problems and achieve preferences
Step 6	Develop management strategies	Specify particular actions that will be undertaken to accomplish objectives Formulate performance criteria for evaluating the effectiveness of the actions Develop a monitoring plan for producing information required to assess performance criteria
Step 7	Implement the plan	Deploy the strategies and conduct monitoring
Step 8	Evaluate implemented strategies and adjust management	Use the information from monitoring to determine whether performance criteria are being met and, if necessary, adjust management

(adapted from U.S. Environmental Protection Agency, 2008)

government agencies, who see value in standardization of protocols from case to case. Such standardization facilitates comparisons among different cases, aids communication, and provides a uniform information base for decision-making. Several different standardized approaches have been developed to meet this need, including the Rosgen Method in the United States (Rosgen, 1996, 2009), the River Styles Framework in Australia (Brierley and Fryirs, 2005), and the Geomorphological Assessment Process in the United Kingdom (Sear et al., 2009). This last method is aimed particularly at addressing conservation and restoration initiatives defined by the Habitat Directive and the Water Framework Directive of the European Commission. Brief overviews are provided of the first two methods, which are outlined in book-length detail and have become popular among assessment practitioners.

16.5.2 What Is the Rosgen Method of River and Watershed Assessment?

The Rosgen Method is the most widely used standardized approach to geomorphological assessment of rivers and watersheds in the United States. This method has been developed primarily by David Rosgen, a former forest hydrologist, who established a consulting business known as Wildland Hydrology, Inc. that specializes in training courses on fluvial geomorphology, watershed assessment and management, river restoration, and monitoring. It consists of a variety of components, including a component aimed at design of river restoration projects, but the centerpiece of the method is a channel classification system that provides the basis for detailed watershed-scale evaluations of river dynamics (Rosgen, 1994). Both the design and assessment components have become the most widely adopted standards of practice within the river restoration community in the United States (Lave, 2009).

16.5.2.1 What Is the Role of Stream Classification in the Rosgen Method?

The river classification system for the Rosgen Method has several objectives: to predict a river's behavior based on characteristics of its form, to associate hydraulic and sediment characteristics with distinctive channel forms, to allow extrapolation of site-specific conditions to other settings, to establish a framework for identifying equilibrium states and departures from these states, and to provide a consistent set of categories for communication within and across disciplines about stream behavior and characteristics (Rosgen, 2009). These objectives are consistent with the general goal of classification in environmental management, which is to establish classes of river types under the assumption that all rivers within a class are comparable and can be managed similarly (Tadaki et al., 2014).

The delineative criteria for the Rosgen river classification scheme include channel planform and bar-element characteristics; entrenchment ratio; channel width/depth ratio; channel sinuosity; channel slope; and geographic context. Major classes of channel types, designated by capital letters, are distinguished on the basis of these criteria (Table 16.6). Additional distinctions are defined within each major class based on the type of boundary materials of the channels, designated by numbers, and differences in slope ranges, designated by lowercase letters. The entrenchment ratio (*ER*) is the width of the valley bottom that is prone to flooding when the water level reaches a depth two times the maximum water depth in the bankfull channel divided by the width of the bankfull channel. When this ratio is small, the stream channel is confined within a narrow valley, whereas when it is large, the valley bottom adjacent to the stream is broad (Figure 16.4).

Classification involves two levels of analysis. In the first level, stream reaches of interest are assigned to preliminary types through evaluation of channel and valley characteristics using secondary data sources, such as aerial photographs and maps, supplemented by reconnaissance field assessments. In the second level, data on channel morphology and materials collected in the field at representative locations are used to determine entrenchment ratio, bankfull width–depth ratio, channel slope, sinuosity, bed features, and size of bed material or presence of bedrock. Based on the results of the field analysis, the reaches are assigned to major types and subclasses of these types (Table 16.6).

The selection of representative locations for collection of field data to characterize types necessitates the identification of reference reaches – reaches that are representative of all reaches of that type. Because collection of data in all reaches usually is not possible, reference reaches provide the basis for extrapolating the characteristics of a certain type of stream to other reaches of that type where data have not been collected. These reaches should represent the unimpaired, stable state of the stream at particular positions within a drainage basin with hydrological conditions characteristic of those positions. Highly modified streams or channels that are actively evolving

should not be selected as reference reaches. Within reference reaches, the identification of the stage or elevation of the bankfull discharge is the single most important factor for second-level analysis (Rosgen, 1996). This stage is needed to determine the bankfull width and bankfull mean depth of the channel, which are used to compute entrenchment and channel width–depth ratios.

Classification of streams provides the basis for assessment of stream conditions. This assessment is based on the assumption that natural streams adjust to prevailing environmental conditions to attain a condition of equilibrium or stability (see Chapter 6). Thus, a stream that has been severely disturbed has been displaced from its potential state or best proper functioning condition in relation to its environmental setting (Rosgen, 1996). Such a stream is unstable, characterized by net aggradation or degradation and an evolving channel form. The Rosgen Method is aimed at providing an assessment that informs how to manage an unstable stream to reestablish equilibrium. Thus, the motto of Wildland Hydrology: "Your job is to help the river be what it wants to be."

16.5.2.2 How Are River Conditions Assessed in the Rosgen Method?

The formal procedure for assessing stream condition consists of a three-tiered methodology known as the Watershed Assessment of River Stability and Sediment Supply (WARSSS) (Rosgen, 2009). At each level the methodology considers hillslope, channel, and hydrologic processes. Reconnaissance-level assessment focuses on problem verification in areas of suspected problems related to erosional and depositional processes in river systems. Once problems are verified, a second, more detailed level of assessment, Rapid Resource Inventory for Sediment and Stability Consequence, is conducted to determine whether problem areas are of moderate or high risk in relation to sediment-related concerns. This level of analysis provides the basis for management recommendations for areas of moderate risk. For areas of high risk, a third, highly detailed level of analysis, Prediction Level Assessment (PLA), is performed to quantify how impaired channel, hillslope, and hydrological processes contribute to sediment problems relative to unimpaired, or reference, conditions. The goal is to establish a predictive-level understanding of impaired-condition contributions so that process-based management strategies can be developed and implemented to mitigate these contributions. The PLA analysis is quite involved and includes extensive collection of field data and predictive modeling of sediment fluxes based on these field data. Channel stability assessments are conducted as part of the PLA analysis.

The Rosgen stream classification scheme provides a baseline for channel stability assessments. When quantities

Table 16.6. General stream type descriptions, quantitative criteria, and landscape setting in Rosgen stream classification.

Stream type[a]	General description	ER	W_{bk}/D_{bk} ratio	Sinuosity	Slope	Common landscape setting
Aa+	Very steep, deeply entrenched, debris transport, torrent streams. Vertical steps with deep scour pools, waterfalls.	<1.4	<12	1.0–1.1	>0.10	Very high-relief landscapes with debris flow potential. Narrow, V-shaped valleys with steep side slopes.
A	Steep, entrenched, cascading, step-pool streams. High energy/debris transport associated with depositional soils. Very stable if bedrock- or boulder-dominated channel. Frequently spaced, deep pools in associated step-pool bed morphology.	<1.4	<12	1.0–1.2	0.04–0.10	High-relief landscapes. Narrow, V-shaped valleys with steep side slopes. Debris cones or alluvial fans may be present.
B	Moderately entrenched, moderate gradient, riffle-dominated channel with infrequently spaced pools. Very stable plan and profile. Stable banks. Rapids predominate with scour pools.	1.4–2.2	>12	>1.2	0.02–0.039 0.04–0.099 = a <0.02 = c	Moderate-relief landscapes. Narrow valleys with moderate side slopes. Debris cones or alluvial fans may be present.
C	Low gradient, meandering, point bar, riffle-pool, alluvial channels with broad well-defined floodplains.	>2.2	>12	>1.2	0.001–0.02 0.02–0.039 = b <0.001 = c	Broad alluvial valleys, possible terraces.
D	Braided channels with longitudinal or transverse bars. Very wide channel with eroding banks. Active lateral adjustment, abundant bedload, and aggradational processes. Convergence and divergence of bed features.	n/a	>40	n/a	0.001–0.02 0.02–0.039 = b <0.001 = c	Broad, moderate-gradient alluvial valleys with locally steeper sections on alluvial fans or glacial debris. Abundance of sediment supply.
DA	Anastomosing, laterally stable, narrow, deep channels. Well-vegetated floodplains and associated wetlands consisting of fine-grained deposits. Very stable streambanks. Very low bedload, high wash load.	>2.2	Highly variable	Highly variable	<0.005	Wide, low-gradient valleys with fine alluvium and/or lacustrine soils.
E	Low gradient, highly sinuous meandering riffle/pool stream with low width–depth ratio, little deposition, and stable channel with well-vegetated banks.	>2.2	<12	>1.5	<0.02 0.02–0.039 = b	Wide alluvial valleys/meadows with floodplains.

Table 16.6. (cont.)

Stream type	General description	ER	W$_{bk}$/D$_{bk}$ ratio	Sinuosity	Slope	Common landscape setting
F	Low gradient, entrenched meandering channel with riffle/pool morphology with high width–depth ratio. Active meandering with high lateral migration and bank erosion rates.	<1.4	>12	>1.2	<0.02 0.02–0.039 = b	Gentle-gradient canyons, gorges, and narrow valleys incised into bedrock or highly weathered material.
G	Entrenched gullies with step-pool morphology, moderate slopes, low width–depth ratios. High rates of erosion of the channel bed and banks.	<1.4	<12	>1.2	0.02–0.039 < 0.02 = c	Found in many valley settings where channels have deeply incised into alluvial or colluvial materials.

[a] Stream types can be subcategorized by type of channel materials: 1 Bedrock, 2 Boulders, 3 Cobble, 4 Gravel, 5 Sand, 6 Silt and Clay. Slope designations, if appropriate, follow the material designation. For example, a cobble-bed type C stream with a slope of 0.03 is designated C3b.

ER: entrenchment ratio.

(adapted from Rosgen, 1994)

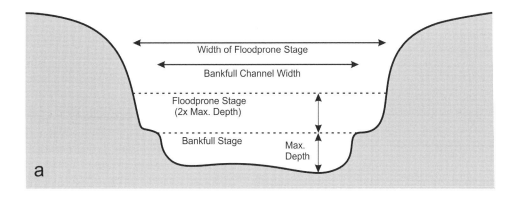

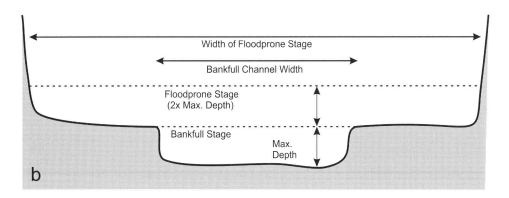

Figure 16.4. Entrenchment ratio (*ER*) is the ratio of the width of the floodprone stage to the bankfull channel width. (a) *ER* ≈ 1.25; (b) *ER* ≈ 2.25.

of morphological metrics for streams or stream reaches in a region of an expected channel type deviate from ranges of values associated with stable reference conditions, these streams or reaches may be in a state of instability that results in morphological evolution to a new stream type. The channel evolution model, which describes the morphological response of streams to an increase in erosional energy due to channelization, is an example of such a sequence of adjustments (see

Figure 15.19). In the Rosgen classification scheme, the sensitivity of streams to disturbance varies by stream type (Table 16.7). Also, the Rosgen Method recognizes a variety of possible channel evolution scenarios besides the classic sequence associated with the channel evolution model (Figure 16.5). The WARSSS PLA determines the channel adjustment sequence associated with a particular stream type and the current state of the stream within the stream sequence (Rosgen, 2009). The potential for erosional instability is high when a stable channel is disturbed to the extent that it begins to evolve into an evolutionary sequence that results in channel incision and enlargement with associated excavation of large amounts of sediment (e.g., B→G, C→D, C→F, C→Gc, E→A, E→Gc, or D→G), whereas the potential is low when it is in a state that will likely evolve into a stable end-state form in the sequence (e.g., B→D, C→E, D→C, F→C, F→Bc, Fb→B, or G→B).

The Rosgen Method represents the development of a design-science tool to meet specific needs of society. In this case, the method has contributed to the growing demand for assessment tools to evaluate aspects of river dynamics deemed by society to be problematic. The main focus of the method is on assessment of channel instability, set within the context of equilibrium river dynamics, and on the production of sediment from channel erosion that leads to impaired water quality. Although the method can be integrated with biological assessments, it does not directly evaluate the quality of physical habitat in relation to stream organisms.

16.5.2.3 What Concerns Have Been Raised about the Rosgen Method?

Within the geomorphological community, the method has been criticized on several grounds (Miller and Ritter, 1996; Juracek and Fitzpatrick, 2003; Simon et al., 2007). All classification schemes are approximate at best, but the use of morphological characteristics only to determine river type ignores the dynamics of rivers and the important of processes in determining the morphological character of rivers (Simon et al., 2007). The concept of equilibrium, on which the Rosgen Method is based, has been challenged by recognition of nonlinear dynamical behavior of some rivers, especially those that are highly sensitive to change due to individual hydrological events (see Chapter 6). A key component of the Rosgen Method is that it supposedly provides the basis for "predicting a river's behavior from its appearance" (Rosgen, 1994, p. 170). By classifying a stream reach into a type, supposedly one can assess its present stability, past conditions, and probable trend if left undisturbed (Small and Doyle, 2012). However, the sequential evolution of river forms of WARSSS (Figure 16.5) relies mainly on documented cases and has not been extensively tested or quantitatively related to

mechanistic processes that control channel adjustment. Current understanding of river dynamics does not provide the basis for highly reliable predictions of river adjustments in response to disturbance.

Determination of bankfull stage, a critical component of the method, can be highly subjective, and not all streams, especially those that have recently undergone incision, have well-defined banks shaped by a distinct channel-forming discharge. Tests of the classification scheme indicate that identification of stream type is sensitive to variation among observers in identification of bankfull stage, which affects estimates of the entrenchment ratio – a critical variable in classification (Roper et al., 2008). Identification of reference reaches, those presumed to be undisturbed, is somewhat subjective, particularly as direct and indirect impacts of human actions on rivers increase in heavily populated landscapes. Moreover, no accommodation is made in the classification scheme for rivers that do not conform readily to the established types. Thus, the scheme is universal in the sense that when it is applied, all reaches of interest must be assigned to one of the existing types, which supposedly encompass all rivers. Although it does include many types of rivers recognized by fluvial geomorphologists, such as braided, meandering, anastomosing, and step-pool systems, the classification scheme does not account directly for some recognized types, such as wandering gravel-bed and anabranching rivers. Nonetheless, the scheme does perform consistently in relation to other geomorphological classification schemes (Kasprak et al., 2016).

Despite its limitations, the Rosgen Method has called attention to the importance of geomorphological information in the environmental management of rivers. It has also fueled the development of professional practice of river assessment based on geomorphological principles. The method has laid the foundation for growing contributions of geomorphological knowledge in decision-making about river-related environmental concerns of public interest.

16.5.3 What Is the River Styles Framework of Assessment?

The River Styles Framework is a coherent set of procedural guidelines to effectively describe the current form and dynamics of rivers, to explain factors that control the current form and dynamics as well as factors that have led to evolution of the river into its current state, and to predict future form and dynamics (Brierley and Fryirs, 2005). The scheme recognizes that each river system is in important ways unique – a product of general fluvial processes as well as site-specific conditions. In this sense, it embraces diversity of fluvial forms and processes. Understanding a particular system must be

Table 16.7. Characteristics of different stream types in relation to management concerns.

Stream type	Sensitivity to disturbance	Recovery potential	Sediment supply	Streambank erosion potential	Vegetation controlling influence
A1	very low	excellent	very low	very low	negligible
A2	very low	excellent	very low	very low	negligible
A3	very high	very poor	very high	very high	negligible
A4	extreme	very poor	very high	very high	negligible
A5	extreme	very poor	very high	very high	negligible
A6	high	poor	high	high	negligible
B1	very low	excellent	very low	very low	negligible
B2	very low	excellent	very low	very low	negligible
B3	low	excellent	low	low	moderate
B4	moderate	excellent	moderate	low	moderate
B5	moderate	excellent	moderate	moderate	moderate
B6	moderate	excellent	moderate	low	moderate
C1	low	very good	very low	low	moderate
C2	low	very good	low	low	moderate
C3	moderate	good	moderate	moderate	very high
C4	very high	good	high	very high	very high
C5	very high	fair	very high	very high	very high
C6	very high	good	high	high	very high
D3	very high	poor	very high	very high	moderate
D4	very high	poor	very high	very high	moderate
D5	very high	poor	very high	very high	moderate
D6	high	poor	high	high	moderate
DA4	moderate	good	very low	low	very high
DA5	moderate	good	low	low	very high
DA6	moderate	good	very low	very low	very high
E3	high	good	low	moderate	very high
E4	very high	good	moderate	high	very high
E5	very high	good	moderate	high	very high
E6	very high	good	low	moderate	very high
F1	low	fair	low	moderate	low
F2	low	fair	moderate	moderate	low
F3	moderate	poor	very high	very high	moderate
F4	extreme	poor	very high	very high	moderate

Table 16.7. (cont.)

Stream type	Sensitivity to disturbance	Recovery potential	Sediment supply	Streambank erosion potential	Vegetation controlling influence
F5	very high	poor	very high	very high	moderate
F6	very high	fair	high	very high	moderate
G1	low	good	low	low	low
G2	moderate	fair	moderate	moderate	low
G3	very high	poor	very high	very high	high
G4	extreme	very poor	very high	very high	high
G5	extreme	very poor	very high	very high	high
G6	very high	poor	high	high	high

(from Rosgen, 1994, 2009)

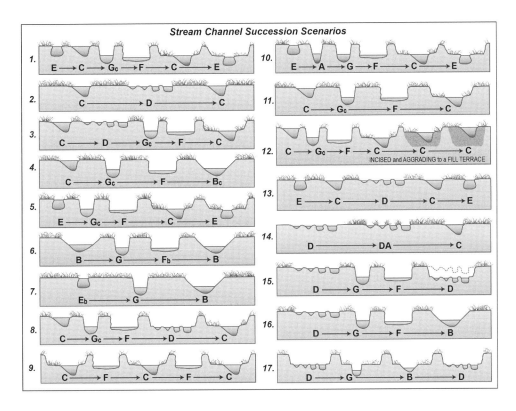

Figure 16.5. Evolutionary succession sequences associated with the Rosgen Method of channel stability assessment (Rosgen, 2014).

based on general knowledge along with detailed investigation of the particular environmental setting and evolutionary history of the system. The flexibility of the approach allows each river to "speak for itself" with regard to its specific physical characteristics within the continuum of river geomorphological diversity. The framework also is targeted toward producing information that allows management to work with natural tendencies of the river. It promotes catchphrases of "know your watershed" (Brierley and Fryirs, 2005) and "don't fight the site" (Brierley and Fryirs, 2009) in watershed-scale management of rivers based on geomorphological information. The scheme has been mainly applied to rivers in Australia, where it was developed, but has been used in other parts of the world as well (Kasprak et al., 2016).

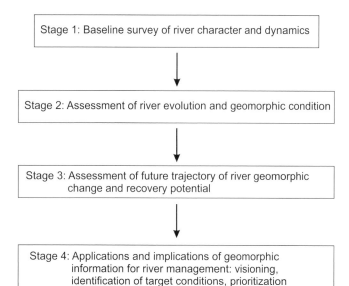

Figure 16.6. Stages in the River Styles Framework (adapted from Brierley and Fryirs, 2005).

Although the River Styles Framework is geared toward geomorphological assessment, it provides the starting point for comprehensive geoecological evaluations of interacting biological and physical processes within river and watershed systems (Brierley and Fryirs, 2005). The framework adopts the perspective that geomorphological processes and forms configure physical habitat for organisms – an important component of stream ecosystems. The geomorphological diversity of rivers shapes the abundance of physical habitat. While the framework acknowledges that physical habitat alone does not determine ecological conditions within a river, it also maintains that failure to consider the relation between geomorphology and ecology can introduce uncertainty and diminish the effectiveness of ecosystem-based management.

The River Styles Framework (RSF) consists of several stages (Brierley and Fryirs, 2005) (Figure 16.6). Characterization of existing river conditions is the focus of Stage 1. The RSF fully recognizes the hierarchical structure of river systems in a manner similar to that presented in this book (Figure 1.7). The conceptualization differs somewhat, but the framework emphasizes that river character and dynamics at reach scales (comparable to planform scale) consist of components at the scale of geomorphic units (similar to bar-unit and bar-element scales) but are controlled largely by processes operating at the scales of landscape units (link scale) and catchments (stream-network or drainage-basin scale).

16.5.3.1 What Role Do Geomorphic Units Play in Assessment?

River styles, defined at reach scales, are constructed from components at smaller spatial scales and controlled by processes at larger spatial scales. Characterization of river styles is based on associations between geomorphic units and channel planform, the type of valley setting, and attributes of bed material. Identification and mapping of geomorphic units is a critical component of this process because these units are viewed as the building blocks of river form (Wheaton et al., 2015). Interpretation of assemblages of geomorphic units and process interactions among units determines the character and dynamics of a reach (Fryirs and Brierley, 2018). A wide variety of geomorphic units have been identified as part of the RSF and include both instream units and floodplain units (Brierley and Fryirs, 2005, 2012). Examples of instream geomorphic units include various types of bars (alternate, point, scroll, longitudinal, transverse, medial, and compound), pools, riffles, concave-bank benches, and step-pool sequences. Floodplain geomorphic units are identified separately from in-channel units. These units consist of the active floodplain, which may be continuous or discontinuous, as well as a variety of other features, such as levees, crevasse splays, ridge and swale topography, oxbow lakes, secondary channels, backswamps, and terraces.

16.5.3.2 How Are River Styles Determined?

A hierarchical, procedural decision tree that depends on valley setting is used to determine the river style (Figure 16.7). The qualitative classification procedure is flexible, can be applied to a broad range of rivers, and involves expert judgment, particularly regarding identification and interpretation of relevant geomorphic units. When applied to a particular river system at the scale of a drainage basin, it provides named styles of various reaches throughout the basin. A style name consists of a sequence of terms that defines a reach identity based on its salient characteristics (Fryirs and Brierley, 2018). The name has a hierarchical structure that reflects the sequential steps in the procedural decision tree (Table 16.8). Once styles are determined, the spatial distribution of these styles can be mapped, and the relation of the styles to controlling factors at watershed scales, including the longitudinal profile, can be examined. Such analysis is often useful for identifying connections between spatial variation in river styles and spatial variation in watershed conditions.

Once river styles are identified, mapped, and related to watershed controlling factors, subsequent steps in the RSF involve determining whether reaches of a specific style depart from the expected state for that style because of human impacts, evaluating inherent evolutionary trajectories of certain styles, and predicting future river conditions based on recovery potential. These aspects of the RSF are somewhat subjective and prone to uncertainty. Assessment of departures from an expected state depends on specification of a reference condition. The RSF is relatively flexible in regard to the definition of a reference condition, including within the scope of this condition human-impacted reaches that have undergone irreversible change. Nevertheless, it relies on the concept of equilibrium or steady state by associating a morphological configuration that is compatible with existing fluxes of water and sediment as the theoretical basis for the reference condition.

16.5.3.3 Why Is Analysis of Geomorphic Change Important in the River Styles Assessment Framework?

Because rivers are dynamic, geomorphic change and evolution cannot be ignored. The second stage of the RSF

examines inherent evolutionary tendencies of a river system along with evolution of the system under the influence of human activity to determine whether change related to human impact is reversible or irreversible. Such determinations are not easy and depend on the availability of appropriate historical information on changes in watershed conditions, hydrological response, and channel form. Of particular value is information on planform change over time acquired from historical maps, aerial photographs, and satellite imagery, as well as historical photographs or surveys of channels that can be used to document channel incision or aggradation. In addition, detailed fieldwork often is required to reconstruct past channel environments using paleo-sedimentological methods. Ergodic methods, where reaches at different stages of evolutionary development at various locations in a watershed are used to infer past and future temporal evolution at each location, can supplement direct documentation of change over time (Fryirs et al., 2012). Capacities for adjustment must be evaluated within the context of particular river styles, given that different styles exhibit different inherent dynamics. Attempts to determine the environmental history of a river system can involve considerable investments in data collection and analysis that may exceed the budget and scope of many river management efforts. Moreover, given the diversity of geomorphological responses of rivers to disturbance, attempts to define distinct categories of adjustment in relation to process domains or disturbance type for the purpose of informing management can be challenging (Lisenby and Fryirs, 2016).

If enough information is available to reconstruct channel changes over time, reference conditions, evolutionary trajectories, the capacity for adjustment, sensitivity to future change, and management objectives, the foci of stages 2 through 4 of the RSF (Figure 16.6), can be framed within the context of historical range of variability (HRV) (Wohl, 2011a). The basic idea underlying HRV is that when a river system is in a steady state, or

Table 16.8. Examples of river styles.

Confined, bedrock-margin controlled, canyon, gravel bed
Partly confined, bedrock margin controlled, discontinuous floodplain, gravel bed
Laterally unconfined, continuous channel, meandering, sand bed
Partly confined, planform controlled, braided, terrace constrained, discontinuous floodplain, gravel bed
Laterally unconfined, continuous channel, braided, gravel bed
Laterally unconfined, continuous channel, anastomosing, fine-grained bed
Laterally unconfined, continuous channel, anabranching, fine-grained bed
Laterally unconfined, continuous channel, wandering, gravel bed

(from Fryirs and Brierley, 2018)

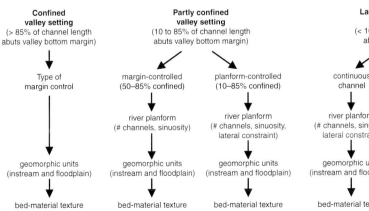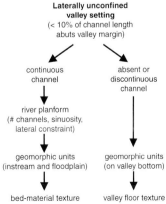

Figure 16.7. Decision tree in River Styles Framework river classification procedure (from Fryirs and Brierley, 2018 modified from Brierley and Fryirs, 2005).

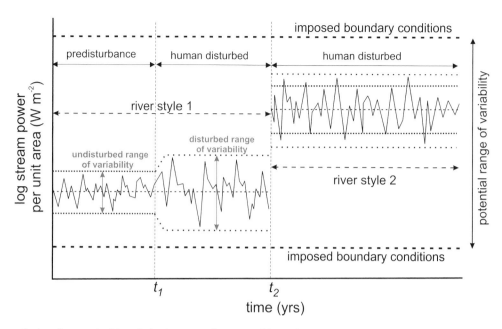

Figure 16.8. River evolution diagram in River Styles Framework. Imposed boundary conditions limit potential range of variability in stream power. Under predisturbance conditions, prior to time t_1, stream power associated with river style 1 fluctuates (solid line) around mean state (dashed line) within undisturbed range of variability. After human disturbance at t_1, the range of variability of stream power increases, but the river style remains unchanged. After human disturbance at time t_2, a new river style (river style 2) develops as disturbed flux boundary conditions (widely spaced dotted lines) shift upward. The change in river style is irreversible without addressing the shift in flux boundary conditions. Management may seek to establish the undisturbed flux boundary conditions associated with River Style 2 (closely spaced dotted lines) (adapted from Brierley and Fryirs, 2005).

state of equilibrium with prevailing environmental conditions, it fluctuates around a mean state over time (see Chapter 6). In other words, fluvial forms and processes vary over time, but within a range of restricted variability. A reference condition can be associated with this historical range of variability, so that the reference condition is viewed as dynamic, not static, but that dynamism is constrained. Changes in environmental conditions, such as those associated with direct and indirect human impacts (see Chapter 15), can result in changes in the range of variability around the mean state as well as abrupt changes in the mean state and its associated range of variability.

The River Styles Framework uses a river evolution diagram to examine the range of variability and its relation to reference conditions (Figure 16.8) (Brierley and Fryirs, 2005). This diagram depicts changes in river energy characteristics, defined on the basis of stream power per unit area, over time in relation to flux boundary conditions and imposed boundary conditions. Valley confinement, valley slope, upstream drainage area, and caliber of supplied sediment impose watershed-scale constraints that define the maximum range of formative energy conditions within which a range of river styles can possibly develop for a specific reach of a river system (Brierley and Fryirs, 2012). Flux boundary conditions, determined by the prevailing hydrologic and sediment regimes, define the range of conditions over which stream power varies for a river style

adjusted to prevailing environmental conditions. Human disturbance can alter flux boundary conditions, and thus the prevailing range of stream power, without altering the river style. Nevertheless, the dynamics of the river system will change as a result of this disturbance, because the system will exhibit enhanced or reduced variability in stream power. For example, a human-induced increase in the variability of stream power of a meandering stream likely will result in greater variability in rates of bank erosion and lateral migration. Given that the river style has not changed, the effects of the disturbance may be reversible through management practices that mitigate the change in flux boundary conditions. The reference condition becomes the undisturbed state, and the goal often is to restore the system to this state (Wohl, 2011a). In such cases, management that seeks to establish a river style that requires a fundamentally different historical range of variability is likely to fail (Woelfle-Erskine et al., 2012).

In other cases, human disturbance may be so severe, frequent, and outside the HRV that the domain of flux boundary conditions shifts, resulting in a fundamental transformation of river style (Rathburn et al., 2013). For the example of a meandering river, if stream power is increased beyond a certain threshold, the river may become braided. This evolution of river style from one type to another is irreversible unless mitigation of human impact on water and sediment fluxes occurs at the watershed scale. Human disturbance may produce flux boundary conditions for the transformed style that differ from the undisturbed, or reference, flux boundary conditions for this style. In this case, management may be directed at mitigating the effects of these disturbed flux boundary conditions or at reestablishing the reference flux boundary conditions. In other cases, slow recovery may gradually return the disturbed system to its predisturbance style (Rathburn et al., 2013), but recovery times may greatly exceed ordinary management timescales, which are usually only on the order of several decades. In such evolving river systems, long-term management may have to be directed at reference conditions that are moving targets (Brierley and Fryirs, 2016).

16.6 What Is the Role of Geomorphology in Implementation of Management Strategies?

Besides providing useful information and analysis for developing appropriate management strategies to address problems related to river dynamics, fluvial geomorphology can contribute directly to implementation of these strategies. A variety of management strategies are possible, ranging from doing nothing to undertaking various forms of restoration, naturalization, mitigation, or even creation. Except in the case of doing nothing, existing river forms and processes usually are actively reconfigured to achieve more desirable conditions. Implementation of such strategies involves the formulation of environmental designs so that input from the applied design science of fluvial geomorphology becomes highly relevant at this stage of management. Although scientific and technical information, including geomorphological information, usually is taken into account in implementing management strategies, decisions about implementation also involve social, political, economic, and aesthetic considerations.

16.6.1 When Is It Appropriate to Give the River Room to Operate?

Given that society values land along rivers, encroachment onto floodplains is common. A well-known consequence of floodplain occupation is increased flood risk. Floodplains are also active areas of erosion and deposition (see Chapter 15), leading to higher risk of harm associated with these fluvial processes. One approach to managing these risks is to leave the river alone and simply provide it with enough room to operate hydrologically and geomorphologically. This passive management strategy recognizes that rivers are dynamic features and that floodplains are integral components of river systems. The concept has been referred to using labels such as erodible river corridor (Piegay

et al., 2005), freedom space (Biron et al., 2014; Buffin-Belanger et al., 2015), and *espace de liberté* (Kondolf, 2012).

The strategy of letting rivers operate unimpeded to preserve existing characteristics, or "heal themselves" if impaired, is best accomplished where stream power and sediment supply are sufficiently high that erosional and depositional processes can dynamically sustain existing complexity or reshape human-simplified channels into natural forms over timescales of a few decades (Kondolf, 2011). This strategy has been proposed mainly for active meandering or wandering gravel-bed rivers, although it should also be applicable to braided rivers in appropriate contexts. From a practical perspective, the strategy requires space; it is not feasible where the river corridor is substantially constrained by human infrastructure. The amount of necessary space can be evaluated using historical evidence on past channel migration as well as simulation modeling of future migration (Piegay et al., 2005). From an ecological perspective, dynamic, evolving rivers produce a diverse array of habitat types that support a wide range of floral and faunal species. Such rivers have greater ecological richness than dynamically constrained, morphologically simplified rivers modified by humans (Kondolf, 2006).

In the United States, Vermont has adopted a river corridor approach to river management that relies on geomorphic data to define the land area necessary for meandering rivers to migrate across their floodplains within a naturally established meander belt (Kline and Cahoon, 2010). River corridors are protected through land acquisition programs and zoning regulations that restrict development, providing dynamic rivers with space to evolve and change. This passive approach to river management has arisen largely in response to constructed restoration projects that failed to adequately accommodate fluvial processes.

16.6.2 What Is Natural Channel Design?

Natural channel design (NCD) is an approach to river restoration that seeks to establish self-sustaining and self-regulating physical, chemical, and biological functions of a river system and to emulate the natural stable form within watershed-scale constraints (Rosgen, 2011). The approach is based on the Rosgen Method of watershed assessment and extends this method by using information derived from watershed assessments to formulate designs of restored rivers. Although the method considers chemical and biological aspects of river systems, its main focus is on physical functions. For this reason, it relies heavily on geomorphological principles and information. As with the Rosgen Method, NCD draws upon the interrelated concepts of equilibrium, steady state, and stability as foundational principles underlying the aim of river management. The goal is to reestablish for unstable

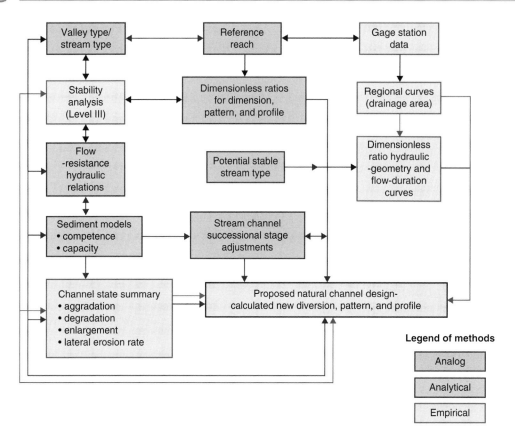

Figure 16.9. Methodology of natural channel design showing relations among analog, analytical, and empirical components (from Rosgen, 2007). (A black and white version of this figure will appear in some formats. For the color version, please refer to the plate section.)

reaches physical stability that integrates the processes responsible for creating and maintaining the characteristic pattern, geometry, and longitudinal profile of a specific type of river channel (Rosgen, 2011).

The overall approach involves 10 phases (Rosgen, 2011). The development of restoration objectives (Phase I) is guided by the goal of reestablishing stability. Phases II and III overlap with and generally are accomplished as part of the WARSSS in the Rosgen Method. Phases IV–VIII determine whether passive or active management is required to meet restoration goals and, if active restoration is required, include the development of a natural channel design. In Phase IX, the restoration project is implemented, and Phase X encompasses monitoring and maintenance of the implemented project.

Natural channel design utilizes analog, empirical, and analytical methods (Rosgen, 2007) (Figure 16.9). Empirical methods include collection and analysis of hydrological and geomorphological data on existing watershed conditions as well as the characteristics of reference and impaired stream reaches. A key part of empirical analysis involves the development of regional curves relating bankfull discharge, bankfull channel width, bankfull channel depth, and bankfull channel cross-sectional area to drainage area (Bieger et al., 2015). These curves provide an empirical basis for sizing the channel dimensions of a restored stream in relation to its position within a watershed.

Reference reaches serve as the foundation for analog methods of design. These reaches provide examples of what an impaired, unstable reach can become if stabilized. References reaches ideally should represent the same potential stream type (Rosgen classification), valley type, flow regime, sediment regime, stream bank type, and riparian vegetation community as the impaired reach to be restored. Characteristics of channel morphology in the reference reach are used to establish dimensionless relations that define the stable channel dimensions, channel pattern, and bed profile for a particular stream type and valley type. Information from empirical and analog methods can then be used to determine channel characteristics for restoration. For example, the bankfull channel width of an impaired reach targeted for restoration ($W_{bk(res)}$) is computed as (Rosgen, 2007)

$$W_{bk(res)} = \left[\frac{W_{bk(ref)}}{D_{bk(ref)}} A_{bk(res)} \right]^{0.5} \tag{16.1}$$

where $W_{bk(ref)}$ is the bankfull channel width of the reference reach, $D_{bk(ref)}$ is the bankfull mean channel depth of the reference reach, and $A_{bk(res)}$ is the bankfull channel cross-sectional area of the reach targeted for restoration. The value of $A_{bk(res)}$ is derived from a regional curve relating the bankfull channel cross-sectional area of reference reaches of an appropriate potential stream type to drainage area. In both the analog and empirical methods, reference reaches serve as analogs for restoration. In this sense, reference reaches are critical to Natural Channel Design. Analytical methods are employed in NCD to evaluate the competence of a restored stream to transport the largest-size particles on its bed and to determine whether the sediment-transport capacity of this stream is sufficient to convey the sediment supplied to it from the watershed. The complete NCD procedure involves 40 steps (Rosgen, 2007) and the prediction of 67 form variables (Rosgen, 2011). It yields estimates of channel morphology (geometry, planform characteristics, and slope) for a restored stream that conform to conditions of a stable reference reach.

Actual implementation of NCD restoration designs often involves substantial amounts of earthmoving to reconfigure an unstable stream into the desired stable form. Reshaping of channels can require the use of heavy construction equipment and removal of vegetation to allow this equipment access to the channel. Thus, many NCD sites are quite barren until vegetation can become reestablished (Lave, 2014). In some designs, not all aspects of stability can be fully accommodated simply by reshaping and revegetating constructed channels. River stabilization and "enhancement" structures may be required to reduce erosion before stabilizing vegetation becomes established, to reduce accelerated bank erosion, to dissipate excess erosional energy, and to protect critical infrastructure along streams (Figure 16.10). Cross vanes and w-weirs are grade control structures designed to dissipate excess erosional energy by creating hydraulic drops and jumps along the river (Rosgen, 2007). Engineered logjams can also be used for this purpose (Abbe and Brooks, 2011). Streambank erosion-control structures include J-hooks, bendway weirs, stream barbs, log or rock vanes, and spur dikes, all of which disrupt high-velocity flow near the outer banks of bends (Yochum, 2018). Banks may be shielded directly from erosion using gabions, riprap, and stone toe protection. Bioengineering techniques for streambank protection include live plantings as well as brush layers and mattresses (Harman and Starr, 2011). Besides enhancing channel stability, some engineered structures may also enhance habitat for fish or macroinvertebrates, but efforts to enhance aquatic habitat often seek to avoid rock structures, instead preferring arrangements of logs or wood that have a more natural appearance (Abbe and Brooks, 2011; Yochum, 2018).

Natural Channel Design has become a popular method for stream restoration among consulting firms as well as local, state, and federal agencies in the United States, including the Natural Resource Conservation Service (Rosgen, 2007), the U.S. Forest Service (Yochum, 2018), the U.S. Fish and Wildlife Service (Harman and Starr, 2011), and the Bureau of Land Management (Lave,

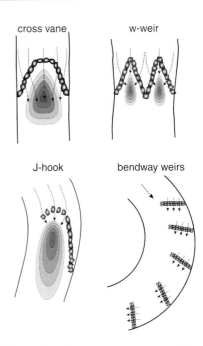

Figure 16.10. Examples of instream structures used in natural channel design and stream restoration.

2014). The method has been used widely to accommodate requirements for stream mitigation under section 404 of the Clean Water Act administered jointly by the U.S. Army Corps of Engineers and the U.S. Environmental Protection Agency. The need to address federal regulatory issues related to stream mitigation has fueled the growth of stream restoration (Doyle and Shields, 2012; Doyle et al., 2013). To meet demand, Natural Channel Design has been largely responsible for privatizing stream restoration or establishing it as a marketable commodity within the private sector (Lave et al., 2010).

As with the Rosgen Method, NCD has not been immune to controversy and criticism (Lave, 2012). Major critiques from a geomorphological perspective center around its one-size-fits all, or "cookbook," methodology for restoration, which is allegedly applicable to any river; its emphasis on form versus process in channel design; its inadequate consideration of geographical context and the place-specific geomorphological history of river systems; its uncertainty related to identification of stable reference reaches to serve as morphological analogs in restoration design; its insistence on channel stability, especially stability in the sense that restored channels should not change at all in response to fluxes of water and sediment, and its heavy reliance on bankfull discharge for channel design, rather than consideration of the full spectrum of channel-shaping events (Simon et al., 2007; Lave, 2009; Wilcock, 2012). The fixed-channel stability criticism, which has been referred to as "stability" (Lave, 2009), is not consistent with the goal of NCD to allow channel adjustment within a stable range (Rosgen, 2011) but has been leveled at implemented projects that have tried to achieve fixed conditions (Lave, 2009). A limitation of the NCD methodology is that it focuses primarily on restoration of single-thread meandering rivers and is not readily adapted to braided, anastomosing, or anabranching rivers (Kondolf, 2006). Failures of restoration projects by practitioners trained in NCD have been documented where project goals were not accomplished because reconfigured channels or control structures were not sustained by subsequent flows (Kondolf et al., 2001; Smith and Prestegaard, 2005; Miller and Kochel, 2010).

In a more general sense, the commodification of stream restoration by NCD has shaped the way in which restoration practice has evolved. Restoration has become big business within the private sector (Bernhardt et al., 2005; Lave et al., 2010) and to some extent occupies a niche within the contemporary river management arena similar to that occupied in the past by traditional river engineering. As new industries arise to meet the demands of society, the best interests of these industries are served by the safeguarding of viewpoints consistent with the aims of the industries. The best interests of the stream restoration industry are served by the perspective that river systems are in need of repair by this industry to undo damage produced by humans. Natural Channel Design is not directly responsible for promoting such an attitude, but the conflation of its focus on geomorphic stability as the ideal condition of a river with the notion that stable rivers should not erode at all contributes to the perception that eroding channels are abnormal and need to be fixed. This perception in turn can lead to implementation of "restoration" projects to control any erosion that may be occurring along a river (Biron et al., 2018), when river mobility associated with bank erosion is often beneficial to the development of aquatic habitat (Chone and Biron, 2016).

The recent shift toward an emphasis on achieving ecological goals in restoration (Palmer et al., 2005), rather than channel stability, has also led to criticism of NCD (Palmer et al., 2014). Implicit in NCD is the assumption that once stable physical habitat is established, improvement in ecological functions will follow (Lave, 2014). Not only is physical habitat just one component of ecological function, but the act of implementing construction-intensive restoration designs itself represents a form of ecological disturbance. Thus, successful ecological restoration is in no way guaranteed by efforts to achieve channel stability.

16.6.3 What Is the Scientific Basis for Process-Based Stream Restoration?

Recent efforts within the field of stream restoration have increasingly emphasized the need to fully integrate social,

ecological, geomorphological, and water-quality considerations to achieve multifarious goals of restoration (Roni and Beechie, 2013). What has become known as the process-based approach to stream restoration has largely arisen in response to concerns about the adequacy of Natural Channel Design for achieving these goals. The emphasis in NCD on channel stability and the use of morphological relations derived from stable reference reaches to guide restoration design has led to characterization of it as primarily a hydrogeomorphic, form-based approach to restoration (Simon et al., 2007). A concern is that it fails to account fully for the role of fluvial processes in river dynamics. The process-based approach aims to rectify this shortcoming by prioritizing river processes or functions in restoration, rather than only river form (Wohl et al., 2015). In contrast to NCD, the process-based approach has emerged from a diverse spectrum of sources, primarily centered within the scientific community. This approach is not highly standardized, and the design science underpinning it remains under development (Palmer et al., 2014; Wohl et al., 2015). Although some general themes can be identified, the approach primarily consists of a loosely organized array of knowledge, principles, and practices intended to provide general guidance on restoration rather than a prescribed, uniform methodology (Roni and Beechie, 2013).

In general, process-based restoration seeks to reestablish in human-impacted streams normative rates and magnitudes of physical, chemical, and biological processes that create and sustain river and floodplain ecosystems (Beechie et al., 2010). This approach highlights full integration of the biotic and abiotic aspects of river systems and the connectivity between rivers and floodplains (Vaughan et al., 2009; Wohl et al., 2015). Geomorphic processes are an important aspect of this approach but cannot be considered in isolation from biogeochemical and ecological processes (Figure 16.11). Ecological research has shown that improving physical habitat alone does not guarantee success in restoring stream biodiversity (Palmer et al., 2010b).

The concept of a reference condition is relevant to the process-based approach to restoration but does not limit this condition to a pristine state unaffected by humans (Beechie et al., 2013b). In other words, process-based restoration does not necessarily attempt to return the river to exactly what it was like before it was disturbed by human actions. Instead, the reference condition refers to a useful baseline for assessing degradation and for defining the geomorphological and ecological aims of restoration. Beyond this general statement, the concept of reference in process-based restoration is somewhat fuzzy. To a large extent, the notion of reference depends on what is sustainable. If watershed-scale fluxes of water and sediment have not been altered substantially, pristine conditions, if still intact at places within the watershed, may serve as a reference. If human activity has changed watershed-scale fluxes to some

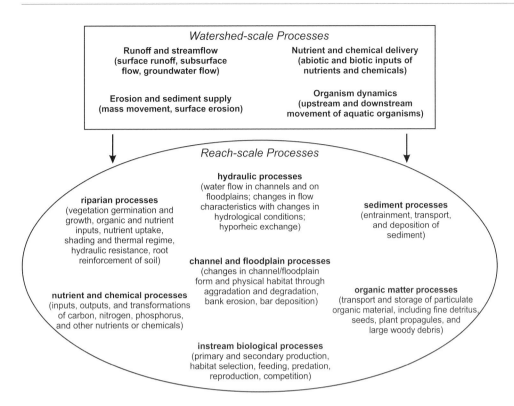

Figure 16.11. Key watershed-scale and integrated reach-scale processes considered in process-based restoration (based on Beechie et al., 2013a).

degree, resulting in relatively minor changes in river dynamics, least-altered examples of river reaches may be desirable reference conditions (Wohl, 2011b). On the other hand, if change in watershed-scale fluxes has been substantial, and nothing close to the pristine state exists or is sustainable, a reference condition adjusted to and sustainable by the current range of variability in undisturbed flux boundary conditions may be most appropriate (Figure 16.8). Identifying examples of reference conditions along existing rivers is valuable for guiding the objectives of restoration and for confirming that such conditions are sustainable within the contemporary environmental setting.

The process-based approach pursues restoration strategies that work with dynamic processes of adjustment to and recovery from disturbance rather than employing highly engineered practices to create artificial, static forms. Natural fluvial processes, such as bank erosion and lateral channel migration, which often become targets of erosion-control treatments in traditional river engineering or even in some Natural Channel Design projects, are viewed as integral to the proper functioning of river systems (Florsheim et al., 2008). These processes produce mosaics of physical habitat that change, or shift, dynamically over time at bar-element, bar-unit, and planform scales (Pringle et al., 1988), thereby contributing to ecosystem heterogeneity as well as ecological diversity and productivity both locally and at watershed scales (Thorp et al., 2006; Brennan et al., 2019). Because it recognizes strong process linkages across scales, the approach seeks not just to target actions at the reach scale – the locus of many

restoration projects – but to address underlying watershed-scale causes of river degradation (Beechie et al., 2013b). From a geomorphological perspective, local problems related to channel erosion or deposition are often connected to changes in fluxes of water and sediment at the watershed scale (see Chapter 15).

Attempts to engage in watershed-scale restoration aimed at modifying sediment supply and runoff from hillslopes are typically difficult to accomplish where land is privately owned and government agencies have only limited authority to regulate land use or land management practices. Also, the lag between watershed-scale management efforts, even when successfully implemented, and local changes in fluvial processes may be on the order of several decades, particularly with regard to sediment dynamics. Nevertheless, reach-scale projects may be unsustainable if changes in watershed conditions are not adequately accommodated or addressed (Gregory and Downs, 2008; Beechie et al. 2013b). Assessments that seek to identify appropriate reference conditions, and the watershed-scale factors that support these conditions, can help to inform sustainable reach-scale restoration.

The process-based approach also stresses the importance of geographical context and environmental history in restoration. It fully recognizes that the dynamics of river systems are a product of both general processes and place-based contingencies that determine specific relations between process and form. To highlight the importance of issues of scale and

Table 16.9. Four principles of process-based stream restoration.

Principle	Description
Target root causes of habitat and ecosystem change	Restoration actions that target root causes of degradation rely on assessments of processes that drive habitat conditions, and actions are designed to correct human alterations to those driving processes.
Tailor restoration actions to local potential	Each reach in a river network has a relatively narrow range of channel and riparian conditions that match its physiographic and climatic setting, and understanding processes controlling restoration outcomes helps design restoration actions that redirect channel and habitat conditions into that range.
Match the scale of restoration to the scale of the problem	When disrupted processes causing degradation are at the reach scale (e.g., channel modification, levees, removal of riparian vegetation), restoration actions at individual sites can effectively address root causes. When causes of degradation are at the watershed scale (e.g., increased erosion, increased runoff due to impervious surfaces), many individual site-scale actions are required to address root causes.
Be explicit about expected outcomes	Process-based restoration is a long-term endeavor, and there are often long lag times between implementation and recovery. Ecosystem features will also continuously change through natural dynamics, and biota may not improve dramatically with any single individual action. Hence, quantifying the restoration outcome is critical to setting appropriate expectations for river restoration.

(from Beechie et al., 2010)

geographical context, process-based restoration is guided by four basic principles: target root causes of degradation, tailor restoration activities to local potential, adequately consider the scale of processes, and establish realistic expectations about outcomes (Table 16.9).

One possible way to accomplish the goal of process-based restoration is to allow a river system to recover passively from disturbance (see Section 16.6.1). In many cases, however, strategies for process-based restoration aim to achieve "assisted recovery" so that processes within the disturbed system adjust quickly to, or at least toward, the targeted reference condition. The extent to which human intervention occurs to achieve this goal varies widely. Assisted recovery generally involves undoing human-imposed or human-induced constraints that limit the operation of river processes. Every effort is made in process-based restoration to accomplish restoration goals through activities that involve minimal intrusion and avoid the need for construction-intensive approaches (Table 16.10). Nevertheless, the approach does encompass interventionist instream techniques, such as remeandering of straightened channels; implementation of structures to enhance the complexity of physical habitat, including pools, riffles, and areas of cover; and augmentation of coarse substrate material to improve spawning habitat (Roni et al., 2013a).

Many process-based restoration strategies focus on reestablishment of connectivity that has been disrupted by human modification of river systems. Connectivity can be disrupted along the river (longitudinal connectivity), between the channel and adjacent floodplain (lateral connectivity), and between the river and its channel bed (vertical connectivity). Barriers to connectivity change hydrological, geomorphological, and ecological processes by altering fluxes of water, sediment, nutrients, chemicals, and organisms throughout the river system. Changes in these fluxes result in loss or deterioration of process linkages that are vital for sustaining the physical and ecological characteristics of the river system. Efforts to reestablish connectivity seek to restore these vital process linkages. Examples of strategies to improve connectivity include removal of dams to restore fish passage, reconnecting flow in the main channel to flow through artificially disconnected secondary channels or to flow into floodplain wetlands, setting back or breaching artificial levees, and removing fine sediment that blocks hyporheic exchange through the bed (Kondolf et al., 2006).

Connectivity between the channel and the floodplain includes consideration in process-based restoration of the riparian corridor and its relation to geomorphological and ecological processes (Gonzalez et al., 2015). In many forested streams, wood loading is an important component of river dynamics and habitat heterogeneity (Wohl et al., 2017). Geomorphologically, instream wood increases hydraulic resistance, promotes storage of sediment and organic matter, enhances local erosion and deposition of the channel bed and banks, and can direct water out of the channel onto the floodplain (Wohl and Scott, 2017).

Table 16.10. Examples of process-based approaches to stream restoration.

Target	Purpose	Examples
Longitudinal, lateral, or vertical connectivity	reestablish migration pathways for organisms allow natural transport of nutrients and sediments reconnect floodplain and main-channel habitats allow natural channel migration	remove or breach dams set back or remove levees reconnect main channel to floodplain sloughs and wetlands augment gravel in bed material to enhance hyporheic exchange promote aggradation in incised channels use engineered logjams to direct in-channel flow onto floodplains
Sediment and water supply	reduce or modify sediment supply increase or decrease runoff and hydrology improve water quality provide adequate flows for aquatic biota	implement best management practices within the watershed to reduce runoff and soil erosion change land use stabilize runoff and erosion from unpaved roads employ pervious surfaces to reduce rates of runoff
Riparian vegetation	change vegetation composition of riparian zone to include more native species limit bank erosion of rivers increase or decrease shading to regulate stream thermal regime provide buffers along stream to filter sediment and nutrients from hillslope runoff	plant trees and vegetation thin or remove understory and invasive species reduce grazing establish riparian corridors
Instream habitat	improve habitat diversity, abundance, stability, and complexity	remeander straightened streams create or enhance pools and riffles introduce wood or engineered logjams create structural habitat features stabilize eroding streams to allow geomorphic processes to create and sustain habitat elements add spawning gravel provide room for the river to migrate laterally and dynamically create habitat reintroduce beaver and the attendant construction of beaver dams
Nutrients and productivity	boost productivity of system to improve biotic production compensate for reduced nutrient levels from lack of anadromous fishes reduce nutrient levels	add organic and inorganic nutrients use streamside wetlands to filter nutrients break subsurface tile lines

(adapted from Roni et al., 2013a)

Overall, it contributes to geomorphological complexity and channel stability. Ecologically, wood retains nutrients and organic matter, increases the diversity of physical habitat, serves as a food source, and enhances biodiversity (Wohl, 2015b). Downed wood on floodplains provides habitat for terrestrial and aquatic organisms and germination sites for plants; by influencing the hydraulics of overbank flows, it also affects patterns of floodplain erosion and deposition (Lininger et al., 2017). Despite its geomorphological and ecological benefits, wood is often perceived negatively (Chin et al., 2008) and removed from rivers (Wohl, 2014).

Process-based restoration promotes riparian management strategies that preserve or enhance forested corridors and avoid removal of instream wood (Roni et al., 2013a). Such strategies include, where appropriate, reintroduction of beavers to river systems and the use of beaver dams in restoration (Pollock et al., 2014). Typically, beaver dams are seen as unnecessary impediments to flow, and beavers themselves are viewed as nuisance rodents; however, the geomorphological and ecological benefits of beaver dams are similar to those of large wood accumulations (Wohl, 2015b). More interventionist approaches to wood management involve implementing designed wood structures known as engineered logjams into streams (Abbe and Brooks, 2011). Through proper design and placement, these engineered structures can serve dual roles of erosion control and habitat enhancement.

16.6.4 How Does Dam Removal Affect Rivers?

Although controversial (Fox et al., 2016; Sneddon et al., 2017), the removal of dams from rivers has become increasingly common. Over the past three decades, more than 1500 dams have been removed from streams and rivers throughout the United States (American Rivers, 2019). The vast majority of these structures have been low-head dams or run-of-river dams. The desire to restore the physical and ecological connectivity of rivers is an important reason why dams are being removed as part of process-based river restoration (Magilligan et al., 2016), but diminished function, costly maintenance, and safety concerns have been the main motivating factors (Foley et al., 2017a).

An important contribution of geomorphological research in relation to the practice of dam removal has been to shed light on how rivers respond to this practice so that negative consequences can be anticipated and avoided. Dam removal changes river hydraulics and can mobilize stored sediment following dam removal, resulting in geomorphological adjustments of rivers (Foley et al., 2017b; Major et al., 2017). These adjustments, if pronounced, can impact ecological conditions. As long as the severity and duration of instability do not substantially limit ecological recovery, the perceived long-term ecological gains of dam removal typically are favored relative to the short-term costs (Stanley and Doyle, 2003). However, few long-term studies of change in ecological conditions resulting from dam removal have been conducted, and a critical need exists for more research to confirm the ecological value of this restoration strategy (Foley et al., 2017b).

The geomorphological response of rivers to removal generally is the opposite of the response to dam construction (see Chapter 15): stored sediment upstream is eroded and moves downstream where, depending on the transport capacity of the river system, it may be deposited (Pizzuto, 2002). The magnitude and rate of response vary depending on the volume of stored material, the texture of this material, the rate of release, and the background sediment flux. The volume of stored material is largely determined by the height of the dam and corresponding size of the reservoir, whereas the rate of release of stored material is influenced by the method of dam removal (Major et al., 2017; Foley et al., 2017b). Sudden, complete removals release sediment

at higher rates than phased, partial removals. Background sediment flux is related to the transport capacity of the river and the rate of sediment supply from the watershed, both of which reflect broadly the geographic setting of the river.

16.6.4.1 What Geomorphic Responses Occur Upstream of the Removal Site?

Removal of small run-of-river dams with little or no stored sediment within the channel upstream (e.g., Csiki and Rhoads, 2014) may not trigger detectable geomorphological responses. On the other hand, sudden removal of a high dam can result in a wave of channel incision migrating upstream as a knickpoint or series of knickpoints followed by subsequent bank collapse and channel widening (Doyle et al., 2003). Aggradation may ensue within the widened channel as eroded sediment from the upstream-migrating wave of incision is flushed downstream. This sequence is essentially identical to the channel evolution model associated with channelization or other mechanisms that locally decrease base level (Figure 16.12). Phased removals produce less dramatic responses, particularly if the wave of incision associated with each phase is allowed to progress far upstream before a subsequent phase of removal is initiated (Foley et al., 2017b).

The release of sediment associated with incision tends to decrease nonlinearly with the large initial pulse followed by steadily decreasing amounts as both the rate of upstream migration of incision slows and the height of the knickpoint decreases (Foley et al., 2017b). Lateral migration contributes to sustained excavation of reservoir deposits as bars develop within the incised channel. Following the sudden breaching of a coffer dam constructed to remove the 15 m high Marmot Dam on Sandy Creek in Oregon, a knickpoint several meters high migrated upstream about 150 m in the first 50 minutes (Major et al., 2012). Rapid channel incision and widening resulted in about 17% of the total sediment volume of the reservoir being removed within the first 60 hours and 40% of this volume being removed over eight weeks (Major et al., 2012). The height of the knickpoint and its rate of migration decreased progressively over time, producing a corresponding nonlinear decrease in the rate of eroded reservoir deposits. After two years, about 60% of the total sediment volume had been removed (Figure 16.13).

Over the long term, continued excavation of reservoir deposits also depends on grain size (Foley et al., 2017b). In general, cohesive fine-grained deposits that rapidly become vegetated following subaerial exposure stabilize quickly and are difficult to excavate completely. The persistence of fine-grained deposits behind breached mill dams that failed decades or even centuries ago attests to the resistance of these deposits to mobilization (see Chapter 15). On the other hand,

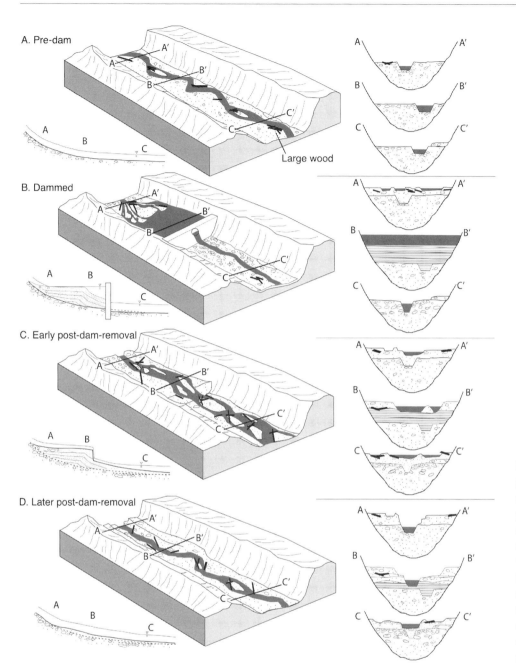

A. Pre-dam

Large wood

B. Dammed

C. Early post-dam-removal

D. Later post-dam-removal

Figure 16.12. Conceptual model of geomorphological adjustments of a gravel-bed river to dam placement and removal showing changes in longitudinal profile (left of block diagrams) and changes in valley/channel characteristics (right) at locations upstream (A), near (B), and downstream (C) of the dam location. (a) Predam, (b) dammed, (c) early post–dam removal, (d) later post–dam removal (from East et al., 2018).

deposits consisting of noncohesive sand and fine gravel remain erodible over long periods and may contribute to the sediment load of a river for many years or decades. As the input of sediment to the incised river upstream of the removed dam gradually declines, morphological characteristics of the channel and floodplain should become adjusted to the background sediment flux. This adjustment can occur fairly quickly; in the case of the Marmot Dam removal, channel dimensions and gradient upstream of the removed dam exhibited little change over time after two to three years (Major et al., 2012). Terraces consisting either of stabilized, unexcavated reservoir deposits or of reworked alluvium deposited during successive waves of

incision, widening, and aggradation may flank the channel-floodplain system (Figure 16.12).

16.6.4.2 What Geomorphic Responses Occur Downstream of the Removal Site?

The downstream response to dam removal typically involves channel aggradation, bar development or growth, filling of pools, and changes in channel width, gradient, and bed-material texture (Major et al., 2017). The downstream effect of dam removal depends strongly on the texture of eroded reservoir deposits. If these materials consist of fine-grained silts and clays that are readily transported as wash load,

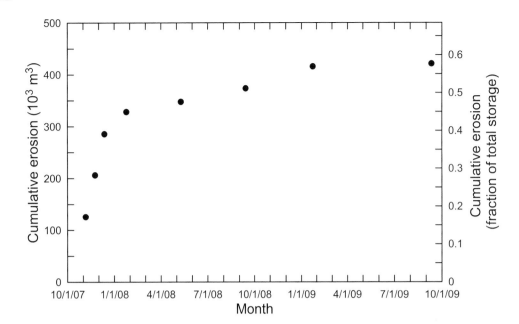

Figure 16.13. Cumulative erosion of reservoir sediment following the breaching of an earthen coffer dam on October 19, 2007 during the removal of Marmot Dam (from Major et al., 2012).

downstream effects are generally minimal. On the other hand, the flushing of large amounts of sand and gravel downstream as bedload or coarse suspended bed-material load can promote channel aggradation, bar formation or growth, pool infilling, embedding of gravel matrices with sand, or burial of static armor layers produced by sediment-starved pre-removal flows (Major et al., 2017; East et al., 2018). If aggradation and bar formation are pronounced, the channel may become braided and widen through bank erosion (East et al., 2018). Channel slope adjustments downstream occur in conjunction with gradient change caused by incision through the reservoir reach to produce a uniform longitudinal profile. Major adjustment typically occurs upstream of the removal site where a new channel incises into the relatively flat reservoir deposits. Slope adjustments downstream occur mainly in response to accumulation of eroded sediment. Large accumulations of eroded sand and gravel downstream of the removed dam can produce an increase in the channel gradient to sustain transport of this coarse material.

Short-term fluvial adjustments downstream of a removed dam will reflect the influence of mobilized reservoir sediment on channel form. In cases where eroded reservoir deposits are transported as bed-material load, the magnitude of downstream channel change and the timescale of these adjustments depend on the volume of eroded material introduced to the river relative to the capacity of postremoval flows to transport this additional material (Major et al., 2017). When the volume of introduced material is small relative to prevailing transport capacity, channel change may be pronounced immediately following removal, but recovery occurs quickly, often within a few years (Major et al., 2012). When the volume of introduced material is large relative to transport capacity, abundant material is likely to accumulate downstream, producing major channel change. Removal of material may require many years or even decades (East et al., 2018). As downstream sediment is gradually mobilized and moved downstream, channel characteristics below the removed dam should evolve toward relatively stable morphological characteristics adjusted to the background sediment flux. During recovery, net aggradation may occur immediately downstream of the removed dam, particularly if interruption of bed-material transport by the dam induced downstream incision (Major et al., 2012).

The distance downstream over which adjustments occur is contingent on the degree to which postremoval flows can entrain and transport different-size fractions of the introduced material along the river. When large proportions of the introduced material are relatively fine (e.g., sand) and can be transported substantial distances downstream, effects are generally diffuse but can extend far from the removal site. When large proportions of the introduced material are relatively coarse (e.g., gravel) and mobility is limited, effects are typically concentrated a short distance downstream from the removal site (Major et al., 2017).

The growing interest in removing dams has prompted the need to predict morphodynamic responses and changes in sediment load following removal. The capacity to predict these responses is useful for formulating removal strategies as well as anticipating and managing negative impacts. Predictive models have been developed to address this need (Cui et al., 2006, 2017; Cantelli et al., 2007; Ding and Langendoen, 2018); however, model calibration and testing are limited by the small number of dam removal cases where high-quality data have been collected. Additional postremoval studies are needed to evaluate and refine the predictive accuracy of dam-removal models.

16.6.5 What Are the Challenges of Stream Naturalization?

Challenges to environmental management of rivers are considerable in geographical settings in which entire landscapes have been substantially altered by humans at the watershed scale. Examples of such settings include intensively managed agricultural landscapes (Kumar et al., 2018), such as those found throughout the midwestern United States, and expansive urban environments, such as those associated with major cities. In these settings, what can be considered the "natural" environment now and into the future consists of human-modified landscapes that differ dramatically from preexisting unmodified landscapes. Runoff and sediment regimes produced by changes in land cover and subsurface drainage differ dramatically from those that existed prior to human modification. Changes in land cover have involved widespread clearing of native vegetation and replacement of this vegetation with cropland or paved surfaces. Subsurface drainage modifications include installation of drainage tiles underlying agricultural fields or storm sewer networks underlying urban areas. Streams have been extensively channelized both to accommodate changes in hydrological conditions and to maximize the amount of land, which is of high value, available for human use. Water quality often has been radically altered through the input of particulates, contaminants, or nutrients into streams. Moreover, given the need for persistent, intensive management of human-dominated landscapes to maintain the economic value of these landscapes, human interaction with the geomorphological, hydrological, ecological, and chemical processes of rivers and streams has, for all intents and purposes, become part of the natural dynamics of these landscapes. In other words, social, political, and economic processes now strongly influence and are influenced by river biophysical processes. Thus, consideration of how rivers function in such environments must explicitly be embedded in a societal context (Petts, 2007).

Although the term "stream restoration" is sometimes used to describe efforts to address environmental-quality concerns of streams in human-dominated landscapes (Bernhardt et al., 2007), use of this term in these contexts, where the return to a predisturbance condition that existed in the absence of humans is practically impossible, can contribute to false impressions of what is feasible (Cockerill and Anderson, 2014). The concept of stream naturalization has therefore been proposed as an alternative to describe attempts to environmentally manage streams in these challenging settings (Rhoads et al., 1999; Wade et al., 2002). Naturalization explicitly embraces both the embedding of social processes within stream dynamics and the need to manage without contemporary examples or historical information to guide the selection of appropriate reference conditions. It also accepts that reference conditions, even if they could be identified, are likely not sustainable in the contemporary environmental setting (Rhoads and Herricks, 1996). This situation has led to the call for environmental management of rivers in human-dominated environments to focus on desired ecosystem services rather than on establishing or reestablishing a particular reference condition (Dufour and Piegay, 2009).

16.6.5.1 How Are Ecosystem Services Relevant to Naturalization?

Four broad categories of ecosystem services can be defined: provisioning, regulating, supporting, and cultural (Millennium Ecosystem Assessment, 2005) (Table 16.11). The challenge of a services-oriented approach to the management of ecosystems, including rivers, is to balance considerations among these different services so that critical supporting and regulating services maintain the functionality of the river system to meet demands placed on provisioning and cultural services by human societies (Yeakley et al., 2016). Meeting this challenge involves scientific understanding of how processes underlying the different services interact. Thus, the need to develop an applied design science as the foundation for an ecosystem-services approach to river management remains crucial, as it does in a reference-based approach to management.

Streams in human-dominated environments often have relatively low-quality supporting and regulating services, which are governed mainly by hydrological, geomorphological, and ecological processes. This quality assessment is based on public perceptions that these services are not operating as desired to maintain preferred provisioning and cultural services. In many urban settings, flows are flashy, erosive, and polluted. Channels are straight and featureless. The threat of flooding looms large because little or no undeveloped space is available to accommodate overbank flows. Few aquatic organisms live in urban streams, and those that do must be highly tolerant of environmental stress. Such conditions are ever more frequently being viewed as undesirable in relation to public preferences about provisioning or cultural services. The desire for clean water, an uncontaminated environment, attractive landscape features, educational and recreational opportunities, and healthy, biologically diverse streams can motivate human actions that seek to undo consequences of past actions and enhance ecosystem services. The goal of enhancing ecosystem services

Table 16.11. Ecosystem services and rivers.

Service	Definition	Examples
Provisioning	Benefits from products provided by ecosystems	Drinking water Domestic use (e.g., sanitation) Industrial use (e.g., manufacturing) Agricultural use (e.g. irrigation) Transportation/navigation Building material (e.g., sand and gravel mining) Food (e.g., fish) Energy production (hydropower)
Regulating	Benefits from regulation of ecosystem processes	Flood attenuation by floodplain storage Sediment attenuation by floodplain storage Dilution of point source inputs Riparian control of river erosion potential Mutual adjustments of process and form following disturbance
Supporting	Benefits obtained from natural processes that support ecosystem development and formation	Soil production on floodplains Nutrient cycling Food web and predator/prey dynamics Physical habitat development and maintenance Hyporheic exchange
Cultural	Nonmaterial benefits obtained from ecosystems through value-laden personal or group experiences	Recreation Aesthetics Inspiration Sense of place Education Community pride

can be accomplished in myriad ways, but achieving it should be guided by underlying knowledge that helps to determine what is feasible. The desire to move the system away from its existing condition with limited ecosystem services toward a condition that increases the scope and benefits of these services is entirely consistent with the goal of stream naturalization (Rhoads et al., 1999).

Attempts to improve ecosystem services of headwater streams in human-dominated agricultural environments of the midwestern United States are greatly constrained by land-use practices that reflect valuation of some cultural and provisioning ecosystem services over others. In east-central Illinois and other areas of the Midwest, these headwater channels consist largely of drainage ditches that either have been excavated to extend the original drainage network headward or represent channelized streams that existed prior to European settlement (Rhoads and Herricks, 1996). The ditches provide outlets for tile drainage systems underlying farm fields. Thus, the original drainage system has been modified because of the high value placed on the provisioning service of food production and the need for adequate land drainage to support this service. Ditches are maintained on a regular basis to remove accumulated sediment and preserve straight trapezoidal channels with flat bottoms.

A major consequence of widespread channel modification and maintenance is the elimination of physical habitat, particularly geomorphological structure at the planform, bar-unit, and bar-element scales. A rich body of historical information on fisheries in east-central Illinois indicates that the lack of physical habitat is the most important factor limiting fish community characteristics (Frothingham et al., 2002). Because hydrological and water-quality conditions remain satisfactory for most fish species, improvement of species richness depends mainly on increasing the diversity of physical habitat. A wide range of possible strategies could be adopted to engage in naturalization of drainage ditches to increase habitat diversity (Figure 16.3); however, the value placed on the need to preserve the efficiency of land drainage restricts the possibilities. The total biomass of fish and the diversity of fish species are greater in meandering channels than in straight trapezoidal ditches (Frothingham et al., 2001; Rhoads et al., 2003), but allowing channels to meander is viewed as an unacceptable practice that will compromise drainage efficiency and also result in loss of farmable land.

16.6.5.2 What Are Some Examples of Stream Naturalization?

When channelization produces relatively wide ditches, inefficient transport of coarse-bed material load at low flow eventually leads to the development of vegetated depositional benches and a meandering low-flow channel within the ditch (Figure 15.22). Characteristics of physical habitat and fish communities in such ditches are of similar quality to those found in meandering streams (Rhoads et al., 2003; Rhoads and Massey, 2012). One common approach to naturalizing drainage ditches is to build or reconfigure these channels into a two-stage form with an inset low-flow channel flanked by a bench or floodplain on the bottom of a deep, wide ditch

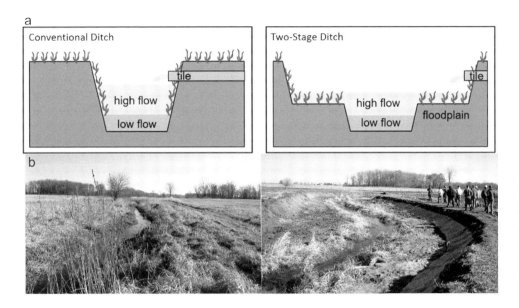

Figure 16.14. (a) Diagram illustrating the difference between a conventional and a two-stage drainage ditch. (b) Shatto drainage ditch in Indiana before (left) and after (right) conversion into a two-stage ditch (courtesy of Jennifer Tank).

(Powell et al., 2007) (Figure 16.14). Following implementation, the inset channel and benches are allowed to adjust through erosion and deposition caused by subsequent flows. Postproject appraisals indicate that this approach to stream naturalization increases the availability of physical habitat and the species richness of fish (DeZiel et al., 2019), enhances removal of nutrients and sediment from streamflow (Davis et al., 2015; Hanrahan et al., 2018), and does not compromise the overall geomorphological stability of ditches (D'Ambrosio et al., 2015; Krider et al., 2017). This naturalization strategy represents a viable method for enhancing the hydraulic, geomorphological, and ecological diversity of ditches, thereby improving ecosystem services while maintaining ditch functionality for land drainage.

In urban settings, constraints on naturalization are somewhat similar, but for slightly different reasons. Many urban channels have straight, trapezoidal forms with flat bottoms. Reconfiguring the planform of these human-modified channels is often constrained by built infrastructure that extends to the margins of the channels. The need to connect urban streams to storm-sewer outfalls also limits the capacity to create nonlinear channel planforms. Naturalization of a 900 m reach of a straight urban stream in the suburbs of Chicago involved creating pools and riffles on the channel bed to enhance physical habitat, regrading eroding channel banks, and revegetating the banks (Figure 16.15). The design for the pool-riffle structures was based on field evaluation of existing hydrological, hydraulic, geomorphological, and ecological conditions, as well as physical and numerical modeling of hydraulic performance of the channel with and without the pool-riffle structures (Wade et al., 2002; Rhoads et al., 2008). Postproject evaluation indicated that the project locally increased fish abundance, biomass, and diversity for two years following implementation, but the improvement was modest and did not compare favorably with fish characteristics in nonurban streams (Schwartz and Herricks, 2007). On the other hand, the pool-riffle structures were self-sustaining and remain relatively unchanged seven years after implementation of the project (Rhoads et al., 2011).

The limited ecological benefits obtained from local habitat-improvement projects highlight the challenge of enhancing ecological conditions in urban streams (Stranko et al., 2012). Other factors, such as poor water quality, lack of refugia from extreme in-channel hydraulic stresses during stormwater runoff events, and barriers to movement that disconnect habitats from sources of organisms, can completely undermine local efforts to improve the ecology of urban streams by enhancing physical habitat. Watershed-scale approaches to urban stream naturalization may be required in these contexts, but such efforts constitute complicated political endeavors that require coordination of goals and commitment of resources among a diverse array of stakeholders – a level of cooperation that realistically is difficult to achieve.

16.6.6 What Is the Role of Uncertainty in Environmental Management of Rivers?

The general goal of reestablishing important geomorphological, hydrological, hydraulic, and ecological functions within a river system where those functions have been lost or severely impaired by human activity is not an easy endeavor. In many ways, efforts to achieve this goal involve the opposite of traditional engineering, which has typically sought to simplify systems to achieve some measure of control over rivers to produce predictable, desired types of responses. The construction of dams, for example, allows river flow to be regulated

Figure 16.15. West Fork of the North Branch Chicago River in downtown Northbrook Illinois before (a) and after (b) stream naturalization.

and controlled to produce hydroelectric power or to facilitate navigability. In contrast to traditional river engineering, environmental management practices such as restoration, naturalization, and mitigation represent attempts to undo constraints imposed on river processes by past human practices and in this sense constitute forms of deengineering of rivers. The goal is to increase complexity of the river system rather than simplify it. This approach to river management seeks to replicate, at least to some extent, integration among biophysical processes that would develop within the river system if left undisturbed. Successfully achieving this goal is fraught with uncertainty, but uncertainty is not unique to environmental management of rivers. It is typical of all efforts to manage environmental systems and needs to be embraced and accepted.

From a scientific perspective, uncertainty in environmental management of rivers is normally associated with limited knowledge and understanding, which constrain the capacity to predict outcomes (Graf, 2008; Wheaton et al., 2008). This type of uncertainty is potentially reducible through continued research on river systems that produces new knowledge and understanding. The degree to which uncertainty of this type can be reduced depends on the pace of knowledge acquisition and also on the extent to which the dynamics of river systems are characterized by inherent randomness or deterministic chaos (Phillips, 1996), which may limit predictability even with nearly perfect or complete knowledge.

Within an applied design-science context, predictive models often serve as useful tools for guiding management that seeks to establish a priori how certain management strategies will yield certain outcomes. Natural Channel Design represents one such predictive model. Considerable potential exists to move beyond NCD using predictive process-based models,

including rational regime models (Millar and Eaton, 2011), one-dimensional channel evolution models (Langendoen, 2011), and two-dimensional morphodynamic models (Schuurman et al., 2016; Williams et al., 2016a). At present, most reach-scale, process-based approaches to design simply aim to predict the channel size, dimensions, and gradient required to transport the supplied sediment with the available water, given information on the sediment and water supply (Shields et al., 2003; Wilcock, 2012) – an approach consistent with rational regime models. Most modeling of this type relies on the concept of dominant discharge (see Chapter 6) to predict analytically characteristics of a stable channel with a mobile bed (Doyle et al., 2007; Soar and Thorne, 2011). On the other hand, process-based models capable of simulating channel dynamics may be especially useful for evaluating uncertainty related to inherent river dynamism, particularly where the goal of management is to allow fluvial processes to operate relatively unconstrained. Inclusion of effects of vegetation (van Oorschot et al., 2016) and flow variability (Williams et al., 2016b) into modeling scenarios can provide insight into how geomorphological processes interact with hydrological and biotic processes. Models are subject to uncertainty, particularly when applied in specific contexts to predict exact morphodynamic responses of particular rivers. Models often are treated with reverence and accorded unreasonable confidence (Stewardson and Rutherford, 2008). Instead, sources of uncertainty in modeling should be carefully considered and model predictions interpreted cautiously based on evaluations of model uncertainty.

An important way to reduce uncertainty related to knowledge limitations is to engage in analysis of implemented projects to evaluate performance in relation to project goals (Downs and Kondolf, 2002). Postproject appraisals are most informative when data collection also occurs prior to implementation. The collection of baseline data should be targeted toward aspects of the project immediately relevant to project goals. The baseline data should also be consistent with data collected during the postproject appraisal to provide the basis for before and after comparisons. Even more rigor can be added by monitoring a nearby control site that is not treated to ensure that any changes detected before and after project implementation are not the result of environmental change rather than a treatment effect (Roni et al., 2013b).

Rigorous, detailed postproject appraisals can yield data that inform how well a particular project achieved its intended goals. The feedback between design and implementation is an important element in the adaptive process of refining science and technology to support successful approaches to environmental management of rivers (Figure 16.1). The effectiveness of different approaches cannot be adequately discerned without such appraisals. Lessons learned from postproject appraisals provide the basis for assessing how well current scientific understanding can be translated into effective a priori management plans. Through such lessons, the knowledge base for restoration can be refined to reduce uncertainty. Unfortunately, many implemented projects do not include postproject appraisals (Bernhardt et al., 2007). When postproject appraisals are conducted, the focus is often on short-term performance. Given that rivers are dynamic systems, various components of implemented projects are likely to have finite lifespans (Schwindt et al., 2019) and the benefits of the projects will likely change over time (Kail et al., 2015). Monitoring over many years or even decades is necessary to evaluate long-term benefits. A long-term perspective is especially important in the light of potential changes in environmental conditions associated with human-induced climate change.

Besides limitations in knowledge and understanding, a variety of human factors add to uncertainty in river management projects. Effective communication is vital for all aspects of these projects, from initial formulation of the need for action to implementation and long-term monitoring. Breakdowns in communication add to uncertainty through failures to convey important information throughout the chain of decision-making (Graf, 2008). As a result, uninformed decisions may be made that can compromise the success of the project by not designing or implementing it in an appropriate manner to achieve project goals. Of particular relevance to the scientist or technical expert is the importance of effectively communicating with decision-makers during project planning about important technical aspects of a project that will facilitate the accomplishment of project goals. A breakdown at this stage can prevent important technical information from being properly incorporated into the project. Another common point of potential miscommunication is at the stage of project implementation, which often requires interaction between a contractor and technical experts responsible for the project design. Miscommunication at this stage can result in improper implementation of the project, with an increase in uncertainty that the project will perform as expected.

Social, cultural, political, and economic factors also add to uncertainty. Decision-makers need to account for public opinion, cultural preferences, political forces, and cost when weighing options about details of projects. Competing factions and voices can greatly complicate how various aspects of a project are conceived, deliberated, and finalized (McDonald et al., 2004). Scientific and technical information sometimes plays only a minor role in the outcomes of this complicated, primarily social process. Disconnect can occur between what is desired and technical aspects of how this desire can be met, or even whether it is feasible to fulfill this desire based on an understanding of river dynamics. As a result, projects may be implemented that lack an adequate scientific and technical basis to achieve objectives, or, even worse, the implemented project may have objectives that cannot be realistically achieved. Differences in opinion about outcomes constitute another source of uncertainty. Evaluations of the success of a project are prone to human subjectivity (Bernhardt et al., 2007; Jaehnig et al., 2011). What may be considered successful based on improved appearance may not be judged successful based on ecological criteria.

Appendix A Power Functions in Fluvial Geomorphology

Many quantitative relationships between variables in fluvial geomorphology are expressed as power functions. A power function relates a dependent variable (y) to an independent variable (x), where x is multiplied by a constant a and raised to a power b:

$$y = ax^b \qquad (A.1)$$

The concept of allometry provides a useful framework for examining power-function relationships (Mark and Church, 1980). Allometry examines the relative rates of change of two variables. For x and y, let dy/dt be the rate of change of variable y over time and dx/dt be the rate of change of variable x over time. The relative rates of change describing the rates of change relative to the magnitudes of each variable are $(dy/dt)/y$ and $(dx/dt)/x$. Assuming the relative rate of change in y is a constant proportion of the relative rate of change of x, then

$$(dy/dt)/y = b(dx/dt)/x \qquad (A.2)$$

where b is a constant of proportionality. Multiplying both sides by dt yields

$$(1/y)dy = b(1/x)dx \qquad (A.3)$$

Integrating both sides:

$$\int (1/y)dy = b\int (1/x)dx \qquad (A.4)$$

gives

$$\ln(y) + c_1 = b\ln(x) + c_2 \qquad (A.5)$$

where c_1 and c_2 are constants of integration. This equation can also be written as

$$\ln(y) = \ln(a) + b\ln(x) \qquad (A.6)$$

where $\ln(a) = c_2 - c_1$. Taking antilogs of this equation yields Eq. (A.1). This derivation shows that a power function expresses the relative rate of change of y to the relative rate of change in x. Although the derivation considers relative rates of change in y and x over time, it also applies to relative rates of change of two variables over space.

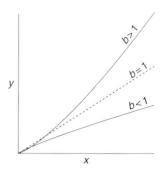

Figure A.1. Form of power-function relations on arithmetic x and y axes for different values of the exponent b.

The exponent b describes the nature of relative rates of change (Figure A.1). If $b > 1$, the value of y increases at an increasing rate with increasing x. If $b < 1$, the value of y increases at a decreasing rate with increases in x. If $b = 1$, y increases as a constant (linear) proportion of x. Thus, for non-zero values of b, unless $b = 1$, power functions define nonlinear relations between two variables. When a bivariate relationship conforms to a power function with $b \neq 1$, data for the two variables plotted in relation to arithmetic x and y axes will exhibit curvilinear patterns. The same data plotted on a graph with log-log axes will exhibit a linear pattern.

When the value of the exponent in a power function balances the units on each side of the equation so that the coefficient a is dimensionless, the relation is referred to as isometric. If x and y both have dimensions of length, this condition is achieved when the exponent has a value of 1; in other words, when the relationship is linear. On the other hand, if x is measured in units of area (L^2) and y is measured in units of length (L), an exponent of 0.5 is required for the relationship to be isometric ($(L^2)^{0.5} = L$). In an isometric relationship, x and y change in linear proportion to one another. Such a relationship is scale invariant and defines a state of self-similarity. Under these conditions, the proportionality, or ratio, of x and y remains constant as the magnitude of x changes.

Most relations between fluvial geomorphic variables can be expressed as power functions because these variables have log-normal rather than normal probability

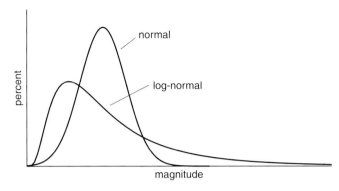

Figure A.2. Normal versus log-normal probability distributions. Note the long tail of extreme positive values for the log-normal curve, which is typical for geomorphic variables that have lower bounds of zero (e.g., flood discharges can be very large in size but cannot have negative values – the lowest rate of flow in a river is zero).

distributions (Figure A.2). A log-normal distribution is the probability distribution of any random variable whose logarithm is normally distributed. This type of distribution arises because many geomorphic variables have definite lower bounds of zero (i.e., their values cannot be negative) yet have less restrictive upper bounds. Thus, the distributions of these variables are skewed. By computing log-transformed values of two variables with log-normal distributions, the variables can be related to one another in linear form (Eq. (A.6)). Statistically, linear regression analysis or other linear best-fit procedures can be applied to log-transformed values of two log-normally distributed variables (Eq. (A.6)) to determine values of a and b in the corresponding power function (Eq. (A.1)).

Appendix B Characterization of Fluvial Sediment

Samples of fluvial sediments, whether obtained from the bed, banks, or floodplain of a river system, contain a mixture of different grain sizes. Collectively, these different sizes of particles define a grain-size distribution of the sample. This distribution is expressed in the form of the frequency of occurrence, either by weight or by number, plotted against the size of the particles. A histogram consists of vertical bars that show the percentage of particles (by weight or number) in intervals of grain size (Figure B.1). A cumulative frequency curve is formed by summing the percentages by interval depicted in the histogram; it indicates the percentage of particles finer than a given diameter (Figure B.2). The mode is the most commonly occurring grain size, or the peak of the histogram. If the distribution has two or more peaks, it is multimodal (e.g., bimodal or trimodal) (Figure B.1). The median particle size divides the cumulative frequency distribution into two equal parts: 50% of the particles are coarser than this size and 50% are finer than this size. Values of percentile grain sizes are determined easily from the cumulative frequency distribution by reading from the appropriate percentile over to the curve and down to the corresponding grain size. The sizes associated with specific percentiles are symbolized as d_i, where i is the percentile. For example, d_{50} is the grain size corresponding to the 50th percentile, or the median particle size, and d_{84} is the grain size corresponding to the 84th percentile; i.e., 84% of the particles by weight or number are finer than this grain size (Figure B.2).

Many fluvial samples have log-normal, rather than normal, distributions (Figure B.3). A normal distribution has a probability density function, a continuous representation of the histogram, that is symmetrical. A log-normal distribution exhibits skewness toward either large or small values. A log-normal distribution is easily converted into a normal distribution by transforming the data into logarithmic values. A commonly used transformation is the phi grain size measure (Folk, 1980):

$$d_\phi = -\log_2 d \tag{B.1}$$

$$\log_2 d = (\log_{10} d / \log_{10} 2) \tag{B.2}$$

where d is the grain size in mm. Note that this measure is the log base 2 transformation and that negative values correspond to coarse sediments. The phi system is a geometric progression with a reference value of 0 at 1 mm (i.e., $-\log_2(1) = 0$).

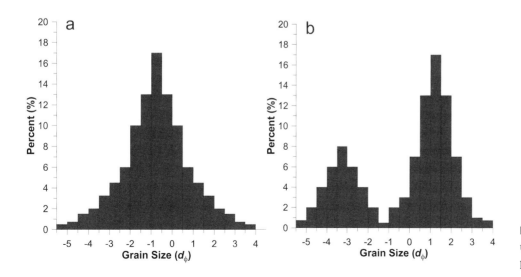

Figure B.1. Histograms for unimodal (a) and bimodal (b) particle-size distributions.

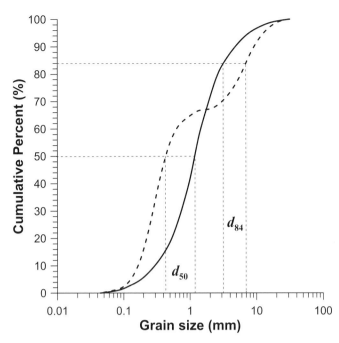

Figure B.2. Cumulative curves for the unimodal (solid line) and bimodal (heavy dashed line) particle-size distributions in Figure B.1. Light dashed lines illustrate values of d_{50} and d_{84} for each distribution.

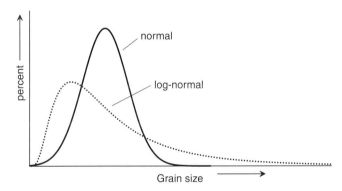

Figure B.3. Normal versus log-normal grain-size distributions.

The phi scale corresponds to the descriptive Wentworth (1922) size classification (Table B.1) and variants on this scheme (Blott and Pye, 2012).

The fraction by number in specific size intervals typically is determined for samples consisting entirely of coarse particles (8 mm and larger) because the amount of material by weight required to adequately represent all size fractions of this type of sediment is prohibitively large (Church et al., 1987; Gale and Hoare, 1992). Sampling is conducted at low flow or when the bed is exposed. A grid is established over the area of interest, and particles are sampled from the bed at grid nodes. This method is often referred to as pebble counts,

because it focuses on collecting a number of particles and measuring the size of the particles using an appropriate measuring device, such as a caliper. The size of each particle is determined based on a measurement of the diameter of its intermediate (*b*) axis, and all the individual particles are assigned to size classes based on the measured diameters. Particles can also be sampled using areal methods, such as collecting all particles on the surface within a defined area of the bed or assessing particle sizes using photographs of the bed. In some cases, conversion factors are required to transform grain-size distribution data generated by different methods so that results of various procedures are comparable. Details on methods of sampling, sample size, sampling strategy, conversion factors, and particle analysis of river sediments can be found in Bunte and Abt (2001) and Kondolf and Lisle (2016).

Bulk samples of fine gravel and smaller material can be obtained using scoops, shovels, or various types of corers (Smith, 1984; Edwards and Glysson, 1999; Singer, 2008b; Smith and Elmore, 2014; Shchepetkina et al., 2015). The frequency in percentage by weight of fine gravel and sand is determined by sieve analysis (Guy, 1969). Mesh sizes of sieves generally correspond to size intervals in the phi scheme (Table B.1); thus, the weight of particles retained on each sieve provides data on the weight of particles in each size interval. The fraction by weight in each interval is computed by dividing the retained weight by the total weight of the sample. Particle-size analysis of fine sediments (<0.0625 mm) is performed using sedimentation (Guy, 1969; Gee and Bauder, 1986) or laser-diffraction techniques (Eshel et al., 2004; Yang et al., 2015b).

In the past, many statistics for grain-size distributions were calculated graphically from cumulative frequency curves (Folk and Ward, 1957; Folk, 1980). Today, these statistics can be readily calculated using spreadsheet software, obviating the need for graphical methods, which at best are approximate. Most commonly, geometric statistical moments are computed, including the mean, standard deviation, skewness, and kurtosis, given the tendency for fluvial sediment to exhibit log-normal, rather than normal, distributions.

For samples of coarse material involving measurements of particle diameters for a total number of particles (*n*), the geometric mean ($d_{\phi m}$) (phi units) is

$$d_{\phi m} = \sum_{j=1}^{n} d_{\phi j}/n \qquad (B.3)$$

where $d_{\phi j}$ is the diameter of particle *j* in phi units and *n* = the number of particles. The geometric mean (d_m) in mm is

Table B.1. Particle size and the Wentworth classification.

Sieve mesh (mm)	mm	μm	Phi	Wentworth (1922) size class	Main size fractions	Blott and Pye (2012)
	4096		−12			
	2048		−11			Megaclast
	1024		−10			Very large boulder
	512		−9			Large boulder
	256		−8	Boulder		Medium boulder
	181		−7.5			
	128		−7			Small boulder
90.0	90.5		−6.5			
63.0	64.0		−6	Cobble		Very small boulder
45.0	45.3		−5.5			
31.5	32.0		−5			Very coarse gravel
22.4	22.6		−4.5			
16.0	16.0		−4		Gravel	Coarse gravel
11.2	11.3		−3.5			
8.0	8.0		−3			Medium gravel
5.6	5.7		−2.5			
4.0	4.0		−2	Pebble		Fine gravel
2.8	2.8		−1.5			
2.0	2.0		−1	Granule		Very fine gravel
1.4	1.4		−0.5			
1.0	1.0	1000	0.0	Very coarse sand		Very coarse sand
0.71	0.71	710	0.5			
0.50	0.50	500	1.0	Coarse sand		Coarse sand
0.35	0.35	350	1.5			
0.25	0.25	250	2.0	Medium sand	Sand	Medium sand
0.18	0.18	180	2.5			
0.125	0.125	125	3.0	Fine sand		Fine sand
0.090	0.088	88	3.5			
0.063	0.063	63	4.0	Very fine sand		Very fine sand
	0.044	44	4.5			
	0.031	31	5.0	Coarse silt		Very coarse silt
	0.022	22	5.5			
	0.0156	15.6	6.0	Medium silt	Silt	Coarse silt
	0.0078	7.8	7.0	Fine silt		Medium silt
	0.0039	3.9	8.0	Very fine silt		Fine silt
	<0.0039	<3.9	<8.0	Clay		Very fine silt (8–9 phi)
						Clay (< 9 phi)

$d_m = 2^{-d_{\phi m}}$. Unbiased estimates in phi units of the sample standard deviation (σ_ϕ), skewness (sk_ϕ), and kurtosis (ku_ϕ) are computed as

$$\sigma_\phi = \sqrt{\sum_{j=1}^{n}(d_{\phi j} - d_{\phi m})^2/(n-1)} \tag{B.4}$$

$$sk_\phi = \frac{n}{(n-1)(n-2)}\sum_{j=1}^{n}\left(\frac{(d_{\phi j} - d_{\phi m})}{\sigma_\phi}\right)^3 \tag{B.5}$$

$$ku_\phi = \left[\frac{n(n+1)}{(n-1)(n-2)(n-3)}\sum_{j=1}^{n}\left(\frac{d_{\phi j} - d_{\phi m}}{\sigma_\phi}\right)^4\right] - \left(\frac{3(n-1)^2}{(n-2)(n-3)}\right) \tag{B.6}$$

Table B.2. Qualitative phi classes for sorting, skewness, and kurtosis.

Sorting (σ_ϕ)		Skewness (sk_ϕ)		Kurtosis (ku_ϕ)	
Very well sorted	< 0.35	Very fine skewed	> 1.30		
Well sorted	0.35 to 0.50	Fine skewed	0.43 to 1.30	Platykurtic	< 0
Moderately well sorted	0.50 to 0.70	Symmetrical	−0.43 to 0.43	Mesokurtic	≈ 0
Moderately sorted	0.70 to 1.00	Coarse skewed	−0.43 to −1.30	Leptokurtic	> 0
Poorly sorted	1.00 to 2.00	Very coarse skewed	< −1.30		
Very poorly sorted	2.00 to 4.00				
Extremely poorly sorted	> 4.00				

(adapted from Blott and Pye, 2001)

The mean in phi units for samples by weight is

$$d_{\phi m} = \sum_{i=1}^{j} f_i d_{\phi mdi} \tag{B.7}$$

where f_i is the fraction of the total weight of the sample in grain-size interval i (weight in interval i divided by the total weight of the sample), j is the total number of grain-size intervals, and $d_{\phi mdi}$ is the grain size in phi units of the mid-point of grain-size interval i. Values of sorting, skewness, and kurtosis are estimated as (Blott and Pye, 2001)

$$\sigma_\phi = \sqrt{\sum_{i=1}^{j} f_i (d_{\phi mdi} - d_{\phi m})^2} \tag{B.8}$$

$$sk_\phi = \frac{\sum_{i=1}^{j} f_i (d_{\phi mdi} - d_{\phi m})^3}{\sigma_\phi{}^3} \tag{B.9}$$

$$ku_\phi = \left[\frac{\sum_{i=1}^{j} f_i (d_{\phi mdi} - d_{\phi m})^4}{\sigma^4} \right] - 3 \tag{B.10}$$

The standard deviation provides information on the spread of the particle-size distribution and is often viewed as a measure of the degree of sorting of fluvial sediment by running water, where sorting refers to the extent to which deposition of transported sediment promotes uniformity of sediment size. Small standard deviations indicate well-sorted sediment in which the material is all of similar size, whereas large standard deviations are characteristic of poorly sorted sediment in which the deposit contains particles of many different sizes (Table B.2).

Skewness measures the symmetry of the particle-size distribution. Symmetrical distributions have a skewness close to zero (Table B.2). When particle size is expressed in phi units, distributions with negative skewness have long tails that extend toward coarse particle sizes (coarse-skewed), whereas those with positive skewness have tails that extend toward fine particle sizes (fine-skewed). Note that this pattern of skewness reflects the nature of the phi-unit system, wherein phi values decrease as particle size increases. If particle size is expressed in millimeters, positive skewness indicates a distribution with a long tail that extends toward large particle sizes.

Kurtosis refers to the extent to which the grain-size distribution is excessively peaked or flattened relative to a normal distribution. A distribution that conforms to the shape of a normal distribution is mesokurtic and has a kurtosis close to zero (Table B.2). When the kurtosis greatly exceeds zero, the distribution is excessively peaked or leptokurtic. An excessively flattened distribution has a kurtosis less than 1 and is referred to as platykurtic.

Shape is another important characteristic of grains that influences how fast a particle will fall through a fluid at rest, the drag exerted on the flow by the particle, and how easily the particle is entrained and transported by the flow. Measurements of the three major axes of a particle are

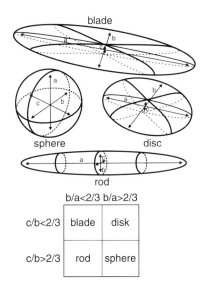

	b/a<2/3	b/a>2/3
c/b<2/3	blade	disk
c/b>2/3	rod	sphere

Figure B.4. Zingg classification of particle shape. Letters indicate subscripts for different particle axes.

used to define its shape, including d_a – the long axis, d_b – the intermediate axis, and d_c – the short axis. In fluvial studies, measurements of shape most often are obtained for coarse particles (gravels). The Zingg (1935) scheme is commonly used to classify particles by shape. This scheme uses ratios of d_b/d_a and d_c/d_b to categorize particles into four categories: spheres, rods, discs, and blades (Figure B.4). Quantitatively, grain shape can be evaluated as

$$SF = \frac{d_c}{\sqrt{d_a d_b}} \tag{B.11}$$

where SF is the shape factor. For a perfectly spherical particle $d_a = d_b = d_c$; thus, SF = 1. For natural sands, SF is approximately 0.7 (Dietrich, 1982).

Appendix C Measuring Discharge and Velocities in Rivers

The measurement of discharge (Q) in a river involves two basic components: 1) determining the geometry of the flow area at a cross section of the river and 2) determining the mean velocity of the flow through the cross section (Turnipseed and Sauer, 2010). These two components are integrated in the velocity-area method:

$$Q = \sum_{i=1}^{n} a_i \bar{u}_i \tag{C.1}$$

where a_i is the area of partial section i of the cross section, $\bar{u}_i$ is the depth-averaged mean velocity of this partial section, and n is the number of partial sections. This approach is implemented using the mid-section method (Figure C.1).

Endpoints of a cross section are established on both sides of a river, and the measurements proceed from one side of the river to the other across the flow. Measurements are obtained at observation points, or verticals (i), that constitute midpoints of rectangular partial sections. Generally, about 20–30 verticals are established and spaced so that no partial section contains more than 10% of the

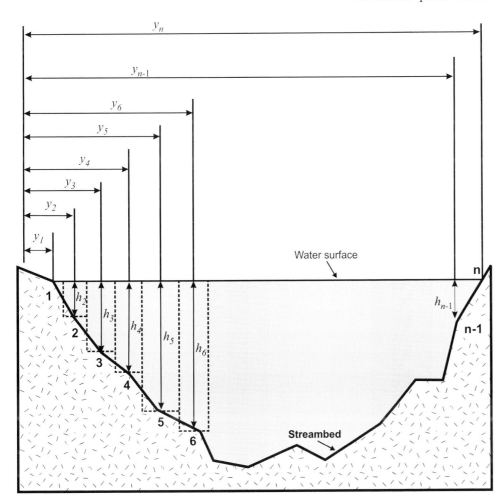

Figure C.1. Diagram illustrating mid-section method of discharge computation.

total discharge. Ideally, no partial section should contain more than 5% of the total discharge. The spacing of the verticals varies with depth and velocity, with the spacing decreasing with increasing depth and velocity. Horizontal distances (y_i) are measured from the starting endpoint to the position of each vertical, with verticals always located at each edge of water. At each vertical, the local depth of the flow (h_i) and the mean velocity over the flow depth ($\bar{u}_i$) are measured. Mean velocities over depth can be determined by 1) measuring numerous velocities throughout the water column, multiplying the velocities by the portion of the total depth each velocity represents, summing these products, and dividing the sum by the total depth; 2) measuring velocities at 2/10th and 8/10th of the total flow depth and computing the average of these two measurements; or 3) measuring a single velocity at 6/10th of the depth. The accuracy of the methods decreases as the number of measured velocities at each vertical decreases. Also, the two-point method and the 6/10th-depth method are based on the assumption that the velocity profile varies logarithmically over depth (Dingman, 2009) (see Chapter 4). These two methods can yield inaccurate estimates of depth-averaged velocities if the profile varies substantially from a logarithmic form.

Partial discharges for the first ($i = 1$) and last ($i = n$) partial sections are computed as

$$q_1 = \bar{u}_1 \left(\frac{y_2 - y_1}{2}\right) h_1 \tag{C.2}$$

$$q_n = \bar{u}_n \left(\frac{y_n - y_{n-1}}{2}\right) h_n \tag{C.3}$$

In many cases, h_1 and h_n will be zero, in which case q_1 and q_n will equal zero. If the edge of water is a vertical face, such as a concrete wall, measurements of $\bar{u}$ and h are obtained immediately adjacent to this face to produce a nonzero partial discharge. Eq. (C.2) and (C.3) also are used whenever water occurs on only one side of an observation point, such as at edges of islands. For verticals with water on both sides of the partial section defined by the vertical, the partial discharge is computed as

$$q_i = \bar{u}_i \left(\frac{y_{i+1} - y_{i-1}}{2}\right) h_i \tag{C.4}$$

The partial discharges for all partial sections are then summed to determine the total discharge for the cross section. The surface width of the flow (W) is $W = y_n - y_1$, assuming no islands occur within the cross section. The area of the flow (A) equals the sum of the partial areas of the sections. These partial areas equal the products of partial widths (second term) and depths (third term) on the right-hand side of Eq. (C.2)–(C.4). Mean depth (D) equals A/W, and mean velocity (U) equals

Q/A. The wetted perimeter of the cross section can be computed as

$$P = \sum_{i=2}^{n} \left\{ \frac{y_i - y_{i-1}}{\cos\left[\tan^{-1}\left(\frac{|h_i - h_{i-1}|}{y_i - y_{i-1}}\right)\right]} \right\} \tag{C.5}$$

and $R = A/P$.

Measuring tapes or surveying instruments, such as total stations or global positioning systems (GPS), are used to measure distances (y) to verticals. Depths (h) are determined using wading rods, survey rods, sounding lines and weights, or hydroacoustic sensors (fathometers).

A variety of instruments are available to measure velocities in rivers (Figure C.2). Mechanical current meters measure velocity through rotation of cups or blades. The meters have either vertical or horizontal axes of rotation. For vertical-axis meters, the flow rotates a set of cups, whereas for horizontal-axis meters, the flow turns a propeller consisting of a set of blades. Often these meters are connected electronically to a digital display that provides information on average velocity at the measurement location over a specific sampling duration.

Electromagnetic current meters operate on Faraday's Law. Water, an electrical conductor, moves through a magnetic field generated by the instrument, and the voltage induced by the flowing water is proportional to its velocity. These meters have no moving parts and are relatively immune to effects of small amounts of debris within the water.

Hydroacoustic instruments, including acoustic Doppler current profilers (ADCPs) and acoustic Doppler velocimeters (ADVs), are now used extensively in measurement of river flow. Both instruments transmit sound waves with a known frequency. These sound waves interact with particles in the flow and are reflected back to a receiver. Changes in the phase lag of the returns of multiple pulses of sound waves transmitted at a known time interval reflect changes in the movement of particles toward or away from the instrument. The change in phase lag over the time interval of transmission is proportional to the flow velocity, assuming that the particles move at the same rate as the flow. Many acoustic instruments have multiple transducers, which allow the velocity signal to be resolved into streamwise, lateral, and vertical components. These types of sensors provide information on three-dimensional fluid motion.

ADCPs transmit sound waves in a spreading pattern throughout the water column, and the resulting velocity data are arranged into discrete bins that are stacked into columns over depth. These columns of bins, which represent profiles of three-dimensional velocity components over the flow depth, are known as ensembles. As an ADCP moves across a river, it progressively produces ensembles, yielding abundant three-

Figure C.2. (a) An acoustic Doppler current profiler at the bottom of a boat mount with a GPS unit at the top of the mount. (b) An acoustic Doppler velocimeter. (c) A vertical-axis mechanical current meter. (d) A horizontal-axis mechanical current meter.

dimensional velocity data over an entire flow cross section. Measurements cannot be obtained close to the sensor head, a region known as the blanking distance, or near the bed of the river, where interference between acoustic signals degrades the velocity measurements. Software for the instruments contains algorithms for interpolating velocities in these regions for discharge calculations. A major advantage of ADCPs is that velocity, depth, and distance data received from an instrument mounted on a boat or on a small catamaran pulled across the river can be collated as it is received and the discharge calculated in real time using the velocity-area method. The acoustic signals generated by ADCPs provide information on flow depth and distance along a transect, which can be determined using a GPS mounted above the instrument. In this sense, ADCPs have revolutionized discharge measurements in rivers, which in the past, when mechanical current meters were used, was a time-consuming, labor-intensive endeavor.

In contrast to ADCPs, ADVs focus sound waves on a small sampling volume (≤ 0.125 cm^3) a short distance from the sensor head (5 to 10 cm). These instruments are capable of operating at high sampling rates (typically $\geq$ 25 Hz) necessary for studies of turbulence in rivers. Information on three-dimensional velocities can be used to characterize Reynolds stresses and turbulent kinetic energy (see Chapter 4). The high sampling rate is also useful for investigating the time-dependent behavior of turbulent flow in rivers.

Appendix D Measurement of Sediment Transport in Rivers

The measurement of sediment transport in rivers has involved separate sampling methods for suspended load versus bedload (Hicks and Gomez, 2016). In each case, techniques of measurement include direct and indirect sampling methods. Direct methods collect actual samples of the sediment in transport. Indirect methods use sensors that detect movement of solid material in transport but do not obtain samples of sediment.

Manual samplers used in direct sampling of suspended sediment include depth-integrating and point-integrating samplers (Figure D.1). Both these samplers consist of a streamlined metal housing containing a glass sample bottle. When lowered into the water either by hand or using a cable and reel system, a mixture of water and sediment enters an inlet nozzle at the front of the sampler and is captured in the bottle. On point-integrating samplers, the nozzle can be opened and closed to collect individual samples at particular depths in the flow. On depth-integrating samplers, the inlet nozzle cannot be closed. Individual samples are obtained by raising and lowering the sampler gradually until the bottle is nearly filled when it is moved down and up over the entire flow depth.

Automated pump samplers consist of a housing containing a pump, an intake tube that extends from the pump to the stream, and numerous bottles on a carousel within the housing. Samples of water and sediment are pumped from the stream into the bottles automatically through programmable controls on the sampler. These samplers are useful for measuring suspended sediment during storm events.

Processing of direct samples requires filtration of the water–sediment mixtures, drying of the filtered sediment samples, and determination of the weight of the dried sediment on a preweighed filter. The concentration of sediment is determined as the weight of the sediment per unit volume of water. Loads are determined by multiplying the mean concentration throughout the cross section of a river by the discharge at the cross section at the time of sampling. Thus, suspended-sediment concentrations should be representative of the mean concentration throughout the cross section of the river at which a discharge is measured.

Instruments used to indirectly measure suspended sediment in streams include optical, acoustic, and laser sensors (Voichick and Topping, 2014). Optical sensors determine the extent to which sediment in the water interferes with the transmission of light. These sensors are also known as turbidity sensors. Acoustic and laser sensors determine how emitted acoustic or laser signals are attenuated by particles within the water. The use of all three of these sensors requires calibration of the signals obtained from the sensors against manual measurements of suspended-sediment concentrations to generate rating functions that relate magnitudes of the signals to suspended sediment concentrations. Acoustic sensors include acoustic Doppler current profilers (ADCPs), which, although developed to measure water velocities, also have been used to estimate suspended-sediment concentrations (Latosinski et al., 2014).

Direct sampling of bedload has proved to be a difficult problem, given the potential for a sampler placed on the bed of the river to disrupt flow around it, producing a local effect that yields unrepresentative samples of bedload. Manual samplers consist of a metal housing with a rectangular inlet and a flared rear section to which a mesh bag is attached. Known as a pressure-difference sampler, this design maintains an entrance velocity equal to the ambient flow velocity and also produces a decrease in pressure at the rear of the sampler. Small handheld versions can be deployed by wading, whereas large versions require the use of a cable and reel system. The sampler is positioned on the bed for a fixed amount of time and bedload is collected in the mesh bag. The material is dried and weighed to determine the transport rate ($M\ T^{-1}$).

Bedload traps consist of lined trenches sunk into the streambed, the tops of which are flush with the bed. Openings allow material transported on the bed to fall into the trench. The material can be removed from the trap manually after a period of time, or, in sophisticated installations, pressure sensors can be used to weigh the material within the trench as it accumulates, or the material can be conveyed to a weighing device on the banks using pumps or a conveyor belt.

Figure D.1. Examples of a depth-integrating suspended sediment sampler (a), a pressure-differential bedload sampler (b), an optical turbidity probe (c), and an automatic pump sampler (d). Panel (e) shows the pump sampler installed by a stream with a solar panel for power.

Particle tracing determines the movement of individual particles as bedload and can be used to determine differences in the mobility of different-size clasts in a river (see Chapter 5). Estimates of bedload-transport rate are also possible if information on scour and fill is combined with clast tracing (Brenna et al., 2019). This method is most suitable where the size of bed material consists of coarse gravel.

Efforts to indirectly measure bedload transport have included impact sensors, where rates of particle motion are determined by impacts of particles on sensors, and passive acoustic sensors, where rates of particle motion are determined by noise generated by particles colliding with one another (Rigby et al., 2016). A promising indirect method is to use ADCPs to estimate bedload-transport rates (Latosinski et al., 2017). An ADCP determines motion of the sensor relative to the bed through a bottom tracking function, whereas the actual velocity of the sensor can be determined using an accurate differential global positioning system. When particles on the bed of the river are moving, the difference between the actual velocity of the sensor and the velocity determined by bottom tracking represents the velocity of the moving particles.

Symbols

Symbol	Definition	Units (SI) or Numeric Value
a	generic coefficient in power function or linear relation; coefficient in hydraulic-geometry relation for width	units dependent on function; $m^{1-3b}\ s^b$ in hydraulic-geometry relation
A	area of flow at a river cross section	m^2
A_{bk}	bankfull cross-sectional area	m^2
$A_{bk(res)}$	bankfull channel area of a reach to be restored	m^2
A_{br}	bar area	m^2
a_c	coefficient for scaling grain roughness length scale	dimensionless
A_d	drainage-basin area	km^2
$\overline{A}_{d\Upsilon}$	average drainage area of streams in an order Υ	m^2 or km^2
A_{ex}	excess cross-sectional area of deepest tributary channel at a confluence	m^2
A_f	amplification factor of bed perturbation	dimensionless
A_{fmax}	maximum amplification factor in stability analysis	dimensionless
a_i	partial area of flow at a river cross section	m^2
A_m	meander bend amplitude	m
A_o	reference drainage basin area	km^2
a_q	coefficient relating Manning's n to discharge	$s^2\ m^{-3.33}$
A_s	cross-sectional area of a particle	m^2
A_{st}	step amplitude of step-pool unit	m
A_T	total cross-sectional area of deepest tributary channel at a confluence	m^2
A'	scour factor in meander migration model	dimensionless
b	generic exponent in power function; exponent in hydraulic-geometry relation for width	dimensionless
b_{1D}	coefficient representing influence of sediment type and vegetation on channel depth	dimensionless
b_{1W}	coefficient representing influence of sediment type and vegetation on channel width	dimensionless
B_i	braiding index	dimensionless
BI	bench index	dimensionless
B_{ia}	active braiding index	dimensionless

Symbol	Definition	Units (SI) or Numeric Value
b_q	exponent relating Manning's n to discharge	dimensionless
c	coefficient in power function for depth hydraulic-geometry relation	$m^{1-3f}\,s^f$
C	Chezy coefficient	$m^{0.5}\,s^{-1}$
C^*	dimensionless channel curvature	dimensionless
C_c	mean curvature of a meander bend	m^{-1}
C_{cs}	local curvature of a meandering channel	m^{-1}
C_d	circularity of a drainage basin	dimensionless
C_D	particle drag coefficient	dimensionless
C_{dm}	dimensionless Chezy coefficient	dimensionless
c_e	dimensionless bank erosion coefficient	dimensionless
c_f	frictional resistance coefficient	dimensionless
c_f'	component of frictional resistance coefficient associated with grain resistance	dimensionless
c_f''	component of frictional resistance coefficient associated with bedform resistance	dimensionless
C_F	crop management factor in USLE	dimensionless
C_L	particle lift coefficient	dimensionless
C_{LC}	chord length of a meander bend	m
C_m	constant of channel maintenance	$km^2\,km^{-1}$
C_o	surface concentration of cosmogenic radionuclide	atoms g^{-1}
C_r	particle resultant resistance coefficient	dimensionless
C_s	sediment concentration	$mg\,l^{-1}$
C_{sb}	dimensionless volumetric bedload concentration	dimensionless
C_{sbm}	concentration of transported bed material	parts per million
C_{sbv}	volumetric bedload concentration	$m^2\,s^{-1}$
C_{svmd}	volumetric suspended-sediment concentration at mid-depth	dimensionless
C_{svrf}	reference volumetric suspended-sediment concentration	dimensionless
C_{svz}	local mean volumetric suspended-sediment concentration	dimensionless
C'_{svz}	fluctuating component of local mean volumetric suspended-sediment concentration	dimensionless
C_{sz}	concentration of suspended sediment at level z in water column	$mg\,l^{-1}$
C_{vg}	extent of vegetation cover on floodplain	%
d	particle diameter	m
D	mean flow depth at a river cross section	m
D'	depth related to grain resistance	m
D''	depth related to form resistance	m

Symbol	Definition	Units (SI) or Numeric Value
d_{50}	median grain size	mm or m
d_{50srf}	median grain size of surface bed material	mm or m
d_{50sub}	median grain size of subsurface bed material	mm or m
d_{84}	particle size corresponding to 84th percentile of size distribution	mm or m
d_{90}	particle size corresponding to 90th percentile of size distribution	mm or m
d_a	diameter of longest particle axis	mm or m
D_{af}	depth of accreting fines on floodplain	cm
d_b	diameter of intermediate particle axis	mm or m
d_{bd}	diameter of substrate particles in grain pivoting	m
D_{bk}	bankfull mean depth of a river channel	m
$\widetilde{D}_{bk}$	dimensionless bankfull depth	dimensionless
$D_{bk(ref)}$	bankfull channel depth of a reference reach	m
d_c	diameter of shortest particle axis	mm or m
D_d	drainage density	km km^{-2}
D_{dd}	dimensionless drainage density	dimensionless
D_e	soil diffusivity coefficient	m^2 s^{-1}
D_f	fractal dimension of drainage basin	dimensionless
d_g	grain size of gravel	m
d_i	particle diameter corresponding to a specific cumulative grain-size percentile i	mm or m
D_l	fractal dimension of river lengths	dimensionless
d_m	geometric mean grain size	mm or m
d_o	particle-size scaling factor	mm
d_p	diameter of a pivoting particle	m
d_{pd}	diameter of a pipe	m
d_r	reference grain size	m
d_s	grain size of sand	m
D_{sv}	volumetric rate of deposition of particles per unit area of the bed	m s^{-1}
d_ϕ	grain size in phi units	$-\log_2(d)$, d in mm
$d_{\phi m}$	geometric mean of grain-size distribution in phi units	phi units
$d_{\phi mdi}$	grain size at the mid-point of grain-size interval i	phi units
E	water-surface elevation	m
E_{af}	spatial extent of accreting fines on floodplain	%
e_b	bedload-transport efficiency	dimensionless
e_g	glacial erosion factor	dimensionless

Symbol	Definition	Units (SI) or Numeric Value
ER	entrenchment ratio	dimensionless
e_s	suspended-sediment transport efficiency	dimensionless
e_{se}	human-induced soil erosion factor	dimensionless
E_{sv}	volumetric rate of entrainment of bed material per unit area of the bed	m s^{-1}
e_t	error term	units of dependent variable
E_T	total energy per unit area of a hydrologic event within a river system above a threshold for channel change	J
ΔE_n	cross-stream change in water-surface elevation	m
ΔE_s	downstream change in water-surface elevation	m
ΔE_{st}	step drop height of step-pool unit	m
f	exponent in hydraulic-geometry relation for depth	dimensionless
F	Froude number	dimensionless
F	force	N
F_b	shear force of a body of water in a river on the bed	N
F_D	drag force acting on a particle on a river bed	N
$\mathbf{F}_D$	densimetric Froude number	dimensionless
f_f	Darcy-Weisbach friction factor	dimensionless
f_f'	component of Darcy–Weisbach friction factor associated with grain resistance	dimensionless
f_f''	component of Darcy–Weisbach friction factor associated with bedform resistance	dimensionless
F_g	gravitational force acting on a particle on a river bed	N
F_g'	modified gravitational force acting on a particle	N
f_i	proportion or fraction of material by weight in size interval i of bed material	dimensionless
F_L	lift force acting on a particle on a river bed	N
F_r	resultant force acting on a particle	N
F_s	factor of safety for bank failure	dimensionless
f_{ssz}	net vertical flux of suspended sediment	m s^{-1}
F_x	body force in x direction	N m^{-3}
F_y	body force in y direction	N m^{-3}
F_z	body force in z direction	N m^{-3}
g	gravitational acceleration	m s^{-2}
g'	reduced gravitational acceleration	m s^{-2}
G	floodplain diffusivity function	dimensionless

Symbol	Definition	Units (SI) or Numeric Value
$G(o_i)$	bedload transport function	dimensionless
h	local depth of overland or river flow	m
H	hole size in quadrant analysis of turbulence	dimensionless
h_b	channel-bank height	m
H_b	bar-unit height	m
h_{bc}	critical channel-bank height	m
h_{cl}	channel centerline depth	m
h_f	frictional head loss	m
h_{max}	maximum water depth at a cross section	m
h_{md}	half of the depth ($h/2$)	m
h_n	local depth normal to the transverse slope of the channel perimeter	m
H_r	riffle height	m
H_s	step height of step-pool unit	m
h_{sd}	thickness of sediment deposited on floodplain	m
h_{sdo}	thickness of sediment deposited at top of the channel banks	m
H_t	total energy head of flow	m
I	rainfall intensity	m s^{-1}
Im	impulse	N s
I_r	incision rate	m s^{-1}
j	time step	Δt
k	coefficient in power function relating velocity to discharge	m^{1-3m} s^{m-1}
k_d	coefficient in drainage area–distance relation	m$^{2-\gamma}$
k_{di}	diffusion coefficient in diffusive model of river incision	m^2 s^{-1}
k_e	bank erosion coefficient	m^3 N^{-1} s^{-1}
$\hat{k}_e$	dimensionless bank erosion coefficient	dimensionless
K_F	soil erodibility factor in USLE	ton hour J^{-1} mm^{-1}
k_i	erosion coefficient in stream power incision model	m^{1-2m} s^{-1}
k_s	roughness length scale associated with grain resistance	m
k_{sb}	coefficient of bedload transport	dimensionless
k_{si}	coefficient in slope–drainage area relation (steepness index)	m$^{-2\Phi}$
k_y	transverse mixing coefficient	(units of tracer)^2 t^{-1}
ku_ϕ	kurtosis of grain-size distribution in phi units	phi units
L	length of a stream	m or km
l_a	length of pivoting arm of a grain resting on the bed	m

Symbol	Definition	Units (SI) or Numeric Value
L_{br}	bar length	m
L_d	length of a drainage basin	km
L_f	lithology erosion factor	dimensionless
L_F	slope length factor in USLE	dimensionless
L_l	length of a braid channel	m
l_o	estimate of the average hillslope length required to support a unit length of channel	m
L_{pd}	path length of particle displacement during a transport episode	m
$\overline{L}_{pd}$	mean path length of particle displacement	m
L_r	rill or gully length	m
L_R	river reach length	m or km
L_s	step spacing of step-pool unit	m
L_t	total length of streams in a drainage network	km
L_V	valley length	m or km
$\overline{L}_\Upsilon$	average length of streams in an order Υ	m or km
m	exponent for velocity in hydraulic-geometry relation, for q in hillslope sediment-transport model, or for drainage area in stream power incision model	dimensionless
M	number of exterior links in a river network	integer value
m_{br}	bar mode	dimensionless
m_r	rank of discharge in flood-frequency analysis	integer value
M_r	momentum flux ratio of flows at a confluence	dimensionless
M_{tl}	turbulent lateral transport of streamwise momentum over the flow depth normal to the channel boundary	kg s^{-2}
ΔM_{st}	annual change in material mass in a drainage basin	ton yr^{-1}
n	Manning coefficient, or	s m$^{-0.33}$
	exponent for slope in hillslope sediment-transport model and in stream power incision model	dimensionless
n'	component of Manning's n associated with grain resistance	s m$^{-0.33}$
n''	component of Manning's n associated with bedform resistance	s m$^{-0.33}$
n_b	number of boxes intersecting a stream segment	integer
N_b	number of particles moving in the bed per unit area	integer m^{-2}
n_d	number of days in a hydrologic record	integer
N_l	number of channels along a transect	integer
n_t	number of floods in partial-duration flood series	integer
N_{xs}	number of transects along a channel	integer
n_y	number of floods in an annual-maximum flood series	integer

Symbol	Definition	Units (SI) or Numeric Value
N_Υ	number of streams in an order Υ	integer
$\overline{N}_{\Upsilon,\Psi}$	average number of side streams of a specific order Ψ joining absorbing streams of a specific order Υ	real number
$\overline{p}$	pressure	N m^{-2}
P	wetted perimeter	m
$p(instability)$	probability a channel will incise	dimensionless
P_b	power of river flow	W
P_{br}	bar perimeter	m
P_d	perimeter of a drainage basin	km
Pe	Peclet number	dimensionless
P_F	erosion control practice factor in USLE	dimensionless
p_i	fraction of material by weight in size interval i of bedload	dimensionless
P_L	perpendicular distance from a meander chord to the bend apex	m
P_{ma}	mean annual precipitation	mm
p_o	porosity of the active layer of the channel bed	dimensionless
P_o	surface production rate of cosmogenic radionuclide	atoms g^{-1}
p_p	plotting position of a flood in flood-frequency analysis	dimensionless
P_s	proportion of sand in the particle-size distribution	dimensionless
q	discharge per unit width	m^2 s^{-1}
Q	discharge (volumetric water flux)	m^3 s^{-1} or km^3 yr^{-1}
$\overline{Q}$	mean discharge	m^3 s^{-1}
$Q_{1.5}$	discharge with a recurrence interval of 1.5 years, on average, or a probability of 0.67	m^3 s^{-1}
Q_{100}	discharge with a recurrence interval of 100 years, on average, or a probability of 0.01	m^3 s^{-1}
Q_2	discharge with a recurrence interval of two years, on average, or a probability of 0.50	m^3 s^{-1}
Q_5	discharge with a recurrence interval of five years, on average, or a probability of 0.2	m^3 s^{-1}
Q_{bk}	bankfull discharge	m^3 s^{-1}
$\hat{Q}_{bk}$	dimensionless bankfull discharge	dimensionless
q_c	critical unit discharge of sediment entrainment	m^2 s^{-1}
Q_c	critical discharge for sediment movement	m^3 s^{-1}
Q_{dd}	dominant discharge	m^3 s^{-1}
Q_{ef}	effective discharge	m^3 s^{-1}
Q_{efb}	effective discharge for bedload transport	m^3 s^{-1}

Symbol	Definition	Units (SI) or Numeric Value
Q_{efs}	effective discharge for suspended-sediment transport	$m^3\,s^{-1}$
Q_{fe}	functionally equivalent discharge	$m^3\,s^{-1}$
Q_{hl}	half-load discharge	$m^3\,s^{-1}$
Q_i	flood of a specific recurrence interval i	$m^3\,s^{-1}$
q_i	partial discharge at a cross section	$m^3\,s^{-1}$
Q_{in}	floodplain-inundating discharge	$m^3\,s^{-1}$
Q_m	mean annual discharge	$m^3\,s^{-1}$
Q_{maf}	mean annual flood	$m^3\,s^{-1}$
Q_n	cross-stream discharge	$m^3\,s^{-1}$
Q_r	discharge ratio at a confluence	dimensionless
Q_{rd}	reversal discharge in pool-riffle maintenance	$m^3\,s^{-1}$
q_s	volumetric sediment flux per unit width	$m^2\,s^{-1}$
Q_s	sediment load	$Mt\,yr^{-1}$, $kt\,yr^{-1}$, or $t\,yr^{-1}$
q_{sbdw}	bedload transport rate by dry weight per unit width	$kg\,m^{-1}\,s^{-1}$
Q_{sbk}	volumetric bankfull sediment load	$m^3\,s^{-1}$
q_{sbl}	lateral volumetric flux of bedload	$m^2\,s^{-1}$
q_{sbsw}	bedload transport rate by submerged weight per unit width	$kg\,m^{-1}\,s^{-1}$
Q_{sbv}	volumetric flux of bed material	$m^3\,s^{-1}$
q_{sbv}	volumetric rate of bedload transport per unit width	$m^2\,s$
q_{sbv}^*	dimensionless volumetric rate of bedload transport per unit width	dimensionless
q_{sbvn}	volumetric rate of bedload transport per unit width in the n direction	$m^2\,s^{-1}$
q_{sbvs}	volumetric rate of bedload transport per unit width in the s direction	$m^2\,s^{-1}$
Q_{sc}	channel erosion rate per unit drainage area	$ton\,yr^{-1}\,km^{-2}$
Q_{se}	hillslope erosion rate per unit drainage area	$ton\,yr^{-1}\,km^{-2}$
Q_{si}	annual material mass influx to a drainage basin	$ton\,yr^{-1}$
Q_{sp}	total sediment erosion per unit drainage area	$ton\,yr^{-1}\,km^{-2}$
q_{ss}	suspended-sediment load per unit width	$kg\,m^{-1}\,s^{-1}$
Q_{ss}	suspended-sediment load	$kg\,s^{-1}$ or $ton\,day^{-1}$
q_{ssl}	vertically integrated volumetric lateral flux of suspended sediment	$m^2\,s^{-1}$
q_{sssw}	suspended-sediment transport rate by submerged weight per unit width	$kg\,m^{-1}\,s^{-1}$
Q_{st}	sediment storage rate per unit drainage area	$ton\,yr^{-1}\,km^{-2}$
q_{stsw}	total sediment transport rate by submerged weight per unit width	$kg\,m^{-1}\,s^{-1}$
Q_{sv}	volumetric sediment flux	$m^3\,yr^{-1}$
Q_{svc}	volumetric sediment-transport capacity	$m^3\,s^{-1}$
Q_{sy}	sediment yield	$Mton$, $kton$, or $ton\,yr^{-1}\,km^{-2}$

Symbol	Definition	Units (SI) or Numeric Value
Q_t	discharge at time t	$m^3 \, s^{-1}$
Q_{th}	threshold discharge for channel change	$m^3 \, s^{-1}$
Q_v	ratio of Q_5/Q_m	dimensionless
$\overline{\ln Q}$	mean of logarithms of discharge	$\ln(m^3 \, s^{-1})$
r	particle radius	m
R	hydraulic radius	m
$\mathbf{R}$	Reynolds number	dimensionless
R_A	drainage-area ratio	dimensionless
R_b	bifurcation ratio	dimensionless
$\mathbf{R}_b$	boundary Reynolds number	dimensionless
R_{bk}	bankfull hydraulic radius	m
r_c	radius of curvature of a bend	m
R_F	rainfall erosivity factor in USLE	$J \, mm \, hour^{-1} \, km^{-2} \, yr^{-1}$
R_L	stream length ratio	dimensionless
R_o	Rouse number	dimensionless
$R_o{'}$	modified Rouse number	dimensionless
$\mathbf{R}_p$	particle Reynolds number	dimensionless
R_r	relief ratio of a watershed	dimensionless
R_{ra}	average basin relief	m
R_{rm}	maximum basin relief	km
R_s	resistance to erosion of hillslope material	$N \, m^{-2}$
S	slope or gradient of a hillslope or channel	$m \, m^{-1}$
s,n	coordinates parallel and perpendicular to channel centerline in a meandering river	m
$\hat{s}$	dimensionless position along the centerline of a meandering river	dimensionless
s'	distance upstream along a meandering river	m
$\hat{s}'$	dimensionless distance upstream along a meandering channel	dimensionless
s_b	box size of fractal analysis	dimensionless
SC_{bd}	percentage of silt and clay in the channel beds	%
SC_{bk}	percentage of silt and clay in the channel banks	%
SC_{wt}	weighted percentage of silt and clay in the channel perimeter	%
S_d	mean drainage-basin slope	dimensionless
SDR	sediment delivery ratio	dimensionless
S_e	slope of energy grade line (energy gradient)	dimensionless
S_F	slope gradient factor in USLE	dimensionless

Symbol	Definition	Units (SI) or Numeric Value
SF	shape factor for particles	dimensionless
S_I	channel sinuosity	dimensionless
S_{IT}	total sinuosity of a braided reach	dimensionless
S_v	valley slope or gradient	dimensionless
S_{ws}	water-surface slope	m m^{-1}
sk_ϕ	skewness of grain-size distribution in phi units	phi units
t	time	s, hour, yr
t_a	timescale of avulsion	s
T_{da}	drainage basin–averaged temperature	°C
T_e	reservoir trapping efficiency factor	dimensionless
t_f	time interval between formative, or channel-shaping events in a river system	years
TKE	turbulent kinetic energy	$\text{m}^2\,\text{s}^{-2}$
t_m	timescale of lateral migration	s
t_r	timespan for recovery of a river system from change produced by formative events	years
T_r	estimated recurrence interval of a flood in flood-frequency analysis	years
t_{sr}	residence time of sediment	years
t_{tmin}	minimum timescale of turbulence	s
Δt_c	time during an event that flow exceeds the threshold of motion for a particle	s
$\mathbf{u}$	magnitude of velocity vector at a position in a river flow	m s^{-1}
u	mean downstream component of velocity	m s^{-1}
U	mean velocity at a river cross section	m s^{-1}
$\hat{u}$	instantaneous downstream velocity	m s^{-1}
u'	fluctuating component of downstream velocity	m s^{-1}
u_*	shear velocity	m s^{-1}
u_*'	shear velocity associated with grain friction	m s^{-1}
$\bar{u}$	depth-averaged downstream velocity	m s^{-1}
u_b	near-bed velocity	m s^{-1}
u_b'	fluctuating component of near-bed downstream velocity	m s^{-1}
u_{*bk}	bankfull shear velocity	m s^{-1}
U_{bk}	bankfull mean velocity of flow in a river channel	m s^{-1}
u_{*bnk}	shear velocity acting on the bank	m s^{-1}
u_{*cbnk}	critical shear velocity of bank material	m s^{-1}

Symbol	Definition	Units (SI) or Numeric Value
u_c	characteristic velocity of flow	m s^{-1}
U_c	critical velocity for entrainment	m s^{-1}
u_{*c}	critical shear velocity	m s^{-1}
u_{*c}'	critical bank shear velocity	m s^{-1}
$\bar{u}_{nb}$	depth-averaged velocity near the outer bank	m s^{-1}
$\bar{u}_{nb}'$	excess depth-averaged near-bank velocity	m s^{-1}
u_o	freestream velocity of flow unaffected by a boundary	m s^{-1}
u_p	primary velocity components in depth-averaged velocity frame of reference	m s^{-1}
U_R	velocity ratio of flows at a confluence	dimensionless
U_r	tectonic uplift rate	m s^{-1}
U_{ro}	coefficient in uplift–distance relation	m$^{1-\vartheta}$ s^{-1}
u_s	surface flow velocity	m s^{-1}
u_{sb}	mean bedload grain velocity	m s^{-1}
U_{ss}	velocity of suspended sediment	m s^{-1}
U_{vs}	virtual velocity of a particle during transport	m s^{-1}
$\mathbf{u_{xy}}$	local two-dimensional velocity vector magnitude	m s^{-1}
$\mathbf{U_{xy}}$	depth-averaged two-dimensional velocity vector magnitude	m s^{-1}
u_z	velocity at height z above the bed	m s^{-1}
v	mean cross-stream component of velocity	m s^{-1}
$\hat{v}$	instantaneous cross-stream velocity	m s^{-1}
v'	fluctuating component of cross-stream velocity	m s^{-1}
$\bar{v}$	depth-averaged cross-stream velocity	m s^{-1}
V_{cov}	percentage of vegetation cover	%
v_{ef}	effective fall velocity	m s^{-1}
v_f	fall velocity of a particle	m s^{-1}
V_p	volume of a particle	m^3
$\overline{V_p}$	mean volume of a group of particles	m^3
V_r	volume of rill or ephemeral gully	m^3
v_s	secondary velocity component in depth-averaged velocity frame of reference	m s^{-1}
V_s	volume of sediment	m^3 or km^3
V_{sin}	volume of sediment input	m^3
V_{sout}	volume of sediment output	m^3
ΔV_s	change in sediment volume	m^3

Symbol	Definition	Units (SI) or Numeric Value
W	water surface width at a cross section	m
w	mean vertical component of velocity	m s^{-1}
$\hat{w}$	instantaneous vertical velocity	m s^{-1}
w'	fluctuating component of vertical velocity	m s^{-1}
w'_b	fluctuating component of near-bed vertical velocity	m s^{-1}
W_{bk}	bankfull width of a river channel	m
$\widetilde{W}_{bk}$	dimensionless bankfull width	dimensionless
$W_{bk(eq)}$	equilibrium bankfull channel width	m
$W_{bk(ref)}$	bankfull channel width of a reference reach	m
$W_{bk(res)}$	bankfull channel width of a reach to be restored	m
W_{br}	bar width	m
W_F	width function of a network (number of channel at a given distance from the basin outlet)	dimensionless
W_{fp}	distance from channel margin to valley wall	m
w_g	weighting function in dynamic channel-geometry model	dimensionless
Wt	weight of water	N
$(W/D)_c$	critical width/depth ratio for braiding	dimensionless
x	distance along a hillslope or river	m or km
X	length of a body of water in a river	m
x_{bl}	distance to base level of a watershed	m or km
x_c	critical hillslope distance for erosion	m
x_h	distance over which characteristic particle size decreases by 0.5	km
y	lateral distance	m
y_d	dimensionless lateral distance from the channel across the floodplain	dimensionless
y_g	generic variable for some aspect of channel geometry (width, depth, or slope)	units specific to y_g
z	vertical distance	m
Z	elevation head or elevation of the river-channel bed	m
$z^{\star}$	depth of production of cosmogenic radionuclides	cm
Z_b	elevation at mouth of a river	m
Z_{ci}	variable in reference-level suspended-sediment function	dimensionless
Z_h	elevation of river channel at base level	m
z_o	roughness length	m
z_{rf}	reference height above the channel bed for suspended sediment	m
z_t	thickness of viscous sublayer and buffer layer	m
z_t^+	dimensionless thickness of viscous sublayer and buffer layer	dimensionless

Symbol	Definition	Units (SI) or Numeric Value
z_v	thickness of viscous sublayer	m
z_v^+	dimensionless thickness of viscous sublayer	dimensionless
ℓ_{emin}	minimum length scale of turbulence	m
ℓ	length scale	m or km
ℓ_c	critical length scale	m or km
ℓ_e	length scale of turbulent eddies	m
α_1, α_2	angles of the channel path relative to the downvalley direction on each side of a meander loops	degrees
$\alpha_{\mathbf{u}1}$	horizontal angle of the velocity vector	degrees
$\alpha_{\mathbf{u}2}$	vertical angle of the velocity vector	degrees
α_a	coefficient in dimensionless width–discharge relation	dimensionless
α_c	coefficient in dimensionless depth–discharge relation	dimensionless
α_e	energy coefficient	dimensionless
α_g	coefficient in dimensionless slope–discharge relation	dimensionless
β_c	suspended-sediment diffusivity correction coefficient	dimensionless
γ	exponent in the drainage area–distance power function	dimensionless
δ	downstream fining rate exponent	km^{-1}
δ_a	downstream fining rate associated with abrasion	km^{-1}
δ_b	thickness of active layer of the channel bed	m
δ_c	critical stress depth	m
δ_d	denudation rate	$mm\ kyr^{-1}$
δ_L	boundary-layer thickness	m
δ_p	stress depth	m
δ_s	downstream fining rate associated with sorting	km^{-1}
ε	eddy viscosity	$kg\ m^{-1}\ s^{-1}$
ε_c	diffusion coefficient for transported particles	$m^2\ s^{-1}$
ε_d	turbulent dissipation	$m^2\ s^{-3}$
ε_{sv}	vertical suspended-sediment diffusivity	$m^2\ s^{-1}$
ε_{sy}	lateral sediment diffusivity	$m^2\ s^{-1}$
ζ_a	vertical aggradation rate	$m\ s^{-1}$
η_b	exponent in dimensionless width–discharge relation	dimensionless
η_f	exponent in dimensionless depth–discharge relation	dimensionless
η_{fc}	grain friction correction factor for bedload transport	dimensionless
η_s	composite erosion factor	dimensionless
η_{sw}	sidewall correction for bedload transport	dimensionless

Symbol	Definition	Units (SI) or Numeric Value
η_z	exponent in dimensionless slope–discharge relation	dimensionless
θ	slope angle of the channel bed or a hillslope	degrees
θ'	angle of the channel bank	degrees
θ_e	slope angle of energy grade line	degrees
θ_j	junction angle	degrees
$\bar{\theta}_j$	mean junction angle	degrees
θ_l	transverse slope of the channel perimeter	degrees
$\theta_{\mathbf{u}}$	angle of local two-dimensional velocity vector	degrees
θ_w	slope angle of the water surface	degrees
Θ	dimensionless bed shear stress	dimensionless
Θ_{bk}	dimensionless bed shear stress for the bankfull discharge	dimensionless
Θ_c	dimensionless critical bed shear stress	dimensionless
Θ_c'	critical dimensionless bed shear stress associated with grain resistance	dimensionless
Θ_{ci}	critical dimensionless bed shear stress corresponding to grain-size fraction i	dimensionless
Θ_{cpb}	critical dimensionless shear stress of a particle of size d_p resting on particles of size d_b	dimensionless
Θ_{cps}	critical dimensionless shear stress of a particle of size d_p on a smooth bed	dimensionless
Θ_{cr}	critical dimensionless shear stress of a reference grain size	dimensionless
κ	von Karman's constant	dimensionless
λ	grain sorting parameter in suspended-sediment reference-level function	dimensionless
λ_b	bar-unit wavelength	m
λ_c	decay rate of cosmogenic radionuclides	yr^{-1}
λ_{cs}	wavelength of lateral channel spacing on a hillslope	m
λ_H	wavelength of horizontal coherent structures	m
λ_m	meander wavelength	m
λ_p	wavelength of hillslope perturbation	m
λ_{pr}	pool-to-pool or riffle-to-riffle spacing	m
λ_s	step wavelength of step-pool unit	m
λ_v	wavelength of valley spacing	m or km
Λ	cosmic ray absorption free path	g cm^{-2}
μ	dynamic viscosity	$\text{kg m}^{-1}\,\text{s}^{-1}$
μ'	ratio of critical bank shear stress to critical bed shear stress	dimensionless
$\hat{\mu}$	bank strength impact factor	dimensionless
ν	kinematic viscosity	$\text{m}^2\,\text{s}^{-1}$

Symbol	Definition	Units (SI) or Numeric Value
ξ_{ex}	vertically integrated excess sediment concentration	m
ξ_m	bank erosion or lateral migration rate	m s^{-1}
$\hat{\xi}_m$	dimensionless bank erosion or channel migration rate	dimensionless
o_i	ratio of dimensionless bed shear stress to critical dimensionless bed shear stress for grain-size fraction i	dimensionless
ρ	density of water	kg m^{-3}
ρ_s	density of sediment	kg m^{-3}
σ_{dx}^2	mean squared displacement (variance) of moved particles over time	m^2
$\sigma_y{}^2$	variance in concentration or value of a tracer	(units of tracer)^2
$\sigma_{yus}{}^2$	variance in concentration or value of a tracer upstream of a confluence	(units of tracer)^2
σ_{lnQ}	standard deviation of the logarithms of discharge	ln(m^3 s^{-1})
σ_s	sediment sorting (standard deviation of grain size distribution)	m or phi units (see σ_ϕ)
$\sigma_{u'}$	standard deviation of streamwise velocity fluctuations	m s^{-1}
$\sigma_{w'}$	standard deviation of vertical velocity fluctuations	m s^{-1}
σ_z	standard deviation of detrended bed elevations	m
σ_ϕ	standard deviation (sorting) of grain-size distribution in phi units	phi units
ς	exponent defining sensitivity to antecedent conditions in partial adjustment model of channel morphodynamics	dimensionless
ϑ	exponent in uplift–distance power function	dimensionless
Υ	stream order	integer
Υ_m	stream order at basin outlet	integer
τ	shear stress	N m^{-2}
τ_b	bed shear stress	N m^{-2}
τ_b'	component of bed shear stress associated with grain resistance	N m^{-2}
τ_b''	component of bed shear stress associated with form resistance	N m^{-2}
$\tau_{b(cl)}$	bed shear stress at the center of the river channel	N m^{-2}
τ_{bc}	critical bed shear stress	N m^{-2}
τ_{bci}	critical bed shear stress for particle diameter corresponding to cumulative grain-size percentile i	N m^{-2}
τ_{bcm}	critical bed shear stress for the geometric mean grain size of the particle-size distribution	N m^{-2}
τ_{bn}	bed shear stress in n direction	N m^{-2}
τ_{bnk}	bank shear stress	N m^{-2}
τ_{bs}	bed shear stress in s direction	N m^{-2}
τ_{cbnk}	critical bank shear stress	N m^{-2}
τ_{cg}	critical bed shear stress for gravel	N m^{-2}

Symbol	Definition	Units (SI) or Numeric Value
τ_{cs}	critical bed shear stress for sand	N m^{-2}
τ_{fp}	boundary shear stress on the floodplain	N m^{-2}
τ_g	gravitational stress	N m^{-2}
τ_{ma}	bed shear stress at which mobile armor begins to develop	N m^{-2}
τ_{na}	bed shear stress above which armor layer does not exist	N m^{-2}
τ_{sa}	bed shear stress at which static armor begins to develop	N m^{-2}
τ_t	turbulent shear stress of flowing water	N m^{-2}
τ_{txx}	Reynolds normal stress in streamwise direction	N m^{-2}
τ_{txy}	Reynolds shear stress in x-y plane	N m^{-2}
τ_{txz}	Reynolds shear stress in x-z plane	N m^{-2}
τ_{Txz}	total shear stress in x-z plane	N m^{-2}
τ_{tyy}	Reynolds normal stress in cross-stream direction	N m^{-2}
τ_{tyz}	Reynolds shear stress in y-z plane	N m^{-2}
τ_{tzz}	Reynolds normal stress in vertical direction	N m^{-2}
τ_v	viscous shear stress of flowing water	N m^{-2}
τ_{vxz}	viscous shear stress in the x-z plane	N m^{-2}
ϕ_1, ϕ_2	subangles of the junction angle related to the major (ϕ_1) and minor (ϕ_2) tributaries	degrees
ϕ'	angle of repose of bank sediment (bank friction angle for unconsolidated sediment)	degrees
ϕ_b	modified bank friction angle	degrees
ϕ_o	pivoting angle of grain resting on the bed	degrees
$\phi_\mathbf{U}$	angle of local 2D depth-averaged velocity vector	degrees
Φ	exponent in slope-drainage area relation (concavity index)	dimensionless
χ	Standardized distance metric for longitudinal profiles	m
ψ_{LE}	lithologic erodibility	dimensionless
ψ_m	mixing metric at confluences	dimensionless
Ψ	order of side streams	integer
ω	power per unit area of flow	W m^{-2}
$\omega^\star$	dimensionless stream power per unit area	dimensionless
ω_{bk}	bankfull stream power per unit area	W m^{-2}
ω_c	critical stream power per unit area for sediment entrainment	W m^{-2}
ω_p	Potential stream power per unit area	W m^{-2}
ω_{pth}	threshold potential stream power per unit area	W m^{-2}
ω_{th}	threshold stream power per unit area	W m^{-2}
Ω	stream power per unit length of flow	W m^{-1}

References

Aalto, R., Dunne, T., Guyot, J. L., 2006. Geomorphic controls on Andean denudation rates. *Journal of Geology*, **114**(1), 85–99.

Abad, J. D., Garcia, M. H., 2009. Experiments in a high-amplitude Kinoshita meandering channel: 2. Implications of bend orientation on bed morphodynamics. *Water Resources Research*, **45**. 10.1029/2008wr007017.

Abbado, D., Slingerland, R., Smith, N. D., 2005. Origin of anastomosis in the upper Columbia River, British Columbia, Canada. In: M. D. Blum, S. B. Marriott, S. F. LeClair (eds.), *Fluvial Sedimentology VII*, International Association of Sedimentologists Special Publication No. 35, pp. 3–15.

Abban, B., Papanicolaou, A. N., Cowles, M. K., et al., 2016. An enhanced Bayesian fingerprinting framework for studying sediment source dynamics in intensively managed landscapes. *Water Resources Research*, **52**(6), 4646–4673. 10.1002/2015wr018030.

Abbe, T., Brooks, A., 2011. Geomorphic, engineering, and ecological considerations when using wood in river restoration. In: A. Simon, S. J. Bennett, J. M. Castro (eds.), *Stream Restoration in Dynamic Fluvial Systems: Scientific Approaches, Analyses, and Tools*. American Geophysical Union, Washington, DC, pp. 419–451.

Abernethy, B., Rutherfurd, I. D., 2000. Stabilising stream banks with riparian vegetation. *Natural Resource Management*, **3**(2), 2–9.

Abrahams, A. D., 1977. Factor of relief in evolution of channel networks in mature drainage basins. *American Journal of Science*, **277**(5), 626–646.

Abrahams, A. D., 1984. Channel networks: a geomorphological perspective. *Water Resources Research*, **20**(2), 161–188.

Abrahams, A. D., Li, G., Atkinson, J. F., 1995. Step-pool streams – adjustment to maximum flow resistance. *Water Resources Research*, **31**(10), 2593–2602.

Abrams, D. M., Lobkovsky, A. E., Petroff, A. P., et al., 2009. Growth laws for channel networks incised by groundwater flow. *Nature Geoscience*, **2**(3), 193–196.

Adami, L., Bertoldi, W., Zolezzi, G., 2016. Multidecadal dynamics of alternate bars in the Alpine Rhine River. *Water Resources Research*, **52**(11), 8921–8938. 10.1002/2015wr018228.

Adamowski, K., 1981. Plotting formula for flood frequency. *Water Resources Bulletin*, **17**(2), 197–202.

Adams, P. N., Slingerland, R. L., Smith, N. D., 2004. Variations in natural levee morphology in anastomosed channel flood plain complexes. *Geomorphology*, **61**(1–2), 127–142.

Agouridis, C. T., Edwards, D. R., Workman, S. R., et al., 2005. Streambank erosion associated with grazing practices in the humid region. *Transactions of the Association of Agricultural Engineers*, **48**(1), 181–190.

Ahadiyan, J., Adeli, A., Bahmanpouri, F., Gualtieri, C., 2018. Numerical simulation of flow and scour in a laboratory junction. *Geosciences*, **8**(5), 162. 10.3390/geosciences8050162.

Aharonson, O., Zuber, M. T., Rothman, D. H., Schorghofer, N., Whipple, K. X., 2002. Drainage basins and channel incision on Mars. *Proceedings of the National Academy of Sciences of the United States of America*, **99**(4), 1780–1783.

Ahilan, S., O'Sullivan, J. J., Bruen, M., Brauders, N., Healy, D., 2013. Bankfull discharge and recurrence intervals in Irish rivers. *Proceedings of the Institution of Civil Engineers-Water Management*, **166**(7), 381–393.

Ahmed, J., Constantine, J. A., Dunne, T., 2019. The role of sediment supply in the adjustment of channel sinuosity across the Amazon Basin. *Geology*, **47**(9), 807–810. 10.1130 /G46319.1.

Ahnert, F., 1970. Functional relationships between denudation, relief, and uplift in large mid-latitude drainage basins. *American Journal of Science*, **268**(3), 243–263.

Aich, V., Zimmermann, A., Elsenbeer, H., 2014. Quantification and interpretation of suspended-sediment discharge hysteresis patterns: how much data do we need? *Catena*, **122**, 120–129.

Aksoy, H., Kavvas, M. L., 2005. A review of hillslope and watershed scale erosion and sediment transport models. *Catena*, **64**(2–3), 247–271. 10.1016/j.catena.2005.08.008.

Allen, G. H., Pavelsky, T. M., 2018. Global extent of rivers and streams. *Science*, **361**(6402), 585–587.

Allmendinger, N. E., Pizzuto, J. E., Moglen, G. E., Lewicki, M., 2007. A sediment budget for an urbanizing watershed, 1951–1996, Montgomery County, Maryland, U.S.A. *Journal of the American Water Resources Association*, **43**(6), 1483–1498.

Allred, T. M., Schmidt, J. C., 1999. Channel narrowing by vertical accretion along the Green River near Green River, Utah. *Geological Society of America Bulletin*, **111**(12), 1757–1772.

Alonso, C. V., 1980. Selecting a formula to estimate sediment transport in non-vegetated channels. Chapter 5. In: W. G. Knisel (ed.), *CREAMS – A Field Scale Model for Chemicals, Runoff, and Erosion from Agriculture Management Systems, Conservation Research Report No. 26*. U.S. Department of Agriculture, Washington, DC.

Alonso, C. V., Bennett, S. J., Stein, O. R., 2002. Predicting head cut erosion and migration in concentrated flows typical of upland areas. *Water Resources Research*, **38**(12), 39-1–39-15. 10.1029/2001wr001173.

American Rivers, 2019. Dam Removal Database, https://figshare.com/articles/_/5234068 (accessed May 14, 2019).

An, H.-P., Chen, S.-C., Chan, H.-C., Hsu, Y., 2013. Dimension and frequency of bar formation in a braided river. *International Journal of Sediment Research*, **28**(3), 358–367.

Anderson, D. G., 1970. Effects of urban development on floods in Northern Virginia. U.S. Geological Survey Water-Supply Paper 2001-C, U.S. Government Printing Office, Washington, DC.

Anderson, R. S., Anderson, S. P., 2010. *Geomorphology: The Mechanics and Chemistry of Landscape.* Cambridge University Press, New York.

Andrews, E. D., 1980. Effective and bankfull discharges of streams in the Yampa River basin, Colorado and Wyoming. *Journal of Hydrology*, **46**(3–4), 311–330.

Andrews, E. D., 1983. Entrainment of gravel from naturally sorted riverbed material. *Geological Society of America Bulletin*, **94**(10), 1225–1231.

Andrews, E. D., 1984. Bed-material entrainment and hydraulic geometry of gravel-bed rivers in Colorado. *Geological Society of America Bulletin*, **95**(3), 371–378.

Andrews, E. D., Erman, D. C., 1986. Persistence in the size distribution of surficial bed material during an extreme snowmelt flood. *Water Resources Research*, **22**(2), 191–197.

Anisimov, O., Vandenberghe, J., Lobanov, V., Kondratiev, A., 2008. Predicting changes in alluvial channel patterns in North-European Russia under conditions of global warming. *Geomorphology*, **98**(3–4), 262–274.

Annable, W. K., Watson, C. C., Thompson, P. J., 2012. Quasi-equilibrium conditions of urban gravel-bed stream channels in southern Ontario, Canada. *River Research and Applications*, **28**(3), 302–325.

Ansari, K., Morvan, H. P., Hargreaves, D. M., 2011. Numerical investigation into secondary currents and wall shear in trapezoidal channels. *Journal of Hydraulic Engineering*, **137**(4), 432–440.

Anthony, D. J., Harvey, M. D., 1991. Stage-dependent cross section adjustments in a meandering reach of Fall River, Colorado. *Geomorphology*, **4**(3–4), 187–203.

Arcement, G. J., Schneider, V. R., 1989. Guide for selecting Manning's roughness coefficients for natural channels and flood plains. U.S. Geological Survey Water-Supply Paper 2339, U.S. Government Printing Office, Washington, DC.

Arnell, N. W., Gosling, S. N., 2016. The impacts of climate change on river flood risk at the global scale. *Climatic Change*, **134**(3), 387–401.

Arnold, C. L., Boison, P. J., Patton, P. C., 1982. Sawmill Brook: an example of rapid geomorphic change related to urbanization. *Journal of Geology*, **90**(2), 155–166. 10.1086/628660.

Arrospide, F., Mao, L., Escauriaza, C., 2018. Morphological evolution of the Maipo River in central Chile: influence of instream gravel mining. *Geomorphology*, **306**, 182–197.

Asahi, K., Shimizu, Y., Nelson, J., Parker, G., 2013. Numerical simulation of river meandering with self-evolving banks. *Journal of*

Geophysical Research – Earth Surface, **118**(4), 2208–2229. 10.1002/jgrf.20150.

Ashmore, P. E., 1982. Laboratory modeling of gravel braided stream morphology. *Earth Surface Processes and Landforms*, **7**(3), 201–225.

Ashmore, P. E., 1991a. How do gravel-bed rivers braid? *Canadian Journal of Earth Sciences*, **28**(3), 326–341.

Ashmore, P. E., 1991b. Channel morphology and bedload pulses in braided, gravel-bed streams. *Geografiska Annaler Series A Physical Geography*, **73**, 37–52.

Ashmore, P. E., 1993. Anabranch confluence kinetics and sedimentation processes in gravel-braided streams. In: J. L. Best, C. S. Bristow (eds.), *Braided Rivers*. Special Publication No. 75, Geological Society of London, London, pp. 129–146.

Ashmore, P. E., 2009. Intensity and characteristic length of braided channel patterns. *Canadian Journal of Civil Engineering*, **36**(10), 1656–1666.

Ashmore, P. E., 2013. Morphology and dynamics of braided rivers. In: J. Shroder (ed.), *Treatise on Geomorphology, vol. 9, Fluvial Geomorphology*, E. Wohl (vol. ed.). Academic Press, San Diego, CA, pp. 289–312.

Ashmore, P. E., 2015. Towards a sociogeomorphology of rivers. *Geomorphology*, **251**, 149–156.

Ashmore, P. E., Church, M., 1998. Sediment transport and river morphology: a paradigm for study. In: P. C. Klingeman, R. L. Beschta, P. D. Komar, J. B. Bradley (eds.), *Gravel-Bed Rivers in the Environment*. Water Resources Publications, Highland Ranch, CO, pp. 115–148.

Ashmore, P. E., Day, T. J., 1988. Effective discharge for suspended sediment transport in streams of the Saskatchewan River basin. *Water Resources Research*, **24**(6), 864–870.

Ashmore, P. E., Gardner, J. T., 2008. Unconfined confluences in braided rivers. In: S. P. Rice, A. G. Roy, B. L. Rhoads (eds.), *River Confluences, Tributaries and the Fluvial Network*. Wiley, Chichester, UK, pp. 119–147.

Ashmore, P. E., Parker, G., 1983. Confluence scour in coarse braided streams. *Water Resources Research*, **19**(2), 392–402.

Ashmore, P. E., Sauks, E., 2006. Prediction of discharge from water surface width in a braided river with implications for at-a-station hydraulic geometry. *Water Resources Research*, **42**(3). 10.1029/2005wr003993.

Ashmore, P. E., Ferguson, R. I., Prestegaard, K. L., Ashworth, P. J., Paola, C., 1992. Secondary flow in anabranch confluences of a braided, gravel-bed stream. *Earth Surface Processes and Landforms*, **17**(3), 299–311. 10.1002/esp.3290170308.

Ashmore, P. E., Bertoldi, W., Gardner, J. T., 2011. Active width of gravel-bed braided rivers. *Earth Surface Processes and Landforms*, **36**(11), 1510–1521. 10.1002/esp.2182.

Ashworth, P. J., 1996. Mid-channel bar growth and its relationship to local flow strength and direction. *Earth Surface Processes and Landforms*, **21**(2), 103–123. 10.1002/(sici)1096–9837(199602)21:2<103::aid-esp569>3.0.co;2-o.

Ashworth, P. J., Ferguson, R. I., 1989. Size-selective entrainment of bed load in gravel bed streams. *Water Resources Research*, **25**(4), 627–634. 10.1029/WR025i004p00627.

Ashworth, P. J., Lewin, J., 2012. How do big rivers come to be different? *Earth-Science Reviews*, **114**(1–2), 84–107. 10.1016/j.earscirev.2012.05.003.

Ashworth, P. J., Ferguson, R. I., Ashmore, P. E., et al., 1992. Measurements in a braided river chute and lobe: 2. Sorting of bed load during entrainment, transport, and deposition. *Water Resources Research*, **28**(7), 1887–1896. 10.1029/92wr00702.

Ashworth, P. J., Best, J. L., Roden, J. E., Bristow, C. S., Klaassen, G. J., 2000. Morphological evolution and dynamics of a large, sand braid-bar, Jamuna River, Bangladesh. *Sedimentology*, **47**(3), 533–555. 10.1046/j.1365–3091.2000.00305.x.

Ashworth, P. J., Smith, G. H. S., Best, J. L., et al., 2011. Evolution and sedimentology of a channel fill in the sandy braided South Saskatchewan River and its comparison to the deposits of an adjacent compound bar. *Sedimentology*, **58**(7), 1860–1883.

Aslan, A., Autin, W. J., 1999. Evolution of the Holocene Mississippi River floodplain, Ferriday, Louisiana: insights on the origin of fine-grained floodplains. *Journal of Sedimentary Research*, **69**(4), 800–815. 10.2110/jsr.69.800.

Asselman, N. E. M., 2000. Fitting and interpretation of sediment rating curves. *Journal of Hydrology*, **234**(3–4), 228–248. 10.1016/s0022-1694(00)00253–5.

Asselman, N. E. M., Middelkoop, H., 1995. Floodplain sedimentation – quantities, patterns, and processes. *Earth Surface Processes and Landforms*, **20**(6), 481–499. 10.1002/esp.3290200602.

Attal, M., Lave, J., 2009. Pebble abrasion during fluvial transport: experimental results and implications for the evolution of the sediment load along rivers. *Journal of Geophysical Research – Earth Surface*, **114**. 10.1029/2009jf001328.

Ayles, C. P., Church, M., 2015. Downstream channel gradation in the regulated Peace River. In: M. Church (ed.), *Regulation of Peace River: A Case Study for River Management*. Wiley, Chichester, UK, pp. 39–66.

Baartman, J. E. M., Jetten, V. G., Ritsema, C. J., de Vente, J., 2012. Exploring effects of rainfall intensity and duration on soil erosion at the catchment scale using openLISEM: Prado catchment, SE Spain. *Hydrological Processes*, **26**(7), 1034–1049. 10.1002/hyp.8196.

Bagherimiyab, F., Lemmin, U., 2013. Shear velocity estimates in rough-bed open-channel flow. *Earth Surface Processes and Landforms*, **38**(14), 1714–1724. 10.1002/esp.3421.

Bagnold, R. A., 1966. An approach to the sediment transport problem from general physics. U.S. Geological Survey Professional Paper 422-I, U.S. Government Printing Office, Washington, DC.

Bagnold, R. A., 1973. Nature of saltation and of "bed-load" transport in water. *Proceedings of the Royal Society of London, Series A – Mathematical Physical and Engineering Sciences*, **332**(1591), 473–504. 10.1098/rspa.1973.0038.

Bagnold, R. A., 1980. An empirical correlation of bedload transport rates in flumes and natural rivers. *Proceedings of the Royal Society of London, Series A – Mathematical Physical and Engineering Sciences*, **372**(1751), 453–473. 10.1098/rspa.1980.0122.

Bai, S., Li, J., 2013. Sediment wash-off from an impervious urban land surface. *Journal of Hydrologic Engineering*, **18**(5), 488–498. 10.1061/(asce)he.1943–5584.0000654.

Bailey, B. B., 1967. *James Hutton – The Founder of Modern Geology*. Elsevier, New York.

Bak, P., 1996. *How Nature Works: The Science of Self-Organized Criticality*. Copernicus, New York.

Baker, V. R., 1977. Stream channel response to floods, with examples from central Texas. *Geological Society of America Bulletin*, **88**(8), 1057–1071. 10.1130/0016–7606(1977)88<1057:srtfwe>2.0.co;2.

Baker, V. R., Costa, J., 1987. Flood power. In: L. Mayer, D. B. Nash (eds.), *Catastrophic Flooding*. Allen and Unwin, London, pp. 1–22.

Baker, V. R., Kochel, R. C., 1988. Flood sedimentation in bedrock fluvial systems. In: V. R. Baker, R. C. Kochel, P. C. Patton (eds.), *Flood Geomorphology*. Wiley and Sons, New York, pp. 123–137.

Bakke, P. D., Basdekas, P. O., Dawdy, D. R., Klingeman, P. C., 1999. Calibrated Parker-Klingeman model for gravel transport. *Journal of Hydraulic Engineering*, **125**(6), 657–660. 10.1061/(asce)0733–9429(1999)125:6(657).

Bakke, P. D., Sklar, L. S., Dawdy, D. R., Wang, W. C., 2017. The design of a site-calibrated Parker-Klingeman gravel transport model. *Water*, **9**(6), 441; 10.3390/w9060441.

Baranya, S., Olsen, N. R. B., Jozsa, J., 2015. Flow analysis of a river confluence with field measurements and RANS model with next grid approach. *River Research and Applications*, **31**(1), 28–41. 10.1002/rra.2718.

Barman, B., Kumar, B., Sarma, A. K., 2018. Turbulent flow structures and geomorphic characteristics of a mining affected alluvial channel. *Earth Surface Processes and Landforms*, **43**(9), 1811–1824. 10.1002/esp.4355.

Barnard, R. S., Melhorn, W. N., 1982. Morphologic and morphometric response to channelization: the case of history of the Big Pine Creek Ditch, Benton County, Iowa. In: R. G. Craig, J. L. Craft (eds.), *Applied Geomorphology*. George Allen and Unwin, London, pp. 224–239.

Barnes, H. H., 1967. Roughness characteristics of natural channels. U.S. Geological Survey Water-Supply Paper 1849, U.S. Government Printing Office, Washington, DC.

Barry, J. J., Buffington, J. M., Goodwin, P., King, J. G., Emmett, W. W., 2008. Performance of bed-load transport equations relative to geomorphic significance: predicting effective discharge and its transport rate. *Journal of Hydraulic Engineering*, **134**(5), 601–615. 10.1061/(asce)0733–9429(2008)134:5(601).

Barth, N. A., Villarini, G., Nayak, M. A., White, K., 2017. Mixed populations and annual flood frequency estimates in the western United States: The role of atmospheric rivers. *Water Resources Research*, **53**(1), 257–269. 10.1002/2016wr019064.

Basso, S., Frascati, A., Marani, M., Schirmer, M., Botter, G., 2015. Climatic and landscape controls on effective discharge. *Geophysical Research Letters*, **42**(20), 8441–8447. 10.1002/2015gl066014.

Bates, B. C., 1990. A statistical log piecewise linear model of at-station hydraulic geometry. *Water Resources Research*, **26**(1), 109–118. 10.1029/WR026i001p00109.

Bathurst, J. C., 1985. Flow resistance estimation in mountain rivers. *Journal of Hydraulic Engineering*, **111**(4), 625–643.

Bathurst, J. C., 1987. Critical conditions for bed material movement in steep, boulder-bed streams. In: R. L. Beschta, T. Blinn, G. E. Grant, G. G. Ice, F. J. Swanson (eds.), *Erosion and Sedimentation in the Pacific Rim*. IAHS Publication No. 165, IAHS Press, Wallingford, UK, pp. 309–318.

Bathurst, J. C., 2013. Critical conditions for particle motion in coarse bed materials of nonuniform size distribution. *Geomorphology*, **197**, 170–184.

Bathurst, J. C., Thorne, C. R., Hey, R. D., 1977. Direct measurements of secondary currents in river bends. *Nature*, **269**(5628), 504–506.

Bathurst, J. C., Thorne, C. R., Hey, R. D., 1979. Secondary flow and shear stress at river bends. *Journal of the Hydraulics Division – ASCE*, **105**(10), 1277–1295.

Bathurst, J. C., Benson, I. A., Valentine, E. M., Nalluri, C., 2002. Overbank sediment deposition patterns for straight and meandering flume channels. *Earth Surface Processes and Landforms*, **27**(6), 659–665.

Bauer, B. O., Sherman, D. J., Wolcott, J. F., 1992. Sources of uncertainty in shear stress and roughness length estimates derived from velocity profiles. *Professional Geographer*, **44**(4), 453–464.

Bayat, E., Rodriguez, J. F., Saco, P. M., et al., 2017. A tale of two riffles: using multidimensional, multifractional, time-varying sediment transport to assess self-maintenance in pool-riffle sequences. *Water Resources Research*, **53**(3), 2095–2113. 10.1002/2016wr019464.

Baynes, E. R. C., Lague, D., Attal, M., et al., 2018a. River self-organisation inhibits discharge control on waterfall migration. *Scientific Reports*, **8**. 10.1038/s41598-018-20767-6.

Baynes, E. R. C., van de Lageweg, W. I., McLelland, S. J., et al., 2018b. Beyond equilibrium: Re-evaluating physical modelling of fluvial systems to represent climate changes. *Earth-Science Reviews*, **181**, 82–97.

Beach, T., 1994. The fate of eroded soil – sediment sink and sediment budgets of agrarian landscapes in southern Minnesota, 1851–1988. *Annals of the Association of American Geographers*, **84**(1), 5–28.

Beaumont, C., Kooi, H., Willet, S., 2000. Coupled tectonic-surface process models with applications to rifted margins and collision orogens. In: M. A. Summerfield (ed.), *Geomorphology and Global Tectonics*. Wiley and Sons, Chichester, UK, pp. 29–55.

Beauvais, A. A., Montgomery, D. R., 1997. Are channel networks statistically self-similar? *Geology*, **25**(12), 1063–1066.

Beechie, T., Imaki, H., 2014. Predicting natural channel patterns based on landscape and geomorphic controls in the Columbia River basin, USA. *Water Resources Research*, **50**(1), 39–57. 10.1002/2013wr013629.

Beechie, T. J., Liermann, M., Pollock, M. M., Baker, S., Davies, J., 2006. Channel pattern and river-floodplain dynamics in forested mountain river systems. *Geomorphology*, **78**(1–2), 124–141.

Beechie, T. J., Sear, D. A., Olden, J. D., et al., 2010. Process-based principles for restoring river ecosystems. *Bioscience*, **60**(3), 209–222.

Beechie, T., Richardson, J. S., Gurnell, A. M., Negishi, J., 2013a. Watershed processes, human impacts, and process-based restoration. In: P. Roni, T. Beechie (eds.), *Stream and Watershed Restoration: A Guide to Restoring Riverine Processes and Habitats*. Wiley-Blackwell, Chichester, UK, pp. 11–49.

Beechie, T., Pess, G.R., Morley, S., et al., 2013b. Watershed assessments and identification of restoration needs. In: T. Beechie, P. Roni (eds.), *Stream and Watershed Restoration: A Guide to Restoring Riverine Processes and Habitats*. Wiley-Blackwell, Chichester, UK, pp. 50–113.

Beer, T., Borgas, M., 1993. Horton Laws and the fractal nature of streams. *Water Resources Research*, **29**(5), 1475–1487.

Beeson, C. E., Doyle, P. F., 1995. Comparison of bank erosion at vegetated and non-vegetated channel bends. *Water Resources Bulletin*, **31**(6), 983–990.

Begin, Z. B., Meyer, D. F., Schumm, S. A., 1981. Development of longitudinal profiles of alluvial channels in response to base-level lowering. *Earth Surface Processes and Landforms*, **6**(1), 49–68.

Belmont, P., Willenbring, J. K., Schottler, S. P., et al., 2014. Toward generalizable sediment fingerprinting with tracers that are conservative and nonconservative over sediment routing timescales. *Journal of Soils and Sediments*, **14**(8), 1479–1492.

Benaichouche, A., Stab, O., Tessier, B., Cojan, I., 2016. Evaluation of a landscape evolution model to simulate stream piracies: Insights from multivariable numerical tests using the example of the Meuse basin, France. *Geomorphology*, **253**, 168–180.

Benda, L., Poff, N. L., Miller, D., et al., 2004a. The network dynamics hypothesis: How channel networks structure riverine habitats. *Bioscience*, **54**(5), 413–427.

Benda, L., Andras, K., Miller, D., Bigelow, P., 2004b. Confluence effects in rivers: Interactions of basin scale, network geometry, and disturbance regimes. *Water Resources Research*, **40**(5). 10.1029/2003wr002583.

Bendixen, M., Best, J. L., Hackney, C., Lonsmann Iverson, L., 2019. Time is running out for sand. *Nature*, **571**, 29–31.

Benedetti, M. M., 2003. Controls on overbank deposition in the upper Mississippi River. *Geomorphology*, **56**(3–4), 271–290.

Bennett, S. J., 1999. Effect of slope on the growth and migration of headcuts in rills. *Geomorphology*, **30**(3), 273–290.

Bennett, S. J., Alonso, C. V., 2005. Kinematics of flow within headcut scour holes on hillslopes. *Water Resources Research*, **41**(9). 10.1029/2004wr003752.

Bennett, S. J., Liu, R. J., 2016. Basin self-similarity, Hack's law, and the evolution of experimental rill networks. *Geology*, **44**(1), 35–38.

Bennett, S. J., Hou, Y. T., Atkinson, J. F., 2014. Turbulence suppression by suspended sediment within a geophysical flow. *Environmental Fluid Mechanics*, **14**(4), 771–794.

Bennett, S. J., Gordon, L. M., Neroni, V., Wells, R. R., 2015. Emergence, persistence, and organization of rill networks on a soil-mantled experimental landscape. *Natural Hazards*, **79**, S7–S24.

Benson, M. A., Thomas, D. M., 1966. A definition of dominant discharge. *International Association of Hydrological Sciences Bulletin*, **11**, 76–80.

Bent, G. C., Waite, A. M., 2013. Equations for estimating bankfull channel geometry and discharge for streams in Massachusetts. U.S. Geological Survey Scientific Investigations Report 2013–5155.

Bentley, P. J., Gulbrandsen, M., Kyvik, S., 2015. The relationship between basic and applied research in universities. *Higher Education*, **70**(4), 689–709.

Bergeron, N. E., Abrahams, A. D., 1992. Estimating shear velocity and roughness length from velocity profiles. *Water Resources Research*, **28**(8), 2155–2158.

Berhanu, M., Petroff, A., Devauchelle, O., Kudrolli, A., Rothman, D. H., 2012. Shape and dynamics of seepage erosion in a horizontal granular bed. *Physical Review E*, **86**(4), 9. 10.1103/PhysRevE.86.041304.

Berlin, M. M., Anderson, R. S., 2007. Modeling of knickpoint retreat on the Roan Plateau, western Colorado. *Journal of Geophysical Research – Earth Surface*, **112**(F3). 10.1029/2006jf000553.

Bernhardt, E. S., Palmer, M. A., Allan, J. D., et al., 2005. Synthesizing US river restoration efforts. *Science*, **308**(5722), 636–637.

Bernhardt, E. S., Sudduth, E. B., Palmer, M. A., et al., 2007. Restoring rivers one reach at a time: results from a survey of US river restoration practitioners. *Restoration Ecology*, **15**(3), 482–493.

Bertoldi, W., 2012. Life of a bifurcation in a gravel-bed braided river. *Earth Surface Processes and Landforms*, **37**(12), 1327–1336.

Bertoldi, W., Tubino, M., 2005. Bed and bank evolution of bifurcating channels. *Water Resources Research*, **41**(7). 10.1029/2004wr003333.

Bertoldi, W., Tubino, M., 2007. River bifurcations: experimental observations on equilibrium configurations. *Water Resources Research*, **43**(10). 10.1029/2007wr005907.

Bertoldi, W., Amplatz, T., Miori, T., Zanoni, L., Tubino, M., 2006. Bed load fluctuations and channel processes in a braided network laboratory model. In: R. Ferreira, C. Alves, G. Leal, A. Cardoso (eds.), *River Flow 2006*, Vols. 1 & 2. Taylor & Francis, London, pp. 937–945.

Bertoldi, W., Zanoni, L., Tubino, M., 2009a. Planform dynamics of braided streams. *Earth Surface Processes and Landforms*, **34**(4), 547–557.

Bertoldi, W., Ashmore, P., Tubino, M., 2009b. A method for estimating the mean bed load flux in braided rivers. *Geomorphology*, **103**(3), 330–340. 10.1016/j.geomorph.2008.06.014.

Bertoldi, W., Zanoni, L., Miori, S., Repetto, R., Tubino, M., 2009c. Interaction between migrating bars and bifurcations in gravel bed rivers. *Water Resources Research*, **45**. 10.1029/2008wr007086.

Bertoldi, W., Zanoni, L., Tubino, M., 2010. Assessment of morphological changes induced by flow and flood pulses in a gravel bed braided river: the Tagliamento River (Italy). *Geomorphology*, **114**(3), 348–360.

Bertoldi, W., Gurnell, A. M., Drake, N. A., 2011a. The topographic signature of vegetation development along a braided river: results of a combined analysis of airborne lidar, color air photographs, and ground measurements. *Water Resources Research*, **47**. 10.1029/2010wr010319.

Bertoldi, W., Drake, N. A., Gurnell, A. M., 2011b. Interactions between river flows and colonizing vegetation on a braided river: exploring spatial and temporal dynamics in riparian vegetation cover using satellite data. *Earth Surface Processes and Landforms*, **36**(11), 1474–1486.

Bertoldi, W., Welber, M., Gurnell, A. M., et al., 2015. Physical modelling of the combined effect of vegetation and wood on river morphology. *Geomorphology*, **246**, 178–187.

Beschta, R. L., 1978. Long-term patterns of sediment production following road construction and logging in the Oregon Coast Range. *Water Resources Research*, **14**(6), 1011–1016.

Beschta, R. L., Pyles, M. R., Skaugset, A. E., Surfleet, C. G., 2000. Peakflow responses to forest practices in the western cascades of Oregon, USA. *Journal of Hydrology*, **233**(1–4), 102–120.

Best, J. L., 1987. Flow dynamics at river channel confluences: Implications for sediment transport and bed morphology. In: F. G. Ethridge, R. M. Flores, M. D. Harvey (eds.), *Recent Developments in Fluvial Sedimentology*, Special Publication 39. Society of Economic Paleontologists and Mineralogists, Tulsa, OK, pp. 27–35.

Best, J. L., 1988. Sediment transport and bed morphology at river channel confluences. *Sedimentology*, **35**, 481–498.

Best, J. L., 1992. On the entrainment of sediment and initiation of bed defects – insights from recent developments within turbulent boundary-layer research. *Sedimentology*, **39**(5), 797–811.

Best, J. L., 2005. The fluid dynamics of river dunes: a review and some future research directions. *Journal of Geophysical Research – Earth Surface*, **110**(F4). 10.1029/2004jf000218.

Best, J. L., 2019. Anthropogenic stresses on the world's big rivers. *Nature Geoscience*, **12**(1), 7–21.

Best, J. L., Ashworth, P. J., 1997. Scour in large braided rivers, and the recognition of sequence stratigraphic boundaries. *Nature*, **387**(6630), 275–277.

Best, J. L., Reid, I., 1984. Separation zone at open-channel junctions. *Journal of Hydraulic Engineering*, **110**(11), 1588–1594.

Best, J. L., Rhoads, B. L., 2008. Sediment transport, bed morphology, and the sedimentology of river channel confluences. In: S. P. Rice, A. G. Roy, B. L. Rhoads (eds.), *River Confluences, Tributaries and the Fluvial Network*. Wiley, Chichester, UK, pp. 45–72.

Best, J. L., Roy, A. G., 1991. Mixing layer distortion at the confluence of channels of different depth. *Nature*, **350**(6317), 411–413.

Bettis, E. A., Benn, D. W., Hajic, E. R., 2008. Landscape evolution, alluvial architecture, environmental history, and the archaeological record of the Upper Mississippi River Valley. *Geomorphology*, **101**(1–2), 362–377.

Bettis, R., White, W. R., 1987. Extremal hypotheses applied to river regime. In: C. R. Thorne, J. C. Bathurst, R. D. Hey (eds.), *Sediment Transport in Gravel-Bed Rivers*. Wiley, Chichester, UK, pp. 767–789.

Bevan, V., MacVicar, B., Chapuis, M., et al., 2018. Enlargement and evolution of a semi-alluvial creek in response to urbanization. *Earth Surface Processes and Landforms*, **43**(11), 2295–2312.

Beven, K., 1981. The effect of ordering on the geomorphic effectiveness of hydrologic events. In: T. R. H. Davies, A. J. Pearce (eds.), *Erosion and Sediment Transport in the Pacific Rim Steeplands*. IAHS Publication No. 132, IAHS Press, Wallingford, UK, pp. 510–526.

Bezak, N., Brilly, M., Sraj, M., 2014. Comparison between the peaks-over-threshold method and the annual maximum method for flood frequency analysis. *Hydrological Sciences Journal*, **59**(5), 959–977.

Bhowmik, N. G., Demissie, M., 1982. Bed material sorting in pools and riffles. *Journal of the Hydraulics Division – ASCE*, **108**(10), 1227–1231.

Biedenharn, D. S., Copeland, R. R., Thorne, C. R., et al., 2000. *Effective discharge calculation: A practical guide*. U.S. Army Corps of Engineers Research and Development Center, Washington, DC.

Bieger, K., Rathjens, H., Allen, P. M., Arnold, J. G., 2015. Development and evaluation of bankfull hydraulic geometry relationships for the physiographic regions of the United States. *Journal of the American Water Resources Association*, **51**(3), 842–858.

Billi, P., Preciso, E., Salemi, E., 2014. Field investigation on step-pool morphology and processes in steep mountain streams. *Agriculture and Forestry*, **60**(3), 7–28.

Bingner, R. L., Theurer, F. D., Yuan, Y., 2015. *AnnAGNPS Technical Processes*. National Sedimentation Laboratory, Agricultural Research Service, U.S. Department of Agriculture, Oxford, Mississippi.

Bird, G., Brewer, P. A., Macklin, M. G., et al., 2008. River system recovery following the Novat-Rosu tailings dam failure, Maramures County, Romania. *Applied Geochemistry*, **23**(12), 3498–3518.

Biron, P., Roy, A. G., Best, J. L., Boyer, C. J., 1993. Bed morphology and sedimentology at the confluence of unequal depth channels. *Geomorphology*, **8**, 115–129.

Biron, P., Best, J. L., Roy, A. G., 1996a. Effects of bed discordance on flow dynamics at open channel confluences. *Journal of Hydraulic Engineering*, **122**(12), 676–682.

Biron, P., Boy, A. G., Best, J. L., 1996b. Turbulent flow structure at concordant and discordant open-channel confluences. *Experiments in Fluids*, **21**(6), 437–446.

Biron, P. M., Richer, A., Kirkbride, A. D., Roy, A. G., Han, S., 2002. Spatial patterns of water surface topography at a river confluence. *Earth Surface Processes and Landforms*, **27**(9), 913–928.

Biron, P. M., Robson, C., Lapointe, M. F., Gaskin, S. J., 2004a. Comparing different methods of bed shear stress estimates in simple and complex flow fields. *Earth Surface Processes and Landforms*, **29**(11), 1403–1415.

Biron, P. M., Ramamurthy, A. S., Han, S., 2004b. Three-dimensional numerical modeling of mixing at river confluences. *Journal of Hydraulic Engineering*, **130**(3), 243–253.

Biron, P. M., Buffin-Belanger, T., Larocque, M., et al., 2014. Freedom space for rivers: A sustainable management approach to enhance river resilience. *Environmental Management*, **54**(5), 1056–1073.

Biron, P. M., Buffin-Belanger, T., Masse, S., 2018. The need for river management and stream restoration practices to integrate hydrogeomorphology. *Canadian Geographer*, **62**(2), 288–295.

Bishop, P., 2007. Long-term landscape evolution: Linking tectonics and surface processes. *Earth Surface Processes and Landforms*, **32**(3), 329–365.

Bishop, P., Hoey, T. B., Jansen, J. D., Artza, I. L., 2005. Knickpoint recession rate and catchment area: The case of uplifted rivers in Eastern Scotland. *Earth Surface Processes and Landforms*, **30**(6), 767–778.

Blainey, J. B., Webb, R. H., Moss, M. E., Baker, V. R., 2002. Bias and information content of paleoflood data in flood-frequency analysis. In: P. K. House, R. H. Webb, V. R. Baker, D. R. Levish (eds.), *Ancient Floods, Modern Hazards: Principles and Applications of Paleoflood Hydrology*. American Geophysical Union, Washington, DC, pp. 161–174.

Blanckaert, K., 2009. Saturation of curvature-induced secondary flow, energy losses, and turbulence in sharp open-channel bends: Laboratory experiments, analysis, and modeling. *Journal of Geophysical Research – Earth Surface*, **114**. 10.1029/2008jf001137.

Blanckaert, K., 2010. Topographic steering, flow recirculation, velocity redistribution, and bed topography in sharp meander bends. *Water Resources Research*, **46**. 10.1029/2009wr008303.

Blanckaert, K., 2011. Hydrodynamic processes in sharp meander bends and their morphological implications. *Journal of Geophysical Research – Earth Surface*, **116**. 10.1029/2010jf001806.

Blanckaert, K., 2015. Flow separation at convex banks in open channels. *Journal of Fluid Mechanics*, **779**, 432–467.

Blanckaert, K., de Vriend, H. J., 2004. Secondary flow in sharp open-channel bends. *Journal of Fluid Mechanics*, **498**, 353–380.

Blanckaert, K., de Vriend, H. J., 2010. Meander dynamics: a nonlinear model without curvature restrictions for flow in open-channel bends. *Journal of Geophysical Research – Earth Surface*, **115**. 10.1029/2009jf001301.

Blanckaert, K., Graf, W. H., 2004. Momentum transport in sharp open-channel bends. *Journal of Hydraulic Engineering*, **130**(3), 186–198.

Blanckaert, K., Duarte, A., Chen, Q., Schleiss, A. J., 2012. Flow processes near smooth and rough (concave) outer banks in curved open channels. *Journal of Geophysical Research – Earth Surface*, **117**. 10.1029/2012jf002414.

Blanckaert, K., Duarte, A., Schleiss, A. J., 2010. Influence of shallowness, bank inclination and bank roughness on the variability of flow patterns and boundary shear stress due to secondary currents in straight open-channels. *Advances in Water Resources*, 33(9), 1062–1074. 10.1016/j.advwatres.2010.06.012.

Blanckaert, K., Kleinhans, M. G., McLelland, S. J., et al., 2013. Flow separation at the inner (convex) and outer (concave) banks of constant-width and widening open-channel bends. *Earth Surface Processes and Landforms*, **38**(7), 696–716.

Bledsoe, B. P., Watson, C. C., 2001a. Logistic analysis of channel pattern thresholds: Meandering, braiding, and incising. *Geomorphology*, **38**(3–4), 281–300.

Bledsoe, B. P., Watson, C. C., 2001b. Effects of urbanization on channel instability. *Journal of the American Water Resources Association*, **37**(2), 255–270.

Bledsoe, B. P., Watson, C. C., Biedenharn, D. S., 2002. Quantification of incised channel evolution and equilibrium. *Journal of the American Water Resources Association*, **38**(3), 861–870.

Blettler, M. C. M., Amsler, M. L., Ezcurra de Drago, I., et al., 2015. The impact of significant input of fine sediment on benthic fauna at tributary junctions: A case study of the Bermejo-Paraguay River confluence, Argentina. *Ecohydrology*, **8**(2), 340–352.

Blom, A., Viparelli, E., Chavarrias, V. C., 2016. The graded alluvial river: Profile concavity and downstream fining. *Geophysical Research Letters*, **43**(12), 6285–6293. 10.1002/2016gl068898.

Blom, A., Arkesteijn, L., Chavarrias, V., Viparelli, E., 2017a. The equilibrium alluvial river under variable flow and its channel-forming discharge. *Journal of Geophysical Research – Earth Surface*, **122**(10), 1924–1948.

Blom, A., Chavarrias, V., Ferguson, R. I., Viparelli, E., 2017b. Advance, retreat, and halt of abrupt gravel-sand transitions in alluvial rivers. *Geophysical Research Letters*, **44**(19), 9751–9760. 10.1002/2017gl074231.

Blondeaux, P., Seminara, G., 1985. A unified bar bend theory of river meanders. *Journal of Fluid Mechanics*, **157**(AUG), 449–470.

Bloschl, G., Gaal, L., Hall, J., et al., 2015. Increasing river floods: fiction or reality? *Water*, **2**(4), 329–344.

Blott, S. J., Pye, K., 2001. GRADISTAT: A grain size distribution and statistics package for the analysis of unconsolidated sediments. *Earth Surface Processes and Landforms*, **26**(11), 1237–1248.

Blott, S. J., Pye, K., 2012. Particle size scales and classification of sediment types based on particle size distributions: Review and recommended procedures. *Sedimentology*, **59**(7), 2071–2096.

Bluck, B. J., 1987. Bed forms and clast size changes in gravel-bed rivers. In: K. S. Richards (ed.), *River Channels: Environment and Process*. Blackwell, Oxford, UK, pp. 159–178.

Blue, B., 2018. What's wrong with healthy rivers? Promise and practice in the search for a guiding ideal for freshwater management. *Progress in Physical Geography*, **42**(4), 462–477.

Blue, B., Brierley, G., 2016. "But what do you measure?" Prospects for a constructive critical physical geography. *Area*, **48**(2), 190–197.

Blum, M. D., Guccione, M. J., Wysocki, D. A., Robnett, P. C., Rutledge, E. M., 2000. Late Pleistocene evolution of the lower Mississippi River valley, southern Missouri to Arkansas. *Geological Society of America Bulletin*, **112**(2), 221–235.

Bolla Pittaluga, M., Repetto, R., Tubino, M., 2003. Channel bifurcation in braided rivers: Equilibrium configurations and stability. *Water Resources Research*, **39**(3). 10.1029/2001wr001112.

Bolla Pittaluga, M., Coco, G., Kleinhans, M. G., 2015. A unified framework for stability of channel bifurcations in gravel and sand fluvial systems. *Geophysical Research Letters*, **42**(18), 7521–7536.

Bomhof, J., Rennie, C. D., Jenkinson, R. W., 2015. Use of local soil and vegetation classifications to improve regional downstream hydraulic geometry relations. *Journal of Hydraulic Engineering*, **141**(5). 10.1061/(asce)hy.1943–7900.0000978.

Booker, D. J., Sear, D. A., Payne, A. J., 2001. Modelling three-dimensional flow structures and patterns of boundary shear stress in a natural pool-riffle sequence. *Earth Surface Processes and Landforms*, **26**(5), 553–576.

Booth, D. B., Fischenich, C. J., 2015. A channel evolution model to guide sustainable urban stream restoration. *Area*, **47**(4), 408–421.

Booth, D. B., Roy, A. H., Smith, B., Capps, K. A., 2016. Global perspectives on the urban stream syndrome. *Freshwater Science*, **35**(1), 412–420.

Borrelli, P., Robinson, D. A., Fleischer, L. R., et al., 2017. An assessment of the global impact of 21st century land use change on soil erosion. *Nature Communications*, **8**. 10.1038/s41467-017-02142-7.

Bouchez, J., Lajeunesse, E., Gaillardet, J., et al., 2010. Turbulent mixing in the Amazon River: The isotopic memory of confluences. *Earth and Planetary Science Letters*, **290**(1–2), 37–43.

Boyer, C., Roy, A. G., Best, J. L., 2006. Dynamics of a river channel confluence with discordant beds: Flow turbulence, bed load sediment transport, and bed morphology. *Journal of Geophysical Research – Earth Surface*, **111**(F4). 10.1029/2005JF000458.

Bracken, L. J., Wainwright, J., 2006. Geomorphological equilibrium: myth and metaphor? *Transactions of the Institute of British Geographers*, **31**(2), 167–178.

Bradbrook, K. F., Lane, S. N., Richards, K. S., 2000. Numerical simulation of three-dimensional, time-averaged flow structure at river confluences. *Water Resources Research*, **36** (9), 2731–2746.

Bradbrook, K. F., Lane, S. N., Richards, K. S., Biron, P. M., Roy, A. G., 2001. Role of bed discordance at asymmetrical river confluences. *Journal of Hydraulic Engineering*, **127**(5), 351–368.

Brandt, S. A., 2000. Classification of geomorphological effects downstream of dams. *Catena*, **40**(4), 375–401.

Brasington, J., Rumsby, B. T., McVey, R. A., 2000. Monitoring and modelling morphological change in a braided gravel-bed river using high resolution GPS-based survey. *Earth Surface Processes and Landforms*, **25**(9), 973–990.

Brasington, J., Langham, J., Rumsby, B., 2003. Methodological sensitivity of morphometric estimates of coarse fluvial sediment transport. *Geomorphology*, **53**(3–4), 299–316.

Braudrick, C. A., Dietrich, W. E., Leverich, G. T., Sklar, L. S., 2009. Experimental evidence for the conditions necessary to sustain meandering in coarse-bedded rivers. *Proceedings of the National Academy of Sciences, of the United States of America* **106**(40), 16936–16941.

Braun, J., Sambridge, M., 1997. Modelling landscape evolution on geological time scales: A new method based on irregular spatial discretization. *Basin Research*, **9**(1), 27–52.

Bray, D. I., 1982. Regime equations for gravel-bed rivers. In: R. D. Hey, J. C. Bathurst, C. R. Thorne (eds.), *Gravel-Bed Rivers*. Wiley, Chichester, UK, pp. 517–542.

Brayshaw, A. C., 1984. Characteristics and origin of cluster bedforms in coarse-grained alluvial channels. In: E. H. Koster, R. J. Steele (eds.), Sedimentology of Gravels and Conglomerates. Canadian Society of Petroleum Geologists Memoir 10, pp. 77–85.

Brebner, A., Wilson, K. C., 1967. Derivation of regime equations from relationships for pressurized flow by use of minimum energy degradation rate. *Proceedings of the Institution of Civil Engineers*, **36**(JAN), 47–62.

Brenna, A., Surian, N., Mao, L., 2019. Virtual velocity approach for estimating bed material transport in gravel-bed rivers: Key factors and significance. *Water Resources Research*, **55**(2), 1651–1674. 10.1029/2018wr023556.

Brennan, S. R., Schindler, D. E., Cline, T. J., et al., 2019. Shifting habitat mosaics and fish production across river basins. *Science*, **364**(6442), 783–786.

Bressan, F., Papanicolaou, A. N., Abban, B., 2014. A model for knickpoint migration in first- and second-order streams. *Geophysical Research Letters*, **41**(14), 4987–4996. 10.1002/2014gl060823.

Brewer, P. A., Lewin, J., 1998. Planform cyclicity in an unstable reach: Complex fluvial response to environmental change. *Earth Surface Processes and Landforms*, **23**(11), 989–1008.

Brewer, P. A., Leeks, G. J. L., Lewin, J., 1992. Direct measurement of in-channel abrasion processes. In: J. Bogen, D. E. Walling, T. J. Day (eds.), *Erosion and Sediment Transport Monitoring Programmes in River Basins*. IAHS Publication No. 210, IAHS Press, Wallingford, UK, pp. 21–29.

Bridge, J. S., 1985. Paleochannel patterns inferred from alluvial deposits – a critical evaluation. *Journal of Sedimentary Petrology*, **55**(4), 579–589.

Bridge, J. S., 1993. The interaction between channel geometry, water flow, sediment transport and deposition in braided rivers. In: J. L. Best, C. S. Bristow (eds.), *Braided Rivers*. Special Publication No. 75. Geological Society of London, London, pp. 13–71.

Bridge, J. S., 2003. *Rivers and Floodplains: Form, Process and Sedimentary Record*. Blackwell Publishing, Oxford, UK.

Bridge, J. S., Bennett, S. J., 1992. A model for the entrainment and transport of sediment grains of mixed sizes, shapes, and densities. *Water Resources Research*, **28**(2), 337–363.

Bridge, J. S., Gabel, S. L., 1992. Flow and sediment dynamics in a low sinuosity braided river – Calamus River, Nebraska Sandhills. *Sedimentology*, **39**(1), 125–142.

Bridge, J. S., Lunt, I. A., 2006. Depositional models of braided rivers. In: G. H. S. Smith, J. L. Best, C. S. Bristow, G. E. Petts (eds.), *Braided Rivers: Process, Deposits, Ecology and Management*. Special Publications of the International Association of Sedimentologists, 36, Blackwell, Malden, MA, pp. 11–50.

Bridge, J. S., Smith, N. D., Trent, F., Gabel, S. L., Bernstein, P., 1986. Sedimentology and morphology of a low-sinuosity river – Calamus River, Nebraska Sand Hills. *Sedimentology*, **33**(6), 851–870.

Bridge, J. S., Alexander, J., Collier, R. E. L., Gawthorpe, R. L., Jarvis, J., 1995. Ground-penetrating radar and coring used to study the large-scale structure of point-bar deposits in 3 dimensions. *Sedimentology*, **42**(6), 839–852.

Brierley, G. J., 1989. River planform facies models – the sedimentology of braided, wandering and meandering reaches of the Squamish River, British Columbia. *Sedimentary Geology*, **61**(1–2), 17–35.

Brierley, G. J., 1991a. Bar sedimentology of the Squamish River, British Columbia – definition and application of morphostratigraphic units. *Journal of Sedimentary Petrology*, **61**(2), 211–225.

Brierley, G. J., 1991b. Floodplain sedimentology of the Squamish River, British Columbia – relevance of element analysis. *Sedimentology*, **38**(4), 735–750.

Brierley, G. J., 2010. Landscape memory: The imprint of the past on contemporary landscape forms and processes. *Area*, **42**(1), 76–85.

Brierley, G. J., Hickin, E. J., 1985. The downstream gradation of particle size in the Squamish River, British Columbia. *Earth Surface Processes and Landforms*, **10**(6), 597–606.

Brierley, G. J., Hickin, E. J., 1991. Channel planform as a non-controlling factor in fluvial sedimentology – the case of the Squamish River floodplain, British Columbia. *Sedimentary Geology*, **75**(1–2), 67–83.

Brierley, G. J., Hickin, E. J., 1992. Floodplain development based on selective preservation of sediments, Squamish River, British Columbia. *Geomorphology*, **4**(6), 381–391.

Brierley, G., Fryirs, K., 1998. A fluvial sediment budget for upper Wolumla Creek, south coast, New South Wales, Australia. *Australian Geographer*, **29**(2), U1-111.

Brierley, G., Fryirs, K., 2005. *Geomorphology and River Management*. Blackwell, Malden, MA.

Brierley, G., Fryirs, K., 2009. Don't fight the site: Three geomorphic considerations in catchment-scale river rehabilitation planning. *Environmental Management*, **43**(6), 1201–1218.

Brierley, G., Fryirs, K., 2012. *Geomorphic Analysis of Rivers: An Approach to Reading the Landscape*. Wiley, Chichester, UK.

Brierley, G. J., Fryirs, K. A., 2016. The use of evolutionary trajectories to guide "moving targets" in the management of river futures. *River Research and Applications*, **32**(5), 823–835.

Brierley, G. J., Ferguson, R. J., Woolfe, K. J., 1997. What is a fluvial levee? *Sedimentary Geology*, **114**(1–4), 1–9.

Bristow, C. S., Skelly, R. L., Ethridge, F. G., 1999. Crevasse splays from the rapidly aggrading, sand-bed, braided Niobrara River, Nebraska: Effect of base-level rise. *Sedimentology*, **46**(6), 1029–1047.

Brizga, S. O., Finlayson, B. L., 1990. Channel avulsion and river metamorphosis – the case of the Thomson River, Victoria, Australia. *Earth Surface Processes and Landforms*, **15**(5), 391–404.

Brookes, A., 1985. River channelization: Traditional engineering methods, physical consequences and alternative practices. *Progress in Physical Geography*, **9**(1), 44–73.

Brookes, A., 1987a. The distribution and management of channelized streams in Denmark. *Regulated Rivers*, **1**, 3–16.

Brookes, A., 1987b. River channel adjustments downstream from channelization works in England and Wales. *Earth Surface Processes and Landforms*, **12**(4), 337–351.

Brookes, A., 1987c. Restoring the sinuosity of artificially straightened stream channels. *Environmental Geology and Water Sciences*, **10** (1), 33–41.

Brookes, A., 1988. *Channelized Rivers: Perspectives for Environmental Management*. Wiley, Chichester, UK.

Brookes, A., 1995. Challenges and objectives for geomorphology in UK river management. *Earth Surface Processes and Landforms*, **20**(7), 593–610.

Brooks, A. P., Brierley, G. J., 1997. Geomorphic responses of lower Bega River to catchment disturbance, 1851–1926. *Geomorphology*, **18**(3–4), 291–304.

Brooks, A. P., Brierley, G. J., Millar, R. G., 2003. The long-term control of vegetation and woody debris on channel and flood-plain evolution: Insights from a paired catchment study in southeastern Australia. *Geomorphology*, **51**(1–3), 7–29.

Browand, F. K., 1986. The structure of the turbulent mixing layer. *Physica D*, **18**(1–3), 135–148.

Brown, A. G., Carey, C., Erkens, G., et al., 2009. From sedimentary records to sediment budgets: Multiple approaches to catchment sediment flux. *Geomorphology*, **108**(1–2), 35–47.

Brown, A. G., Tooth, S., Chiverrell, R. C., et al., 2013. The Anthropocene: Is there a geomorphological case? *Earth Surface Processes and Landforms*, **38**(4), 431–434.

Brown, A. G., Petit, F., James, L. A., 2016. Archeology and human artefacts. In: G. M. Kondolf, H. Piegay (eds.), *Tools in Fluvial Geomorphology*, 2nd Edition. Wiley, Chichester, UK, pp. 40–55.

Brown, A. G., Tooth, S., Bullard, J. E., et al., 2017. The geomorphology of the Anthropocene: Emergence, status and implications. *Earth Surface Processes and Landforms*, **42**(1), 71–90.

Brune, G. M., 1953. Trap efficiency of reservoirs. *Transactions, American Geophysical Union*, **34**(3), 407–418.

Brunsden, D., 2001. A critical assessment of the sensitivity concept in geomorphology. *Catena*, **42**(2–4), 99–123.

Brunsden, D., Thornes, J. B., 1979. Landscape sensitivity and change. *Transactions of the Institute of British Geographers*, **4**(4), 463–484.

Brush, L. M., Jr., 1961. Drainage basins, channels, and flow characteristics of selected streams in central Pennsylvania U.S. Geological Survey Professional Paper 282-F. U.S. Government Printing Office, Washington, DC.

Bryan, R. B., Jones, J. A. A., 1997. The significance of soil piping processes: inventory and prospect. *Geomorphology*, **20**(3–4), 209–218.

Buffin-Belanger, T., Roy, A. G., 1998. Effects of a pebble cluster on the turbulent structure of a depth-limited flow in a gravel-bed river. *Geomorphology*, **25**(3–4), 249–267.

Buffin-Belanger, T., Roy, A. G., Kirkbride, A. D., 2000. On large-scale flow structures in a gravel-bed river. *Geomorphology*, **32**(3–4), 417–435.

Buffin-Belanger, T., Biron, P. M., Larocque, M., et al., 2015. Freedom space for rivers: An economically viable river management concept in a changing climate. *Geomorphology*, **251**, 137–148.

Buffington, J. M., 1999. The legend of A. F. Shields. *Journal of Hydraulic Engineering*, **125**(4), 376–387.

Buffington, J. M., Montgomery, D. R., 1997. A systematic analysis of eight decades of incipient motion studies, with special reference to gravel-bedded rivers. *Water Resources Research*, **33**(8), 1993–2029.

Buffington, J. M., Montgomery, D. R., 2013. Geomorphic classification of rivers. In: J. W. Shroder (ed.), *Treatise on Geomorphology*, Vol. 9, Fluvial geomorphology, E. Wohl (vol. ed.). Academic Press, San Diego, CA, pp. 730–767.

Buffington, J. M., Dietrich, W. E., Kirchner, J. W., 1992. Friction angle measurements on a naturally formed gravel streambed – implications for critical boundary shear stress. *Water Resources Research*, **28**(2), 411–425.

Buffington, J. M., Lisle, T. E., Woodsmith, R. D., Hilton, S., 2002. Controls on the size and occurrence of pools in coarse-grained forest rivers. *River Research and Applications*, **18**(6), 507–531.

Bull, L. J., 1997. Relative velocities of discharge and sediment waves for the River Severn, UK. *Hydrological Sciences Journal*, **42**(5), 649–660.

Bull, L. J., Kirkby, M. J., 1997. Gully processes and modelling. *Progress in Physical Geography*, **21**(3), 354–374.

Bull, L. J., Lawler, D. M., Leeks, G. J. L., Marks, S., 1995. Downstream changes in suspended sediment fluxes in the River Severn, UK. In: W. R. Osterkamp (ed.), *Effects of Scale on Interpretation and Management of Sediment and Water Quality*. IAHS Publication 226, IAHS Press, Wallingford, UK, pp. 27–37.

Bull, W. B., Scott, K. M., 1974. Impact of mining gravel from urban streambeds in the southwestern United States. *Geology*, **2**, 171–174.

Bunte, K., Abt, S. R., 2001. Sampling Surface and Subsurface Particle-Size Distributions in Wadable Gravel- and Cobble-Bed Streams for Analyses in Sediment Transport, Hydraulics, and Streambed Monitoring. General Technical Report RMRS-GTR-74, Rocky Mountain Research Station, Forest Service, U.S. Department of Agriculture, Fort Collins, CO.

Bunte, K., Abt, S. R., Swingle, K. W., Cenderelli, D. A., 2014. Effective discharge in Rocky Mountain headwater streams. *Journal of Hydrology*, **519**, 2136–2147.

Buraas, E. M., Renshaw, C. E., Magilligan, F. J., Dade, W. B., 2014. Impact of reach geometry on stream channel sensitivity to extreme floods. *Earth Surface Processes and Landforms*, **39**(13), 1778–1789.

Burckhardt, J. C., Todd, B. L., 1998. Riparian forest effect on lateral stream channel migration in the glacial till plains. *Journal of the American Water Resources Association*, **34**(1), 179–184.

Burge, L. M., 2005. Wandering Miramichi rivers, New Brunswick, Canada. *Geomorphology*, **69**(1–4), 253–274. 10.1016/j.geomorph.2005.01.010.

Burge, L. M., 2006. Stability, morphology and surface grain size patterns of channel bifurcation in gravel-cobble bedded anabranching rivers. *Earth Surface Processes and Landforms*, **31** (10), 1211–1226.

Burge, L. M., Lapointe, M. F., 2005. Understanding the temporal dynamics of the wandering Renous River, New Brunswick, Canada. *Earth Surface Processes and Landforms*, **30**(10), 1227–1250.

Burkham, D. E., 1972. Channel changes of the Gila River in the Safford Valley, Arizona 1846–1970. U.S. Geological Survey Professional Paper 655-G, U.S. Government Printing Office Washington, DC.

Buscombe, D., Conley, D. C., 2012. Effective shear stress of graded sediments. *Water Resources Research*, **48**. 10.1029/2010wr010341.

Bussi, G., Dadson, S. J., Bowes, M. J., Whitehead, P. G., 2017. Seasonal and interannual changes in sediment transport identified through sediment rating curves. *Journal of Hydrologic Engineering*, **22**(2). 10.1061/(asce)he.1943–5584.0001466.

Caamano, D., Goodwin, P., Buffington, J. M., Liou, J. C. P., Daley-Laursen, S., 2009. Unifying criterion for the velocity reversal hypothesis in gravel-bed rivers. *Journal of Hydraulic Engineering*, **135**(1), 66–70.

Caamano, D., Goodwin, P., Buffington, J. M., 2012. Flow structure through pool-riffle sequences and a conceptual model for their sustainability in gravel-bed rivers. *River Research and Applications*, **28**(3), 377–389.

Cabezas, A., Angulo-Martinez, M., Gonzalez-Sanchis, M., Jimenez, J. J., Comin, F. A., 2010. Spatial variability in floodplain sedimentation: The use of generalized linear mixed-effects models. *Hydrology and Earth System Sciences*, **14**(8), 1655–1668.

Cahoon, D. R., White, D. A., Lynch, J. C., 2011. Sediment infilling and wetland formation dynamics in an active crevasse splay of the Mississippi River delta. *Geomorphology*, **131**(3–4), 57–68.

Call, B. C., Belmont, P., Schmidt, J. C., Wilcock, P. R., 2017. Changes in floodplain inundation under nonstationary hydrology for an adjustable, alluvial river channel. *Water Resources Research*, **53**(5), 3811–3834.

Calle, M., Alho, P., Benito, G., 2017. Channel dynamics and geomorphic resilience in an ephemeral Mediterranean river affected by gravel mining. *Geomorphology*, **285**, 333–346.

Calver, A., Anderson, M. G., 2004. Conceptual framework for the persistence of flood-initiated geomorphological features. *Transactions of the Institute of British Geographers*, **29**(1), 129–137.

Camacho, R., Yen, B. C., 1991. Nonlinear resistance relationships for alluvial channels. In: B. C. Yen (ed.), *Channel Flow Resistance: Centennial of Manning's Formula*. Water Resources Publications, Highlands, Colorado, pp. 186–194.

Campbell, A. J., Sidle, R. C., 1985. Bedload transport in a pool-riffle sequence of a coastal Alaska stream. *Water Resources Bulletin*, **21**(4), 579–590.

Campodonico, V. A., Garcia, M. G., Pasquini, A. I., 2015. The dissolved chemical and isotopic signature downflow the confluence of two large rivers: The case of the Parana and Paraguay rivers. *Journal of Hydrology*, **528**, 161–176.

Camporeale, C., Perona, P., Porporato, A., Ridolfi, L., 2007. Hierarchy of models for meandering rivers and related morphodynamic processes. *Reviews of Geophysics*, **45**(1). 10.1029/2005rg000185.

Camporeale, C., Perucca, E., Ridolfi, L., 2008. Significance of cutoff in meandering river dynamics. *Journal of Geophysical Research – Earth Surface*, **113**(F1). 10.1029/2006jf000694.

Canovaro, F., Solari, L., 2007. Dissipative analogies between a schematic macro-roughness arrangement and step-pool morphology. *Earth Surface Processes and Landforms*, **32**(11), 1628–1640.

Cant, D. J., Walker, R. G., 1978. Fluvial processes and facies sequences in sandy braided south Saskatchewan River, Canada. *Sedimentology*, **25**(5), 625–648.

Cantelli, A., Muto, T., 2014. Multiple knickpoints in an alluvial river generated by a single instantaneous drop in base level: experimental investigation. *Earth Surface Dynamics*, **2**(1), 271–278.

Cantelli, A., Wong, M., Parker, G., Paola, C., 2007. Numerical model linking bed and bank evolution of incisional channel created by dam removal. *Water Resources Research*, **43**(7). 10.1029/2006wr005621.

Cao, S. Y., Knight, D. W., 1997. Entropy-based design approach of threshold alluvial channels. *Journal of Hydraulic Research*, **35**(4), 505–524.

Cao, S. Y., Knight, D. W., 1998. Design for hydraulic geometry of alluvial channels. *Journal of Hydraulic Engineering*, **124**(5), 484–492.

Cao, Z. X., Egashira, S. J., Carling, P. A., 2003. Role of suspended-sediment particle size in modifying velocity profiles in open channel flows. *Water Resources Research*, **39**(2). 10.1029/2001wr000934.

Capra, A., Di Stefano, C., Ferro, V., Scicolone, B., 2009. Similarity between morphological characteristics of rills and ephemeral gullies in Sicily, Italy. *Hydrological Processes*, **23**(23), 3334–3341.

Carling, P. A., 1983. Threshold of coarse sediment transport in broad and narrow natural streams. *Earth Surface Processes and Landforms*, **8**(1), 1–18.

Carling, P. A., 1984. Deposition of fine and coarse sand in an open-work gravel bed. *Canadian Journal of Fisheries and Aquatic Sciences*, **41**(2), 263–280.

Carling, P., 1988. The concept of dominant discharge applied to two gravel-bed streams in relation to channel stability thresholds. *Earth Surface Processes and Landforms*, **13**(4), 355–367.

Carling, P. A., 1991. An appraisal of the velocity-reversal hypothesis for stable pool riffle sequences in the River Severn, England. *Earth Surface Processes and Landforms*, **16**(1), 19–31.

Carling, P. A., 1999. Subaqueous gravel dunes. *Journal of Sedimentary Research*, **69**(3), 534–545.

Carling, P. A., Orr, H. G., 2000. Morphology of riffle-pool sequences in the River Severn, England. *Earth Surface Processes and Landforms*, **25**(4), 369–384.

Carling, P. A., Reader, N. A., 1982. Structure, composition and bulk properties of upland stream gravels. *Earth Surface Processes and Landforms*, **7**(4), 349–365.

Carling, P. A., Wood, N., 1994. Simulation of flow over pool-riffle topography – a consideration of the velocity reversal hypothesis. *Earth Surface Processes and Landforms*, **19**(4), 319–332.

Carling, P. A., Kelsey, A., Glaister, M. S. 1992. Effect of bed roughness, particle shape and orientation on initial motion criteria. In: P. Billi, R. D. Hey, C. R. Thorne, P. Tacconi (eds.), *Dynamics of Gravel-Bed Rivers*. Wiley, Chichester, UK, pp. 23–38.

Carling, P. A., Cao, Z. X., Holland, M. J., Ervine, D. A., Babaeyan-Koopaei, K., 2002. Turbulent flow across a natural compound channel. *Water Resources Research*, **38**(12). 10.1029/2001wr000902.

Carling, P., Jansen, J., Meshkova, L., 2014. Multichannel rivers: Their definition and classification. *Earth Surface Processes and Landforms*, **39**(1), 26–37.

Carling, P. A., Gupta, N., Atkinson, P. M., He, H. Q., 2016. Criticality in the planform behavior of the Ganges River meanders. *Geology*, **44**(10), 859–862.

Carling, P. A., Perillo, M., Best, J., Garcia, M. H., 2017. The bubble bursts for cavitation in natural rivers: Laboratory experiments reveal minor role in bedrock erosion. *Earth Surface Processes and Landforms*, **42**(9), 1308–1316.

Carlston, C. W., 1969. Downstream variations in hydraulic geometry of streams – special emphasis on mean velocity. *American Journal of Science*, **267**(4), 499–509.

Carretier, S., Poisson, B., Vassallo, R., Pepin, E., Farias, M., 2009. Tectonic interpretation of transient stage erosion rates at different spatial scales in an uplifting block. *Journal of Geophysical Research – Earth Surface*, **114**, 19. 10.1029/2008jf001080.

Carson, M. A., 1984a. The meandering-braided river threshold – a reappraisal. *Journal of Hydrology*, **73**(3–4), 315–334. 10.1016/0022–1694(84)90006–4.

Carson, M. A., 1984b. Observations on the meandering-braided transition, the Canterbury Plains, New Zealand. Part One. *New Zealand Geographer*, **40**, 12–17.

Carson, M. A., 1986. Characteristics of high-energy "meandering" rivers – the Canterbury Plains, New Zealand. *Geological Society of America Bulletin*, **97**(7), 886–895.

Carson, M. A., Griffiths, G. A., 1989. Gravel transport in the braided Waimakariri River – mechanisms, measurements, and predictions. *Journal of Hydrology*, **109**(3–4), 201–220.

Carson, M. A., Lapointe, M. F., 1983. The inherent asymmetry of river meander planform. *Journal of Geology*, **91**(1), 41–55.

Cartigny, M. J. B., Ventra, D., Postma, G., van den Berg, J. H., 2014. Morphodynamics and sedimentary structures of bedforms under supercritical-flow conditions: new insights from flume experiments. *Sedimentology*, 61(3), 712–748. 10.1111/sed.12076.

Castelltort, S., Simpson, G., 2006. River spacing and drainage network growth in widening mountain ranges. *Basin Research*, **18**(3), 267–276.

Castillo, C., Gomez, J. A., 2016. A century of gully erosion research: Urgency, complexity and study approaches. *Earth-Science Reviews*, **160**, 300–319.

Castillo, M., Bishop, P., Jansen, J. D., 2013. Knickpoint retreat and transient bedrock channel morphology triggered by base-level fall in small bedrock river catchments: The case of the Isle of Jura, Scotland. *Geomorphology*, **180**, 1–9.

Castro, J. M., Jackson, P. L., 2001. Bankfull discharge recurrence intervals and regional hydraulic geometry relationships: Patterns in the Pacific Northwest, USA. *Journal of the American Water Resources Association*, **37**(5), 1249–1262.

Cazanacli, D., Smith, N. D., 1998. A study of morphology and texture of natural levees – Cumberland Marshes, Saskatchewan, Canada. *Geomorphology*, **25**(1–2), 43–55.

Celik, A. O., Diplas, P., Dancey, C. L., 2013. Instantaneous turbulent forces and impulse on a rough bed: Implications for initiation of bed material movement. *Water Resources Research*, **49**(4), 2213–2227. 10.1002/wrcr.20210.

Chang, H. H., 1979. Minimum stream power and river channel patterns. *Journal of Hydrology*, **41**, 303–327.

Chang, H. H., 1980. Geometry of gravel streams. *Journal of the Hydraulics Division – ASCE*, **106**(9), 1443–1456.

Chang, H. H., 1988a. *Fluvial Processes in River Engineering*. Wiley, New York.

Chang, H. H., 1988b. On the cause of river meandering. In: W. R. White (ed.), *International Conference on River Regime*. Wiley, New York, pp. 83–93.

Chang, H. Y., Simons, D. B., Woolhiser, D. A., 1971. Flume experiments on alternate bar formation. *Journal of the Waterways, Harbors and Coastal Engineering Division, ASCE*, **97**, 155–165.

Chartrand, S. M., Whiting, P. J., 2000. Alluvial architecture in headwater streams with special emphasis on step-pool topography. *Earth Surface Processes and Landforms*, **25**(6), 583–600.

Chartrand, S. M., Jellinek, M., Whiting, P. J., Stamm, J., 2011. Geometric scaling of step-pools in mountain streams: observations and implications. *Geomorphology*, **129**(1–2), 141–151. 10.1016/j.geomorph.2011.01.020.

Chartrand, S. M., Jellinek, A. M., Hassan, M. A., Ferrer-Boix, C., 2018. Morphodynamics of a width-variable gravel bed stream: new insights on pool-riffle formation from physical experiments. *Journal of Geophysical Research – Earth Surface*, **123**(11), 2735–2766. 10.1029/2017jf004533.

Chen, A., Darbon, J., Morel, J.-M., 2014. Landscape evolution models: A review of their fundamental equations. *Geomorphology*, **219**, 68–86. 10.1016/j.geomorph.2014.04.037.

Chen, D., Jirka, G. H., 1995. Experimental study of plane turbulent wakes in a shallow water layer. *Fluid Dynamics Research*, **16**, 11–41.

Chen, X., Zhu, D. Z., Steffler, P. M., 2017. Secondary currents induced mixing at channel confluences. *Canadian Journal of Civil Engineering*, **44**(12), 1071–1083. 10.1139/cjce-2016-0228.

Cheng, C., Song, Z. Y., Wang, Y. G., Zhang, J. S., 2013. Parameterized expressions for an improved Rouse equation. *International Journal of Sediment Research*, **28**(4), 523–534.

Cheng, L., Yaeger, M., Viglione, A., et al., 2012. Exploring the physical controls of regional patterns of flow duration curves – Part 1: insights from statistical analyses. *Hydrology and Earth System Sciences*, **16**(11), 4435–4446.

Cheng, N. S., 1997. Simplified settling velocity formula for sediment particle. *Journal of Hydraulic Engineering*, **123**(2), 149–152.

Chew, L. C., Ashmore, P. E., 2001. Channel adjustment and a test of rational regime theory in a proglacial braided stream. *Geomorphology*, **37**(1–2), 43–63.

Chin, A., 1989. Step pools in stream channels. *Progress in Physical Geography*, **13**(3), 390–407.

Chin, A., 1998. On the stability of step-pool mountain streams. *Journal of Geology*, **106**(1), 59–69.

Chin, A., 1999a. On the origin of step-pool sequences in mountain streams. *Geophysical Research Letters*, **26**(2), 231–234.

Chin, A., 1999b. The morphologic structure of step-pools in mountain streams. *Geomorphology*, **27**(3–4), 191–204.

Chin, A., 2002. The periodic nature of step-pool mountain streams. *American Journal of Science*, **302**(2), 144–167.

Chin, A., 2003. The geomorphic significance of step-pools in mountain streams. *Geomorphology*, **55**(1–4), 125–137.

Chin, A., 2006. Urban transformation of river landscapes in a global context. *Geomorphology*, **79**(3–4), 460–487.

Chin, A., Phillips, J. D., 2007. The self-organization of step-pools in mountain streams. *Geomorphology*, **83**(3–4), 346–358.

Chin, A., Wohl, E., 2005. Toward a theory for step pools in stream channels. *Progress in Physical Geography*, **29**(3), 275–296.

Chin, A., Daniels, M. D., Urban, M. A., et al., 2008. Perceptions of wood in rivers and challenges for stream restoration in the United States. *Environmental Management*, **41**(6), 893–903.

Chin, A., Gidley, R., Tyner, L., Gregory, K., 2017. Adjustment of dryland stream channels over four decades of urbanization. *Anthropocene*, **20**, 24–36.

Chiu, C. L., Lin, G. F., 1983. Computation of 3-D flow and shear in open channels. *Journal of Hydraulic Engineering*, **109**(11), 1424–1440.

Chone, G., Biron, P. M., 2016. Assessing the relationship between river mobility and aquatic habitat. *River Research and Applications*, **32**(4), 528–539.

Chorley, R. J., Beckinsdale, R. P., 1964. *The History of the Study of Landforms or The Development of Geomorphology. Volume One: Geomorphology before Davis.* Methuen, London.

Chorley, R. J., Beckinsdale, R. P., Dunn, A. J., 1973. *The History of the Study of Landforms or the Development of Geomorphology, Volume 2. The Life and Work of William Morris Davis.* Methuen, London.

Chorley, R. J., Schumm, S. A., Sugden, D. E., 1984. *Geomorphology.* Methuen, London.

Chow, V. T., 1959. *Open-Channel Hydraulics.* McGraw-Hill, New York.

Chu, V. H., Babarutsi, S., 1988. Confinement and bed-friction effects in shallow turbulent mixing layers. *Journal of Hydraulic Engineering*, **114**, 1257–1274.

Chuang, S.-C., Chen, H., Lin, G.-W., Lin, C.-W., Chang, C.-P., 2009. Increase in basin sediment yield from landslides in storms following major seismic disturbance. *Engineering Geology*, **103**(1–2), 59–65.

Church, M., 1972. *Baffin Island Sandurs: A Study of Arctic Fluvial Processes. (Bulletin of the Geological Survey of Canada 216).* Ottawa, Canada.

Church, M., 1983. Pattern of instability in a wandering gravel bed channel. In: J. D. Collinson, J. Lewin (eds.), *Modern and Ancient Fluvial Systems*, International Association of Sedimentologists Special Publication No. 6. Blackwell, Oxford, UK, pp. 169–180.

Church, M., 1995. Geomorphic response to river flow regulation – case studies and time-scales. *Regulated Rivers-Research & Management*, **11**(1), 3–22. 10.1002/rrr.3450110103.

Church, M., 2006. Bed material transport and the morphology of alluvial river channels. *Annual Review of Earth and Planetary Sciences*, **34**, 325–354.

Church, M., 2011. Observations and experiments. In: K. J. Gregory, A. S. Goudie (eds.), *The Sage Handbook of Geomorphology.* Sage Publications, London, pp. 121–141.

Church, M., Ferguson, R. I., 2015. Morphodynamics: rivers beyond steady state. *Water Resources Research*, **51**(4), 1883–1897. 10.1002/2014wr016862.

Church, M., Haschenburger, J. K., 2017. What is the "active layer"? *Water Resources Research*, **53**(1), 5–10. 10.1002/2016wr019675.

Church, M., Hassan, M. A., 1992. Size and distance of travel of unconstrained clasts on a streambed. *Water Resources Research*, **28**(1), 299–303.

Church, M., Kellerhals, R., 1978. Statistics of grain-size variation along a gravel river. *Canadian Journal of Earth Sciences*, **15**(7), 1151–1160.

Church, M., Rice, S. P., 2009. Form and growth of bars in a wandering gravel-bed river. *Earth Surface Processes and Landforms*, **34**(10), 1422–1432.

Church, M., Slaymaker, O., 1989. Disequilibrum of Holocene sediment yield in glaciated British Columbia. *Nature*, **337**(6206), 452–454.

Church, M., Zimmermann, A., 2007. Form and stability of step-pool channels: research progress. *Water Resources Research*, **43**(3). 10.1029/2006wr005037.

Church, M. A., McLean, D. G., Wolcott, J. F., 1987. River bed gravels: sampling and analysis. In: C. R. Thorne, J. C. Bathurst, R. D. Hey (eds.), *Sediment Transport in Gravel-Bed Rivers.* Wiley, Chichester, UK, pp. 43–48.

Church, M., Ham, D., Hassan, M., Slaymaker, O., 1999. Fluvial clastic sediment yield in Canada: Scaled analysis. *Canadian Journal of Earth Sciences*, **36**(8), 1267–1280.

Citterio, A., Piegay, H., 2009. Overbank sedimentation rates in former channel lakes: Characterization and control factors. *Sedimentology*, **56**(2), 461–482.

Claude, N., Rodrigues, S., Bustillo, V., et al., 2012. Estimating bedload transport in a large sand-gravel bed river from direct sampling, dune tracking and empirical formulas. *Geomorphology*, **179**, 40–57.

Clayton, J. A., Pitlick, J., 2007. Spatial and temporal variations in bed load transport intensity in a gravel bed river bend. *Water Resources Research*, **43**(2). 10.1029/2006wr005253.

Clifford, N. J., 1993. Differential bed sedimentology and the maintenance of riffle-pool sequences. *Catena*, **20**(5), 447–468.

Clifford, N. J., Richards, K. S., 1992. The reversal hypothesis and the maintenance of riffle-pool sequences: A review and field appraisal. In: P. A. Carling, G. E. Petts (eds.), *Lowland Floodplain Rivers: Geomorphological Perspectives.* Wiley, Chichester, UK, pp. 43–70.

Clifford, N. J., Hardisty, J., French, J. R., Hart, S., 1993. Downstream variation in bed material characteristics: A turbulence-controlled form-process feedback mechanism In: J. L. Best, C. S. Bristow (eds.), *Braided Rivers*. Geological Society of London Special Publication 75, Geological Society, London, pp. 89–104.

Clubb, F. J., Mudd, S. M., Milodowski, D. T., Hurst, M. D., Slater, L. J., 2014. Objective extraction of channel heads from high-resolution topographic data. *Water Resources Research*, **50**(5), 4283–4304. 10.1002/2013wr015167.

Cockerill, K., Anderson, W. P., Jr., 2014. Creating false images: stream restoration in an urban setting. *Journal of the American Water Resources Association*, **50**, 468–482.

Cohen, S., Kettner, A. J., Syvitski, J. P. M., 2014. Global suspended sediment and water discharge dynamics between 1960 and 2010: continental trends and intra-basin sensitivity. *Global and Planetary Change*, **115**, 44–58.

Cohen, Y., Devauchelle, O., Seybold, H. F., et al., 2015. Path selection in the growth of rivers. *Proceedings of the National Academy of Sciences of the United States of America*, **112**(46), 14132–14137.

Colberg, J. S., Anders, A. M., 2014. Numerical modeling of spatially-variable precipitation and passive margin escarpment evolution. *Geomorphology*, **207**, 203–212.

Collins, A. L., Walling, D. E., 2004. Documenting catchment suspended sediment sources: Problems, approaches and prospects. *Progress in Physical Geography*, **28**(2), 159–196.

Collins, A. L., Walling, D. E., Leeks, G. J. L., 1997. Source type ascription for fluvial suspended sediment based on a quantitative composite fingerprinting technique. *Catena*, **29**(1), 1–27.

Collins, A. L., Walling, D. E., Leeks, G. J. L., 1998. Use of composite fingerprints to determine the provenance of the contemporary suspended sediment load transported by rivers. *Earth Surface Processes and Landforms*, **23**(1), 31–52.

Collins, A. L., Walling, D. E., Webb, L., King, P., 2010. Apportioning catchment scale sediment sources using a modified composite fingerprinting technique incorporating property weightings and prior information. *Geoderma*, **155**(3–4), 249–261.

Collins, B. D., Dunne, T., 1989. Gravel transport, gravel harvesting, and channel-bed degradation in rivers draining the southern Olympic Mountains, Washington, USA. *Environmental Geology and Water Sciences*, **13**(3), 213–224.

Collins, B. D., Montgomery, D. R., Fetherston, K. L., Abbe, T. B., 2012. The floodplain large-wood cycle hypothesis: a mechanism for the physical and biotic structuring of temperate forested alluvial valleys in the North Pacific coastal ecoregion. *Geomorphology*, **139**, 460–470.

Collinson, J. D., 1996. Alluvial sedimentation. In: H. G. Reading (ed.), *Sedimentary Environments: Processes, Facies, and Stratigraphy*. Blackwell Science, Oxford, UK, pp. 37–82.

Colombini, M., 1993. Turbulence-driven secondary flows and formation of sand ridges. *Journal of Fluid Mechanics*, **254**, 701–719.

Colombini, M., Parker, G., 1995. Longitudinal streaks. Journal of Fluid Mechanics, **304**, 161–183.

Colombini, M., Seminara, G., Tubino, M., 1987. Finite-amplitude alternate bars. *Journal of Fluid Mechanics*, **181**, 213–232.

Colosimo, M. F., Wilcock, P. R., 2007. Alluvial sedimentation and erosion in an urbanizing watershed, Gwynns Falls, Maryland. *Journal of the American Water Resources Association*, **43**(2), 499–521.

Comiti, F., Andreoli, A., Lenzi, M. A., 2005. Morphological effects of local scouring in step-pool streams. *Earth Surface Processes and Landforms*, **30**(12), 1567–1581.

Comiti, F., Mao, L., Wilcox, A., Wohl, E. E., Lenzi, M. A., 2007. Field-derived relationships for flow velocity and resistance in high-gradient streams. *Journal of Hydrology*, **340**(1–2), 48–62.

Constantine, C. R., Mount, M. F., Florsheim, J. L., 2003. The effects of longitudinal differences in gravel mobility on the downstream fining pattern in the Cosumnes River, California. *Journal of Geology*, **111**(2), 233–241.

Constantine, C. R., Dunne, T., Hanson, G. J., 2009. Examining the physical meaning of the bank erosion coefficient used in meander migration modeling. *Geomorphology*, **106**(3–4), 242–252.

Constantine, J. A., Dunne, T., 2008. Meander cutoff and the controls on the production of oxbow lakes. *Geology*, **36**(1), 23–26.

Constantine, J. A., McLean, S. R., Dunne, T., 2010a. A mechanism of chute cutoff along large meandering rivers with uniform floodplain topography. *Geological Society of America Bulletin*, **122**(5–6), 855–869.

Constantine, J. A., Dunne, T., Piegay, H., Kondolf, G. M., 2010b. Controls on the alluviation of oxbow lakes by bed-material load along the Sacramento River, California. *Sedimentology*, **57**(2), 389–407.

Constantine, J. A., Dunne, T., Ahmed, J., Legleiter, C., Lazarus, E. D., 2014. Sediment supply as a driver of river meandering and floodplain evolution in the Amazon Basin. *Nature Geoscience*, **7**(12), 899–903.

Constantinescu, G., Koken, M., Zeng, J., 2011a. The structure of turbulent flow in an open channel bend of strong curvature with deformed bed: Insight provided by detached eddy simulation. *Water Resources Research*, **47**. 10.1029/2010wr010114.

Constantinescu, G., Miyawaki, S., Rhoads, B., Sukhodolov, A., Kirkil, G., 2011b. Structure of turbulent flow at a river confluence with momentum and velocity ratios close to 1: Insight provided by an eddy-resolving numerical simulation. *Water Resources Research*, **47**. 10.1029/2010wr010018.

Constantinescu, G., Miyawaki, S., Rhoads, B., Sukhodolov, A., 2012. Numerical analysis of the effect of momentum ratio on the dynamics and sediment-entrainment capacity of coherent flow structures at a stream confluence. *Journal of Geophysical Research – Earth Surface*, **117**. 10.1029/2012jf002452.

Constantinescu, G., Miyawaki, S., Rhoads, B., Sukhodolov, A., 2014. Numerical evaluation of the effects of planform geometry and inflow conditions on flow, turbulence structure, and bed shear velocity at a stream confluence with a concordant bed. *Journal of Geophysical Research – Earth Surface*, **119**(10), 2079–2097. 10.1002/2014jf003244.

Constantinescu, G., Miyawaki, S., Rhoads, B., Sukhodolov, A., 2016. Influence of planform geometry and momentum ratio on thermal mixing at a stream confluence with a concordant bed. *Environmental Fluid Mechanics*, **16**(4), 845–873.

Contos, J., Tripcevich, N., 2014. Correct placement of the most distant source of the Amazon River in the Mantaro River drainage. *Area*, **46**(1), 27–39.

Cook, K. L., Turowski, J. M., Hovius, N., 2013. A demonstration of the importance of bedload transport for fluvial bedrock erosion and knickpoint propagation. *Earth Surface Processes and Landforms*, **38**(7), 683–695.

Coon, W. F., 1998. Estimates of roughness coefficients for selected natural stream channels with vegetated banks in New York. U.S. Geological Survey Open-file Report 93–161, U.S. Geological Survey, Denver, CO.

Cooper, A. H., Brown, T. J., Price, S. J., Ford, J. R., Waters, C. N., 2018. Humans are the most significant global geomorphological driving force of the 21st century. *Anthropocene Review*, **5**(3), 222–229.

Coopersmith, E., Yaeger, M. A., Ye, S., Cheng, L., Sivapalan, M., 2012. Exploring the physical controls of regional patterns of flow duration curves – Part 3: a catchment classification system based on regime curve indicators. *Hydrology and Earth System Sciences*, **16**(11), 4467–4482.

Costa, J. E., 1974. Response and recovery of a Piedmont watershed from tropical storm Agnes, June 1972. *Water Resources Research*, **10**(1), 106–112.

Costa, J. E., 1975. Effects of agriculture on erosion and sedimentation in the Piedmont Province, Maryland. *Geological Society of America Bulletin*, **86**(9), 1281–1286.

Costa, J. E., O'Connor, J. E., 1995. Geomorphically effective floods. In: J. E. Costa, A. J. Miller, K. P. Potter, P. R. Wilcock (eds.), *Natural and Anthropogenic Influences in Fluvial Geomorphology (The Wolman Volume)*. American Geophysical Union, Washington, DC, pp. 45–56.

Costigan, K. H., Gerken, J. E., 2016. Channel morphology and flow structure of an abandoned channel under varying stages. *Water Resources Research*, **52**(7), 5458–5472. 10.1002/2015wr017601.

Costigan, K. H., Daniels, M. D., Perkin, J. S., Gido, K. B., 2014. Longitudinal variability in hydraulic geometry and substrate characteristics of a Great Plains sand-bed river. *Geomorphology*, **210**, 48–58.

Coulthard, T. J., 2005. Effects of vegetation on braided stream pattern and dynamics. *Water Resources Research*, **41**(4). 10.1029/2004wr003201.

Coulthard, T. J., van de Wiel, M. J., 2013a. Numerical modeling in fluvial geomorphology. In: J. W. Shroder (ed.), *Treatise on Geomorphology, Vol. 9, Fluvial Geomorphology*, E. Wohl (vol. ed.). Academic Press, San Diego, CA, pp. 694–710.

Coulthard, T. J., Van de Wiel, M. J., 2013b. Climate, tectonics or morphology: what signals can we see in drainage basin sediment yields? *Earth Surface Dynamics*, **1**(1), 13–27.

Coulthard, T. J., Macklin, M. G., Kirkby, M. J., 2002. A cellular model of Holocene upland river basin and alluvial fan evolution. *Earth Surface Processes and Landforms*, **27**(3), 269–288.

Coulthard, T. J., Hancock, G. R., Lowry, J. B. C., 2012. Modelling soil erosion with a downscaled landscape evolution model. *Earth Surface Processes and Landforms*, **37**(10), 1046–1055.

Covault, J. A., Craddock, W. H., Romans, B. W., Fildani, A., Gosai, M., 2013. Spatial and temporal variations in landscape evolution: Historic and longer-term sediment flux through global catchments. *Journal of Geology*, **121**(1), 35–56.

Cowan, W. L., 1956. Estimating hydraulic roughness coefficients. *Agricultural Engineering*, **37**, 473–475.

Cox, N. J., Warburton, J., Armstrong, A., Holliday, V. J., 2008. Fitting concentration and load rating curves with generalized linear models. *Earth Surface Processes and Landforms*, **33**(1), 25–39.

Creelle, S., Schindfessel, L., De Mulder, T., 2017. Modelling of the tributary momentum contribution to predict confluence head losses. *Journal of Hydraulic Research*, **55**(2), 175–189.

Crosato, A., Mosselman, E., 2009. Simple physics-based predictor for the number of river bars and the transition between meandering and braiding. *Water Resources Research*, **45**. 10.1029/2008wr007242.

Crosato, A., Mosselman, E., Desta, F. B., Uijttewaal, W. S. J., 2011. Experimental and numerical evidence for intrinsic nonmigrating bars in alluvial channels. *Water Resources Research*, **47**. 10.1029/2010wr009714.

Crosato, A., Desta, F. B., Cornelisse, J., Schuurman, F., Uijttewaal, W. S. J., 2012. Experimental and numerical findings on the long-term evolution of migrating alternate bars in alluvial channels. *Water Resources Research*, **48**. 10.1029/2011wr011320.

Crosby, B. T., Whipple, K. X., 2006. Knickpoint initiation and distribution within fluvial networks: 236 waterfalls in the Waipaoa River, North Island, New Zealand. *Geomorphology*, **82**(1–2), 16–38.

Crosby, B. T., Whipple, K. X., Gasparini, N. M., Wobus, C. W., 2007. Formation of fluvial hanging valleys: Theory and simulation. *Journal of Geophysical Research – Earth Surface*, **112**(F3). 10.1029/2006jf000566.

Crowder, D. W., Knapp, H. V., 2005. Effective discharge recurrence intervals of Illinois streams. *Geomorphology*, **64**(3–4), 167–184.

Crowder, D. W., Demissie, M., Markus, M., 2007. The accuracy of sediment loads when log-transformation produces nonlinear sediment load-discharge relationships. *Journal of Hydrology*, **336**(3–4), 250–268.

Cruise, J. F., Laymon, C. A., Al-Hamdan, O. Z., 2010. Impact of 20 years of land-cover change on the hydrology of streams in the southeastern United States. *Journal of the American Water Resources Association*, **46**(6), 1159–1170.

Crutzen, P. J., 2002. Geology of mankind. *Nature*, **415**(6867), 23–23.

Csiki, S., Rhoads, B. L., 2010. Hydraulic and geomorphological effects of run-of-river dams. *Progress in Physical Geography*, **34**(6), 755–780.

Csiki, S. J. C., Rhoads, B. L., 2014. Influence of four run-of-river dams on channel morphology and sediment characteristics in Illinois, USA. *Geomorphology*, **206**, 215–229.

Cui, Y. T., Parker, G., 1998. The arrested gravel front: Stable gravel-sand transitions in rivers – Part 2: General numerical solution. Journal of Hydraulic Research, **36**(2), 159–182.

Cui, Y. T., Parker, G., Paola, C., 1996. Numerical simulation of aggradation and downstream fining. *Journal of Hydraulic Research*, **34**(2), 185–204.

Cui, Y. T., Parker, G., Braudrick, C., Dietrich, W. E., Cluer, B., 2006. Dam removal express assessment models (DREAM). Part 1: model development and validation. *Journal of Hydraulic Research*, **44**(3), 291–307.

Cui, Y., Booth, D. B., Monschke, J., et al., 2017. Analyses of the erosion of fine sediment deposit for a large dam-removal project: An empirical approach. *International Journal of River Basin Management*, **15**(1), 103–114.

Curran, J. C., 2007. Step-pool formation models and associated step spacing. *Earth Surface Processes and Landforms*, **32**(11), 1611–1627. 10.1002/esp.1589.

Curran, J. C., Wilcock, P. R., 2005. Characteristic dimensions of the step-pool bed configuration: an experimental study. *Water Resources Research*, **41**(2). 10.1029/2004wr003568.

Curran, J. H., Wohl, E. E., 2003. Large woody debris and flow resistance in step-pool channels, Cascade Range, Washington. *Geomorphology*, **51**(1–3), 141–157.

Curtis, K. E., Renshaw, C. E., Magilligan, F. J., Dade, W. B., 2010. Temporal and spatial scales of geomorphic adjustments to reduced competency following flow regulation in bedload-dominated systems. *Geomorphology*, **118**(1–2), 105–117.

Czegledi, I., Saly, P., Takacs, P., et al., 2016. The scales of variability of stream fish assemblages at tributary confluences. *Aquatic Sciences*, **78**(4), 641–654.

Czuba, J. A., Foufoula-Georgiou, E., 2015. Dynamic connectivity in a fluvial network for identifying hotspots of geomorphic change. *Water Resources Research*, **51**(3), 1401–1421. 10.1002/2014wr016139.

Czuba, J. A., David, S. R., Edmonds, D. A., Ward, A. S., 2019. Dynamics of surface-water connectivity in a low-gradient meandering river floodplain. *Water Resources Research*, **55**(3), 1849–1870. 10.1029/2018wr023527.

da Silva, A. M. F., 2006. On why and how do rivers meander. *Journal of Hydraulic Research*, **44**(5), 579–590.

da Silva, A. M. F., 2009. On the stable geometry of self-formed alluvial channels: theory and practical application. *Canadian Journal of Civil Engineering*, **36**(10), 1667–1679.

da Silva, A. M. F., Ahmari, H., 2009. Size and effect on the mean flow of large-scale horizontal coherent structures in open-channel flows: An experimental study. *Canadian Journal of Civil Engineering*, **36**(10), 1643–1655.

Dade, W. B., Friend, P. F., 1998. Grain-size, sediment-transport regime, and channel slope in alluvial rivers. *Journal of Geology*, **106**(6), 661–675.

Dadson, S. J., Hovius, N., Chen, H., et al., 2004. Earthquake-triggered increase in sediment delivery from an active mountain belt. *Geology*, **32**(8), 733–736.

Dai, A., 2016. Historical and future changes in streamflow and continental runoff: A review. In: Q. Tang, T. Oki (eds.), *Terrestrial Water Cycle and Climate Change: Natural and Human-induced Impacts*. American Geophysical Union, Washington, DC, pp. 17–37.

Dalrymple, T., 1960. Flood Frequency Analyses. Manual of Hydrology, Part 3. Flood-Flow Techniques, U.S. Geological Survey Water-supply Paper 1543-A. U.S. Government Printing Office, Washington, DC.

D'Ambrosio, J. L., Ward, A. D., Witter, J. D., 2015. Evaluating geomorphic change in constructed two-stage ditches. *Journal of the American Water Resources Association*, **51**(4), 910–922.

Daniels, M. D., Rhoads, B. L., 2004. Effect of large woody debris configuration on three-dimensional flow structure in two low-energy meander bends at varying stages. *Water Resources Research*, **40**(11). 10.1029/2004wr003181.

Daniels, R. B., 1960. Entrenchment of the Willow drainage ditch, Harrison County, Iowa. *American Journal of Science*, **258**(3), 161–176.

Darby, S. E., Thorne, C. R., 1992. Impact of channelization on the Mimmshall Brook, Hertfordshire, UK. *Regulated Rivers Research and Management*, **7**(2), 193–204.

Darby, S. E., Thorne, C. R., 1996. Development and testing of riverbank-stability analysis. *Journal of Hydraulic Engineering, ASCE*, **122**(8), 443–454. 10.1061/(asce)0733-9429(1996) 122:8(443).

Darby, S. E., Gessler, D., Thorne, C. R., 2000. Computer program for stability analysis of steep, cohesive riverbanks. *Earth Surface Processes and Landforms*, **25**(2), 175–190.

Darby, S. E., Alabyan, A. M., Van de Wiel, M. J., 2002. Numerical simulation of bank erosion and channel migration in meandering rivers. *Water Resources Research*, **38**(9), 2-1-2–12.

Darby, S. E., Rinaldi, M., Dapporto, S., 2007. Coupled simulations of fluvial erosion and mass wasting for cohesive river banks. *Journal of Geophysical Research – Earth Surface*, **112**(F3). 10.1029/2006jf000722.

David, G. C. L., Wohl, E., Yochum, S. E., Bledsoe, B. P., 2010. Controls on at-a-station hydraulic geometry in steep headwater streams, Colorado, USA. *Earth Surface Processes and Landforms*, **35**(15), 1820–1837. 10.1002/esp.2023.

David, S. R., Edmonds, D. A., Letsinger, S. L., 2017. Controls on the occurrence and prevalence of floodplain channels in meandering rivers. *Earth Surface Processes and Landforms*, **42**(3), 460–472. 10.1002/esp.4002.

Davidson, G. R., Carnley, M., Lange, T., Galicki, S. J., Douglas, A., 2004. Changes in sediment accumulation rate in an oxbow lake following late 19th century clearing of land for agricultural use: A Pb-210, Cs-137, and C-14 study in Mississippi, USA. *Radiocarbon*, **46**(2), 755–764.

Davidson, S. K., Hey, R. D., 2011. Regime equations for natural meandering cobble- and gravel-bed rivers. *Journal of Hydraulic Engineering*, **137**(9), 894–910.

Davidson, S. L., Eaton, B. C., 2018. Beyond regime: A stochastic model of floods, bank erosion, and channel migration. *Water Resources Research*, **54**(9), 6282–6298. 10.1029/2017wr022059.

Davies, T. R. H., Sutherland, A. J., 1983. Extremal hypotheses for river behavior. *Water Resources Research*, **19**(1), 141–148.

Davies, T. R. H., Tinker, C. C., 1984. Fundamental characteristics of stream meanders. *Geological Society of America Bulletin*, **95**(5), 505–512.

Davies-Vollum, K. S., Smith, N. D., 2008. Factors affecting the accumulation of organic-rich deposits in a modern avulsive floodplain: Examples from the Cumberland Marshes, Saskatchewan, Canada. *Journal of Sedimentary Research*, **78**(9–10), 683–692.

Davis, R. T., Tank, J. L., Mahl, U. H., Winikoff, S. G., Roley, S. S., 2015. The influence of two-stage ditches with constructed floodplains on water column nutrients and sediments in agricultural streams. *Journal of the American Water Resources Association*, **51**(4), 941–955.

Davis, W. M., 1889. The rivers and valleys of Pennsylvania. *National Geographic Magazine*, **1**, 11–26.

Davis, W. M., 1899. The geographical cycle. *Geographical Journal*, **14**, 481–504.

Davis, W. M., 1902. Base level, grade and peneplain. *Journal of Geology*, **10**, 77–111.

Davis, W. M., 1903. The development of river meanders. *The Geological Magazine, New Series, Decade IV*, **Vol. X**, 145–148.

Davoren, A., Mosley, M. P., 1986. Observations of bedload movement, bar development and sediment supply in the braided Ohua River. *Earth Surface Processes and Landforms*, **11**(6), 643–652.

Davy, B. W., Davies, T. R. H., 1979. Entropy concepts in fluvial geomorphology – re-evaluation. *Water Resources Research*, **15**(1), 103–106.

Day, S. S., Gran, K. B., Belmont, P., Wawrzyniec, T., 2013. Measuring bluff erosion part 2: Pairing aerial photographs and terrestrial laser scanning to create a watershed scale sediment budget. *Earth Surface Processes and Landforms*, **38**(10), 1068–1082.

de Almeida, G. A. M., Rodriguez, J. F., 2011. Understanding pool-riffle dynamics through continuous morphological simulations. *Water Resources Research*, **47**. 10.1029/2010wr009170.

de Almeida, G. A. M., Rodriguez, J. F., 2012. Spontaneous formation and degradation of pool-riffle morphology and sediment sorting using a simple fractional transport model. *Geophysical Research Letters*, **39**. 10.1029/2012gl051059.

de Azeredo Freitas, H. R., Freitas, C. d. C., Rosim, S., de Freitas Oliveira, J. R., 2016. Drainage networks and watersheds delineation derived from TIN-based digital elevation models. *Computers & Geosciences*, **92**, 21–37.

De Cacqueray, N., Hargreaves, D. M., Morvan, H. P., 2009. A computational study of shear stress in smooth rectangular channels. *Journal of Hydraulic Research*, **47**(1), 50–57.

De Cesare, G., Schleiss, A., Hermann, F., 2001. Impact of turbidity currents on reservoir sedimentation. *Journal of Hydraulic Engineering*, **127**(1), 6–16.

De Serres, B., Roy, A. G., 1990. Flow direction and branching geometry at junctions in dendritic river networks. *Professional Geographer*, **42**(2), 194–201.

De Serres, B., Roy, A. G., Biron, P. M., Best, J. L., 1999. Three-dimensional structure of flow at a confluence of river channels with discordant beds. *Geomorphology*, **26**(4), 313–335.

de Vente, J., Poesen, J., Arabkhedri, M., Verstraeten, G., 2007. The sediment delivery problem revisited. *Progress in Physical Geography*, **31**(2), 155–178.

de Vente, J., Poesen, J., Verstraeten, G., et al., 2013. Predicting soil erosion and sediment yield at regional scales: Where do we stand? *Earth-Science Reviews*, **127**, 16–29.

Dean, D. J., Schmidt, J. C., 2013. The geomorphic effectiveness of a large flood on the Rio Grande in the Big Bend region: Insights on geomorphic controls and post-flood geomorphic response. *Geomorphology*, **201**, 183–198.

Dearing, J. A., Jones, R. T., 2003. Coupling temporal and spatial dimensions of global sediment flux through lake and marine sediment records. *Global and Planetary Change*, **39**(1–2), 147–168.

Dedkov, A., 2004. The relationship between sediment yield and drainage basin area. In: V. Golosov, V. Belyaev, D. E. Walling (eds.), *Sediment Transfer through the Fluvial System*. IAHS Publication No. 288, IAHS Press, Wallingford, UK, pp. 197–204.

Dedkov, A. P., Gusarov, A. V., 2006. Suspended sediment yield from continents into the World Ocean: spatial and temporal changeability. In: J. S. Rowan, R. W. Duck, A. Werritty (eds.), *Sediment Dynamics and the Hydromorphology of Fluvial Systems*. IAHS Publication No. 306, IAHS Press, Wallingford, UK, pp. 3–11.

Dedkov, A., Moszherin, V. I., 1992. Erosion and sediment yields in mountain regions of the world. In: D. E. Walling, T. R. Davies, B. Hasholt (eds.), *Erosion, Debris Flows and Environment in Mountain Regions*. IAHS Publication No. 209, IAHS Press, Wallingford, UK, pp. 29–36.

Defina, A., 2003. Numerical experiments on bar growth. *Water Resources Research*, **39**(4). 10.1029/2002wr001455.

Delmas, M., Pak, L. T., Cerdan, O., et al., 2012. Erosion and sediment budget across scale: A case study in a catchment of the European loess belt. *Journal of Hydrology*, **420**, 255–263.

Desloges, J. R., Church, M., 1987. Channel and floodplain facies in a wandering gravel-bed river. In: F. G. Ethridge, R. M. Flores, M. D. Harvey (eds.), *Recent Developments in Fluvial Sedimentology*, SEPM Special Publication No. 39. SEPM, Tulsa, OK, pp. 99–109.

Desloges, J. R., Church, M. A., 1989. Canadian landform examples – 13: Wandering gravel-bed rivers. *Canadian Geographer*, **33**(4), 360–364.

Devauchelle, O., Petroff, A. P., Seybold, H. F., Rothman, D. H., 2012. Ramification of stream networks. *Proceedings of the National Academy of Sciences of the United States of America*, **109**(51), 20832–20836.

DeVries, P., 2003. Bedload layer thickness and disturbance depth in gravel bed streams. *Journal of Hydraulic Engineering*, **128**, 983–991.

Dey, A. K., Tsujimoto, T., Kitamura, T., 2007. Experimental investigations on different modes of headcut migration. *Journal of Hydraulic Research*, **45**(3), 333–346.

Dey, S., 2014. *Fluvial Hydrodynamics: Hydrodynamic and Sediment Transport Phenomena*. Springer, Berlin.

Dey, S., Sarkar, S., Solari, L., 2011. Near-bed turbulence characteristics at the entrainment threshold of sediment beds. *Journal of Hydraulic Engineering*, **137**(9), 945–958.

DeZiel, B. A., Krider, L., Hansen, B., et al., 2019. Habitat improvements and fish community response associated with an agricultural two-stage ditch in Mower County, Minnesota. *Journal of the American Water Resources Association*, **55**(1), 154–188.

D'Haen, K., Verstraeten, G., Degryse, P., 2012. Fingerprinting historical fluvial sediment fluxes. *Progress in Physical Geography*, **36**(2), 154–186.

Dieras, P. L., Constantine, J. A., Hales, T. C., Piegay, H., Riquier, J., 2013. The role of oxbow lakes in the off-channel storage of bed material along the Ain River, France. *Geomorphology*, **188**, 110–119.

Dietrich, J. T., 2016. Riverscape mapping with helicopter-based Structure-from-Motion photogrammetry. *Geomorphology*, **252**, 144–157.

Dietrich, W. E., 1982. Settling velocity of natural particles. *Water Resources Research*, **18**(6), 1615–1626.

Dietrich, W. E., 1987. Mechanics of flow and sediment transport in river bends. In: K. S. Richards (ed.), *River Channels: Environment and Process*. Basil Blackwell, London, pp. 179–227.

Dietrich, W. E., Dunne, T., 1978. Sediment budget for a small catchment in mountainous terrain. *Zeitschrift fur Geomorphologie*, **Supplementband 29**, 191–206.

Dietrich, W. E., Dunne, T., 1993. Channel heads. In: M. J. Kirkby, K. Beven (eds.), *Channel Network Hydrology*. Wiley, New York, pp. 175–219.

Dietrich, W. E., Smith, J. D., 1983. Influence of the point bar on flow through curved channels. *Water Resources Research*, **19**(5), 1173–1192.

Dietrich, W. E., Smith, J. D., 1984. Bed load transport in a river meander. *Water Resources Research*, **20**(10), 1355–1380.

Dietrich, W. E., Whiting, P. J., 1989. Boundary shear stress and sediment transport in river meanders of sand and gravel. In: S. Ikeda, G. Parker (eds.), *River Meandering*. American Geophysical Union, Washington, DC., pp. 1–50.

Dietrich, W. E., Smith, J. D., Dunne, T., 1979. Flow and sediment transport in a sand bedded meander. *Journal of Geology*, **87**(3), 305–315.

Dietrich, W. E., Kirchner, J. W., Ikeda, H., Iseya, F., 1989. Sediment supply and the development of the coarse surface-layer in gravel-bedded rivers. *Nature*, **340**(6230), 215–217.

Dietrich, W. E., Wilson, C. J., Montgomery, D. R., McKean, J., Bauer, R., 1992. Erosion thresholds and land surface morphology. *Geology*, **20**(8), 675–679.

Dietrich, W. E., Wilson, C. J., Montgomery, D. R., McKean, J., 1993. Analysis of erosion thresholds, channel networks, and landscape morphology using a digital terrain model. *Journal of Geology*, **101**(2), 259–278.

Dietrich, W. E., Bellugi, G. E., Sklar, L. S., et al., 2003. Geomorphic transport laws for predicting landscape form and dynamics. In: P. R. Wilcock, R. M. Iverson (eds.), *Prediction in Geomorphology*. American Geophysical Union, Washington, DC, pp. 103–132.

Diffenbaugh, N. S., Singh, D., Mankin, J. S., et al., 2017. Quantifying the influence of global warming on unprecedented extreme climate events. *Proceedings of the National Academy of Sciences of the United States of America*, **114**(19), 4881–4886.

Dinehart, R. L., 1992. Evolution of coarse gravel bed forms – field measurements at flood stage. *Water Resources Research*, **28**(10), 2667–2689.

Ding, Y., Langendoen, E. J., 2018. Simulation and control of sediment transport due to dam removal. *Journal of Applied Water Engineering and Research*, **6**(2), 95–108.

Dingman, S. L., 1984. *Fluvial Hydrology*. W. H. Freeman and Co., New York.

Dingman, S. L., 2007. Analytical derivation of at-a-station hydraulic-geometry relations. *Journal of Hydrology*, **334**(1–2), 17–27.

Dingman, S. L., 2009. *Fluvial Hydraulics*. Oxford University Press, New York.

Diplas, P., 1987. Bedload transport in gravel-bed streams. *Journal of Hydraulic Engineering*, **113**(3), 277–292.

Diplas, P., 1990. Characteristics of self-formed straight channels. *Journal of Hydraulic Engineering*, **116**(5), 707–728.

Diplas, P., Vigilar, G., 1992. Hydraulic geometry of threshold channels. *Journal of Hydraulic Engineering*, **118**(4), 597–614.

Diplas, P., Dancey, C. L., Celik, A. O., et al., 2008. The role of impulse on the initiation of particle movement under turbulent flow conditions. *Science*, **322**(5902), 717–720.

Dixon, S. J., Sambrook Smith, G. H., Best, J. L., et al., 2018. The planform mobility of river channel confluences: Insights from analysis of remotely sensed imagery. *Earth-Science Reviews*, **176**, 1–18.

Dodds, P. S., Rothman, D. H., 2000. Scaling, universality, and geomorphology. *Annual Review of Earth and Planetary Sciences*, **28**, 571–610.

Dodds, P. S., Rothman, D. H., 2001a. Geometry of river networks. I. Scaling, fluctuations, and deviations. *Physical Review E*, **63**(1), 016115.

Dodds, P. S., Rothman, D. H., 2001b. Geometry of river networks. II. Distributions of component size and number. *Physical Review E*, **63**(1), 016116.

Dodds, P. S., Rothman, D. H., 2001c. Geometry of river networks. III. Characterization of component connectivity. *Physical Review E*, **63**(1), 016117.

Dodov, B., Foufoula-Georgiou, E., 2004. Generalized hydraulic geometry: Derivation based on a multiscaling formalism. *Water Resources Research*, **40**(6). 10.1029/2003wr002082.

Doeschl, A. B., Ashmore, P. E., Davison, M., 2006. Methods for assessing exploratory computational models of braided rivers. In: G. H. S. Smith, J. L. Best, C. S. Bristow, G. E. Petts (eds.), *Braided Rivers: Process, Deposits, Ecology and Management*. Special Publications of the International Association of Sedimentologists, 36, Blackwell, Malden, MA, pp. 177–197.

Doeschl-Wilson, A. B., Ashmore, P. E., 2005. Assessing a numerical cellular braided-stream model with a physical model. *Earth Surface Processes and Landforms*, **30**(5), 519–540.

Domokos, G., Jerolmack, D. J., Sipos, A. A., Toeroek, A., 2014. How river rocks round: Resolving the shape-size paradox. *PLoS ONE*, **9**(2). 10.1371/journal.pone.0088657.

Dooge, J. C. I., 1991. The Manning formula in context. In: B. C. Yen (ed.), *Channel Flow Resistance: Centennial of Manning's*

Formula. Water Resources Publications, Highland Ranch, Colorado, pp. 136–185.

Dordevic, D., 2013. Numerical study of 3D flow at right-angled confluences with and without upstream planform curvature. *Journal of Hydroinformatics*, **15**(4), 1073–1088.

Dotterweich, M., 2013. The history of human-induced soil erosion: geomorphic legacies, early descriptions and research, and the development of soil conservation – a global synopsis. *Geomorphology*, **201**, 1–34.

Douglas, I., 1967. Man, vegetation and the sediment yields of rivers. *Nature*, **215**(5104), 925–928.

Downs, P. W., Gregory, K. J., 1995. Approaches to river channel sensitivity. *Professional Geographer*, **47**(2), 168–175.

Downs, P. W., Kondolf, G. M., 2002. Post-project appraisals in adaptive management of river channel restoration. *Environmental Management*, **29**(4), 477–496.

Downs, P. W., Soar, P. J., Taylor, A., 2016. The anatomy of effective discharge: The dynamics of coarse sediment transport revealed using continuous bedload monitoring in a gravel-bed river during a very wet year. *Earth Surface Processes and Landforms*, **41**(2), 147–161.

Doyle, M. W., Shields, C. A., 2008. An alternative measure of discharge effectiveness. *Earth Surface Processes and Landforms*, **33**(2), 308–316.

Doyle, M. W., Shields, F. D., 2012. Compensatory mitigation for streams under the Clean Water Act: Reassessing science and redirecting policy. *Journal of the American Water Resources Association*, **48**(3), 494–509.

Doyle, M. W., Stanley, E. H., Harbor, J. M., 2003. Channel adjustments following two dam removals in Wisconsin. *Water Resources Research*, **39**(1). 10.1029/2002wr001714.

Doyle, M. W., Shields, D., Boyd, K. F., Skidmore, P. B., Dominick, D., 2007. Channel-forming discharge selection in river restoration design. *Journal of Hydraulic Engineering*, **133**(7), 831–837.

Doyle, M. W., Lave, R., Robertson, M. M., Ferguson, J., 2013. River federalism. *Annals of the Association of American Geographers*, **103**(2), 290–298.

Doyle, M. W., Singh, J., Lave, R., Robertson, M. M., 2015. The morphology of streams restored for market and nonmarket purposes: Insights from a mixed natural-social science approach. *Water Resources Research*, **51**(7), 5603–5622. 10.1002/2015wr017030.

Drake, T. G., Shreve, R. L., Dietrich, W. E., Whiting, P. J., Leopold, L. B., 1988. Bedload transport of fine gravel observed by motion-picture photography. *Journal of Fluid Mechanics*, **192**, 193–217.

Duan, J. G., Julien, P. Y., 2010. Numerical simulation of meandering evolution. *Journal of Hydrology*, **391**(1–2), 36–48.

Duckson, D. W., Duckson, L. J., 1995. Morphology of bedrock step pool systems. *Water Resources Bulletin*, **31**(1), 43–51.

Duckson, D. W., Duckson, L. J., 2001. Channel bed steps and pool shapes along Soda Creek, Three Sisters Wilderness, Oregon. *Geomorphology*, 38(3–4), 267–279. 10.1016/s0169-555x(00)00098-2.

Dufour, S., Piégay, H., 2009. From the myth of a lost paradise to targeted river restoration: Forget natural references and focus on human benefits. *River Research and Applications*, **25**(5), 568–581.

Dunne, T., 1980. Formation and controls of channel networks. *Progress in Physical Geography*, **4**, 211–239.

Dunne, T., 1990. Hydrology, mechanics, and geomorphic implications of erosion by subsurface flow. In: C. G. Higgins, D. R. Coates (eds.), Groundwater Geomorphology: The Role of Subsurface Water in Earth-Surface Processes and Landforms, Geological Society of America Special Paper 252, pp. 1–28.

Dunne, T., Aubry, B. F., 1986. Evaluation of Horton's theory of sheetwash and rill erosion on the basis of field experiments. In: A. D. Abrahams (ed.), *Hillslope Processes*. Allen and Unwin, Winchester, MA, pp. 31–53.

Dunne, T., Leopold, L. B., 1978. *Water in Environmental Planning*. Freeman, San Francisco, CA.

Dunne, T., Mertes, L. A. K., Meade, R. H., Richey, J. E., Forsberg, B. R., 1998. Exchanges of sediment between the flood plain and channel of the Amazon River in Brazil. *Geological Society of America Bulletin*, **110**(4), 450–467.

Durkin, P. R., Hubbard, S. M., Holbrook, J., Boyd, R., 2018. Evolution of fluvial meander-belt deposits and implications for the completeness of the stratigraphic record. *Geological Society of America Bulletin*, **130**(5–6), 721–739.

Dury, G. H., 1964. Principles of underfit streams. U.S. Geological Survey Professional Paper 452-A. U.S. Government Printing Office, Washington, DC.

Dury, G. H., 1966. The concept of grade. In: G. H. Dury (ed.), *Essays in Geomorphology*. Heinemann, London, pp. 211–234.

Dyhouse, G. R., 1985. Stage-frequency analysis at a major river junction. *Journal of Hydraulic Engineering*, **111**(4), 565–583.

Dzubakova, K., Piegay, H., Riquier, J., Trizna, M., 2015. Multi-scale assessment of overflow-driven lateral connectivity in floodplain and backwater channels using LiDAR imagery. *Hydrological Processes*, **29**(10), 2315–2330.

East, A. E., Logan, J. B., Mastin, M. C., et al., 2018. Geomorphic evolution of a gravel-bed river under sediment-starved versus sediment-rich conditions: river response to the world's largest dam removal. *Journal of Geophysical Research – Earth Surface*, **123**(12), 3338–3369. 10.1029/2018jf004703.

Eaton, B. C., 2006. Bank stability analysis for regime models of vegetated gravel bed rivers. *Earth Surface Processes and Landforms*, **31**(11), 1438–1444.

Eaton, B. C., Church, M., 2004. A graded stream response relation for bed load-dominated streams. *Journal of Geophysical Research – Earth Surface*, **109**(F3), 18. 10.1029/2003jf000062.

Eaton, B. C., Church, M., 2007. Predicting downstream hydraulic geometry: a test of rational regime theory. *Journal of Geophysical Research – Earth Surface*, **112**(F3). 10.1029/2006jf000734.

Eaton, B. C., Millar, R. G., 2017. Predicting gravel bed river response to environmental change: The strengths and limitations of a regime-based approach. *Earth Surface Processes and Landforms*, **42**(6), 994–1008.

Eaton, B. C., Church, M., Millar, R. G., 2004. Rational regime model of alluvial channel morphology and response. *Earth Surface Processes and Landforms*, **29**(4), 511–529.

Eaton, B. C., Church, M., Davies, T. R. H., 2006. A conceptual model for meander initiation in bedload-dominated streams. *Earth Surface Processes and Landforms*, **31**(7), 875–891.

Eaton, B. C., Millar, R. G., Davidson, S., 2010. Channel patterns: braided, anabranching, and single-thread. *Geomorphology*, **120**(3–4), 353–364.

Eden, S., Tunstall, S. M., Tapsell, S. M., 2000. Translating nature: river restoration as nature culture. *Environment and Planning D-Society & Space*, **18**(2), 257–273.

Edwards, T. K., Glysson, G. D., 1999. Field methods for measurement of fluvial sediment. U.S. Geological Survey Techniques of Water Resources Investigations, Book 3, Chapter C2. Reston, VA, pp. 1–89.

Eekhout, J. P. C., Hoitink, A. J. F., Mosselman, E., 2013. Field experiment on alternate bar development in a straight sand-bed stream. *Water Resources Research*, **49**(12), 8357–8369. 10.1002/2013wr014259.

Egozi, R., Ashmore, P., 2008. Defining and measuring braiding intensity. *Earth Surface Processes and Landforms*, **33**(14), 2121–2138.

Egozi, R., Ashmore, P., 2009. Experimental analysis of braided channel pattern response to increased discharge. *Journal of Geophysical Research – Earth Surface*, **114**. 10.1029/2008jf001099.

Einstein, H. A., 1942. Formulas for the transportation of bedload. *Transactions of the American Society of Civil Engineers*, **117**, 561–597.

Einstein, H. A., 1950. The bedload function for bedload transportation in open channel flows. Technical Bulletin No. 1026, U.S.D.A., Soil Conservation Service.

Einstein, H. A., Chien, N., 1953. Can the rate of wash load be predicted from the bed-load function? *Transactions of the American Geophysical Union*, **34**, 876–882.

Einstein, H. A., Li, H., 1958. Secondary currents in straight channels. *Transactions, American Geophysical Union*, **39**, 1085–1088.

Einstein, H. A., Shen, H. W., 1964. Study on meandering in straight alluvial channels. *Journal of Geophysical Research*, **69**(24), 5239–5247.

Eke, E., Parker, G., Shimizu, Y., 2014. Numerical modeling of erosional and depositional bank processes in migrating river bends with self-formed width: Morphodynamics of bar push and bank pull. *Journal of Geophysical Research – Earth Surface*, **119**(7), 1455–1483. 10.1002/2013jf003020.

Ellery, W. N., Ellery, K., Rogers, K. H., McCarthy, T. S., 1995. The role of *Cyperus papyrus* L. In channel blockage and abandonment in the northeastern Okavango Delta, Botswana. *African Journal of Ecology*, **33**(1), 25–49.

Ellis, E. R., Church, M., 2005. Hydraulic geometry of secondary channels of lower Fraser River, British Columbia, from acoustic Doppler profiling. *Water Resources Research*, **41**(8). 10.1029/2004wr003777.

Ely, L. L., Enzel, Y., Baker, V. R., Cayan, D. R., 1993. A 5000-year record of extreme floods and climate change in the Southwestern United States. *Science*, **262**(5132), 410–412.

Emerson, J. W., 1971. Channelization – a case study. *Science*, **173**(3994), 325–326.

Emery, S. B., Perks, M. T., Bracken, L. J., 2013. Negotiating river restoration: The role of divergent reframing in environmental decision-making. *Geoforum*, **47**, 167–177.

Emmett, W. W., 1975. The channels and waters of the upper Salmon River area, Idaho. U.S. Geological Survey Professional Paper 870-A. U.S. Government Printing Office, Washington, DC.

Emmett, W. W., Wolman, M. G., 2001. Effective discharge and gravel-bed rivers. *Earth Surface Processes and Landforms*, **26**(13), 1369–1380.

Endreny, T. A., 2007. Estimation of channel bankfull occurrence from instantaneous discharge data. *Journal of Hydrologic Engineering*, **12**(5), 524–531.

Engel, F. L., Rhoads, B. L., 2012. Interaction among mean flow, turbulence, bed morphology, bank failures and channel planform in an evolving compound meander loop. *Geomorphology*, **163**, 70–83.

Engel, F. L., Rhoads, B. L., 2016. Three-dimensional flow structure and patterns of bed shear stress in an evolving compound meander bend. *Earth Surface Processes and Landforms*, **41**(9), 1211–1226.

Engel, F. L., Rhoads, B. L., 2017. Velocity profiles and the structure of turbulence at the outer bank of a compound meander bend. *Geomorphology*, **295**, 191–201.

Engelund, F., Skovgaard, O., 1973. Origin of meandering and braiding in alluvial streams. *Journal of Fluid Mechanics*, **57**(FEB6), 289–302.

England, J. F., Jr., Cohn, T. A., Faber, B. A., et al., 2018. Guidelines for determining flood flow frequency – Bulletin 17C. U.S. Geological Survey Techniques and Methods Book 4 Chapter B5.

Ergenzinger, P., Schmidt, K. H., 1990. Stochastic elements of bed load transport in a step-pool mountain river. In: R. O. Sinninger, M. Monbaron (eds.), *Hydrology in Mountainous Regions. II – Artificial Reservoirs; Water and Slopes*. IAHS Publication No. 194, IAHS Press, Wallingford, UK, pp. 39–46.

Ergenzinger, P., Schmidt, K. H., Busskamp, R., 1989. The pebble transmitter system (PETS) – 1st results of a technique for studying coarse material erosion, transport and deposition. *Zeitschrift fur Geomorphologie*, **33**(4), 503–508.

Erskine, W. D., 2011. Geomorphic controls on historical channel planform changes on the lower Pages River, Hunter Valley, Australia. *Australian Geographer*, **42**(3), 289–307. 10.1080/00049182.2011.595768.

Erskine, W., McFadden, C., Bishop, P., 1992. Alluvial cutoffs as indicators of former channel conditions. *Earth Surface Processes and Landforms*, **17**(1), 23–37.

Eshel, G., Levy, G. J., Mingelgrin, U., Singer, M. J., 2004. Critical evaluation of the use of laser diffraction for particle-size distribution analysis. *Soil Science Society of America Journal*, **68**(3), 736–743.

Evans, R., 1998. The erosional impacts of grazing animals. *Progress in Physical Geography*, **22**(2), 251–268.

Everitt, B., 1993. Channel responses to declining flow on the Rio Grande between Ft. Quitman and Presidio, Texas. *Geomorphology*, **6**(3), 225–242.

Fagan, S. D., Nanson, G. C., 2004. The morphology and formation of floodplain-surface channels, Cooper Creek, Australia. *Geomorphology*, **60**(1–2), 107–126.

Fahnestock, R. K., 1963. Morphology and hydrology of a glacial stream – White River, Mt. Rainier, Washington. U.S. Geological Survey Professional Paper 422-A. U.S. Government Printing Office, Washington, DC.

Farnsworth, K. L., Milliman, J. D., 2003. Effects of climatic and anthropogenic change on small mountainous rivers: The Salinas River example. *Global and Planetary Change*, **39**(1–2), 53–64.

Fathel, S., Furbish, D., Schmeeckle, M., 2016. Parsing anomalous versus normal diffusive behavior of bedload sediment particles. *Earth Surface Processes and Landforms*, **41**(12), 1797–1803.

Faulkner, D. J., 1998. Spatially variable historical alluviation and channel incision in west-central Wisconsin. *Annals of the Association of American Geographers*, **88**(4), 666–685.

Faulkner, D. J., Larson, P. H., Jol, H. M., et al., 2016. Autogenic incision and terrace formation resulting from abrupt late-glacial base-level fall, lower Chippewa River, Wisconsin, USA. *Geomorphology*, **266**, 75–95.

Federici, B., Paola, C., 2003. Dynamics of channel bifurcations in noncohesive sediments. *Water Resources Research*, **39**(6). 10.1029/2002wr001434.

Federici, B., Seminara, R., 2003. On the convective nature of bar instability. *Journal of Fluid Mechanics*, **487**, 125–145.

Fencl, J. S., Mather, M. E., Costigan, K. H., Daniels, M. D., 2015. How big of an effect do small dams have? Using geomorphological footprints to quantify spatial impact of low-head dams and identify patterns of across-dam variation. *PLoS ONE*, **10**(11), 22. 10.1371/journal.pone.0141210.

Ferdowsi, B., Ortiz, C. P., Houssais, M., Jerolmack, D. J., 2017. River-bed armouring as a granular segregation phenomenon. *Nature Communications*, **8**. 10.1038/s41467-017-01681-3.

Ferguson, R. I., 1973. Sinuosity of supraglacial streams. *Geological Society of America Bulletin*, **84**(1), 251–255.

Ferguson, R. I., 1981. Channel form and channel changes. In: J. Lewin (ed.), *British Rivers*. Allen and Unwin, London, pp. 90–125.

Ferguson, R. I., 1986a. River loads underestimated by rating curves. *Water Resources Research*, **22**(1), 74–76.

Ferguson, R. I., 1986b. Hydraulics and hydraulic geometry. *Progress in Physical Geography*, **10**(1), 1–31.

Ferguson, R. I., 1987a. Accuracy and precision of methods for estimating river loads. *Earth Surface Processes and Landforms*, **12**(1), 95–104.

Ferguson, R., 1987b. Hydraulic and sedimentary controls of channel pattern. In: K. Richards (ed.), *River Channels: Environment and Process*. Blackwell, New York, pp. 129–158.

Ferguson, R. I., 1993. Understanding braiding processes in gravel-bed rivers: Progress and unsolved problems. In: J. L. Best, C. S. Bristow (eds.), *Braided Rivers*. Geological Society of London Special Publication No. 75, Geological Society, London, pp. 73–87.

Ferguson, R. I., 1994. Critical discharge for entrainment of poorly sorted gravel. *Earth Surface Processes and Landforms*, **19**(2), 179–186.

Ferguson, R. I., 2003. Emergence of abrupt gravel to sand transitions along rivers through sorting processes. *Geology*, **31**(2), 159–162.

Ferguson, R. I., 2005. Estimating critical stream power for bedload transport calculations in gravel-bed rivers. *Geomorphology*, **70**(1–2), 33–41.

Ferguson, R., 2007. Flow resistance equations for gravel- and boulder-bed streams. *Water Resources Research*, **43**(5). 10.1029/2006WR005422.

Ferguson, R. I., 2012. River channel slope, flow resistance, and gravel entrainment thresholds. *Water Resources Research*, **48**. 10.1029/2011wr010850.

Ferguson, R., Ashworth, P., 1991. Slope-induced changes in channel character along a gravel-bed stream – the Allt Dubhaig, Scotland. *Earth Surface Processes and Landforms*, **16**(1), 65–82.

Ferguson, R. J., Brierley, G. J., 1999. Levee morphology and sedimentology along the lower Tuross River, south-eastern Australia. *Sedimentology*, **46**(4), 627–648.

Ferguson, R. I., Church, M., 2004. A simple universal equation for grain settling velocity. *Journal of Sedimentary Research*, **74**(6), 933–937. 10.1306/051204740933.

Ferguson, R. I., Hoey, T. B., 2002. Long-term slowdown of river tracer pebbles: Generic models and implications for interpreting short-term tracer studies. *Water Resources Research*, **38**(8). 10.1029/2001WR000637.

Ferguson, R. I., Wathen, S. J., 1998. Tracer-pebble movement along a concave river profile: Virtual velocity in relation to grain size and shear stress. *Water Resources Research*, **34**(8), 2031–2038.

Ferguson, R., Werritty, A., 1983. Bar development and channel changes in the gravelly River Feshie, Scotland. In: J. D. Collinson, J. Lewin (eds.), *Modern and Ancient Fluvial Systems*. Special Publication of the International Association of Sedimentologists No. 6, Blackwell, Oxford, UK, pp.181–193.

Ferguson, R. I., Ashmore, P. E., Ashworth, P. J., Paola, C., Prestegaard, K. L., 1992. Measurements in a braided river chute and lobe. 1. Flow pattern, sediment transport, and channel change. *Water Resources Research*, **28**(7), 1877–1886. 10.1029/92wr00700.

Ferguson, R. I., Kirkbride, A. D., Roy, A. G., 1996a. Markov analysis of velocity fluctuations in gravel-bed rivers. In: P. J. Ashworth, S. J. Bennett, J. Best, S. McLelland (eds.), *Coherent Flow Structures in Open Channels*. John Wiley, New York, pp. 165–183.

Ferguson, R., Hoey, T., Wathen, S., Werritty, A., 1996b. Field evidence for rapid downstream fining of river gravels through selective transport. *Geology*, **24**(2), 179–182.

Ferguson, R. I., Bloomer, D. J., Hoey, T. B., Werritty, A., 2002. Mobility of river tracer pebbles over different timescales. *Water Resources Research*, **38**(5). 10.1029/2001wr000254.

Ferguson, R. I., Parsons, D. R., Lane, S. N., Hardy, R. J., 2003. Flow in meander bends with recirculation at the inner bank. *Water Resources Research*, **39**(11). 10.1029/2003wr001965.

Ferguson, R. I., Cudden, J. R., Hoey, T. B., Rice, S. P., 2006. River system discontinuities due to lateral inputs: generic styles and controls. *Earth Surface Processes and Landforms*, **31**(9), 1149–1166.

Ferguson, R. I., Bloomer, D. J., Church, M., 2011. Evolution of an advancing gravel front: Observations from Vedder Canal, British Columbia. *Earth Surface Processes and Landforms*, **36**(9), 1172–1182.

Ferguson, R. I., Church, M., Rennie, C. D., Venditti, J. G., 2015. Reconstructing a sediment pulse: Modeling the effect of placer mining on Fraser River, Canada. *Journal of Geophysical Research – Earth Surface*, **120**(7), 1436–1454. 10.1002/2015jf003491.

Ferguson, R. I., Sharma, B. P., Hodge, R. A., Hardy, R. J., Warburton, J., 2017. Bed load tracer mobility in a mixed bedrock/alluvial channel. *Journal of Geophysical Research – Earth Surface*, **122**(4), 807–822.

Fernandez Luque, R., Van Beek, R., 1976. Erosion and transport of bed sediment. *Journal of Hydraulic Research*, **14**, 127–144.

Ferreira, R. M. L., 2015. The von Karman constant for flows over rough mobile beds. Lessons learned from dimensional analysis and similarity. *Advances in Water Resources*, **81**, 19–32.

Ferrer-Boix, C., Chartrand, S. M., Hassan, M. A., Martin-Vide, J. P., Parker, G., 2016. On how spatial variations of channel width influence river profile curvature. *Geophysical Research Letters*, **43**(12), 6313–6323. 10.1002/2016gl069824.

Ferro, V., Porto, P., 2012. Identifying a dominant discharge for natural rivers in southern Italy. *Geomorphology*, **139**, 313–321.

Filgueira-Rivera, M., Smith, N. D., Slingerland, R. L., 2007. Controls on natural levee development in the Columbia river, British Columbia, Canada. *Sedimentology*, **54**(4), 905–919.

Finnegan, N. J., Roe, G., Montgomery, D. R., Hallet, B., 2005. Controls on the channel width of rivers: Implications for modeling fluvial incision of bedrock. *Geology*, **33**(3), 229–232.

Fischer, H. B., Imberger, J., List, E. J., Koh, R. C. Y., Brooks, N. H., 1979. *Mixing in Inland and Coastal Waters*. Academic Press, San Diego, CA.

Fisk, H. N., 1947. Fine-grained alluvial deposits and their effects on Mississippi River activity. Mississippi River Commission, U.S. Waterways Experiment Station, Vols. 1 & 2.

FitzHugh, T. W., Vogel, R. M., 2011. The impact of dams on flood flows in the United States. *River Research and Applications*, **27**(10), 1192–1215.

Flint, J. J., 1973. Experimental development of headward growth of channel networks. *Geological Society of America Bulletin*, **84**(3), 1087–1093.

Flint, J. J., 1980. Tributary arrangements in fluvial systems. *American Journal of Science*, **280**(1), 26–45.

Flores-Cervantes, J. H., Istanbulluoglu, E., Bras, R. L., 2006. Development of gullies on the landscape: a model of headcut retreat resulting from plunge pool erosion. *Journal of Geophysical Research – Earth Surface*, **111**(F1). 10.1029/2004jf000226.

Florsheim, J. L., Mount, J. F., 2002. Restoration of floodplain topography by sand-splay complex formation in response to intentional levee breaches, Lower Cosumnes River, California. *Geomorphology*, **44**(1–2), 67–94.

Florsheim, J. L., Mount, J. F., Chin, A., 2008. Bank erosion as a desirable attribute of rivers. *Bioscience*, **58**(6), 519–529.

Flynn, K. M., Kirby, W. H., Hummel, P. R., 2006. User's Manual for PeakFQ Flood Frequency Analysis for Program Peak FQ Flood Frequency Analysis using Bulletin 17B Guidelines. U.S. Geological Survey, Techniques and Methods Book4, Chapter B4, Reston, VA.

Fola, M. E., Rennie, C. D., 2010. Downstream hydraulic geometry of clay-dominated cohesive bed rivers. *Journal of Hydraulic Engineering*, **136**(8), 524–527.

Foley, M. M., Magilligan, F. J., Torgersen, C. E., et al., 2017a. Landscape context and the biophysical response of rivers to dam removal in the United States. *PLoS ONE*, **12**(7). 10.1371/journal.pone.0180107.

Foley, M. M., Bellmore, J. R., O'Connor, J. E., et al., 2017b. Dam removal: Listening in. *Water Resources Research*, **53**(7), 5229–5246. 10.1002/2017wr020457.

Folk, R. L., 1980. *Petrology of Sedimentary Rocks*. Hemphill Publishing Co., Austin, TX.

Folk, R. L., Ward, W. C., 1957. Brazos River bar: A study in the significance of grain size parameters. *Journal of Sedimentary Petrology*, **27**, 3–26.

Fonstad, M. A., Marcus, W. A., 2010. High resolution, basin extent observations and implications for understanding river form and process. *Earth Surface Processes and Landforms*, **35**(6), 680–698.

Forte, A. M., Whipple, K. X., 2018. Criteria and tools for determining drainage divide stability. *Earth and Planetary Science Letters*, **493**, 102–117.

Foufoula-Georgiou, E., Sapozhnikov, V. B., 1998. Anisotropic scaling in braided rivers: An integrated theoretical framework and results from application to an experimental river. *Water Resources Research*, **34**(4), 863–867.

Foufoula-Georgiou, E., Sapozhnikov, V., 2001. Scale invariances in the morphology and evolution of braided rivers. *Mathematical Geology*, **33**(3), 273–291.

Fournier, F., 1960. *Climate et Erosion*. Presses Universitaires de France, Paris.

Fox, C. A., Magilligan, F. J., Sneddon, C. S., 2016. "You kill the dam, you are killing a part of me": Dam removal and the environmental politics of river restoration. *Geoforum*, **70**, 93–104.

Fox, G. A., Felice, R. G., 2014. Bank undercutting and tension failure by groundwater seepage: predicting failure mechanisms. *Earth Surface Processes and Landforms*, **39**(6), 758–765.

Fox, G. A., Wilson, G. V., Simon, A., et al., 2007. Measuring streambank erosion due to ground water seepage: Correlation to bank pore water pressure, precipitation and stream stage. *Earth Surface Processes and Landforms*, **32**(10), 1558–1573.

Fox, J. F., Papanicolaou, A. N., 2008. An un-mixing model to study watershed erosion processes. *Advances in Water Resources*, **31**(1), 96–108.

Franca, M. J., Lemmin, U., 2015. Detection and reconstruction of large-scale coherent flow structures in gravel-bed rivers. *Earth Surface Processes and Landforms*, **40**(1), 93–104.

Francalanci, S., Solari, L., Toffolon, M., Parker, G., 2012. Do alternate bars affect sediment transport and flow resistance in gravel-bed rivers? *Earth Surface Processes and Landforms*, **37**(8), 866–875.

Frankel, K. L., Pazzaglia, F. J., Vaughn, J. D., 2007. Knickpoint evolution in a vertically bedded substrate, upstream-dipping terraces, and Atlantic slope bedrock channels. *Geological Society of America Bulletin*, **119**(3–4), 476–486.

Frascati, A., Lanzoni, S., 2010. Long-term river meandering as a part of chaotic dynamics? A contribution from mathematical modelling. *Earth Surface Processes and Landforms*, **35**(7), 791–802.

Fredsoe, J., 1978. Meandering and braiding of rivers. *Journal of Fluid Mechanics*, **84**(FEB), 609–624.

Frias, C. E., Abad, J. D., Mendoza, A., et al., 2015. Planform evolution of two anabranching structures in the Upper Peruvian Amazon River. *Water Resources Research*, **51**(4), 2742–2759.

Friedkin, J. F., 1945. *A laboratory study of the meandering of alluvial rivers*. U.S. Waterways Experiment Station, Vicksburg, MS.

Frings, R. M., 2008. Downstream fining in large sand-bed rivers. *Earth-Science Reviews*, **87**(1–2), 39–60.

Frings, R. M., Ottevanger, W., Sloff, K., 2011. Downstream fining processes in sandy lowland rivers. *Journal of Hydraulic Research*, **49**(2), 178–193.

Frothingham, K. M., Rhoads, B. L., 2003. Three-dimensional flow structure and channel change in an asymmetrical compound meander loop, Embarras River, Illinois. *Earth Surface Processes and Landforms*, **28**(6), 625–644.

Frothingham, K. M., Rhoads, B. L., Herricks, E. E., 2001. Stream geomorphology and fish community structure in channelized and meandering reaches of an agricultural stream. In: J. M. Dorava, D. R. Montgomery, B. B. Palcsak, F. A. Fitzpatrick (eds.), *Geomorphic Processes and Riverine Habitat*, American Geophysical Union, Washington, DC, pp. 105–117.

Frothingham, K. M., Rhoads, B. L., Herricks, E. E., 2002. A multiscale conceptual framework for integrated ecogeomorphological research to support stream naturalization in the agricultural Midwest. *Environmental Management*, **29**(1), 16–33.

Fryirs, K., 2013. (Dis)Connectivity in catchment sediment cascades: A fresh look at the sediment delivery problem. *Earth Surface Processes and Landforms*, **38**(1), 30–46.

Fryirs, K. A., 2017. River sensitivity: A lost foundation concept in fluvial geomorphology. *Earth Surface Processes and Landforms*, **42**(1), 55–70.

Fryirs, K., Brierley, G. J., 2001. Variability in sediment delivery and storage along river courses in Bega catchment, NSW, Australia: Implications for geomorphic river recovery. *Geomorphology*, **38**(3–4), 237–265.

Fryirs, K. A., Brierley, G. J., 2018. What's in a name? A naming convention for geomorphic river types using the River Styles Framework. *PLoS ONE*, **13**(9). 10.1371/journal.pone.0201909.

Fryirs, K. A., Brierley, G. J., Preston, N. J., Kasai, M., 2007. Buffers, barriers and blankets: The (dis)connectivity of catchment-scale sediment cascades. *Catena*, **70**(1), 49–67.

Fryirs, K., Spink, A., Brierley, G., 2009. Post-European settlement response gradients of river sensitivity and recovery across the upper Hunter catchment, Australia. *Earth Surface Processes and Landforms*, **34**(7), 897–918.

Fryirs, K., Brierley, G. J., Erskine, W. D., 2012. Use of ergodic reasoning to reconstruct the historical range of variability and evolutionary trajectory of rivers. *Earth Surface Processes and Landforms*, **37**(7), 763–773.

Fryirs, K., Lisenby, P., Croke, J., 2015. Morphological and historical resilience to catastrophic flooding: The case of Lockyer Creek, SE Queensland, Australia. *Geomorphology*, **241**, 55–71.

Fryirs, K. A., Wheaton, J. M., Brierley, G. J., 2016. An approach for measuring confinement and assessing the influence of valley setting on river forms and processes. *Earth Surface Processes and Landforms*, **41**(5), 701–710.

Fujita, Y., 1989. Bar and channel formation in braided streams. In: S. Ikeda, G. Parker (eds.), *River Meandering*. American Geophysical Union, Washington, DC, pp. 417–462.

Fuller, I. C., Large, A. R. G., Milan, D. J., 2003. Quantifying channel development and sediment transfer following chute cutoff in a wandering gravel-bed river. *Geomorphology*, **54**(3–4), 307–323.

Furbish, D. J., 1991. Spatial autoregressive structure in meander evolution. *Geological Society of America Bulletin*, **103**(12), 1576–1589.

Furbish, D. J., 1997. *Fluid Physics in Geology*. Oxford University Press, New York.

Furbish, D. J., Haff, P. K., Roseberry, J. C., Schmeeckle, M. W., 2012. A probabilistic description of the bed load sediment flux: 1. theory. *Journal of Geophysical Research – Earth Surface*, **117**. 10.1029/2012jf002352.

Furbish, D. J., Schmeeckle, M., Fathel, S., 2017. Particle motions and bedload theory: The entrainment forms of the flux and the Exner equation. In: D. Tsutsumi, J. B. Laronne (eds.), *Gravel-Bed Rivers: Processes and Disasters*. Wiley, Chichester, UK, pp. 97–120.

Furey, P. R., Troutman, B. M., 2008. A consistent framework for Horton regression statistics that leads to a modified Hack's law. *Geomorphology*, **102**(3–4), 603–614.

Gagliano, S. M., Howard, P. C., 1984. The neck cutoff oxbow lake cycle along the lower Mississippi River. In: C. M. Elliott (ed.), *River Meandering*. American Society of Civil Engineers, New York, pp. 147–158.

Galay, V. J., 1983. Causes of river bed degradation. *Water Resources Research*, **19**(5), 1057–1090.

Gale, S. J., Hoare, P. G., 1992. Bulk sampling of coarse clastic sediments for particle size analysis. *Earth Surface Processes and Landforms*, **17**(7), 729–733.

Galster, J. C., Pazzaglia, F. J., Germanoski, D., 2008. Measuring the impact of urbanization on channel widths using historic aerial photographs and modern surveys. *Journal of the American Water Resources Association*, **44**(4), 948–960.

Ganti, V., Meerschaert, M. M., Foufoula-Georgiou, E., Viparelli, E., Parker, G., 2010. Normal and anomalous diffusion of gravel tracer particles in rivers. *Journal of Geophysical Research – Earth Surface*, **115**. 10.1029/2008jf001222.

Gao, P., 2008. Understanding watershed suspended sediment transport. *Progress in Physical Geography*, **32**(3), 243–263.

Gao, P., 2013. Rill and gully development processes. In: J. W. Schroder (ed.), *Treatise on Geomorphology*, Vol. 7, Hillslope Processes, M. Stoffel, R. Marston (vol. eds.). Wiley, New York, pp. 122–131.

Gao, P., Josefson, M., 2012a. Temporal variations of suspended sediment transport in Oneida Creek watershed, central New York. *Journal of Hydrology*, **426**, 17–27.

Gao, P., Josefson, M., 2012b. Event-based suspended sediment dynamics in a central New York watershed. *Geomorphology*, **139**, 425–437.

Gao, P., Nearing, M. A., Commons, M., 2013. Suspended sediment transport at the instantaneous and event time scales in semiarid watersheds of southeastern Arizona, USA. *Water Resources Research*, **49**(10), 6857–6870. 10.1002/wrcr.20549.

Garcia, M. H., 2008. Sediment transport and morphodynamics. In: M. H. Garcia (ed.), *Sedimentation Engineering: Processes, Measurements, Modeling, and Practice*. American Society of Civil Engineers, Reston, VA, pp. 21–163.

Garcia, M., Nino, Y., 1993. Dynamics of sediment bars in straight and meandering channels – experiments on the resonance phenomenon. *Journal of Hydraulic Research*, **31**(6), 739–761.

Garcia, M., Parker, G., 1991. Entrainment of bed sediment into suspension. *Journal of Hydraulic Engineering*, **117**(4), 414–435.

Garcia-Flores, M., Maza-Alvarez, J. A., 1997. Inicio de movimiento y acorazamiento, Capitulo 8 del Manual de Ingeniería de Ríos, Series del Instituto de Ingeniería 592, UNAM, Mexico (in Spanish).

Gardner, T. W., 1983. Experimental study of knickpoint and longitudinal profile evolution in cohesive, homogenous material. *Geological Society of America Bulletin*, **94**(5), 664–672.

Gardner, T., Ashmore, P., Leduc, P., 2018. Morpho-sedimentary characteristics of proximal gravel braided river deposits in a Froude-scaled physical model. *Sedimentology*, **65**(3), 877–896.

Garrels, R. M., MacKenzie, F. T., 1971. *Evolution of Sedimentary Rocks*. Norton, New York.

Gasparini, N. M., Tucker, G. E., Bras, R. L., 1999. Downstream fining through selective particle sorting in an equilibrium drainage network. *Geology*, **27**(12), 1079–1082.

Gaudet, J. M., Roy, A. G., 1995. Effect of bed morphology on flow mixing length at river confluences. *Nature*, **373**(6510), 138–139.

Gaudio, R., Miglio, A., Dey, S., 2010. Non-universality of von Karman's κ in fluvial streams. *Journal of Hydraulic Research*, **48**(5), 658–663.

Gaurav, K., Tandon, S. K., Devauchelle, O., Sinha, R., Metivier, F., 2017. A single width-discharge regime relationship for individual threads of braided and meandering rivers from the Himalayan Foreland. *Geomorphology*, **295**, 126–133.

Gay, G. R., Gay, H. H., Gay, W. H., et al., 1998. Evolution of cutoffs across meander necks in Powder River, Montana, USA. *Earth Surface Processes and Landforms*, **23**(7), 651–662.

Gee, G. W., Bauder, J. W., 1986. Particle-size analysis. In: A. Klute (ed.), *Methods of Soil Analysis*. Part 1, 2nd ed., Agronomy Monograph 9. Soil Science Society of America and American Society of Agronomy, Madison, WI, pp. 383–411.

Gellis, A. C., 2013. Factors influencing storm-generated suspended-sediment concentrations and loads in four basins of contrasting land use, humid-tropical Puerto Rico. *Catena*, **104**, 39–57.

Gellis, A. C., Myers, M. K., Noe, G. B., et al., 2017. Storms, channel changes, and a sediment budget for an urban-suburban stream, Difficult Run, Virginia, USA. *Geomorphology*, **278**, 128–148.

Germanoski, D., Schumm, S. A., 1993. Changes in braided river morphology resulting from aggradation and degradation. *Journal of Geology*, **101**(4), 451–466.

Ghinassi, M., Ielpi, A., Aldinucci, M., Fustic, M., 2016. Downstream-migrating fluvial point bars in the rock record. *Sedimentary Geology*, **334**, 66–96.

Ghoshal, S., James, L. A., Singer, M. B., Aalto, R., 2010. Channel and floodplain change analysis over a 100-year period: Lower Yuba River, California. *Remote Sensing*, **2**(7), 1797–1825.

Giachetta, E., Refice, A., Capolongo, D., Gasparini, N. M., Pazzaglia, F. J., 2014. Orogen-scale drainage network evolution and response to erodibility changes: insights from numerical experiments. *Earth Surface Processes and Landforms*, **39**(9), 1259–1268.

Giere, R. N., 1988. *Explaining Science: A Cognitive Approach*. University of Chicago Press, Chicago.

Gilbert, G. K., 1877. Report on the geology of the Henry Mountains. U.S. Geographical and Geological Survey of the Rocky Mountains Region. U.S. Government Printing Office, Washington, DC.

Gilbert, G. K., 1907. Rate of Recession of Niagara Falls, U.S. Geological Survey Bulletin No. 36. U.S. Government Printing Office, Washington, DC.

Gilbert, G. K., 1914. The transportation of debris by running water. U.S. Geological Survey Professional Paper 86. U.S. Government Printing Office, Washington, DC.

Gilbert, G. K., 1917. Hydraulic-mining debris in the Sierra Nevada. U.S. Geological Survey Professional Paper 105, U.S. Government Printing Office, Washington, DC.

Gilvear, D., Bryant, R., 2016. Analysis of remotely sensed data for fluvial geomorphology and river science. In: G. M. Kondolf, H. Piegay (eds.), *Tools in Fluvial Geomorphology*, 2nd Edition. Wiley, Chichester, UK, pp. 103–132.

Gleason, C. J., Smith, L. C., 2014. Toward global mapping of river discharge using satellite images and at-many-stations hydraulic geometry. *Proceedings of the National Academy of Sciences of the United States of America*, **111**(13), 4788–4791.

Gleason, C. J., Wang, J., 2015. Theoretical basis for at-many-stations hydraulic geometry. *Geophysical Research Letters*, **42**(17), 7107–7114. 10.1002/2015gl064935.

Gleason, C. J., Smith, L. C., Lee, J., 2014. Retrieval of river discharge solely from satellite imagery and at-many-stations hydraulic geometry: Sensitivity to river form and optimization parameters. *Water Resources Research*, **50**(12), 9604–9619. 10.1002/2014wr016109.

Glock, W. S., 1931. The development of drainage systems: A synoptic view. *Geographical Review*, **21**(3), 475–482. 10.2307/209434.

Glymph, L. M., 1954. Studies of sediment yields from watersheds. IAHS Publication 36, IAHS Press, Wallingford, UK, 178–191.

Goff, J. R., Ashmore, P., 1994. Gravel transport and morphological change in braided Sunwapta River, Alberta, Canada. *Earth Surface Processes and Landforms*, **19**(3), 195–212.

Goldrick, G., Bishop, P., 2007. Regional analysis of bedrock stream long profiles: evaluation of Hack's SL form, and formulation and assessment of an alternative (the DS form). *Earth Surface Processes and Landforms*, **32**(5), 649–671.

Gomez, B., 1983. Temporal variations in the particle-size distribution of surficial bed material – the effect of progressive bed armoring. *Geografiska Annaler Series A Physical Geography*, **65**(3–4), 183–192.

Gomez, B., 1984. Typology of segregated (armored paved) surfaces – some comments. *Earth Surface Processes and Landforms*, **9**(1), 19–24.

Gomez, B., 1995. Bedload transport and changing grain size distributions. In: A. Gurnell, G. Petts (eds.), *Changing River Channels*. Wiley, Chichester, UK, pp. 177–199.

Gomez, B., Church, M., 1989. An assessment of bedload sediment transport formulas for gravel bed rivers. *Water Resources Research*, **25**(6), 1161–1186.

Gomez, B., Mullen, V. T., 1992. An experimental study of sapped drainage network development. *Earth Surface Processes and Landforms*, **17**(5), 465–476.

Gomez, B., Mertes, L. A. K., Phillips, J. D., Magilligan, F. J., James, L. A., 1995. Sediment characteristics of an extreme flood – 1993 upper Mississippi River Valley. *Geology*, **23**(11), 963–966.

Gomez, B., Phillips, J. D., Magilligan, F. J., James, L. A., 1997. Floodplain sedimentation and sensitivity: summer 1993 flood, upper Mississippi River valley. *Earth Surface Processes and Landforms*, **22**(10), 923–936.

Gomez, B., Eden, D. N., Peacock, D. H., Pinkney, E. J., 1998. Floodplain construction by recent, rapid vertical accretion: Waipaoa River, New Zealand. *Earth Surface Processes and Landforms*, **23**(5), 405–413.

Gomez, B., Coleman, S. E., Sy, V. W. K., Peacock, D. H., Kent, M., 2007. Channel change, bankfull and effective discharges on a vertically accreting, meandering, gravel-bed river. *Earth Surface Processes and Landforms*, **32**(5), 770–785. 10.1002/esp.1424.

Gomez, J. A., Darboux, F., Nearing, M. A., 2003. Development and evolution of rill networks under simulated rainfall. *Water Resources Research*, **39**(6), 14. 10.1029/2002wr001437.

Gomi, T., Sidle, R. C., Woodsmith, R. D., Bryant, M. D., 2003. Characteristics of channel steps and reach morphology in headwater streams, southeast Alaska. *Geomorphology*, **51**(1–3), 225–242.

Gonzalez, E., Sher, A. A., Tabacchi, E., Masip, A., Poulin, M., 2015. Restoration of riparian vegetation: A global review of implementation and evaluation approaches in the international, peer-reviewed literature. *Journal of Environmental Management*, **158**, 85–94.

Gonzalez, R. L., Pasternack, G. B., 2015. Reenvisioning cross-sectional at-a-station hydraulic geometry as spatially explicit hydraulic topography. *Geomorphology*, **246**, 394–406.

Goode, J. R., Wohl, E., 2007. Relationships between land-use and forced-pool characteristics in the Colorado Front Range. *Geomorphology*, **83**(3–4), 249–265.

Goodwell, A. E., Zhu, Z. D., Dutta, D., et al., 2014. Assessment of floodplain vulnerability during extreme Mississippi River flood 2011. *Environmental Science & Technology*, **48**(5), 2619–2625.

Goodwin, P., 2004. Analytical solutions for estimating effective discharge. *Journal of Hydraulic Engineering-ASCE*, **130**(8), 729–738.

Gorrick, S., Rodriguez, J. F., 2014. Flow and force-balance relations in a natural channel with bank vegetation. *Journal of Hydraulic Research*, **52**(6), 794–810.

Gorycki, M. A., 1973. Hydraulic drag – a meander-initiating mechanism. *Geological Society of America Bulletin*, **84**(1), 175–186.

Grabowski, R. C., Gurnell, A. M., 2016. Using historical data in fluvial geomorphology. In: G. M. Kondolf, H. Piegay (eds.), *Tools in Fluvial Geomorphology*, 2nd Edition. Wiley, Chichester, UK, pp. 56–75.

Graf, W. H., Cellino, M., 2002. Suspension flows in open channels; experimental study. *Journal of Hydraulic Research*, **40**(4), 435–447.

Graf, W. L., 1975. Impact of suburbanization on fluvial geomorphology. *Water Resources Research*, **11**(5), 690–692.

Graf, W. L., 1977a. Rate law in fluvial geomorphology. *American Journal of Science*, **277**(2), 178–191.

Graf, W. L., 1977b. Network characteristics in suburbanizing streams. *Water Resources Research*, **13**(2), 459–463.

Graf, W. L., 1983a. Downstream changes in stream power in the Henry Mountains, Utah. *Annals of the Association of American Geographers*, **73**(3), 373–387.

Graf, W. L., 1983b. Flood-related change in an arid-region river. *Earth Surface Processes and Landforms*, **8**(2), 125–139.

Graf, W. L., 1983c. The arroyo problem: Paleohydrology and paleohydraulics in the short term. In: K. Gregory (ed.), *Background to Paleohydrology*. Wiley, New York, pp. 279–302.

Graf, W. L., 1988a. *Fluvial Processes in Dryland Rivers*. Springer-Verlag, Berlin.

Graf, W. L., 1988b. Definition of flood plains along arid-region rivers. In: V. R. Baker, R. C. Kochel, P. C. Patton (eds.), *Flood Geomorphology*. Wiley, New York, pp. 231–242.

Graf, W. L., 1990. Fluvial dynamics of Thorium-230 in the Church Rock Event, Puerco River, New Mexico. *Annals of the Association of American Geographers*, **80**(3), 327–342.

Graf, W. L., 1993. *Landscapes, Commodities, and Ecosystems: The Relationship between Science and Policy for American Rivers, Sustaining Our Water Resources*. National Academy Press, Washington, DC, pp. 11–42.

Graf, W. L., 1999. Dam nation: A geographic census of American dams and their large-scale hydrologic impacts. *Water Resources Research*, **35**(4), 1305–1311.

Graf, W. L., 2006. Downstream hydrologic and geomorphic effects of large dams on American rivers. *Geomorphology*, **79**(3–4), 336–360.

Graf, W. L., 2008. Sources of uncertainty in river restoration research. In: S. E. Darby, D. A. Sear (eds.), *River Restoration: Managing the Uncertainty in Restoring Physical Habitat*. Wiley, Chichester, UK, pp. 15–19.

Grams, P. E., Schmidt, J. C., 2002. Streamflow regulation and multi-level flood plain formation: Channel narrowing on the aggrading Green River in the eastern Uinta Mountains, Colorado and Utah. *Geomorphology*, **44**(3–4), 337–360.

Grams, P. E., Schmidt, J. C., 2005. Equilibrium or indeterminate? Where sediment budgets fail: Sediment mass balance and adjustment of channel form, Green River downstream from Flaming Gorge Dam, Utah and Colorado. *Geomorphology*, **71**(1–2), 156–181.

Grams, P. E., Schmidt, J. C., Topping, D. J., 2007. The rate and pattern of bed incision and bank adjustment on the Colorado River in Glen Canyon downstream from Glen Canyon Dam, 1956–2000. *Geological Society of America Bulletin*, **119**(5–6), 556–575.

Gran, K. B., Czuba, J. A., 2017. Sediment pulse evolution and the role of network structure. *Geomorphology*, **277**, 17–30.

Gran, K. B., Paola, C., 2001. Riparian vegetation controls on braided stream dynamics. *Water Resources Research*, **37**(12), 3275–3283.

Gran, K. B., Tal, M., Wartman, E. D., 2015. Co-evolution of riparian vegetation and channel dynamics in an aggrading braided river system, Mount Pinatubo, Philippines. *Earth Surface Processes and Landforms*, **40**(8), 1101–1115. 10.1002/esp.3699.

Granger, D. E., Schaller, M., 2014. Cosmogenic nuclides and erosion at the watershed scale. *Elements*, **10**(5), 369–373.

Grant, G. E., Swanson, F. J., Wolman, M. G., 1990. Pattern and origin of stepped-bed morphology in high-gradient streams, western Cascades, Oregon. *Geological Society of America Bulletin*, **102**(3), 340–352.

Grant, G. E., Schmidt, J. C., Lewis, S., 2003. A geological framework for interpreting downstream effects of dams on rivers. In: J. E. O'Connor, G. E. Grant (eds.), *A Peculiar River*. American Geophysical Union, Washington, DC., pp. 203–219.

Grant, G. E., O'Connor, J. E., Wolman, M. G., 2013. A river runs through it: conceptual models in fluvial geomorphology. In: J. W. Schroder (ed.), *Treatise on Geomorphology*, Vol. 9, Fluvial geomorphology, E. Wohl (vol. ed.). Academic Press, San Diego, CA, pp. 6–21.

Gray, A. B., Warrick, J. A., Pasternack, G. B., Watson, E. B., Goni, M. A., 2014. Suspended sediment behavior in a coastal dry-summer subtropical catchment: Effects of hydrologic preconditions. *Geomorphology*, **214**, 485–501.

Gray, R. J., Glysson, G. D., Turcios, L. M., Schwartz, G. E., 2000. Comparability of suspended-sediment concentration and total suspended solids data. U.S. Geological Survey Water-Resources Investigation Report 00–4191. U.S. Geological Survey, Reston, VA.

Gregoretti, C., 2008. Inception sediment transport relationships at high slopes. *Journal of Hydraulic Engineering*, **134**(11), 1620–1629.

Gregory, K. J., 1976. Drainage networks and climate. In: E. Derbyshire (ed.), *Geomorphology and Climate*. Wiley and Sons, Chichester, UK, pp. 289–315.

Gregory, K. J., 2006. The human role in changing river channels. *Geomorphology*, **79**(3–4), 172–191.

Gregory, K. J., Downs, P., 2008. The sustainability of restored rivers: catchment-scale perspectives on long-term response. In: S. E. Darby, D. A. Sear (eds.), *River Restoration: Managing the Uncertainty in Restoring Physical Habitat*. Wiley, Chichester, UK, pp. 253–286.

Gregory, K. J., Gurnell, A. M., Hill, C. T., Tooth, S., 1994. Stability of the pool riffle sequence in changing river channels. *Regulated Rivers-Research & Management*, **9**(1), 35–43.

Grenfell, M., Aalto, R., Nicholas, A., 2012. Chute channel dynamics in large, sand-bed meandering rivers. *Earth Surface Processes and Landforms*, **37**(3), 315–331.

Grenfell, M. C., Nicholas, A. P., Aalto, R., 2014. Mediative adjustment of river dynamics: The role of chute channels in tropical sand-bed meandering rivers. *Sedimentary Geology*, **301**, 93–106.

Griffiths, G. A., 1984. Extremal hypotheses for river regime – an illusion of progress. *Water Resources Research*, **20**(1), 113–118.

Grill, G., Lehner, B., Thieme, M., et al., 2019. Mapping the world's free-flowing rivers. *Nature*, **569**(7755), 215–221. 10.1038/s41586-019-1111-9.

Grimaldi, S., Petroselli, A., Nardi, F., 2012. A parsimonious geomorphological unit hydrograph for rainfall-runoff modelling in small ungauged basins. *Hydrological Sciences Journal*, **57**(1), 73–83. 10.1080/02626667.2011.636045.

Grimaud, J. L., Paola, C., Voller, V., 2016. Experimental migration of knickpoints: Influence of style of base-level fall and bed lithology. *Earth Surface Dynamics*, **4**(1), 11–23.

Grimley, D. A., Anders, A. M., Bettis, E. A., III, et al., 2017. Using magnetic fly ash to identify post-settlement alluvium and its record of atmospheric pollution, central USA. *Anthropocene*, **17**, 84–98.

Grissinger, E. H., 1996. Rill and gullies erosion. In: M. Agassi (ed.), *Soil Erosion, Conservation and Rehabilitation*. Marcel Dekker, New York, pp. 153–167.

Grissinger, E. H., Murphey, J. B., 1984. Morphometric evolution of man-modified channels. In: C. M. Elliott (ed.), *River Meandering*. American Society of Civil Engineers, New York, pp. 273–283.

Grudzinski, B. P., Daniels, M. D., 2018. Bison and cattle grazing impacts on grassland stream morphology in the Flint Hills of Kansas. *Rangeland Ecology & Management*, **71**(6), 783–791.

Grudzinski, B. P., Daniels, M. D., Anibas, K., Spencer, D., 2016. Bison and cattle grazing management, bare ground coverage, and links to suspended sediment concentrations in grassland streams. *Journal of the American Water Resources Association*, **52**(1), 16–30.

Gualtieri, C., Filizola, N., de Oliveira, M., Santos, A. M., Ianniruberto, M., 2018. A field study of the confluence between Negro and Solimoes Rivers. Part 1: Hydrodynamics and sediment transport. *Comptes Rendus Geoscience*, **350**(1–2), 31–42.

Guillen-Ludena, S., Franca, M. J., Cardoso, A. H., Schleiss, A. J., 2015. Hydro-morphodynamic evolution in a 90 degrees movable bed discordant confluence with low discharge ratio. *Earth Surface Processes and Landforms*, **40**(14), 1927–1938.

Guillen-Ludena, S., Franca, M. J., Cardoso, A. H., Schleiss, A. J., 2016. Evolution of the hydromorphodynamics of mountain river confluences for varying discharge ratios and junction angles. *Geomorphology*, **255**, 1–15.

Guillen-Ludena, S., Franca, M. J., Alegria, F., Schleiss, A. J., Cardoso, A. H., 2017a. Hydromorphodynamic effects of the width ratio and local tributary widening on discordant confluences. *Geomorphology*, **293**, 289–304.

Guillen-Ludena, S., Cheng, Z., Constantinescu, G., Franca, M. J., 2017b. Hydrodynamics of mountain-river confluences and its relationship to sediment transport. *Journal of Geophysical Research – Earth Surface*, **122**(4), 901–924. 10.1002/2016jf004122.

Guinoiseau, D., Bouchez, J., Gelabert, A., et al., 2016. The geochemical filter of large river confluences. *Chemical Geology*, **441**, 191–203.

Guneralp, I., Marston, R. A., 2012. Process-form linkages in meander morphodynamics: Bridging theoretical modeling and real world complexity. *Progress in Physical Geography*, **36**(6), 718–746.

Guneralp, I., Rhoads, B. L., 2008. Continuous characterization of the planform geometry and curvature of meandering rivers. *Geographical Analysis*, **40**(1), 1–25.

Guneralp, I., Rhoads, B. L., 2009. Empirical analysis of the planform curvature-migration relation of meandering rivers. *Water Resources Research*, **45**. 10.1029/2008wr007533.

Guneralp, I., Rhoads, B. L., 2010. Spatial autoregressive structure of meander evolution revisited. *Geomorphology*, **120**(3–4), 91–106.

Guneralp, I., Rhoads, B. L., 2011. Influence of floodplain erosional heterogeneity on planform complexity of meandering rivers. *Geophysical Research Letters*, **38**. 10.1029/2011gl048134.

Guo, J., 2002. Hunter Rouse and Shields diagram advances in hydraulics and water engineering, Proceedings of the 13th IAHR-APD Congress. World Scientific Singapore, pp. 1096–1098.

Guo, J. K., 2013. Modified log-wake law for smooth rectangular open channel flow. Proceedings of the 35th IAHR World Congress, Vols I and II, 2010–2019.

Guo, J. K., Julien, P. Y., 2001. Turbulent velocity profiles in sediment-laden flows. *Journal of Hydraulic Research*, **39**(1), 11–23.

Guo, J., Julien, P. Y., 2005. Shear stress in smooth rectangular open-channel flows. *Journal of Hydraulic Engineering*, **131**(1), 30–37.

Gupta, N., Kleinhans, M. G., Addink, E. A., Atkinson, P. M., Carling, P. A., 2014. One-dimensional modeling of a recent Ganga avulsion: Assessing the potential effect of tectonic subsidence on a large river. *Geomorphology*, **213**, 24–37.

Gurnell, A. M., Petts, G. E., Hannah, D. M., et al., 2001. Riparian vegetation and island formation along the gravel-bed Fiume Tagliamento, Italy. *Earth Surface Processes and Landforms*, **26**(1), 31–62.

Gurnell, A., Lee, M., Souch, C., 2007. Urban rivers: hydrology, geomorphology, ecology and opportunities for change. *Geography Compass*, **1**, 1118–1137.

Gurnell, A., Surian, N., Zanoni, L., 2009. Multi-thread river channels: a perspective on changing European alpine river systems. *Aquatic Sciences*, **71**(3), 253–265.

Gurram, S. K., Karki, K. S., Hager, W. H., 1997. Subcritical junction flow. *Journal of Hydraulic Engineering*, **123**(5), 447–455.

Guy, H. P., 1969. Laboratory theory and methods for sediment analysis, U.S. Geological Survey Techniques of Water Resources Investigations, Book 5, Chapter C1. Reston, VA, pp. 1–58.

Habersack, H. M., 2001. Radio-tracking gravel particles in a large braided river in New Zealand: A field test of the stochastic theory of bed load transport proposed by Einstein. *Hydrological Processes*, **15**(3), 377–391.

Hack, J. T., 1957. Studies of longitudinal profiles in Virginia and Maryland. U.S. Geological Survey Professional Paper 294-B. U.S. Government Printing Office, Washington, DC.

Hack, J. T., 1960. Interpretation of erosional topography in humid temperate regions. *American Journal of Science*, **258**, 80–97.

Hackney, C., Best, J., Leyland, J., et al., 2015. Modulation of outer bank erosion by slump blocks: disentangling the protective and destructive role of failed material on the three-dimensional flow structure. *Geophysical Research Letters*, **42**(24), 10663–10670. 10.1002/2015gl066481.

Hackney, C. R., Darby, S. E., Parsons, D. R., et al., 2018. The influence of flow discharge variations on the morphodynamics of a diffluence-confluence unit on a large river. *Earth Surface Processes and Landforms*, **43**(2), 349–362.

Hadley, R. F., Schumm, S. A., 1961. Sediment sources and drainage basin characteristics in the upper Cheyenne River basin, Wyoming. U.S. Geological Survey Water-Supply Paper 1531-B. U.S. Government Printing Office, Washington, DC.

Hagerty, D. J., 1991. Piping sapping erosion. 1. Basic considerations. *Journal of Hydraulic Engineering*, **117**(8), 991–1008.

Halbert, K., Nguyen, C. C., Payrastre, O., Gaume, E., 2016. Reducing uncertainty in flood frequency analyses: a comparison of local and regional approaches involving information on extreme historical floods. *Journal of Hydrology*, **541**, 90–98.

Hall, P., 2004. Alternating bar instabilities in unsteady channel flows over erodible beds. *Journal of Fluid Mechanics*, **499**, 49–73.

Hammer, T. R., 1972. Stream channel enlargement due to urbanization. *Water Resources Research*, **8**, 1530–1540.

Han, B., Endreny, T. A., 2014. Detailed river stage mapping and head gradient analysis during meander cutoff in a laboratory river. *Water Resources Research*, **50**(2), 1689–1703. 10.1002/2013wr013580.

Hancock, G. R., Evans, K. G., 2006. Channel head location and characteristics using digital elevation models. *Earth Surface Processes and Landforms*, **31**(7), 809–824.

Hancock, G., Willgoose, G., 2001a. The interaction between hydrology and geomorphology in a landscape simulator experiment. *Hydrological Processes*, **15**(1), 115–133.

Hancock, G., Willgoose, G., 2001b. Use of a landscape simulator in the validation of the SIBERIA catchment evolution model: Declining equilibrium landforms. *Water Resources Research*, **37**(7), 1981–1992.

Hancock, G. R., Willgoose, G. R., 2002. The use of a landscape simulator in the validation of the SIBERIA landscape evolution model: Transient landforms. *Earth Surface Processes and Landforms*, **27**(12), 1321–1334.

Hancock, G. R., Crawter, D., Fityus, S. G., Chandler, J., Wells, T., 2008. The measurement and modelling of rill erosion at angle of repose slopes in mine spoil. *Earth Surface Processes and Landforms*, **33**(7), 1006–1020.

Hancock, G. R., Willgoose, G. R., Lowry, J., 2014a. Transient landscapes: gully development and evolution using a landscape evolution model. *Stochastic Environmental Research and Risk Assessment*, **28**(1), 83–98.

Hancock, G. J., Wilkinson, S. N., Hawdon, A. A., Keen, R. J., 2014b. Use of fallout tracers 7Be, 210Pb and 137Cs to distinguish the form of sub-surface soil erosion delivering sediment to rivers in large catchments. *Hydrological Processes*, **28**(12), 3855–3874.

Hanrahan, B. R., Tank, J. L., Dee, M. M., et al., 2018. Restored floodplains enhance denitrification compared to naturalized floodplains in agricultural streams. *Biogeochemistry*, **141**(3), 419–437.

Happ, S. C., 1945. Sedimentation in South Carolina Piedmont valleys. *American Journal of Science*, **243**(3), 113–126.

Happ, S. C., Rittenhouse, G., Dobson, G. C., 1940. Some principles of accelerated stream and valley sedimentation. U.S. Department of Agriculture Technical Bulletin 695. U.S. Government Printing Office, Washington, DC.

Harden, C. P., 2013. Impacts of vegetation clearance on channel change: Historical perspective. In: J. C. Shroder (ed.), *Treatise on Geomorphology*, Vol. 13, Geomorphology of Human Disturbances, Climate Change, and Natural Hazards, L. A. James, C. P. Harden, J. J. Clague (vol. eds.). Academic Press, San Diego, CA, pp. 14–27.

Harden, D. R., 1990. Controlling factors in the distribution and development of incised meanders in the central Colorado Plateau. *Geological Society of America Bulletin*, **102**(2), 233–242.

Hardy, R. J., Bates, P. D., Anderson, M. G., 2000. Modelling suspended sediment deposition on a fluvial floodplain using a two-dimensional dynamic finite element model. *Journal of Hydrology*, **229**(3–4), 202–218.

Hardy, R. J., Best, J. L., Lane, S. N., Carbonneau, P. E., 2009. Coherent flow structures in a depth-limited flow over a gravel surface: The role of near-bed turbulence and influence of Reynolds number. *Journal of Geophysical Research – Earth Surface*, **114**. 10.1029/2007jf000970.

Hardy, R. J., Best, J. L., Parsons, D. R., Marjoribanks, T. I., 2016. On the evolution and form of coherent flow structures over a gravel bed: Insights from whole flow field visualization and measurement. *Journal of Geophysical Research – Earth Surface*, **121**(8), 1472–1493. 10.1002/2015jf003753.

Harman, C., Stewardson, M., DeRose, R., 2008. Variability and uncertainty in reach bankfull hydraulic geometry. *Journal of Hydrology*, **351**(1–2), 13–25.

Harman, W., Starr, R., 2011. Natural Channel Design Review Checklist. US Fish and Wildlife Service, Chesapeake Bay Field Office, US Environmental Protection Agency, Office of Wetlands, Oceans, and Watersheds, Wetlands Division, Washington, DC.

Harrison, L. R., Keller, E. A., 2007. Modeling forced pool-riffle hydraulics in a boulder-bed stream, southern California. *Geomorphology*, **83**(3–4), 232–248.

Harrison, L. R., Dunne, T., Fisher, G. B., 2015. Hydraulic and geomorphic processes in an overbank flood along a meandering, gravel-bed river: Implications for chute formation. *Earth Surface Processes and Landforms*, **40**(9), 1239–1253.

Harvey, A. M., 2002. Effective timescales of coupling within fluvial systems. *Geomorphology*, **44**(3–4), 175–201.

Harwood, K., Brown, A. G., 1993. Fluvial processes in a forested anastomosing river – flood partitioning and change flow patterns. *Earth Surface Processes and Landforms*, **18**(8), 741–748.

Haschenburger, J. K., 2011. Vertical mixing of gravel over a long flood series. *Earth Surface Processes and Landforms*, **36**(8), 1044–1058.

Haschenburger, J. K., 2013a. Bedload kinematics and fluxes. In: J. Schroder (ed.), *Treatise on Geomorphology*, Vol. 9, Fluvial Geomorphology, E. Wohl (vol. ed.) Academic Press, San Diego, CA, pp. 103–123.

Haschenburger, J. K., 2013b. Tracing river gravels: insights into dispersion from a long-term field experiment. *Geomorphology*, **200**, 121–131.

Haschenburger, J. K., 2017. Streambed disturbance over a long flood series. *River Research and Applications*, **33**(5), 753–765.

Haschenburger, J. K., Church, M., 1998. Bed material transport estimated from the virtual velocity of sediment. *Earth Surface Processes and Landforms*, **23**(9), 791–808.

Haschenburger, J. K., Cowie, M., 2009. Floodplain stages in the braided Ngaruroro River, New Zealand. *Geomorphology*, **103**(3), 466–475.

Hassan, M. A., Bradley, D. N., 2017. Geomorphic controls on tracer particle dispersion in gravel-bed rivers. In: D. Tsutsumi, J. B. Laronne (eds.), *Gravel-Bed Rivers: Processes and Disasters*. Wiley, Chichester, UK, pp. 159–184.

Hassan, M. A., Church, M., 1994. Vertical mixing of coarse particles in gravel-bed rivers – a kinematic model. *Water Resources Research*, **30**(4), 1173–1185. 10.1029/93wr03351.

Hassan, M. A., Reid, I., 1990. The influence of microform bed roughness elements on flow and sediment transport in gravel bed rivers. *Earth Surface Processes and Landforms*, **15**(8), 739–750.

Hassan, M. A., Roy, A. G., 2016. Coarse particle tracing in fluvial geomorphology. In: G. M. Kondolf, H. Piegay (eds.), *Tools in Fluvial Geomorphology*, 2nd Edition. Wiley and Sons, Chichester, UK, pp. 306–323.

Hassan, M. A., Schick, A. P., Laronne, J. B., 1984. The recovery of flood dispersed coarse sediment particles – a 3-dimensional magnetic tracing method. *Catena*, 153–162.

Hassan, M. A., Church, M., Ashworth, P. J., 1992. Virtual rate and mean distance of travel of individual clasts in gravel-bed rivers. *Earth Surface Processes and Landforms*, **17**(6), 617–627.

Hassan, M. A., Brayshaw, D., Alila, Y., Andrews, E., 2014. Effective discharge in small formerly glaciated mountain streams of

British Columbia: Limitations and implications. *Water Resources Research*, **50**(5), 4440–4458. 10.1002/2013wr014529.

Hauer, C., Habersack, H., 2009. Morphodynamics of a 1000-year flood in the Kamp River, Austria, and impacts on floodplain morphology. *Earth Surface Processes and Landforms*, **34**(5), 654–682.

Haviv, I., Enzel, Y., Whipple, K. X., et al., 2010. Evolution of vertical knickpoints (waterfalls) with resistant caprock: Insights from numerical modeling. *Journal of Geophysical Research – Earth Surface*, **115**. 10.1029/2008jf001187.

Haw, M. D., 2002. Colloidal suspensions, Brownian motion, molecular reality: A short history. *Journal of Physics – Condensed Matter*, **14**(33), 7769–7779.

Hawley, R. J., 2018. Making stream restoration more sustainable: a geomorphically, ecologically, and socioeconomically principled approach to bridge the practice with the science. *Bioscience*, **68**(7), 517–528.

Hawley, R. J., Bledsoe, B. P., 2011. How do flow peaks and durations change in suburbanizing semi-arid watersheds? A southern California case study. *Journal of Hydrology*, **405**(1–2), 69–82.

Hawley, R. J., Bledsoe, B. P., 2013. Channel enlargement in semiarid suburbanizing watersheds: A southern California case study. *Journal of Hydrology*, **496**, 17–30.

Hawley, R. J., Bledsoe, B. P., Stein, E. D., Haines, B. E., 2012. Channel evolution model of semiarid stream response to urban-induced hydromodification. *Journal of the American Water Resources Association*, **48**(4), 722–744.

Hawley, R. J., MacMannis, K. R., Wooten, M. S., 2013. Bed coarsening, riffle shortening, and channel enlargement in urbanizing watersheds, northern Kentucky, USA. *Geomorphology*, **201**, 111–126.

Hayakawa, Y., Matsukura, Y., 2003. Recession rates of waterfalls in Boso Peninsula, Japan, and a predictive equation. *Earth Surface Processes and Landforms*, **28**(6), 675–684.

Hayakawa, Y. S., Oguchi, T., 2014. Spatial correspondence of knickzones and stream confluences along bedrock rivers in Japan: implications for hydraulic formation of knickzones. *Geografiska Annaler Series A Physical Geography*, **96**(1), 9–19.

He, L., Wilkerson, G. V., 2011. Improved bankfull channel geometry prediction using two-year return-period discharge. *Journal of the American Water Resources Association*, **47**(6), 1298–1316.

Heidel, S. G., 1956. The progressive lag of sediment concentration with flood waves. *Transactions, American Geophysical Union*, **37**, 56–66.

Hejazi, M. I., Markus, M., 2009. Impacts of urbanization and climate variability on floods in northeastern Illinois. *Journal of Hydrologic Engineering*, **14**(6), 606–616.

Henck, A. C., Montgomery, D. R., Huntington, K. W., Liang, C., 2010. Monsoon control of effective discharge, Yunnan and Tibet. *Geology*, **38**(11), 975–978.

Henderson, F. M., 1966. *Open Channel Flow*. Macmillan, New York.

Henkle, J. E., Wohl, E., Beckman, N., 2011. Locations of channel heads in the semiarid Colorado Front Range, USA. *Geomorphology*, **129**(3–4), 309–319.

Heritage, G., Hetherington, D., 2007. Towards a protocol for laser scanning in fluvial geomorphology. *Earth Surface Processes and Landforms*, **32**(1), 66–74.

Herrero, H. S., Garcia, C. M., Pedocchi, F., et al., 2016. Flow structure at a confluence: Experimental data and the bluff body analogy. *Journal of Hydraulic Research*, **54**(3), 263–274.

Herrero H. S., Diaz Lozada, J. M., Garcia, C. M., et al., 2018. The influence of tributary flow density differences on the hydrodynamic behavior of a confluent meander bend and implications for flow mixing. *Geomorphology*, **304**, 99–112.

Hey, R. D., 1978. Determinate hydraulic geometry of river channels. *Journal of the Hydraulics Division – ASCE*, **104**(6), 869–885.

Hey, R. D., Thorne, C. R., 1986. Stable channels with mobile gravel beds. *Journal of Hydraulic Engineering*, **112**(8), 671–689.

Hickin, E. J., 1979. Concave-bank benches on the Squamish River, British Columbia, Canada. *Canadian Journal of Earth Sciences*, **16**(1), 200–203.

Hickin, E. J., 1986. Concave-bank benches in the floodplains of Muskwa and Fort Nelson Rivers, British Columbia. *Canadian Geographer*, **30**(2), 111–122.

Hickin, E. J., Nanson, G. C., 1984. Lateral migration rates of river bends. *Journal of Hydraulic Engineering*, **110**(11), 1557–1567.

Hickin, A. S., Kerr, B., Barchyn, T. E., Paulen, R. C., 2009. Using ground-penetrating radar and capacitively coupled resistivity to investigate 3-D fluvial architecture and grain-size distribution of a gravel floodplain in northeast British Columbia, Canada. *Journal of Sedimentary Research*, **79**(5–6), 457–477.

Hicks, D. M., Gomez, B., 2016. Sediment transport. In: G. M. Kondolf, H. Piegay (eds.), *Tools in Fluvial Geomorphology*, 2nd Edition. Wiley, Chichester, UK, pp. 324–356.

Hicks, D. M., Mason, P. D., 1991. *Roughness Characteristics of New Zealand Rivers*. National Institute of Water and Atmospheric Research, Wellington, New Zealand.

Hicks, D. M., Hill, J., Shankar, U., 1996. Variation of suspended sediment yields around New Zealand: The relative importance of rainfall and geology. In: D. E. Walling and B. E. Webb (eds.), *Erosion and Sediment Yield: Global and Regional Perspectives*. IAHS Publication No. 236, IAHS Press, Wallingford, UK, pp. 149–156.

Hicks, D. M., Duncan, M. J., Lane, S. N., Tal, M., Westaway, R., 2008. Contemporary morphological change in braided gravel-bed rivers: New developments from field and laboratory studies, with particular reference to the influence of riparian vegetation. In: H. Habersack, H. Piegay, M. Rinaldi (eds.), *Gravel-Bed Rivers VI: From Process Understanding to River Restoration*, Vol. 11. Elsevier, Amsterdam, the Netherlands, pp. 557–586.

Hinderer, M., 2012. From gullies to mountain belts: A review of sediment budgets at various scales. *Sedimentary Geology*, **280**, 21–59.

Hirsch, P. J., Abrahams, A. D., 1981. The properties of bed sediments in pools and riffles. *Journal of Sedimentary Petrology*, **51**(3), 757–760.

Hobbs, W. H., 1921. Studies of the cycle of glaciation. *Journal of Geology*, **29**, 370–386.

Hodge, R. A., Hoey, T. B., Sklar, L. S., 2011. Bed load transport in bedrock rivers: The role of sediment cover in grain entrainment, translation, and deposition. *Journal of Geophysical Research – Earth Surface*, **116**. 10.1029/2011jf002032.

Hodge, R. A., Sear, D. A., Leyland, J., 2013. Spatial variations in surface sediment structure in riffle-pool sequences: a preliminary test of the Differential Sediment Entrainment Hypothesis (DSEH). *Earth Surface Processes and Landforms*, **38**(5), 449–465.

Hoegh-Guldberg, O., Jacob, D., Taylor, M., et al., 2018. Impacts of 1.5°C global warming on natural and human systems. In: V. Masson-Delmotte, P. Zhai, H.-O. Pörtner, et al. (eds.), Global Warming of 1.5°C. Intergovernmental Panel On Climate Change, www.ipcc.ch/sr15/chapter/chapter-3/.

Hoey, T. B., Bluck, B. J., 1999. Identifying the controls over downstream fining of river gravels. *Journal of Sedimentary Research*, **69**(1), 40–50.

Hoey, T. B., Ferguson, R., 1994. Numerical simulation of downstream fining by selective transport in gravel bed rivers – model development and illustration. *Water Resources Research*, **30**(7), 2251–2260.

Hoey, T. B., Ferguson, R. I., 1997. Controls of strength and rate of downstream fining above a river base level. *Water Resources Research*, **33**(11), 2601–2608.

Hoey, T. B., Sutherland, A. J., 1991. Channel morphology and bedload pulses in braided river – a laboratory study. *Earth Surface Processes and Landforms*, **16**(5), 447–462.

Hoffmann, T., 2015. Sediment residence time and connectivity in non-equilibrium and transient geomorphic systems. *Earth-Science Reviews*, **150**, 609–627.

Holeman, J. N., 1968. Sediment yield of major rivers of the world. *Water Resources Research*, **4**(4), 737–747.

Hollis, G. E., 1975. The effect of urbanization on floods of different recurrence interval. *Water Resources Research*, **11**(3), 431–435.

Holtan, H. N., Kirkpatrick, M. H., 1950. Rainfall, infiltration, and hydraulics of flow in runoff computation. *Transactions, American Geophysical Union*, **31**(5), 771–779.

Hong, L. B., Davies, T. R. H., 1979. A study of stream braiding – summary. *Geological Society of America Bulletin*, **90**(12), 1094–1095.

Hooke, J. M., 1977. The distribution and nature of changes in river channel patterns. In: K. J. Gregory (ed.), *River Channel Changes*. Wiley, Chichester, UK, pp. 265–280.

Hooke, J. M., 1979. Analysis of the processes of river bank erosion. *Journal of Hydrology*, **42**(1–2), 39–62.

Hooke, J. M., 1980. Magnitude and distribution of rates of river bank erosion. *Earth Surface Processes and Landforms*, **5**(2), 143–157.

Hooke, J. M., 1986. Applicable and applied geomorphology of rivers. *Geography*, **71**(310), 1–13.

Hooke, J. M., 1995a. Processes of channel planform change on meandering channels in the UK. In: A. M. Gurnell, K. J. Gregory (eds.), *Changing River Channels*. Wiley, Chichester, UK, pp. 87–115.

Hooke, J. M., 1995b. River channel adjustment to meander cutoffs on the River Bollin and River Dane, northwest England. *Geomorphology*, **14**(3), 235–253.

Hooke, J. M., 2003a. Coarse sediment connectivity in river channel systems: a conceptual framework and methodology. *Geomorphology*, **56**(1–2), 79–94.

Hooke, J. M., 2003b. River meander behaviour and instability: a framework for analysis. *Transactions of the Institute of British Geographers*, **28**(2), 238–253.

Hooke, J. M., 2004a. Analysis of coarse sediment connectivity in semiarid river channels. In: V. Golosov, V. Belyaev, D. E. Walling (eds.), *Sediment Transfer through the Fluvial System*. IAHS Publication No. 288, IAHS Press, Wallingford, UK, pp. 269–275.

Hooke, J. M., 2004b. Cutoffs galore!: Occurrence and causes of multiple cutoffs on a meandering river. *Geomorphology*, **61**(3–4), 225–238.

Hooke, J. M., 2007. Complexity, self-organisation and variation in behaviour in meandering rivers. *Geomorphology*, **91**(3–4), 236–258.

Hooke, J. M., 2016a. Morphological impacts of flow events of varying magnitude on ephemeral channels in a semiarid region. *Geomorphology*, **252**, 128–143.

Hooke, J. M., 2016b. Geomorphological impacts of an extreme flood in SE Spain. *Geomorphology*, **263**, 19–38.

Hooke, J. M., Harvey, M. D., 1983. Meander changes in relation to bed morphology and secondary flows. In: J. D. Collinson, J. Lewin (eds.), *Modern and Ancient Fluvial Systems*. International Association of Sedimentologists, Blackwell, Oxford, UK, pp. 121–132.

Hooke, J. M., Mant, J., 2015. Morphological and vegetation variations in response to flow events in rambla channels of SE Spain. In: A. P. Dykes, M. Mulligan, J. Wainwright (eds.), *Monitoring and Modeling of Dynamic Environments*. Wiley, Chichester, UK, pp. 61–96.

Hooke, R. L. B., 1975. Distribution of sediment transport and shear stress in a meander bend. *Journal of Geology*, **83**(5), 543–565.

Hooke, R. L., 1994. On the efficacy of humans as geomorphic agents. *GSA Today*, **4**, 217, 224–225.

Hooke, R. L., 2000. On the history of humans as geomorphic agents. *Geology*, **28**(9), 843–846.

Hooke, R. L., 2012. Land transformation by humans: A review. *GSA Today*, **22**, 4–10.

Hooshyar, M., Singh, A., Wang, D., 2017. Hydrologic controls on junction angle of river networks. *Water Resources Research*, **53**(5), 4073–4083.

Hopkinson, L. C., Wynn-Thompson, T. M., 2016. Comparison of direct and indirect boundary shear stress measurements along vegetated streambanks. *River Research and Applications*, **32**(8), 1755–1764.

Horn, J. D., Joeckel, R. M., Fielding, C. R., 2012. Progressive abandonment and planform changes of the central Platte River in Nebraska, central USA, over historical timeframes. *Geomorphology*, **139**, 372–383.

Horowitz, A. J., 2003. An evaluation of sediment rating curves for estimating suspended sediment concentrations for subsequent flux calculations. *Hydrological Processes*, **17**(17), 3387–3409.

Horton, A. J., Constantine, J. A., Hales, T. C., et al., 2017. Modification of river meandering by tropical deforestation. *Geology*, **45**(6), 511–514.

Horton, R. E., 1945. Erosional development of streams and their drainage basins: A hydrophysical approach to quantitative morphology. *Geological Society of America Bulletin*, **56**, 275–370.

Hovius, N., 1996. Regular spacing of drainage outlets from linear mountain belts. *Basin Research*, **8**(1), 29–44.

Hovius, N., Meunier, P., Ching-Weei, L., et al., 2011. Prolonged seismically induced erosion and the mass balance of a large earthquake. *Earth and Planetary Science Letters*, **304**(3–4), 347–355.

Howard, A. D., 1967. Drainage analysis in geologic interpretation: A summation. *American Association of Petroleum Geologists Bulletin*, **51**, 2246–2259.

Howard, A. D., 1971a. Simulation of stream networks by headward growth and branching. *Geographical Analysis*, **3**(1), 29–50.

Howard, A. D., 1971b. Simulation model of stream capture. *Geological Society of America Bulletin*, **82**(5), 1355–1375.

Howard, A. D., 1971c. Optimal angles of stream junction: Geometric, stability to capture, and minimum power criteria. *Water Resources Research*, **7**(4), 863–873.

Howard, A. D., 1982. Equilibrium and time scales in geomorphology – application to sand-bed alluvial streams. *Earth Surface Processes and Landforms*, **7**(4), 303–325.

Howard, A. D., 1990. Theoretical model of optimal drainage networks. *Water Resources Research*, **26**(9), 2107–2117.

Howard, A. D., 1992. Modelling channel migration and floodplain sedimentation in meandering streams. In: P. A. Carling, G. E. Petts (eds.), *Lowland Floodplain Rivers*. Wiley, New York, pp. 1–41.

Howard, A. D., 1994. A detachment-limited model of drainage basin evolution. *Water Resources Research*, **30**(7), 2261–2285.

Howard, A., Dolan, R., 1981. Geomorphology of the Colorado River in the Grand Canyon. *Journal of Geology*, **89**(3), 269–298.

Howard, A. D., Knutson, T. R., 1984. Sufficient conditions for river meandering – a simulation approach. *Water Resources Research*, **20**(11), 1659–1667. 10.1029/WR020i011p01659.

Howard, A. D., McLane, C. F., 1988. Erosion of cohesionless sediment by groundwater seepage. *Water Resources Research*, **24**(10), 1659–1674. 10.1029/WR024i010p01659.

Hsu, C.-C., Lee, W.-J., Chang, C.-H., 1998a. Subcritical open-channel junction flow. *Journal of Hydraulic Engineering*, **24**, 847–855.

Hsu, C.-C., Wu, F.-S., Lee, W.-J., 1998b. Flow at 90° equal-width open-channel junction. *Journal of Hydraulic Engineering*, **124**, 186–191.

Huang, H. Q., 2010. Reformulation of the bed load equation of Meyer-Peter and Muller in light of the linearity theory for alluvial channel flow. *Water Resources Research*, **46**. 10.1029/2009wr008974.

Huang, H. Q., Nanson, G. C., 1997. Vegetation and channel variation; a case study of four small streams in southeastern Australia. *Geomorphology*, **18** (3–4), 237–249. 10.1016/s0169-555x(96)00028–1.

Huang, H. Q., Nanson, G. C., 1998. The influence of bank strength on channel geometry: An integrated analysis of some observations. *Earth Surface Processes and Landforms*, **23**(10), 865–876.

Huang, H. Q., Nanson, G. C., 2000. Hydraulic geometry and maximum flow efficiency as products of the principle of least action. *Earth Surface Processes and Landforms*, **25**(1), 1–16.

Huang, H. Q., Nanson, G. C., 2002. A stability criterion inherent in laws governing alluvial channel flow. *Earth Surface Processes and Landforms*, **27**(9), 929–944.

Huang, H. Q., Nanson, G. C., 2007. Why some alluvial rivers develop an anabranching pattern. *Water Resources Research*, **43**(7). 10.1029/2006wr005223.

Huang, H. Q., Warner, R. F., 1995. The multivariate controls of hydraulic geometry – a causal investigation in terms of boundary shear distribution. *Earth Surface Processes and Landforms*, **20**(2), 115–130.

Huang, H., Chang, H. H., Nanson, G. C., 2004. Minimum energy as the general form of critical flow and maximum flow efficiency and for explaining variations in river channel pattern. *Water Resources Research*, **40**(4), W04502. 10.1029/2003wr002539.

Huang, J. C., Weber, L. J., Lai, Y. G., 2002. Three-dimensional numerical study of flows in open-channel junctions. *Journal of Hydraulic Engineering*, **128**(3), 268–280.

Huang, S. Y., Cheng, S. J., Wen, J. C., Lee, J. H., 2008. Identifying peak-imperviousness-recurrence relationships on a growing-impervious watershed, Taiwan. *Journal of Hydrology*, **362**(3–4), 320–336.

Huckleberry, G., 1994. Contrasting channel response to floods on the middle Gila River, *Arizona Geology*, **22**(12), 1083–1086.

Hudson, P. F., 2002. Pool-riffle morphology in an actively migrating alluvial channel: The Lower Mississippi River. *Physical Geography*, **23**(2), 154–169.

Hudson, P. F., Heitmuller, F. T., 2003. Local- and watershed-scale controls on the spatial variability of natural levee deposits in a large fine-grained floodplain: Lower Panuco Basin, Mexico. *Geomorphology*, **56**(3–4), 255–269.

Hudson-Edwards, K. A., Schell, C., Macklin, M. G., 1999. Mineralogy and geochemistry of alluvium contaminated by metal mining in the Rio Tinto area, southwest Spain. *Applied Geochemistry*, **14** (8), 1015–1030.

Hundey, E. J., Ashmore, P. E., 2009. Length scale of braided river morphology. *Water Resources Research*, **45**. 10.1029/2008wr007521.

Hung, C.-L.J., James, L. A., Carbone, G. J., 2018. Impacts of urbanization on stormflow magnitudes in small catchments in the Sandhills of South Carolina, USA. *Anthropocene*, **23**, 17–28.

Hupp, C. R., 1992. Riparian vegetation recovery patterns following stream channelization – a geomorphic perspective. *Ecology*, **73** (4), 1209–1226.

Hupp, C. R., Simon, A., 1991. Bank accretion and the development of vegetation depositional surfaces along modified alluvial channels. *Geomorphology*, **4**(2), 111–124.

Hurther, D., Lemmin, U., 2000. Shear stress statistics and wall similarity analysis in turbulent boundary layers using a high-resolution 3-D ADVP. *IEEE Journal of Oceanic Engineering*, **25**(4), 446–457.

Hutton, J., 1795. *Theory of the Earth, with Proofs and Illustrations.* Vol. 1&2. Edinburgh; London: Printed for Cadell and Davies; William Creech.

Hyde, K. D., Wilcox, A. C., Jencso, K., Woods, S., 2014. Effects of vegetation disturbance by fire on channel initiation thresholds. *Geomorphology*, **214**, 84–96.

Ianniruberto, M., Trevethan, M., Pinheiro, A., et al., 2018. A field study of the confluence between Negro and Solimoes Rivers. Part 2: Bed morphology and stratigraphy. *Comptes Rendus Geoscience*, **350**(1–2), 43–54.

Ibbitt, R. P., Willgoose, G. R., Duncan, M. J., 1999. Channel network simulation models compared with data from the Ashley River, New Zealand. *Water Resources Research*, **35**(12), 3875–3890.

Ijjasz-Vasquez, E. J., Bras, R. L., 1995. Scaling regimes of local slope versus contributing area in digital elevation models. *Geomorphology*, **12**(4), 299–311.

Ijjasz-Vasquez, E. J., Bras, R. L., Rodriguez-Iturbe, I., 1993. Hack's relation and optimal channel networks – the elongation of river basins as a consequence of energy minimization. *Geophysical Research Letters*, **20**(15), 1583–1586.

Ikeda, H., 1973. A study on the formation of sand bars in an experimental flume. *Geographical Review of Japan*, **46**–7, 435–451.

Ikeda, H., 1989. Sedimentary controls on channel migration and origin of point bars in sand-bedded meandering rivers. In: S. Ikeda, G. Parker (eds.), *River Meandering*. Water Resources Monograph 12. American Geophysical Union, Washington, DC, pp. 51–68.

Ikeda, S., 1984a. Flow and bed topography in channels with alternate bars. In: C. M. Elliott (ed.), *River Meandering*. American Society of Civil Engineers, New York, pp. 733–746.

Ikeda, S., 1984b. Prediction of alternate bar wavelength and height. *Journal of Hydraulic Engineering*, **110**, 371–386.

Ikeda, S., Izumi, N., 1990. Width and depth of self-formed straight gravel rivers with bank vegetation. *Water Resources Research*, **26**(10), 2353–2364.

Ikeda, S., Parker, G., Sawai, K., 1981. Bend theory of river meanders. 1. Linear development. *Journal of Fluid Mechanics*, **112**(NOV), 363–377.

Ikeda, S., Parker, G., Kimura, Y., 1988. Stable width and depth of straight gravel rivers with heterogeneous bed materials. *Water Resources Research*, **24**(5), 713–722.

Imaizumi, F., Hattanji, T., Hayakawa, Y. S., 2010. Channel initiation by surface and subsurface flows in a steep catchment of the Akaishi Mountains, Japan. *Geomorphology*, **115**(1–2), 32–42.

Imhoff, K. S., Wilcox, A. C., 2016. Coarse bedload routing and dispersion through tributary confluences. *Earth Surface Dynamics*, **4**(3), 591–605.

Inoue, T., Parker, G., Stark, C. P., 2017. Morphodynamics of a bedrock-alluvial meander bend that incises as it migrates outward: approximate solution of permanent form. *Earth Surface Processes and Landforms*, **42**(9), 1342–1354.

Ishii, Y., Hori, K., 2016. Formation and infilling of oxbow lakes in the Ishikari lowland, northern Japan. *Quaternary International*, **397**, 136–146.

Istanbulluoglu, E., Bras, R. L., 2005. Vegetation-modulated landscape evolution: effects of vegetation on landscape processes, drainage density, and topography. *Journal of Geophysical Research – Earth Surface*, **110**(F2), 19. 10.1029/2004jf000249.

Istanbulluoglu, E., Bras, R. L., 2006. On the dynamics of soil moisture, vegetation, and erosion: Implications of climate variability and change. *Water Resources Research*, **42**(6). 10.1029/2005wr004113.

Istanbulluoglu, E., Tarboton, D. G., Pack, R. T., Luce, C., 2002. A probabilistic approach for channel initiation. *Water Resources Research*, **38**(12). 10.1029/2001wr000782.

Izenberg, N. R., Arvidson, R. E., Brackett, R. A., et al., 1996. Erosional and depositional patterns associated with the 1993 Missouri river floods inferred from SIR-C and TOPSAR radar data. *Journal of Geophysical Research – Planets*, **101**(E10), 23149–23167.

Izumi, N., Parker, G., 1995. Inception of channelization and drainage-basin formation – upstream-driven theory. *Journal of Fluid Mechanics*, **283**, 341–363.

Izumi, N., Parker, G., 2000. Linear stability analysis of channel inception: Downstream-driven theory. *Journal of Fluid Mechanics*, **419**, 239–262.

Jackson, C. R., Martin, J. K., Leigh, D. S., West, L. T., 2005. A southeastern piedmont watershed sediment budget: Evidence for a multi-millennial agricultural legacy. *Journal of Soil and Water Conservation*, **60**(6), 298–310.

Jackson, R. G., 1976. Largescale ripples of the lower Wabash River. *Sedimentology*, **23**(5), 593–623.

Jackson, W. L., Beschta, R. L., 1982. A model of 2-phase bedload transport in an Oregon Coast Range stream. *Earth Surface Processes and Landforms*, **7**(6), 517–527.

Jacobson, R. B., Femmer, S. R., McKenney, R. A., 2001. Land use changes and the physical habitat of streams – a review with emphasis on studies within the U.S. Geological Survey federal-state cooperative program. U.S. Geological Survey Circular 1175, Reston, VA.

Jacobson, R. B., O'Connor, J. E., Oguchi, T., 2016. Surficial geological tools in fluvial geomorphology. In: G. M. Kondolf, H. Piegay (eds.), *Tools in Fluvial Geomorphology*, 2nd Edition. Wiley, Chichester, UK, pp. 15–39.

Jaeger, K. L., Montgomery, D. R., Bolton, S. M., 2007. Channel and perennial flow initiation in headwater streams: management implications of variability in source-area size. *Environmental Management*, **40**(5), 775–786.

Jaeggi, M. N. R., 1984. Formation and effects of alternate bars. *Journal of Hydraulic Engineering*, **110**, 142–156.

Jaehnig, S. C., Lorenz, A. W., Hering, D., et al., 2011. River restoration success: A question of perception. *Ecological Applications*, **21**(6), 2007–2015.

Jain, S. C., 1990. Armor or pavement. *Journal of Hydraulic Engineering*, **116**(3), 436–440.

Jain, V., Fryirs, K., Brierley, G., 2008. Where do floodplains begin? The role of total stream power and longitudinal profile form on floodplain initiation processes. *Geological Society of America Bulletin*, **120**(1–2), 127–141.

James, C. S., 1985. Sediment transfer to overbank sections. *Journal of Hydraulic Research*, **23**(5), 435–452.

James, C. S., 1990. Prediction of entrainment conditions for nonuniform, noncohesive sediments. *Journal of Hydraulic Research*, **28**(1), 25–41.

James, L. A., 1989. Sustained storage and transport of hydraulic gold mining sediment in the Bear River, California. *Annals of the Association of American Geographers*, **79**(4), 570–592.

James, L. A., 1991. Incision and morphological evolution of an alluvial channel recovering from hydraulic mining sediment. *Geological Society of America Bulletin*, **103**(6), 723–736.

James, L. A., 1997. Channel incision on the lower American river, California, from streamflow gage records. *Water Resources Research*, **33**(3), 485–490.

James, L. A., 2004. Tailings fans and valley-spur cutoffs created by hydraulic mining. *Earth Surface Processes and Landforms*, **29**(7), 869–882.

James, L.A., 2006. Bed waves at the basin scale: Implications for river management and restoration. *Earth Surface Processes and Landforms*, **31**(13), 1692–1706.

James, L. A., 2013. Impact of early agriculture and deforestation on geomorphic systems. In: J. C. Shroder (ed.), *Treatise on Geomorphology*, Vol. 13, Geomorphology of Human Disturbances, Climate Change, and Natural Hazards, L. A. James, C. P. Harden, J. J. Clague (vol. eds.). Academic Press, San Diego, CA, pp. 48–67.

James, L. A., 2018. Ten conceptual models of large-scale legacy sedimentation – a review. *Geomorphology*, **317**, 199–217.

James, L. A., Lecce, S. A., 2013. Impacts of land-use and land-cover change on river systems. In: J. C. Shroder (ed.), *Treatise on Geomorphology*, Vol. 9, Fluvial Geomorphology, E. Wohl (vol. ed.). Academic Press, San Diego, CA, pp. 768–793.

James, L. A., Phillips, J. D., Lecce, S. A., 2017. A centennial tribute to GK Gilbert's Hydraulic Mining Debris in the Sierra Nevada. *Geomorphology*, **294**, 4–19.

James, L. A., Monohan, C., Ertis, B., 2019. Long-term hydraulic mining sediment budgets: Connectivity as a management tool. *Science of the Total Environment*, **651**, 2024–2035.

James, W. R., Krumbein, W. C., 1969. Frequency distributions of stream link lengths. *Journal of Geology*, **77**(5), 544–565.

Jansen, J. D., Nanson, G. C., 2004. Anabranching and maximum flow efficiency in Magela Creek, northern Australia. *Water Resources Research*, **40**(4). 10.1029/2003wr002408.

Jansen, J. D., Nanson, G. C., 2010. Functional relationships between vegetation, channel morphology, and flow efficiency in an alluvial (anabranching) river. *Journal of Geophysical Research – Earth Surface*, **115**. 10.1029/2010jf001657.

Jansen, J. M. L., Painter, R. B., 1974. Predicting sediment yield from climate and topography. *Journal of Hydrology*, **21**, 371–380.

Jansson, M., 1985. A comparison of detransformed logarithmic regressions and power function regressions. *Geografiska Annaler Series A Physical Geography*, **67**(1–2), 61–70.

Jansson, M. B., 1996. Estimating a sediment rating curve of the Reventazon River at Palomo using logged mean loads within discharge classes. *Journal of Hydrology*, **183**(3–4), 227–241.

Jarrett, R. D., 1984. Hydraulics of high-gradient streams. *Journal of Hydraulic Engineering*, **110**(11), 1519–1539.

Jarvis, N., Koestel, J., Messing, I., Moeys, J., Lindahl, A., 2013. Influence of soil, land use and climatic factors on the hydraulic conductivity of soil. *Hydrology and Earth System Sciences*, **17**(12), 5185–5195.

Javemick, L., Brasington, J., Caruso, B., 2014. Modeling the topography of shallow braided rivers using structure-from-motion photogrammetry. *Geomorphology*, **213**, 166–182.

Jefferson, A. J., McGee, R. W., 2013. Channel network extent in the context of historical land use, flow generation processes, and landscape evolution in the North Carolina Piedmont. *Earth Surface Processes and Landforms*, **38**(6), 601–613.

Jeffries, R., Darby, S. E., Sear, D. A., 2003. The influence of vegetation and organic debris on flood-plain sediment dynamics: Case study of a low-order stream in the New Forest, England. *Geomorphology*, **51**(1–3), 61–80.

Jeong, A., Dorn, R. I., 2019. Soil erosion from urbanization processes in the Sonoran Desert, Arizona, USA. *Land Degradation & Development*, **30**(2), 226–238.

Jerolmack, D. J., Brzinski, T. A., III, 2010. Equivalence of abrupt grain-size transitions in alluvial rivers and eolian sand seas: a hypothesis. *Geology*, **38**(8), 719–722.

Jerolmack, D. J., Mohrig, D., 2007. Conditions for branching in depositional rivers. *Geology*, **35**(5), 463–466.

Jha, S., Bombardelli, F., 2009. Two-phase modeling of turbulence in dilute sediment-laden, open-channel flows. *Environmental Fluid Mechanics*, **9**(2), 237–266.

Jia, Y. F., 1990. Minimum Froude number and the equilibrium of alluvial sand rivers. *Earth Surface Processes and Landforms*, **15**(3), 199–209.

Joeckel, R. M., Tucker, S. T., McMullin, J. D., 2016. Morphosedimentary features from a major flood on a small, lower-sinuosity, single-thread river: The unknown quantity of overbank deposition, historical-change context, and comparisons with a multichannel river. *Sedimentary Geology*, **343**, 18–37.

Johannesson, H., Parker, G., 1989. Linear theory of river meanders. In: S. Ikeda, G. Parker (eds.), *River Meandering*. Water Resources Monograph 12. American Geophysical Union, Washington, DC, pp. 181–212.

Johnson, J. P. L., Aronovitz, A. C., Kim, W., 2015. Coarser and rougher: Effects of fine gravel pulses on experimental step-pool channel morphodynamics. *Geophysical Research Letters*, **42**(20), 8432–8440. 10.1002/2015gl066097.

Johnson, P. A., Heil, T. M., 1996. Uncertainty in estimating bankfull conditions. *Water Resources Bulletin*, **32**(6), 1283–1291.

Johnson, W. C., 1994. Woodland expansion in the Platte River, Nebraska – patterns and causes. *Ecological Monographs*, **64**(1), 45–84.

Johnson, W.C., 1997. Equilibrium response of riparian vegetation to flow regulation in the Platte River, Nebraska. *Regulated Rivers – Research & Management*, **13**(5), 403–415.

Johnston, C. E., Andrews, E. D., Pitlick, J., 1998. In situ determination of particle friction angles of fluvial gravels. *Water Resources Research*, **34**(8), 2017–2030.

Jones, J. A., Grant, G. E., 1996. Peak flow responses to clear-cutting and roads in small and large basins, western Cascades, Oregon. *Water Resources Research*, **32**(4), 959–974.

Jones, L. S., Humphrey, N. F., 1997. Weathering-controlled abrasion in a coarse-grained, meandering reach of the Rio Grande: implications for the rock record. *Geological Society of America Bulletin*, **109**(9), 1080–1088.

Jones, L. S., Schumm, S. A., 1999. Causes of avulsion: an overview. In: N.D. Smith, J. Rodgers (eds.), *Fluvial Sedimentology VI*, International Association of Sedimentologists Special Publication 28, Blackwell, Malden, MA, pp.171–178.

Jones, O. T., 1924. The Upper Towry drainage system. *Quarterly Journal of the Geological Society of London*, **80**, 560–609.

Jordan, B. A., Annable, W. K., Watson, C. C., Sen, D., 2010. Contrasting stream stability characteristics in adjacent urban watersheds: Santa Clara Valley, California. *River Research and Applications*, **26**(10), 1281–1297.

Jothiprakash, V., Garg, V., 2008. Re-look to conventional techniques for trapping efficiency estimation of a reservoir. *International Journal of Sediment Research*, **23**(1), 76–84.

Joy, D. M., Townsend, R. D., 1981. Improved flow characteristics at a 90° channel confluence, 5th Canadian Hydrotechnical Conference. Canadian Society of Civil Engineers, Fredericton, New Brunswick, pp. 781–799.

Julian, J. P., Elmore, A. J., Guinn, S. M., 2012. Channel head locations in forested watersheds across the mid-Atlantic United States: A physiographic analysis. *Geomorphology*, **177**, 194–203.

Jung, K., Marpu, P. R., Ouarda, T., 2015. Improved classification of drainage networks using junction angles and secondary tributary lengths. *Geomorphology*, **239**, 41–47.

Junker, B., Buchecker, M., Mueller-Boeker, U., 2007. Objectives of public participation: Which actors should be involved in the decision making for river restorations? *Water Resources Research*, **43**(10). 10.1029/2006WR005584.

Juracek, K. E., 2004. Historical channel-bed elevation change as a result of multiple disturbances, Soldier Creek, Kansas. *Physical Geography*, **25**(4), 269–290.

Juracek, K. E., Fitzpatrick, F. A., 2003. Limitations and implications of stream classification. *Journal of the American Water Resources Association*, **39**(3), 659–670.

Kabiri-Samani, A., Farshi, F., Chamani, M. R., 2013. Boundary shear stress in smooth trapezoidal open channel flows. *Journal of Hydraulic Engineering*, **139**(2), 205–212.

Kail, J., Brabec, K., Poppe, M., Januschke, K., 2015. The effect of river restoration on fish, macroinvertebrates and aquatic macrophytes: A meta-analysis. *Ecological Indicators*, **58**, 311–321.

Karcz, I., 1966. Secondary currents and the configuration of a natural stream bed. *Journal of Geophysical Research*, **71**, 3109–3117.

Karlstrom, L., Gajjar, P., Manga, M., 2013. Meander formation in supraglacial streams. *Journal of Geophysical Research – Earth Surface*, **118**(3), 1897–1907.

Karr, J. R., 1981. Assessment of biotic integrity using fish communities. *Fisheries*, **6**(6), 21–27.

Kasprak, A., Magilligan, F. J., Nislow, K. H., et al., 2013. Differentiating the relative importance of land cover change and geomorphic processes on fine sediment sequestration in a logged watershed. *Geomorphology*, **185**, 67–77.

Kasprak, A., Wheaton, J. M., Ashmore, P. E., Hensleigh, J. W., Peirce, S., 2015. The relationship between particle travel distance and channel morphology: Results from physical models of braided rivers. *Journal of Geophysical Research – Earth Surface*, **120**(1), 55–74.

Kasprak, A., Hough-Snee, N., Beechie, T., et al., 2016. The blurred line between form and process: A comparison of stream channel classification frameworks. *PLoS ONE*, **11**(3), 31. 10.1371/journal.pone.0150293.

Kasvi, E., Vaaja, M., Kaartinen, H., et al., 2015. Sub-bend scale flow-sediment interaction of meander bends – a combined approach of field observations, close-range remote sensing and computational modelling. *Geomorphology*, **238**, 119–134.

Kasvi, E., Hooke, J., Kurkela, M., et al., 2017. Modern empirical and modelling study approaches in fluvial geomorphology to elucidate sub-bend-scale meander dynamics. *Progress in Physical Geography*, **41**(5), 533–569.

Kean, J. W., Smith, J. D., 2004. Flow and boundary shear stress in channels with woody bank vegetation. In: S. J. Bennett, A. Simon (eds.), *Riparian Vegetation and Fluvial Geomorphology*. American Geophysical Union, Washington, DC, pp. 237–252.

Kean, J. W., Smith, J. D., 2006a. Form drag in rivers due to small-scale natural topographic features: 1. regular sequences. *Journal of Geophysical Research – Earth Surface*, **111**(F4). 10.1029/2006jf000467.

Kean, J. W., Smith, J. D., 2006b. Form drag in rivers due to small-scale natural topographic features: 2. irregular sequences. *Journal of Geophysical Research – Earth Surface*, **111**(F4). 10.1029/2006jf000490.

Kean, J. W., Kuhnle, R. A., Smith, J. D., Alonso, C. V., Langendoen, E. J., 2009. Test of a method to calculate near-bank velocity and boundary shear stress. *Journal of Hydraulic Engineering*, **135**(7), 588–601.

Kearsley, L. H., Schmidt, J. C., Warren, K. D., 1994. Effects of Glen Canyon Dam on Colorado River sand deposits used as campsites in Grand Canyon National Park, USA. *Regulated Rivers-Research & Management*, **9**(3), 137–149.

Keast, D., Ellison, J., 2013. Magnitude frequency analysis of small floods using the annual and partial series. *Water*, **5**(4), 1816–1829.

Keller, E. A., 1971. Areal sorting of bed-load material: the hypothesis of velocity reversal. *Geological Society of America Bulletin*, **82**, 753–756.

Keller, E. A., 1972. Development of alluvial stream channels – a five-stage model. *Geological Society of America Bulletin*, **83**(5), 1531–1536.

Keller, E. A., 1975. Channelization – search for a better way. *Geology*, **3**(5), 246–248.

Keller, E. A., Florsheim, J. L., 1993. Velocity-reversal hypothesis – a model approach. *Earth Surface Processes and Landforms*, **18**(8), 733–740.

Keller, E. A., Melhorn, W. N., 1978. Rhythmic spacing and origin of pools and riffles. *Geological Society of America Bulletin*, **89**(5), 723–730.

Kelly, S., 2006. Scaling and hierarchy in braided rivers and their deposits: examples and implications for reservoir modelling. In: G. H. Sambrook Smith, J. B. Best, C. S. Bristow, G. E. Petts (eds.), *Braided Rivers: Process, Deposits, Ecology and Management*, International Association of Sedimentologists Special Publication 36. Blackwell, Malden, MA, pp. 75–106.

Kelly, S. A., Takbiri, Z., Belmont, P., Foufoula-Georgiou, E., 2017. Human amplified changes in precipitation-runoff patterns in large river basins of the midwestern United States. *Hydrology and Earth System Sciences*, **21**(10), 5065–5088.

Kemp, J., 2010. Downstream channel changes on a contracting, anabranching river: The Lachlan, southeastern Australia. *Geomorphology*, **121**(3–4), 231–244.

Kennedy, B. A., 1984. On Playfair's Law of accordant junctions. *Earth Surface Processes and Landforms*, **9**(2), 153–173.

Kennedy, J. F., 1995. The Albert Shields story. *Journal of Hydraulic Engineering*, **121**(11), 766–772.

Kenworthy, S. T., Rhoads, B. L., 1995. Hydrologic control of spatial patterns of suspended sediment concentration at a stream confluence. *Journal of Hydrology*, **168**(1–4), 251–263.

Kesel, R. H., Yodis, E. G., 1992. Some effects of human modifications on sand-bed channels in southwestern Mississippi, USA. *Environmental Geology and Water Sciences*, **20**(2), 93–104.

Kesseli, T. E., 1941. The concept of the graded river. *Journal of Geology*, **49**(6), 561–588.

Keylock, C. J., 2015. Flow resistance in natural, turbulent channel flows: the need for a fluvial fluid mechanics. *Water Resources Research*, **51**(6), 4374–4390. 10.1002/2015wr016989.

Keylock, C. J., Constantinescu, G., Hardy, R. J., 2012. The application of computational fluid dynamics to natural river channels: Eddy resolving versus mean flow approaches. *Geomorphology*, **179**, 1–20.

Keylock, C. J., Singh, A., Foufoula-Georgiou, E., 2013. The influence of migrating bed forms on the velocity-intermittency structure of turbulent flow over a gravel bed. *Geophysical Research Letters*, **40**(7). 10.1002/grl.50337.

Keylock, C. J., Singh, A., Venditti, J. G., Foufoula-Georgiou, E., 2014a. Robust classification for the joint velocity-intermittency structure of turbulent flow over fixed and mobile bedforms. *Earth Surface Processes and Landforms*, **39**(13), 1717–1728.

Keylock, C. J., Lane, S. N., Richards, K. S., 2014b. Quadrant/octant sequencing and the role of coherent structures in bed load sediment entrainment. *Journal of Geophysical Research – Earth Surface*, **119**(2), 264–286. 10.1002/2012jf002698.

Khodashenas, S. R., Paquier, A., 1999. A geometrical method for computing the distribution of boundary shear stress across irregular straight open channels. *Journal of Hydraulic Research*, **37**(3), 381–388.

Kidson, R., Richards, K. S., 2005. Flood frequency analysis: assumptions and alternatives. *Progress in Physical Geography*, **29**(3), 392–410.

Kiffney, P. M., Greene, C. M., Hall, J. E., Davies, J. R., 2006. Tributary streams create spatial discontinuities in habitat, biological productivity, and diversity in mainstem rivers. *Canadian Journal of Fisheries and Aquatic Sciences*, **63**(11), 2518–2530.

Kim, S. C., Friedrichs, C. T., Maa, J. P. Y., Wright, L. D., 2000. Estimating bottom stress in tidal boundary layer from Acoustic Doppler Velocimeter data. *Journal of Hydraulic Engineering*, **126**(6), 399–406.

Kincey, M., Warburton, J., Brewer, P., 2018. Contaminated sediment flux from eroding abandoned historical metal mines: spatial and temporal variability in geomorphological drivers. *Geomorphology*, **319**, 199–215.

King, L. C., 1953. Canons of landscape evolution. *Geological Society of America Bulletin*, **64**(7), 721–752.

Kinnell, P. I. A., 2004. Sediment delivery ratios: a misaligned approach to determining sediment delivery from hillslopes. *Hydrological Processes*, **18**(16), 3191–3194.

Kinnell, P. I. A., 2008a. Sediment delivery from hillslopes and the Universal Soil Loss Equation: some perceptions and misconceptions. *Hydrological Processes*, **22**(16), 3168–3175.

Kinnell, P. I. A., 2008b. Discussion: misrepresentation of the USLE in "Is sediment delivery a fallacy?". *Earth Surface Processes and Landforms*, **33**(10), 1627–1629.

Kirby, E., Whipple, K., 2001. Quantifying differential rock-uplift rates via stream profile analysis. *Geology*, **29**(5), 415–418.

Kirchner, J. W., 1993. Statistical inevitability of Horton's Laws and the apparent randomness of stream channel networks. *Geology*, **21**(7), 591–594.

Kirchner, J. W., Dietrich, W. E., Iseya, F., Ikeda, H., 1990. The variability of critical shear stress, friction angle, and grain protrusion in water-worked sediments. *Sedimentology*, **37**(4), 647–672.

Kirchner, J. W., Finkel, R. C., Riebe, C. S., et al., 2001. Mountain erosion over 10 yr, 10 k.y., and 10 m.y. time scales. *Geology*, **29**(7), 591–594.

Kirkby, M. J., 1976. Tests of random network model and its application to basin hydrology. *Earth Surface Processes and Landforms*, **1**(3), 197–212.

Kirkby, M. J., 1980. The streamhead as a significant geomorphic threshold. In: D. R. Coates, J. D. Vitek (eds.), *Thresholds in Geomorphology*. Allen and Unwin, Boston, MA, pp. 53–73.

Kirkby, M. J., 1996. A role for theoretical models in geomorphology? In: B. L. Rhoads, C. E. Thorn (eds.), *The Scientific Nature of Geomorphology*. Wiley and Sons, Chichester, UK, pp. 257–272.

Kirkby, M. J., Bracken, L. J., 2009. Gully processes and gully dynamics. *Earth Surface Processes and Landforms*, **34**(14), 1841–1851.

Kisling-Moller, J., 1992. Lateral sediment transport by bedforms in a meander bend. *Earth Surface Processes and Landforms*, **17**(5), 501–513.

Kitanidis, P. K., Kennedy, J. F., 1984. Secondary current and river-meander formation. *Journal of Fluid Mechanics*, **144**(JUL), 217–229.

Klavon, K., Fox, G., Guertault, L., et al., 2017. Evaluating a process-based model for use in streambank stabilization: Insights on the Bank Stability and Toe Erosion Model (BSTEM). *Earth Surface Processes and Landforms*, **42**(1), 191–213.

Klein, M., 1984. Anticlockwise hysteresis in suspended sediment concentration during individual storms – Holbeck Catchment –

Yorkshire, England. *Catena*, 11(2–3), 251–257. 10.1016/0341-8162(84)90014-6.

Kleinhans, M. G., 2004. Sorting in grain flows at the lee side of dunes. *Earth-Science Reviews*, **65**(1–2), 75–102.

Kleinhans, M. G., van den Berg, J. H., 2011. River channel and bar patterns explained and predicted by an empirical and a physics-based method. *Earth Surface Processes and Landforms*, **36**(6), 721–738.

Kleinhans, M. G., Jagers, H. R. A., Mosselman, E., Sloff, C. J., 2008. Bifurcation dynamics and avulsion duration in meandering rivers by one-dimensional and three-dimensional models. *Water Resources Research*, **44**(8). 10.1029/2007wr005912.

Kleinhans, M. G., de Haas, T., Lavooi, E., Makaske, B., 2012. Evaluating competing hypotheses for the origin and dynamics of river anastomosis. *Earth Surface Processes and Landforms*, **37** (12), 1337–1351.

Kleinhans, M. G., Ferguson, R. I., Lane, S. N., Hardy, R. J., 2013. Splitting rivers at their seams: bifurcations and avulsion. *Earth Surface Processes and Landforms*, **38**(1), 47–61.

Kleinhans, M. G., van Dijk, W. M., van de Lageweg, W. I., et al., 2014. Quantifiable effectiveness of experimental scaling of river-and delta morphodynamics and stratigraphy. *Earth-Science Reviews*, **133**, 43–61.

Kleinhans, M. G., de Vries, B., Braat, L., van Oorschot, M., 2018. Living landscapes: Muddy and vegetated floodplain effects on fluvial pattern in an incised river. *Earth Surface Processes and Landforms*, **43**(14), 2948–2963. 10.1002/esp.4437.

Kline, M., Cahoon, B., 2010. Protecting river corridors in Vermont. *Journal of the American Water Resources Association*, **46**(2), 227–236. 10.1111/j.1752–1688.2010.00417.x.

Klonsky, L., Vogel, R. M., 2011. Effective measures of "effective" discharge. *Journal of Geology*, **119**(1), 1–14.

Knaapen, M. A. F., Hulscher, S., De Vriend, H. J., Van Harten, A., 2001. Height and wavelength of alternate bars in rivers: Modelling vs. laboratory experiments. *Journal of Hydraulic Research*, **39**(2), 147–153.

Knight, D. W., Shiono, K., 1990. Turbulence measurements in a shear-layer region of a compound channel. *Journal of Hydraulic Research*, **28**(2), 175–196.

Knighton, A. D., 1975. Variations in at-a-station hydraulic geometry. *American Journal of Science*, **275**(2), 186–218.

Knighton, A. D., 1980. Longitudinal changes in size and sorting of stream-bed material in four English rivers. *Geological Society of America Bulletin*, **91**(1), 55–62.

Knighton, A. D., 1982a. Asymmetry of river channel cross-sections. 2. Mode of development and local variation. *Earth Surface Processes and Landforms*, 7(2), 117–131.

Knighton, A. D., 1982b. Longitudinal changes in the size and shape of stream bed material: Evidence of variable transport conditions. *Catena*, **9**, 25–34.

Knighton, A. D., 1987. River channel adjustment – the downstream dimension. In: K. S. Richards (ed.), *River Channels: Environment and Process*. Blackwell, New York, pp. 95–128.

Knighton, A. D., 1989. River adjustment to changes in sediment load – the effects of tin mining on the Ringarooma River,

Tasmania, 1875–1984. *Earth Surface Processes and Landforms*, **14**(4), 333–359.

Knighton, A. D., 1991. Channel bed adjustment along mine-affected rivers of northeast Tasmania. *Geomorphology*, **4**(3–4), 205–219.

Knighton, A. D., 1998. *Fluvial Forms and Processes*. Arnold, London.

Knighton, A. D., 1999a. Downstream variation in stream power. *Geomorphology*, **29**(3–4), 293–306.

Knighton, A. D., 1999b. The gravel-sand transition in a disturbed catchment. *Geomorphology*, **27**(3–4), 325–341.

Knighton, A. D., Nanson, G. C., 1993. Anastomosis and the continuum of channel pattern. *Earth Surface Processes and Landforms*, **18**(7), 613–625.

Knox, J. C., 1972. Valley alluviation in southwestern Wisconsin. *Annals of the Association of American Geographers*, **62**(3), 401–410.

Knox, J. C., 1976. Concept of the graded stream. In: W. N. Melhorn, R. C. Flemal (eds.), *Theories of Landform Development*. Binghamton State University of New York, Binghamton, NY, pp. 169–198.

Knox, J. C., 1977. Human impacts on Wisconsin stream channels. *Annals of the Association of American Geographers*, **67**(3), 323–342.

Knox, J. C., 1987. Historical valley floor sedimentation in the upper Mississippi Valley. *Annals of the Association of American Geographers*, **77**(2), 224–244.

Knox, J. C., 1993. Large increases in flood magnitude in response to modest changes in climate. *Nature*, **361**(6411), 430–432.

Knox, J. C., 2000. Sensitivity of modern and Holocene floods to climate change. *Quaternary Science Reviews*, **19**(1–5), 439–457.

Knox, J. C., 2001. Agricultural influence on landscape sensitivity in the Upper Mississippi River Valley. *Catena*, **42**(2–4), 193–224.

Knox, J. C., 2006. Floodplain sedimentation in the Upper Mississippi Valley: Natural versus human accelerated. *Geomorphology*, **79** (3–4), 286–310.

Kochel, R. C., 1988. Geomorphic impact of large floods: Review and new perspectives on frequency and magnitude. In: V. R. Baker, R. C. Kochel, P. C. Patton (eds.), *Flood Geomorphology*. Wiley, New York, pp. 169–187.

Kochel, R. C., Piper, J. F., 1986. Morphology of large valleys on Hawaii – evidence for groundwater sapping and comparisons with Martian valleys. *Journal of Geophysical Research – Solid Earth and Planets*, **91**(B13), E175–E192. 10.1029/JB091iB13p0E175.

Kochel, R. C., Howard, A. D., McLane, C. F., 1985. Channel networks developed in fine-grained sediments: analogs to Martian valleys. In: M. J. Woldenberg (ed.), *Models in Geomorphology*. Allen and Unwin, London, pp. 313–341.

Kochel, R. C., Simmons, D. W., Piper, J. F., 1988. Groundwater sapping experiments in weakly consolidated layered sediments: A summary. In: A. D. Howard, R. C. Kochel, H. E. Holt (eds.), *Sapping Features of the Colorado Plateau: A Comparative Planetary Geology Field Guide*. National Aeronautics and Space Administration, Washington, DC, pp. 84–93.

Kodama, Y., 1994a. Experimental study of abrasion and its role in producing downstream fining in gravel-bed rivers. *Journal of Sedimentary Research*, **64**(1), 76–85.

Kodama, Y., 1994b. Downstream changes in the lithology and grain size of fluvial gravels, the Watarase River, Japan – evidence of the role of abrasion in downstream fining. *Journal of Sedimentary Research*, **64**(1), 68–75.

Koiter, A. J., Owens, P. N., Petticrew, E. L., Lobb, D. A., 2013. The behavioural characteristics of sediment properties and their implications for sediment fingerprinting as an approach for identifying sediment sources in river basins. *Earth-Science Reviews*, **125**, 24–42.

Komar, P. D., 1983. Shapes of streamlined islands on Earth and Mars – experiments and analyses of the minimum-drag form. *Geology*, **11**(11), 651–654.

Komar, P. D., Li, Z., 1986. Pivoting analysis of the selective entrainment of sediments by shape and size with application to gravel threshold. *Sedimentology*, **33**(3), 425–436.

Komar, P. D., Li, Z., 1988. Applications of grain-pivoting and sliding analyses to selective entrainment of gravel and to flow-competence evaluations. *Sedimentology*, **35**(4), 681–695.

Komar, P. D., Wang, C., 1984. Processes of selective grain transport and the formation of placers on beaches. *Journal of Geology*, **92**(6), 637–655.

Kondolf, G. M., 1994. Geomorphic and environmental effects of instream gravel mining. *Landscape and Urban Planning*, **28**(2–3), 225–243.

Kondolf, G. M., 1997. Hungry water: effects of dams and gravel mining on river channels. *Environmental Management*, **21**(4), 533–551.

Kondolf, G. M., 2000. Assessing salmonid spawning gravel quality. *Transactions of the American Fisheries Society*, **129**(1), 262–281.

Kondolf, G. M., 2006. River restoration and meanders. *Ecology and Society*, **11**(2). www.ecologyandsociety.org/vol11/iss12/art42/.

Kondolf, G. M., 2011. Setting goals in river restoration: when and where can the river "heal itself"? In: A. Simon, S. J. Bennett, J. M. Castro (eds.), *Stream Restoration in Dynamic Fluvial Systems: Scientific Approaches, Analyses, and Tools*. American Geophysical Union, Washington, DC, pp. 29–43.

Kondolf, G. M., 2012. The *Espace de Liberte* and restoration of fluvial process: when can the river restore itself and when must we intervene? In: P. J. Boon, P. J. Raven (eds.), *River Conservation and Management*. Wiley, Chichester, UK, pp. 225–241.

Kondolf, G. M., Lisle, T. E., 2016. Measuring bed sediment. In: G. M. Kondolf, H. Piegay (eds.), *Tools in Fluvial Geomorphology*. Wiley, Chichester, UK, pp. 278–305.

Kondolf, G. M., Matthews, W. V. G., 1991. Unmeasured residuals in sediment budgets – a cautionary note. *Water Resources Research*, **27**(9), 2483–2486.

Kondolf, G. M., Piegay, H. (eds.), 2016. *Tools in Fluvial Geomorphology*. Wiley, Chichester, UK.

Kondolf, G. M., Swanson, M. L., 1993. Channel adjustments to reservoir construction and gravel extraction along Stony Creek, California. *Environmental Geology*, **21**(4), 256–269.

Kondolf, G. M., Wilcock, P. R., 1996. The flushing flow problem: defining and evaluating objectives. *Water Resources Research*, **32**(8), 2589–2599.

Kondolf, G. M., Smeltzer, M. W., Railsback, S. F., 2001. Design and performance of a channel reconstruction project in a coastal California gravel-bed stream. *Environmental Management*, **28**(6), 761–776.

Kondolf, G. M., Boulton, A. J., O'Daniel, S., et al., 2006. Process-based ecological river restoration: Visualizing three-dimensional connectivity and dynamic vectors to recover lost linkages. *Ecology and Society*, **11** (2). www.ecologyandsociety.org/vol11/iss12/art15.

Kondolf, G. M., Piegay, H., Schmitt, L., Montgomery, D. R., 2016. Geomorphic classification of rivers and streams. In: G. M. Kondolf, H. Piegay (eds.), *Tools in Fluvial Geomorphology*, 2nd Edition. Wiley, Chichester, UK, pp. 133–158.

Kong, D., Miao, C., Wu, J., et al., 2017. Environmental impact assessments of the Xiaolangdi Reservoir on the most hyperconcentrated laden river, Yellow River, China. *Environmental Science and Pollution Research*, **24**(5), 4337–4351.

Konrad, C. P., Booth, D. B., Burges, S. J., 2005. Effects of urban development in the Puget Lowland, Washington, on interannual streamflow patterns: Consequences for channel form and streambed disturbance. *Water Resources Research*, **41**(7), 1–15. 10.1029/2005WR004097.

Konsoer, K. M., Rhoads, B. L., 2014. Spatial-temporal structure of mixing interface turbulence at two large river confluences. *Environmental Fluid Mechanics*, **14**(5), 1043–1070.

Konsoer, K. M., Rhoads, B. L., Langendoen, E. J., et al., 2016a. Spatial variability in bank resistance to erosion on a large meandering, mixed bedrock-alluvial river. *Geomorphology*, **252**, 80–97.

Konsoer, K. M., Rhoads, B. L., Best, J. L., et al., 2016b. Three-dimensional flow structure and bed morphology in large elongate meander loops with different outer bank roughness characteristics. *Water Resources Research*, **52**(12), 9621–9641. 10.1002/2016wr019040.

Konsoer, K., Rhoads, B., Best, J., et al., 2017. Length scales and statistical characteristics of outer bank roughness for large elongate meander bends: The influence of bank material properties, floodplain vegetation and flow inundation. *Earth Surface Processes and Landforms*, **42**(13), 2024–2037.

Kooi, H., Beaumont, C., 1996. Large-scale geomorphology: Classical concepts reconciled and integrated with contemporary ideas via a surface processes model. *Journal of Geophysical Research – Solid Earth*, **101**(B2), 3361–3386.

Koppes, M. N., Montgomery, D. R., 2009. The relative efficacy of fluvial and glacial erosion over modern to orogenic timescales. *Nature Geoscience*, **2**(9), 644–647.

Korup, O., 2012. Earth's portfolio of extreme sediment transport events. *Earth-Science Reviews*, **112**(3–4), 115–125.

Krider, L., Magner, J., Hansen, B., et al., 2017. Improvements in fluvial stability associated with two-stage ditch construction in Mower County, Minnesota. *Journal of the American Water Resources Association*, **53**(4), 886–902.

Krumbein, W. C., 1941. The effects of abrasion on the size, shape and roundness of rock fragments. *Journal of Geology*, **49**(5), 482–520.

Kuenen, P. H., 1956. Experimental abrasion of pebbles. 2. Rolling by current. *Journal of Geology*, **64**(4), 336–368.

Kuhnle, R. A., 1993a. Incipient motion of sand-gravel sediment mixtures. *Journal of Hydraulic Engineering*, **119**(12), 1400–1415.

Kuhnle, R. A., 1993b. Fluvial transport of sand and gravel mixtures with bimodal size distributions. *Sedimentary Geology*, **85**(1–4), 17–24.

Kumar, P., Le, P. V. V., Papanicolaou, A. N. T., et al., 2018. Critical transition in critical zone of intensively managed landscapes. *Anthropocene*, **22**, 10–19.

La Barbera, P., Rosso, R., 1989. On the fractal dimension of stream networks. *Water Resources Research*, **25**(4), 735–741.

La Barbera, P., Rosso, R., 1990. On the fractal dimension of stream networks – reply. *Water Resources Research*, **26**(9), 2245–2248.

Lacey, R. W. J., Roy, A. G., 2008. The spatial characterization of turbulence around large roughness elements in a gravel-bed river. *Geomorphology*, **102**(3–4), 542–553.

Lackey, R. T., 2001. Values, policy, and ecosystem health. *Bioscience*, **51**(6), 437–443.

Lague, D., 2014. The stream power river incision model: Evidence, theory and beyond. *Earth Surface Processes and Landforms*, **39**(1), 38–61.

Laity, J. E., Malin, M. C., 1985. Sapping processes and the development of theater-headed valley networks on the Colorado Plateau. *Geological Society of America Bulletin*, **96**(2), 203–217.

Lajeunesse, E., Malverti, L., Charru, F., 2010. Bed load transport in turbulent flow at the grain scale: experiments and modeling. *Journal of Geophysical Research – Earth Surface*, **115**. 10.1029/2009jf001628.

Lam, D., Thompson, C., Croke, J., Sharma, A., Macklin, M., 2017. Reducing uncertainty with flood frequency analysis: the contribution of paleoflood and historical flood information. *Water Resources Research*, **53**(3), 2312–2327. 10.1002/2016wr019959.

Lam, M. Y., Ghidaoui, M. S., Kolyshkin, A. A., 2016. The roll-up and merging of coherent structures in shallow mixing layers. *Physics of Fluids*, **28**(9). 10.1063/1.4960391.

Lamarre, H., Roy, A. G., 2008. The role of morphology on the displacement of particles in a step-pool river system. *Geomorphology*, **99**(1–4), 270–279.

Lamarre, H., MacVicar, B., Roy, A. G., 2005. Using passive integrated transponder (PIT) tags to investigate sediment transport in gravel-bed rivers. *Journal of Sedimentary Research*, **75**(4), 736–741.

Lamb, M. P., Dietrich, W. E., 2009. The persistence of waterfalls in fractured rock. *Geological Society of America Bulletin*, **121**(7–8), 1123–1134.

Lamb, M. P., Venditti, J. G., 2016. The grain size gap and abrupt gravel-sand transitions in rivers due to suspension fallout. *Geophysical Research Letters*, **43**(8), 3777–3785. 10.1002/2016gl068713.

Lamb, M. P., Howard, A. D., Johnson, J., et al., 2006. Can springs cut canyons into rock? *Journal of Geophysical Research-Planets*, **111**(E7), 18. 10.1029/2005je002663.

Lamb, M. P., Dietrich, W. E., Aciego, S. M., DePaolo, D. J., Manga, M., 2008a. Formation of Box Canyon, Idaho, by megaflood: implications for seepage erosion on Earth and Mars. *Science*, **320**(5879), 1067–1070.

Lamb, M. P., Dietrich, W. E., Sklar, L. S., 2008b. A model for fluvial bedrock incision by impacting suspended and bed load sediment. *Journal of Geophysical Research-Earth Surface*, **113**(F3). 10.1029/2007jf000915.

Lamb, M. P., Dietrich, W. E., Venditti, J. G., 2008c. Is the critical Shields stress for incipient sediment motion dependent on channel-bed slope? *Journal of Geophysical Research-Earth Surface*, **113**(F2). 10.1029/2007jf000831.

Lana-Renault, N., Regues, D., 2009. Seasonal patterns of suspended sediment transport in an abandoned farmland catchment in the Central Spanish Pyrenees. *Earth Surface Processes and Landforms*, **34**(9), 1291–1301.

Lancaster, S. T., Underwood, E. F., Frueh, W. T., 2010. Sediment reservoirs at mountain stream confluences: dynamics and effects of tributaries dominated by debris-flow and fluvial processes. *Geological Society of America Bulletin*, **122**(11–12), 1775–1786.

Landemaine, V., Gay, A., Cerdan, O., Salvador-Blanes, S., Rodrigues, S., 2015. Morphological evolution of a rural headwater stream after channelisation. *Geomorphology*, **230**, 125–137.

Landwehr, K., Rhoads, B. L., 2003. Depositional response of a headwater stream to channelization, east central Illinois, USA. *River Research and Applications*, **19**(1), 77–100.

Lane, E. W., 1954. The Importance of Fluvial Morphology in Hydraulic Engineering. Bureau of Reclamation, Engineering Laboratory Division, Hydraulic Laboratories Report No. 372, U.S. Department of Interior, Denver, CO.

Lane, S. N., 2006. Approaching the system-scale understanding of braided river behaviour. In: G. H. S. Smith, J. L. Best, C. S. Bristow, G. E. Petts (eds.), *Braided Rivers: Process, Deposits, Ecology and Management*. Special Publications of the International Association of Sedimentologists, 36, Blackwell, Malden, MA, pp. 107–135.

Lane, S. N., Chandler, J. H., Richards, K. S., 1994. Developments in monitoring and modeling small-scale river bed topography. *Earth Surface Processes and Landforms*, **19**(4), 349–368.

Lane, S. N., Richards, K. S., Chandler, J. H., 1995. Morphological estimation of the time-integrated bed-load transport rate. *Water Resources Research*, **31**(3), 761–772.

Lane, S. N., Westaway, R. M., Hicks, D. M., 2003. Estimation of erosion and deposition volumes in a large, gravel-bed, braided river using synoptic remote sensing. *Earth Surface Processes and Landforms*, **28**(3), 249–271.

Lane, S. N., Parsons, D. R., Best, J. L., *et al.*, 2008. Causes of rapid mixing at a junction of two large rivers: Rio Parana and Rio Paraguay, Argentina. *Journal of Geophysical Research-Earth Surface*, **113**(F2). 10.1029/2006jf000745.

Lang, M., Ouarda, T., Bobee, B., 1999. Towards operational guidelines for over-threshold modeling. *Journal of Hydrology*, **225** (3–4), 103–117.

Lang, A., Bork, H. R., Mackel, R., et al., 2003. Changes in sediment flux and storage within a fluvial system: Some examples from the Rhine catchment. *Hydrological Processes*, **17**(16), 3321–3334.

Langbein, W. B., 1949. Annual floods and the partial duration series. *Transactions, American Geophysical Union*, **30**(6), 879–881.

Langbein, W. B., 1964. Geometry of river channels. *American Society of Civil Engineers Proceedings*, **90**, 301–312.

Langbein, W. B., Leopold, L. B., 1964. Quasi-equilibrium states in channel morphology. *American Journal of Science*, **262**(6), 782–794.

Langbein, W. B., Leopold, L. B., 1966. River meanders – theory of minimum variance. U.S. Geological Survey Professional Paper 422-H. U.S. Government Printing Office, Washington, DC.

Langbein, W. B., Schumm, S. A., 1958. Yield of sediment in relation to mean annual precipitation. *Transactions, American Geophysical Union*, **39**(6), 1076–1084.

Langendoen, E. J., 2011. Application of the CONCEPTS channel evolution model in stream restoration strategies. In: A. Simon, S. J. Bennett, J. M. Castro (eds.), *Stream Restoration in Dynamic Fluvial Systems: Scientific Approaches, Analyses, and Tools*. American Geophysical Union, Washington, DC, pp. 487–502.

Langendoen, E. J., Alonso, C. V., 2008. Modeling the evolution of incised streams: I. Model formulation and validation of flow and streambed evolution components. *Journal of Hydraulic Engineering*, **134**(6), 749–762.

Lanz, B., Dietz, S., Swanson, T., 2018. The expansion of modern agriculture and global biodiversity decline: An integrated assessment. *Ecological Economics*, **144**, 260–277.

Lanzoni, S., 2000a. Experiments on bar formation in a straight flume 1. Uniform sediment. *Water Resources Research*, **36**(11), 3337–3349.

Lanzoni, S., 2000b. Experiments on bar formation in a straight flume 2. Graded sediment. *Water Resources Research*, **36**(11), 3351–3363.

Lanzoni, S., Seminara, G., 2006. On the nature of meander instability. *Journal of Geophysical Research-Earth Surface*, **111**(F4). 10.1029/2005jf000416.

Lara Pinto Coelho, M. M., 2015. Experimental determination of free surface levels at open-channel junctions. *Journal of Hydraulic Research*, **53**(3), 394–399.

Laraque, A., Guyot, J. L., Filizola, N., 2009. Mixing processes in the Amazon River at the confluences of the Negro and Solimoes Rivers, Encontro das Aguas, Manaus, Brazil. *Hydrological Processes*, **23**(22), 3131–3140.

Laronne, J. B., Reid, I., Yitshak, Y., Frostick, L. E., 1994. The non-layering of gravel streambeds under ephemeral flood regimes. *Journal of Hydrology*, **159**(1–4), 353–363.

Larsen, I. J., Montgomery, D. R., Greenberg, H. M., 2014. The contribution of mountains to global denudation. *Geology*, **42**(6), 527–530.

Latosinski, F. G., Szupiany, R. N., Garcia, C. M., Guerrero, M., Amsler, M. L., 2014. Estimation of concentration and load of suspended bed sediment in a large river by means of acoustic Doppler technology. *Journal of Hydraulic Engineering*, **140**(7). 10.1061/(asce)hy.1943-7900.0000859.

Latosinski, F. G., Szupiany, R. N., Guerrero, M., Amsler, M. L., Vionnet, C., 2017. The ADCP's bottom track capability for bedload prediction: Evidence on method reliability from sandy river applications. *Flow Measurement and Instrumentation*, **54**, 124–135.

Latrubesse, E. M., 2008. Patterns of anabranching channels: the ultimate end-member adjustment of mega rivers. *Geomorphology*, **101**(1–2), 130–145.

Latrubesse, E. M., Franzinelli, E., 2005. The late Quaternary evolution of the Negro River, Amazon, Brazil: Implications for island and floodplain formation in large anabranching tropical systems. *Geomorphology*, **70**(3–4), 372–397.

Latrubesse, E. M., Amsler, M. L., de Morais, R. P., Aquino, S., 2009. The geomorphologic response of a large pristine alluvial river to tremendous deforestation in the South American tropics: the case of the Araguaia River. *Geomorphology*, **113**(3–4), 239–252.

Latrubesse, E. M., Arima, E. Y., Dunne, T., et al., 2017. Damming the rivers of the Amazon basin. *Nature*, **546**(7658), 363–369.

Lau, J. K., Lauer, T. E., Weinman, M. L., 2006. Impacts of channelization on stream habitats and associated fish assemblages in east central Indiana. *American Midland Naturalist*, **156**(2), 319–330.

Lauer, J. W., Parker, G., 2008. Net local removal of floodplain sediment by river meander migration. *Geomorphology*, **96**(1–2), 123–149.

Lave, R., 2009. The controversy over natural channel design: substantive explanations and potential avenues for resolution. *Journal of the American Water Resources Association*, **45**(6), 1519–1532.

Lave, R., 2012. Bridging political ecology and STS: A field analysis of the Rosgen wars. *Annals of the Association of American Geographers*, **102**(2), 366–382.

Lave, R., 2014. Freedom and constraint: Generative expectations in the US stream restoration field. *Geoforum*, **52**, 236–244.

Lave, R., 2016. Stream restoration and the surprisingly social dynamics of science. *Wiley Interdisciplinary Reviews-Water*, **3**(1), 75–81.

Lave, R., 2018. Stream mitigation banking. *Wiley Interdisciplinary Reviews-Water*, **5**(3). 10.1002/wat2.1279.

Lave, R., Doyle, M., Robertson, M., 2010. Privatizing stream restoration in the US. *Social Studies of Science*, **40**(5), 677–703.

Lavelle, J. W., Mofjeld, H. O., 1987. Do critical stresses for incipient motion and erosion really exist? *Journal of Hydraulic Engineering*, **113**(3), 370–385.

Lawler, D. M., 1993. Needle ice processes and sediment mobilization on river banks – the River Ilston, West-Glamorgan, UK. *Journal of Hydrology*, **150**(1), 81–114.

Le Coz, J., Michalkova, M., Hauet, A., et al., 2010. Morphodynamics of the exit of a cutoff meander: Experimental findings from field and laboratory studies. *Earth Surface Processes and Landforms*, **35**(3), 249–261.

Lecce, S. A., 1997a. Nonlinear downstream changes in stream power on Wisconsin's Blue River. *Annals of the Association of American Geographers*, **87**(3), 471–486.

Lecce, S. A., 1997b. Spatial patterns of historical overbank sedimentation and floodplain evolution: Blue River, Wisconsin. *Geomorphology*, **18**(3–4), 265–277.

Lecce, S. A., Pease, P. P., Gares, P. A., Wang, J. Y., 2006. Seasonal controls on sediment delivery in a small coastal plain watershed, North Carolina, USA. *Geomorphology*, **73**(3–4), 246–260.

Lecce, S. A., Pavlowsky, R. T., Bassett, G. S., Martin, D. J., 2011. Metal contamination from gold mining in the Cid District, North Carolina. *Physical Geography*, **32**(5), 469–495.

Leclerc, R. F., Hickin, E. J., 1997. The internal structure of scrolled floodplain deposits based on ground-penetrating radar, North Thompson River, British Columbia. *Geomorphology*, **21**(1), 17–38.

Leddy, J. O., Ashworth, P. J., Best, J. L., 1993. Mechanisms of anabranch avulsion within gravel-bed braided rivers: observations from a scaled physical model. In: J. L. Best, C. S. Bristow (eds.), *Braided Rivers*. Geological Society of London Special Publication No. 75. Geological Society of London, London, pp. 119–127.

Lee, A. J., Ferguson, R. I., 2002. Velocity and flow resistance in step-pool streams. *Geomorphology*, **46**(1–2), 59–71.

Lee, H. G., Kim, J., 2015. Two-dimensional Kelvin-Helmholtz instabilities of multi-component fluids. *European Journal of Mechanics B-Fluids*, **49**, 77–88.

Lee, H. Y., Fu, D. T., Song, M. H., 1993. Migration of rectangular mining pit composed of uniform sediment. *Journal of Hydraulic Engineering*, **119**(1), 64–80.

Lee, J.-S., Julien, P. Y., 2006. Downstream hydraulic geometry of alluvial channels. *Journal of Hydraulic Engineering*, **132**(12), 1347–1352.

Leeder, M. R., Bridges, P. H., 1975. Flow separation in meander bends. *Nature*, **253**(5490), 338–339.

Legg, N., Heimburg, C., Collins, B. D., Olsen, P. L., 2014. *The Channel Migration Toolbox*. Washington State Department of Ecology, Olympia, WA.

Legleiter, C. J., Harrison, L. R., Dunne, T., 2011. Effect of point bar development on the local force balance governing flow in a simple, meandering gravel bed river. *Journal of Geophysical Research-Earth Surface*, **116**, 29. 10.1029/2010jf001838.

Leigh, D. S., 1994. Mercury contamination and floodplain sedimentation from former gold mines in north Georgia. *Water Resources Bulletin*, **30**(4), 739–748.

Leigh, D. S., 2008. Late Quaternary climates and river channels of the Atlantic Coastal Plain, Southeastern USA. *Geomorphology*, **101**(1–2), 90–108.

Leighly, J., 1932. Toward a theory of the morphological significance of turbulence in the flow of streams. *University of California Publications in Geography*, **6**(1), 1–22.

Leite Ribeiro, M., Blanckaert, K., Roy, A. G., Schleiss, A. J., 2012a. Flow and sediment dynamics in channel confluences. *Journal of Geophysical Research-Earth Surface*, **117**. 10.1029/2011jf002171.

Leite Ribeiro, M., Blanckaert, K., Roy, A. G., Schleiss, A. J., 2012b. Hydromorphological implications of local tributary widening for river rehabilitation. *Water Resources Research*, **48**. 10.1029/2011wr011296.

Lenzi, M. A., 2001. Step-pool evolution in the Rio Cordon, northeastern Italy. *Earth Surface Processes and Landforms*, **26**(9), 991–1008.

Lenzi, M. A., Mao, L., Comiti, F., 2006. Effective discharge for sediment transport in a mountain river: computational approaches and geomorphic effectiveness. *Journal of Hydrology*, **326**(1–4), 257–276.

Leopold, L. B., 1953. Downstream change of velocity in rivers. *American Journal of Science*, **251**(8), 606–624.

Leopold, L. B., 1973. River channel change with time – an example. *Geological Society of America Bulletin*, **84**(6), 1845–1860.

Leopold, L. B., 1977. A reverence for rivers. *Geology*, **5**(7), 429–430.

Leopold, L. B., 1990. Lag times for small drainage basins. *Catena*, **18**, 157–171.

Leopold, L. B., Bull, W. B., 1979. Base level, aggradation, and grade. *Proceedings of the American Philosophical Society*, **123**(3), 168–202.

Leopold, L. B., Langbein, W. B., 1962. The concept of entropy in landscape evolution. U.S. Geological Survey Professional Paper 500-A. U.S. Government Printing Office, Washington, DC.

Leopold, L. B., Maddock, T., 1953. The hydraulic geometry of stream channels and some physiographic implications. U.S. Geological Survey Professional Paper 252. U.S. Government Printing Office, Washington, DC.

Leopold, L. B., Miller, J., 1956. Ephemeral streams – hydraulic factors and their relation to the drainage net. U.S. Geological Survey Professional Paper 282-A. U.S. Government Printing Office, Washington, DC.

Leopold, L. B., Wolman, M. G., 1957. River channel patterns: braided, meandering and straight. U.S. Geological Survey Professional Paper 282-B. U.S. Government Printing Office, Washington, DC.

Leopold, L. B., Wolman, M. G., 1960. River meanders. *Geological Society of America Bulletin*, 71, 769–794.

Leopold, L., Wolman, M. G., Miller, J. P., 1964. *Fluvial Processes in Geomorphology*. Freeman, San Francisco, CA.

Leopold, L. B., Huppman, R., Miller, A., 2005. Geomorphic effects of urbanization in forty-one years of observation. *Proceedings of the American Philosophical Society*, **149**(3), 349–371.

Lewin, J., 1976. Initiation of bed forms and meanders in coarse-grained sediment. *Geological Society of America Bulletin*, **87**(2), 281–285.

Lewin, J., 1978. Meander development and floodplain sedimentation: a case study from mid-Wales. *Geological Journal*, 13(1), 25–36.

Lewin, J., Ashworth, P. J., 2014a. Defining large river channel patterns: alluvial exchange and plurality. *Geomorphology*, **215**, 83–98.

Lewin, J., Ashworth, P. J., 2014b. The negative relief of large river floodplains. *Earth-Science Reviews*, **129**, 1–23.

Lewin, J., Brewer, P. A., 2001. Predicting channel patterns. *Geomorphology*, **40**(3–4), 329–339.

Lewin, J., Brewer, P. A., 2002. Laboratory simulation of clast abrasion. *Earth Surface Processes and Landforms*, **27**(2), 145–164.

Lewin, J., Brewer, P. A., 2003. Reply to Van den Berg and Bledsoe's comment on Lewin and Brewer (2001) "Predicting Channel Patterns". *Geomorphology*, **53**(3–4), 339–342.

Lewin, J., Macklin, M. G., 2010. Floodplain catastrophes in the UK Holocene: messages for managing climate change. *Hydrological Processes*, **24**(20), 2900–2911.

Lewin, J., Macklin, M. G., 2014. Marking time in geomorphology: should we try to formalise an Anthropocene definition? *Earth Surface Processes and Landforms*, **39**(1), 133–137.

Lewin, J., Ashworth, P. J., Strick, R. J. P., 2017. Spillage sedimentation on large river floodplains. *Earth Surface Processes and Landforms*, **42**(2), 290–305.

Lewis, G. W., Lewin, J., 1983. Alluvial cutoffs in Wales and the Borderlands. In: J.D. Collinson, J. Lewin (eds.), *Modern and Ancient Fluvial Systems*. International Association of Sedimentologists Special Publications 6, Blackwell, Oxford, UK, pp. 145–154.

Lewis, Q. W., Rhoads, B. L., 2015. Rates and patterns of thermal mixing at a small stream confluence under variable incoming flow conditions. *Hydrological Processes*, **29**(20), 4442–4456.

Lewis, Q. W., Rhoads, B. L., 2018. LSPIV measurements of two-dimensional flow structure in streams using small unmanned aerial systems: 2. hydrodynamic mapping at river confluences. *Water Resources Research*, **54**(10), 7981–7999. 10.1029/2018wr022551.

Lewis, S. L., Maslin, M. A., 2015. Defining the Anthropocene. *Nature*, **519**(7542), 171–180.

Li, C., Czapiga, M. J., Eke, E. C., Viparelli, E., Parker, G., 2015a. Variable Shields number model for river bankfull geometry: bankfull shear velocity is viscosity-dependent but grain size-independent. *Journal of Hydraulic Research*, **53**(1), 36–48.

Li, J., Bristow, C. S., 2015. Crevasse splay morphodynamics in a dryland river terminus: Rio Colorado in Salar de Uyuni Bolivia. *Quaternary International*, **377**, 71–82.

Li, J., Bristow, C. S., Luthi, S. M., Donselaar, M. E., 2015b. Dryland anabranching river morphodynamics: Rio Capilla, Salar de Uyuni, Bolivia. *Geomorphology*, **250**, 282–297.

Li, M. Z., Komar, P. D., 1992. Selective entrainment and transport of mixed size and density sands – flume experiments simulating the formation of black-sand placers. *Journal of Sedimentary Petrology*, **62**(4), 584–590.

Li, Z., Komar, P. D., 1986. Laboratory measurements of pivoting angles for applications to selective entrainment of gravel in a current. *Sedimentology*, **33**, 413–423.

Liebault, F., Bellot, H., Chapuis, M., Klotz, S., Deschatres, M., 2012. Bedload tracing in a high-sediment-load mountain stream. *Earth Surface Processes and Landforms*, **37**(4), 385–399.

Limerinos, J. T., 1970. Determination of the Manning coefficient from measured bed roughness in natural channels. U.S. Geological Survey Water-Supply Paper 1898-B. U.S. Government Printing Office, Washington, DC.

Lin, J. D., Soong, H. K., 1979. Junction losses in open channel flows. *Water Resources Research*, **15**, 414–418.

Lin, Z., Oguchi, T., 2004. Drainage density, slope angle, and relative basin position in Japanese bare lands from high-resolution DEMs. *Geomorphology*, **63**(3–4), 159–173.

Lindow, N., Fox, G. A., Evans, R. O., 2009. Seepage erosion in layered stream bank material. *Earth Surface Processes and Landforms*, **34**(12), 1693–1701.

Lininger, K. B., Wohl, E., Sutfin, N. A., Rose, J. R., 2017. Floodplain downed wood volumes: a comparison across three biomes. *Earth Surface Processes and Landforms*, **42**(8), 1248–1261.

Lisenby, P. E., Fryirs, K. A., 2016. Catchment- and reach-scale controls on the distribution and expectation of geomorphic channel adjustment. *Water Resources Research*, **52**(5), 3408–3427. 10.1002/2015wr017747.

Lisenby, P. E., Croke, J., Fryirs, K. A., 2018. Geomorphic effectiveness: a linear concept in a non-linear world. *Earth Surface Processes and Landforms*, **43**(1), 4–20.

Lisle, T. E., 1979. A sorting mechanism for a riffle-pool sequence. *Geological Society of America Bulletin*, Part II, **90**(7), 1142–1157.

Lisle, T. E., 1986. Stabilization of a gravel channel by large streamside obstructions and bedrock bends, Jacoby Creek, northwestern California. *Geological Society of America Bulletin*, **97**(8), 999–1011.

Lisle, T. E., Hilton, S., 1992. The volume of fine sediment in pools – an index of sediment supply in gravel-bed streams. *Water Resources Bulletin*, **28**(2), 371–383.

Lisle, T. E., Ikeda, H., Iseya, F., 1991. Formation of stationary alternate bars in a steep channel with mixed-size sediment – a flume experiment. *Earth Surface Processes and Landforms*, **16**(5), 463–469.

Lisle, T. E., Iseya, F., Ikeda, H., 1993. Response of a channel with alternate bars to a decrease in supply of mixed-size bedload – a flume experiment. *Water Resources Research*, **29**(11), 3623–3629.

Lisle, T. E., Cui, Y. T., Parker, G., Pizzuto, J. E., Dodd, A. M., 2001. The dominance of dispersion in the evolution of bed material waves in gravel-bed rivers. *Earth Surface Processes and Landforms*, **26**(13), 1409–1420.

Liu, T.-h., Chen, L., Fan, B.-l., 2012. Experimental study on flow pattern and sediment transportation at a 90° open-channel confluence. *International Journal of Sediment Research*, **27**(2), 178–187.

Lobkovsky, A. E., Jensen, B., Kudrolli, A., Rothman, D. H., 2004. Threshold phenomena in erosion driven by subsurface flow. *Journal of Geophysical Research-Earth Surface*, **109**(F4). 10.1029/2004jf000172.

Lobkovsky, A. E., Smith, B. E., Kudrolli, A., Mohrig, D. C., Rothman, D. H., 2007. Erosive dynamics of channels incised by subsurface water flow. *Journal of Geophysical Research-Earth Surface*, **112**(F3). 10.1029/2006jf000517.

Loewenherz, D. S., 1991. Stability and the initiation of channelized surface drainage – a reassessment of the short wavelength limit. *Journal of Geophysical Research-Solid Earth and Planets*, **96**(B5), 8453–8464.

Loewenherz-Lawrence, D. S., 1994. Hydrodynamic description for advective sediment transport processes and rill initiation. *Water Resources Research*, **30**(11), 3203–3212.

Lofthouse, C., Robert, A., 2008. Riffle-pool sequences and meander morphology. *Geomorphology*, **99**(1–4), 214–223.

Loget, N., Van Den Driessche, J., 2009. Wave train model for knickpoint migration. *Geomorphology*, **106**(3–4), 376–382.

Long, D. G. F., 2011. Architecture and depositional style of fluvial systems before land plants: a comparison of Precambrian, early Paleozoic, and modern river deposits. In: S. K. Davidson, S. Leleu, C. P. North (eds.), *From River to Rock Record: The Preservation of Fluvial Sediments and Their Subsequent Interpretation*. Society for Sedimentary Geology Special Publication, SEPM, Tulsa, OK, pp. 37–61.

Lopez, F., Garcia, M. H., 1999. Wall similarity in turbulent open-channel flow. *Journal of Engineering Mechanics*, **125**(7), 789–796.

Lopez-Tarazon, J. A., Estrany, J., 2017. Exploring suspended sediment delivery dynamics of two Mediterranean nested catchments. *Hydrological Processes*, **31**(3), 698–715.

Lotsari, E., Vaaja, M., Flener, C., et al., 2014. Annual bank and point bar morphodynamics of a meandering river determined by high-accuracy multitemporal laser scanning and flow data. *Water Resources Research*, **50**(7), 5532–5559. 10.1002/2013wr014106.

Lotsari, E., Thorndycraft, V., Alho, P., 2015. Prospects and challenges of simulating river channel response to future climate change. *Progress in Physical Geography*, **39**(4), 483–513.

Lu, H., Moran, C. J., Sivapalan, M., 2005. A theoretical exploration of catchment-scale sediment delivery. *Water Resources Research*, **41**(9). 10.1029/2005wr004018.

Lu, S. S., Willmarth, W. W., 1973. Measurements of structure of Reynolds stress in a turbulent boundary layer. *Journal of Fluid Mechanics*, **60**(SEP18), 481–511.

Lubowe, J. K., 1964. Steam junction angles in the dendritic drainage pattern. *American Journal of Science*, **262**(3), 325–339.

Lucas, R. W., Baker, T. T., Wood, M. K., Allison, C. D., VanLeeuwen, D. M., 2009. Streambank morphology and cattle grazing in two montane riparian areas in western New Mexico. *Journal of Soil and Water Conservation*, **64**(3), 183–189.

Ludwig, W., Probst, J. L., 1998. River sediment discharge to the oceans: present-day controls and global budgets. *American Journal of Science*, **298**(4), 265–295.

Lunt, I. A., Bridge, J. S., Tye, R. S., 2004. A quantitative, three-dimensional depositional model of gravelly braided rivers. *Sedimentology*, **51**(3), 377–414.

Luo, H., Fytanidis, D. K., Schmidt, A. R., Garcia, M. H., 2018. Comparative 1D and 3D numerical investigation of open-channel junction flows and energy losses. *Advances in Water Resources*, **117**, 120–139. 10.1016/j.advwatres.2018.05.012.

Lyubimova, T., Lepikhin, A., Konovalov, V., Parshakova, Y., Tiunov, A., 2014. Formation of the density currents in the zone of confluence of two rivers. *Journal of Hydrology*, **508**, 328–342.

Ma, Y., Huang, H. Q., Xu, J., Brierley, G. J., Yao, Z., 2010. Variability of effective discharge for suspended sediment transport in a large semi-arid river basin. *Journal of Hydrology*, **388**(3–4), 357–369.

MacFarlane, W. A., Wohl, E., 2003. Influence of step composition on step geometry and flow resistance in step-pool streams of the Washington Cascades. *Water Resources Research*, **39**(2), 10.1029/2001wr001238.

Mackay, J. R., 1970. Lateral mixing of the Liard and Mackenzie rivers downstream from their confluence. *Canadian Journal of Earth Sciences*, **7**, 111–124.

Mackin, J. H., 1948. Concept of the graded river. *Geological Society of America Bulletin*, **59**(5), 463–511.

Mackin, J., 1963. Methods of investigation in geology. In: C. C. Albritton (ed.), *The Fabric of Geology*. Addison-Wesley, Reading, MA, pp. 135–163.

Macklin, M. G., Lewin, J., Woodward, J. C., 2012. The fluvial record of climate change. *Philosophical Transactions of the Royal Society A: Mathematical, Physical and Engineering Sciences*, **370**(1966), 2143–2172.

Macklin, M. G., Lewin, J., Jones, A. F., 2014. Anthropogenic alluvium: an evidence-based meta-analysis for the UK Holocene. *Anthropocene*, **6**, 26–38.

MacVicar, B., Best, J., 2013. A flume experiment on the effect of channel width on the perturbation and recovery of flow in straight pools and riffles with smooth boundaries. *Journal of Geophysical Research-Earth Surface*, **118**(3), 1850–1863.

MacVicar, B. J., Roy, A. G., 2007a. Hydrodynamics of a forced riffle pool in a gravel bed river: 1. mean velocity and turbulence intensity. *Water Resources Research*, **43**(12). 10.1029/2006wr005272.

MacVicar, B. J., Roy, A. G., 2007b. Hydrodynamics of a forced riffle pool in a gravel bed river: 2. scale and structure of coherent turbulent events. *Water Resources Research*, **43**(12). 10.1029/2006wr005274.

MacVicar, B. J., Roy, A. G., 2011. Sediment mobility in a forced riffle-pool. *Geomorphology*, **125**(3), 445–456.

MacWilliams, M. L., Jr., Wheaton, J. M., Pasternack, G. B., Street, R. L., Kitanidis, P. K., 2006. Flow convergence routing hypothesis for pool-riffle maintenance in alluvial rivers. *Water Resources Research*, **42**(10). 10.1029/2005wr004391.

Maddock, T., Jr., 1970. Indeterminate hydraulics of alluvial channels. *Journal of the Hydraulics Division – ASCE*, **96**(HY11), 2309–2323.

Madej, M. A., Ozaki, V., 1996. Channel response to sediment wave propagation and movement, Redwood Creek, California, USA. *Earth Surface Processes and Landforms*, **21**(10), 911–927.

Madej, M. A., Ozaki, V., 2009. Persistence of effects of high sediment loading in a salmon-bearing river, northern California. In: L. A. James, S. L. Rathburn, G. R. Whittecar (eds.), Management and Restoration of Fluvial Systems with Broad Historical Changes and Human Impacts. Geological Society of America Special Papers, 451, pp. 43–55.

Magilligan, F. J., 1985. Historical floodplain sedimentation in the Galena River Basin, Wisconsin and Illinois. *Annals of the Association of American Geographers*, **75**(4), 583–594.

Magilligan, F. J., 1992a. Thresholds and the spatial variability of flood power during extreme floods. *Geomorphology*, **5**(3–5), 373–390.

Magilligan, F. J., 1992b. Sedimentology of a fine-grained aggrading floodplain. *Geomorphology*, **4**(6), 393–408.

Magilligan, F. J., Nislow, K. H., 2005. Changes in hydrologic regime by dams. *Geomorphology*, **71**(1–2), 61–78.

Magilligan, F. J., Phillips, J. D., James, L. A., Gomez, B., 1998. Geomorphic and sedimentological controls on the effectiveness of an extreme flood. *Journal of Geology*, **106**(1), 87–95.

Magilligan, F. J., Nislow, K. H., Graber, B. E., 2003. Scale-independent assessment of discharge reduction and riparian disconnectivity following flow regulation by dams. *Geology*, **31**(7), 569–572.

Magilligan, F. J., Buraas, E. M., Renshaw, C. E., 2015. The efficacy of stream power and flow duration on geomorphic responses to catastrophic flooding. *Geomorphology*, **228**, 175–188.

Magilligan, F. J., Graber, B. E., Nislow, K. H., et al., 2016. River restoration by dam removal: enhancing connectivity at watershed scales. *Elementa-Science of the Anthropocene*, **4**, 1–14.

Mahowald, N. M., Baker, A. R., Bergametti, G., et al., 2005. Atmospheric global dust cycle and iron inputs to the ocean. *Global Biogeochemical Cycles*, **19**(4). 10.1029/2004gb002402.

Major, J. J., O'Connor, J. E., Podolak, C. J., et al., 2012. Geomorphic response of the Sandy River, Oregon, to removal of the Marmot Dam. U.S. Geological Survey Professional Paper 1792, Reston, VA.

Major, J. J., East, A. E., O'Connor, J. E., et al., 2017. Geomorphic responses to dam removal in the United States – a two-decade perspective. In: D. Tsutsumi, J. B. Laronne (eds.), *Gravel-Bed Rivers: Processes and Disasters*. Wiley, Chichester, UK, pp. 355–381.

Makaske, B., 2001. Anastomosing rivers: a review of their classification, origin and sedimentary products. *Earth-Science Reviews*, **53**(3–4), 149–196.

Makaske, B., Smith, D. G., Berendsen, H. J. A., 2002. Avulsions, channel evolution and floodplain sedimentation rates of the anastomosing upper Columbia River, British Columbia, Canada. *Sedimentology*, **49**(5), 1049–1071.

Makaske, B., Smith, D. G., Berendsen, H. J. A., et al., 2009. Hydraulic and sedimentary processes causing anastomosing morphology of the upper Columbia River, British Columbia, Canada. *Geomorphology*, **111** (3–4),194–205.

Makaske, B., Lavooi, E., de Haas, T., Kleinhans, M. G., Smith, D. G., 2017. Upstream control of river anastomosis by sediment overloading, upper Columbia River, British Columbia, Canada. *Sedimentology*, **64**(6), 1488–1510.

Malamud, B. D., Turcotte, D. L., 2006. The applicability of power-law frequency statistics to floods. *Journal of Hydrology*, **322**(1–4), 168–180.

Maloney, E. M., 2019. How do we take the pulse of an aquatic ecosystem? Current and historical approaches to measuring ecosystem integrity. *Environmental Toxicology and Chemistry*, **38**(2), 289–301.

Mandelbrot, B., 1982. *The Fractal Geometry of Nature*. W.H. Freeman and Company, San Francisco, CA.

Mannering, J. V., Fenster, C. R., 1983. What is conservation tillage? *Journal of Soil and Water Conservation*, **38**(3), 141–143.

Mao, L., Picco, L., Lenzi, M. A., Surian, N., 2017. Bed material transport estimate in large gravel-bed rivers using the virtual velocity approach. *Earth Surface Processes and Landforms*, **42**(4), 595–611.

Marchi, L., Cavalli, M., Amponsah, W., Borga, M., Crema, S., 2016. Upper limits of flash flood stream power in Europe. *Geomorphology*, **272**, 68–77.

Marcinkowski, P., Grabowski, R. C., Okruszko, T., 2017. Controls on anastomosis in lowland river systems: towards process-based solutions to habitat conservation. *Science of the Total Environment*, **609**, 1544–1555.

Mark, D. M., Church, M., 1980. On size and scale in geomorphology. *Progress in Physical Geography*, **4**, 342–390.

Markham, A. J., Thorne, C. R., 1992. Geomorphology of gravel-bed river bends. In: P. Billi, R. D. Hey, C. R. Thorne, P. Taconi (eds.), *Dynamics of Gravel-Bed Rivers*. Wiley and Sons, Chichester, UK, pp. 433–456.

Maroulis, J. C., Nanson, G. C., 1996. Bedload transport of aggregated muddy alluvium from Cooper Creek, central Australia: a flume study. *Sedimentology*, **43**(5), 771–790.

Marra, W. A., Parsons, D. R., Kleinhans, M. G., Keevil, G. M., Thomas, R. E., 2014. Near-bed and surface flow division patterns in experimental river bifurcations. *Water Resources Research*, **50**(2), 1506–1530.

Marra, W. A., McLelland, S. J., Parsons, D. R., et al., 2015. Groundwater seepage landscapes from distant and local sources in experiments and on Mars. *Earth Surface Dynamics*, **3**(3), 389–408.

Marriott, S., 1992. Textural analysis and modeling of a flood deposit – River Severn, UK. *Earth Surface Processes and Landforms*, **17**(7), 687–697.

Marsh, W., Dozier, J., 1981. *Landscape: An Introduction to Physical Geography*. Addison-Wesley, Reading, MA.

Marston, R. A., 1982. The geomorphic significance of log steps in forest streams. *Annals of the Association of American Geographers*, **72**(1), 99–108.

Marston, R. A., 1983. Supraglacial stream dynamics on the Juneau Icefield. *Annals of the Association of American Geographers*, **73**(4), 597–608.

Martin, R. L., Jerolmack, D. J., Schumer, R., 2012. The physical basis for anomalous diffusion in bed load transport. *Journal of Geophysical Research-Earth Surface*, **117**. 10.1029/2011jf002075.

Martin, Y., Church, M., 1995. Bed-material transport estimated from channel surveys – Vedder River, British Columbia. *Earth Surface Processes and Landforms*, **20**(4), 347–361.

Martin, Y., Church, M., 2004. Numerical modelling of landscape evolution: geomorphological perspectives. *Progress in Physical Geography*, **28**(3), 317–339.

Martin-Vide, J. P., Ferrer-Boix, C., Ollero, A., 2010. Incision due to gravel mining: modeling a case study from the Gallego River, Spain. *Geomorphology*, **117**(3–4), 261–271.

Martin-Vide, J. P., Plana-Casado, A., Sambola, A., Capape, S., 2015. Bedload transport in a river confluence. *Geomorphology*, **250**, 15–28.

Masek, J. G., Turcotte, D. L., 1993. A diffusion-limited aggregation model for the evolution of drainage networks. *Earth and Planetary Science Letters*, **119**(3), 379–386.

Masteller, C. C., Finnegan, N. J., 2017. Interplay between grain protrusion and sediment entrainment in an experimental flume. *Journal of Geophysical Research-Earth Surface*, **122**(1), 274–289. 10.1002/2016jf003943.

Matsubara, Y., Howard, A. D., 2014. Modeling planform evolution of a mud-dominated meandering river: Quinn River, Nevada, USA. *Earth Surface Processes and Landforms*, **39**(10), 1365–1377.

Matsubara, Y., Howard, A. D., Burr, D. M., et al., 2015. River meandering on Earth and Mars: a comparative study of Aeolis Dorsa meanders, Mars and possible terrestrial analogs of the Usuktuk River, AK, and the Quinn River, NV. *Geomorphology*, **240**, 102–120.

Mattingly, R. L., Herricks, E. E., Johnston, D. M., 1993. Channelization and levee construction in Illinois – review and implications for management. *Environmental Management*, **17**(6), 781–795.

McCarthy, T. S., Stanistreet, I. G., Cairncross, B., 1991. The sedimentary dynamics of active fluvial channels on the Okavango Fan, Botswana. *Sedimentology*, **38**(3), 471–487.

McCarthy, T. S., Ellery, W. N., Stanistreet, I. G., 1992. Avulsion mechanisms on the Okavango Fan, Botswana – the control of a fluvial system by vegetation. *Sedimentology*, **39**(5), 779–795.

McCarthy, T. S., Ellery, W. N., Bloem, A., 1998. Some observations on the geomorphological impact of hippopotamus (*Hippopotamus amphibius* L) in the Okavango Delta, Botswana. *African Journal of Ecology*, **36**(1), 44–56.

McCormick, J., 1989. *Reclaiming Paradise: The Global Environmental Movement*. Indiana University Press, Bloomington, IN.

McDonald, A., Lane, S. N., Haycock, N. E., Chalk, E. A., 2004. Rivers of dreams: on the gulf between theoretical and practical aspects of an upland river restoration. *Transactions of the Institute of British Geographers*, **29**(3), 257–281.

McEwan, I., Heald, J., 2001. Discrete particle modeling of entrainment from flat uniformly sized sediment beds. *Journal of Hydraulic Engineering*, **127**(7), 588–597.

McGuire, L. A., Pelletier, J. D., 2016. Controls on valley spacing in landscapes subject to rapid base-level fall. *Earth Surface Processes and Landforms*, **41**(4), 460–472.

McGuire, L. A., Pelletier, J. D., Gomez, J. A., Nearing, M. A., 2013. Controls on the spacing and geometry of rill networks on hillslopes: rain splash detachment, initial hillslope roughness, and the competition between fluvial and colluvial transport. *Journal of Geophysical Research-Earth Surface*, **118**(1), 241–256. 10.1002/jgrf.20028.

McLean, D. G., Church, M., 1999. Sediment transport along lower Fraser River – 2. Estimates based on the long-term gravel budget. *Water Resources Research*, **35**(8), 2549–2559.

McLean, D. G., Church, M., Tassone, B., 1999. Sediment transport along lower Fraser River – 1. Measurements and hydraulic computations. *Water Resources Research*, **35**(8), 2533–2548. 10.1029/1999wr900101.

McLelland, S. J., Ashworth, P. J., Best, J. L., 1996. The origin and downstream development of coherent flow structures at channel junctions. In: P. J. Ashworth, S. J. Bennett, J. L. Best, S. J. McLelland (eds.), *Coherent Flow Structures in Open Channels*. John Wiley and Sons, Chichester, UK, pp. 459–490.

McNamara, J. P., Borden, C., 2004. Observations on the movement of coarse gravel using implanted motion-sensing radio transmitters. *Hydrological Processes*, **18**(10), 1871–1884.

McNamara, J. P., Ziegler, A. D., Wood, S. H., Vogler, J. B., 2006. Channel head locations with respect to geomorphologic thresholds derived from a digital elevation model: a case study in northern Thailand. *Forest Ecology and Management*, **224**(1–2), 147–156.

McParland, D., Eaton, B., Rosenfeld, J., 2016. At-a-station hydraulic geometry simulator. *River Research and Applications*, **32**(3), 399–410.

Meade, R. H., 1982. Sources, sinks, and storage of river sediment in the Atlantic Drainage of the United States. *Journal of Geology*, **90**(3), 235–252.

Mejia, A. I., Niemann, J. D., 2008. Identification and characterization of dendritic, parallel, pinnate, rectangular, and trellis networks based on deviations from planform self-similarity. *Journal of Geophysical Research-Earth Surface*, **113**(F2), 21. 10.1029/2007jf000781.

Melton, M. A., 1957. An analysis of the relation among elements of climate, surface properties, and geomorphology. Office of Naval Research Technical Report No. 11, Columbia University, New York.

Melton, M. A., 1961. Discussion: the effect of sediment type on the shape and stratification of some modern fluvial deposits. *American Journal of Science*, 259, 231–233.

Merritt, D. M., Wohl, E. E., 2003. Downstream hydraulic geometry and channel adjustment during a flood along an ephemeral, arid-region drainage. *Geomorphology*, **52**(3–4), 165–180.

Merz, B., Aerts, J., Arnbjerg-Nielsen, K., et al., 2014. Floods and climate: emerging perspectives for flood risk assessment and management. *Natural Hazards and Earth System Sciences*, **14**(7), 1921–1942.

Messager, M. L., Lehner, B., Grill, G., Nedeva, I., Schmitt, O., 2016. Estimating the volume and age of water stored in global lakes using a geostatistical approach. *Nature Communications*, 7. 7:13603 | 10.1038/ncomms13603.

Metivier, F., Gaudemer, Y., 1999. Stability of output fluxes of large rivers in South and East Asia during the last 2 million years: implications on floodplain processes. *Basin Research*, **11**(4), 293–303.

Metivier, F., Paola, C., Kozarek, J. L., Tai, M., 2016. Experimental studies and practical challenges in fluvial geomorphology. In: G. M. Kondolf, H. Piegay (eds.), *Tools in Fluvial Geomorphology*, 2nd Edition. Wiley, London, pp. 454–475.

Meybeck, M., 2003. Global analysis of river systems: from Earth system controls to Anthropocene syndromes. *Philosophical Transactions of the Royal Society B-Biological Sciences*, **358**(1440), 1935–1955.

Meyer, J. L., 1997. Stream health: incorporating the human dimension to advance stream ecology. *Journal of the North American Benthological Society*, **16**(2), 439–447.

Meyer-Peter, E., Muller, R., 1948. Formulas for bedload transport. Proceedings of the Second Congress, International Association for Hydraulic Research Stockholm, pp. 39–64.

Miall, A. D., 2006. *The Geology of Fluvial Deposits.* 4th Corrected Edition. Springer Verlag, Berlin.

Miao, C., Borthwick, A. G. L., Liu, H., Liu, J., 2015. China's policy on dams at the crossroads: removal or further construction? *Water*, **7**(5), 2349–2357.

Micheli, E. R., Larsen, E. W., 2011. River channel cutoff dynamics, Sacramento River, California, USA. *River Research and Applications*, **27**(3), 328–344.

Middleton, G. V., Wilcock, P. R., 1994. *Mechanics in the Earth and Environmental Sciences.* Cambridge University Press, Cambridge, UK.

Mignot, E., Vinkovic, I., Doppler, D., Riviere, N., 2014. Mixing layer in open-channel junction flows. *Environmental Fluid Mechanics*, **14**(5), 1027–1041.

Milan, D. J., 2013a. Virtual velocity of tracers in a gravel-bed river using size-based competence duration. *Geomorphology*, **198**, 107–114.

Milan, D.J., 2013b. Sediment routing hypothesis for pool-riffle maintenance. *Earth Surface Processes and Landforms*, **38**(14), 1623–1641.

Milan, D. J., Heritage, G. L., Large, A. R. G., Charlton, M. E., 2001. Stage dependent variability in tractive force distribution through a riffle-pool sequence. *Catena*, **44**(2), 85–109.

Milan, D. J., Heritage, G. L., Hetherington, D., 2007. Application of a 3D laser scanner in the assessment of erosion and deposition volumes and channel change in a proglacial river. *Earth Surface Processes and Landforms*, **32**(11), 1657–1674.

Milan, D. J., Heritage, G. L., Large, A. R. G., Fuller, I. C., 2011. Filtering spatial error from DEMs: implications for morphological change estimation. *Geomorphology*, **125**(1), 160–171.

Millar, R.G., 2000. Influence of bank vegetation on alluvial channel patterns. *Water Resources Research*, **36**(4), 1109–1118.

Millar, R. G., 2005. Theoretical regime equations for mobile gravel-bed rivers with stable banks. *Geomorphology*, **64**(3–4), 207–220.

Millar, R. G., Eaton, B. C., 2011. Bank vegetation, bank strength, and application of the University of British Columbia regime model to stream restoration. In: A. Simon, S. J. Bennett, J. M. Castro (eds.), *Stream Restoration in Dynamic Fluvial Systems.* American Geophysical Union, Washington, DC, pp. 475–485.

Millar, R. G., Quick, M. C., 1993. Effect of bank stability on geometry of gravel rivers. *Journal of Hydraulic Engineering*, **119**(12), 1343–1363.

Millar, R. G., Quick, M. C., 1998. Stable width and depth of gravel-bed rivers with cohesive banks. *Journal of Hydraulic Engineering*, **124**(10), 1005–1013.

Millard, C., Hajek, E., Edmonds, D. A., 2017. Evaluating controls on crevasse-splay size: implications for floodplain-basin filling. *Journal of Sedimentary Research*, **87**(7), 722–739.

Millennium Ecosystem Assessment, 2005. *Ecosystems and Human Well-being: A Framework for Assessment.* Island Press, Washington, DC.

Miller, A. J., 1990. Flood hydrology and geomorphic effectiveness in the central Appalachians. *Earth Surface Processes and Landforms*, **15**(2), 119–134.

Miller, J. R., 1991b. Development of anastomosing channels in south-central Indiana. *Geomorphology*, **4**(3–4), 221–229.

Miller, J. R., 1991c. The influence of bedrock geology on knickpoint development and channel-bed degradation along downcutting streams in south-central Indiana. *Journal of Geology*, **99**(4), 591–605.

Miller, J. R., 1997. The role of fluvial geomorphic processes in the dispersal of heavy metals from mine sites. *Journal of Geochemical Exploration*, **58**(2–3),101–118.

Miller, J. R., 2017. Causality of historic arroyo incision in the southwestern United States. *Anthropocene*, **18**, 69–75. 10.1016/j.ancene.2017.06.003.

Miller, J. R., Kochel, R. C., 2010. Assessment of channel dynamics, in-stream structures and post-project channel adjustments in North Carolina and its implications to effective stream restoration. *Environmental Earth Sciences*, **59**(8), 1681–1692. 10.1007/s12665-009–0150-1.

Miller, J. R., Ritter, J. B., 1996. An examination of the Rosgen classification of natural rivers. *Catena*, **27**(3–4), 295–299. 10.1016/0341–8162(96)00017–3.

Miller, J. R., Lechler, P. J., Desilets, M., 1998. The role of geomorphic processes in the transport and fate of mercury in the Carson River basin, west-central Nevada. *Environmental Geology*, **33**(4), 249–262. 10.1007/s002540050244.

Miller, K. L., Szabo, T., Jerolmack, D. J., Domokos, G., 2014. Quantifying the significance of abrasion and selective transport for downstream fluvial grain size evolution. *Journal of Geophysical Research-Earth Surface*, **119**(11), 2412–2429. 10.1002/2014jf003156.

Miller, M. C., McCave, I. N., Komar, P. D., 1977. Threshold of sediment motion under unidirectional currents. *Sedimentology*, **24**(4), 507–527.

Miller, R. L., Byrne, R. J., 1966. The angle of repose for a single grain on a fixed rough bed. *Sedimentology*, **6**, 303–314.

Miller, S. O., Ritter, D. F., Kochel, R. C., Miller, J. R., 1993. Fluvial responses to land-use changes and climatic variations within the Drury Creek watershed, southern Illinois. *Geomorphology*, **6**(4), 309–329.

Miller, T. K., 1984. A system model of stream-channel shape and size. *Geological Society of America Bulletin*, **95**(2), 237–241.

Miller, T. K., 1991a. A model of stream channel adjustment – assessment of Rubey's hypothesis. *Journal of Geology*, **99**(5), 699–710.

Miller, T. K., Onesti, L. J., 1979. Relationship between channel shape and sediment characteristics in the channel perimeter. *Geological Society of America Bulletin*, **90**(3), 301–304. 10.1130/0016–7606(1979)90<301:trbcsa>2.0.co;2.

Milliman, J. D., Farnsworth, K. L., 2011. *River Discharge to the Coastal Ocean: A Global Synthesis.* Cambridge University Press, Cambridge, UK.

Milliman, J. D., Meade, R. H., 1983. World-wide delivery of river sediment to the oceans. *Journal of Geology*, **91**(1), 1–21.

Milliman, J. D., Syvitski, J. P. M., 1992. Geomorphic/tectonic control of sediment discharge to the ocean – the importance of small mountainous rivers. *Journal of Geology*, **100**(5), 525–544.

Milne, J. A., 1982. Bed-material size and the riffle-pool sequence. *Sedimentology*, **29**(2), 267–278.

Milzow, C., Molnar, P., McArdell, B. W., Burlando, P., 2006. Spatial organization in the step-pool structure of a steep mountain stream (Vogelbach, Switzerland). *Water Resources Research*, **42**(4). 10.1029/2004WR003870.

Miori, S., Repetto, R., Tubino, M., 2006. A one-dimensional model of bifurcations in gravel bed channels with erodible banks. *Water Resources Research*, **42**(11). 10.1029/2006wr004863.

Miori, S., Hardy, R. J., Lane, S. N., 2012. Topographic forcing of flow partition and flow structures at river bifurcations. *Earth Surface Processes and Landforms*, **37**(6), 666–679.

Mohajeri, S. H., Righetti, M., Wharton, G., Romano, G. P., 2016. On the structure of turbulent gravel bed flow: implications for sediment transport. *Advances in Water Resources*, **92**, 90–104.

Molnar, P., Anderson, R. S., Kier, G., Rose, J., 2006. Relationships among probability distributions of stream discharges in floods, climate, bed load transport, and river incision. *Journal of Geophysical Research-Earth Surface*, **111**(F2). 10.1029/2005jf000310.

Molnar, P., Densmore, A. L., McArdell, B. W., Turowski, J. M., Burlando, P., 2010. Analysis of changes in the step-pool morphology and channel profile of a steep mountain stream following a large flood. *Geomorphology*, **124**(1–2), 85–94.

Montgomery, D. R., 2007a. *Dirt: The Erosion of Civilizations*. The University of California Press, Berkeley, CA.

Montgomery, D. R., 2007b. Soil erosion and agricultural sustainability. *Proceedings of the National Academy of Sciences of the United States of America*, **104**(33), 13268–13272.

Montgomery, D. R., Buffington, J. M., 1997. Channel-reach morphology in mountain drainage basins. *Geological Society of America Bulletin*, **109**(5), 596–611.

Montgomery, D. R., Dietrich, W. E., 1988. Where do channels begin? *Nature*, **336**(6196), 232–234.

Montgomery, D. R., Dietrich, W. E., 1989. Sources area, drainage density, and channel initiation. *Water Resources Research*, **25**(8), 1907–1918.

Montgomery, D. R., Dietrich, W. E., 1992. Channel initiation and the problem of landscape scale. *Science*, **255**(5046), 826–830.

Montgomery, D.R., Dietrich, W.E., 1994. Landscape dissection and drainage area-slope thresholds. In: M.J. Kirkby (ed.), *Process Models and Theoretical Geomorphology*. Wiley, Chichester, UK, pp. 221–246.

Montgomery, D. R., Foufoula-Georgiou, E., 1993. Channel network source representation using digital elevation models. *Water Resources Research*, **29**(12), 3925–3934.

Montgomery, D. R., Gran, K. B., 2001. Downstream variations in the width of bedrock channels. *Water Resources Research*, **37**(6), 1841–1846.

Montgomery, D. R., Buffington, J. M., Smith, R. D., Schmidt, K. M., Pess, G., 1995. Pool spacing in forest channels. *Water Resources Research*, **31**(4), 1097–1105.

Moody, J. A., Troutman, B. M., 2000. Quantitative model of the growth of floodplains by vertical accretion. *Earth Surface Processes and Landforms*, **25**(2), 115–133.

Moody, J. A., Pizzuto, J. E., Meade, R. H., 1999. Ontogeny of a flood plain. *Geological Society of America Bulletin*, **111**(2), 291–303.

Morgan, R. P. C., Quinton, J. N., Smith, R. E., et al., 1998. The European Soil Erosion Model (EUROSEM): a dynamic approach for predicting sediment transport from fields and small catchments. *Earth Surface Processes and Landforms*, **23**(6), 527–544.

Morisawa, M. E., 1962. Quantitative geomorphology of some watersheds in the Appalachian Plateau. *Geological Society of America Bulletin*, **73**(9), 1025–1046.

Morisawa, M., 1964. Development of drainage systems on upraised lake floor. *American Journal of Science*, **262**(3), 340–354.

Morisawa, M., 1985. *Rivers: Form and Process*. Longman, London.

Moron, S., Edmonds, D. A., Amos, K., 2017. The role of floodplain width and alluvial bar growth as a precursor for the formation of anabranching rivers. *Geomorphology*, **278**, 78–90.

Morozova, G. S., Smith, N. D., 1999. Holocene avulsion history of the lower Saskatchewan fluvial system, Cumberland Marshes, Saskatchewan-Manitoba, Canada. In: N. D. Smith, J. Rogers (eds.), *Fluvial Sedimentology VI*, International Association of Sedimentologists Special Publication No. 28, Blackwell, Malden, MA, pp. 231–249.

Morozova, G. S., Smith, N. D., 2000. Holocene avulsion styles and sedimentation patterns of the Saskatchewan River, Cumberland Marshes, Canada. *Sedimentary Geology*, **130**(1–2), 81–105.

Morris, G. L., Annandale, G., Hotchkiss, R., 2008. Reservoir sedimentation. In: M. H. Garcia (ed.), *Sedimentation Engineering: Processes, Measurements, Modeling, and Practice*. American Society of Civil Engineers, New York, pp. 579–612.

Morris, P. H., Williams, D. J., 1999a. Worldwide correlations for subaerial aqueous flows with exponential longitudinal profiles. *Earth Surface Processes and Landforms*, **24**(10), 867–879.

Morris, P. H., Williams, D. J., 1999b. A worldwide correlation for exponential bed particle size variation in subaerial aqueous flows. *Earth Surface Processes and Landforms*, **24**(9), 835–847.

Mosher, S.-J., Martini, I. P., 2002. Coarse-grained flood bars formed at the confluence of two subarctic rivers affected by hydroelectric dams, Ontario, Canada. In: I. P. Martini, V. R. Baker, G. Garzon (eds.), *Flood and Megaflood Deposits: Recent and Ancient Examples*. International Association of Sedimentologists Special Publication 32, Blackwell, Oxford, UK, pp. 213–231.

Mosley, M. P., 1976. An experimental study of channel confluences. *Journal of Geology*, **84**, 535–562.

Mosley, M. P., 1982. Analysis of the effect of changing discharge on channel morphology and instream uses in a braided river, Ohua River, New Zealand. *Water Resources Research*, **18**(4), 800–812.

Moss, J. H., Kochel, R. C., 1978. Unexpected geomorphic effects of Hurricane Agnes storm and flood, Conestoga Drainage Basin, southeastern Pennsylvania. *Journal of Geology*, **86**(1), 1–11.

Mossa, J., McLean, M., 1997. Channel planform and land cover changes on a mined river floodplain. *Applied Geography*, **17**, 43–54.

Mosselman, E., 2005. Basic equations for sediment transport in CFD for fluvial morphodynamics. In: P. D. Bates, S. N. Lane, R. I. Ferguson (eds.), *Computational Fluid Dynamics: Applications in Environmental Hydraulics*. Wiley, Chichester, UK, pp. 71–89.

Motta, D., Abad, J. D., Langendoen, E. J., Garcia, M. H., 2012a. A simplified 2D model for meander migration with physically-based bank evolution. *Geomorphology*, **163**, 10–25.

Motta, D., Abad, J. D., Langendoen, E. J., Garcia, M. H., 2012b. The effects of floodplain soil heterogeneity on meander planform shape. *Water Resources Research*, **48**. 10.1029/2011wr011601.

Motta, D., Langendoen, E. J., Abad, J. D., Garcia, M. H., 2014. Modification of meander migration by bank failures. *Journal of Geophysical Research-Earth Surface*, **119**(5), 1026–1042. 10.1002/2013jf002952.

Mueller, E. R., Pitlick, J., 2013. Sediment supply and channel morphology in mountain river systems: 1. relative importance of lithology, topography, and climate. *Journal of Geophysical Research-Earth Surface*, **118**(4), 2325–2342. 10.1002/2013jf002843.

Mueller, E. R., Pitlick, J., 2014. Sediment supply and channel morphology in mountain river systems: 2. single thread to braided transitions. *Journal of Geophysical Research-Earth Surface*, **119**(7), 1516–1541. 10.1002/2013jf003045.

Mulder, T., Syvitski, J. P. M., 1996. Climatic and morphologic relationships of rivers: implications of sea-level fluctuations on river loads. *Journal of Geology*, **104**(5), 509–523.

Murphy, P. J., Randle, T. J., Fotherby, L. M., Daraio, J. A., 2004. *The Platte River Channel: History and Restoration*. U.S. Department of Interior, Bureau of Reclamation, Denver, CO.

Murray, A. B., Paola, C., 1994. A cellular model of braided rivers. *Nature*, **371**(6492), 54–57. 10.1038/371054a0.

Murray, A. B., Paola, C., 1997. Properties of a cellular braided-stream model. *Earth Surface Processes and Landforms*, **22**(11), 1001–1025.

Murray, A. B., Paola, C., 2003. Modelling the effect of vegetation on channel pattern in bedload rivers. *Earth Surface Processes and Landforms*, **28**(2), 131–143.

Muste, M., Yu, K., Fujita, I., Ettema, R., 2005. Two-phase versus mixed-flow perspective on suspended sediment transport in turbulent channel flows. *Water Resources Research*, **41**(10). 10.1029/2004wr003595.

Naden, P. S., 1992. Spatial variability in flood estimation for large catchments – the exploitation of channel network structure. *Hydrological Sciences Journal*, **37**(1), 53–71.

Naiman, R. J., Turner, M. G., 2000. A future perspective on North America's freshwater ecosystems. *Ecological Applications*, **10**(4), 958–970.

Nakato, T., 1990. Tests of selected sediment transport formulas. *Journal of Hydraulic Engineering*, **116**(3), 362–379.

Nanson, G. C., 1980. Point bar and floodplain formation of the meandering Beatton River, northeastern British Columbia, Canada. *Sedimentology*, **27**(1), 3–29.

Nanson, G. C., 1986. Episodes of vertical accretion and catastrophic stripping – a model of disequilibrium floodplain development. *Geological Society of America Bulletin*, **97**(12), 1467–1475.

Nanson, G., 2013. Anabranching and anastomosing rivers. In: J. Schroder (ed.), *Treatise on Geomorphology*, Vol. 9, Fluvial Geomorphology, E. Wohl (vol. ed.). Academic Press, San Diego, CA, pp. 330–345.

Nanson, G. C., Croke, J. C., 1992. A genetic classification of floodplains. *Geomorphology*, **4**(6), 459–486.

Nanson, G. C., Hickin, E. J., 1983. Channel migration and incision on the Beatton River. *Journal of Hydraulic Engineering, ASCE*, **109**(3), 327–337.

Nanson, G. C., Hickin, E. J., 1986. A statistical analysis of bank erosion and channel migration in western Canada. *Geological Society of America Bulletin*, **97**(4), 497–504.

Nanson, G. C., Huang, H. Q., 2008. Least action principle, equilibrium states, iterative adjustment and the stability of alluvial channels. *Earth Surface Processes and Landforms*, **33**(6), 923–942.

Nanson, G. C., Huang, H. Q., 2017. Self-adjustment in rivers: evidence for least action as the primary control of alluvial-channel form and process. *Earth Surface Processes and Landforms*, **42**(4), 575–594.

Nanson, G. C., Huang, H. Q., 2018. A philosophy of rivers: equilibrium states, channel evolution, teleomatic change and least action principle. *Geomorphology*, **302**, 3–19.

Nanson, G. C., Knighton, A. D., 1996. Anabranching rivers: their cause, character and classification. *Earth Surface Processes and Landforms*, **21**(3), 217–239.

Nanson, G. C., Page, K., 1983. Lateral accretion of fine-grained concave benches on meandering rivers. *Special Publication of the International Association of Sedimentologists*, **6**, 133–143.

Nanson, G. C., Rust, B. R., Taylor, G., 1986. Coexistent mud braids and anastomosing channels in an arid-zone river – Cooper Creek, central Australia. *Geology*, **14**(2), 175–178.

Nanson, R. A., 2010. Flow fields in tightly curving meander bends of low width-depth ratio. *Earth Surface Processes and Landforms*, **35**(2), 119–135. 10.1002/esp.1878.

Naot, D., Rodi, W., 1982. Calculation of secondary currents in channel flow. *Journal of the Hydraulics Division – ASCE*, **108**, 948–968.

Naqshband, S., McElroy, B., Mahon, R. C., 2017. Validating a universal model of particle transport lengths with laboratory measurements of suspended grain motions. *Water Resources Research*, **53**(5), 4106–4123. 10.1002/2016wr020024.

Nash, D. B., 1994. Effective sediment-transporting discharge from magnitude-frequency analysis. *Journal of Geology*, **102**(1), 79–95.

National Research Council, 1992. *Restoration of Aquatic Ecosystems*. National Academy Press, Washington, DC.

Navratil, O., Albert, M. B., 2010. Non-linearity of reach hydraulic geometry relations. *Journal of Hydrology*, **388**(3–4), 280–290.

Navratil, O., Albert, M. B., Herouin, E., Gresillon, J. M., 2006. Determination of bankfull discharge magnitude and frequency: comparison of methods on 16 gravel-bed river reaches. *Earth Surface Processes and Landforms*, **31**(11), 1345–1363.

Navratil, O., Breil, P., Schmitt, L., Grosprêtre, L., Albert, M. B., 2013. Hydrogeomorphic adjustments of stream channels disturbed by urban runoff (Yzeron River basin, France). *Journal of Hydrology*, **485**, 24–36.

Naylor, L. A., Spencer, T., Lane, S. N., et al., 2017. Stormy geomorphology: geomorphic contributions in an age of climate extremes. *Earth Surface Processes and Landforms*, **42**(1), 166–190.

Nazari-Giglou, A., Jabbari-Sahebari, A., Shakibaeinia, A., Borghei, S. M., 2016. An experimental study of sediment transport in channel confluences. *International Journal of Sediment Research*, **31**(1), 87–96.

Nearing, M. A., Nichols, M. H., Stone, J. J., Renard, K. G., Simanton, J. R., 2007. Sediment yields from unit-source semi-arid watersheds at Walnut Gulch. *Water Resources Research*, **43**(6). 10.1029/2006wr005692.

Nearing, M. A., Xie, Y., Liu, B., Ye, Y., 2017. Natural and anthropogenic rates of soil erosion. *International Soil and Water Conservation Research*, **5**(2), 77–84.

Neely, A. B., Bookhagen, B., Burbank, D. W., 2017. An automated knickzone selection algorithm (KZ-Picker) to analyze transient landscapes: calibration and validation. *Journal of Geophysical Research – Earth Surface*, **122**(6), 1236–1261. 10.1002/2017jf004250.

Neitsch, S. L., Arnold, J. G., Kiniry, J. R., Williams, J. R., 2009. Soil and Water Assessment Tool: Theoretical Documentation Version 2009. Texas Water Resources Institute Technical Report No. 406, Texas A&M University College Station, Texas.

Nelson, A. D., Church, M., 2012. Placer mining along the Fraser River, British Columbia: the geomorphic impact. *Geological Society of America Bulletin*, **124**(7–8), 1212–1228.

Nelson, J. M., 1990. The initial instability and finite-amplitude stability of alternate bars in straight channels. *Earth-Science Reviews*, **29**(1–4), 97–115.

Nelson, J. M., Smith, J. D., 1989. Evolution and stability of erodible channel beds. In: S. Ikeda, G. Parker (eds.), *River Meandering*, Water Resources Monograph 12. American Geophysical Union, Washington, DC, pp. 321–377.

Nelson, J. M., Shreve, R. L., McLean, S. R., Drake, T. G., 1995. Role of near-bed turbulence structure in bed load transport and bed form mechanics. *Water Resources Research*, **31**(8), 2071–2086.

Nelson, J. M., McDonald, R. R., Shimizu, Y., et al., 2016. Modelling flow, sediment transport and morphodynamics in rivers. In: G. M. Kondolf, H. Piegay (eds.), *Tools in Fluvial Geomorphology*, 2nd Edition. Wiley, Chichester, UK, pp. 412–441.

Nelson, P. A., Smith, J. A., Miller, A. J., 2006. Evolution of channel morphology and hydrologic response in an urbanizing drainage basin. *Earth Surface Processes and Landforms*, **31**(9), 1063–1079. 10.1002/esp.1308.

Newson, M. D., Large, A. R. G., 2006. "Natural" rivers, "hydromorphological quality" and river restoration: a challenging new agenda for applied fluvial geomorphology. *Earth Surface Processes and Landforms*, **31**(13), 1606–1624.

Nezu, I., Nakagawa, H., 1993. *Turbulence in Open-Channel Flows*. Balkema, Rotterdam.

Nezu, I., Nakagawa, H., Rodi, W., 1989. Significant difference between secondary currents in closed channels and narrow open channels, Proceedings of the 28th Congress, International Association of Hydraulic Research, Ottawa, Canada, pp. A-125–A132.

Nezu, I., Tominaga, A., Nakagawa, H., 1993. Field measurements of secondary currents in straight rivers. *Journal of Hydraulic Engineering*, **119**(5), 598–614.

Nicholas, A., 2013. Morphodynamic diversity of the world's largest rivers. *Geology*, **41**(4), 475–478.

Nicholas, A. P., Mitchell, C. A., 2003. Numerical simulation of overbank processes in topographically complex floodplain environments. *Hydrological Processes*, **17**(4), 727–746.

Nicholas, A. P., Sambrook Smith, G. H., 1999. Numerical simulation of three-dimensional flow hydraulics in a braided channel. *Hydrological Processes*, **13**(6), 913–929.

Nicholas, A. P., Walling, D. E., 1997. Modelling flood hydraulics and overbank deposition on river floodplains. *Earth Surface Processes and Landforms*, **22**(1), 59–77.

Nicholas, A. P., Walling, D. E., 1998. Numerical modelling of floodplain hydraulics and suspended sediment transport and deposition. *Hydrological Processes*, **12**(8), 1339–1355.

Nicholas, A. P., Thomas, R., Quine, T. A., 2006a. Cellular modelling of braided river form and process. In: G. H. S. Smith, J. L. Best, C. S. Bristow, G. E. Petts (eds.), *Braided Rivers: Process, Deposits, Ecology and Management*. Special Publications of the International Association of Sedimentologists, 36, Blackwell, Malden, MA, pp. 137–151.

Nicholas, A. P., Walling, D. E., Sweet, R. J., Fang, X., 2006b. Development and evaluation of a new catchment-scale model of floodplain sedimentation. *Water Resources Research*, **42**(10), 13. 10.1029/2005wr004579.

Nicholas, A. P., Ashworth, P. J., Smith, G. H. S., Sandbach, S. D., 2013. Numerical simulation of bar and island morphodynamics in anabranching megarivers. *Journal of Geophysical Research-Earth Surface*, **118**(4), 2019–2044. 10.1002/jgrf.20132.

Nichols, A. L., Viers, J. H., 2017. Not all breaks are equal: variable hydrologic and geomorphic responses to intentional levee breaches along the lower Cosumnes River, California. *River Research and Applications*, **33**(7), 1143–1155.

Nichols, M. H., Nearing, M. A., Polyakov, V. O., Stone, J. J., 2013. A sediment budget for a small semiarid watershed in southeastern Arizona, USA. *Geomorphology*, **180**, 137–145.

Nichols, M. H., Nearing, M., Hernandez, M., Polyakov, V. O., 2016. Monitoring channel head erosion processes in response to an artificially induced abrupt base level change using time-lapse photography. *Geomorphology*, **265**, 107–116.

Nickolotsky, A., Pavlowsky, R. T., 2007. Morphology of step-pools in a wilderness headwater stream: the importance of standardizing geomorphic measurements. *Geomorphology*, **83**(3–4), 294–306.

Niemann, J. D., Bras, R. L., Veneziano, D., Rinaldo, A., 2001. Impacts of surface elevation on the growth and scaling properties of simulated river networks. *Geomorphology*, **40**(1–2), 37–55.

Nienhuis, J. H., Tornqvist, T. E., Esposito, C. R., 2018. Crevasse splays versus avulsions: a recipe for land building with levee breaches. *Geophysical Research Letters*, **45**(9), 4058–4067.

Niiniluoto, I., 1993. The aim and structure of applied research. *Erkenntnis*, **38**, 1–21.

Niiniluoto, I., 2014. Values in design sciences. *Studies in History and Philosophy of Science*, **46**, 11–15.

Nijssen, B., O'Donnell, G. M., Hamlet, A. F., Lettenmaier, D. P., 2001. Hydrologic sensitivity of global rivers to climate change. *Climatic Change*, **50**(1–2), 143–175.

Nikora, V., Goring, D., 2000. Flow turbulence over fixed and weakly mobile gravel beds. *Journal of Hydraulic Engineering*, **126**(9), 679–690.

Nikora, V., Habersack, H., Huber, T., McEwan, I., 2002. On bed particle diffusion in gravel bed flows under weak bed load transport. *Water Resources Research*, **38**(6). 10.1029/2001wr000513.

Nilsson, C., Reidy, C. A., Dynesius, M., Revenga, C., 2005. Fragmentation and flow regulation of the world's large river systems. *Science*, **308**(5720), 405–408.

Nino, Y., Garcia, M., 1998. Using Lagrangian particle saltation observations for bedload sediment transport modelling. *Hydrological Processes*, **12**(8), 1197–1218.

Nino, Y., Garcia, M., Ayala, L., 1994. Gravel saltation: 1. experiments. *Water Resources Research*, **30**(6), 1907–1914.

Nino, Y., Lopez, F., Garcia, M., 2003. Threshold for particle entrainment into suspension. *Sedimentology*, **50**(2), 247–263.

Noble, C. A., Palmquist, R. C., 1968. Meander growth in artificially straightened streams. *Proceedings of the Iowa Academy of Science*, **75**, 234–242.

Nolan, K. M., Kelsey, H. M., Marron, D. C. (eds.), 1995. Geomorphic processes and aquatic habitat in the Redwood Creek basin, northwestern California. U.S. Geological Survey Professional Paper 1454. U.S. Government Printing Office, Washington, DC.

Notebaert, B., Broothaerts, N., Verstraeten, G., 2018. Evidence of anthropogenic tipping points in fluvial dynamics in Europe. *Global and Planetary Change*, **164**, 27–38.

Nunnally, N. R., 1978. Stream renovation – an alternative to channelization. *Environmental Management*, **2**(5), 403–411.

Nykanen, D. K., Foufoula-Georgiou, E., Sapozhnikov, V. B., 1998. Study of spatial scaling in braided river patterns using synthetic aperture radar imagery. *Water Resources Research*, **34**(7), 1795–1807.

O'Brien, P. E., Wells, A. T., 1986. A small, alluvial crevasse splay. *Journal of Sedimentary Petrology*, **56**(6), 876–879.

Odoni, N. A., Lane, S. N., 2011. The significance of models in geomorphology: from concepts to experiments. In: K. J. Gregory, A. S. Goudie (eds.), *The Sage Handbook on Geomorphology*. Sage, Los Angeles, CA, pp. 154–173.

Oeurng, C., Sauvage, S., Sanchez-Perez, J.-M., 2010. Dynamics of suspended sediment transport and yield in a large agricultural catchment, southwest France. *Earth Surface Processes and Landforms*, **35**(11), 1289–1301.

Ohmori, H., 1991. Change in the mathematical function type describing the longitudinal profile of a river through an evolutionary process. *Journal of Geology*, **99**(1), 97–110.

Onda, Y., 1994. Seepage erosion and its implication to the formation of amphitheater valley heads – a case study at Obara, Japan. *Earth Surface Processes and Landforms*, **19**(7), 627–640.

O'Neill, M. P., Abrahams, A. D., 1984. Objective identification of pools and riffles. *Water Resources Research*, **20**(7), 921–926.

Orme, A. R., 2013. The scientific roots of geomorphology before 1830. In: J. W. Shroder (ed.), *Treatise on Geomorphology*, Vol. 1, The Foundations of Geomorphology, A.R. Orme, D. Sack (vol. eds.). Elsevier, New York, pp. 11–36.

Osman, A. M., Thorne, C. R., 1988. Riverbank stability analysis. 1: theory. *Journal of Hydraulic Engineering*, **114**(2), 134–150.

Osterkamp, W. R., Costa, J. E., 1987. Changes accompanying an extraordinary flood on a sand-bed stream. In: L. Mayer, D. B. Nash (eds.), *Catastrophic Flooding*. Allen and Unwin, London, pp. 201–224.

Ottevanger, W., Blanckaert, K., Uijttewaal, W. S. J., 2012. Processes governing the flow redistribution in sharp river bends. *Geomorphology*, **163**, 45–55.

Ottevanger, W., Blanckaert, K., Uijttewaal, W. S. J., de Vriend, H. J., 2013. Meander dynamics: a reduced-order nonlinear model without curvature restrictions for flow and bed morphology. *Journal of Geophysical Research – Earth Surface*, **118**(2), 1118–1131. 10.1002/jgrf.20080.

Oudin, L., Salavati, B., Furusho-Percot, C., Ribstein, P., Saadi, M., 2018. Hydrological impacts of urbanization at the catchment scale. *Journal of Hydrology*, **559**, 774–786.

Owen, L. A., 2013. Tectonic geomorphology: a perspective. In: J. W. Schroder (ed.), *Treatise on Geomorphology*, Vol. 5, Tectonic Geomorphology, J. A. Owen (vol. ed.). Elsevier, New York, pp. 4–12.

Owens, P. N., Batalla, R. J., Collins, A. J., et al., 2005. Fine-grained sediment in river systems: environmental significance and management issues. *River Research and Applications*, **21**(7), 693–717.

Page, K., Nanson, G., 1982. Concave-bank benches and associated floodplain formation. *Earth Surface Processes and Landforms*, 7 (6), 529–543.

Page, K. J., Nanson, G. C., Frazier, P. S., 2003. Floodplain formation and sediment stratigraphy resulting from oblique accretion on the Murrumbidgee River, Australia. *Journal of Sedimentary Research*, **73**(1), 5–14.

Paiement-Paradis, G., Marquis, G., Roy, A., 2011. Effects of turbulence on the transport of individual particles as bedload in a gravel-bed river. *Earth Surface Processes and Landforms*, **36**(1), 107–116.

Pal, D., Ghoshal, K., 2016. Effect of particle concentration on sediment and turbulent diffusion coefficients in open-channel turbulent flow. *Environmental Earth Sciences*, **75**(18). 10.1007/s12665-016-6045-z.

Palmer, M. A., Hondula, K. L., 2014. Restoration as mitigation: analysis of stream mitigation for coal mining impacts in Southern Appalachia. *Environmental Science & Technology*, **48** (18), 10552–10560.

Palmer, M. A., Bernhardt, E. S., Allan, J. D., et al., 2005. Standards for ecologically successful river restoration. *Journal of Applied Ecology*, **42**(2), 208–217.

Palmer, M. A., Bernhardt, E. S., Schlesinger, W. H., et al., 2010a. Mountaintop mining consequences. *Science*, **327**(5962), 148–149.

Palmer, M. A., Menninger, H. L., Bernhardt, E., 2010b. River restoration, habitat heterogeneity and biodiversity: a failure of theory or practice? *Freshwater Biology*, **55**, 205–222.

Palmer, M. A., Hondula, K. L., Koch, B. J., 2014. Ecological restoration of streams and rivers: shifting strategies and shifting goals. In: D. J. Futuyma (ed.), *Annual Review of Ecology, Evolution, and Systematics*, 45, 247–269.

Paola, C., 2001. Modelling stream braiding over a range of scales. In: M. P. Mosley (ed.), *Gravel Bed Rivers V*. New Zealand Hydrological Society, Wellington, pp. 11–46.

Paola, C., Seal, R., 1995. Grain-size patchiness as a cause of selective deposition and downstream fining. *Water Resources Research*, **31**(5), 1395–1407.

Paola, C., Parker, G., Seal, R., et al., 1992. Downstream fining by selective deposition in a laboratory flume. *Science*, **258**(5089), 1757–1760.

Paola, C., Straub, K., Mohrig, D., Reinhardt, L., 2009. The "unreasonable effectiveness" of stratigraphic and geomorphic experiments. *Earth-Science Reviews*, **97**(1–4), 1–43.

Papangelakis, E., Hassan, M. A., 2016. The role of channel morphology on the mobility and dispersion of bed sediment in a small gravel-bed stream. *Earth Surface Processes and Landforms*, **41**(15), 2191–2206.

Papanicolaou, A. N., Diplas, P., Dancey, C. L., Balakrishnan, M., 2001. Surface roughness effects in near-bed turbulence: implications to sediment entrainment. *Journal of Engineering Mechanics*, **127**(3), 211–218.

Park, C. C., 1977. World-wide variations in hydraulic geometry exponents of stream channels – analysis and some observations. *Journal of Hydrology*, **33**(1–2), 133–146.

Park, C. C., 1981. Man, river systems, and environmental impact. *Progress in Physical Geography*, **5**, 1–31.

Park, E., Latrubesse, E. M., 2015. Surface water types and sediment distribution patterns at the confluence of mega rivers: the Solimoes-Amazon and Negro Rivers junction. *Water Resources Research*, **51**(8), 6197–6213. 10.1002/2014wr016757.

Parker, C., Clifford, N. J., Thorne, C. R., 2011. Understanding the influence of slope on the threshold of coarse grain motion: revisiting critical stream power. *Geomorphology*, **126**(1–2), 51–65. 10.1016/j.geomorph.2010.10.027.

Parker, G., 1976. Cause and characteristic scales of meandering and braiding in rivers. *Journal of Fluid Mechanics*, **76**(AUG11), 457–480.

Parker, G., 1978a. Self-formed straight rivers with equilibrium banks and mobile bed. Part 1. The sand-silt river. *Journal of Fluid Mechanics*, **89**(NOV), 109–125. 10.1017/s0022112078002499.

Parker, G., 1978b. Self-formed straight rivers with equilibrium banks and mobile bed. Part 2. The gravel river. *Journal of Fluid Mechanics*, **89**(NOV), 127–146.

Parker, G., 1979. Hydraulic geometry of active gravel rivers. *Journal of the Hydraulics Division – ASCE*, **105**(9), 1185–1201.

Parker, G., 1990. Surface-based bedload transport relation for gravel rivers. *Journal of Hydraulic Research*, **28**(4), 417–436.

Parker, G., 1991a. Selective sorting and abrasion of river gravel. 1. Theory. *Journal of Hydraulic Engineering*, **117**(2), 131–149.

Parker, G., 1991b. Selective sorting and abrasion of river gravel. 2. Applications. *Journal of Hydraulic Engineering*, **117**(2), 150–171.

Parker, G., 2008. Transport of gravel and sediment mixtures. In: M. H. Garcia (ed.), *Sedimentation Engineering: Processes, Measurements, Modeling, and Practice*. American Society of Civil Engineers, Reston, VA, pp. 165–251.

Parker, G., Andres, D., 1976. Detrimental effects of river channelization, Rivers '76. American Society of Civil Engineers, New York, pp. 1248–1266.

Parker, G., Andrews, E. D., 1986. On the time development of meander bends. *Journal of Fluid Mechanics*, **162**, 139–156. 10.1017/s0022112086001970.

Parker, G., Cui, Y. T., 1998. The arrested gravel front: stable gravel-sand transitions in rivers – Part 1: simplified analytical solution. *Journal of Hydraulic Research*, **36**(1), 75–100.

Parker, G., Johannesson, H., 1989. Observations on some recent theories of resonance and overdeepening in meandering channels. In: S. Ikeda, G. Parker (eds.), *River Meandering*. Water Resources Monograph 12. American Geophysical Union, Washington, DC, pp. 379–415.

Parker, G., Klingeman, P. C., 1982. On why gravel bed streams are paved. *Water Resources Research*, **18**(5), 1409–1423. 10.1029/WR018i005p01409.

Parker, G., Sutherland, A. J., 1990. Fluvial armor. *Journal of Hydraulic Research*, **28**(5), 529–544.

Parker, G., Toro-Escobar, C. M., 2002. Equal mobility of gravel in streams: the remains of the day. *Water Resources Research*, **38**(11), 46/41–46/48. 10.1029/2001wr000669.

Parker, G., Klingeman, P. C., McLean, D. G., 1982a. Bedload and size distribution in paved gravel-bed streams. *Journal of the Hydraulics Division-ASCE*, **108**(4), 544–571.

Parker, G., Dhamotharan, S., Stefan, H., 1982b. Model experiments on mobile, paved gravel bed streams. *Water Resources Research*, **18**(5), 1395–1408.

Parker, G., Sawai, K., Ikeda, S., 1982c. Bend theory of river meanders. 2. Nonlinear deformation of finite-amplitude bends. *Journal of Fluid Mechanics*, **115**(FEB), 303–314.

Parker, G., Diplas, P., Akiyama, J., 1983. Meander bends of high amplitude. *Journal of Hydraulic Engineering*, **109**(10), 1323–1337.

Parker, G., Wilcock, P. R., Paola, C., Dietrich, W. E., Pitlick, J., 2007. Physical basis for quasi-universal relations describing bankfull hydraulic geometry of single-thread gravel bed rivers. *Journal of Geophysical Research-Earth Surface*, **112**(F4). 10.1029/2006jf000549.

Parker, G., Hassan, M. A., Wilcock, P. R., 2008. Adjustment of the bed surface size distribution of gravel-bed rivers in response to cycled hydrographs. In: H. Habersack, H. Piegay, M. Rinaldi (eds.), *Gravel-bed Rivers VI: From Process Understanding to River Restoration*. Elsevier, Amsterdam, the Netherlands, pp. 241–285.

Parker, G., Shimizu, Y., Wilkerson, G. V., et al., 2011. A new framework for modeling the migration of meandering rivers. *Earth Surface Processes and Landforms*, **36**(1), 70–86.

Parker, R. S., Schumm, S. A., 1982. Experimental study of drainage networks. In: R. B. Bryan, A. Yair (eds.), *Badland: Geomorphology and Piping*. Geobooks, Norwich, UK, pp. 153–168.

Parsons, A. J., 2012. How useful are catchment sediment budgets? *Progress in Physical Geography*, **36**(1), 60–71.

Parsons, A. J., Wainwright, J., Brazier, R. E., Powell, D. M., 2006. Is sediment delivery a fallacy? *Earth Surface Processes and Landforms*, **31**(10), 1325–1328.

Parsons, A. J., Wainwright, J., Brazier, R. E., Powell, D. M., 2008a. Is sediment delivery a fallacy? Reply. *Earth Surface Processes and Landforms*, **33**(10), 1630–1631.

Parsons, D. R., Best, J. L., Orfeo, O., et al., 2005. Morphology and flow fields of three-dimensional dunes, Rio Parana, Argentina: results from simultaneous multibeam echo sounding and acoustic Doppler current profiling. *Journal of Geophysical Research-Earth Surface*, **110**(F4). 10.1029/2004jf000231.

Parsons, D. R., Best, J. L., Lane, S. N., et al., 2007. Form roughness and the absence of secondary flow in a large confluence-diffluence, Rio Parana, Argentina. *Earth Surface Processes and Landforms*, **32**(1), 155–162.

Parsons, D. R., Best, J. L., Lane, S. N., et al., 2008b. Large river channel confluences. In: S. P. Rice, A. G. Roy, B. L. Rhoads (eds.), *River Confluences, Tributaries and the Fluvial Network*. Wiley, Chichester, UK, pp. 73–91.

Pasternack, G. B., Ellis, C. R., Marr, J. D., 2007. Jet and hydraulic jump near-bed stresses below a horseshoe waterfall. *Water Resources Research*, **43**(7), 14. 10.1029/2006wr005774.

Pasternack, G. B., Bounrisavong, M. K., Parikh, K. K., 2008. Backwater control on riffle-pool hydraulics, fish habitat quality, and sediment transport regime in gravel-bed rivers. *Journal of Hydrology*, **357**(1–2), 125–139.

Patil, S., Sivapalan, M., Hassan, M. A., et al., 2012. A network model for prediction and diagnosis of sediment dynamics at the watershed scale. *Journal of Geophysical Research – Earth Surface*, **117**. 10.1029/2012jf002400.

Patton, P. C., 1988. Geomorphic response of streams to floods in the glaciated terrain of southern New England. In: V. R. Baker, R. C. Kochel, P. C. Patton (eds.), *Flood Geomorphology*. Wiley, New York, pp. 261–277.

Pavlowsky, R. T., Lecce, S. A., Owen, M. R., Martin, D. J., 2017. Legacy sediment, lead, and zinc storage in channel and floodplain deposits of the Big River, Old Lead Belt Mining District, Missouri, USA. *Geomorphology*, **299**, 54–75.

Payne, B. A., Lapointe, M. F., 1997. Channel morphology and lateral stability: effects on distribution of spawning and rearing habitat for Atlantic salmon in a wandering cobble-bed river. *Canadian Journal of Fisheries and Aquatic Sciences*, **54**(11), 2627–2636.

Peakall, J., Ashworth, P., Best, J., 1996. Physical modeling in fluvial geomorphology: principles, applications and unresolved issues. In: B. L. Rhoads, C. E. Thorn (eds.), *The Scientific Nature of Geomorphology*. Wiley and Sons, Chichester, UK, pp. 221–253.

Peakall, J., Ashworth, P. J., Best, J. L., 2007. Meander-bend evolution, alluvial architecture, and the role of cohesion in sinuous river channels: a flume study. *Journal of Sedimentary Research*, **77**(3–4), 197–212.

Pearson, A. J., Pizzuto, J., 2015. Bed load transport over run-of-river dams, Delaware, USA. *Geomorphology*, **248**, 382–395.

Peckham, S. D., 1995. New results for self-similar trees with applications to river networks. *Water Resources Research*, **31**(4), 1023–1029.

Peckham, S. D., Gupta, V. K., 1999. A reformulation of Horton's laws for large river networks in terms of statistical self-similarity. *Water Resources Research*, **35**(9), 2763–2777.

Pelletier, J. D., 1999. Self-organization and scaling relationships of evolving river networks. *Journal of Geophysical Research – Solid Earth*, **104**(B4), 7359–7375.

Pelletier, J. D., 2003. Drainage basin evolution in the Rainfall Erosion Facility: dependence on initial conditions. *Geomorphology*, **53**(1–2), 183–196.

Pelletier, J. D., 2004. Persistent drainage migration in a numerical landscape evolution model. *Geophysical Research Letters*, **31**(20). 10.1029/2004gl020802.

Pelletier, J. D., 2012a. Fluvial and slope-wash erosion of soil-mantled landscapes: detachment- or transport-limited? *Earth Surface Processes and Landforms*, **37**(1), 37–51.

Pelletier, J. D., 2012b. A spatially distributed model for the long-term suspended sediment discharge and delivery ratio of drainage basins. *Journal of Geophysical Research-Earth Surface*, **117**. 10.1029/2011jf002129.

Pelletier, J. D., 2013. A robust, two-parameter method for the extraction of drainage networks from high-resolution digital elevation models (DEMs): evaluation using synthetic and real-world DEMs. *Water Resources Research*, **49**(1), 75–89.

Pelosi, A., Schumer, R., Parker, G., Ferguson, R. I., 2016. The cause of advective slowdown of tracer pebbles in rivers: implementation of Exner-Based Master Equation for coevolving streamwise and vertical dispersion. *Journal of Geophysical Research-Earth Surface*, **121**(3), 623–637. 10.1002/2015jf003497.

Penck, W., 1972. *Morphological Analysis of Land Forms: A Contribution to Physical Geology*. Translated by Hella Czech and Cumming Boswell. Hafner Pub. Co., New York.

Penna, N., De Marchis, M., Canelas, O. B., et al., 2018. Effect of the junction angle on turbulent flow at a hydraulic confluence. *Water*, **10**(4). 10.3390/w10040469.

Penning-Rowsell, E. C., Townshend, J. R. G., 1978. Influence of scale on the factors affecting stream channel slope. *Transactions of the Institute of British Geographers*, **3**(4), 395–415.

Perkins, H. J., 1970. The formation of vorticity in turbulent flow. *Journal of Fluid Mechanics*, **44**, 721–740.

Perron, J. T., Fagherazzi, S., 2012. The legacy of initial conditions in landscape evolution. *Earth Surface Processes and Landforms*, **37**(1), 52–63.

Perron, J. T., Royden, L., 2013. An integral approach to bedrock river profile analysis. *Earth Surface Processes and Landforms*, **38**(6), 570–576.

Perron, J. T., Kirchner, J. W., Dietrich, W. E., 2008a. Spectral signatures of characteristic spatial scales and nonfractal structure in landscapes. *Journal of Geophysical Research – Earth Surface*, **113** (F4). 10.1029/2007jf000866.

Perron, J. T., Dietrich, W. E., Kirchner, J. W., 2008b. Controls on the spacing of first-order valleys. *Journal of Geophysical Research – Earth Surface*, **113**(F4). 10.1029/2007jf000977.

Perron, J. T., Kirchner, J. W., Dietrich, W. E., 2009. Formation of evenly spaced ridges and valleys. *Nature*, **460**(7254), 502–505.

Perron, J. T., Richardson, P. W., Ferrier, K. L., Lapotre, M., 2012. The root of branching river networks. *Nature*, **492**(7427), 100–103.

Persendt, F. C., Gomez, C., 2016. Assessment of drainage network extractions in a low-relief area of the Cuvelai Basin (Namibia) from multiple sources: LiDAR, topographic maps, and digital aerial orthophotographs. *Geomorphology*, **260**, 32–50.

Perucca, E., Camporeale, C., Ridolfi, L., 2005. Nonlinear analysis of the geometry of meandering rivers. *Geophysical Research Letters*, **32**(3). 10.1029/2004gl021966.

Perucca, E., Camporeale, C., Ridolfi, L., 2007. Significance of the riparian vegetation dynamics on meandering river morphodynamics. *Water Resources Research*, **43**(3). 10.1029/2006wr005234.

Petit, F., 1994. Dimensionless critical shear stress evaluation from flume experiments using different gravel beds. *Earth Surface Processes and Landforms*, **19**(6), 565–576.

Petit, F., Pauquet, A., 1997. Bankfull discharge recurrence interval in gravel-bed rivers. *Earth Surface Processes and Landforms*, **22**(7), 685–693.

Petit, F., Gob, F., Houbrechts, G., Assani, A. A., 2005. Critical specific stream power in gravel-bed rivers. *Geomorphology*, **69**(1–4), 92–101.

Petit, F., Houbrechts, G., Peeters, A., et al., 2015. Dimensionless critical shear stress in gravel-bed rivers. *Geomorphology*, **250**, 308–320.

Petroff, A. P., Devauchelle, O., Seybold, H., Rothman, D. H., 2013. Bifurcation dynamics of natural drainage networks. *Philosophical Transactions of the Royal Society A – Mathematical Physical and Engineering Sciences*, **371**(2004). 10.1098/rsta.2012.0365.

Petts, G. E., Gurnell, A. M., 2005. Dams and geomorphology: research progress and future directions. *Geomorphology*, **71** (1–2), 27–47.

Petts, G., Gurnell, A., 2013. Hydrogeomorphic effects of reservoirs, dams and diversions. In: J. C. Shroder (ed.), *Treatise on Geomorphology*, Vol. 13, Geomorphology of Human Disturbances, Climate Change and Natural Hazards, L. A. James, C. P. Harden, C. C. Clague (vol. eds). Academic Press, San Diego, CA, pp. 96–114.

Petts, G. E., Thoms, M. C., 1987. Morphology and sedimentology of a tributary confluence bar in a regulated river – North Tyne, UK. *Earth Surface Processes and Landforms*, **12**(4), 433–440.

Petts, J., 2007. Learning about learning: lessons from public engagement and deliberation on urban river restoration. *Geographical Journal*, **173**, 300–311.

Pfeiffer, A. M., Finnegan, N. J., Willenbring, J. K., 2017. Sediment supply controls equilibrium channel geometry in gravel rivers. *Proceedings of the National Academy of Sciences of the United States of America*, **114**(13), 3346–3351.

Phillips, C. B., Jerolmack, D. J., 2014. Dynamics and mechanics of bed-load tracer particles. *Earth Surface Dynamics*, **2**(2), 513–530.

Phillips, C. B., Jerolmack, D. J., 2016. Self-organization of river channels as a critical filter on climate signals. *Science*, **352** (6286), 694–697.

Phillips, C. B., Scatena, F. N., 2013. Reduced channel morphological response to urbanization in a flood-dominated humid tropical environment. *Earth Surface Processes and Landforms*, **38**(9), 970–982.

Phillips, C. B., Martin, R. L., Jerolmack, D. J., 2013. Impulse framework for unsteady flows reveals superdiffusive bed load transport. *Geophysical Research Letters*, **40**(7). 10.1002/grl.50323.

Phillips, J. D., 1990. The instability of hydraulic geometry. *Water Resources Research*, **26**(4), 739–744.

Phillips, J. D., 1991. Fluvial sediment budgets in the North Carolina Piedmont. *Geomorphology*, **4**(3–4), 231–241.

Phillips, J. D., 1992. The end of equilibrium? *Geomorphology*, **5**(3–5), 195–201.

Phillips, J. D., 1993. Interpreting the fractal dimension of river networks. In: N. S. Lam, L. De Cola (eds.), *Fractals in Geography*. Prentice Hall, Englewood Cliffs, NJ, pp. 142–157.

Phillips, J. D., 1996. Deterministic complexity, explanation, and predictability in geomorphic systems. In: B. L. Rhoads, C. E. Thorn (eds.), *The Scientific Nature of Geomorphology*. Wiley, Chichester, UK, pp. 315–335.

Phillips, J. D., 2002. Erosion, isostatic response, and the missing peneplains. *Geomorphology*, **45**(3–4), 225–241.

Phillips, J. D., 2003. Alluvial storage and the long-term stability of sediment yields. *Basin Research*, **15**(2), 153–163.

Phillips, J. D., 2006a. Evolutionary geomorphology: thresholds and nonlinearity in landform response to environmental change. *Hydrology and Earth System Sciences*, **10**(5), 731–742.

Phillips, J. D., 2006b. Deterministic chaos and historical geomorphology: a review and look forward. *Geomorphology*, **76**(1–2), 109–121.

Phillips, J. D., 2010. The job of the river. *Earth Surface Processes and Landforms*, **35**(3), 305–313.

Phillips, J. D., 2012. Log-jams and avulsions in the San Antonio River Delta, Texas. *Earth Surface Processes and Landforms*, **37**(9), 936–950.

Phillips, J. D., 2014. Anastomosing channels in the lower Neches River valley, Texas. *Earth Surface Processes and Landforms*, **39** (14), 1888–1899.

Phillips, J. D., Slattery, M. C., 2006. Sediment storage, sea level, and sediment delivery to the ocean by coastal plain rivers. *Progress in Physical Geography*, **30**(4), 513–530.

Phillips, J. D., Van Dyke, C., 2016. Principles of geomorphic disturbance and recovery in response to storms. *Earth Surface Processes and Landforms*, **41**(7), 971–979.

Phillips, J. D., Slattery, M. C., Musselman, Z. A., 2005. Channel adjustments of the lower Trinity River, Texas, downstream of Livingston Dam. *Earth Surface Processes and Landforms*, **30**(11), 1419–1439.

Pickup, G., Rieger, W. A., 1979. Conceptual model of the relationship between channel characteristics and discharge. *Earth Surface Processes and Landforms*, **4**(1), 37–42.

Pickup, G., Warner, R. F., 1976. Effects of hydrological regime on magnitude and frequency of dominant discharge. *Journal of Hydrology*, **29**(1–2), 51–75.

Piegay, H., Vaudor, L., 2016. Statistics and fluvial geomorphology. In: G. M. Kondolf, H. Piegay (eds.), *Tools in Fluvial Geomorphology*, 2nd Edition. Wiley, Chichester, UK, pp. 476–506.

Piegay, H., Darby, S. E., Mosselman, E., Surian, N., 2005. A review of techniques available for delimiting the erodible river corridor: a sustainable approach to managing bank erosion. *River Research and Applications*, **21**(7), 773–789.

Piegay, H., Grant, G., Nakamura, F., Trustrum, N., 2006. Braided river management: from assessment of river behavior to improved sustainable management. In: G. H. Sambrook Smith, J. L. Best, C. S. Bristow, G. E. Petts (eds.), *Braided Rivers: Process, Deposits, Ecology and Management*, International Association of Sedimentologists Special Publication 36. Blackwell, Malden, MA, pp. 257–274.

Piegay, H., Hupp, C. R., Citterio, A., et al., 2008. Spatial and temporal variability in sedimentation rates associated with cutoff channel infill deposits: Ain River, France. *Water Resources Research*, **44** (5). 10.1029/2006wr005260.

Pierce, A. R., King, S. L., 2008. Spatial dynamics of overbank sedimentation in floodplain systems. *Geomorphology*, **100**(3–4), 256–268.

Pieri, D. C., 1984. Junction angles in drainage networks. *Journal of Geophysical Research*, **89**(NB8), 6878–6884.

Pierson, T. C., 2005. Hyperconcentrated flow – transitional process between water flow and debris flow. In: M. Jakob, O. Hungr (eds.), *Debris-Flow Hazards and Related Phenomena*. Springer, Berlin, pp. 159–202.

Pietsch, T. J., Nanson, G. C., 2011. Bankfull hydraulic geometry; the role of in-channel vegetation and downstream declining discharges in the anabranching and distributary channels of the Gwydir distributive fluvial system, southeastern Australia. *Geomorphology*, **129**(1–2), 152–165.

Pitlick, J., 1992. Flow resistance under conditions of intense gravel transport. *Water Resources Research*, **28**(3), 891–903.

Pittaluga, M. B., Nobile, G., Seminara, G., 2009. A nonlinear model for river meandering. *Water Resources Research*, **45**. 10.1029/2008wr007298.

Pizzuto, J. E., 1984a. An evaluation of methods for calculating the concentration of suspended bed material in rivers. *Water Resources Research*, **20**(10), 1381–1389.

Pizzuto, J. E., 1984b. Bank erodibility of shallow sandbed streams. *Earth Surface Processes and Landforms*, **9**(2), 113–124.

Pizzuto, J. E., 1987. Sediment diffusion during overbank flows. *Sedimentology*, **34**(2), 301–317.

Pizzuto, J. E., 1995. Downstream fining in a network of gravel-bedded rivers. *Water Resources Research*, **31**(3), 753–759.

Pizzuto, J. E., 2002. Effects of dam removal on river form and process. *Bioscience*, **52**(8), 683–691.

Pizzuto, J. E., 2016. Modelling fluvial morphodynamics. In: G. M. Kondolf, H. Piegay (eds.), *Tools in Fluvial Geomorphology*, 2nd Edition. Wiley, Chichester, UK, pp. 442–455.

Pizzuto, J. E., Meckelnburg, T. S., 1989. Evaluation of a linear bank erosion equation. *Water Resources Research*, **25**(5), 1005–1013.

Pizzuto, J. E., Hession, W. C., McBride, M., 2000. Comparing gravel-bed rivers in paired urban and rural catchments of southeastern Pennsylvania. *Geology*, **28**(1), 79–82.

Placzkowska, E., Gornik, M., Mocior, E., et al., 2015. Spatial distribution of channel heads in the Polish Flysch Carpathians. *Catena*, **127**, 240–249.

Playfair, J., 1802. *Illustrations of the Huttonian Theory of the Earth*. Edinburgh; London: Printed for Cadell and Davies; William Creech. *Nineteenth Century Collections Online*, http://tinyurl.gale.com/tinyurl/CPVmk4 (accessed December 1, 2019).

Plumb, B. D., Annable, W. K., Thompson, P. J., Hassan, M. A., 2017. The impact of urbanization on temporal changes in sediment transport in a gravel bed channel in southern Ontario, Canada. *Water Resources Research*, **53**(10), 8443–8458. 10.1002/2016WR020288.

Poesen, J., 2018. Soil erosion in the Anthropocene: research needs. *Earth Surface Processes and Landforms*, **43**(1), 64–84.

Poff, N. L., Olden, J. D., Merritt, D. M., Pepin, D. M., 2007. Homogenization of regional river dynamics by dams and global biodiversity implications. *Proceedings of the National Academy of Sciences of the United States of America*, **104**(14), 5732–5737.

Pollen, N., 2007. Temporal and spatial variability in root reinforcement of streambanks: accounting for soil shear strength and moisture. *Catena*, **69**(3), 197–205.

Pollen, N., Simon, A., 2005. Estimating the mechanical effects of riparian vegetation on stream bank stability using a fiber bundle model. *Water Resources Research*, **41**(7). 10.1029/2004wr003801.

Pollock, M. M., Beechie, T. J., Wheaton, J. M., et al., 2014. Using beaver dams to restore incised stream ecosystems. *Bioscience*, **64**(4), 279–290.

Polvi, L. E., Wohl, E., Merritt, D. M., 2014. Modeling the functional influence of vegetation type on streambank cohesion. *Earth Surface Processes and Landforms*, **39**, 1245–1258.

Pornprommin, A., Izumi, N., 2010. Inception of stream incision by seepage erosion. *Journal of Geophysical Research – Earth Surface*, **115**. 10.1029/2009jf001369.

Pornprommin, A., Takei, Y., Wubneh, A. M., Izumi, N., 2010. Channel inception in cohesionless sediment by seepage erosion. *Journal of Hydro-Environment Research*, **3**(4), 232–238.

Portenga, E. W, Bierman, P.A., 2011. Understanding Earth's eroding surface with [10]Be. *GSA Today*, **21**, 4–10.

Portenga, E. W., Westaway, K. E., Bishop, P., 2016. Timing of post-European settlement alluvium deposition in SE Australia: a legacy of European land-use in the Goulburn Plains. *Holocene*, **26**(9), 1472–1485.

Potter, K. W., 1991. Hydrological impacts of changing land management practices in a moderate-sized agricultural catchment. *Water Resources Research*, **27**(5), 845–855.

Powell, D. M., 1998. Patterns and processes of sediment sorting in gravel-bed rivers. *Progress in Physical Geography*, **22**(1), 1–32.

Powell, D. M., Ashworth, P. J., 1995. Spatial pattern of flow competence and bed load transport in a divided gravel bed river. *Water Resources Research*, **31**(3), 741–752.

Powell, D. M., Ockelford, A., Rice, S. P., et al., 2016. Structural properties of mobile armors formed at different flow strengths in gravel-bed rivers. *Journal of Geophysical Research – Earth Surface*, **121**(8), 1494–1515. 10.1002/2015jf003794.

Powell, G. E., Ward, A. D., Mecklenburg, D. E., Jayakaran, A. D., 2007. Two-stage channel systems: Part 1, a practical approach for sizing agricultural ditches. *Journal of Soil and Water Conservation*, **62**(4), 277–286.

Powell, J. W., 1875. Exploration of the Colorado River of the West and its tributaries. Smithsonian Institution, U.S. Government Printing Office, Washington, DC.

Prancevic, J. P., Lamb, M. P., 2015a. Unraveling bed slope from relative roughness in initial sediment motion. *Journal of Geophysical Research – Earth Surface*, **120**(3), 474–489. 10.1002/2014jf003323.

Prancevic, J. P., Lamb, M. P., 2015b. Particle friction angles in steep mountain channels. *Journal of Geophysical Research – Earth Surface*, **120**(2), 242–259. 10.1002/2014jf003286.

Praskievicz, S., 2015. A coupled hierarchical modeling approach to simulating the geomorphic response of river systems to anthropogenic climate change. *Earth Surface Processes and Landforms*, **40**(12), 1616–1630.

Prestegaard, K. L., 1983. Variables influencing water-surface slopes in gravel-bed streams at bankfull stage. *Geological Society of America Bulletin*, **94**(5), 673–678.

Pringle, C. M., Naiman, R. J., Bretschko, G., et al., 1988. Patch dynamics in lotic systems – the stream as a mosaic. *Journal of the North American Benthological Society*, **7**(4), 503–524.

Pritchard, D., Roberts, G. G., White, N. J., Richardson, C. N., 2009. Uplift histories from river profiles. *Geophysical Research Letters*, **36**. 10.1029/2009gl040928.

Pyne, S. J., 1980. *Grove Karl Gilbert: A Great Engine of Research*. University of Texas Press, Austin, TX.

Pyrce, R. S., Ashmore, P. E., 2003a. The relation between particle path length distributions and channel morphology in gravel-bed streams: a synthesis. *Geomorphology*, **56**(1–2), 167–187.

Pyrce, R. S., Ashmore, P. E., 2003b. Particle path length distributions in meandering gravel-bed streams: results from physical models. *Earth Surface Processes and Landforms*, **28**(9), 951–966.

Pyrce, R. S., Ashmore, P. E., 2005. Bedload path length and point bar development in gravel-bed river models. *Sedimentology*, **52**(4), 839–857.

Qian, H., Cao, Z., Liu, H., Pender, G., 2017. Numerical modelling of alternate bar formation, development and sediment sorting in straight channels. *Earth Surface Processes and Landforms*, **42**(4), 555–574.

Qin, J., Zhong, D. Y., Wang, G. Q., Ng, S. L., 2012. On characterization of the imbrication of armored gravel surfaces. *Geomorphology*, **159**, 116–124.

Qing-Yuan, Y., Xian-Ye, W., Wei-Zhen, L., Xie-Kang, W., 2009. Experimental study on characteristics of separation zone in confluence zones in rivers. *Journal of Hydrologic Engineering*, **14**(2), 166–171.

Quraishy, M. S., 1944. The origin of curves in rivers. *Current Science*, **13**, 36–39.

Radoane, M., Radoane, N., Dumitriu, D., Miclaus, C., 2008. Downstream variation in bed sediment size along the East Carpathian rivers: evidence of the role of sediment sources. *Earth Surface Processes and Landforms*, **33**(5), 674–694.

Ramamurthy, A. S., Carballada, L. B., Tran, D. M., 1988. Combining open channel flow at right angled junctions. *Journal of Hydraulic Engineering*, **114**, 1449–1460.

Ramon, C. L., Hoyer, A. B., Armengol, J., Dolz, J., Rueda, F. J., 2013. Mixing and circulation at the confluence of two rivers entering a meandering reservoir. *Water Resources Research*, **49**(3), 1429–1445.

Ramon, C. L., Armengol, J., Dolz, J., Prats, J., Rueda, F. J., 2014. Mixing dynamics at the confluence of two large rivers undergoing weak density variations. *Journal of Geophysical Research – Oceans*, **119**(4), 2386–2402.

Ramon, C. L., Prats, J., Rueda, F. J., 2016. The influence of flow inertia, buoyancy, wind, and flow unsteadiness on mixing at the asymmetrical confluence of two large rivers. *Journal of Hydrology*, **539**, 11–26.

Rantz, S. E., and others, 1982. Measurement and computation of streamflow: Vol. 2. Computation of discharge, U.S. Geological Survey Water Supply Paper 2175. U.S. Government Printing Office, Washington, DC.

Rathbun, R. E., Rostad, C. E., 2004. Lateral mixing in the Mississippi River below the confluence with the Ohio River. *Water Resources Research*, **40**(5). 10.1029/2003wr002381.

Rathburn, S. L., Rubin, Z. K., Wohl, E. E., 2013. Evaluating channel response to an extreme sedimentation event in the context of historical range of variability: Upper Colorado River, USA. *Earth Surface Processes and Landforms*, **38**(4), 391–406.

Recking, A., 2009. Theoretical development on the effects of changing flow hydraulics on incipient bed load motion. *Water Resources Research*, **45**. 10.1029/2008wr006826.

Recking, A., 2016. A generalized threshold model for computing bed load grain size distribution. *Water Resources Research*, **52**(12), 9274–9289.

Recking, A., Liebault, F., Peteuil, C., Jolimet, T., 2012. Testing bed-load transport equations with consideration of time scales. *Earth Surface Processes and Landforms*, **37**(7), 774–789.

Refice, A., Giachetta, E., Capolongo, D., 2012. SIGNUM: a Matlab, TIN-based landscape evolution model. *Computers & Geosciences*, **45**, 293–303.

Reid, D. A., Hassan, M. A., Floyd, W., 2016. Reach-scale contributions of road-surface sediment to the Honna River, Haida Gwaii, BC. *Hydrological Processes*, **30**(19), 3450–3465.

Reid, D. E., Hickin, E. J., Babakaiff, S. C., 2010. Low-flow hydraulic geometry of small, steep mountain streams in southwest British Columbia. *Geomorphology*, **122**(1–2), 39–55.

Reid, I., Laronne, J. B., 1995. Bedload sediment transport in an ephemeral stream and a comparison with seasonal and perennial counterparts. *Water Resources Research*, **31**(3), 773–781.

Reid, L. M., Dunne, T., 1984. Sediment production from forest road surfaces. *Water Resources Research*, **20**(11), 1753–1761.

Reid, L. M., Dunne, T., 2016. Sediment budgets as an organizing framework in fluvial geomorphology. In: G. M. Kondolf, H. Piegay (eds.), *Tools in Fluvial Geomorphology*. Wiley and Sons, Chichester, UK, pp. 357–380.

Reinfelds, I., Nanson, G., 1993. Formation of braided river floodplains, Waimakiriri River, New Zealand. *Sedimentology*, **40**(6), 1113–1127.

Renard, K. G., Foster, G. R., Weesies, G. A., Porter, J. P., 1991. RUSLE – revised universal soil loss equation. *Journal of Soil and Water Conservation*, **46**(1), 30–33.

Renschler, C. S., 2003. Designing geo-spatial interfaces to scale process models: the GeoWEPP approach. *Hydrological Processes*, **17**(5), 1005–1017.

Renwick, W. H., 1992. Equilibrium, disequilibrium, and nonequilibrium landforms in the landscape. *Geomorphology*, **5**(3–5), 265–276.

Renwick, W. H., Andereck, Z. D., 2006. Reservoir sedimentation trends in Ohio, USA: sediment delivery and response to land-use change. In: J. S. Rowan, R. W. Duck, A. Werritty (eds.), *Sediment Dynamics and the Hydromorphology of Fluvial Systems*. IAHS Publication No. 306, IAHS Press, Wallingford, UK, pp. 341–347.

Rhoads, B. L., 1987a. Stream power terminology. *Professional Geographer*, **39**(2), 189–195.

Rhoads, B. L., 1987b. Changes in stream channel characteristics at tributary junctions. *Physical Geography*, **8**, 346–361.

Rhoads, B. L., 1988. Mutual adjustments between process and form in a desert mountain fluvial system. *Annals of the Association of American Geographers*, **78**(2), 271–287.

Rhoads, B. L., 1989. Longitudinal variations in the size and sorting of bed material along six arid-region mountain streams. In: A. Yair, S. M. Berkowicz (eds.), *Arid and Semiarid Environments – Geomorphological and Pedological Aspects*, Catena Supplement 14. Catena Verlag, Cremlingen-Destadt, W. Germany, pp. 87–105.

Rhoads, B. L., 1990a. Hydrologic characteristics of a small desert mountain stream: implications for short-term magnitude and frequency of bedload transport. *Journal of Arid Environments*, **18**, 151–163.

Rhoads, B. L., 1990b. The impact of stream channelization on the geomorphic stability of an arid-region river. *National Geographic Research*, **6**, 157–177.

Rhoads, B. L., 1991a. A continuously varying parameter model of downstream hydraulic geometry. *Water Resources Research*, **27**(8), 1865–1872.

Rhoads, B. L., 1991b. Multicollinearity and parameter estimation in simultaneous-equation models of fluvial systems. *Geographical Analysis*, **23**(4), 346–361.

Rhoads, B. L., 1991c. Impact of agricultural development on regional drainage in the lower Santa Cruz Valley, Arizona, USA. *Environmental Geology and Water Sciences*, **18**(2), 119–135.

Rhoads, B. L., 1992. Statistical models of fluvial systems. *Geomorphology*, **5**(3–5), 433–455.

Rhoads, B. L., 1995. Stream power: a unifying theme for urban fluvial geomorphology. In: E. E. Herricks (ed.), *Stormwater Runoff and Receiving Systems: Impact, Monitoring, and Assessment*. Lewis Publishers, Boca Raton, FL, pp. 65–75.

Rhoads, B. L., 1996. Mean structure of transport-effective flows at an asymmetrical confluence when the main stream is dominant. In: P. J. Ashworth, S. J. Bennett, J. L. Best, S. J. McLelland (eds.), *Coherent Flow Structures in Open Channels*. John Wiley & Sons Ltd, Chichester, UK, pp. 491–517.

Rhoads, B. L., 1999. Beyond pragmatism: the value of philosophical discourse for physical geography. *Annals of the Association of American Geographers*, **89**(4), 760–771.

Rhoads, B. L., 2006a. The dynamic basis of geomorphology reenvisioned. *Annals of the Association of American Geographers*, **96**(1), 14–30.

Rhoads, B. L., 2006b. Scaling of confluence dynamics in river systems: some general considerations. In: G. Parker, M. H. Garcia (eds.), *River, Coastal and Estuarine Morphodynamics, RCEM 2005*. Taylor and Francis, London, pp. 379–387.

Rhoads, B. L., 2013. Process in geomorphology. In: J. W. Shroder (ed.), *Treatise on Geomorphology*, Vol. 1, The Foundations of Geomorphology, A. R. Orme and D. Sack (vol. eds.). Elsevier, New York, pp. 190–204.

Rhoads, B. L., Cahill, R. A., 1999. Geomorphological assessment of sediment contamination in an urban stream system. *Applied Geochemistry*, **14**(4), 459–483.

Rhoads, B. L., Herricks, E. E., 1996. Naturalization of headwater streams in Illinois: challenges and possibilities. In: F. D. Shields, Jr., A. Brookes (eds.), *River Channel Restoration*. Wiley, Chichester, UK, pp. 331–367.

Rhoads, B. L., Johnson, K. K., 2018. Three-dimensional flow structure, morphodynamics, suspended sediment, and thermal mixing at an asymmetrical river confluence of a straight tributary and curving main channel. *Geomorphology*, **323**, 51–69. 10.1016/j.geomorph.2018.09.009.

Rhoads, B. L., Kenworthy, S. T., 1995. Flow structure at an asymmetrical stream confluence. *Geomorphology*, **11**, 273–293.

Rhoads, B. L., Kenworthy, S. T., 1998. Time-averaged flow structure in the central region of a stream confluence. *Earth Surface Processes and Landforms*, **23**, 171–191.

Rhoads, B. L., Kenworthy, S. T., 1999. On secondary circulation, helical motion and Rozovskii-based analysis of time-averaged two-dimensional velocity fields at confluences. *Earth Surface Processes and Landforms*, **24**(4), 369–375.

Rhoads, B. L., Massey, K. D., 2012. Flow structure and channel change in a sinuous grass-lined stream within an agricultural drainage ditch: implications for ditch stability and aquatic habitat. *River Research and Applications*, **28**(1), 39–52.

Rhoads, B. L., Miller, M. V., 1991. Impact of flow variability on the morphology of a low-energy meandering river. *Earth Surface Processes and Landforms*, **16**(4), 357–367.

Rhoads, B. L., Sukhodolov, A. N., 2001. Field investigation of three-dimensional flow structure at stream confluences: 1.

thermal mixing and time-averaged velocities. *Water Resources Research*, **37**(9), 2393–2410.

Rhoads, B. L., Sukhodolov, A. N., 2004. Spatial and temporal structure of shear layer turbulence at a stream confluence. *Water Resources Research*, **40**(6). 10.1029/2003WR002811.

Rhoads, B. L., Sukhodolov, A. N., 2008. Lateral momentum flux and the spatial evolution of flow within a confluence mixing interface. *Water Resources Research*, **44**(8). 10.1029/2007wr006634.

Rhoads, B. L., Thorn, C. E., 1993. Geomorphology as science – the role of theory. *Geomorphology*, **6**(4), 287–307.

Rhoads, B. L., Thorn, C. E., 1996. Toward a philosophy of geomorphology. In: B. L. Rhoads, C. E. Thorn (eds.), *The Scientific Nature of Geomorphology*. Wiley, Chichester, UK, pp. 115–143.

Rhoads, B. L., Thorn, C. E., 2011. The role and character of theory in geomorphology. In: K. J. Gregory, A. S. Goudie (eds.), *The Sage Handbook of Geomorphology*. Sage, Los Angeles, CA, pp. 59–77.

Rhoads, B. L., Welford, M. R., 1991. Initiation of river meandering. *Progress in Physical Geography*, **15**(2), 127–156.

Rhoads, B. L., Wilson, D., Urban, M., Herricks, E. E., 1999. Interaction between scientists and nonscientists in community-based watershed management: emergence of the concept of stream naturalization. *Environmental Management*, **24**(3), 297–308.

Rhoads, B. L., Schwartz, J. S., Porter, S., 2003. Stream geomorphology, bank vegetation, and three-dimensional habitat hydraulics for fish in midwestern agricultural streams. *Water Resources Research*, **39**(8). 10.1029/2003WR002294

Rhoads, B .L., Garcia, M. H., Rodriguez, J., et al., 2008. Methods for evaluating the geomorphological performance of naturalized rivers: examples from the Chicago Metropolitan Area. In: S. E. Darby, D. A. Sear (eds.), *River Restoration: Managing for Uncertainty in Restoring Physical Habitat*. Wiley, Chichester, UK, pp. 209–228.

Rhoads, B. L., Riley, J. D., Mayer, D. R., 2009. Response of bed morphology and bed material texture to hydrological conditions at an asymmetrical stream confluence. *Geomorphology*, **109**(3–4), 161–173.

Rhoads, B. L., Engel, F. L., Abad, J. D., 2011. Pool-riffle design based on geomorphic principles for naturalizing straight channels. In: A. Simon, S. J. Bennett, J. M. Castro (eds.), *Stream Restoration in Dynamic Fluvial Systems: Scientific Approaches, Analyses, and Tools*. American Geophysical Union, Washington, DC, pp. 367–384.

Rhoads, B. L., Lewis, Q. W., Andresen, W., 2016. Historical changes in channel network extent and channel planform in an intensively managed landscape: natural versus human-induced effects. *Geomorphology*, **252**, 17–31. 10.1016/j.geomorph.2015.04.021.

Rhodes, D. D., 1977. The b-f-m diagram – graphical representation and interpretation of at-a-station hydraulic geometry. *American Journal of Science*, **277**(1), 73–96.

Rhodes, D. D., 1978. World-wide variations in hydraulic geometry exponents of stream channels – analysis and some observations – comments. *Journal of Hydrology*, **39**(1–2), 193–197.

Rhodes, D. D., 1987. The b-f-m diagram for downstream hydraulic geometry. *Geografiska Annaler Series A Physical Geography*, **69**(1), 147–161.

Rice, S., 1998. Which tributaries disrupt downstream fining along gravel-bed rivers? *Geomorphology*, **22**(1), 39–56.

Rice, S., 1999. The nature and controls on downstream fining within sedimentary links. *Journal of Sedimentary Research*, **69**(1), 32–39.

Rice, S. P., 2017. Tributary connectivity, confluence aggradation and network biodiversity. *Geomorphology*, **277**, 6–16.

Rice, S., Church, M., 1998. Grain size along two gravel-bed rivers: statistical variation, spatial pattern and sedimentary links. *Earth Surface Processes and Landforms*, **23**(4), 345–363.

Rice, S. P., Church, M., 2001. Longitudinal profiles in simple alluvial systems. *Water Resources Research*, **37**(2), 417–426.

Rice, S. P., Church, M., 2010. Grain-size sorting within river bars in relation to downstream fining along a wandering channel. *Sedimentology*, **57**(1), 232–251.

Rice, S. P., Kiffney, P., Greene, C., Pess, G. R., 2008. The ecological importance of tributaries and confluences. In: S. P. Rice, A. G. Roy, B. L. Rhoads (eds.), *River Confluences, Tributaries and the Fluvial Network*. Wiley, Chichester, UK, pp. 209–242.

Rice, S.P., Roy, A.G., Rhoads, B.L. (eds.), 2008. *River Confluences, Tributaries, and the Fluvial Network*. Wiley, Chichester, UK.

Rice, S. P., Church, M., Wooldridge, C. L., Hickin, E. J., 2009. Morphology and evolution of bars in a wandering gravel-bed river; lower Fraser river, British Columbia, Canada. *Sedimentology*, **56**(3), 709–736.

Richards, D., 2018. Three-dimensional flow, morphologic change, and sediment deposition and distribution of actively evolving neck cutoffs located on the White River, Arkansas. PhD Dissertation, Louisiana State University.

Richards, K. S., 1973. Hydraulic geometry and channel roughness – a nonlinear system. *American Journal of Science*, **273**(10), 877–896.

Richards, K. S., 1976a. Morphology of pool-riffle sequences. *Earth Surface Processes and Landforms*, **1**(1), 71–88.

Richards, K. S., 1976b. Channel width and the riffle pool sequence. *Geological Society of America Bulletin*, **87**(6), 883–890.

Richards, K. S., 1978. Channel geometry in the riffle-pool sequence. *Geografiska Annaler Series A Physical Geography*, **60**(1–2), 23–27.

Richards, K. S., 1980. A note on changes in channel geometry at tributary junctions. *Water Resources Research*, **16**(1), 241–244.

Richards, K. S., 1993. Sediment delivery and the drainage network. In: K. Beven, M. J. Kirkby (eds.), *Channel Network Hydrology*. Wiley and Sons, Chichester, UK, pp. 221–254.

Richards, K. S., 1996. Samples and cases: generalisation and explanation in geomorphology. In: B. L. Rhoads, C. E. Thorn (eds.), *The Scientific Nature of Geomorphology*. Wiley and Sons, Chichester, UK, pp. 171–190.

Richards, K. S., Clifford, N., 1991. Fluvial geomorphology – structured beds in gravelly rivers. *Progress in Physical Geography*, **15**(4), 407–422.

Richardson, W. R., Thorne, C. R., 1998. Secondary currents around braid bar in Brahmaputra River, Bangladesh. *Journal of Hydraulic Engineering*, **124**(3), 325–328.

Richardson, W. R., Thorne, C. R., 2001. Multiple thread flow and channel bifurcation in a braided river: Brahmaputra-Jamuna River, Bangladesh. *Geomorphology*, **38**(3–4), 185–196.

Rickenmann, D., Recking, A., 2011. Evaluation of flow resistance in gravel-bed rivers through a large field data set. *Water Resources Research*, **47**. 10.1029/2010wr009793.

Ridenour, G. S., Giardino, J. R., 1991. The statistical study of hydraulic geometry – a new direction for compositional data analysis. *Mathematical Geology*, **23**(3), 349–366.

Ridenour, G. S., Giardino, J. R., 1995. Logratio linear modeling of hydraulic geometry using indexes of flow resistance as covariates. *Geomorphology*, **14**(1), 65–72.

Rigby, J. R., Wren, D. G., Kuhnle, R. A., 2016. Passive acoustic monitoring of bed load for fluvial applications. *Journal of Hydraulic Engineering*, **142**(9). 10.1061/(asce)hy.1943-7900.0001122.

Righini, M., Surian, N., Wohl, E., et al., 2017. Geomorphic response to an extreme flood in two Mediterranean rivers (northeastern Sardinia, Italy): analysis of controlling factors. *Geomorphology*, **290**, 184–199.

Rigon, R., Rodriguez-Iturbe, I., Maritan, A., et al., 1996. On Hack's law. *Water Resources Research*, **32**(11), 3367–3374.

Riley, J. D., Rhoads, B. L., 2012. Flow structure and channel morphology at a natural confluent meander bend. *Geomorphology*, **163**, 84–98.

Riley, J. D., Rhoads, B. L., Parsons, D. R., Johnson, K. K., 2015. Influence of junction angle on three-dimensional flow structure and bed morphology at confluent meander bends during different hydrological conditions. *Earth Surface Processes and Landforms*, **40**(2), 252–271.

Riley, S. J., 1972. A comparison of morphometric measures of bankfull. *Journal of Hydrology*, **17**, 23–31.

Rinaldi, M., Wyzga, B., Surian, N., 2005. Sediment mining in alluvial channels: physical effects and management perspectives. *River Research and Applications*, **21**(7), 805–828.

Rinaldi, M., Mengoni, B., Luppi, L., Darby, S. E., Mosselman, E., 2008. Numerical simulation of hydrodynamics and bank erosion in a river bend. *Water Resources Research*, **44**(9). 10.1029/2008wr007008.

Rinaldi, M., Surian, N., Comiti, F., Bussettini, M., 2013. A method for the assessment and analysis of the hydromorphological condition of Italian streams: the Morphological Quality Index (MQI). *Geomorphology*, 180, 96–108. 10.1016/j.geomorph.2012.09.009.

Rinaldo, A., Dietrich, W. E., Rigon, R., Vogel, G. K., Rodriguez-Iturbe, I., 1995. Geomorphological signatures of varying climate. *Nature*, **374**(6523), 632–635.

Rinaldo, A., Rigon, R., Banavar, J. R., Maritan, A., Rodriguez-Iturbe, I., 2014. Evolution and selection of river networks: statics, dynamics, and complexity. *Proceedings of the National Academy of Sciences of the United States of America*, **111**(7), 2417–2424.

Riquier, J., Piegay, H., Lamouroux, N., Vaudor, L., 2017. Are restored side channels sustainable aquatic habitat features? Predicting the potential persistence of side channels as aquatic habitats based on their fine sedimentation dynamics. *Geomorphology*, **295**, 507–528.

Rittenour, T. M., Blum, M. D., Goble, R. J., 2007. Fluvial evolution of the lower Mississippi River valley during the last 100 k.y. glacial cycle: response to glaciation and sea-level change. *Geological Society of America Bulletin*, **119** (5–6),586–608.

Ritter, D. F., 1975. Stratigraphic implications of coarse-grained gravel deposited as overbank sediment, southern Illinois. *Journal of Geology*, **83**(5), 645–650.

Ritter, D. F., 1986. *Process Geomorphology*, 2nd Edition. W.C. Brown, Dubuque, IA.

Riviere, N., Wei, C., Kouyi, G. L., Momplot, A., Mignot, E., 2015. Mixing downstream of a 90° open channel junction. Proceedings of the 36th IAHR World Congress: Deltas of the Future and What Happens Upstream. International Association of Hydro-Environment Engineering and Research.

Robert, A., 2003. *River Processes: An Introduction to Fluvial Dynamics*. Arnold, London.

Robert, A., Roy, A. G., 1990. On the fractal interpretation of the mainstream length-drainage area relationship. *Water Resources Research*, **26**(5), 839–842.

Robertson, J. M., Augspurger, C. K., 1999. Geomorphic processes and spatial patterns of primary forest succession on the Bogue Chitto River, USA. *Journal of Ecology*, **87**(6), 1052–1063.

Robertson-Rintoul, M. S. E., Richards, K. S., 1993. Braided-channel pattern and palaeohydrology using an index of total sinuosity. In: J. L. Best, C. S. Bristow (eds.), *Braided Rivers*. Geological Society of London Special Publication No. 75, Geological Society, London, pp. 113–118.

Rodrigues, S., Mosselman, E., Claude, N., Wintenberger, C. L., Juge, P., 2015. Alternate bars in a sandy gravel bed river: generation, migration and interactions with superimposed dunes. *Earth Surface Processes and Landforms*, **40**(5), 610–628.

Rodriguez, J. F., Garcia, M. H., 2008. Laboratory measurements of 3-D flow patterns and turbulence in straight open channel with rough bed. *Journal of Hydraulic Research*, **46**(4), 454–465.

Rodriguez-Iturbe, I., 1993. The geomorphological unit hydrograph. In: K. Beven, M. J. Kirkby (eds.), *Channel Network Hydrology*. Wiley, Chichester, UK, pp. 43–68.

Rodriguez-Iturbe, I., Rinaldo, A., 1997. *Fractal River Basins*. Cambridge University Press, Cambridge, UK.

Rodriguez-Iturbe, I., Valdes, J. B., 1979. Geomorphologic structure of hydrologic response. *Water Resources Research*, **15**(6), 1409–1420.

Roehl, J. E., 1962. Sediment source areas, delivery ratios and influencing morphological factors. IAHS Publication No. 59, IAHS Press, Wallingford, UK, pp. 202–213.

Rogers, M. M., Moser, R. D., 1992. The 3-dimensional evolution of a plane mixing layer – the Kelvin-Helmholtz rollup. *Journal of Fluid Mechanics*, **243**, 183–226.

Roll-Hansen, N., 2017. A historical perspective on the distinction between basic and applied science. *Journal for General Philosophy of Science*, **48**(4), 535–551.

Roni, P., Beechie, T. (eds.), 2013. *Stream and Watershed Restoration: A Guide to Restoring Riverine Processes and Habitats*. Wiley-Blackwell, Chichester, UK.

Roni, P., Pess, G. R., Hanson, K., Pearsons, M., 2013a. Selecting appropriate stream and watershed restoration techniques. In:

P. Roni, T. Beechie (eds.), *Stream and Watershed Restoration: A Guide to Restoring Riverine Processes and Habitats*. Wiley-Blackwell, Chichester, UK, pp. 144–188.

Roni, P., Liermann, M., Muhar, S., Schmutz, S., 2013b. Monitoring and evaluation of restoration actions. In: P. Roni, T. Beechie (eds.), *Stream and Watershed Restoration: A Guide to Restoring Riverine Processes and Habitats*. Wiley-Blackwell, Chichester, UK, pp. 254–279.

Roodsari, B. K., Chandler, D. G., 2017. Distribution of surface imperviousness in small urban catchments predicts runoff peak flows and stream flashiness. *Hydrological Processes*, **31**(17), 2990–3002.

Roper, B. B., Buffington, J. M., Archer, E., Moyer, C., Ward, M., 2008. The role of observer variation in determining Rosgen stream types in northeastern Oregon mountain streams. *Journal of the American Water Resources Association*, **44**(2), 417–427.

Rosburg, T. T., Nelson, P. A., Bledsoe, B. P., 2017. Effects of urbanization on flow duration and stream flashiness: a case study of Puget Sound streams, western Washington, USA. *Journal of the American Water Resources Association*, **53**(2), 493–507.

Rosgen, D. L., 1994. A classification of natural rivers. *Catena*, **22**(3), 169–199.

Rosgen, D. L., 1996. *Applied River Morphology*. Wildland Hydrology, Pagosa Springs, CO.

Rosgen, D. L., 2007. Rosgen geomorphic channel design (Chapter 11). Natural Resources Conservation Service National Engineering Handbook Part 654 Stream Restoration Design, U.S. Department of Agriculture, Washington, DC.

Rosgen, D. L., 2009. *Watershed Assessment of River Stability and Sediment Supply (WARSSS)*, 2nd Edition. Wildland Hydrology, Pagosa Springs, CO.

Rosgen, D. L., 2011. Natural channel design: fundamental concepts, assumptions, and methods. In: A. Simon, S. J. Bennett, J. M. Castro (eds.), *Stream Restoration in Dynamic Fluvial Systems: Scientific Approaches, Analyses, and Tools*. American Geophysical Union, Washington, DC, pp. 69–93.

Rosgen, D. L., 2014. *River Stability Field Guide*, 2nd edition. Wildland Hydrology, Fort Collins, CO.

Rosso, R., 1984. Nash model relation to Horton order ratios. *Water Resources Research*, **20**(7), 914–920.

Rouse, H., 1939. Discussion of "Laboratory investigation of flume traction and transportation". *Transactions of the American Society of Civil Engineers*, **104**, 1303–1308.

Rowland, J. C., Lepper, K., Dietrich, W. E., Wilson, C. J., Sheldon, R., 2005. Tie channel sedimentation rates, oxbow formation age and channel migration rate from optically stimulated luminescence (OSL) analysis of floodplain deposits. *Earth Surface Processes and Landforms*, **30**(9), 1161–1179.

Rowland, J. C., Dietrich, W. E., Day, G., Parker, G., 2009. Formation and maintenance of single-thread tie channels entering floodplain lakes: observations from three diverse river systems. *Journal of Geophysical Research-Earth Surface*, **114**. 10.1029/2008jf001073.

Roy, A. G., 1983. Optimal angular geometry models of river branching. *Geographical Analysis*, **15**(2), 87–96.

Roy, A. G., Bergeron, N., 1990. Flow and particle paths in a natural river confluence with coarse bed material. *Geomorphology*, **3**, 99–112.

Roy, A. G., Roy, R., 1988. Short communication: changes in channel size at river confluences with coarse bed material. *Earth Surface Processes and Landforms*, **13**, 77–84.

Roy, A. G., Woldenberg, M. J., 1986. A model for changes in channel form at a river confluence. *Journal of Geology*, **94**(3), 402–411. 10.1086/629038.

Roy, A. G., Roy, R., Bergeron, N., 1988. Hydraulic geometry and changes in flow velocity at a river confluence with coarse bed material. *Earth Surface Processes and Landforms*, **13**, 583–598.

Roy, A. G., Buffin-Belanger, T., Deland, S., 1996. Scales of coherent turbulent flow structures in a gravel-bed river. In: P. J. Ashworth, S. J. Bennett, J. L. Best, S. McLelland (eds.), *Coherent Flow Structures in Open Channels*. John Wiley, New York, pp. 147–164.

Roy, A. G., Buffin-Belanger, T., Lamarre, H., Kirkbride, A. D., 2004. Size, shape and dynamics of large-scale turbulent flow structures in a gravel-bed river. *Journal of Fluid Mechanics*, **500**, 1–27.

Roy, N. G., Sinha, R., 2014. Effective discharge for suspended sediment transport of the Ganga River and its geomorphic implication. *Geomorphology*, **227**, 18–30.

Royall, D., 2013. Land use impacts on the hydrogeomorphology of small watersheds. In: J. C. Shroder (ed.), *Treatise on Geomorphology*, Vol. 13, Geomorphology of Human Disturbances, Climate Change, and Natural Hazards, L. A. James, C. P. Harden, J. J. Clague (vol. eds.). Academic Press, San Diego, CA, pp. 28–47.

Royden, L., Perron, J. T., 2013. Solutions of the stream power equation and application to the evolution of river longitudinal profiles. *Journal of Geophysical Research-Earth Surface*, **118**(2), 497–518. 10.1002/jgrf.20031.

Rozovskii, I.L., 1957. *Flow of Water in Bends of Open Channels*. Academy of Sciences of the Ukrainian S.S.R., Kiev. (Translated from Russian by Y. Prushansky, Israel Program for Scientific Translations, 1961.)

Rubey, W. W., 1933a. Settling velocities of gravel, sand, and silt particles. *American Journal of Science*, **25**(148), 325–338.

Rubey, W. W., 1933b. Equilibrium-conditions in debris-laden streams. *Transactions, American Geophysical Union*, **14**, 497–505.

Rubey, W. W., 1952. Geology and mineral resources of the Hardin and Brussels quadrangles (in Illinois). U.S. Geological Survey Professional Paper 218. U.S. Government Printing Office, Washington, DC.

Ruhe, R. V., 1952. Topographic discontinuities of the Des Moines Lobe. *American Journal of Science*, **250**(1), 46–56.

Russell, K. L., Vietz, G. J., Fletcher, T. D., 2017. Global sediment yields from urban and urbanizing watersheds. *Earth-Science Reviews*, **168**, 73–80.

Russell, K. L., Vietz, G. J., Fletcher, T. D., 2018. Urban catchment runoff increases bedload sediment yield and particle size in stream channels. *Anthropocene*, **23**, 53–66.

Rust, B. R., 1981. Sedimentation in an arid-zone anastomosing fluvial system – Cooper's Creek, central Australia. *Journal of Sedimentary Petrology*, **51**(3), 745–755.

Ruther, N., Olsen, N. R. B., 2007. Modelling free-forming meander evolution in a laboratory channel using three-dimensional computational fluid dynamics. *Geomorphology*, **89**(3–4), 308–319.

Rutherford, J. C., 1994. *River Mixing*. Wiley, Chichester, UK.

Sack, D., 2013. Geomorphology and nineteenth century explorations of the American West In: J. W. Shroder (ed.), *Treatise on Geomorphology*, Vol. 1, The Foundations of Geomorphology, A.R. Orme, D. Sack (vol. eds.). Elsevier, New York, pp. 53–63.

Saletti, M., Molnar, P., Zimmermann, A., Hassan, M. A., Church, M., 2015. Temporal variability and memory in sediment transport in an experimental step-pool channel. *Water Resources Research*, **51**(11), 9325–9337. 10.1002/2015wr016929.

Saletti, M., Molnar, P., Hassan, M. A., Burlando, P., 2016. A reduced-complexity model for sediment transport and step-pool morphology. *Earth Surface Dynamics*, **4**(3), 549–566.

Sambrook Smith, G. H., Ferguson, R. I., 1995. The gravel sand transition along river channels. *Journal of Sedimentary Research Section a – Sedimentary Petrology and Processes*, **65**(2), 423–430.

Sambrook Smith, G. H., Ferguson, R. I., 1996. The gravel-sand transition: flume study of channel response to reduced slope. *Geomorphology*, **16**(2), 147–159.

Sambrook Smith, G. H., Ashworth, P. J., Best, J. L., Woodward, J., Simpson, C. J., 2005. The morphology and facies of sandy braided rivers: some considerations of scale invariance. In: M. D. Blum, S. B. Marriott, S. F. Leclair (eds.), *Fluvial Sedimentology VII*, International Association of Sedimentologists Special Publication No. 35. Blackwell, Malden, MA, pp. 145–158.

Sambrook Smith, G. H., Ashworth, P. J., Best, J. L., Woodward, J., Simpson, C. J., 2006. The sedimentology and alluvial architecture of the sandy braided South Saskatchewan River, Canada. *Sedimentology*, **53**(2), 413–434.

Sambrook Smith, G. H., Best, J. L., Leroy, J. Z., Orfeo, O., 2016. The alluvial architecture of a suspended sediment dominated meandering river: the Rio Bermejo, Argentina. *Sedimentology*, **63**(5), 1187–1208.

Sandercock, P. J., Hooke, J. M., 2011. Vegetation effects on sediment connectivity and processes in an ephemeral channel in SE Spain. *Journal of Arid Environments*, **75**(3), 239–254.

Sangireddy, H., Carothers, R. A., Stark, C. P., Passalacqua, P., 2016a. Controls of climate, topography, vegetation, and lithology on drainage density extracted from high resolution topography data. *Journal of Hydrology*, **537**, 271–282.

Sangireddy, H., Stark, C. P., Kladzyk, A., Passalacqua, P., 2016b. GeoNet: an open source software for the automatic and objective extraction of channel heads, channel network, and channel morphology from high resolution topography data. *Environmental Modelling & Software*, **83**, 58–73.

Sapozhnikov, V., Foufoula-Georgiou, E., 1996. Self-affinity in braided rivers. *Water Resources Research*, **32**(5), 1429–1439.

Sapozhnikov, V. B., Foufoula-Georgiou, E., 1999. Horizontal and vertical self-organization of braided rivers toward a critical state. *Water Resources Research*, **35**(3), 843–851.

Saucier, R. T., 1994a. Evidence of late glacial runoff in the lower Mississippi Valley. *Quaternary Science Reviews*, **13**(9–10), 973–981.

Saucier, R. T., 1994b. *Geomorphology and Quaternary geologic history of the lower Mississippi Valley*. U.S. Army Corps of Engineers, Vicksburg, MS.

Sawyer, A. M., Pasternack, G. B., Moir, H. J., Fulton, A. A., 2010. Riffle-pool maintenance and flow convergence routing observed on a large gravel-bed river. *Geomorphology*, **114**(3), 143–160.

Scheidegger, A. E., 1967. A stochastic model for drainage patterns into an intramontane trench. *Bulletin for the Association of Scientific Hydrology*, **12**, 15–20.

Scheingross, J. S., Lamb, M. P., 2017a. A mechanistic model of waterfall plunge pool erosion into bedrock. *Journal of Geophysical Research-Earth Surface*, **122**(11), 2079–2104.

Scheingross, J. S., Lo, D. Y., Lamb, M. P., 2017b. Self-formed waterfall plunge pools in homogeneous rock. *Geophysical Research Letters*, **44**(1), 200–208.

Schick, A. P., Hassan, M. A., Lekach, J., 1987. A vertical exchange model for coarse bedload movement – numerical considerations. *Catena Supplement*, **10**, 73–83.

Schiefer, E., Hassan, M. A., Menounos, B., Pelpola, C. P., Slaymaker, O., 2010. Interdecadal patterns of total sediment yield from a montane catchment, southern Coast Mountains, British Columbia, Canada. *Geomorphology*, **118**(1–2), 207–212.

Schielen, R., Doelman, A., De Swart, H. E., 1993. On the nonlinear dynamics of free bars in straight channels. *Journal of Fluid Mechanics*, **252**, 325–356.

Schindfessel, L., Creelle, S., De Mulder, T., 2015. Flow patterns in an open channel confluence with increasingly dominant tributary inflow. *Water*, **7**(9), 4724–4751.

Schindfessel, L., Creelle, S., De Mulder, T., 2017. How different cross-sectional shapes influence the separation zone of an open-channel confluence. *Journal of Hydraulic Engineering*, **143**(9). 10.1061/(asce)hy.1943–7900.0001336.

Schlichting, H., Gersten, K., 2016. *Boundary-Layer Theory*. Springer, Berlin.

Schmeeckle, M. W., Nelson, J. M., Shreve, R. L., 2007. Forces on stationary particles in near-bed turbulent flows. *Journal of Geophysical Research-Earth Surface*, **112**(F2). 10.1029/2006jf000536.

Schmidt, J. C., Wilcock, P. R., 2008. Metrics for assessing the downstream effects of dams. *Water Resources Research*, **44**(4). 10.1029/2006wr005092.

Schmidt, K. H., Ergenzinger, P., 1992. Bedload entrainment, travel lengths, step lengths, rest periods – studied with passive (iron, magnetic) and active (radio) tracer techniques. *Earth Surface Processes and Landforms*, **17**(2), 147–165.

Schneider, A., Gerke, H. H., Maurer, T., Nenov, R., 2013. Initial hydro-geomorphic development and rill network evolution in an artificial catchment. *Earth Surface Processes and Landforms*, **38**(13), 1496–1512.

Schorghofer, N., Jensen, B., Kudrolli, A., Rothman, D. H., 2004. Spontaneous channelization in permeable ground: theory, experiment, and observation. *Journal of Fluid Mechanics*, **503**, 357–374.

Schroder, R., 1991. Test of Hack's slope to bed material relationship in the southern Eifel Uplands, Germany. *Earth Surface Processes and Landforms*, **16**(8), 731–736.

Schumann, R. R., 1989. Morphology of Red Creek, Wyoming, an arid-region anastomosing channel system. *Earth Surface Processes and Landforms*, **14**(4), 277–288.

Schumm, S. A., 1956. Evolution of drainage systems and slopes in badlands at Perth Amboy, New Jersey. *Geological Society of America Bulletin*, **67**(5), 597–646.

Schumm, S. A., 1960. The shape of alluvial channels in relation to sediment type. U.S. Geological Survey Professional Paper 352-B, U.S. Government Printing Office, Washington, DC.

Schumm, S. A., 1968. River adjustment to altered hydrologic regime – Murrumbidgee River and paleochannels. U.S. Geological Survey Professional Paper 598. U.S. Government Printing Office, Washington, DC.

Schumm, S. A., 1969. River metamorphosis. *Journal of the Hydraulics Division – ASCE*, HY1, 255–273.

Schumm, S. A., 1971. Fluvial geomorphology: channel adjustment and river metamorphosis. In: H. W. Shen (ed.), *River Mechanics*, Vol. 1. H. W. Shen, Ft. Collins, CO, pp. 5.1–5.22.

Schumm, S. A., 1977. *The Fluvial System*. Wiley, New York.

Schumm, S. A., 1979. Geomorphic thresholds – concepts and its application. *Transactions of the Institute of British Geographers*, **4**(4), 485–515.

Schumm, S. A., 1981. Evolution and response of the fluvial system, sedimentological implications. Society of Economic Paleontologists and Mineralogists Special Publication No. 31, 19–29.

Schumm, S.A., 1985. Patterns of alluvial rivers. *Annual Review of Earth and Planetary Sciences*, **13**, 5–27.

Schumm, S. A., 1993. River response to baselevel change – implications for sequence stratigraphy. *Journal of Geology*, **101**(2), 279–294.

Schumm, S. A., Khan, H. R., 1972. Experimental study of channel patterns. *Geological Society of America Bulletin*, **83**(6), 1755–1770.

Schumm, S. A., Lichty, R. W., 1963. Channel widening and flood-plain construction along the Cimarron River in southwestern Kansas. U.S. Geological Survey Professional Paper 352-D. U.S. Government Printing Office, Washington, DC.

Schumm, S. A., Lichty, R. W., 1965. Time, space, and causality in geomorphology. *American Journal of Science*, **263**(2), 110–119.

Schumm, S. A., Rea, D. K., 1995. Sediment yield from disturbed earth systems. *Geology*, **23**(5), 391–394.

Schumm, S. A., Stevens, M. A., 1973. Abrasion in place: a mechanism for rounding and size reduction of coarse sediment in rivers. *Geology*, **1**, 37–40.

Schumm, S. A., Harvey, M. D., Watson, C. C., 1984. *Incised Channels: Morphology, Dynamics and Controls*. Water Resources Publications, Littleton, CO.

Schumm, S. A., Mosley, M. P., Weaver, W. E., 1987. *Experimental Fluvial Geomorphology*. Wiley, Chichester, UK.

Schumm, S. A., Boyd, K. F., Wolff, C. G., Spitz, W. J., 1995. A groundwater sapping landscape in the Florida panhandle. *Geomorphology*, **12**(4), 281–297.

Schumm, S. A., Erskine, W. D., Tilleard, J. W., 1996. Morphology, hydrology, and evolution of the anastomosing Owens and King Rivers, Victoria, Australia. *Geological Society of America Bulletin*, **108**(10), 1212–1224.

Schuurman, F., Kleinhans, M. G., 2015. Bar dynamics and bifurcation evolution in a modelled braided sand-bed river. *Earth Surface Processes and Landforms*, **40**(10), 1318–1333.

Schuurman, F., Marra, W. A., Kleinhans, M. G., 2013. Physics-based modeling of large braided sand-bed rivers: bar pattern formation, dynamics, and sensitivity. *Journal of Geophysical Research-Earth Surface*, **118**(4), 2509–2527. 10.1002/2013jf002896.

Schuurman, F., Shimizu, Y., Iwasaki, T., Kleinhans, M. G., 2016. Dynamic meandering in response to upstream perturbations and floodplain formation. *Geomorphology*, **253**, 94–109.

Schwanghart, W., Scherler, D., 2017. Bumps in river profiles: uncertainty assessment and smoothing using quantile regression techniques. *Earth Surface Dynamics*, **5**(4), 821–839.

Schwartz, J. S., Herricks, E. E., 2007. Evaluation of pool-riffle naturalization structures on habitat complexity and the fish community in an urban Illinois stream. *River Research and Applications*, **23**(4), 451–466.

Schwendel, A. C., Nicholas, A. P., Aalto, R. E., Smith, G. H. S., Buckley, S., 2015. Interaction between meander dynamics and floodplain heterogeneity in a large tropical sandbed river: the Rio Beni, Bolivian Amazon. *Earth Surface Processes and Landforms*, **40**(15), 2026–2040.

Schwendel, A., Aalto, R., Nicholas, A. P., Parsons, D., 2018. Fill characteristics of abandoned river channels and resulting stratigraphy of a mobile sand-bed river floodplain. In: M. Ghinassi, L. Columbera, N. P. Mountney, A. J. H. Reesink (eds.), *Fluvial Meanders and their Sedimentary Products in the Rock Record*, International Association of Sedimentologists Special Publication 48. Wiley and Sons, Hoboken, NJ, pp. 251–272.

Schwenk, J., Foufoula-Georgiou, E., 2016. Meander cutoffs nonlocally accelerate upstream and downstream migration and channel widening. *Geophysical Research Letters*, **43**(24), 12437–12445.

Schwenk, J., Lanzoni, S., Foufoula-Georgiou, E., 2015. The life of a meander bend: connecting shape and dynamics via analysis of a numerical model. *Journal of Geophysical Research – Earth Surface*, **120**(4), 690–710. 10.1002/2014jf003252.

Schwenk, J., Khandelwal, A., Fratkin, M., Kumar, V., Foufoula-Georgiou, E., 2017. High spatiotemporal resolution of river planform dynamics from Landsat: the RivMAP toolbox and results from the Ucayali River. *Earth and Space Science*, **4**(2), 46–75. 10.1002/2016ea000196.

Schwindt, S., Pasternack, G. B., Bratovich, P. M., Rabone, G., Simodynes, D., 2019. Hydro-morphological parameters generate lifespan maps for stream restoration management. *Journal of Environmental Management*, **232**, 475–489.

Scorpio, V., Zen, S., Bertoldi, W., et al., 2018. Channelization of a large alpine river: what is left of its original morphodynamics? *Earth Surface Processes and Landforms*, **43** (5), 1044–1062.

Scown, M. W., Thoms, M. C., De Jager, N. R., 2015. Floodplain complexity and surface metrics: influences of scale and geomorphology. *Geomorphology*, **245**, 102–116.

Scown, M. W., Thoms, M. C., De Jager, N. R., 2016. An index of floodplain surface complexity. *Hydrology and Earth System Sciences*, **20**(1), 431–441.

Seal, R., Paola, C., 1995. Observations of downstream fining on the North Fork Toutle River near Mount St. Helens, Washington. *Water Resources Research*, **31**(5), 1409–1419.

Seal, R., Paola, C., Parker, G., Southard, J. B., Wilcock, P. R., 1997. Experiments on downstream fining of gravel. 1. Narrow-channel runs. *Journal of Hydraulic Engineering*, **123** (10), 874–884.

Sear, D. A., 1994. River restoration and geomorphology. *Aquatic Conservation – Marine and Freshwater Ecosystems*, **4**(2), 169–177.

Sear, D. A., 1996. Sediment transport processes in pool-riffle sequences. *Earth Surface Processes and Landforms*, **21**(3), 241–262.

Sear, D., Newson, M., Hill, C., Old, J., Branson, J., 2009. A method for applying fluvial geomorphology in support of catchment-scale river restoration planning. *Aquatic Conservation – Marine and Freshwater Ecosystems*, **19**(5), 506–519.

Sear, D. A., Millington, C. E., Kitts, D. R., Jeffries, R., 2010. Logjam controls on channel: floodplain interactions in wooded catchments and their role in the formation of multi-channel patterns. *Geomorphology*, **116**(3–4), 305–319.

Searcy, J. K., 1959. Flow-duration curves. U.S. Geological Survey Water-supply Paper 1542-A. U.S. Government Printing Office, Washington, DC.

Selby, M., 1985. *Earth's Changing Surface: An Introduction to Geomorphology*. Clarendon Press, Oxford, UK.

Selley, R. C., 2000. *Applied Sedimentology*. Academic Press, San Diego, CA.

Seminara, G., Tubino, M., 1989. Alternate bars and meandering: free, forced and mixed interactions. In: S. Ikeda, G. Parker (eds.), *River Meandering*. Water Resources Monograph 12. American Geophysical Union, Washington, DC, pp. 267–320.

Seminara, G., Tubino, M., 1992. Weakly nonlinear theory of regular meanders. *Journal of Fluid Mechanics*, **244**, 257–288.

Seminara, G., Zolezzi, G., Tubino, M., Zardi, D., 2001. Downstream and upstream influence in river meandering. Part 2. Planimetric development. *Journal of Fluid Mechanics*, **438**, 213–230.

Sennatt, K. M., Salant, N. L., Renshaw, C. E., Magilligan, F. J., 2006. Assessment of methods for measuring embeddedness: application to sedimentation in flow regulated streams. *Journal of the American Water Resources Association*, **42**(6), 1671–1682.

Seto, K. C., Fragkias, M., Guneralp, B., Reilly, M. K., 2011. A meta-analysis of global urban land expansion. *PLoS ONE*, **6**(8). 10.1371/journal.pone.0023777.

Seybold, H., Rothman, D. H., Kirchner, J. W., 2017. Climate's watermark in the geometry of stream networks. *Geophysical Research Letters*, **44**(5), 2272–2280. 10.1002/2016gl072089.

Shabayek, S., Steffler, P., Hicks, F., 2002. Dynamic model for subcritical combining flows in channel junctions. *Journal of Hydraulic Engineering*, **128**, 821–828.

Shakibainia, A., Tabatabai, M. R. M., Zarrati, A. R., 2010. Three-dimensional numerical study of flow structure in channel confluences. *Canadian Journal of Civil Engineering*, **37**(5), 772–781.

Shchepetkina, A., Gingras, M. K., Pemberton, S. G., 2015. The removal-cap suction corer: an inexpensive and durable device to extract unconsolidated, wet sediments. *Journal of Sedimentary Research*, **85**(12), 1431–1437. 10.2110/jsr.2015.91.

Shelef, E., Hilley, G. E., 2013. Impact of flow routing on catchment area calculations, slope estimates, and numerical simulations of landscape development. *Journal of Geophysical Research-Earth Surface*, **118**(4), 2105–2123. 10.1002/jgrf.20127.

Shen, C., Wang, S., Liu, X., 2016. Geomorphological significance of at-many-stations hydraulic geometry. *Geophysical Research Letters*, **43**(8), 3762–3770. 10.1002/2016gl068364.

Shen, H., Zheng, F., Wen, L., Lu, J., Jiang, Y., 2015a. An experimental study of rill erosion and morphology. *Geomorphology*, **231**, 193–201.

Shen, Z., Tornqvist, T. E., Mauz, B., et al., 2015b. Episodic overbank deposition as a dominant mechanism of floodplain and delta-plain aggradation. *Geology*, **43**(10), 875–878.

Shepherd, R. G., 1985. Regression analysis of river profiles. *Journal of Geology*, **93**(3), 377–384.

Sherriff, S. C., Rowan, J. S., Fenton, O., et al., 2016. Storm event suspended sediment-discharge hysteresis and controls in agricultural watersheds: implications for watershed scale sediment management. *Environmental Science & Technology*, **50**(4), 1769–1778.

Shields, F. D., Jr., Abt, S. R., 1989. Sediment deposition in cutoff meander bends and implications for effective management. *Regulated Rivers Research and Management*, **4**(4), 381–396.

Shields, F. D., Jr., Copeland, R. R., Klingeman, P. C., Doyle, M. W., Simon, A., 2003. Design for stream restoration. *Journal of Hydraulic Engineering*, **129**(8), 575–584.

Shields, F. D., Jr., Copeland, R. R., Klingeman, P. C., Doyle, M. W., Simon, A., 2008. Stream restoration. In: M. H. Garcia (ed.), *Sedimentation Engineering: Processes, Measurements, Modeling, and Practice*. American Society of Civil Engineers, New York, pp. 461–503.

Shih, W., Diplas, P., Celik, A. O., Dancey, C., 2017. Accounting for the role of turbulent flow on particle dislodgement via a coupled quadrant analysis of velocity and pressure sequences. *Advances in Water Resources*, **101**, 37–48.

Shiklomanov, I.A., 1993. World fresh water resources. In: P. Gleick (ed.), *Water in Crisis*. Oxford University Press, New York, pp. 13–23.

Shiono, K., Knight, D. W., 1991. Turbulent open channel flows with variable depth across the channel. *Journal of Fluid Mechanics*, **222**, 617–646.

Sholtes, J. S., Bledsoe, B. P., 2016. Half-yield discharge: process-based predictor of bankfull discharge. *Journal of Hydraulic Engineering*, **142**(8). 10.1061/(asce)hy.1943-7900.0001137.

Sholtes, J., Werbylo, K., Bledsoe, B., 2014. Physical context for theoretical approaches to sediment transport magnitude-frequency analysis in alluvial channels. *Water Resources Research*, **50**(10), 7900–7914. 10.1002/2014wr015639.

Shreve, R. L., 1966. Statistical law of stream numbers. *Journal of Geology*, **74**(1), 17–37.

Shreve, R. L., 1967. Infinite topologically random channel networks. *Journal of Geology*, **75**(2), 178–186.

Shulits, S., 1941. Rational equation of river-bed profile. *Transactions, American Geophysical Union*, **22**, 622–630.

Shvidchenko, A. B., Pender, G., 2000. Flume study of the effect of relative depth on the incipient motion of coarse uniform sediments. *Water Resources Research*, **36**(2), 619–628.

Shvidchenko, A. B., Pender, G., 2001b. Macroturbulent structure of open-channel flow over gravel beds. *Water Resources Research*, **37**(3), 709–719.

Shvidchenko, A. B., Pender, G., Hoey, T. B., 2001a. Critical shear stress for incipient motion of sand/gravel streambeds. *Water Resources Research*, **37**(8), 2273–2283.

Sichingabula, H. M., 1999. Magnitude-frequency characteristics of effective discharge for suspended sediment transport, Fraser River, British Columbia, Canada. *Hydrological Processes*, **13**(9), 1361–1380.

Simoes, F. J. M., 2014. Shear velocity criterion for incipient motion of sediment. *Water Science and Engineering*, 7, 183–193.

Simon, A., 1989a. A model of channel response in disturbed alluvial channels. *Earth Surface Processes and Landforms*, **14**(1), 11–26.

Simon, A., 1989b. The discharge of sediment in channelized alluvial streams. *Water Resources Bulletin*, **25**(6), 1177–1188.

Simon, A., 1992. Energy, time, and channel evolution in catastrophically disturbed fluvial systems. *Geomorphology*, **5**(3–5), 345–372.

Simon, A., Collison, A. J. C., 2002. Quantifying the mechanical and hydrologic effects of riparian vegetation on streambank stability. *Earth Surface Processes and Landforms*, **27**(5), 527–546.

Simon, A., Rinaldi, M., 2000. Channel instability in the loess area of the midwestern United States. *Journal of the American Water Resources Association*, **36**(1), 133–150.

Simon, A., Rinaldi, M., 2006. Disturbance, stream incision, and channel evolution: the roles of excess transport capacity and boundary materials in controlling channel response. *Geomorphology*, **79**(3–4), 361–383.

Simon, A., Rinaldi, M., 2013. Incised channels: disturbance, evolution and the roles of excess transport capacity and boundary materials in controlling channel response. In: J. C. Shroder (ed.), *Treatise on Geomorphology*, Vol. 9, Fluvial Geomorphology, E. Wohl (vol. ed.), Academic Press, San Diego, CA, pp. 574–594.

Simon, A., Dickerson, W., Heins, A., 2004. Suspended-sediment transport rates at the 1.5-year recurrence interval for ecoregions of the United States: transport conditions at the bankfull and effective discharge. *Geomorphology*, **58** (1–4),243–262.

Simon, A., Doyle, M., Kondolf, M., et al., 2007. Critical evaluation of how the Rosgen classification and associated "natural channel design" methods fail to integrate and quantify fluvial processes and channel response. *Journal of the American Water Resources Association*, **43**(5), 1117–1131.

Simons, D. B., Richardson, E. V., 1966. Resistance to flow in alluvial channels. U.S. Geological Survey Professional Paper 422-J. U.S. Government Printing Office, Washington, DC.

Simpson, G., Schlunegger, F., 2003. Topographic evolution and morphology of surfaces evolving in response to coupled fluvial and hillslope sediment transport. *Journal of Geophysical Research – Solid Earth*, **108**(B6), 10.1029/2002jb002162.

Singer, M. B., 2008a. Downstream patterns of bed material grain size in a large, lowland alluvial river subject to low sediment supply. *Water Resources Research*, **44**(12). 10.1029/2008wr007183.

Singer, M.B., 2008b. A new sampler for extracting bed material sediment from sand and gravel beds in navigable rivers. *Earth Surface Processes and Landforms*, **33**(14), 2277–2284.

Singh, U., Crosato, A., Giri, S., Hicks, M., 2017. Sediment heterogeneity and mobility in the morphodynamic modelling of gravel-bed braided rivers. *Advances in Water Resources*, **104**, 127–144.

Singh, V. P., 2003. On the theories of hydraulic geometry. *International Journal of Sediment Research*, **18**, 196–218.

Singh, V. P., Zhang, L., 2008a. At-a-station hydraulic geometry relations, 1: theoretical development. *Hydrological Processes*, **22**(2), 189–215.

Singh, V.P., Zhang, L., 2008b. At-a-station hydraulic geometry relations, 2: calibration and testing. *Hydrological Processes*, **22**(2), 216–228.

Singh, V. P., Yang, C. T., Deng, Q.-Z., 2003. Downstream hydraulic geometry relations: 2. calibration and testing. *Water Resources Research*, **12**. 10.1029/2003WR002498.

Sinha, S. K., Parker, G., 1996. Causes of concavity in longitudinal profiles of rivers. *Water Resources Research*, **32**(5), 1417–1428.

Siviglia, A., Stecca, G., Vanzo, D., et al., 2013. Numerical modelling of two-dimensional morphodynamics with applications to river bars and bifurcations. *Advances in Water Resources*, **52**, 243–260.

Skalak, K., Pizzuto, J., Hart, D. D., 2009. Influence of small dams on downstream channel characteristics in Pennsylvania and Maryland: implications for the long-term geomorphic effects of dam removal. *Journal of the American Water Resources Association*, **45**(1), 97–109.

Sklar, L., Dietrich, W. E., 1998. River longitudinal profiles and bedrock incision models: stream power and the influence of sediment supply. In: K. Tinkler, E. E. Wohl (eds.), *Rivers over Rock: Fluvial Process in Bedrock Channels*, 107. American Geophysical Union, Washington, DC, pp. 237–260.

Sklar, L. S., Dietrich, W. E., 2001. Sediment and rock strength controls on river incision into bedrock. *Geology*, **29**(12), 1087–1090.

Sklar, L. S., Dietrich, W. E., 2004. A mechanistic model for river incision into bedrock by saltating bed load. *Water Resources Research*, **40**(6). 10.1029/2003wr002496.

Sklar, L. S., Dietrich, W. E., 2008. Implications of the saltation-abrasion bedrock incision model for steady-state river longitudinal profile relief and concavity. *Earth Surface Processes and Landforms*, **33**(7), 1129–1151.

Sklar, L. S., Dietrich, W. E., 2012. Correction to "A mechanistic model for river incision into bedrock by saltating bed load". *Water Resources Research*, **48**. 10.1029/2012wr012267.

Skolasinska, K., 2014. Inquiry of levee formation by grain size analysis – a case study from the Warta River (central Poland). *Catena*, **122**, 103–110.

Slater, L. J., Singer, M. B., 2013. Imprint of climate and climate change in alluvial riverbeds: continental United States, 1950–2011. *Geology*, **41**(5), 595–598.

Slattery, M. C., Bryan, R. B., 1992. Hydraulic conditions for rill incision under simulated rainfall – a laboratory experiment. *Earth Surface Processes and Landforms*, **17**(2), 127–146.

Slaymaker, O., 2003. The sediment budget as conceptual framework and management tool. *Hydrobiologia*, **494**(1–3), 71–82.

Slaymaker, O., 2006. Towards the identification of scaling relations in drainage basin sediment budgets. *Geomorphology*, **80**(1–2), 8–19.

Slingerland, R., Smith, N. D., 1998. Necessary conditions for a meandering-river avulsion. *Geology*, **26**(5), 435–438.

Slowik, M., 2018. The formation of an anabranching planform in a sandy floodplain by increased flows and sediment load. *Earth Surface Processes and Landforms*, **43**(3), 623–638.

Small, M. J., Doyle, M. W., 2012. Historical perspectives on river restoration design in the USA. *Progress in Physical Geography*, **36**(2), 138–153.

Smart, J. S., 1968. Statistical properties of stream lengths. *Water Resources Research*, **4**(5), 1001–1014.

Smith, B., Clifford, N. J., Mant, J., 2014. The changing nature of river restoration. *Wiley Interdisciplinary Reviews – Water*, **1**(3), 249–261.

Smith, C. E., 1998. Modeling high sinuosity meanders in a small flume. *Geomorphology*, **25**(1–2), 19–30.

Smith, D. G., 1976. Effect of vegetation on lateral migration of anastomosed channels of a glacier meltwater river. *Geological Society of America Bulletin*, **87**(6), 857–860.

Smith, D. G., 1983. Anastomosed fluvial deposits: examples from western Canada. In: J. D. Collinson, J. Lewin (eds.), *Modern and Ancient Fluvial Systems*, Special Publication of the International Association of Sedimentologists No. 6, Blackwell, Oxford, UK, pp. 155–168.

Smith, D. G., 1984. Vibracoring fluvial and deltaic sediments – tips on improving penetration and recovery. *Journal of Sedimentary Petrology*, **54**(2), 660–663.

Smith, D. G., 1986. Anastomosing river deposits, sedimentation rates and basin subsidence, Magdalena River, northwestern Columbia, South America. *Sedimentary Geology*, **46**(3–4), 177–196.

Smith, D., Elmore, A. C., 2014. A modification of freeze-core technology for collecting granular fluvial sediment samples. *Environmental Earth Sciences*, **71**(9), 4149–4156.

Smith, D. G., Smith, N. D., 1980. Sedimentation in anastomosed river systems: examples from alluvial valleys near Banff, Alberta. *Journal of Sedimentary Petrology*, **50**, 157–164.

Smith, D. G., Hubbard, S. M., Leckie, D. A., Fustic, M., 2009a. Counter point bar deposits: lithofacies and reservoir significance in the meandering modern Peace River and ancient McMurray Formation, Alberta, Canada. *Sedimentology*, **56**(6), 1655–1669. 10.1111/j.1365–3091.2009.01050.x.

Smith, L. C., Isacks, B. L., Bloom, A. L., Murray, A. B., 1996. Estimation of discharge from three braided rivers using synthetic aperture radar satellite imagery: potential application to ungaged basins. *Water Resources Research*, **32**(7), 2021–2034.

Smith, N. D., 1974. Sedimentology and bar formation in upper Kicking Horse River, a braided outwash stream. *Journal of Geology*, **82**(2), 205–223.

Smith, N. D., Perez-Arlucea, M., 1994. Fine-grained splay deposition in the avulsion belt of the lower Saskatchewan River, Canada. *Journal of Sedimentary Research Section B – Stratigraphy and Global Studies*, **64**(2), 159–168.

Smith, N. D., Perez-Arlucea, M., 2008. Natural levee deposition during the 2005 flood of the Saskatchewan River. *Geomorphology*, **101**(4), 583–594.

Smith, N. D., Smith, D. G., 1984. William River – an outstanding example of channel widening and braiding caused by bedload addition. *Geology*, **12**(2), 78–82.

Smith, N. D., Cross, T. A., Dufficy, J. P., Clough, S. R., 1989. Anatomy of an avulsion. *Sedimentology*, **36**(1), 1–23.

Smith, N. D., McCarthy, T. S., Ellery, W. N., Merry, C. L., Ruther, H., 1997. Avulsion and anastomosis in the panhandle region of the Okavango Fan, Botswana. *Geomorphology*, **20**(1–2), 49–65.

Smith, N. D., Slingerland, R. L., Perez-Arlucea, M., Morozova, G. S., 1998. The 1870s avulsion of the Saskatchewan River. *Canadian Journal of Earth Sciences*, **35**(4), 453–466. 10.1139/e97-113.

Smith, N. D., Perez-Arlucea, M., Edmonds, D. A., Slingerland, R. L., 2009b. Elevation adjustments of paired natural levees during flooding of the Saskatchewan River. *Earth Surface Processes and Landforms*, **34**(8), 1060–1068. 10.1002/esp.1792.

Smith, S. M., Prestegaard, K. L., 2005. Hydraulic performance of a morphology-based stream channel design. *Water Resources Research*, **41**(11). 10.1029/2004wr003926.

Smith, T. R., 2010. A theory for the emergence of channelized drainage. *Journal of Geophysical Research – Earth Surface*, **115**. 10.1029/2008jf001114.

Smith, T. R., Bretherton, F. P., 1972. Stability and the conservation of mass in drainage basin evolution. *Water Resources Research*, **8**, 1506–1529.

Sneddon, C. S., Magilligan, F. J., Fox, C. A., 2017. Science of the dammed: expertise and knowledge claims in contested dam removals. *Water Alternatives-an Interdisciplinary Journal on Water Politics and Development*, **10**(3), 677–696.

Snow, R. S., Slingerland, R. L., 1987. Mathematical modeling of graded river profiles. *Journal of Geology*, **95**(1), 15–33.

Snyder, N. P., Whipple, K. X., Tucker, G. E., Merritts, D. J., 2003. Importance of a stochastic distribution of floods and erosion thresholds in the bedrock river incision problem. *Journal of Geophysical Research – Solid Earth*, **108**(B2). 10.1029/2001jb001655.

Soar, P. J., Thorne, C. R., 2011. Design discharge for river restoration. In: A. Simon, S. J. Bennett, J. M. Castro (eds.), *Stream Restoration in Dynamic Fluvial Systems: Scientific Approaches, Analyses, and Tools*. American Geophysical Union, Washington, DC, pp. 123–149.

Song, X., Bai, Y., 2015. A new empirical river pattern discriminant method based on flow resistance characteristics. *Catena*, **135**, 163–172.

Soong, D. T., Prater, C. D., Halfar, T. M., Wobig, L. A., 2012. Manning's roughness coefficients for Illinois streams. U.S. Geological Survey Data Series 668. U.S. Geological Survey, Reston, VA.

Spink, A., Hillman, M., Fryirs, K., Brierley, G., Lloyd, K., 2010. Has river rehabilitation begun? Social perspectives from the Upper Hunter catchment, New South Wales, Australia. *Geoforum*, **41** (3), 399–409.

Stanistreet, I. G., McCarthy, T. S., 1993. The Okavango Fan and the classification of subaerial fan systems. *Sedimentary Geology*, **85** (1–4), 115–133.

Stanistreet, I. G., Cairncross, B., McCarthy, T. S., 1993. Low sinuosity and meandering bedload rivers of the Okavango Fan: channel confinement by vegetated levees. *Sedimentary Geology*, **85**(1–4), 135–156.

Stanley, E. H., Doyle, M. W., 2003. Trading off: the ecological removal effects of dam removal. *Frontiers in Ecology and the Environment*, **1**(1), 15–22.

Stark, C. P., 1991. An invasion percolation model of drainage network evolution. *Nature*, **352**(6334), 423–425.

Stark, C. P., Barbour, J. R., Hayakawa, Y. S., et al., 2010. The climatic signature of incised river meanders. *Science*, **327**(5972), 1497–1501. 10.1126/science.1184406.

Stecca, G., Measures, R., Hicks, D. M., 2017. A framework for the analysis of noncohesive bank erosion algorithms in morphodynamic modeling. *Water Resources Research*, **53**(8), 6663–6686. 10.1002/2017wr020756.

Steffen, W., Persson, A., Deutsch, L., et al., 2011. The Anthropocene: from global change to planetary stewardship. *Ambio*, **40**(7), 739–761.

Steffen, W., Broadgate, W., Deutsch, L., Gaffney, O., Ludwig, C., 2015. The trajectory of the Anthropocene: the Great Acceleration. *Anthropocene Review*, **2**(1), 81–98.

Stein, O. R., Julien, P. Y., 1993. Criterion delineating the mode of the headcut migration. *Journal of Hydraulic Engineering*, **119**(1), 37–50.

Stein, O. R., LaTray, D. A., 2002. Experiments and modeling of head cut migration in stratified soils. *Water Resources Research*, **38** (12). 10.1029/2001wr001166.

Sternberg, H., 1875. Untersuchen uber Langen- und Querprofile geschiebefuhrende Flusse. *Zeitschrift Fur Bauwesen*, **25**, 483–506.

Stevaux, J. C., Souza, I. A., 2004. Floodplain construction in an anastomosed river. *Quaternary International*, **114**, 55–65.

Stevens, A. J., Clarke, D., Nicholls, R. J., 2016. Trends in reported flooding in the UK: 1884–2013. *Hydrological Sciences Journal*, **61**(1), 50–63.

Stevens, M. A., Simons, D. B., Richardson, E. V., 1975. Nonequilibrium river form. *Journal of the Hydraulics Division-ASCE*, **101**(NHY5), 557–566.

Stewardson, M., 2005. Hydraulic geometry of stream reaches. *Journal of Hydrology*, **306**(1–4), 97–111.

Stewardson, M., Rutherford, I., 2008. Conceptual and mathematical modelling in river restoration: do we have unreasonable confidence? In: S. E. Darby, D. A. Sear (eds.), *River Restoration: Managing for Uncertainty in Restoring Physical Habitat*. Wiley, Chichester, UK, pp. 61–78.

Stolum, H. H., 1996. River meandering as a self-organization process. *Science*, **271**(5256), 1710–1713.

Stout, J. C., Belmont, P., 2014. TerEx Toolbox for semi-automated selection of fluvial terrace and floodplain features from lidar. *Earth Surface Processes and Landforms*, **39**(5), 569–580.

Stout, J. C., Belmont, P., Schottler, S. P., Willenbring, J. K., 2014. Identifying sediment sources and sinks in the Root River, southeastern Minnesota. *Annals of the Association of American Geographers*, **104**(1), 20–39.

Stouthamer, E., Berendsen, H. J. A., 2001. Avulsion frequency, avulsion duration, and interavulsion period of Holocene channel belts in the Rhine-Meuse Delta, the Netherlands. *Journal of Sedimentary Research*, **71**(4), 589–598.

Stouthamer, E., Berendsen, H. J. A., 2007. Avulsion: the relative roles of autogenic and allogenic processes. *Sedimentary Geology*, **198** (3–4), 309–325.

Strahler, A. N., 1952a. Dynamic basis of geomorphology. *Geological Society of America Bulletin*, **63**, 923–938.

Strahler, A. N., 1952b. Hypsometric (area-altitude) analysis of erosional topography. *Geological Society of America Bulletin*, **63** (11), 1117–1142.

Strahler, A. N., 1965. *Introduction to Physical Geography*. Wiley, New York.

Stranko, S. A., Hilderbrand, R. H., Palmer, M. A., 2012. Comparing the fish and benthic macroinvertebrate diversity of restored urban streams to reference streams. *Restoration Ecology*, **20**(6), 747–755.

Strick, R. J. P., Ashworth, P. J., Awcock, G., Lewin, J., 2018. Morphology and spacing of river meander scrolls. *Geomorphology*, **310**, 57–68.

Strom, K. B., Papanicolaou, A. N., 2008. Morphological characterization of cluster microforms. *Sedimentology*, **55**(1), 137–153.

Stubblefield, A. P., Reuter, J. E., Goldman, C. R., 2009. Sediment budget for subalpine watersheds, Lake Tahoe, California, USA. *Catena*, **76**(3), 163–172.

Sui, B., Huang, S.-h., 2017. Numerical analysis of flow separation zone in a confluent meander bend channel. *Journal of Hydrodynamics*, **29**(4), 716–723.

Sukhodolov, A. N., 2015. Field based research in fluvial hydraulics: potential, paradigms and challenges. *Journal of Hydraulic Research*, **53**(1), 1–19.

Sukhodolov, A. N., Rhoads, B. L., 2001. Field investigation of three-dimensional flow structure at stream confluences 2. Turbulence. *Water Resources Research*, **37**(9), 2411–2424.

Sukhodolov, A. N., Sukhodolova, T. A., 2019. Dynamics of flow at concordant gravel bed river confluences: effects of junction angle and momentum flux ratio. *Journal of Geophysical Research – Earth Surface*, **124**(2), 588–615. 10.1029/2018jf004648.

Sukhodolov, A., Thiele, M., Bungartz, H., 1998. Turbulence structure in a river reach with sand bed. *Water Resources Research*, **34**(5), 1317–1334.

Sukhodolov, A. N., Schnauder, I., Uijttewaal, W. S. J., 2010. Dynamics of shallow lateral shear layers: experimental study in a river with a sandy bed. *Water Resources Research*, **46**. 10.1029/2010wr009245.

Sukhodolov, A. N., Krick, J., Sukhodolova, T. A., et al., 2017. Turbulent flow structure at a discordant river confluence: asymmetric jet dynamics with implications for channel morphology. *Journal of Geophysical Research – Earth Surface*, **122**(6), 1278–1293. 10.1002/2016jf004126.

Summerfield, M. A., Hulton, N. J., 1994. Natural controls of fluvial denudation rates in major world drainage basins. *Journal of Geophysical Research – Solid Earth*, **99**(B7), 13871–13883.

Sun, T., Meakin, P., Jossang, T., Schwarz, K., 1996. A simulation model for meandering rivers. *Water Resources Research*, **32**(9), 2937–2954.

Sun, T., Meakin, P., Jossang, T., 2001. A computer model for meandering rivers with multiple bed load sediment sizes 2. Computer simulations. *Water Resources Research*, **37**(8), 2243–2258.

Surian, N., 2006. Effects of human impact on braided river morphology: examples from northern Italy. In: G. H. Sambrook Smith, J. L. Best, C. S. Bristow, G. E. Petts (eds.), *Braided Rivers: Process, Deposits, Ecology and Management*, International Association of Sedimentologists Special Publication 36. Blackwell, Oxford, UK, pp. 327–338.

Sutherland, A. J., 1987. Static armor layers by progressive erosion. In: C. R. Thorne, J.C. Bathurst, and R.D. Hey (eds.), *Sediment Transport in Gravel-bed Rivers*. Wiley, Chichester, UK, pp. 243–260.

Swanson, B. J., Meyer, G., 2014. Tributary confluences and discontinuities in channel form and sediment texture: Rio Chama, NM. *Earth Surface Processes and Landforms*, **39**(14), 1927–1943.

Swanson, M. L., Kondolf, G. M., Boison, P. J., 1989. An example of rapid gully initiation and extension by subsurface erosion: coastal San Mateo County, California. *Geomorphology*, **2**(4), 393–403.

Sweet, W. V., Geratz, J. W., 2003. Bankfull hydraulic geometry relationships and recurrence intervals for North Carolina's Coastal Plain. *Journal of the American Water Resources Association*, **39**(4), 861–871.

Sylvester, Z., Durkin, P., Covault, J. A., 2019. High curvatures drive river meandering. *Geology*, **47**(3), 263–266.

Syvitski, J. P. M., 2003. Sediment fluxes and rates of sedimentation. In: G. Middleton (ed.), *Encyclopedia of Sediments and Sedimentary Rocks*. Kluwer Academic Publishers, Dordrecht, Netherlands, pp. 600–606.

Syvitski, J. P. M., Kettner, A., 2011. Sediment flux and the Anthropocene. *Philosophical Transactions of the Royal Society A – Mathematical Physical and Engineering Sciences*, **369**(1938), 957–975.

Syvitski, J. P. M., Milliman, J. D., 2007. Geology, geography, and humans battle for dominance over the delivery of fluvial sediment to the coastal ocean. *Journal of Geology*, **115**(1), 1–19.

Syvitski, J. P., Morehead, M. D., 1999. Estimating river-sediment discharge to the ocean: application to the Eel margin, northern California. *Marine Geology*, **154**(1–4), 13–28.

Syvitski, J. P. M., Saito, Y., 2007. Morphodynamics of deltas under the influence of humans. *Global and Planetary Change*, **57**(3–4), 261–282.

Syvitski, J. P. M., Peckham, S. D., Hilberman, R., Mulder, T., 2003. Predicting the terrestrial flux of sediment to the global ocean: a planetary perspective. *Sedimentary Geology*, **162**(1–2), 5–24.

Syvitski, J. P. M., Vorosmarty, C. J., Kettner, A. J., Green, P., 2005. Impact of humans on the flux of terrestrial sediment to the global coastal ocean. *Science*, **308**(5720), 376–380.

Syvitski, J. P. M., Kettner, A. J., Overeem, I., et al., 2009. Sinking deltas due to human activities. *Nature Geoscience*, **2**(10), 681–686.

Syvitski, J. P., Kettner, A., Overeem, I., Brakenridge, G. R., Cohen, S., 2019. Latitudinal controls on siliciclastic sediment production and transport, Latitudinal Controls on Stratigraphic Models and Sedimentary Concepts, SEPM Special Publication 108, 14–28.

Szupiany, R. N., Amsler, M.L., Parsons, D. R., Best, J. L., 2009. Morphology, flow structure, and suspended bed sediment transport at two large braid-bar confluences. *Water Resources Research*, **45**. 10.1029/2008wr007428.

Szupiany, R. N., Amsler, M. L., Hernandez, J., et al., 2012. Flow fields, bed shear stresses, and suspended bed sediment dynamics in bifurcations of a large river. *Water Resources Research*, **48**. 10.1029/2011wr011677.

Tabata, K. K., Hickin, E. J., 2003. Interchannel hydraulic geometry and hydraulic efficiency of the anastomosing Columbia River, southeastern British Columbia, Canada. *Earth Surface Processes and Landforms*, **28**(8), 837–852.

Tadaki, M., Brierley, G., Cullum, C., 2014. River classification: theory, practice, politics. *Wiley Interdisciplinary Reviews – Water*, **1**(4), 349–367.

Tal, M., Paola, C., 2007. Dynamic single-thread channels maintained by the interaction of flow and vegetation. *Geology*, **35**(4), 347–350.

Tal, M., Paola, C., 2010. Effects of vegetation on channel morphodynamics: results and insights from laboratory experiments. *Earth Surface Processes and Landforms*, **35**(9), 1014–1028.

Tal, M., Gran, K., Murray, A. B., Paola, C., Hicks, D. M., 2004. Riparian vegetation as a primary control on channel characteristics in multi-thread rivers. In: S. J. Bennett, A. Simon (eds.), *Riparian Vegetation and Fluvial Geomorphology*, 8, pp. 43–58.

Talbot, T., Lapointe, M., 2002a. Numerical modeling of gravel bed river response to meander straightening: the coupling between the evolution of bed pavement and long profile. *Water Resources Research*, **38**(6). 10.1029/2001wr000330.

Talbot, T., Lapointe, M., 2002b. Modes of response of a gravel bed river to meander straightening: the case of the Sainte-Marguerite River, Saguenay Region, Quebec, Canada. *Water Resources Research*, **38**(6). 10.1029/2001wr000324.

Talling, P.J., Sowter, M.J., 1999. Drainage density on progressively tilted surfaces with different gradients, Wheeler Ridge, California. *Earth Surface Processes and Landforms*, **24**(9), 809–824.

Tang, H., Zhang, H., Yuan, S., 2018. Hydrodynamics and contaminant transport on a degraded bed at a 90-degree channel confluence. *Environmental Fluid Mechanics*, **18**(2), 443–463.

Taniguchi, K. T., Biggs, T. W., 2015. Regional impacts of urbanization on stream channel geometry: a case study in semiarid southern California. *Geomorphology*, **248**, 228–236.

Tanner, W. F., 1960. Helicoidal flow, a possible cause of meandering. *Journal of Geophysical Research*, **65**, 993–995.

Tanner, W. F., 1971. The river profile. *Journal of Geology*, **79**(4), 482–492.

Tarboton, D. G., Bras, R. L., Rodriguez-Iturbe, I., 1988. The fractal nature of river networks. *Water Resources Research*, **24**(8), 1317–1322.

Tarboton, D. G., Bras, R. L., Rodriguez-Iturbe, I., 1990. On the fractal dimension of stream networks – comment. *Water Resources Research*, **26**(9), 2243–2244.

Taylor, C. F. H., 1999. The role of overbank flow in governing the form of an anabranching river: the Fitzroy River, northwestern Australia. In: N. D. Smith, J. Rogers (eds.), *Fluvial Sedimentology VI*, International Association of Sedimentologists Special Publication No. 28, Blackwell, Oxford, UK, pp. 77–91.

Taylor, E. H., 1944. Flow characteristics at rectangular open-channel junctions. *Transactions, American Society of Civil Engineers*, **109**, 893–912.

Tennekes, H., Lumley, J. L., 1972. *A First Course in Turbulence*. MIT Press, Cambridge, MA.

Termini, D., 2009. Experimental observations of flow and bed processes in large-amplitude meandering flume. *Journal of Hydraulic Engineering*, **135**(7), 575–587.

Tessler, Z. D., Voeroesmarty, C. J., Grossberg, M., et al., 2015. Profiling risk and sustainability in coastal deltas of the world. *Science*, **349**(6248), 638–643.

Thayer, J. B., 2017. Downstream regime relations for single-thread channels. *River Research and Applications*, **33**, 182–186.

Thomas, R., Nicholas, A. P., 2002. Simulation of braided river flow using a new cellular routing scheme. *Geomorphology*, **43**(3–4), 179–195.

Thomas, R., Nicholas, A. P., Quine, T. A., 2007. Cellular modelling as a tool for interpreting historic braided river evolution. *Geomorphology*, **90**(3–4), 302–317.

Thomas, R. B., Megahan, W. F., 1998. Peak flow responses to clear-cutting and roads in small and large basins, western Cascades, Oregon: a second opinion. *Water Resources Research*, **34**(12), 3393–3403.

Thomas, R. E., Parsons, D. R., Sandbach, S. D., et al., 2011. An experimental study of discharge partitioning and flow structure at symmetrical bifurcations. *Earth Surface Processes and Landforms*, **36**(15), 2069–2082.

Thompson, A., 1986. Secondary flows and the pool-riffle unit – a case study of the processes of meander development. *Earth Surface Processes and Landforms*, **11**(6), 631–641.

Thompson, C., Croke, J., 2008. Channel flow competence and sediment transport in upland streams in southeast Australia. *Earth Surface Processes and Landforms*, **33**(3), 329–352.

Thompson, C., Croke, J., 2013. Geomorphic effects, flood power, and channel competence of a catastrophic flood in confined and unconfined reaches of the upper Lockyer valley, southeast Queensland, Australia. *Geomorphology*, **197**, 156–169.

Thompson, D. M., 2001. Random controls on semi-rhythmic spacing of pools and riffles in constriction-dominated rivers. *Earth Surface Processes and Landforms*, **26**(11), 1195–1212.

Thompson, D. M., 2004. The influence of pool length on local turbulence production and energy slope: a flume experiment. *Earth Surface Processes and Landforms*, **29**(11), 1341–1358.

Thompson, D. M., 2006. The role of vortex shedding in the scour of pools. *Advances in Water Resources*, **29**(2), 121–129.

Thompson, D. M., 2007. The characteristics of turbulence in a shear zone downstream of a channel constriction in a coarse-grained forced pool. *Geomorphology*, **83**(3–4), 199–214.

Thompson, D. M., Fixler, S. A., 2017. Formation and maintenance of a forced pool-riffle couplet following loading of large wood. *Geomorphology*, **296**, 74–90.

Thompson, D. M., Hoffman, K. S., 2001. Equilibrium pool dimensions and sediment-sorting patterns in coarse-grained, New England channels. *Geomorphology*, **38**(3–4), 301–316.

Thompson, D. M., McCarrick, C. R., 2010. A flume experiment on the effect of constriction shape on the formation of forced pools. *Hydrology and Earth System Sciences*, **14**(7), 1321–1330.

Thompson, D. M., Wohl, E. E., 2009. The linkage between velocity patterns and sediment entrainment in a forced-pool and riffle unit. *Earth Surface Processes and Landforms*, **34**(2), 177–192.

Thompson, D. M., Wohl, E. E., Jarrett, R. D., 1996. Revised velocity-reversal and sediment-sorting model for a high-gradient, pool-riffle stream. *Physical Geography*, **17**(2), 142–156.

Thompson, D. M., Wohl, E. E., Jarrett, R. D., 1999. Velocity reversals and sediment sorting in pools and riffles controlled by channel constrictions. *Geomorphology*, **27**(3–4), 229–241.

Thoms, M. C., 1987. Channel sedimentation within the urbanized River Tame, UK. *Regulated Rivers Research and Management*, **1**(3), 229–246.

Thomson, J., 1876. On the origin of windings of rivers in alluvial plains. *Proceedings of the Royal Society*, **25**, 5–8.

Thonemann, P., 2011. *The Maeander Valley: a historical geography from antiquity to Byzantium*. Cambridge University Press, Cambridge, UK.

Thorn, C. E., Welford, M. R., 1994. The equilibrium concept in geomorphology. *Annals of the Association of American Geographers*, **84**(4), 666–696.

Thorne, C. R., 1982. Processes and mechanisms of river bank erosion. In: R. D. Hey, J. C. Bathurst, C. R. Thorne (eds.), *Gravel-Bed Rivers*. Wiley, Chichester, UK, pp. 227–259.

Thorne, C. R., 1990. Effects of vegetation on river bank erosion and stability. In: J. B. Thornes (ed.), *Vegetation and Erosion: Processes and Environments*. Wiley, Chichester, UK, pp. 125–144.

Thorne, C. R., Abt, S. R., 1993. Analysis of riverbank stability due to toe scour and lateral erosion. *Earth Surface Processes and Landforms*, **18**(9), 835–843.

Thorne, C. R., Tovey, N. K., 1981. Stability of composite river banks. *Earth Surface Processes and Landforms*, **6**(5), 469–484.

Thorne, C. R., Zevenbergen, L. W., Pitlick, J. C., et al., 1985. Direct measurement of secondary currents in a meandering sand-bed river. *Nature*, **315**(6022), 746–747.

Thorne, S. D., Furbish, D. J., 1995. Influences of coarse bank roughness on flow within a sharply curved river bend. *Geomorphology*, **12**(3), 241–257.

Thorp, J. H., Thoms, M. C., Delong, M. D., 2006. The riverine ecosystem synthesis: biocomplexity in river networks across space and time. *River Research and Applications*, **22**(2), 123–147.

Tinkler, K. J., 1985. *A Short History of Geomorphology*. Barnes and Noble, Totowa, NJ.

Tockner, K., Paetzold, A., Karaus, U., Claret, C., Zettel, J., 2006. Ecology of braided rivers. In: G. H. Sambrook Smith, J. L. Best, C. S. Bristow, G. E. Petts (eds.), *Braided Rivers: Process, Deposits, Ecology and Management*, International Association of Sedimentologists Special Publication 36. Blackwell, Oxford, UK, pp. 339–359.

Toniolo, H., Parker, G., Voller, V., 2007. Role of ponded turbidity currents in reservoir trap efficiency. *Journal of Hydraulic Engineering*, **133**(6), 579–595.

Toonen, W. H. J., Kleinhans, M. G., Cohen, K. M., 2012. Sedimentary architecture of abandoned channel fills. *Earth Surface Processes and Landforms*, **37**(4), 459–472.

Tooth, S., 2005. Splay formation along the lower reaches of ephemeral rivers on the Northern Plains of arid central Australia. *Journal of Sedimentary Research*, **75**(4), 636–649.

Tooth, S., McCarthy, T. S., 2004a. Anabranching in mixed bedrock-alluvial rivers: the example of the Orange River above Augrabies Falls, Northern Cape Province, South Africa. *Geomorphology*, **57**(3–4), 235–262.

Tooth, S., McCarthy, T. S., 2004b. Controls on the transition from meandering to straight channels in the wetlands of the Okavango Delta, Botswana. *Earth Surface Processes and Landforms*, **29**(13), 1627–1649.

Tooth, S., Nanson, G. C., 1999. Anabranching rivers on the Northern Plains of arid central Australia. *Geomorphology*, **29**(3–4), 211–233.

Tooth, S., Nanson, G. C., 2000a. Equilibrium and nonequilibrium conditions in dryland rivers. *Physical Geography*, **21**(3), 183–211.

Tooth, S., Nanson, G. C., 2000b. The role of vegetation in the formation of anabranching channels in an ephemeral river, Northern plains, arid central Australia. *Hydrological Processes*, **14**(16–17), 3099–3117.

Tooth, S., Nanson, G. C., 2004. Forms and processes of two highly contrasting rivers in arid central Australia, and the implications for channel-pattern discrimination and prediction. *Geological Society of America Bulletin*, **116**(7–8), 802–816.

Tooth, S., Jansen, J. D., Nanson, G. C., Coulthard, T. J., Pietsch, T., 2008. Riparian vegetation and the late Holocene development of an anabranching river: Magela Creek, northern Australia. *Geological Society of America Bulletin*, **120**(7–8), 1021–1035.

Topping, D. J., Rubin, D. M., Vierra, L. E., 2000. Colorado River sediment transport – 1. Natural sediment supply limitation and the influence of Glen Canyon Dam. *Water Resources Research*, **36**(2), 515–542.

Torizzo, M., Pitlick, J., 2004. Magnitude-frequency of bed load transport in mountain streams in Colorado. *Journal of Hydrology*, **290**(1–2), 137–151.

Törnqvist, T. E., Bridge, J. S., 2002. Spatial variation of overbank aggradation rate and its influence on avulsion frequency. *Sedimentology*, **49**(5), 891–905.

Törnqvist, T. E., van Dijk, G. J., 1993. Optimizing sampling strategy for radiocarbon dating of Holocene fluvial systems in a vertically aggrading setting. *Boreas*, **22**(2), 129–145.

Torres, A., Brandt, J., Lear, K., Liu, J. G., 2017. A looming tragedy of the sand commons. *Science*, **357**(6355), 970–971.

Trampush, S. M., Huzurbazar, S., McElroy, B., 2014. Empirical assessment of theory for bankfull characteristics of rivers. *Water Resources Research*, **50**, 9211–9220.

Tranmer, A. W., Tonina, D., Benjankar, R., Tiedemann, M., Goodwin, P., 2015. Floodplain persistence and dynamic-equilibrium conditions in a canyon environment. *Geomorphology*, **250**, 147–158.

Trimble, S. W., 1974. *Man-Induced Soil Erosion on the Southern Piedmont, 1700–1970*. Soil Conservation Society of America, Ankeny, IA.

Trimble, S. W., 1975. Denudation studies – can we assume stream steady state? *Science*, **188**(4194), 1207–1208.

Trimble, S. W., 1977. Fallacy of stream equilibrium in contemporary denudation studies. *American Journal of Science*, **277**(7), 876–887.

Trimble, S. W., 1981. Changes in sediment storage in the Coon Creek Basin, Driftless Area, Wisconsin, 1853–1975. *Science*, **214**(4517), 181–183.

Trimble, S. W., 1983. A sediment budget for Coon Creek Basin in the Driftless Area, Wisconsin, 1853–1977. *American Journal of Science*, **283**(5), 454–474.

Trimble, S. W., 1993. The distributed sediment budget model and watershed management in the Paleozoic plateau of the upper midwestern United States. *Physical Geography*, **14**(3), 285–303.

Trimble, S. W., 1994. Erosional effects of cattle on streambanks in Tennessee, USA. *Earth Surface Processes and Landforms*, **19**(5), 451–464.

Trimble, S. W., 1995. Catchment sediment budgets and change. In: A. Gurnell, G. Petts (eds.), *Changing River Channels*. Wiley and Sons, Chichester, UK, pp. 201–215.

Trimble, S. W., 1997. Contribution of stream channel erosion to sediment yield from an urbanizing watershed. *Science*, **278**(5342), 1442–1444.

Trimble, S. W., 1999. Decreased rates of alluvial sediment storage in the Coon Creek Basin, Wisconsin, 1975–93. *Science*, **285**(5431), 1244–1246.

Trimble, S. W., 2008. The use of historical data and artifacts in geomorphology. *Progress in Physical Geography*, **32**(1), 3–29.

Trimble, S. W., 2009. Fluvial processes, morphology and sediment budgets in the Coon Creek Basin, WI, USA, 1975–1993. *Geomorphology*, **108**(1–2), 8–23.

Trimble, S. W., 2013. *Historical Agriculture and Soil Erosion in the Upper Mississippi River Hill Country*. CRC Press, Boca Raton, FL.

Trimble, S. W., Cooke, R. U., 1991. Historical sources for geomorphological research in the United States. *Professional Geographer*, **43**(2), 212–228.

Trimble, S. W., Crosson, P., 2000. U.S. soil erosion rates – myth and reality. *Science*, **289**(5477), 248–250.

Trimble, S. W., Mendel, A. C., 1995. The cow as a geomorphic agent – a critical review. *Geomorphology*, **13**(1–4), 233–253.

Tubino, M., 1991. Growth of alternate bars in unsteady flow. *Water Resources Research*, **27**(1), 37–52.

Tubino, M., Seminara, G., 1990. Free forced interactions in developing meanders and suppression of free bars. *Journal of Fluid Mechanics*, **214**, 131–159.

Tucker, G. E., 2004. Drainage basin sensitivity to tectonic and climatic forcing: implications of a stochastic model for the role of entrainment and erosion thresholds. *Earth Surface Processes and Landforms*, **29**(2), 185–205.

Tucker, G. E., Hancock, G. R., 2010. Modelling landscape evolution. *Earth Surface Processes and Landforms*, **35**(1), 28–50.

Tucker, G. E., Slingerland, R. L., 1994. Erosional dynamics, flexural isostasy, and long-lived escarpments – a numerical modeling study. *Journal of Geophysical Research-Solid Earth*, **99**(B6), 12229–12243. 10.1029/94jb00320.

Tucker, G. E., Lancaster, S., Gasparini, N., Bras, R. L., 2001. The Channel-Hillslope Integrated Landscape Development Model (CHILD) In: R. S. Harmon, W. W. Doe III (eds.), *Landscape Erosion and Evolution Modeling*. Kluwer Academic, New York, pp. 349–388.

Turnipseed, D. P., Sauer, V. B., 2010. Discharge measurements at gaging stations. U.S. Geological Survey Techniques and Methods, Book 3, Chapter A8. http://pubs.usgs.gov/tm/tm3-a8/.

Turowski, J. M., Lague, D., Hovius, N., 2007. Cover effect in bedrock abrasion: a new derivation and its implications for the modeling of bedrock channel morphology. *Journal of Geophysical Research-Earth Surface*, **112**(F4). 10.1029/2006jf000697.

Turowski, J. M., Hovius, N., Wilson, A., Horng, M.-J., 2008. Hydraulic geometry, river sediment and the definition of bedrock channels. *Geomorphology*, **99**(1–4), 26–38.

Turowski, J. M., Yager, E. M., Badoux, A., Rickenmann, D., Molnar, P., 2009. The impact of exceptional events on erosion, bedload transport and channel stability in a step-pool channel. *Earth Surface Processes and Landforms*, **34**(12), 1661–1673.

Tyner, J. S., Yoder, D. C., Chomicki, B. J., Tyagi, A., 2011. A review of construction site best management practices for erosion control. *Transactions of the American Society of Agricultural and Biological Engineers*, **54**(2), 441–450.

Uijttewaal, W. S. J., Booij, R., 2000. Effects of shallowness on the development of free-surface mixing layers. *Physics of Fluids*, **12**(2), 392–402.

Uijttewaal, W. S. J., Tukker, J., 1998. Development of quasi two-dimensional structures in a shallow free-surface mixing layer. *Experiments in Fluids*, **24**(3), 192–200.

Umar, M., Rhoads, B. L., Greenberg, J. A., 2018. Use of multispectral satellite remote sensing to assess mixing of suspended sediment downstream of large river confluences. *Journal of Hydrology*, **556**, 325–338.

United Nations, 2018. World Urbanization Prospects: The 2018 Revision [key facts], United Nations, New York.

Urban, M. A., Rhoads, B. L., 2003. Catastrophic human-induced change in stream-channel planform and geometry in an agricultural watershed, Illinois, USA. *Annals of the Association of American Geographers*, **93**(4), 783–796.

U.S. Army Corps of Engineers, 2010. Illinois Stream Mitigation Guidance. U.S. Army Corps of Engineers. www.mvr.usace.army.mil/Portals/48/docs/regulatory/mitigation/IllinoisMethod.pdf.

U.S. Army Corps of Engineers, 2016. *HEC-RAS River Analysis System, Hydraulic Reference Manual, Version 5.0*. U.S. Army Corps of Engineers Hydrologic Engineering Center, Davis, CA.

U.S. Army Corps of Engineers, 2018. National Inventory of Dams. http://nid.usace.army.mil/ (accessed November 27, 2019).

U.S. Department of Agriculture, 1983. Sediment sources, yields, and delivery ratios, Chapter 6 of Section 3: *Sedimentation, National Engineering Handbook*. U.S. Department of Agriculture, Washington, D.C.

U.S. Department of Agriculture, 1995. U.S.D.A – Water Erosion Prediction Project – Technical Documentation. NSERL Report No. 10, National Soil Erosion Research Laboratory, West Lafayette, IN.

U.S. Department of Agriculture, 2013. Revised Universal Soil Loss Equation Version 2 (RUSLE2) Science Documentation A.R.S. U.S. Department of Agriculture, Washington, DC.

U.S. Department of Agriculture, 2018. Summary Report: 2015 National Resources Inventory. Natural Resources Conservation Service, Washington, DC, Center for Survey Statistics and Methodology, Iowa State University, Ames, IA.

U.S. Environmental Protection Agency, 2008. *Handbook for Developing Watershed Plans to Restore and Protect Waters*. U.S. Environmental Protection Agency, Washington, DC.

Vallé, B. L., Pasternack, G. B., 2006. Submerged and unsubmerged natural hydraulic jumps in a bedrock step-pool mountain channel. *Geomorphology*, **82**(1–2), 146–159.

Valyrakis, M., Diplas, P., Dancey, C. L., Greer, K., Celik, A. O., 2010. Role of instantaneous force magnitude and duration on particle entrainment. *Journal of Geophysical Research – Earth Surface*, **115**. 10.1029/2008jf001247.

Valyrakis, M., Diplas, P., Dancey, C. L., 2011. Entrainment of coarse grains in turbulent flows: an extreme value theory approach. *Water Resources Research*, **47**. 10.1029/2010wr010236.

Valyrakis, M., Diplas, P., Dancey, C. L., 2013. Entrainment of coarse particles in turbulent flows: an energy approach. *Journal of Geophysical Research – Earth Surface*, **118**(1), 42–53. 10.1029/2012jf002354.

van de Lageweg, W. I., van Dijk, W. M., Baar, A. W., Rutten, J., Kleinhans, M. G., 2014. Bank pull or bar push: what drives scroll-bar formation in meandering rivers? *Geology*, **42**(4), 319–322.

van de Wiel, M. J., Darby, S. E., 2007. A new model to analyse the impact of woody riparian vegetation on the geotechnical stability of riverbanks. *Earth Surface Processes and Landforms*, **32** (14), 2185–2198.

van de Wiel, M. J., Rousseau, Y. Y., Darby, S. E., 2016. Models in fluvial geomorphology. In: G. M. Kondolf, H. Piegay (eds.), *Tools in Fluvial Geomorphology*, 2nd Edition. Wiley, Chichester, UK, pp. 383–411.

van den Berg, J. H., 1987. Bedform migration and bed-load transport in some rivers and tidal environments. *Sedimentology*, **34**(4), 681–698.

van den Berg, J. H., 1995. Prediction of alluvial channel pattern of perennial rivers. *Geomorphology*, **12**(4), 259–279.

van den Berg, J. H., Bledsoe, B. P., 2003. Comment on Lewin and Brewer (2001): "Predicting channel patterns". *Geomorphology*, **53**(3–4), 333–337.

van der Mark, C. F., Mosselman, E., 2013. Effects of helical flow in one-dimensional modelling of sediment distribution at river bifurcations. *Earth Surface Processes and Landforms*, **38**(5), 502–511.

van Dijk, W. M., van de Lageweg, W. I., Kleinhans, M. G., 2012. Experimental meandering river with chute cutoffs. *Journal of Geophysical Research-Earth Surface*, **117**. 10.1029/2011jf002314.

van Dijk, W. M., Schuurman, F., van de Lageweg, W. I., Kleinhans, M. G., 2014. Bifurcation instability and chute cutoff development in meandering gravel-bed rivers. *Geomorphology*, **213**, 277–291.

van Maren, D. S., Winterwerp, J. C., Wu, B. S., Zhou, J. J., 2009. Modelling hyperconcentrated flow in the Yellow River. *Earth Surface Processes and Landforms*, **34**(4), 596–612.

van Oorschot, M., Kleinhans, M., Geerling, G., Middelkoop, H., 2016. Distinct patterns of interaction between vegetation and morphodynamics. *Earth Surface Processes and Landforms*, **41** (6), 791–808.

van Rijn, L. C., 1984. Sediment transport, part II: suspended load transport. *Journal of Hydraulic Engineering*, **110**(11), 1613–1641.

van Toorenenburg, K. A., Donselaar, M. E., Weltje, G. J., 2018. The life cycle of crevasse splays as a key mechanism in the aggradation of alluvial ridges and river avulsion. *Earth Surface Processes and Landforms*, **43**(11), 2409–2420.

van Vliet, J., Eitelberg, D. A., Verburg, P. H., 2017. A global analysis of land take in cropland areas and production displacement from urbanization. *Global Environmental Change – Human and Policy Dimensions*, **43**, 107–115.

Vanapalli, S. K., Fredlund, D. G., Pufahl, D. E., Clifton, A. W., 1996. Model for the prediction of shear strength with respect to soil suction. *Canadian Geotechnical Journal*, **33**(3), 379–392.

Vandaele, K., Poesen, J., Govers, G., van Wesemael, B., 1996. Geomorphic threshold conditions for ephemeral gully incision. *Geomorphology*, **16**(2), 161–173.

Vanmaercke, M., Kettner, A. J., van den Ekhaut, M., et al., 2014. Moderate seismic activity affects contemporary sediment yields. *Progress in Physical Geography*, **38**, 145–172.

Vanmaercke, M., Ardizzone, F., Rossi, M., Guzzetti, F., 2017. Exploring the effects of seismicity on landslides and catchment sediment yield: an Italian case study. *Geomorphology*, **278**, 171–183.

Vanoni, V. A., Benedict, P. C., Bondurant, D. C., et al., 1966. Sediment transportation mechanics: initiation of motion. *Journal of the Hydraulics Division – ASCE*, **92**, 291–314.

Vasconselos, P. M., Farley, K. A., Stone, J., Piacentini, T., Fifield, L. K., 2019. Stranded landscapes in the humid tropics: Earth's oldest land surfaces. *Earth and Planetary Science Letters*, **519**, 152–164.

Vaughan, I. P., Diamond, M., Gurnell, A. M., et al., 2009. Integrating ecology with hydromorphology: a priority for river science and management. *Aquatic Conservation – Marine and Freshwater Ecosystems*, **19**(1), 113–125.

Vaughn, D. M., 1990. Flood dynamics of a concrete-lined, urban stream in Kansas City, Missouri. *Earth Surface Processes and Landforms*, **15**(6), 525–537.

Venditti, J. G., 2013. Bedforms in sand-bedded rivers. In: J. L. Schroder (ed.), *Treatise on Geomorphology*, Vol. 9, Fluvial Geomorphology, E. Wohl (vol. ed.), Academic Press, San Diego, CA, pp. 137–162.

Venditti, J. G., Church, M., 2014. Morphology and controls on the position of a gravel-sand transition: Fraser River, British Columbia. *Journal of Geophysical Research – Earth Surface*, **119**(9), 1959–1976.

Venditti, J. G., Nelson, P. A., Minear, J. T., Wooster, J., Dietrich, W. E., 2012. Alternate bar response to sediment supply termination. *Journal of Geophysical Research – Earth Surface*, **117**. 10.1029/2011jf002254.

Venditti, J. G., Hardy, R. J., Church, M., Best, J. L., 2013. What is a coherent flow structure in geophysical flow? In: J. G. Venditti (ed.), *Coherent Flow Structures at Earth's Surface*. Wiley and Sons, Chichester, UK, pp. 1–16.

Venditti, J. G., Domarad, N., Church, M., Rennie, C. D., 2015. The gravel-sand transition: sediment dynamics in a diffuse extension. *Journal of Geophysical Research – Earth Surface*, **120**(6), 943–963. 10.1002/2014jf003328.

Vercruysse, K., Grabowski, R. C., Rickson, R. J., 2017. Suspended sediment transport dynamics in rivers: multi-scale drivers of temporal variation. *Earth-Science Reviews*, **166**, 38–52.

Vericat, D., Wheaton, J. M., Brasington, J., 2017. Revisiting the morphological approach: opportunities and challenges with repeat high resolution topography. In: D. Tsutsumi, J. B. Laronne (eds.), *Gravel-Bed Rivers: Processes and Disasters*. Wiley, Chichester, UK, pp. 121–158.

Vermeulen, B., Hoitink, A. J. F., Labeur, R. J., 2015. Flow structure caused by a local cross-sectional area increase and curvature in a sharp river bend. *Journal of Geophysical Research – Earth Surface*, **120**(9), 1771–1783. 10.1002/2014jf003334.

Verpoorter, C., Kutser, T., Seekell, D. A., Tranvik, L. J., 2014. A global inventory of lakes based on high-resolution satellite imagery.

Geophysical Research Letters, **41**(18), 6396–6402. 10.1002/2014gl060641.

Verstraeten, G., Broothaerts, N., Van Loo, M., et al., 2017. Variability in fluvial geomorphic response to anthropogenic disturbance. *Geomorphology*, **294**, 20–39.

Vetter, T., 2011a. Riffle-pool morphometry and stage-dependent morphodynamics of a large floodplain river (Vereinigte Mulde, Sachsen-Anhalt, Germany). *Earth Surface Processes and Landforms*, **36**(12), 1647–1657.

Vetter, T., 2011b. Analysing riffle-pool dynamics of a large floodplain river with a system-oriented approach. *Zeitschrift fur Geomorphologie*, **55**, 355–372.

Viero, D. P., Lopez Dupon, S., Lanzoni, S., 2018. Chute cutoffs in meandering rivers: formative mechanisms and hydrodynamic forcing. In: M. Ghinassi, L. Columbera, N. P. Mountney, A. J. H. Reesink, M. Bateman (eds.), *Fluvial Meanders and their Sedimentary Products in the Rock Record*. International Association of Sedimentologists Special Publication 48. Wiley and Sons, Hoboken, NJ, pp. 201–229.

Vietz, G. J., Rutherfurd, I. D., Stewardson, M. J., Finlayson, B. L., 2012. Hydrodynamics and sedimentology of concave benches in a lowland river. *Geomorphology*, **147**, 86–101.

Vietz, G. J., Walsh, C. J., Fletcher, T. D., 2016. Urban hydrogeomorphology and the urban stream syndrome: treating the symptoms and causes of geomorphic change. *Progress in Physical Geography*, **40**(3), 480–492.

Vigilar, G. G., Diplas, P., 1997. Stable channels with mobile bed: formulation and numerical solution. *Journal of Hydraulic Engineering*, **123**(3), 189–199.

Vigilar, G. G., Diplas, P., 1998. Stable channels with mobile bed: model verification and graphical solution. *Journal of Hydraulic Engineering*, **124**(11), 1097–1108.

Vogel, R. M., Stedinger, J. R., Hooper, R. P., 2003. Discharge indices for water quality loads. *Water Resources Research*, **39**(10). 10.1029/2002wr001872.

Voichick, N., Topping, D. J., 2014. Extending the turbidity record—making additional use of continuous data from turbidity, acoustic-doppler, and laser diffraction instruments and suspended-sediment samples in the Colorado River in Grand Canyon. U.S. Geological Survey Scientific Investigations Report 2014–5097. Available online only at http://dx.doi.org/10.3133/sir20145097.

Vollmer, S., Kleinhans, M. G., 2007. Predicting incipient motion, including the effect of turbulent pressure fluctuations in the bed. *Water Resources Research*, **43**(5). 10.1029/2006wr004919.

von Blanckenburg, F., 2005. The control mechanisms of erosion and weathering at basin scale from cosmogenic nuclides in river sediment. *Earth and Planetary Science Letters*, **237**(3–4), 462–479.

Vorosmarty, C. J., Meybeck, M., Fekete, B., et al., 2003. Anthropogenic sediment retention: major global impact from registered river impoundments. *Global and Planetary Change*, **39**(1–2), 169–190.

Wade, R. J., Rhoads, B. L., Rodriguez, J., et al., 2002. Integrating science and technology to support stream naturalization near Chicago, Illinois. *Journal of the American Water Resources Association*, **38**(4), 931–944.

Walling, D. E., 1977. Assessing accuracy of suspended sediment rating curves for a small basin. *Water Resources Research*, **13**(3), 530–538.

Walling, D. E., 1983. The sediment delivery problem. *Journal of Hydrology*, **65**(1–3), 209–237.

Walling, D. E., 1999. Linking land use, erosion and sediment yields in river basins. *Hydrobiologia*, **410**, 223–240.

Walling, D. E., 2005. Tracing suspended sediment sources in catchments and river systems. *Science of the Total Environment*, **344**(1–3), 159–184.

Walling, D. E., 2006. Human impact on land-ocean sediment transfer by the world's rivers. *Geomorphology*, **79**(3–4), 192–216.

Walling, D. E., Collins, A. L., 2008. The catchment sediment budget as a management tool. *Environmental Science & Policy*, **11**(2), 136–143.

Walling, D. E., Fang, D., 2003. Recent trends in the suspended sediment loads of the world's rivers. *Global and Planetary Change*, **39**(1–2), 111–126.

Walling, D. E., Gregory, K. J., 1970. The measurement of the effects of building construction on drainage basin dynamics. *Journal of Hydrology*, **11**, 129–144.

Walling, D. E., He, Q., 1997. Investigating spatial patterns of overbank sedimentation on river floodplains. *Water Air and Soil Pollution*, **99**(1–4), 9–20.

Walling, D. E., He, Q., 1998. The spatial variability of overbank sedimentation on river floodplains. *Geomorphology*, **24**(2–3), 209–223.

Walling, D. E., He, Q., 1999. Using fallout lead-210 measurements to estimate soil erosion on cultivated land. *Soil Science Society of America Journal*, **63**(5), 1404–1412.

Walling, D. E., Kleo, A. H. A., 1979. Sediment yields in areas of low precipitation: an overview. In: *The Hydrology of Areas of Low Precipitation*. IAHS Publication No. 128, IAHS Press, Wallingford, UK, pp. 479–493.

Walling, D. E., Webb, B. W., 1983. Patterns of sediment yield. In: K. J. Gregory (ed.), *Background to Palaeohydrology*. Wiley and Sons, New York, pp. 69–100.

Walling, D. E., Webb, B. W., 1996. Erosion and sediment yield: a global overview. In: D. E. Walling, B. W. Webb (eds.), *Erosion and Sediment Yield: Global and Regional Perspectives*. IAHS Publication No. 236, IAHS Press, Wallingford, UK, pp. 3–19.

Walling, D. E., Russell, M. A., Hodgkinson, R. A., Zhang, Y., 2002. Establishing sediment budgets for two small lowland agricultural catchments in the UK. *Catena*, **47**(4), 323–353.

Walling, D. E., Collins, A. L., Jones, P. A., Leeks, G. J. L., Old, G., 2006. Establishing fine-grained sediment budgets for the Pang and Lambourn LOCAR catchments, UK. *Journal of Hydrology*, **330**(1–2), 126–141.

Wallinga, J., 2002. Optically stimulated luminescence dating of fluvial deposits: a review. *Boreas*, **31**(4), 303–322.

Walsh, J., Hicks, D. M., 2002. Braided channels: self-similar or self-affine? *Water Resources Research*, **38**(6). 10.1029/2001wr000749.

Walter, R. C., Merritts, D. J., 2008. Natural streams and the legacy of water-powered mills. *Science*, **319**(5861), 299–304.

Wang, Y., Rhoads, B. L., Wang, D., 2016. Assessment of the flow regime alterations in the middle reach of the Yangtze River associated with dam construction: potential ecological implications. *Hydrological Processes*, **30**(21), 3949–3966.

Wang, Y., Rhoads, B. L., Wang, D., Wu, J., Zhang, X., 2018. Impacts of large dams on the complexity of suspended sediment dynamics in the Yangtze River. *Journal of Hydrology*, **558**, 184–195.

Warner, R. F., 1997. Floodplain stripping: another form of adjustment to secular hydrologic regime change in Southeast Australia. *Catena*, **30**(4), 263–282.

Warrick, J. A., Milliman, J. D., Walling, D. E., et al., 2014. Earth is (mostly) flat: apportionment of the flux of continental sediment over millennial time scales: comment. *Geology*, **42**(1), e316. 10.1130/g34846c.1.

Wathen, S. J., Ferguson, R. I., Hoey, T. B., Werritty, A., 1995. Unequal mobility of gravel and sand in weakly bimodal river sediments. *Water Resources Research*, **31**(8), 2087–2096.

Watson, R. L., 1969. Modified Rubey's law accurately predicts sediment settling velocities. *Water Resources Research*, **5**(5), 1147–1150.

Waylen, P., Woo, M. K., 1982. Prediction of annual floods generated by mixed processes. *Water Resources Research*, **18**(4), 1283–1286.

Waythomas, C., Williams, G. P., 1988. Sediment yield and spurious correlation – toward a better portrayal of the annual suspended-sediment load of rivers. *Geomorphology*, **1**, 309–316.

Webber, N. B., Greated, C. A., 1966. An investigation of flow behavior at the junction of rectangular channels. *Proceedings of the Institute of Civil Engineers*, **34**, 321–334.

Weber, L. J., Schumate, E. D., Mawer, N., 2001. Experiments on flow at a 90° open-channel junction. *Journal of Hydraulic Engineering*, **127**(5), 340–350.

Weichert, R. B., Bezzola, G. R., Minor, H.-E., 2008. Bed morphology and generation of step-pool channels. *Earth Surface Processes and Landforms*, **33**(11), 1678–1692.

Welber, M., Bertoldi, W., Tubino, M., 2012. The response of braided planform configuration to flow variations, bed reworking and vegetation: the case of the Tagliamento River, Italy. *Earth Surface Processes and Landforms*, **37**(5), 572–582.

Welford, M. R., 1993. Field evaluation of empirical equations in straight alluvial channels. *Physical Geography*, **14**(6), 581–598.

Welford, M. R., 1994. A field test of Tubino's (1991) model of alternate bar formation. *Earth Surface Processes and Landforms*, **19**(4), 287–297.

Wellmeyer, J. L., Slattery, M. C., Phillips, J. D., 2005. Quantifying downstream impacts of impoundment on flow regime and channel planform, lower Trinity River, Texas. *Geomorphology*, **69**(1–4), 1–13.

Wende, R., Nanson, G. C., 1998. Anabranching rivers: ridge-form alluvial channels in tropical northern Australia. *Geomorphology*, **22**(3–4), 205–224.

Wentworth, C. K., 1919. A laboratory and field study of cobble abrasion. *Journal of Geology*, **27**(7), 507–521.

Wentworth, C. K., 1922. A scale of grade and class terms for clastic sediments. *Journal of Geology*, **30**, 377–392.

Werritty, A., 1992. Downstream fining in a gravel-bed river in southern Poland – lithologic controls and the role of abrasion. In:

P. Billi, R.D. Hey, C.R. Thorne, P. Tacconi (eds.), *Dynamics of Gravel-Bed Rivers*. Wiley, Chichester, UK, pp. 333–350.

Wheaton, J., Darby, S. E., Sear, D. A., 2008. The scope of uncertainties in river restoration. In: S. E. Darby, D. A. Sear (eds.), *River Restoration: Managing the Uncertainty of Restoring Physical Habitat*. Wiley, Chichester, UK, pp. 21–39.

Wheaton, J. M., Brasington, J., Darby, S. E., Sear, D. A., 2010. Accounting for uncertainty in DEMs from repeat topographic surveys: improved sediment budgets. *Earth Surface Processes and Landforms*, **35**(2), 136–156.

Wheaton, J. M., Brasington, J., Darby, S. E., et al., 2013. Morphodynamic signatures of braiding mechanisms as expressed through change in sediment storage in a gravel-bed river. *Journal of Geophysical Research – Earth Surface*, **118**(2), 759–779.

Wheaton, J. M., Fryirs, K. A., Brierley, G., et al., 2015. Geomorphic mapping and taxonomy of fluvial landforms. *Geomorphology*, **248**, 273–295.

Whipple, K. X., 2001. Fluvial landscape response time: how plausible is steady-state denudation? *American Journal of Science*, **301**(4–5), 313–325.

Whipple, K. X., 2004. Bedrock rivers and the geomorphology of active orogens. *Annual Review of Earth and Planetary Sciences*, **32**, 151–185.

Whipple, K. X., Tucker, G. E., 1999. Dynamics of the stream-power river incision model: implications for height limits of mountain ranges, landscape response timescales, and research needs. *Journal of Geophysical Research-Solid Earth*, **104**(B8), 17661–17674.

Whipple, K. X., DiBiase, R. A., Crosby, B. T., 2013. Bedrock rivers. In: J. Schroder (ed.), *Treatise on Geomorphology*, Vol. 9, Fluvial Geomorphology, E. E. Wohl (vol. ed.). Academic Press, San Diego, CA, pp. 550–573.

Whipple, W., Jr., DiLouie, J., 1981. Coping with increased stream erosion in urbanizing areas. *Water Resources Research*, **17**(5), 1561–1564.

Whitaker, A. C., Potts, D. F., 2007. Analysis of flow competence in an alluvial gravel bed stream, Dupuyer Creek, Montana. *Water Resources Research*, **43**(7). 10.1029/2006wr005289.

White, W. R., Bettess, R., Paris, E., 1982. Analytical approach to river regime. *Journal of the Hydraulics Division – ASCE*, **108**(10), 1179–1193.

Whitehead, A. N., 1925. *Science and the Modern World*. The Free Press, New York.

Whiting, P. J., Dietrich, W. E., 1990. Boundary shear stress and roughness over mobile alluvial beds. *Journal of Hydraulic Engineering*, **116**(12), 1495–1511.

Whiting, P. J., Dietrich, W. E., 1991. Convective accelerations and boundary shear stress over a channel bar. *Water Resources Research*, **27**(5), 783–796.

Whiting, P. J., Dietrich, W. E., 1993a. Experimental studies of bed topography and flow patterns in large-amplitude meanders. 1. Observations. *Water Resources Research*, **29**(11), 3605–3614.

Whiting, P. J., Dietrich, W. E., 1993b. Experimental studies of bed topography and flow patterns in large-amplitude meanders 2. Mechanisms. *Water Resources Research*, **29**(11), 3615–3622.

Whittaker, A. C., Boulton, S. J., 2012. Tectonic and climatic controls on knickpoint retreat rates and landscape response times. *Journal of Geophysical Research – Earth Surface*, **117**. 10.1029/2011jf002157.

Whittaker, J. G., Jaeggi, M. N. R., 1982. Origin of step-pool systems in mountain streams. *Journal of the Hydraulics Division – ASCE*, **108**(6), 758–773.

Wiberg, P. L., Smith, J. D., 1987. Calculations of the critical shear stress for motion of uniform and heterogeneous sediments. *Water Resources Research*, **23**(8), 1471–1480.

Wiberg, P. L., Smith, J. D., 1991. Velocity distribution and bed roughness in high gradient streams. *Water Resources Research*, **27**(5), 825–838.

Wicks, J. M., Bathurst, J. C., 1996. SHESED: a physically based, distributed erosion and sediment yield component for the SHE hydrological modelling system. *Journal of Hydrology*, **175**(1–4), 213–238.

Wilcock, P. R., 1996. Estimating local bed shear stress from velocity observations. *Water Resources Research*, **32**(11), 3361–3366.

Wilcock, P. R., 1997. The components of fractional transport rate. *Water Resources Research*, **33**(1), 247–258.

Wilcock, P. R., 1998. Two-fraction model of initial sediment motion in gravel-bed rivers. *Science*, **280**(5362), 410–412.

Wilcock, P. R., 2012. Stream restoration in gravel-bed rivers. In: M. Church, P. Biron, A. Roy (eds.), *Gravel-Bed Rivers: Processes, Tools, Environments*. Wiley, Chichester, UK, pp. 137–146.

Wilcock, P. R., Crowe, J. C., 2003. Surface-based transport model for mixed-size sediment. *Journal of Hydraulic Engineering*, **129**(2), 120–128.

Wilcock, P. R., DeTemple, B. T., 2005. Persistence of armor layers in gravel-bed streams. *Geophysical Research Letters*, **32**(8). 10.1029/2004gl021772.

Wilcock, P. R., Kenworthy, S. T., 2002. A two-fraction model for the transport of sand/gravel mixtures. *Water Resources Research*, **38**(10). 10.1029/2001wr000684.

Wilcock, P. R., McArdell, B. W., 1993. Surface-based fractional transport rates – mobilization thresholds and partial transport of a sand-gravel mixture. *Water Resources Research*, **29**(4), 1297–1312.

Wilcock, P. R., McArdell, B. W., 1997. Partial transport of a sand/gravel sediment. *Water Resources Research*, **33**(1), 235–245.

Wilcox, A. C., Wohl, E. E., 2006. Flow resistance dynamics in step-pool stream channels: 1. large woody debris and controls on total resistance. *Water Resources Research*, **42**(5). 10.1029/2005wr004277.

Wilcox, A. C., Wohl, E. E., 2007. Field measurements of three-dimensional hydraulics in a step-pool channel. *Geomorphology*, **83**(3–4), 215–231.

Wilcox, A. C., Nelson, J. M., Wohl, E. E., 2006. Flow resistance dynamics in step-pool channels: 2. partitioning between grain, spill, and woody debris resistance. *Water Resources Research*, **42**(5). 10.1029/2005wr004278.

Wilcox, A. C., Wohl, E. E., Comiti, F., Mao, L., 2011. Hydraulics, morphology, and energy dissipation in an alpine step-pool channel. *Water Resources Research*, **47**. 10.1029/2010wr010192.

Wilkerson, G. V., Parker, G., 2011. Physical basis for quasi-universal relationships describing bankfull hydraulic geometry of sand-bed rivers. *Journal of Hydraulic Engineering*, **137**(7), 739–753.

Wilkinson, B. H., McElroy, B. J., 2007. The impact of humans on continental erosion and sedimentation. *Geological Society of America Bulletin*, **119**(1–2), 140–156.

Wilkinson, S. N., Keller, R. J., Rutherfurd, I. D., 2004. Phase-shifts in shear stress as an explanation for the maintenance of pool-riffle sequences. *Earth Surface Processes and Landforms*, **29**(6), 737–753.

Wilkinson, S. N., Prosser, I. P., Hughes, A. O., 2006. Predicting the distribution of bed material accumulation using river network sediment budgets. *Water Resources Research*, **42**(10). 10.1029/2006wr004958.

Wilkinson, S. N., Henderson, A., Chen, Y., Sherman, B., 2008. SedNet User Guide. Client Report, CSIRO Land and Water Canberra.

Wilkinson, S. N., Prosser, I. P., Rustomji, P., Read, A. M., 2009a. Modelling and testing spatially distributed sediment budgets to relate erosion processes to sediment yields. *Environmental Modelling & Software*, **24**(4), 489–501.

Wilkinson, S. N., Wallbrink, P. J., Hancock, G. J., et al., 2009b. Fallout radionuclide tracers identify a switch in sediment sources and transport-limited sediment yield following wildfire in a eucalypt forest. *Geomorphology*, **110**(3–4), 140–151.

Wilkinson, S. N., Dougall, C., Kinsey-Henderson, A. E., et al., 2014. Development of a time-stepping sediment budget model for assessing land use impacts in large river basins. *Science of the Total Environment*, **468**, 1210–1224.

Wilkinson, S. N., Kinsey-Henderson, A. E., Hawdon, A. A., et al., 2018. Grazing impacts on gully dynamics indicate approaches for gully erosion control in northeast Australia. *Earth Surface Processes and Landforms*, **43**(8), 1711–1725.

Willemin, J. H., 2000. Hack's law: sinuosity, convexity, elongation. *Water Resources Research*, **36**(11), 3365–3374.

Willenbring, J. K., Codilean, A. T., McElroy, B., 2013. Earth is (mostly) flat: apportionment of the flux of continental sediment over millennial time scales. *Geology*, **41**(3), 343–346.

Willett, S. D., Brandon, M. T., 2002. On steady states in mountain belts. *Geology*, **30**(2), 175–178.

Willett, S. D., McCoy, S. W., Perron, J. T., Goren, L., Chen, C.-Y., 2014. Dynamic reorganization of river basins. *Science*, **343**(6175). 10.1126/science.1248765.

Willgoose, G., 2005. Mathematical modeling of whole landscape evolution. *Annual Review of Earth and Planetary Sciences*, **33**, 443–459.

Willgoose, G., Bras, R. L., Rodriguez-Iturbe, I., 1991a. A coupled channel network growth and hillslope evolution model 1. theory. *Water Resources Research*, **27**(7), 1671–1684.

Willgoose, G., Bras, R. L., Rodriguez-Iturbe, I., 1991b. A coupled channel network and hillslope evolution model 2. nondimensionalization and applications. *Water Resources Research*, **27**(7), 1685–1696.

Willgoose, G., Bras, R. L., Rodriguez-Iturbe, I., 1991c. Results from a new model of river basin evolution. *Earth Surface Processes and Landforms*, **16**(3), 237–254.

Williams, G. P., 1978a. Bank-full discharge of rivers. *Water Resources Research*, **14**(6), 1141–1154.

Williams, G. P., 1978b. Hydraulic geometry of river cross sections – theory of minimum variance. U.S. Geological Survey Professional Paper 1029. U.S. Government Printing Office, Washington, DC.

Williams, G. P., 1986. River meanders and channel size. *Journal of Hydrology*, **88**, 147–164.

Williams, G. P., 1988. Paleofluvial estimates from dimensions of former channels and meanders. In: V. R. Baker, R. C. Kochel, P. C. Patton (eds.), *Flood Geomorphology*. Wiley, New York, pp. 321–334.

Williams, G. P., Wolman, M. G., 1984. Downstream effects of dams on alluvial rivers. U.S. Geological Survey Professional Paper 1286. U.S. Government Printing Office, Washington, DC.

Williams, J., 1996. Turbulent flow in rivers. In: P. A. Carling, M. R. Dawson (eds.), *Advances in Fluvial Dynamics and Stratigraphy*. Wiley, Chichester, UK, pp. 1–32.

Williams, J. R., Berndt, H. D., 1977. Sediment yield prediction based on watershed hydrology. *Transactions of the American Society of Agricultural Engineers*, **20**(6), 1100–1104.

Williams, R. D., Brasington, J., Vericat, D., Hicks, D. M., 2014. Hyperscale terrain modelling of braided rivers: fusing mobile terrestrial laser scanning and optical bathymetric mapping. *Earth Surface Processes and Landforms*, **39**(2), 167–183.

Williams, R. D., Rennie, C. D., Brasington, J., Hicks, D. M., Vericat, D., 2015. Linking the spatial distribution of bed load transport to morphological change during high-flow events in a shallow braided river. *Journal of Geophysical Research – Earth Surface*, **120**(3), 604–622. 10.1002/2014jf003346.

Williams, R. D., Brasington, J., Hicks, D. M., 2016a. Numerical modelling of braided river morphodynamics: review and future challenges. *Geography Compass*, **10**(3), 102–127.

Williams, R. D., Measures, R., Hicks, D. M., Brasington, J., 2016b. Assessment of a numerical model to reproduce event-scale erosion and deposition distributions in a braided river. *Water Resources Research*, **52**(8), 6621–6642.

Wilson, C. G., Papanicolaou, A. N. T., Denn, K. D., 2012. Partitioning fine sediment loads in a headwater system with intensive agriculture. *Journal of Soils and Sediments*, 12(6), 966–981. 10.1007/s11368-012-0504-2.

Wilson, G., 2011. Understanding soil-pipe flow and its role in ephemeral gully erosion. *Hydrological Processes*, **25**(15), 2354–2364.

Wilson, K. C., 1966. Bedload transport at high shear stresses. *Journal of the Hydraulics Division – ASCE*, **92**, 49–59.

Wilson, L., 1971. Drainage density, length ratios, and lithology in a glaciated area of southern Connecticut. *Geological Society of America Bulletin*, **82**(10), 2955–2956.

Wilson, L., 1973. Variations in mean annual sediment yield as a function of mean annual precipitation. *American Journal of Science*, **273**(4), 335–349.

Winant, C. D., Browand, F. K., 1974. Vortex pairing: the mechanism of turbulent mixing layer growth at moderate Reynolds numbers. *Journal of Fluid Mechanics*, **63**, 237–255.

Winemiller, K. O., McIntyre, P. B., Castello, L., et al., 2016. Balancing hydropower and biodiversity in the Amazon, Congo, and Mekong. *Science*, **351**(6269), 128–129.

Wishart, D., Warburton, J., Bracken, L., 2008. Gravel extraction and planform change in a wandering gravel-bed river: the River Wear, Northern England. *Geomorphology*, **94**(1–2), 131–152.

Wobus, C., Whipple, K. X., Kirby, E., et al., 2006a. Tectonics from topography: procedures, promise, and pitfalls. In: S. D. Willett, N. Hovius, M. T. Brandon, D. M. Fisher (eds.), Tectonics, Climate, and Landscape Evolution. Geological Society of America Special Papers, Special Paper 398. Geological Society of America, Denver, CO, pp. 55–74.

Wobus, C. W., Crosby, B. T., Whipple, K. X., 2006b. Hanging valleys in fluvial systems: controls on occurrence and implications for landscape evolution. *Journal of Geophysical Research – Earth Surface*, **111**(F2). 10.1029/2005jf000406.

Woelfle-Erskine, C., Wilcox, A. C., Moore, J. N., 2012. Combining historical and process perspectives to infer ranges of geomorphic variability and inform river restoration in a wandering gravel-bed river. *Earth Surface Processes and Landforms*, **37**(12), 1302–1312.

Wohl, E., 2004. Limits of downstream hydraulic geometry. *Geology*, **32**(10), 897–900.

Wohl, E., 2008. Review of effects of large floods in resistant-boundary channels. In: H. Habersack, H. Piegay, M. Rinaldi (eds.), *Gravel-Bed Rivers VI: From Process Understanding to River Restoration*. Elsevier, Amsterdam, the Netherlands, pp. 181–212.

Wohl, E., 2011a. What should these rivers look like? Historical range of variability and human impacts in the Colorado Front Range, USA. *Earth Surface Processes and Landforms*, **36**(10), 1378–1390.

Wohl, E., 2011b. Seeing the forest and the trees: wood in stream restoration in the Colorado Front Range, United States. In: A. Simon, S. J. Bennett, J. M. Castro (eds.), *Stream Restoration in Dynamic Fluvial Systems: Scientific Approaches, Analyses, and Tools*. American Geophysical Union, Washington, DC, pp. 399–418.

Wohl, E., 2013a. Field and laboratory experiments in fluvial geomorphology. In: J. W. Shroder (ed.), *Treatise on Geomorphology*, Vol. 9, Fluvial geomorphology, E. Wohl (vol. ed.). Academic Press, San Diego, CA, pp. 679–693.

Wohl, E., 2013b. Migration of channel heads following wildfire in the Colorado Front Range, USA. *Earth Surface Processes and Landforms*, **38**(9), 1049–1053.

Wohl, E., 2014. A legacy of absence: wood removal in US rivers. *Progress in Physical Geography*, **38**(5), 637–663.

Wohl, E., 2015a. Legacy effects on sediments in river corridors. *Earth-Science Reviews*, **147**, 30–53.

Wohl, E., 2015b. Of wood and rivers: bridging the perception gap. *Wiley Interdisciplinary Reviews – Water*, **2**(3), 167–176.

Wohl, E., David, G. C. L., 2008. Consistency of scaling relations among bedrock and alluvial channels. *Journal of Geophysical Research – Earth Surface*, **113**(F4). 10.1029/2008jf000989.

Wohl, E. E., Grodek, T., 1994. Channel bed-steps along Nahael Yael, Negev Desert, Israel. *Geomorphology*, **9**(2), 117–126.

Wohl, E., Legleiter, C. J., 2003. Controls on pool characteristics along a resistant-boundary channel. *Journal of Geology*, **111**(1), 103–114.

Wohl, E., Merritt, D. M., 2007. What is a natural river? *Geography Compass*, **1**, 871–900.

Wohl, E., Merritt, D. M., 2008. Reach-scale channel geometry of mountain streams. *Geomorphology*, **93**(3–4), 168–185.

Wohl, E., Scott, D. N., 2017. Wood and sediment storage and dynamics in river corridors. *Earth Surface Processes and Landforms*, **42**(1), 5–23.

Wohl, E. E., Thompson, D. M., 2000. Velocity characteristics along a small step-pool channel. *Earth Surface Processes and Landforms*, **25**(4), 353–367.

Wohl, E. E., Vincent, K. R., Merritts, D. J., 1993. Pool and riffle characteristics in relation to channel gradient. *Geomorphology*, **6**(2), 99–110.

Wohl, E., Madsen, S., MacDonald, L., 1997. Characteristics of log and clast bed-steps in step-pool streams of northwestern Montana, USA. *Geomorphology*, **20**(1–2), 1–10.

Wohl, E., Cenderelli, D., Mejia-Navarro, M., 2001. Channel change from extreme floods in bedrock canyons. In: D. J. Anthony, M. D. Harvey, D. Laronne, M. P. Mosley (eds.), *Applying Geomorphology to Environmental Management*. Water Resources Publications, LLC, Highlands Ranch, CO, pp. 149–174.

Wohl, E., Angermeier, P. L., Bledsoe, B., et al., 2005. River restoration. *Water Resources Research*, **41**(10). 10.1029/2005wr003985.

Wohl, E., Lane, S. N., Wilcox, A. C., 2015. The science and practice of river restoration. *Water Resources Research*, **51**(8), 5974–5997. 10.1002/2014wr016874.

Wohl, E., Lininger, K. B., Fox, M., Baillie, B. R., Erskine, W. D., 2017. Instream large wood loads across bioclimatic regions. *Forest Ecology and Management*, **404**, 370–380.

Wolman, M. G., 1955. The natural channel of Brandywine Creek Pennsylvania, U.S. Geological Survey Professional Paper 271. U.S. Government Printing Office, Washington, DC.

Wolman, M. G., 1967. A cycle of erosion and sedimentation in urban river channels. *Geografiska Annaler Series A Physical Geography*, **49**, 385–395.

Wolman, M. G., Gerson, R., 1978. Relative scales of time and effectiveness in watershed geomorphology. *Earth Surface Processes and Landforms*, **3**(2), 189–208.

Wolman, M. G., Leopold, L. B., 1957. River floodplains: some observations on their formation, U.S. Geological Survey Professional Paper 282-C. U.S. Government Printing Office, Washington, DC.

Wolman, M. G., Miller, J. P., 1960. Magnitude and frequency of forces in geomorphic processes. *Journal of Geology*, **68**(1), 54–74.

Wolman, M. G., Schick, A. P., 1967. Effects of construction on fluvial sediment, urban and suburban areas of Maryland. *Water Resources Research*, **3**(2), 451–464.

Wolter, A., Ward, B., Millard, T., 2010. Instability in eight sub-basins of the Chilliwack River Valley, British Columbia, Canada: a comparison of natural and logging-related landslides. *Geomorphology*, **120**(3–4), 123–132.

Womble, P., Doyle, M., 2012. The geography of trading ecosystem services: a case study of wetland and stream compensatory mitigation markets. *Harvard Environmental Law Review*, **36**(1), 229–296.

Wong, M., Parker, G., 2006. Reanalysis and correction of bed-load relation of Meyer-Peter and Muller using their own database. *Journal of Hydraulic Engineering*, **132**(11), 1159–1168.

Wong, M., Parker, G., DeVries, P., Brown, T. M., Burges, S. J., 2007. Experiments on dispersion of tracer stones under lower-regime plane-bed equilibrium bed load transport. *Water Resources Research*, **43**(3). 10.1029/2006wr005172.

Wood, A. L., Simon, A., Downs, P. W., Thorne, C. R., 2001. Bank-toe processes in incised channels: the role of apparent cohesion in the entrainment of failed bank materials. *Hydrological Processes*, **15**(1), 39–61.

Wood, P. J., Armitage, P. D., 1997. Biological effects of fine sediment in the lotic environment. *Environmental Management*, **21**(2), 203–217.

Woodget, A. S., Carbonneau, P. E., Visser, F., Maddock, I. P., 2015. Quantifying submerged fluvial topography using hyperspatial resolution UAS imagery and structure from motion photogrammetry. *Earth Surface Processes and Landforms*, **40**(1), 47–64.

Woodget, A. S., Austrums, R., Maddock, I. P., Habit, E., 2017. Drones and digital photogrammetry: from classifications to continuums for monitoring river habitat and hydromorphology. *WIREs Water*, **4**(4), e1222.

Woodyer, K. D., 1968. Bankfull frequency in rivers. *Journal of Hydrology*, **6**(2), 114–142.

Wooldridge, C. L., Hickin, E. J., 2002. Step-pool and cascade morphology, Mosquito Creek, British Columbia: a test of four analytical techniques. *Canadian Journal of Earth Sciences*, **39**(4), 493–503.

Wooldridge, C. L., Hickin, E. J., 2005. Radar architecture and evolution of channel bars in wandering gravel-bed rivers: Fraser and Squamish rivers, British Columbia, Canada. *Journal of Sedimentary Research*, **75**(5), 844–860.

World Energy Council, 2019. Energy Resources: Hydropower. www.worldenergy.org/data/resources/resource/hydropower/ (accessed June 6, 2019).

Worrall, F., Burt, T. P., Howden, N. J. K., Hancock, G. R., 2014. Variation in suspended sediment yield across the UK – a failure of the concept and interpretation of the sediment delivery ratio. *Journal of Hydrology*, **519**, 1985–1996.

Wren, D. G., Davidson, G. R., Walker, W. G., Galicki, S. J., 2008. The evolution of an oxbow lake in the Mississippi alluvial floodplain. *Journal of Soil and Water Conservation*, **63**(3), 129–135.

Wright, K. A., Sendek, K. H., Rice, R. M., Thomas, R. B., 1990. Logging effects on streamflow – storm runoff at Caspar Creek in northwestern California. *Water Resources Research*, **26**(7), 1657–1667.

Wright, S., Parker, G., 2004. Density stratification effects in sand-bed rivers. *Journal of Hydraulic Engineering – ASCE*, **130**(8), 783–795.

Wright, S., Parker, G., 2005. Modeling downstream fining in sand-bed rivers. II: application. *Journal of Hydraulic Research*, **43**(6), 621–631.

Wu, B., Zheng, S., Thorne, C. R., 2012. A general framework for using the rate law to simulate morphological response to disturbance in the fluvial system. *Progress in Physical Geography*, **36**(5), 575–597.

Wyrick, J. R., Pasternack, G. B., 2008. Modeling energy dissipation and hydraulic jump regime responses to channel nonuniformity at river steps. *Journal of Geophysical Research – Earth Surface*, **113**(F3). 10.1029/2007jf000873.

Wyrick, J. R., Pasternack, G. B., 2014. Geospatial organization of fluvial landforms in a gravel-cobble river: beyond the riffle-pool couplet. *Geomorphology*, **213**, 48–65.

Wyzga, B., 1996. Changes in the magnitude and transformation of flood waves subsequent to the channelization of the Raba River, Polish Carpathians. *Earth Surface Processes and Landforms*, **21**(8), 749–763.

Xia, J., Wu, B., Wang, G., Wang, Y., 2010. Estimation of bankfull discharge in the Lower Yellow River using different approaches. *Geomorphology*, **117**(1–2), 66–77.

Xu, J. X., 1996. Channel pattern change downstream from a reservoir: an example of wandering braided rivers. *Geomorphology*, **15**(2), 147–158.

Xu, J. X., 2002. Implication of relationships among suspended sediment size, water discharge and suspended sediment concentration: the Yellow River basin, China. *Catena*, **49**(4), 289–307.

Xu, J. X., 2004. Channel pattern discrimination based on the relationship between channel slope and width. *Zeitschrift fur Geomorphologie*, **48**(3), 391–401.

Xu, J., 2008. Discrimination of channel patterns for gravel-and sand-bed rivers. *Zeitschrift fur Geomorphologie*, **52**(4), 503–523.

Xu, J., Yan., Y., 2005. Scale effects on specific yield in the Yellow River basin and geomorphological explanations. *Journal of Hydrology*, **307**, 219–232.

Yaeger, M., Coopersmith, E., Ye, S., et al., 2012. Exploring the physical controls of regional patterns of flow duration curves – Part 4: a synthesis of empirical analysis, process modeling and catchment classification. *Hydrology and Earth System Sciences*, **16**(11), 4483–4498.

Yager, E. M., Kirchner, J. W., Dietrich, W. E., 2007. Calculating bed load transport in steep boulder bed channels. *Water Resources Research*, **43**(7). 10.1029/2006wr005432.

Yager, E. M., Dietrich, W. E., Kirchner, J. W., McArdell, B. W., 2012. Prediction of sediment transport in step-pool channels. *Water Resources Research*, **48**. 10.1029/2011wr010829.

Yalin, M., 1971. On the formation of dunes and meanders, Proceedings of the 14th Congress, International Association of Hydraulic Research, 3, Paper C13, pp. 1–8.

Yalin, M. S., 2006. Large-scale turbulence and river morphology. In: R. M. L. Ferreira, C. T. L. Alves, G. A. B. Leal, A. H. Cardoso (eds.), *River Flow 2006*, Vols 1 and 2, CRC Press, Boca Raton, FL, pp. 1243–1249.

Yalin, M. S., Da Silva, A. M. F., 2000. Computation of regime channel characteristics on thermodynamic basis. *Journal of Hydraulic Research*, **38**(1), 57–63.

Yalin, M. S., Karahan, E., 1979. Inception of sediment transport. *Journal of the Hydraulics Division – ASCE*, **105**(11), 1433–1443.

Yan, Q., Iwasaki, T., Stumpf, A., et al., 2018. Hydrogeomorphological differentiation between floodplains and terraces. *Earth Surface Processes and Landforms*, **43**(1), 218–228.

Yang, C. T., 1971. On river meanders. *Journal of Hydrology*, **13**, 231–253.

Yang, C. T., 1973. Incipient motion and sediment transport. *Journal of the Hydraulics Division – ASCE*, **99**, 1679–1704.

Yang, C. T., 1984. Unit stream power equation for gravel. *Journal of Hydraulic Engineering*, **110**(12), 1783–1797.

Yang, C. T., 1994. Variational theories in hydrodynamics and hydraulics. *Journal of Hydraulic Engineering*, **120**(6), 737–756.

Yang, C. T., Song, C. C. S., 1979. Theory of minimum rate of energy dissipation. *Journal of the Hydraulics Division – ASCE*, **105**(7), 769–784.

Yang, C. T., Wan, S. G., 1991. Comparisons of selected bed-material load formulas. *Journal of Hydraulic Engineering*, **117**(8), 973–989.

Yang, C. T., Song, C. C. S., Woldenberg, M. J., 1981. Hydraulic geometry and minimum rate of energy dissipation. *Water Resources Research*, **17**(4), 1014–1018.

Yang, K., Cao, S., Knight, D. W., 2007. Flow patterns in compound channels with vegetated floodplains. *Journal of Hydraulic Engineering*, **133**(2), 148–159.

Yang, Q. Y., Liu, T. H., Lu, W. Z., Wang, X. K., 2013. Numerical simulation of confluence flow in open channel with dynamic meshes techniques. *Advances in Mechanical Engineering*. 10.1155/2013/860431.

Yang, S. L., Xu, K. H., Milliman, J. D., Yang, H. F., Wu, C. S., 2015a. Decline of Yangtze River water and sediment discharge: impact from natural and anthropogenic changes. *Scientific Reports*, **5**. 10.1038/srep12581.

Yang, S. Q., 2005. Sediment transport capacity in rivers. *Journal of Hydraulic Research*, **43**(2), 131–138.

Yang, S. Q., Tan, S. K., Lim, S. Y., 2004. Velocity distribution and dip-phenomenon in smooth uniform open channel flows. *Journal of Hydraulic Engineering*, **130**(12), 1179–1186.

Yang, S. Q., Tan, S. K., Wang, X.-K., 2012. Mechanism of secondary currents in open channel flows. *Journal of Geophysical Research – Earth Surface*, **117**. 10.1029/2012jf002510.

Yang, X. L., Zhang, Q. Y., Li, X. Z., et al., 2015b. Determination of soil texture by laser diffraction method. *Soil Science Society of America Journal*, **79**(6), 1556–1566.

Yanites, B. J., 2018. The dynamics of channel slope, width, and sediment in actively eroding bedrock river systems. *Journal of Geophysical Research – Earth Surface*, **123**(7), 1504–1527. 10.1029/2017jf004405.

Yanites, B. J., Tucker, G. E., 2010. Controls and limits on bedrock channel geometry. *Journal of Geophysical Research – Earth Surface*, **115**(F4). 10.1029/2009jf001601.

Yatsu, E., 1955. On the longitudinal profile of the graded river. *Transactions, American Geophysical Union*, **36**(4), 655–663.

Yeakley, J. A., Ervin, D., Chang, H., et al., 2016. Ecosystem services of streams and rivers. In: D. J. Gilvear, M. T. Greenwood, M. C. Thoms, P. J. Woods (eds.), *River Science: Research and Management for the 21st Century*. Wiley, Chichester, UK, pp. 335–352.

Yeh, T. H., Parker, G., 2013. Software for evaluating sediment-induced stratification in open-channel flows. *Computers & Geosciences*, **53**, 94–104.

Yen, B. C., 2002. Open channel flow resistance. *Journal of Hydraulic Engineering*, **128**(1), 20–39.

Yochum, S. E., 2018. Guidance for Stream Restoration. Forest Service: National Stream & Aquatic Ecology Center Technical Note TN-102.4. U.S. Department of Agriculture, Ft. Collins, CO.

Yochum, S. E., Bledsoe, B. P., David, G. C. L., Wohl, E., 2012. Velocity prediction in high-gradient channels. *Journal of Hydrology*, **424**, 84–98.

Yochum, S. E., Comiti, F., Wohl, E., David, G. C. L., Mao, L., 2014. Photographic Guidance for Selecting Flow Resistance Coefficients in High-gradient Channels. Gen. Tech. Rep. RMRS-GTR-323. Forest Service, Rocky Mountain Research Station, U.S. Department of Agriculture, Ft. Collins, CO.

Yochum, S. E., Sholtes, J. S., Scott, J. A., Bledsoe, B. P., 2017. Stream power framework for predicting geomorphic change: the 2013 Colorado Front Range flood. *Geomorphology*, **292**, 178–192.

Yokoo, Y., Sivapalan, M., 2011. Towards reconstruction of the flow duration curve: development of a conceptual framework with a physical basis. *Hydrology and Earth System Sciences*, **15**(9), 2805–2819.

Young, W. J., Davies, T. R. H., 1991. Bedload transport processes in a braided gravel-bed river model. *Earth Surface Processes and Landforms*, **16**(6), 499–511.

Yu, B., Wolman, M. G., 1987. Some dynamic aspects of river geometry. *Water Resources Research*, **23**(3), 501–509.

Yu, M., Rhoads, B. L., 2018. Floodplains as a source of fine sediment in grazed landscapes: tracing the source of suspended sediment in the headwaters of an intensively managed agricultural landscape. *Geomorphology*, **308**, 278–292.

Yuan, S., Tang, H., Xiao, Y., et al., 2016. Turbulent flow structure at a 90-degree open channel confluence: accounting for the distortion of the shear layer. *Journal of Hydro-Environment Research*, **12**, 130–147.

Yuan, S., Tang, H., Xiao, Y., Qui, X., Xia, Y., 2017. Water flow and sediment transport at open-channel confluences: an experimental study. *Journal of Hydraulic Research*, **56**(3). 10.1080/00221686.2017.1354932.

Yuill, B. T., Khadka, A. K., Pereira, J., Allison, M. A., Meselhe, E. A., 2016a. Morphodynamics of the erosional phase of crevasse-splay evolution and implications for river sediment diversion function. *Geomorphology*, **259**, 12–29.

Yuill, B. T., Gaweesh, A., Allison, M. A., Meselhe, E. A., 2016b. Morphodynamic evolution of a lower Mississippi River channel bar after sand mining. *Earth Surface Processes and Landforms*, **41**(4), 526–542.

Zalasiewicz, J., Waters, C. N., Summerhayes, C. P., et al., 2017. The Working Group on the Anthropocene: summary of evidence and interim recommendations. *Anthropocene*, **19**, 55–60.

Zanoni, L., Gurnell, A., Drake, N., Surian, N., 2008. Island dynamics in a braided river from analysis of historical maps and air photographs. *River Research and Applications*, **24**(8), 1141–1159.

Zarfl, C., Lumsdon, A. E., Berlekamp, J., Tydecks, L., Tockner, K., 2015. A global boom in hydropower dam construction. *Aquatic Sciences*, **77**(1), 161–170.

Zen, S., Zolezzi, G., Toffolon, M., Gurnell, A. M., 2016. Biomorphodynamic modelling of inner bank advance in migrating meander bends. *Advances in Water Resources*, **93**, 166–181.

Zhang, C., Xu, M., Hassan, M. A., Chartrand, S. M., Wang, Z., 2018. Experimental study on the stability and failure of individual step-pool. *Geomorphology*, **311**, 51–62.

Zhang, L. T., Li, Z. B., Wang, S. S., 2016b. Spatial scale effect on sediment dynamics in basin-wide floods within a typical agro-watershed: a case study in the hilly loess region of the Chinese Loess Plateau. *Science of the Total Environment*, **572**, 476–486.

Zhang, Y., Slingerland, R., Duffy, C., 2016a. Fully-coupled hydrologic processes for modeling landscape evolution. *Environmental Modelling & Software*, **82**, 89–107.

Zhou, Z., Coco, G., Townend, I., et al., 2017. Is "Morphodynamic Equilibrium" an oxymoron? *Earth-Science Reviews*, **165**, 257–267.

Ziliani, L., Surian, N., Coulthard, T. J., Tarantola, S., 2013. Reduced-complexity modeling of braided rivers: assessing model performance by sensitivity analysis, calibration, and validation. *Journal of Geophysical Research – Earth Surface*, **118**(4), 2243–2262. 10.1002/jgrf.20154.

Zimmermann, A. E., 2013. Step-pool channel features. In: J. Shroder (ed.), *Treatise on Geomorphology*, Vol. 9, Fluvial Geomorphology, E. Wohl (vol. ed.). Academic Press, San Diego, CA, pp. 346–363.

Zimmermann, A. E., Church, M., 2001. Channel morphology, gradient profiles and bed stresses during flood in a step-pool channel. *Geomorphology*, **40**(3–4), 311–327.

Zimmermann, A. E., Church, M., Hassan, M. A., 2008. Identification of steps and pools from stream longitudinal profile data. *Geomorphology*, **102**(3–4), 395–406.

Zimmermann, A. E., Church, M., Hassan, M. A., 2010. Step-pool stability: testing the jammed state hypothesis. *Journal of Geophysical Research – Earth Surface*, **115**. 10.1029/2009jf001365.

Zinger, J. A., Rhoads, B. L., Best, J. L., 2011. Extreme sediment pulses generated by bend cutoffs along a large meandering river. *Nature Geoscience*, **4**(10), 675–678.

Zinger, J. A., Rhoads, B. L., Best, J. L., Johnson, K. K., 2013. Flow structure and channel morphodynamics of meander bend chute cutoffs: a case study of the Wabash River, USA. *Journal of Geophysical Research – Earth Surface*, **118**(4), 2468–2487. 10.1002/jgrf.20155.

Zingg, T., 1935. Beitrage zur Schotteranalyse. *Schweizerische Mineralogische und Petrologische Mitteilungen*, **15**, 39–140.

Zolezzi, G., Seminara, G., 2001. Downstream and upstream influence in river meandering. Part 1. General theory and application to overdeepening. *Journal of Fluid Mechanics*, **438**, 183–211.

Zolezzi, G., Bertoldi, W., Tubino, M., 2006. Morphological analysis and prediction of river bifurcations. In: G. H. S. Smith, J. L. Best, C. S. Bristow, G. E. Petts (eds.), *Braided Rivers: Process, Deposits, Ecology and Management*. Special Publications of the International Association of Sedimentologists, 36, Blackwell, Oxford, UK, pp. 233–256.

Index

abductive reasoning, 10, 145

abrasion, 297, 301, 302, 303, 304, 306, 308
role in downstream fining, 298–299

advective processes, 16, 19, 26, 28, 34, 50, 63, 88, 130, 290, 291, 310, 323, 325, 326

aggradation, 125, 149
and agriculture, 348
and anabranching, 253, 254, 261
and anastomosis, 261, 342
and avulsion, 194, 237, 246, 254, 255
and braiding, 237, 240, 249, 337
and channelization, 359
and complex response, 160
and dam removal, 396, 397, 398
and dams, 364, 366
and downstream fining, 300, 306
and floods, 163
and legacy sediment or post-settlement alluvium, 347
and mega rivers, 266
and mining waste, 351
and river equilibrium, 145, 157, 234, 372
and stream instability, 380
and the longitudinal profile, 297
and timber harvesting, 350
and urbanization, 354
at confluences, 272
channelization, 361

aggradational-degradational episode, 351

agriculture, 371, 375, 399
and soil erosion, 345–346
effects of soil erosion on rivers, 346
effects on river morphology, 347–350
global impact of, 344
impacts on hydrology, 345

alluvial architecture, 335, 337, 340, 341

alluvial fans, 39, 257, 261, 264

alluvium, 144, 225, 257, 264, 314, 320, 321, 331, 332, 333, 334, 335, 347, 349, 357, 397

alternate bars, 193, 198, 200, 201, 204, 235, 237, 321, *See also* bar units
relation to bar units, 198

Amazon River, 3, 48, 55, 60, 266

anabranching rivers, 252, 330, 332, 349, 383, 392
and mega rivers, 195, 266–268
and optimization principles, 252–253
difference from braided rivers, 187
environmental domain of, 194–195, 252
equilibrium of, 253
importance of avulsion in, 255–256
in arid and semi-arid North and South America, 262–264
in Australia, 261–262
on the Okavango Delta, 264–266
processes of anabranching, 253–254
vegetation and bank resistance, 256

anastomosing rivers, 194, 252, 256, 259–261, 264, 266, 267, 325, 330, 383, 392
and equilibrium, 261
environmental domain of, 194

annual maximum series, 138, 141

Anthropocene, 343

antidunes, 84, 314, 315

armor. *See* bed material

arroyo, 149, 357

avulsion, 13, 162, 241, 243, 246, 249, 250, 254, 255–256, 257, 259, 262, 263, 264, 267, 268, 328, 332, 335, 337, 339, 342

backswamp, 336, 386

bank erosion, 98, 99, 100, 102, 104, 214, 221–224, 232, 237, 242, 246, 248, 258, 333, 350, 355, 365, 378, 388, 391, 392, 393
and basal endpoint control, 223
and cattle grazing, 350
and lateral channel migration, 224–225
importance in braided rivers, 248

bank material, 98
and channel geometry, 175, 176

bankfull discharge, 155, 156, 172, 173, 174, 175, 187, 188, 253, 359, 380, 390, 391
determination of, 151
recurrence interval of, 146
relation to dominant discharge, 145–146
relation to effective discharge, 152

bankfull stage, 145, 146–151, 383

complicating factors, 147–151
determination of, 146

bar elements, 12, 13, 208, 209, 210, 211

bar theory
and initiation of braiding, 234–236, 237, 238, 241
of initiation of meandering, 202–203

bar units, 12, 198, 202, 203, 220, 225, 254, 258, 269, 285
and braiding, 234, 235, 236, 237, 238, 239, 241, 245
and confluence-diffluence units, 241, 242
and initiation of meandering, 198–200, 202–205
at confluences, 284–285
forced, 200, 203, 207, 217
in elongate meander bends, 227–228

bar-bend theory
of initiation of meandering, 202, 203–205

bar-element scale, 12, 83, 85, 127, 170, 198, 207, 247, 250, 269, 386, 393, 400

bars, 98, 234, 253, 257, 386
bank-attached, at confluences, 283–284, 285, 286, 289, 292
compound, 237, 245, 250, 258, 337, 339, 340
in braided rivers, 187, 239
mid-channel, at confluences, 285
plug, 328, 329, 330
point. *See* point bar
tributary-mouth, 283, 284, 285
unit, 237, 245, 246, 250, 258, 301, 321, 330, 337, 340

bar-unit scale, 12, 83, 85, 127, 132, 160, 170, 198, 246, 247, 250, 296, 297, 301, 313, 386, 393, 400

base level, 4, 20, 30, 32, 34, 36, 47, 160, 261, 302, 303, 308, 309, 311, 328, 396

baseflow, 135, 143

beaver, in stream restoration, 395

bed material, 92
active layer, 122, 125, 129, 130
and channel geometry, 176
armor, 359, 360, 365, 367, 368
armor, static versus mobile, 125

bed material (cont.)
coarse surface layer, 123–125
downstream fining of, 122, 128, 297–305
winnowing of, 125
bedform scale, 12, 83, 85
bedforms, 12, 13, 84–85, 114, 120, 132, 217, 268, 278, 302, 321
bedload, 100, 116, 119, 122, 152
coarse particle tracing and transport, 129–131
efficiency, 121
Eulerian perspective, 119, 120, 122, 127, 129, 130
global flux, 48
Lagrangian perspective, 129, 130
morphological method of determining transport rate, 132
particle travel distances, 129
role of grain kinematics in transport, 129, 131
sheets, 237, 258, 339, 340
transport and excess boundary shear stress, 120
transport and river equilibrium, 156
transport and stream power, 120–122
transport equations, basic form of, 120
transport intensity, 120
transport, measurement of. See Appendix D
volumetric transport rate, 120, 130, 132
bed-material entrainment, 105–116
and impulse, 112, 115
and thresholds of motion, 108–109
equal mobility, 111, 112, 125, 300–301
in particle mixtures, 109–112, 302
into suspension, 112, 116
size selective, 109, 111, 125
bed-material load, 100, 122, 242, 249, 282, 283, 303, 329, 332, 363, 364, 365, 367, 398, 400
accuracy of predictive models, 122
and changes in channel morphology, 131–133
and equal mobility, 125, 126
and mobile armor layers, 126
equal mobility, 220
fractional transport of, 122–128
fractional transport rates, 125
partial transport, 125
predicting fractional transport rates, 127–128
size-selective transport, 126, 297, 299, 300, 301

suspended, 100, 119, 120, 121, 132
transport and stream power, 120–122
bedrock rivers, 3, 12, 97, 130, 161, 163, 174, 225, 255, 256, 306, 311
erosion of, 307–308
bench index, 148
benches
within river channels, 147–149, 361, 400
bend theory, 203
of initiation of meandering, 202
bifurcation ratio, 41, 137
bifurcations, 237, 241, 246, 248, 250, 258, 268, 269, 329
dynamics of, 242–246
biodiversity, 392, 393, 395, 399, 401
boundary conditions
rough versus smooth, 108
boundary layer
around particle, 116
boundary layer theory, 86
relation to flow resistance and boundary shear stress, 87
boundary layer, turbulent, 87
buffer layer, 87, 91
over a smooth bed, 88
turbulent layer, 88
viscous sublayer, 87
boundary Reynolds number, 106, 107, 183
braided rivers, 234, 256, 260, 261, 267, 268, 285, 321, 330, 332, 367, 383, 389, 392
difference from anabranching rivers, 187
equilibrium of, 248
initiation of braiding, 234–237
morphological attributes of, 237–241
planform dynamics, 246–251
process-form interactions in, 241–246
simulation models of, 249–251
braiding index, 240
active, 246
and bed-material load, 246–247
braiding intensity, 249, 251
braidplain, 187, 338
braidtrain, 187, 338, 339
buoyancy, 291
and mixing at confluences, 292

celerity, 80, 81, 101
centrifugal force, 96, 213, 218, 244, 276, 277
channel classification
and the longitudinal profile, 313–314
in River Styles Framework, 386
in Rosgen Method, 379–380
channel evolution model, 359, 382, 396
Rosgen Method version of, 383

channel formative event, 6, 144, 158, 159, 181, 184, 185, 357
channel geometry, 144, 183, 367, See also rational regime theory, downstream channel geometry
and feedback among variables, 178
models of change over time, 183–185
channel heads, 25–29, 30, 33, 38, 39
and climate, 28
channel infills, 328–331, 335, 336, 337, 339, 340, 342, 354
channel initiation, 27, 28, 36, 45
by overland flow, 16–19
by subsurface flow, 20, 21
stability analysis, 17–20, 21
channel planform, 12, 85, 182, 241, 313, 327, 339, 367, 386
change over time and space, 195–196
differences between meandering and braided rivers, 188–192
factors influencing, 188
importance in river classification, 186
major types of, 186–188
channel slope, 309, 317, 321, 349, 355, 358, 359, 360, 365, 380
and concept of a graded river, 176, 294
defined, 76
dependence on channel planform, 188, 190, 206
equilibrium, 307
interdependence with grain size of bed material, 297, 300, 305
relation to controlling factors, 183, 296
relation to the longitudinal profile, 294
channelization, 367, 371, 375, 382, 396, 399, 400
definition of, purpose, 358
effect on channel stability, 358–359
geomorphic responses of rivers to, 359–361
characteristic form, 12, 13, 145, 158–159, 178, 357
Chezy coefficient, 82, 83, 216
Chezy equation, 82
Clean Water Act, 371, 373, 376, 391
climate, 12, 14, 24, 34, 42, 43, 49, 50, 53, 59, 60, 62, 70, 104, 140, 143, 155, 156, 159, 307, 343, 344
and geomorphic effectiveness of floods, 163
as independent variable in fluvial systems, 12, 15, 24, 49, 308, 357
climate change, 11, 15, 22, 37, 143, 178, 195, 309, 363, 403
effects on rivers, 2, 356–357

Colorado River, 4, 363, 364
competence of flow, 86, 107, 130, 211, 237, 238, 242, 297, 300, 302, 316, 364, 365, 391
    defined, 123
complex response, 160
concave-bank bench, 216, 323, 386
concavity index, 307
confluence hydrodynamic zone, 272
    characteristics of flow within, 273–274
    factors influencing flow within, 274–275
    flow within the mixing interface, 275–276
    helical motion of flow within, 276–278
    influence of bed discordance on flow in, 280–281
    secondary flow in, 278
confluence-diffluence units, 238, 239, 241–242, 246, 268, 269
confluences, 11, 241, 248, 250, 309
    and river ecology, 269
    as fundamental components of river networks, 269
    as planform-scale components of fluvial systems, 269
    bar-unit dynamics in braided rivers, 242
    bed discordance at, 275, 279, 291
    bed morphology of, 281–285
    change in channel form at, 271–272
    changes in energy and depth at, 272–273
    factors affecting junction angle, 270–271
    flow in. See confluence hydrodynamic zone
    in braided and anabranching rivers, 269, 290
    lateral mixing at, 290–293
    planform dynamics of, 290
    planform geometry of, 269–270
    scour at, 282–283, 285, 286
    sediment transport in, 285–287
confluent meander bends, 287–290
connectivity
    and dam removal, 396
    and river restoration, 394–395
    between channels and floodplains, 266, 321, 392
    between sediment sources and sinks, 65
    disruption of, by dams, 362
    ecological, 373
    of hillslopes and rivers, 55, 58
    of hydrological pathways, 103
    of river systems, 13
    of runoff and erosion, 344
    of sediment fluxes, 66, 67, 104, 159, 163, 346, 349, 353
    versus disconnectivity, 67
conservation of mass

and sediment budgets, 65, 71
    as a theoretical principle, 7
    in bed-material transport, 131, 132
constant of channel maintenance, 24
contamination, of bed material, 351, 354
conveyance capacity, 255, 261, 358
Cooper Creek, 261
counterpoint accretion, 323
critical flow, 80
curvature, of meandering channels, 206, 207, 229, 230, 231
cut banks, 207, 208, 210, 221, 225, 332
cutoffs, 226–227, 229, 231, 236, 237, 248, 258, 267, 328, 329, 332, 333, 335, 337
    chute, 329
    neck, 330
cycle of erosion, 5, 8, 15, 29

dam removal, 394
    and sediment flux, 396
    geomorphic responses of rivers to, 396–399
    reasons for, 396
dams, 371, 401
    abundance of, 361–362
    and global sediment flux, 58–59
    effects on river flow, 362–363
    effects on river morphology, 364–366
    impoundment-runoff index, 362
    influence on sediment transport, 363–364
    trap efficiency, 58, 363
Darcy-Weisbach equation, 82
Darcy-Weisbach friction factor, 82, 83, 318
Davis, William Morris, 5
deductive reasoning, 10
degradation, 149, 240, 351
    and channelization, 359, 360
    and dam removal, 396
    and dams, 364
    and river equilibrium, 145, 157, 234, 372
    and stream instability, 380
    and urbanization, 355
    defined, 124
    response to in-channel mining, 367
deltas, 39, 47, 49, 59, 303, 327, 363
denudation rate, 49
    and cosmogenic nuclides, 61–63
    and measured data on sediment load, 60
    relation to controlling factors, 62–63
    relation to relief, 52–53
depth of flow
    local, 77
    mean over a cross section, 73

diffusive processes, 15, 20, 28, 33, 39, 50, 63, 78, 117, 130, 290, 310, 312, 323, 325, 326, 367
    and overbank sedimentation on floodplains, 325
    in bedload transport, 130–131
discharge, 72
    as the basic metric of flow in rivers, 72
    bankfull. See bankfull discharge
    critical unit, 112
    definition of, 72
    dominant. See dominant discharge
    effective. See effective discharge
    functional-equivalent, 155
    half-load, 155
    mean annual, 165, 171, 172, 248
    mean daily, 142, 143, 151, 363
    measurement of, 135, See Appendix C
    peak, 135
disequilibrium, 159, 309, 333
dissolved load, 60, 97
    global flux, 48
dominant discharge, 144–145, 147, 157, 402
    and downstream hydraulic geometry, 172
    relation to bankfull discharge, 145–146
downstream channel geometry, 174, 184, 271, See also rational regime theory
    bedrock rivers, 174–175
    braided rivers, 241
    integrated empirical and analytical approach, 183
    multivariate models of, 175–178
    simultaneous-equation models of, 178
    versus downstream hydraulic geometry, 173–174
downstream fining. See bed material, gravel-sand transitions
drainage area ratio, 41
drainage basin
    and the scale of river systems, 11
    circularity, 32
    definition of, 3
    evolution of, 12, 15, 37–38
    in equilibrium or steady state, 305
    mass balance of, 63
    relation to river dynamics, 4
    stream length-drainage area relation, 44
drainage density, 24, 36, 37
drainage ditch, 345, 361, 400
drainage divide, 3, 21, 25, 26
    migration of, 37–38
drainage-basin scale, 49, 63, 159, 311, 373, 386
dunes, 84, 114, 169, 217, 220, 242, 278, 302, 313, 330, 336, 337

ecology, 2, 123, 142, 197, 269, 345, 366, 369,
    372, 373, 374, 375, 386, 389, 392, 396,
    399, 401, 403
    and river management goals, 372–373
ecosystem, 1, 3, 269, 372, 373, 375, 378, 386,
    392, 393, 399
ecosystem services, 399, 401
    categories of, 399
eddies, in turbulent flow, 79, 88, 89, 90, 91, 99,
    100, 114, 202, 213, 308
effective discharge, 152, 158, 159, 161
    factors affecting, 152–155
    relation to bankfull discharge, 152
embeddedness, 123, 366
energy grade line, 76
energy gradient, 76, 83, 330
energy, of flow in rivers, 75–76
entrenchment ratio, 380, 383
ephemeral streams, 2, 122, 135, 159, 163, 261,
    263, 360, 366
equilibrium, 5, 38, 159
    in relation to steady state and characteristic
        form, 13
    landscape evolution, 34
    models, 34
    of river channels, 144, 145, 152, 156, 172,
        176, 179, 184, 248, 294, 295, 335, 366,
        372, 379, 383, 387, 388, 389
    of river channels, limitations of
        concept, 157
erosion
    detachment-limited, 19, 34, 306, 307
    transport-limited, 19, 34, 307
event timescale, 12, 13, 14, 49, 104, 172, 229,
    241, 245, 246, 269, 369
Exner equation, 131, 132
extremal principles, 170
    and river meandering, 198

fall velocity, 112, 116, 117, 118, 119,
    121
    Stokes Law, 116
flood control, 149, 361, 375
flood frequency analysis, 138–142
    exceedance probability, 140
    peaks over threshold method, 141
    recurrence interval, 138, 140
    recurrence interval, annual maximum
        versus partial duration series, 141
flood frequency curve, 139, 140, 165
    and hydrologic variability, 143
flood power, 161, 163
    and channel change, 161
    duration of, 162
floodbasins, 259, 260

flooding, 1, 319, 380
floodplain, 3, 57, 62, 63, 97, 99, 145, 146, 147,
    163, 186, 221, 254, 256, 260, 261, 262,
    266, 348, 350, 354, 355, 361, 367, 386,
    389, 394
    definition of, 319–321
    deposition. See lateral accretion, vertical
        accretion, overbank sedimentation,
        levee, channel infill
    development of, 321
    erosion, 331–333
    integral component of river systems, 319
    major depositional processes, 321
    major erosional processes, 332
    of anastomosing rivers, 335, 341–342
    of braided rivers, 335, 337–339
    of meandering rivers, 335, 336–337
    of mega rivers, 266
    of wandering gravel-bed rivers, 335,
        339–341
    sedimentology of, 335–342
    shaving of, 332, 333
    stripping of, 332, 334
    typology of, 333–335
floods
    and bifurcations, 268
    and confluences, 269
    flash, 74
    geomorphic effectiveness of, 141, 160–163,
        169, 234, 249, 262, 316, 332, 339,
        348, 349
    impact of climate change on, 357
    impact of dams on, 363
    impact of land-use change on, 348,
        350, 354
    mega, 21
flow duration analysis, 142–143, 151
flow duration curve, 142, 165
    and hydrologic variability, 143
    and runoff-generating mechanisms, 143
flow oscillation theory
    of river meandering, 200–202
flow resistance, 166, 169, 179, 358
    and bedforms, 84–85, 115
    and scales of roughness, 84
    in step-pool channels, 317, 318
    influence of vegetation on, 225
    of sediment grains or particles, 84
    on floodplains, 319
    relation to boundary shear stress, 83
    relation to mean velocity, 82–83
flow separation, 213, 218, 233, 288, 289,
    329, 330
    at confluences, 273, 274, 280, 284
flow stagnation, 216, 288, 323, 329

at confluences, 273, 274, 283
fluvial geomorphology, 2, 9, 29, 97, 157, 343,
    369, 370, 371, 374, 378, 379, 389
    definition of, 2
    growth of, 2
    history of, 4–6
    origin of, 4
    public awareness of, 2
    role in river management, 372
    roots of, 5
    scientific inquiry in, 6–10
    styles of reasoning in, 10
fractal scaling
    and self-similarity, 44
    braided rivers, 248
    definition of, 44
freestream, 86
friction coefficient, 82
Froude number, 80, 191, 314, 315, 317

gaging station, 135
Ganges River, 266, 267
geologic timescale, 12, 49, 67, 231, 261, 266,
    269, 290, 294, 305, 308, 357
geological conditions
    as independent variable in fluvial sys-
        tems, 12
geomorphic effectiveness, 157–163, 185
    and climate change, 357
    major tenets, 157–160
    of floods, 160–163, 349
geomorphic units, 386
geomorphic work, 151–152, 156, 158, 161
geomorphological assessment of rivers and
    watersheds
    goals of, 378–379
Geomorphological Assessment Process, 379
geomorphological instantaneous unit hydro-
    graph, 137
Gilbert, Grove Karl, 4, 107, 144
graded river, 5, 144, 145, 157, 176, 178, 234,
    294, 295, 372
grain scale, 12, 83, 85, 159
gravel-sand transitions
    and downstream fining, 302–304
groundwater, 135
groundwater flow, 72, 97
gullies, 22–24, 28, 34, 70, 157, 227, 255, 332,
    346, 347, 350

habitat, 71, 122, 142, 197, 269, 319, 321, 350,
    354, 358, 366, 372, 373, 378, 383, 386,
    389, 391, 392, 393, 394, 395, 396,
    400, 401
head loss, 76

head, as measure of energy, 75
headcut, 26, 30, 31, 32, 163, 227, 262, 309, 349, 355, 359, 360, 367
    dynamics of, 310–311
helical motion, of flow, 201, 202, 214, 215, 216, 217, 218, 230, 244, 245, 268, 276–278, 282, 286, 289, 290, 292
historical range of variability, 387–389
Horton ratios, 41–43, 44, 45
Horton, Robert E., 5
human impacts on rivers, 6, 14, 371, 388, *See also* agriculture, channelization, climate change, dams, mining, timber harvesting, urbanization
    and changes in land cover, 344
    direct versus indirect, 343–344
    scope of, 343
human impacts on sediment flux, 62, 103, 343
Hutton, James, 4
hydraulic conductivity, 20, 134, 345
hydraulic drop, 81, 391
hydraulic engineering, 144, 181
hydraulic geometry, 164
    at-a-station, 165, 166–171, 183
    at-a-station and extremal principles, 170–171
    at-a-station, of braided rivers, 240
    at-a-station, reach-averaged, 169
    at-many-stations, 170
    defined, 164–166
    downstream, 165, 171–173, *See also* downstream channel geometry
    downstream, for bankfull discharge, 173
    downstream, versus downstream channel geometry, 173–174
hydraulic jump, 81, 318, 391
hydraulic radius, 77, 80, 82, 83, 109, 115, 166, 189, 358
hydraulics, 2, 5, 12, 72, 144, 374, 375
hydrograph, 102, 135
    and river network structure, 137
    factors influencing shape of, 137–138
hydrological cycle, 1, 134
hydrology, 2, 12, 72, 134, 144, 345, 374, 375
hyperconcentrated flow, 99–100
hyporheic exchange, 135, 394
hyporheic zone, 135
hypothesis, 7, 8, 9, 10, 43, 49, 144, 145, 211
    extremal, 179
hysteresis, 168, *See* suspended sediment concentration

imbrication, 123, 316
impulse, 112
    relation to turbulence, 112–113

infiltration capacity, 16, 20, 134, 249, 345, 350, 353, 354
instability, 18, 19, 21, 36, 203, 205, 239, 275, 358
    Kelvin-Helmholtz, 275
    of braided rivers, 234, 241, 337
    of channel banks, 181, 224
    of river channels, 149, 158, 351, 357, 359, 367, 372, 382, 383, 396
    of step-pool units, 316
    of wandering gravel-bed rivers, 257, 258
interception, 20, 49, 134, 345, 350, 353
intermittent streams, 2, 4, 135, 163
islands, 252, 253, 257, 260, 267, 268, 269, 339
    in anabranching rivers, 187
isovels, 95

junction angle, 270
    and channel slope, 271
    and climate, 271
    and stream order, 270

kinematic viscosity, 10
Kinoshita curves, 231
knickpoint, 21, 159, 227, 312, 355, 359, 367, 396
    definition and morphologic characteristics, 309–310
    evolution of, 310–312
Kolmogorov microscales, 10

laminar flow, 78, 86
land drainage, 149, 358, 375, 400
landscape evolution, 5, 15, 29, 97, 269
    models of, 33–39, 305
landscape sensitivity. *See* geomorphic effectiveness
landslides, 27, 60, 71, 104, 350
large woody debris, 67, 82, 210, 225, 256, 257, 314, 318, 339, 349, 358, 394, 395, 396
lateral accretion, 261, 263, 321–323, 335, 336, 339, 340
lateral migration, 221, 224, 225, 229, 230, 232, 242, 245, 246, 249, 255, 256, 259, 263, 327, 332, 334, 335, 348, 349, 350, 367, 388, 396
law of the wall, 115, 117, 119
    and bed shear stress, 92–93
    and logarithmic velocity profile, 90–92
    derivation of, 89–92
    velocity profile equation, 91
legacy sediment, 347–348, 349, 350
levee, 67, 254, 255, 259, 260, 261, 262, 266, 267, 327, 328, 334, 336, 342, 386
    artificial, 327, 358, 394

formation of, 325–327
    front-loading versus backloading, 326
link scale, 12, 269
logging. *See* timber harvesting
longitudinal profile, 12, 164, 248, 294, 313, 364, 386, 390
    adjustments of, 308–312
    changes in channel types along, 313–314
    concave-upward form of, 295, 296, 300, 302, 305, 306, 307, 308
    convex-upward form of, 295, 306
    defined, 294
    dependence on resolution of topographic data, 295–296
    equilibrium form of in alluvial rivers, 305–306
    equilibrium form of in bedrock rivers, 306–308, 309
    functional forms used to depict, 295

management of rivers, 6, 14, 234
    and post-project appraisals, 402–403
    and uncertainty, 401–403
    and watershed planning, 378
    general goals of environmental management, 371
    importance of geomorphology in, 369
    main geomorphological and ecological goals, 372–373
    passive approach to, 389
    relation to river science, 369–371
    role of social processes in, 371–372, 403
Manning equation, 82, 83, 166, 167, 172, 358
Manning's *n*, 82, 169
    and channel geometry, 176
Manning's roughness coefficient, 82, 83
mass movement, 23, 60, 71, 98, 99, 100, 222, 316, 355
mean annual flood, 140, 189, 359
meander belt, 206, 322
meander loops, 207, 228, 231
    compound, 231
    compound form of, 226, 228, 232
meander scar, 328, 332, 336
meander train, 206, 225
meandering rivers, 256, 259, 314, 321, 322, 325, 327, 328, 332, 349, 367, 383, 389, 392, 400
    and extremal hypotheses, 198
    and floodplain shaving, 332
    bank erosion and lateral migration, 224–225
    bed material transport in, 218–221
    bedrock, 225
    flow in, 213–218

meandering rivers (cont.)
  initiation of meandering, 198–205
  lateral migration of, 221, *See also* lateral
    migration
  pattern of bed shear stress, 216, 218
  planform characteristics, 205–207
  planform evolution, 225–233, 337
  sinuosity of, 186, 206
mega rivers, 195, 266–268
mining, 58, 302, 371
  hydraulic, 5, 351–353
  in-channel, effect on rivers, 366–368
  waste, effect on rivers, 351–353
Mississippi River, 3, 160, 266, 322, 325,
    332, 349
mitigation, 370, 376–378, 389, 391, 402
mixed load, 119, 190, 302, 303, 349
mixing interface, 273, 275–276, 280, 289, 290
  versus shear layer, 273
mixing, in rivers, 290–291
model theoretic view (MTV) of scientific
    theory, 7–8
models
  analytical, 8, 21, 38, 152, 231
  cellular automata, 249
  data, 8
  numerical, 8, 37, 83, 95, 96, 132, 203, 205,
    229, 231, 232, 250, 325
  physical, 9, 205, 287, 357
  physics-based, 229, 230, 249, 250, 251
  predictive, 50, 59, 71, 96, 122, 170, 185, 251,
    273, 325, 357, 380, 398, 402
  reduced-complexity, 185, 250
  theoretical, 7, 8, 9, 17, 19, 311
  uncertainty of, 402
modern timescale, 12, 14, 49, 63, 157, 172,
    176, 178, 290, 300, 335, 357, 369
momentum flux, 78, 79, 88, 89, 121, 287
momentum flux ratio, 274, 285, 287, 288, 292

Natural Channel Design, 389–392, 393, 402
naturalization, 370, 375, 378, 389,
    399–401, 402
Navier-Stokes equations, 95
nonequilibrium, 159
nonlinear dynamical behavior, 7, 36, 157,
    231, 346, 369, 383, 402
  of models, 34–35
normal stress
  effective, 224
  turbulent, 88, 115, 201

oblique accretion, 323
Okavango Delta, 264
open-channel hydraulics, 72

and complexity of flow in rivers, 75
classification of flow types, 73–74
dimensionality of fluid motion, 74–75
three-dimensional flow, 95–96, 213
optimality, 7, 182, 183
  and junction angles, 271
optimality principles
  and rational regime theory, 178–181
  and river networks, 46
overbank sedimentation, 302, 321, 323–328,
    335, 339, 342, 348
  mechanisms of, 323
overland flow, 16, 18, 20, 27, 29, 39, 97, 345,
    346, 353, 354
  infiltration excess, 16, 20, 24, 26, 72, 134, 345
  saturation excess, 20, 26, 27, 30, 33, 72, 134,
    135, 143
oxbow lake, 227, 328, 332, 333, 386
  formation of, 328–330
oxbow lakes, 336, 337

partial duration series, 141, 146
  geomorphological importance of, 142
particle path length, 129
particle size
  characterization of. *See* Appendix B
  distributions. *See* Appendix B
  downstream fining. *See* bed material
  phi units. *See* Appendix B
particle Reynolds number, 119, 183
particle step length, 129
path lengths
  of transported particles, 130
percolation, 135
perennial streams, 74, 121, 135, 159, 261, 366
pivoting angle, 105, 110
  relation to grain resistance to motion, 110
  relation to particle size, 110
planform. *See* channel planform
planform scale, 12, 72, 83, 85, 127, 160, 170,
    205, 246, 247, 248, 249, 269, 359, 386,
    393, 400
plate tectonics, 15
Playfair, John, 4
Playfair's Law, 4
point bar, 191, 208, 210, 219, 221, 225, 232,
    258, 288, 289, 321, 335, 336, 337,
    339, 354
  influence on flow through meander bends,
    217–218
  relation to bar units in meandering riv-
    ers, 207
pollution, 1, 71, 99, 142, 351, 371, 373, 399
pools
  infilling of, 350, 397

pools and riffles, 198, 201, 358, 386, 394, 401
  characteristics of, 209–210
  forced, 210–211
  identification of, 209
  maintenance of, 211–213
  relation to bar units in meandering riv-
    ers, 207
  velocity reversal hypothesis, 211–213
post-settlement alluvial. *See* legacy sediment
Powell, John Wesley, 4
power functions. *See* Appendix A
  in fluvial geomorphology, 9
  isometric versus allometric scaling. *See*
    Appendix A
  multivariate, 175
primary channel, 254, 255, 257, 258, 259, 260,
    263, 266, 267, 341
pristine conditions, 372, 374, 375, 392
protection, of rivers, 373–374

rainsplash, 16, 22, 345, 346, 353
rating curve, 136–137, 151, 165
  sediment, 101, 104, 152
rational regime theory, 178–183, 191, 402
  limitations, 182
  numerical/analytical approaches, 181–182
  optimality approaches, 178–181
  threshold approach, 181
  vectorial mechanics, 181
ray-isovel method, 95
reach, of river, 72
reference condition, 372, 374, 380, 387, 388,
    389, 392, 393, 394, 399
reference reach, 380, 383, 390, 391, 392
regime theory, 144, 145, *See also* rational
    regime theory
regulative principles, 7, 145, 157
relative roughness, 314
  and grain resistance, 84
relief, 12, 16, 24, 29, 36, 39, 50, 52, 55, 57, 59,
    62, 159, 270, 294, 304, 344, 350
  as independent variable in fluvial sys-
    tems, 12
relief ratio, 50, 52, 64
reservoirs. *See* dams
restoration, 370, 374–375, 377, 378, 379, 389,
    390, 391, 399, 402
  and Natural Channel Design, 392
  process-based, 392–396
return flow, 135
Reynolds number, 80, 87, 94
Reynolds stresses, 88, 93, 96, 113, 115, 224
ridge and swale topography, 322, 386
rills, 16, 21, 29, 34, 70, 346
  and channel initiation, 16–17, 21

characteristics of, 22
versus gullies, 22–24
ripples, 84, 114, 217, 283, 284, 313, 336, 337
river
definition of, 2–3
river engineering, 223, 369, 371, 392, 393, 402
river network, 2, 3, 4, 5, 11, 12, 15, 29, 33, 34,
39, 63, 269, 308
as example of self-organized criticality, 46
development of, by overland flow, 30–32,
36–37
development of, by subsurface flow, 30,
32–33, 38–39
fractal scaling of, 44–46
geometry of, 39
laws of network composition, 43
method of defining, 39
optimal channel networks, 46
stages of development, 29
topology of, 44, 45
River Styles Framework, 379, 383–389
river system, 3–4
as floodplain and channels, 319
dependent and independent variables,
12–13
heirarchical spatial structure of, 11–12
timescales, 12–13
river-network scale, 11, 33, 269, 297, 306, 359,
373, 386, See also drainage-basin scale
rivers
importance of, 1
Rosgen Method, 379, 389, See also Natural
Channel Design
concerns about, 383
of geomorphic assessment, 379–383
roughness height, 91, 92
Rouse equation, 117, 118, 119
Rouse number, 117, 119

saltation, 100, 116, 307, 308
science
applied design, 370, 371, 374, 389, 392,
399, 402
applied predictive, 370
applied versus basic, 370
scour and fill, 130, 168, 172, 234, 337, 415
scroll bars, 226, 232, 321–322, 323, 330
secondary channels, 254, 255, 257, 258, 259,
260, 263, 266, 267, 321, 332, 341,
386, 394
secondary circulation, 214, 215, 218, 220, 244,
245, 278, 280
secondary flow, 214, 215, 229, 242, 244, 268,
286, 291
and helical motion, 278

sediment budget, 65–67, 132
and connectivity, 66–67
and drainage-basin management, 71
and models of sediment flux, 70–71
and sediment residence time, 67
and sediment tracing, 68
of Coon Creek, 65, 71
sediment delivery problem, 63
sediment delivery ratio, 63–64, 69, 70, 346
and timescale, 67
defined, 63
limitations of, 64–65
sediment fingerprinting. See sediment
tracing
sediment load, 49, 121, See also sediment
yield
and agriculture, 58, 346, 348, 351
and anabranching rivers, 260, 268
and braided rivers, 243
and channel geometry, 178, 179, 181
and channel planform, 188–189
and channel slope, 296
and confluences, 285, 286, 290
and dam removal, 397, 398
and denudation rate, 62
and effective discharge, 151, 152, 155
and human impacts, 58
and longitudinal profile, 296, 305, 306
and mining activities, 352
and runoff, 52
and straight rivers, 192, 193
and timber harvesting, 350
and urbanization, 354
and water quality, 372
factors influencing at drainage-basin
scale, 49
global predictive model, 59
impact of dams on, 58, 364, 365, 366
relation to basin topography, 50, 52–53
relation to climate and vegetation, 49, 60
relation to drainage area, 50
relation to rock and soil erodibility, 50
relative importance of controlling factors,
59–60
versus solid load, 97
sediment production, 55, 63, 64, 67, 104, 346,
350, 351, 355
sediment storage, 54, 55, 62, 63, 65, 104,
248, 321
sediment residence time, 67
sediment tracing, 67–70, 103
sediment transport capacity, 101, 124, 132,
179, 219, 237, 238, 242, 253, 282, 296,
300, 302, 306, 323, 324, 359, 360, 361,
365, 391

sediment yield, 49, 70, 71, 346, 353, See also
sediment load
and earthquakes, 60
and rock erodibility, 57
relation to drainage area, 50, 53–57
relation to precipitation and runoff, 50–52
relation to vegetation, 52
sediment, fluvial
characterization of. See Appendix B
sedimentary links, 304–305
seepage erosion, 20, 21, 26, 27, 28, 30, 32, 33,
38, 224, 346
selective sorting, 297, 299, 304, 306
and downstream fining, 299–302
self-organized criticality, 46, 231, 248
sensitivity, of river systems, 357, See also
geomorphic effectiveness
shear layer, 114, 323, 326
shear stress
acting on grains, 116
bank, 222, 224
bed, 77–78, 82, 83, 87, 113, 115, 116, 123,
125, 181
bed determination of, 92–95
bed, bankfull, in rivers, 182, 183
boundary, of overland flow, 16
boundary, ray-isovel method, 95
critical bank, 225
critical bed, 108, 111, 128, 181, 182, 183,
211, 300, 316
critical bed, for channel initiation, 37
critical bed, of overland flow, 19
critical dimensionless bed, 106, 110, 111,
112, 191, 300
critical dimensionless grain, 116
dimensionless bed, 106, 107, 130, 236,
237, 300
turbulent, 79, 80, 86, 88, 89, 93, 113, 115
viscous, 78, 79, 86, 89
shear velocity, 83, 87, 92, 106, 112, 118, 119,
129, 156
bank, 224
bankfull, 183
critical, 112
critical bank, 224
sheetwash, 16, 346
shingle bars, 227, 228
sinuosity
and channel straightening, 358
and cutoffs, 226
and length of oxbow lakes, 328
and meander evolution, 231
and sediment flux, 229
factors controlling, 206
of braided rivers, 240

sinuosity (cont.)
of single-thread rivers, 186
recovery of, 360
soil creep, 16
soil erosion, 55, 58, 59, 63, 64, 70, 71, 98, 100, 101, 347, 353
impacts of agriculture on, 345–346
solid load, 97
splays, 327–328, 336, 338
bank-top, 327
crevasse, 260, 261, 262, 327, 336, 341, 386
stability, 17, 158, 355
morphological, 314, 316, 357, 374, 401
of river channels, 71, 318, 371, 372, 380, 389, 391, 392, 395
stability analysis, 16, 17, 18, 19, 21, 170, 192, 202, 234, 243
slope, 224
stage, of a river, 135
steady state, 7, 13, 36, 37, 38, 39, 145, 157, 158, 248, 306, 309, 312, 324, 387, 389
steepness index, 307, 309, 310, 312
step-pool channels, 314, 383, 386
flow and sediment transport in, 317–318
formation of, 314–316
morphology of, 316–317
stability of, 316
Sternberg's Law, of downstream fining, 297, 298
storm sewer, 102, 354, 355, 399, 401
Strahler, Arthur, 5
straight rivers
and sinuosity, 186
environmental conditions of, 194
stream capture, 37
stream length ratio, 41
stream links, 12, 39–41, 43, 304
and width function, 137
cis versus trans, 43
stream magnitude, 39–41
stream order, 39–41, 44
stream power, 46, 78, 120, 121, 129, 130, 162, 229, 234, 248, 250, 252, 253, 256, 258, 259, 266, 296, 306, 328, 332, 348, 355, 358, 359, 388
and floodplain characteristics, 333
and floodplain typology, 333–335
critical, 112
dimensionless, 247
meandering versus braided rivers, 189–191
of floods, 161
potential, 190
stream power incision model, 37, 306, 312
stream, versus river, 2

structures for stream stabilization and restoration, 391
subcritical flow, 80, 81, 281, 311, 314, 317
supercritical flow, 81, 281, 314, 317
superelevation, of the water surface, 213, 216, 230, 244, 246, 273, 277, 278, 327
suspended load, 48, 99, 119, 121, 193, 307, 321, 327
suspended sediment, 323
global flux to oceans, 47
transport of, 116–119
transport, measurement of. See Appendix D
suspended sediment concentration, 100, 103, 117, 119, 323, 325, 350
first flush phenomenon, 102
hysteresis, 102–103, 104
relation to discharge and other controlling factors, 100–105
seasonal variations, 104
wave, 101
suspended sediment load, 67, 100, 119, 152, 229, 269, 308
relation to sediment concentration, 101
sustainability, 59, 371, 372, 375, 392, 393
symmetry
of bifurcations, 243
of confluence planform, 270
of confluences versus bifurcations, 270

tectonic uplift, 34, 36, 37, 39, 62, 63, 267, 294, 296, 305, 306, 307, 308, 309, 312
terrace, 149, 162, 266, 272, 349, 365, 366, 386, 397
definition of, 320
thalweg, 98, 188, 189, 192, 193, 194, 200, 201, 202, 256, 321, 336, 337, 360
threshold channel, 181
thresholds
extrinsic, 157
in river systems, 4, 157, 159, 346
intrinsic, 157
throughflow, 72, 97, 134
timber harvesting
effect on rivers, 350–351
topographic steering, of flow, 217, 218, 221, 230
transient conditions, 13, 159, 194, 308, 309, 312
tributaries
effect on downstream fining, 304–305
effect on main rivers, 271–272

turbulence, 87, 88, 112–115, 116, 118, 323
bursting, 114
coherent turbulent structures, 114, 275–276, 291, 364
ejections, 113, 115
horizontal bursting, 202
horizontal coherent structures, 202
quadrant analysis of, 113
sweeps, 113
turbulent dissipation rate, 10
turbulent flow, 10, 78–79, 80, 86, 88, 114
turbulent kinetic energy, 93, 213, 225, 276, 291

urban stream syndrome, 353
urbanization, 353, 371, 375, 399, 401
effects of construction on streams, 353–354
effects of urbanized areas on river channels, 355–356
effects on hydrology, 354–355
effects on sediment supply, 355
phases of, 353

valley slope, 26, 176, 186, 188, 189, 193, 195, 196, 206, 256, 296, 313, 388
valley spacing
in relation to advective and diffusive processes, 35–36
vegetation, 12, 104
and anabranching, 256, 261, 264–266, 268
and bank erosion, 222, 224
and braided river dynamics, 249
and braided rivers, 192, 339
and channel geometry, 175
and channel recovery to disturbance, 359, 361, 365, 366, 396
and lateral migration of rivers, 232
as independent variable in fluvial systems, 12, 357
clearing of, 344, 345, 349, 350, 353, 391, 399
growth on point bars, 233
velocity
fluctuating components of, 88
mean, at a location in the flow, 74
mean, over a cross section, 73
measurement of. See Appendix C
vector, 74
velocity profile, turbulent flow, 91
vertical accretion, 261, 321, 323, 329, 333, 334, 335, 336, 339, 340, 361
self-limiting nature of, 323

virtual velocity
  of transported particles, 129, 130
viscosity, 78, 99
  dynamic, 78
  eddy, 79, 89, 117
  kinematic, 80, 87, 183
viscous sublayer, 108, 114
von Karman's constant, 90, 92
vorticity, 201

wall similarity concept, 94
wandering gravel-bed rivers, 194, 252,
    256–258, 259, 260, 261, 267, 285,
    383, 389
  environmental domain of, 194
wash load, 99, 100, 258,
    303, 397
water quality, 1, 99, 371, 372, 373, 374, 383,
    392, 399, 401

waterfalls, 309, 311
watershed. *See* drainage
    basin
watershed assessment, 378, 379,
    389
watershed planning, 378, 389
wave, of sediment, 351
wetted perimeter, 77
width, of flow, 73